Evolutionary and Molecular Biology
Scientific Perspectives on Divine Action

A Series on "Scientific Perspectives on Divine Action"

First Volume
Quantum Cosmology and the Laws of Nature:
Scientific Perspectives on Divine Action
Edited by Robert John Russell, Nancey Murphy, and C. J. Isham

Second Volume
Chaos and Complexity:
Scientific Perspectives on Divine Action
Edited by Robert John Russell, Nancey Murphy, and Arthur Peacocke

Third Volume
Evolutionary and Molecular Biology:
Scientific Perspectives on Divine Action
Edited by Robert John Russell, William R. Stoeger, S.J., and Francisco J. Ayala

Future Scientific Topics
The Neurosciences and the Person
Quantum Physics and Quantum Field Theory

Jointly published by the Vatican Observatory and
the Center for Theology and the Natural Sciences

Robert John Russell, General Editor of the Series

Supported in part by a grant from
the Wayne and Gladys Valley Foundation

Evolutionary and Molecular Biology
Scientific Perspectives on Divine Action

Robert John Russell
William R. Stoeger, S.J.
Francisco J. Ayala

Editors

Vatican Observatory
Publications,
Vatican City State

Center for Theology and
the Natural Sciences,
Berkeley, California

1998

Robert John Russell (General Editor) is Professor of Theology and Science in Residence at the Graduate Theological Union in Berkeley, California, and Founder and Director of the Center for Theology and the Natural Sciences

William R. Stoeger, S.J. is Staff Astrophysicist at the Vatican Observatory, Vatican Observatory Research Group, Steward Observatory, The University of Arizona, Tucson, Arizona. He is also Adjunct Associate Professor of Astronomy at the University of Arizona.

Francisco J. Ayala is the Donald Bren Professor of Biological Sciences at the University of California, Irvine.

About the cover: the image on the cover of this volume is an evolutionary tree with twenty species derived from the composition of the protein cytochrome-c. The numbers are estimates of the changes in this protein that have occurred during evolution. First developed in the 1960s, this tree is one of the earliest reconstructions of evolution's history using data from molecular evolution. It agrees well with the pattern of phylogenetic relationships worked out by classical techniques of comparative morphology and from the fossil record. See Francisco J. Ayala's article in this volume (p. 52) for a discussion of protein evolution.

Jointly published by the Vatican Observatory and the Center for Theology and the Natural Sciences

Distributed (except in Italy and the Vatican City State) by
    The University of Notre Dame Press
    Notre Dame, Indiana 46556
    U.S.A.

Distributed in Italy and the Vatican City State by
    Libreria Editrice Vaticana
    V-00120 Citta del Vaticano
    Vatican City State

ISBN 0-268-02753-6 (pbk.)

ACKNOWLEDGMENTS

The editors wish to express their gratitude to the Vatican Observatory and the Center for Theology and the Natural Sciences for co-sponsoring this research. Particular appreciation goes to George Coyne and Bill Stoeger, whose leadership and vision made this series of conferences possible.

Editing for this volume began with an initial circulation of papers for critical responses before the main conference in 1996, and continued with extensive interactions between editors and authors after the conference. The editors want to express their gratitude to all the participants for their written responses to pre-conference drafts and for their enthusiastic discussions during the conferences.

Special thanks goes to Kirk Wegter-McNelly, CTNS Editing Coordinator, who devoted meticulous attention and long hours to every stage in the post-conference phases of the production of this volume while pursuing doctoral studies at the GTU. In particular, Kirk worked closely with the editors as the post-conference papers were read and critiqued in light of previous editorial comments and suggestions. He did extensive and detailed copy-editing, including substantial editorial work on some essays, and prepared the two figures in the first essay by Camilo J. Chela-Conde. Kirk was responsible for the layout of the volume, including formatting the book such that an exceptional amount of text could fit within a reasonable overall page length, and he produced the front and back covers. He also designed and implemented both indices. Overall, it was in great measure due to his diligent support that the volume was completed in 1998. Thanks also goes to Jane C. Redmond for her careful copyediting of the French papal statement and to George Coyne for overseeing printing and distribution.

Special thanks goes to Greg Maslowe for organizing a pre-conference held at CTNS in September, 1995, and for preparing and distributing readers, papers, and responses prior to the main conference in June, 1996. Thanks is also offered to the staff of the Vatican Observatory for arranging a second pre-conference in the fall, 1995 and the main conference, both held at Castel Gandolfo.

The editors wish to thank those who have given permission to reprint the following articles in this volume:

"Evolution and the Human Person: The Pope in Dialogue" by George V. Coyne, S.J. From *Science and Theology: The New Consonance*, edited by Ted Peters (Boulder, Colo.: Westview Press, 1998): 153–61. Reprinted by permission of Westview Press, a member of the Perseus Books Group.

"Message to the Vatican Observatory Conference on Evolutionary and Molecular Biology" by Pope John Paul II. Originally printed in *L'Osservatore Romano*, 29 June, 1996. Reprinted by permission of George V. Coyne, S.J.

ACKNOWLEDGMENTS

"Message to the Pontifical Academy of Sciences" by Pope John Paul II. Originally printed in the English edition of the *L'Osservatore Romano*, 30 October, 1996. Reprinted by permission of George V. Coyne, S.J.

"Aux Membres de l'Académie pontificale des Sciences réunis en Assemblée plénière" by Pope John Paul II. Originally printed in *L'Osservatore Romano*, 23 October, 1996. Reprinted by permission of George V. Coyne, S.J.

CONTENTS

CONTENTS

III. RELIGIOUS INTERPRETATIONS OF BIOLOGICAL THEMES

IV. BIOLOGY, ETHICS, AND THE PROBLEM OF EVIL

INTRODUCTION

Robert John Russell

1.1 *Background to the Volume*

Central to most streams of Christian theology is the belief that God creates and redeems the world. As creator, the God of biblical faith is the transcendent source of all that is, including spacetime, the "stuff" of nature, and the laws which science discovers. Yet this radically transcendent God is more immanent to the world than we are to ourselves, bringing order out of chaos and the new out of the old in the interstices of nature, working in, with, under and through the very processes which science describes by its laws. The "mighty acts of God," then, are precisely, though not only, the actions and activities of the natural world, including both ordinary natural events and those extraordinary occasions in which God's creative and providential intentions can be seen as particularly prominent. As immanent redeemer, God suffers with creatures throughout the sweep of life, sharing their pain and transforming their lives into something which, though hidden now, will be eternally wondrous. As transcendent redeemer, God guarantees that this hidden wonder, which we only dimly perceive in the Easter mystery, is in fact grounded in a radically new and coming eschatological future.

Contemporary natural science offers several challenges to these streams in theology which both enhance their vision and call for decisive reconstruction. First science immensely broadens the picture of "the world" which biblical faith understands as God's creation and as the subject of divine redemption. It is increasingly clear to a growing body of scholars that the theological concept of "world" should include not just human history or even just life on earth. Instead it should include the entire fifteen billion year history of the *cosmos*, stretching from the earliest epochs of the expanding universe through the origin and evolution of life on earth—and perhaps elsewhere in the depths of space. Is it now possible to develop afresh a theology of creation and redemption that takes seriously the astounding breadth of the subject of God's love, "the world" (John 3:16)?[1]

Second, science presents an overwhelmingly detailed and lavishly nested series of accounts of the cosmos, whose multiple levels of consistency, coherence and predictive exactitude warrant their claim to joint explanatory power. Beginning with the work of Charles Darwin in the mid-nineteenth century and continuing up through the modern neo-Darwinian synthesis that has been developing throughout the twentieth century, and which since the 1950s includes an increasingly detailed understanding of the molecular foundations of evolution, the biological sciences now present us with an astonishingly precise understanding of the history of life on earth over the past billion years. The evolutionary picture nests within the broader context provided by the physical sciences as they describe the development of first and second generation stars in our galaxy and the production of planetary systems like ours. Their account, in turn, lies within the broad scope of contemporary cosmology and its depiction of an expanding universe. Lurking before this might be the essential singularity ("t=0") of Einstein's Big Bang cosmology or the endless chaotic inflation of a quantum cosmology.

---

[1] It is important to recall that the Greek word translated here as "world" is in fact κόσμος.

This detailed picture requires that theologians go far beyond the comfortable strategy of merely acknowledging what science tells us about the physical universe while continuing to discuss the human sphere in isolation from science. Instead the theological challenge today is to drink deeply from the well of scientific knowledge and to ponder seriously its proffered wisdom—however much this might challenge traditional beliefs and values. But theology also comes with a challenge to science. It increasingly calls for a thorough and critical analysis of the theological, philosophical and valuative assumptions lying within the physical and biological sciences, in order that the relation between theology and science be one of genuine, creative, and mutual respect and interaction. The task, then, is both daunting and invigorating, particularly when it comes to the relevance of science regarding the physical and biological roots of the human person, including our capacities for reason, morality, religion and culture, and to the theological critique of those reductionistic philosophical elements which, when taken into science, undermine the integrity and worth of the human person.

Although it often goes unnoticed by contemporary voices which argue—unfruitfully—that evolution and Christianity are in an irreconcilable conflict,[2] Christian theologians have in fact developed a diversity of responses to Darwin and his scientific descendants beginning in the late nineteenth century and extending to the present. As Claude Welch points out in his illuminating appraisal,[3] two distinct forms of compatibility were explored during the nineteenth century. Some argued for an accommodation between evolution and Christianity, and others for the full-scale integration of evolution into theology. The former included F. D. Maurice, who saw science as a mode of divine revelation; Charles Kingsley, who was an early proponent of the immanence of God in nature; and James McCosh, who interpreted natural selection in terms of supernatural design. The latter included the authors of *Lux Mundi*, who rejected the God of deism (occasional acts by such a God imply an habitual absence) in favor of the biblical God who acts everywhere through evolution; George Wright, who imported Darwinianism into a Calvinistic interpretation of nature; and Henry Drummond, who added to evolution the struggle for the life of others.[4]

---

[2] Richard Dawkins, Francis Crick, Peter Atkins and many others both within and outside of science claim that evolution provides a firm basis for atheism and discredits Christianity—and by implication, world religions in general. Among Christian Fundamentalists, and even among evangelicals and conservatives, this claim—that "evolution equals atheism"—is too often taken without question, and it sets the agenda: attack atheism by attacking evolution, either directly, as with the Scopes Trial of the 1920s, or indirectly, by seeking to replace it with a "scientific" competitor, often described as "creation science" or "Intelligent Design." It is not within the scope and purposes of this volume to consider these approaches; we recommend the variety of responsible texts available to the interested reader. It should be self-evident that the claim to be challenged is that "evolution equals atheism." In its stead, many in this volume will present a Christian interpretation of evolution. Though these will differ significantly, the mere fact of their existence falsifies the atheistic claim *and* challenges the assumption that the best way, let alone the only way, to defend the faith is to attack evolution.

[3] Claude Welch, *Protestant Theology in the Nineteenth Century*, vol. 2 (New Haven: Yale University Press, 1972, 1985), chap. 6; John Hedley Brooke, *Science and Religion: Some Historical Perspectives* (Cambridge: Cambridge University Press, 1991); Arthur Peacocke, *Creation and the World of Science* (Oxford: Clarendon, 1979).

[4] Theological opposition to Darwin is also rooted in the nineteenth century, as Welch and others point out.

These positions continued into the twentieth century while others emerged as well.[5] Many Roman Catholic theologians underscore the importance of science to Christian thought. In his 1950 encyclical, *Humani generis*, Pope Pius XII combined the evolution of the physical body, as discovered by science, with the special creation of the soul, as revealed by Scripture. Ernan McMullin, Karl Rahner, Rosemary Radford Ruether, Karl Schmitz-Moormann, Pierre Teilhard de Chardin, and other Catholics have eloquently thematized Christian faith within broad evolutionary categories. In general, the results of science are welcomed by Catholic theologians because Scripture can be given a metaphorical interpretation. It is in this connection that the current statements by Pope John Paul II (included in this volume) are particularly significant, taking as they do an appreciative view of evolution and a nuanced view of what constitutes the theological distinctiveness of the human person in light of evolution.

Anglican scholars have continued the broad tradition of natural, sacramental, and philosophical theology into the twentieth century, finding a wealth of insight from evolutionary biology. Contributors include William Temple, Lionel Thornton, Austin Farrer, Ian Ramsey, W. H. Austin, Charles Raven, and E. L. Mascall. Although the early central figures in Protestant neo-orthodoxy tended to treat theology and science as dealing with different realms and methods, a detailed analysis would show that some important interactions were considered by them. Beginning in the 1960s with the pioneering work of Ian G. Barbour, however, the intensity and diversity of creative interactions between theologians, philosophers, and scientists have increased exponentially. These scholars include Ralph Burhoe, John Cobb, Jr., John Dillenberger, David R. Griffin, Langdon Gilkey, Charles Hartshorne, Carl Heim, George S. Hendry, Sallie McFague, Jürgen Moltmann, Wolfhart Pannenberg, William Pollard, Holmes Rolston, III, Paul Santmire, and Gerd Theissen. Twentieth-century reaction to Darwin among Evangelical (conservative) Protestant theologians has been mixed. The rise of the Fundamentalist movement in the 1920s, with its sharp opposition to evolutionary theory, tends to obscure the fact that many of the most significant Evangelical scholars crafted creative responses to the new evolutionary world-view. These include A. A. Hodge, B. B. Warfield, A. H. Strong, and more recently Alister McGrath. Bernard Ramm played a role in the Evangelical development of the science-theology dialogue comparable to that of Ian Barbour among the liberals. I am particularly grateful that,

---

[5] For a helpful overview, see Ian G. Barbour, *Issues in Science and Religion* (San Francisco: Harper Torchbook, 1966), chap. 12; idem, *Religion in an Age of Science* (New York: HarperCollins, 1990), chaps.6–9. See also Christopher Kaiser, *Creation and the History of Science* (Grand Rapids, Mich.: Pickering, 1991); John Durant, ed. *Darwinism and Divinity* (Oxford: Basil Blackwell, 1985); Holmes Rolston, III, *Science and Religion: A Critical Survey* (New York: Random House, 1987); Ernan McMullin, ed., *Evolution and Creation* (Notre Dame, Ind.: University of Notre Dame Press, 1985); David C. Lindberg and Ronald L. Numbers eds., *God and Nature: Historical Essays on the Encounter between Christianity and Science* (Berkeley, Calif.: University of California Press, 1986); Svend Andersen and Arthur Peacocke, eds., *Evolution and Creation: A European Perspective* (Aarhus: Aarhus University Press, 1987); John Dillenberger , *Protestant Thought & Natural Science: A Historical Study* (Nashville, Tenn.: Abingdon Press, 1960). For Protestant conservative responses see, for example, Bernard Ramm, *The Christian View of Science and Scripture* (Grand Rapids, Mich.: Eerdmans, 1954) and more recently David N. Livingstone, *Darwin's Forgotten Defenders: The Encounter between Evangelical Theology and Evolutionary Thought* (Grand Rapids, Mich.: Eerdmans, 1987).

in addition to these names, many of the most creative contributors to the dialogue from both Catholic and Protestant communities are represented in this volume.

Meanwhile, during the past several decades the neo-Darwinian synthesis has been extended to include fields ranging from paleontology and comparative anatomy to biogeography, embryology, and molecular genetics. Topics include the question of speciation, gradual versus punctuated evolution, protein evolution, the molecular evolutionary clock, new models of selection (gene, organism, kin, group, species), and possible additional sources of novelty besides mutation and selection. Research on human evolution is focusing on what distinguishes our species from other early hominids: bipedalism, brain size, language, or tools? A variety of scientific theories are also pushing the frontiers of evolutionary and molecular biology from the perspective of physics, including chaos, complexity, and self-organization.

These areas and others are receiving increasing interest by humanities scholars. The biological basis of morality, the biocultural evolution of distinctively human capacities and behaviors, the significance of genetic variation for divine action in evolution, the problem of theodicy in light of "natural evil" (for example, disease, suffering, and death in nature), the challenge to ethics posed by genetic research and engineering, and the possibility and significance of life in the universe, are only some of the many current areas of interest to theologians, ethicists and philosophers of religion. There has also been a small, but growing, interest in exploring the philosophical presuppositions of contemporary science and their grounding in a theological view of the world. Such a case has been made for the historical beginnings of Western science and its continued pursuit is worth considering if, as I have just suggested, the interaction between theology and science is to be genuinely "two-way." Indeed the very diversity in mainstream scientific research can be seen to represent, at least in part, differing philosophical conceptions of nature, functioning here as creative sources for actual scientific research programs.

It is in this broad context of diverse issues including many aspects of the interaction between theology and science that the current volume takes its place.

1.2 Background to the Conference

In June, 1996, twenty-seven scholars, with expertise in physics, cosmology, evolutionary and molecular biology, philosophy of religion, philosophy of science, philosophical and systematic theology, history of theology, and history of science, met for a week-long conference at the Vatican Observatory, Italy. The purpose of the conference was to explore the implications of evolutionary and molecular biology for philosophical and theological issues surrounding the topic of divine action. The resulting papers form the contents of this volume. This conference was the third of five planned for the decade of the 1990s on theology, philosophy, and the natural sciences. The overarching goal is to contribute to constructive theological research as it engages the natural sciences, and to identify and critique the philosophical and theological elements that are present in ongoing scientific research. The conferences are jointly sponsored by the Vatican Observatory and the Center for Theology and the Natural Sciences (CTNS). The Observatory is in the picturesque town of Castel Gandolfo, thirty miles southeast of Rome. Since 1935 it has been the site of basic research in both observational and theoretical astronomy. CTNS, an affiliate of the Graduate Theological Union, Berkeley, California, sponsors and conducts a variety of research, teaching, and public service programs in the interdisciplinary field of theology and science.

This series of conferences grew out of the initiative of Pope John Paul II, who commissioned the Vatican Observatory to hold a major international conference at the Observatory in September, 1987. The resulting publication, *Physics, Philosophy and Theology: A Common Quest for Understanding*,[6] includes a message by the Pope on the relations between the church and the scientific communities, which was reprinted, together with nineteen responses by scientists, philosophers, and theologians, in *John Paul II on Science and Religion: Reflections on the New View From Rome*.[7]

Based on this work, George Coyne, Director of the Observatory, initiated a series of five conferences for the decade of the 1990s. The goal would be to expand upon the research agenda begun in 1987 by moving into additional areas in the physical and biological sciences. Coyne then invited CTNS to co-sponsor the decade of research. The first conference in the series resulted in the publication of *Quantum Cosmology and the Laws of Nature: Scientific Perspectives on Divine Action*.[8] The second conference produced *Chaos and Complexity: Scientific Perspectives on Divine Action*.[9] Future conferences will take up issues in the neurosciences and in quantum physics and quantum field theory.[10]

1.3 *Guiding Theme of the Series of Conferences: Scientific Perspectives on Divine Action*

A major issue in the way research is carried out in the field of theology and science regards the role science ought to play. Too often science tends to set the agenda for the research with little if any initiative taken by theology. From the beginning it was the clear intention of the steering committee that our research expand beyond this format to insure a two-way interaction between scientific and theological research programs. In order to achieve this goal we decided on a two-fold strategy. First, we searched for an overarching theological topic to thematize the entire series of conferences. The topic of divine action, or God's action in and interaction with the world, was quickly singled out as a promising candidate. Clearly it permeates the discussions of theology and science in both philosophical and systematic contexts, and it allows a variety of particular theological and philosophical issues to be pursued under a general umbrella.

Next, we organized a series of individual conferences around specific areas in the natural sciences. We chose quantum cosmology, the origin and status of the laws of nature, and foundational issues in quantum physics and quantum field theory to build on areas of research begun in *Physics, Philosophy and Theology*. We also chose chaos and complexity, biological evolution, molecular biology, genetics, creative self-organization, and the mind-brain problem to extend our research into areas treated less directly in *Physics, Philosophy and Theology*.

---

[6] Robert John Russell, William R. Stoeger, and George V. Coyne, eds. (Vatican City State: Vatican Observatory, 1988).

[7] Robert John Russell, William R. Stoeger, and George V. Coyne, eds. (Vatican City State: Vatican Observatory, 1990).

[8] Robert John Russell, Nancey Murphy, and C. J. Isham, eds. (Vatican City State: Vatican Observatory; Berkeley, Calif.: Center for Theology and the Natural Sciences, 1993).

[9] Robert John Russell, Nancey Murphy and Arthur R. Peacocke, eds. (Vatican City State: Vatican Observatory; Berkeley, Calif.: Center for Theology and the Natural Sciences, 1995).

[10] The former was held in June, 1998 and the latter is scheduled for June, 2000. Publication of the neurosciences conference papers is scheduled for 1999.

The overall research methodology for the series of conferences was considered as well. It was agreed that papers for each conference would be circulated several months in advance of the conference for critical written responses from all participants. Revisions would be read in advance of the conference to maximize the critical discussion of the papers during the event. Post-conference revisions would be carefully reviewed by the volume editors in light of these discussions. We agreed to hold regional pre-conferences in Berkeley and Rome to provide an introduction for participants to relevant technical issues in science, philosophy, and theology and to foster joint research and collaboration among participants prior to the conference. An organizing committee would guide the preparation for, and procedures of each conference, as well as the following editorial process.

Since the topic of God's action in the world was chosen as the guiding theological theme for the conferences, a brief introduction to the topic was provided in the volume on *Chaos and Complexity*.[11] The introduction included a working typology of terms used in the ongoing discussions of divine action by scholars at this series of conferences. Many of these terms also appear in the present volume in light of that typology, and the interested reader is referred to *Chaos and Complexity* for a full discussion of the typology.

1.4 *Summary of the Present Volume*

1.4.1 *Papal Documents and Response*

This volume begins with two recent papal documents and a brief commentary on them by George Coyne.

In a cordial greeting to the 1996 conference out of which this volume emerged, **John Paul II** emphasized that the search by philosophers, theologians, and scientists for a fuller understanding of life in the universe and the role of humanity is consistent with the Church's commitment to intellectual inquiry. Still, science, philosophy and theology can benefit humanity only if they are grounded in truth as found in the works of the Creator and particularly in the human person, created in God's image. They help to clarify the vision of humanity as the focus of creation's dynamism and the supreme object of God's action. Thus science and the betterment of humanity are intimately linked. He closed by reaffirming his support for this series of conferences as they contribute to the exchange between religion and science.

In the following October John Paul II addressed the Plenary Session of the Pontifical Academy of Sciences. He emphasized that the questions of the origins of life and of the nature of humankind as they are explored by the sciences are of deep interest to the Church. Will the scientific conclusions coincide with, or will they appear to contradict, Revelation? Here his response is that truth cannot contradict truth. Moreover, by understanding these results and their potential impact on humankind, the Church will be strengthened in her concern regarding issues of moral conduct. He recalled the position taken by Puis XII in 1950, that evolution and theology are not in opposition regarding humanity and its vocation. He claimed that evolution, though studied as a scientific theory, can also be interpreted philosophically in many ways, and that Revelation must be considered in the process. John Paul II also noted that in 1992, when speaking in regard to Galileo, he had emphasized that exegetes and theologians must understand science as they seek a correct interpretation of Scripture.

---

[11] *Chaos and Complexity*, 1–31; the typology appears in section 3.4, pp. 9–13.

Today the theory of evolution is no longer a mere hypothesis. It is increasingly accepted by scientists and supported by the convergence of research from a variety of fields. What then is its significance? John Paul II addressed this question by first turning to the philosophy of science, commenting on theory construction and its verification by data. A scientific theory such as evolution also draws on natural philosophy, for example, in providing an explanation for the mechanism of evolution. The result is the presence of several theories of evolution based diversely on materialist, reductionist, and spiritualist philosophies. This raises the question of the authentic role of philosophy and of theology in these discussions.

Of direct concern to the Church is the concept of human nature, particularly the *imago Dei*. Being of value *per se* and capable of forming a communion with others and with God, people cannot be subordinated as a means or instrument. John Paul II then reiterated the position stressed by Pius XII: while the human body comes from pre-existent living matter, the spiritual soul is immediately created by God. He rejected a view of mind as epiphenomenal or as emergent from matter as incompatible with the Church's view of the nature and dignity of the human person. Instead humanity represents an ontological difference from the rest of nature. Such a claim is not irreconcilable with the physical continuity pointed to by evolution, since the transition to the spiritual is not observable using scientific methods. Here philosophy is needed to account for self-awareness, moral conscience, freedom, and so on, and theology then seeks out their ultimate meaning. John Paul II concluded by reminding us that the Bible offers us an extraordinary "message of life," calling us to enter into eternal life while speaking of God as the living God.

George Coyne presents an interpretive article for John Paul II's preceding statements on evolution and the human person. Coyne sets the context by starting with the historical background of the Pope's statement which he describes in terms of three approaches to science and religion. During the seventeenth and eighteenth centuries, the Church attempted to appropriate modern science to establish a rational foundation for religious belief. Paradoxically, this led to the corruption of faith and contributed to the rise of modern atheism. The founding of the Vatican Observatory in 1891 signals the second approach. Here the Church attempted to combat anticlericalism by a vigorous, even triumphalistic, agenda. Finally, the twentieth century has seen the Church come to view science as offering rational support for theological doctrine. Coyne cites Pope Pius XII who, in 1952, took Big Bang cosmology as "bearing witness" to the contingency of the universe and to its creation by God.

Still, in three prior statements and in the current one, John Paul II has taken a new approach. Though the Galilean controversy was important to the first two approaches, what Coyne takes to be the key element is John Paul II's call for a genuine and open-ended dialogue in which science and religion, though distinct and marked by their own integrity, can contribute positively to each other. Dialogue sets the context for John Paul II's discussion of evolution.

The discussion is, in fact, mostly scientific, drawing first from research in the life sciences, next from molecular chemistry to life in the evolving universe, and finally to the possibility of early primitive life on Mars and the discovery of extra-solar planets. John Paul II stresses that, though evolution is an established scientific theory, philosophy and theology enter into its formulation, leading to several distinct and competing evolutionary world-views. Some of these—materialism, reductionism, and spiritualism—are "rejected outright." Instead a genuine dialogue begins as the papal message struggles with two views which may or may not be compatible: evolution according to science and the intervention by God to create the human soul.

Thus dialogue risks dissonance between science and religion. Revelation is given an antecedent and primary role compared with scientific discovery. Yet the religious message struggles to remain open, perhaps through a reinterpretation of what science tells us. One possibility would be the body-soul dualism taken by Pius XII. Instead John Paul shifts from an ontological to an epistemological interpretation of the appearance of what he then calls the "spiritual" in humanity. The message closes by indicating that the dialogue should continue. Here Coyne adds that in doing so we think in terms of God's continuous creation through the process of evolution. Rather than intervening, God gives the world freedom to evolve and participates in the process through love. Perhaps this approach can preserve what is special about the emergence of spirit without resort to interventionism.

1.4.2 Section One: Scientific Background

Two papers in section one provide an overview of the scientific background of evolutionary and molecular biology, including the history of the field since Darwin, the facts, processes, and mechanisms of evolution (Ayala), and hominid evolution (Cela-Conde). The third paper discusses the possibility of the evolution of extra-terrestrial life (Chela-Flores).

According to **Francisco Ayala**, the evolution of organisms, that is, their descent with modification from a common origin, is at the core of biology. Though evolution is universally accepted by biologists, its mechanisms are still actively investigated and debated by scientists. Darwin's explanation was essentially correct but incomplete, awaiting the discoveries and power of genetics and molecular biology. Ayala then distinguishes between two questions: whether and how evolution happened.

Ayala briefly traces historical sources and then focuses on Darwin, who proposed natural selection to account for the adaptive organization of living creatures and their apparent purpose or design. Missing in Darwin's work was a theory of inheritance that would account for the preservation of variations on which selection could act. Mendelian genetics eventually provided the "missing link." In addition, Weismann's germ-plasma theory helped counter the Lamarckian alternative to Darwin and contributed to the neo-Darwinian theory that emerged out of the nineteenth century. Further progress came from Dobzhansky in the 1930s. In 1953, Watson and Crick discovered the structure of DNA. In 1968, Kimura's work on "molecular clocks" made possible a reconstruction of the evolutionary history of life with its many branchings. Finally, the recent techniques of DNA cloning and sequencing have provided additional knowledge about evolution.

Next Ayala discusses three related issues: the fact of evolution, the details of evolutionary history in which lineages split, and the mechanisms by which evolution occurs. The first, that organisms are related by common descent with modification, is both fundamental to evolution and heavily supported by the evidence. The second and third are mixed, with some conclusions well established and others less so. Before delving into the details, Ayala briefly comments on the mix of responses to evolution from the religious communities. It can seem incompatible to those holding to a literal interpretation of Genesis, the immortality of the soul, or humans created in the image of God. To others, God is seen as operating through intermediate, natural causes, including those involved in evolution. Here Ayala cites Pope John Paul II's recent comments on evolutionary biology. Ayala then turns to a detailed exposition of the evidence for evolution, drawing on paleontology, comparative anatomy, biogeography, embryology, biochemistry, molecular genetics, and other fields. He focuses on the question of speciation, including models such as adaptive radiation for how reproductive isolation arises. After giving a reconstruction of evolutionary history, Ayala concludes his essay by discussing gradual and punctuated evolution, DNA and protein evolution, the molecular clock of evolution, and human evolution.

It is **Camilo Cela-Conde**'s central claim that "no straight line can be drawn from our ancestors to the modern human species." Instead evolution depicts a much more complex picture of human evolution. A basic question is that of taxonomy: how are we to define a hominid? One way is by discovering an exclusive trait that might serve to distinguish hominids from other primates. Cela-Conde discusses but rejects such candidates as bipedalism, a large brain, an articulated language, a large coefficient of encephalization, the ability to create tools and thus culture, etc. He then takes a different approach, describing in some detail the variety of species that are considered as belonging to the hominid family. He begins with the appearance of early hominids some 4.4 million years ago and points out the many subtleties involved in attempting to classify them. He describes the complex issues surrounding the evolution of *Homo erectus*, *Homo neanderthalensis*, and finally *Homo sapiens*, citing arguments against a direct link between Neanderthals and morphologically modern humans. He concludes his essay with a careful discussion of morphological and genetic studies of the origin of human beings, including two opposite models: multiregional transition and mitochondrial Eve. Although he disagrees with the widespread idea that all humankind shares one ancestral grandmother, he does support the theory of the "out-of-Africa" spread of modern humans.

The focus of **Julian Chela-Flores**' paper is the possibility of the evolution of life elsewhere in our solar system. He first reviews Big Bang cosmology, including its modifications by Guth and Linde. Next he turns to the origin of life on Earth from the 1920s to the present. Although scientists view organic matter as inexorably self-organized according to the laws of physics and chemistry, the complete pathway from the inanimate to life on Earth has not been reproduced experimentally, nor has the importation of organic molecules from space been ruled out. Meanwhile research is now underway in exobiology and bioastronomy via the ongoing space missions. Issues include cross-contamination of either Earth, Mars, or Europa, comparative planetology, the search for extraterrestrial homochirality (SETH), the search for extraterrestrial eukaryotes (SETE), and, since the 1960s, the search for extraterrestrial intelligence (SETI). He concludes this section with speculations on the future of evolution on earth.

Next, Chela-Flores describes recent topics including chemical evolution in the universe, the pathways from precursors to biomolecules, modern taxonomy, the terminology for single-celled organisms, and the evolution of prokaryotic cells in the Precambrian period. He then discusses the evolution of eukaryotes, including the role of oxygen and iron in their first appearance, and the identification of eukaryotes that are morphologically similar to prokaryotes. Next Chela-Flores takes us from eukaryogenesis to the appearance of intelligent life on Earth. Here he presses his case for the inevitable increase of complexity in the transition from bacteria to eukarya. Physics and chemistry imply an "imperative" appearance of life during cosmic evolution which he formulates as a bold, but in principle testable, hypothesis: "once the living process has started, then the cellular plans, or blueprints, are also of universal validity." In short, prokaryotes lead to eukaryotes, and they do so *universally*. Provided that planets have the appropriate volatiles (particularly water and oxygen), Chela-Flores argues that not only life, but eukaryogenesis, is bound to occur. Within the next two decades, a new generation of space missions could test his hypothesis. Moreover, the hypothesis bears on the question whether these missions should search for Earth-like life or something entirely different. Chela-Flores gives various responses to this question, including the relevance of SETE to SETI and the significance of the discoveries of the Murchison and Allan Hills meteorites that originated on Mars.

In closing, Chela-Flores maintains that there is a second environment in our solar system, the Jovian satellite Europa, in which the eukaryogenesis hypothesis may be tested. He first describes other possible sites for extremophiles and other microorganisms, including the atmospheres of Europa, Io, Titan, and Triton, and possible hot springs at the bottom of Europa's (putative) ocean. Then he identifies parameters that may characterize the degree of evolution of Europan biota both at the ice surface and its ocean. He concludes again that a space mission could test these ideas in the near future.

1.4.3 *Section Two: Evolution and Divine Action*

The six papers in section two focus broadly on the question of divine action in light of evolutionary and molecular biology. The first two papers offer philosophical analyses of teleology in light of biology by a scientist (Ayala) and by a theologian (Wildman). The next papers offer assessments of the evidence for teleology from two scientific perspectives (Davies and Stoeger). The final papers develop theological arguments regarding divine action in light of evolution, the first focusing on special providence (Russell), the second working within process theism (Birch).

According to **Francisco Ayala**, Darwin's achievement was to complete the Copernican revolution; biology could now be explained in terms of universal, immanent, natural laws without resorting explicitly to a Creator. The result was to bring biological organisms into the realm of science. Many theologians have seen no contradiction between Darwin and Christian faith, both at the time of Darwin's writings and in the century since. Natural selection is creative: in a "sieve-like" way it retains rare but useful genes. But natural selection is not creative in the Christian sense of *creatio ex nihilo*. Instead it is like a painter mixing pigments on a canvas. It is a non-random process that promotes adaptation, that is, combinations useful to the organisms. By proceeding "stepwise," it produces combinations of genes that otherwise would be highly improbable. It lacks foresight or a preconceived plan, being the consequence of differential reproduction. Thus, though it has the appearance of purposefulness, it does not anticipate the environment of the future. It accounts for the "design" of organisms, since adaptive variations increase relative survival and reproduction. Aquinas and Paley understood that purely random processes will not account for biological nature; but they could not recognize, as Darwin saw, that these processes could be "oriented" by the *functional* design they convey to organisms. In this sense they are not entirely random. Chance is an integral part of evolution, but its random character is counteracted by natural selection which preserves what is useful and eliminates the harmful. Without mutation, evolution could not happen. Without natural selection, mutations would bring disorganization and extinction. Thanks to Darwin we can view the process of evolution as creative though not conscious. The biological world is the result of natural processes governed by natural laws, and this vision has forever changed how we perceive ourselves and our place in the universe.

Ayala next develops a complex conception of teleology. An object or behavior is teleological when it gives evidence of design or appears to be directed toward certain ends. Features of organisms, such as the wings of a bird, are teleological when they are adaptations which originate by natural selection and when they function to increase the reproductive success of their carriers. Inanimate objects and processes, such as a salt molecule or a mountain, are not teleological since they are not directed towards specific ends. Teleological explanations, in turn, account for the existence of teleological features. Ayala then distinguishes between those actions or objects which are purposeful and those which are not. The former exhibit artificial or external teleology. Those resulting from actions which are not purposeful exhibit natural or internal teleology. Bounded natural teleology, in turn, describes an end-state reached in spite of environmental fluctuations, whereas unbounded teleology refers to an end-state that is not specifically predetermined, but results from one of several available alternatives. The adaptations of organisms are teleological in this indeterminate sense. Finally, teleological explanations are fully compatible with efficient causal explanations, and in some cases both are required.

With this in mind Ayala argues that Darwin's theory of evolution and his explanation of design are no more "anti-Christian" than are Newton's laws of motion. Divine action should not be sought in terms of gaps in the scientific account of nature—although the origin of the universe will always remain outside the bounds of scientific explanation.

The essay concludes by acknowledging the success of science as a way of knowing, a major source of economic growth in the United States, a bringer of essential technologies, and a mode of accumulating knowledge that spans generations. Still, science is not the only way of knowing; we also have the arts, common sense, religion, and so on, all of which far predate science. Science is universal in scope but hopelessly incomplete. Much of what is left out, such as meaning and value, may be considered more important than what science includes.

But can we make a connection between the appearance of purposes or ends in nature and the reality of divine action by arguing that such ends indicate genuine teleology and that such teleology is a mode of God's action? **Wesley Wildman** calls this position "the teleological argument for divine action" and provides an extensive analysis of its problematic status. Its more aggressive form as a design argument, as given by William Paley and Alfred Russel Wallace, has been undermined by evolutionary biology, but more modest forms are still possible. Wildman claims that no detailed, supportive argument from biology to theories of divine action can be given, nor can evolution destroy such theories, but biology can influence them to some degree. The principle reason is the "metaphysical ambiguity" which effects each of three major steps in the argument, though it is generally left unnoticed.

Before proceeding, Wildman distinguishes between apparent and real ends, and between open-ended and closed-ended processes. In addition, given the dubious aspects of Aristotle's use of final causes and the equally unnecessary abandonment of teleology in contemporary causal explanation, he includes a four-fold schema regarding teleology. 1) Against Dawkins, who argues that ends are only apparent, Wildman cites Monod's views about teleonomy and chaos. 2) Given that complex systems are open-ended, he asks whether one specific end can be achieved, and by an intentional agent? 3) Moreover, are high-level characteristics of living systems due to the complex external arrangement of their parts, or to the emergence of genuine internal relations? 4) Is teleology expressed in the laws of nature, in chance, or in some basic constituent of nature like mind?

Now Wildman turns to the first stage in the argument. Given that the appearance of ends in nature is ubiquitous, can we establish that real purposes give rise to them? And preliminary to this, are apparent ends in nature only merely apparent? According to Wildman, modern biology has produced the strongest possible reason for answering these affirmatively with its use of efficient causal explanations of apparent ends. Still the conclusion depends on two points which Wildman carefully criticizes: a principle of metaphysical minimalism and a claim that all ends in nature, outside those achieved by agents, can exhaustively be explained solely by means of efficient causes. His position includes scientific and philosophical arguments about the sufficiency of efficient causal explanations.

The second stage requires a metaphysical bridge between real ends in nature, as in the preceding stage, and broad teleological principles which can be connected with a theory of divine action. The problem here is that some teleologies are not amendable to such theories. Foremost is Aristotle's metaphysics, with its unmoved prime mover. Others include the Chinese concept of *li*, the Buddhist view of a purposive nature without God, and theistic mystical theology.

In the third stage Wildman traces the links between metaphysical contextualizations of teleology and theories of divine action, stressing the metaphysical ambiguity which complicates these links. If the locus of the teleological principle is natural law, divine action can only involve the universal determination of natural possibilities and the ontological grounding of nature. If the locus includes chance, divine action can be expressed more directly. If it also includes the constituents of nature, God may offer the initial aims to actual occasions, or the material conditions for the emergence of self-organizing systems.

According to Wildman, then, the teleological argument for divine action is not easily established. There is no unbroken chain of implications from apparent ends in nature to real ends to fundamental teleological principles to the modes of divine action. Additional premises are needed to connect the chain, and none of these is furnished by biological evolution. Moreover, there are profound teleological visions that are antagonistic toward divine action and are equally well supported by evolution. The failure of the teleological argument is located in its underlying metaphysical ambiguity. On the other hand, apparent purposes in nature are not incompatible with teleological theories of divine action; indeed, the implications run more smoothly in this direction. In closing, Wildman offers a final conjecture: if one's premise is that the universe is meaningful, then one is led to affirm that the universe has an overarching teleological sweep. Alternatively, those popular writers on biology who avoid postulating a fundamental teleological principle must either assume that the cosmos is absurd or refuse to consider the implications of their premise.

Paul Davies offers us a modified version of the uniformitarian view of divine action. In selecting the laws of nature, God chooses specific laws which allow not only for chance events but also for the genuine emergence of complexity. He claims that the full gamut of natural complexity cannot be accounted for by neo-Darwinism, relativity, and quantum mechanics; one must also consider nature's inherent powers of self-organization based on, though not reducible to, these laws. Still the emergence of complexity does not require special interventionist divine action.

Davies begins by classifying divine action into three types: interventionist, non-interventionist, and uniform. He rejects the first, since it reduces God to nature and involves theological contradictions. The second is a new possibility which appeals to quantum indeterminism and bottom-up causality or to the mind-body problem and top-down causality. Davies considers several possible objections and responds to them before turning to his own view, a modified form of uniform divine action. This emphasizes God's continuing role in creating the universe each moment though without bringing about particular events which nature "on its own" would not have produced. Davies illustrates this via the game of chess, in which the end of a given game is determined both by the rules and by the specific sequence of moves chosen by each player. Thus God selects the laws of nature; being inherently statistical, they allow for chance events at the quantum or chaos levels as well as for human agency. God need not violate these laws in order to act, and there is room for human freedom and even for inanimate systems to explore novel pathways.

The existence of these very specific laws raises the question of cosmological design. Davies acknowledges that "anthropic" arguments like his might be countered by a cosmic Darwinism, such as the "many worlds" view provided by inflationary cosmology, but he gives several reasons why he rejects these accounts. He then argues that quasi-universal organizing principles will be found to describe self-organizing, complex systems. They will complement the laws of physics, but they would not be reducible to or derivable from physics, nor would they refer to a mystical or vitalistic addition to them.

Davies sees his view of divine action as going beyond ordinary uniformitarianism. Chance in nature is God's bestowal of openness, freedom, and the natural capacity for creativity. The emergence of what he calls the "order of complexity" is a genuine surprise, arising out of the "order of simplicity" described by the laws of physics. He calls this "teleology without teleology." The acid test, according to Davies, is whether we are alone in the universe. If the general trend of matter toward mind and culture is written into the laws of nature, though its form depends on the details of evolution, we would expect that life abounds in the universe. This accounts for the importance of the SETI project. Finally, Davies is open to the possibility of combining his view with a non-interventionist account of divine action.

In his final section, Davies addresses biologists who, he expects, will find the concept of "teleology without teleology" favorable for two reasons. First, biologists have already incorporated elements of self-organization and emergent complexity into the neo-Darwinian account. Second, some biologists see evidence of complexifying trends in biological phenomena. Finally, the theological interpretation he advances would in no way be obligatory on others.

Is there an immanent directionality in nature? If so, can science discover it or must we turn to philosophy and theology to recognize it and its significance? According to **William Stoeger**, some scientists and philosophers conclude from the variety and interrelatedness of nature that there must be a universal plan to it all. Many even assume there must be overarching holistic laws of nature which constrain the universe to behave in ways which clearly manifest purposes or ends. Many other scientists and philosophers, however, report they have found no evidence of an immanent directionality in nature. Often it is even presupposed that there is no overall directionality—much less teleology—in evolution, that complete randomness and uncertainty prevail, presided over only by the laws of physics, chemistry, and biology.

Stoeger's aim is to show first that there is a directionality, perhaps even a teleology, immanent in nature that can be discovered through the natural sciences as they study the emergence of physical and biological structure, complexity, life, and mind. He intentionally stresses this point since so many scientists deny it. Stoeger, however, believes that the discoveries of the natural sciences can be harmonized with an adequate understanding of God's creative action in the world without postulating holistic laws or teleological mechanisms beyond those described by the sciences. The evidence at the scientific level also seems to rule out the necessity, and even the possibility, for divine intervention to complement the principles and processes accessible to science. Finally theology can refer to divine action and teleology, but the results of science should place constraints on the way it describes them. Moreover the laws of nature, as they function within creation, are one of the key ways in which God acts in the universe.

First, Stoeger describes the epistemological and metaphysical assumptions which underlie his approach. He then turns to a lengthy discussion of the scientific account of directionality drawing on such areas as cosmology, astronomy, chemistry, geophysics, biology, self-organization, and Boolean networks. The global cosmic directionality is given by the expanding, cooling universe. More specific sequential focusing of directionality occurs as galaxies and stars form, and they in turn provide stellar and planetary environments in which chemical and biological complexifications may arise. Stoeger goes into considerable scientific detail to show that a definite directionality is established, maintained, and narrowed in the process. Randomness does play an essential role, as do catastrophes—enabling the emergence of variety and diversity—but always within the larger framework of order and regularity.

Thus, the directionality inherent in the evolutionary process is seen in terms of its hierarchically nested character and the way this reflects the structure of the universe as a whole. This means that, for a particular configuration at a given moment, only a certain range of configurations at successive moments are possible. This hierarchical nestedness means further that these directed configurations occur on all levels—with very general types of directionality being characteristic of more global levels (those of the observable universe, or of our own galaxy) and more focused specific directionalities arising on more local levels (those of a given planet, a given organism, or community of organisms).

But in what sense does this directionality constitute a teleology? Stoeger argues that a system can be teleological without necessarily involving a blueprint for a final product. It need only move towards realizing possibilities in an ordered way, partly under the evolving conditions of its ecological environment. The realization of any given possibility typically presupposes the prior realization of other possibilities. The directionalities in nature are flexible and pliable, not fixed. Nor do they indicate consciously directed intention, at least from the point of view of the natural sciences—nor do they rule it out, and this is partly due to their own limitations. Stoeger considers a similar question regarding philosophy before turning to theology. Here he claims that the Christian tradition inevitably involves conscious divine purpose in creation, including the overall process of evolution. He concludes with some thoughts on the difference between "end-resulting," "end-directed," and "goal-seeking" forms of teleology.

Robert Russell works within the context of theistic evolution: biological evolution is God's way of creating life. God is both the transcendent source *ex nihilo* of the universe as a whole, including its sheer existence at each moment and the laws of nature, and the immanent Creator of all physical and biological complexity, acting continuously in, with, under, and through the processes of nature. But can we press the case further and think of God's *special* providence in nature? And can we do so without viewing God's action as an intervention into these processes and a violation of the laws of nature?

To many theologians, the connection between special providence and intervention has seemed unavoidable, leaving them with a forced option. 1) Liberals, attempting to avoid interventionism, reduce special providence to our *subjective* response to what is simply God's uniform action. 2) Conservatives support *objective* special providence and accept its interventionist implications. The purpose of Russell's paper is to move us beyond these options to a new approach: a non-interventionist understanding of objective special providence. This is only possible theologically if nature, at some level, can be interpreted philosophically as ontologically indeterministic in light of contemporary science. Russell's claim is that quantum mechanics provides one such possibility. Moreover, since quantum mechanics underlies the processes of genetic mutation, and since mutation together with natural selection constitute the central features of the neo-Darwinian understanding of evolution, then we can view evolution theologically as genuinely open to objective special providence without being forced into interventionism.

In section two, Russell claims that his project is neither a form of natural theology, of physico-theology, nor an argument from design. Instead it is part of a general constructive trinitarian theology pursued as *fides quaerens intellectum*. He suggests why a non-interventionist view of objective special providence should be important theologically. He argues for an indeterministic interpretation of quantum physics. Finally he address three scientific issues regarding the role of quantum mechanics in genetic mutation and the role of genetic variation in biological evolution.

Section three reviews the history of the project, beginning with the writings of Karl Heim and William Pollard in the 1950s and including recent works by Arthur Peacocke and John Polkinghorne. One key question is whether God acts in all quantum events (as Nancey Murphy claims) or merely in some (as Tom Tracy suggests). Another regards the problem of theodicy when God is taken as acting throughout evolution. Russell closes this section by reflecting on issues raised by these authors.

Section four addresses three caveats. First, Russell's hypothesis is not meant as an explanation of "how" God acts, but merely one domain where the effects of God's special action might occur. Second, it is not meant as either an epistemic or an ontological "gaps" argument. Still quantum mechanics may one day be replaced. Russell's methodology is intentionally designed to handle "gaps" like this by incorporating implications from physics and philosophy into constructive theology while keeping theology open to changes in these implications. Third, Russell's argument is not meant to exclude divine action at other levels in nature or "top-down" and "whole-part" approaches. However, these are unintelligible *without* intervention until the evolution of sufficiently complex phenomena. This leaves a bottom-up approach via quantum mechanics the most reasonable option for the early sweep of evolution.

Section five engages two final challenges. First, "chance" in evolution also challenges the possibility of God achieving a *future* purpose by acting *in the present.* Russell responds that God acts not by foreseeing the future from the present but by eternally *seeing* the future in its own present. In passing Russell comments on a potential conflict with the implications of special relativity regarding this claim. The second challenge is theodicy. Russell notes that suffering, disease and death are conditions required for the evolution of freedom and moral agency. He suggests that we relocate the question of God's action in evolution to a theology of redemption and eschatology if we are to address adequately the problem of theodicy.

According to **Charles Birch**, most biologists now accept neo-Darwinism as the methodological basis of their understanding of biological evolution, supplemented by the concept of self-organization. 1) Research areas in neo-Darwinian evolutionary theory include differences in the object of selection, the problem of deleterious mutations, subtleties concerning the role of chance, the genetic assimilation of environmental effects, influences on natural selection by modification of the environment, and both neutral and punctuated theories of evolution. 2) Self-organization refers to the production of complex order *without* a centralizing agency and is usually invoked to explain the evolution of the pre-biotic world. It may also help to explain complex processes in developmental biology, including cell differentiation. Stuart Kaufman applies the mathematics of chaos theory to some aspects of biological evolution without appealing to natural selection.

Birch acknowledges that both neo-Darwinism and self-organization draw on strictly mechanistic models, but insists that this does not imply that biological entities are in all respects machines. On the one hand, a mechanistic analysis seems to provide all that we need for modern biology from a purely physical perspective. On the other hand, it has little, if anything, to say about the mental experience of biological creatures, namely their experience of freedom, choice, and in the human case at least, self-determination. Such analysis has even less to say about the possibility of divine action in the living world. Birch believes this problem stems from the fact that the organism is treated methodologically by neo-Darwinists as an object and not as a subject. Compounding the problem, the mechanistic methodology has led many Darwinists to argue for an underlying mechanistic metaphysics. As a result, the evolution of mind and consciousness, and the functions which they uniquely serve in nature, have remained an enigma for Darwinism.

In its stead Birch suggests a metaphysics for biological organisms which includes their mental as well as physical aspects. The proposal is drawn from the philosophy of Whitehead, in which all individual entities from protons to people are considered to be subjects. Biological evolution is not simply a matter of change in the external relations of objects, but also one of change in the internal relations of subjects. This includes a subject's relation to its immediate past, analogous to memory, and its relation to its possible future, analogous to anticipation. There is an ever-present urge in life which can be called purpose. Process thought thus posits mentality or experience in some form as an aspect of nature down to the level of fundamental particles. Only at the higher levels of complexity is experience actually conscious. Birch cites David Chalmers, Galen Strawson, and Henry Stapp as non-process scholars who support the validity of experience as universal in nature. The key argument is that mentality did not emerge from the non-mental at some point in the evolutionary sequence.

He then contrasts process philosophy with two alternatives: emergence and reductionism. Emergence involves a category mistake (that the "mental" emerges from the "physical") as well as a scientific problem (drawing the line between sentience and non-sentience). Reductionism, although it is fruitful and represents most scientific analysis, is inadequate because it cannot account for the fact that the whole has properties which the parts do not; moreover, the parts become qualitatively different by being parts of the whole.

In its place, Birch claims that the best answer to the whole-part problem and the strongest argument for rejecting reductionism is the doctrine of internal relations. This process approach is compatible with a lower-to-upper causality, and it has important implications for scientific research, too, offering support for the idea of top-down causation. Theologically, the potentiality of the universe is held in the mind of God. Divine potentiality becomes concrete reality in the universe by means of persuasive love. God interacts with individual entities in three ways. First, the future is open, and God persuasively confronts entities with creative and saving possibilities for their future. Next, the entities of the world are created by God and respond to God's feelings for the world. Finally, God responds to the world with infinite passion, taking actual entities into the divine life.

1.4.4 *Section Three: Religious Interpretations of Biological Themes*

Of the eight papers in section three, the first analyzes arguments drawn uncritically from evolution to support atheism or a science-based religion and offers instead a kenotic-moral theistic interpretation (Ellis). The second revisits Darwin's historical relation to natural theology, evaluates the way metaphors shape Darwinism, and offers a new metaphor from a feminist perspective (Clifford). The third views evolution from a naturalist perspective which takes religion as a way of life (Drees). The remaining five papers present constructive theological arguments involving biological issues. These include the challenge of biocultural evolution and the concept of the created co-creator (Hefner), problems of emergence, propensities, pain, death, and Christology (Peacocke), original sin and saving grace from a trinitarian approach (Edwards), divine kenosis and the power of the future from a process perspective (Haught), and a comparison of models of God in light of philosophical issues raised by evolution (Barbour).

George Ellis analyzes arguments by a number of contemporary scientists that either support atheism or offer a science-based religion. He claims that they are based on scientifically unjustified assumptions and rely on rhetorical or emotional appeal. They ignore the limited scope and method of science, contain an implicit metaphysical agenda, rely on the authority of science while addressing issues outside its scope, and occasionally misrepresent or ignore opposing views.

First, Ellis reminds us that scientific theories are provisional, open, limited in scope, partially supported by evidence, and inherently incomplete. Cosmology in particular takes for granted the laws of physics, but it cannot explain why they exist, why the universe exists, or whether there is an underlying purpose and meaning to the universe. Nor can science provide a foundation for values, although values are essential to the conduct of science. Issues such as the existence of God lie forever outside the competence of science to adjudicate, although the weight of data and experience can influence one's opinion. Ellis acknowledges that the argument from special design has been undermined by evolutionary theory. His concern, however, is with those who construct a "scientific religion" out of either a physically based metaphysics or a scientifically motivated system of values, and who deny the fact that the metaphysical interpretation of science is ambiguous and that the epistemic and ethical scope of science is limited.

To make his case, Ellis turns to detailed critiques of specific authors. He admires the elegance of Carl Sagan when he writes about science, but is concerned that Sagan goes on to deploy a "naturalistic religion" whose metaphysical basis reaches beyond what science can warrant. Such fundamental issues as the existence of the universe and the laws of physics are taken for granted, and Sagan's conclusion about the relative unimportance of human life in the universe is argued using emotion, not logic. Similarly Richard Dawkins ignores these fundamental issues and takes for granted the conditions that make evolution possible. E. O. Wilson attempts to give an exhaustive account of moral behavior as mere genetic programming but ignores the fact that morality presupposes voluntary, intentional action. Daniel Dennett and Jacques Monod ignore the metaphysical issues regarding the existence of the universe, and Monod proposes that science can be a source of universal ethics without recognizing the inadequacies of such an ethics. Finally Peter Atkins dismisses anything outside the scope of science, making reductionism into a dogma and ignoring the metaphysical ambiguity of science. The pay-off for these writers is the hope to achieve absolute certainty and to receive the privileged status of scientific "high priest." Ironically, their exaggerations serve to foster anti-scientific views in the general public. These authors also fail to alert their readers to the speculative status of the scientific theories being considered, such as chaotic inflationary cosmology, cultural "memes," and the arrow of time problem. Ellis is also critical of the specific strategies employed to undercut the opponent's views, such as explaining away religion in functional and evolutionary terms, or appealing to emotion or rhetoric.

In contrast to this, some scientists who deploy scientifically based world views are less dogmatic and more tentative in their approach. Their conclusion is self-critical atheism rather than dogmatic atheism—a position which Ellis finds much more reasonable. He also admits that taking religious experience seriously is problematic, since there is too much data, too much conflict among the data, and much of it involves manifest evil wrought in the name of religion—and here he agrees with the writers he has been criticizing. This leads Ellis to reflect on how the selection of data and the construction and testing of theories in science occur. Perhaps a lesson from the proponents of scientific religion is that tests in religion need to be more seriously developed and widely acknowledged. Can morality and ethics be judged by their fruits: are they life-giving or death-dealing? Ellis believes we might use this process to evaluate the broad spectrum of religions as well as particular sects.

Ellis closes by describing three theories which take the data seriously. 1) The kenotic moral-theistic position attributes an ethical under-pinning to the universe derived from and expressing the self-emptying nature of God. It is reflected in such exemplars as Mahatma Gandhi and Martin Luther King, Jr. With this view one can cut across the lines of religious traditions to retrieve selectively those data and practices that are kenotic. One can also account for the apparent counter-evidence to God's existence, namely evil, as well as the metaphysical ambiguity of nature, since kenosis requires free will, and free will requires genuine metaphysical ambiguity and makes evil deeds a possibility. Moreover, evolution is also entailed; special creation would make the argument from design overwhelming and faith empty. Of course, these insights are not meant as a theistic proof; the proposal provides a viable viewpoint which might or might not be true, and this is, once again, coherent with Ellis' fundamental claim about metaphysical ambiguity.

The second theory is self-critical atheism: conflicting religious data suggest that none can be correct. This view differs from the dogmatic atheism of scientific religions, being open to evidence and aware of both metaphysical ambiguity and the limits of the scientific method. Finally, the evidence may lead to agnosticism. Ellis cautions that neither atheism nor agnosticism provide a solid basis for ethics. Ultimately the choice between these theories is personal, but truth is not irrelevant, logic plays a role, and only indubitable certainty is unattainable. This also means that by ignoring metaphysical and epistemological complications, the arguments for scientific religion bear the marks of pseudo-science rather than true science. Ellis hopes his paper clears the way for scientific world views that are less dogmatic and more open to genuine interaction with alternative views.

Anne Clifford examines Darwin's *The Origin of Species* in relation to nineteenth-century British natural theology. Though the latter was considered a form of science it actually offered a union between science and Christian belief in a creator. Its primary text was nature, not Genesis, and it attempted to provide evidence from nature for God's sovereignty and purposeful design. Clifford warns us not to let the hegemony that Darwin's theory now enjoys undercut our interest in natural theology, partly because we would not fully appreciate what Darwin's revolution accomplished. She sets out to trace that accomplishment, being mindful of the way language in both science and theology, with its metaphorical character, shapes our claims about reality.

Her first move is to challenge the "warfare" model of the relationship between Christianity and Darwin's theory fostered by Andrew Dickson White and John Draper. Wilberforce's attack on evolution was actually based primarily on scientific grounds, not on concerns about biblical revelation. He accepted natural selection as a process that weeded out the unfit within a species, but he felt Darwin had not provided sufficient evidence for the evolution of new species. Wilberforce's argument drew implicitly on Francis Bacon's earlier distinction between the book of revelation and the book of nature. Though God was the author of both books, the distinction provided scientists freedom from forcing their results to conform to biblical texts. Darwin too drew on the two-books tradition and on a close reading of William Paley, who argued from nature to an intelligent designer. Paley went further than Bacon, though, by discussing nature's purpose and by moving from purpose to a personal designer and thus to a personal God. He rejected randomness in nature as well as the extinction of species. The *Bridgewater Treatises* continued this argument, insisting on the fixity of nature and on divine sovereignty which maintains nature and natural laws. These are the actual positions that Darwin's theory of natural selection would reject.

Data gathered from his voyage on the Beagle triggered Darwin's "conversion" from natural theology to his theory of natural selection as an account of the variety and mutability of species. Recent discoveries in geology enhanced his account, including the ancient age of the earth and the possibility of sequencing the fossil record. Also contributing was Darwin's knowledge of animal breeding as well as Malthus' work on population and resources, with its focus on the struggle for existence. Darwin's theory of natural selection and its theme of the survival of the fittest broke with natural theology not only in the concept of God as special designer of each separate species, including their direct creation and their immutability, but also with the benevolence of God. Natural theologians, it seems, had been particularly blind to the abundance of suffering and death in nature.

Clifford then analyzes the role of metaphor in science, drawing on the writings of Janet Soskice, Paul Ricoeur, and Sallie McFague. She focuses on two of Darwin's key metaphors: "the origin of species" and "natural selection." Darwin's theory in effect shifted the meaning of "origins" by describing the emergence of new species while bracketing the question of the origin of life as such. He also transformed the meaning of species; rather than fixed and discrete, they came to be seen as fluid, possessing the capacity to evolve. Darwin's metaphor, "natural selection," combines meanings drawn from animal breeding by humans and from nature in the wild. It suggests that nature "chooses" and, though Darwin rejected vitalism, he has been read as deifying nature. Clifford also points out that Darwin considered his theory compatible with belief in God, though his personal position seems to shift from belief to agnosticism.

According to Clifford, then, Darwin did not intend a warfare against Christianity, only against natural theology, and here only in the form of a highly rationalistic Christian theism coupled to a limited body of scientific data. He challenged Paley's watchmaker analogy that assumed a God of radical sovereignty and a passive and static world. What might we find to replace it? McFague proposes the metaphor of the universe as God's body. Clifford modifies this by suggesting the metaphor of a mother giving birth. It brings together in dynamic tension the reproductive and evolutionary character of nature with the biblical doctrine of God as creator. It is panentheistic, rather than pantheistic, and is, according to Elizabeth Johnson, the "paradigm without equal," drawing on a wealth of biblical texts for God's relation to the world. Finally it is compatible with Darwin's rejection of God as designer, the immutability of species, and it takes up his concern to acknowledge the extent of suffering in nature.

According to **Willem Drees**, at least three issues arise from an evolutionary view of nature. One is the challenge to a literalist understanding of Genesis. Another is that evolution may leave no room for divine action in the world. Finally, evolution can radically modify our understanding of human nature and morality. The latter is the focus of Drees' paper. Rather than seeking an alternative to, or a modification of, evolutionary ideas, Drees intends to stay as close as possible to insights offered and concepts developed in the sciences. He call his position "naturalism" and asks what the consequences are if a naturalist view is correct. His central theses are: 1) upon a sufficiently subtle view of science, evolution can do justice to the richness of experience and of morality, but 2) not to the cognitive concerns of religion; nevertheless 3) there is still room in a naturalist view for religion as a way of life and as a response to limit questions concerning the scientific framework.

Drees first distinguishes between soft or nonreductive naturalism, whose context is ordinary human experience and language, and hard or reductive naturalism. The latter includes epistemological naturalism (a universal application of the scientific method without an ontological commitment) and ontological naturalism (Drees' position). Within ontological naturalism there are three varieties: reductive materialists, who hold for type-type identity, nonreductive materialists, who opt for token-token identity (Drees' position), and eliminative materialists, who would reduce away the higher level of discourse.

In Drees' view, the natural world is all that we know about and interact with; no supernatural realm shows up within the world. All entities are made of the same constituents. Still naturalism (or physicalism) can be non-reductive in the sense that higher level properties may require their own concepts and explanatory schemes. Evolutionary explanations are primarily functional. He argues that such a naturalism need not be atheistic. Instead, physics and cosmology form the boundary of the natural sciences and raise speculative, limit questions about the naturalist view as such, questions about which naturalism can remain agnostic. The integrity, coherence, and completeness of reality as described by science does not imply its self-sufficiency. Contrary to Peter Atkins, Drees sees religious accounts in which the natural world as a whole is dependent on a transcendent Creator as consistent with, though not required by, naturalism. What Drees rejects is a view of God as altering the laws of nature or as acting within the contingencies of nature since, again, nature is complete and the integrity of nature is affirmed.

Next Drees turns to evolutionary explanations of morality. If morality, such as pro-social behavior, is given an evolutionary explanation, can it still be considered "moral"? Drees first argues that evolutionary naturalism as a whole should not be dismissed because of the claims made by those whom Daniel Dennett calls "greedy reductionists." Instead Richard Alexander, David Sloan Wilson, Elliot Sober, Michael Ruse and Francisco Ayala give serious consideration to the importance of cultural and mental aspects in the evolutionary explanation of morality. He describes four reasons why such accounts need not undercut the validity of seeing morality as genuinely "moral." For example, sociobiology undermines the claim that values originate in a supernatural source, but people are still free to choose from among competing values. Morality can go beyond our emotions, as E. O. Wilson argues, and the contingencies of our evolutionary history, as Michael Ruse proposes, to reflect a genuine distinction between is and ought.

But what happens to religion when it becomes the object of scientific study and explanation? Whereas morality and experience seem to survive an evolutionary understanding, the implications for religion are more serious. In effect, a functional and immanent understanding of morality need not be as problematic to moral persons as a similar understanding of religious language may be to believers, since religious language typically refers to transcendent realities. Thus, some of the fears by believers seem warranted. However, Drees holds that the grounds for accepting a naturalistic evolutionary view of reality, including ourselves, are strong. Hence, rather than backing away when a conflict threatens, he prefers to reflect on the options for religion within an evolutionary framework.

According to Drees, naturalism rules out objective reference to divine action in the world and it offers an evolutionary account of how such ideas arose. Thus naturalism renders their cognitive content "extremely unlikely" without claiming absolute proof. Religious traditions can be studied as complex entities and ways of life, each within its own environment. They embody regulative ideals and forms of worship, and they undergird moral and spiritual commitments. Though their cognitive claims may need revision, religions confront and challenges us with these ideals and values, offering a vision for a better world. Moreover, they encourage us to raise limit questions which naturalism alone cannot answer and they in turn offer answers to such questions. The openness expressed in limit questions can induce wonder and gratitude about the world, and this mystical function of religion can be complementary to its more prophetic, functional characteristics. Finally, evolution has bequeathed us the capacity for imagination and thus for transcending any one particular perspective or regulative ideal. This in turn leads us to the notion of divine transcendence.

Philip Hefner begins with the "two-natured" character of the human: the confluence of genetic and cultural information. These co-exist in the central nervous system (CNS) and have co-evolved and co-adapted. The genetic has made the cultural dimension possible; their symbiotic character differentiates humanity from other forms of life. Ralph Burhoe describes them as co-adapted organisms. Though we are conditioned by our evolutionary development and our ecological situation, we are free to consider appropriate behaviors within an environmental and societal matrix of demands, since our freedom serves the interest of the deterministic evolutionary system and is rooted in our "genetically controlled adaptive plasticity." The emergence of conditionedness and freedom are an evolutionary preparation for values and morality; the ought is built into evolution and need not be imported from external sources. Hefner reports that evolutionary psychology and human behavioral ecology have moved beyond their roots in sociobiology. He notes two key issues: how adaptive behaviors are shaped by critical moments in evolutionary history and then transmitted by bundles of adaptations, and how cooperative behaviors, including morality, evolve in the context of genetic, neurobiological, and cultural interactions.

Humans have evolved to seek and to shape meaning, enabled by the CNS, and our survival depends on it. A crucial step is the construction of frameworks and interpretations which are "pre-moral," as Solomon Katz puts it. Moreover, our species, though bounded by evolution, acts within this context, thus inevitably altering the world. Hefner speaks theologically of the human as "created co-creator." We encounter transcendence in several ways: as evolution and the ecosystem transcend themselves when they question their purpose through us; as we act in the non-human world and in culture; and as we open ourselves to our future. Thus the "project" of the human species is also nature's project, and the challenge for us is to discover its content.

Human culture includes diverse strategies for living; its greatest challenge is science and technology. These provide the underlying conditions for our interrelated planetary community but also its pressure on the global ecosystem. They are essential for human life and they thoroughly condition the future of the planet. Culture functions to guide behavior comparably to the role of physico-bio-genetic systems in plants and animals. Its interpretation challenges us intellectually and spiritually. Hefner opts for a non-dualistic interpretation: technology, like all culture, is an emergent form of nature grounded in human neurobiology. Still, as a thoroughly technological civilization we now face a crisis not merely of "tools out of control," but of an all-permeating form of existence that threatens to turn against itself and nature. He calls for "a re-organization of consciousness" adequate to this crisis.

Christian theology, through the doctrine of creation, can provide such perspective. The natural world is vested in meaning by its relation to God as creator *ex nihilo*; nature is entirely God's project, what God intended. The doctrine of continuing creation emphasizes the way in which, at every moment of time, God creates in freedom and love, giving the world its evolutionary character, purpose, and meaning. Our understanding of nature's meaning arises in the context of our scientific experience of the world, including randomness and genetic predisposition. Humanity, as created in the *imago Dei*, becomes a metaphor for the meaning of nature. Human sin represents the epistemic distance between the actual human condition and the primordial intentionality and love that God bequeaths the world. The key question for relating theology to science should be whether we believe that what governs the world at its depths is divine love. The Incarnation and the sacraments are pivotal theological affirmations that nature is capable of being an instrument of God's will and purpose. In Jesus Christ we discover both the normative image of God and the instantiated character of God's freedom, intentionality, and love.

Hefner concludes with six claims regarding the created co-creator as a fully natural creature illuminating the capabilities of nature, the convergence of the human project and the project of nature, and the transcendence and freedom of both. The evolution of the created co-creator reveals that nature's project is also God's project, and the human project must be in the service of nature's project. The *imago Dei*, and particularly Christ, gives content to God's intentionality for this project.

As **Arthur Peacocke** observes, the nineteenth-century theological reaction to Darwin was much more positive, and the scientific reaction much more negative, than many today care to admit. Current theology, however, is far less open—a "churlishness" which Peacocke is committed to rectifying for the sake of the believability and intellectual integrity of Christianity. To do so he turns directly to five broad features of biological evolution and the theological reflections they suggest.

The first is continuity and emergence. Although the seamless web of nature is explained by scientists using strictly natural causes, biological evolution is characterized by genuine emergence and a hierarchy of organization, including new properties, behaviors, and relations. Such emergence entails both epistemic irreducibility and a putative ontology. Emergence, in turn, is God's action as the continuous, ongoing, and immanent Creator in and through the processes of nature.

The second feature is the mechanism of evolution. Although biologists agree on the central role of natural selection, some believe selection alone cannot account for the whole story. Peacocke describes eight approaches to the question which operate entirely within a naturalistic framework, assume a Darwinian perspective, and take chance to mean either: 1) epistemic unpredictability, arising either a) because we cannot accurately determine the initial conditions, or b) because the observed events are the outcome of the crossing of two independent causal chains, or 2) inherent unpredictability as found at the subatomic level. Chance characterizes both mutations in DNA (types 1a and/or 2), and the relation between genetic mutation and the adaptation of progeny (type 1b). Chance in turn was elevated to a metaphysical principle by Jacques Monod, who rejected God's involvement in evolution, but Peacocke disagrees. Instead, chance connotes the many ways in which potential forms of organization are thoroughly explored in nature. Rather than being a sign of irrationality, the interplay of chance and law are creative over time. This fact, for the theist, is one of the God-endowed features of the world reflecting the Creator's intentions.

Next Peacocke raises the question about trends, properties and functions which arise through, and are advantageous in, natural selection. Drawing on G. G. Simpson and Karl Popper, Peacocke claims that there are "propensities" for such properties. Examples include complexity, information-processing and -storage ability, and language. They characterize the gradual evolution of complex organisms and contribute to the eventual existence of persons capable of relating to God. Thus the propensities for these properties can be regarded as the intention of God who continuously creates through the evolutionary processes, though without any *special* action by God at, say, the level of quantum mechanics or genetic mutations.

The fourth feature is the ubiquity of pain, suffering and death in nature. Pain and suffering are the inevitable consequence of possessing systems capable of information processing and storage. Death of the individual and the extinction of species are prerequisites for the creation of biological order. Complex living structures can only evolve in a finite time if they accumulate changes achieved in simpler forms, and are not assembled *de novo*. This includes both the predator-prey cycle, which involves eating pre-formed complex chemical structures, and the modification of existing structures via biological evolution. This, in turn, raises the problem of theodicy. Peacocke stresses that God suffers in and with the suffering of creatures, and cites support from current theologians who reject divine impassibility. God's purpose is to bring about the realm of persons in communion with God and with each other. Moreover, God's suffering with Christ on the cross extends to the whole of nature. Death as the "wages of sin" cannot possibly mean biological death; this requires us to reformulate the classical theology of redemption. The reality of sin must consist in our alienation from God, a falling short of what God intends us to be. It arises because, through evolution, we gain self-consciousness and freedom, and with them, egotism and the possibility of their misuse.

In his final section Peacocke turns to the theological significance of Jesus Christ in an evolutionary perspective. Christ's resurrection shows that such union with God cannot be broken even by death. His invitation to follow him calls us to be transformed by God's act of new creation within human history. But how is this possible for us now? This leads Peacocke to the problem of atonement. Since he rejects objective theories that link biological death to sin and the Fall, the suffering of God and the action of the Holy Spirit in us together must effect our "at-one-ment" with God and enable God to take us into the divine life.

Denis Edwards is concerned with rethinking the doctrine of sin and grace in light of biological evolution. He begins with the insights of Gerd Theissen, Sallie McFague and Philip Hefner. According to Edwards, Theissen argues that the common features of science and theology can be articulated through evolutionary categories. Religion manifests the "central reality," God. Christianity offers the principle of radical solidarity which runs counter to natural selection. The pull in us towards anti-social behavior has a biological foundation, while the work of the Holy Spirit is in the direction of pro-social behavior, helping us see strangers as kin. Theissen supports these points by referring to the three great "mutations" of Christian faith: biblical monotheism, New Testament Christology, and the experience of the Holy Spirit. Edwards then criticizes Theissen's work in terms of both biology and theology: for example, are natural selection and culture, and natural selection and the way of Christ, each so sharply opposed?

Sallie McFague finds the pattern for divine immanence in creation in the story of Jesus: the universe is directed toward inclusive love for all, particularly the oppressed. As Edwards sees it, her "Christic paradigm" extends God's liberating, healing, and inclusive love to non-human creatures. McFague understands nature as the new poor. She finds consonance between natural selection and Christianity, since evolution is not only biological but cultural, and since it is essential that human culture contributes to the welfare of all life on earth. But she finds dissonance between natural selection and Christianity, because neither cultural nor biological evolution includes solidarity with the oppressed. Instead, God suffers with suffering creation, since the world is God's body. But, Edwards asks, is McFague too negative, even moralistic, about natural selection? Is the Christic paradigm opposed to natural selection or does it define God's creative action in and through evolution?

According to Edwards, Philip Hefner sees the human being as a symbiosis of genes and culture. Religion is the central dimension of culture. In view of the ecological crisis we have brought about we need a theology of the human as created co-creator. Hefner views original sin in terms of the discrepancy in the information coming from our genes and our culture, including a clash in us between altruism and genetic selfishness. He also suggests that original sin can be understood in terms of the fallibility and limitation that are essential to human evolution and freedom. Though these are good, they are always accompanied by failure. The religious traditions carry altruistic values, particularly trans-kin altruism, and the biblical commandments ground altruism ultimately in God.

Edwards believes that Hefner's evolutionary insights genuinely illuminate the human condition and the Christian understanding of concupiscence. He argues, however, that discrepancy and fallibility are not in themselves sin. Instead, following Rahner, he distinguishes between the disorder of sin and the disorder that is intrinsic to being human. The former comes from our rejecting God. The latter results from our being both spiritually and bodily finite; it is a form of concupiscence that is morally neutral and not in itself sinful. Our existential state is constituted by both disorders. Edwards finds Hefner's insight as bearing on our natural, but not our sinful, disorder: the structure of the human, though a fallible symbiosis of genes and culture, is not in itself sin. In addition, Edwards suggests that our genetic inheritance can carry messages essential for human life while culture and religion can carry messages of evil. This means that selfishness and sin cannot be identified strictly with our biological side, and unselfish behavior cannot be identified entirely with our cultural side.

With regard to grace, Edwards writes that, while altruism is a radical dimension of divine and human love, it does not express the ultimate vision of that love. Indeed, indiscriminate calls to altruism and self-sacrifice can function to maintain oppression, as feminist theologians have stressed. Moreover, in a trinitarian doctrine of God, love is revealed most radically in mutual, equal, and ecstatic friendship. So, though Hefner sees altruistic love as holding the status of a cosmological and ontological principle, Edwards sees persons-in-mutual-relations as having this status. Drawing on the writings of John Zizioulas, Walter Kasper, and Catherine Mowry LaCugna, Edwards suggests that if the essence of God is relational and if everything that is springs from persons-in-relation, then this points towards an ontology which he calls "being-in-relation." Moreover, such an ontology is partially congruent with evolutionary biology, including its stress on cooperative, coadaptive, symbiotic, and ecological relations. Contrary to Theissen and McFague who tend to oppose natural selection and the Gospel, Edwards wants the "Christic paradigm" to view God as continuously creating through the processes of evolution.

Still the struggle and pain of evolution leads Edwards to face the challenge of theodicy. Following Thomas Tracy, he first suggests that natural selection needs to be considered in non-anthropomorphic and non-moral terms as an objective process in nature, like nucleosynthesis in stars. Theodicy is no more intense a problem for natural selection than it is to all such processes, including death when understood as essential to evolution and life. The trinitarian God who creates through natural selection needs to be understood not only as relational but also as freely accepting the limitations found in loving relationships with creatures. The Incarnation and the Cross point to a conception of God related to natural selection through unthinkable vulnerability and self-limitation. The God of natural selection is thus the liberating, healing, and inclusive God of Jesus. This God is engaged with and suffers with creation; at the same time, creatures participate in God's being and trinitarian relationships.

According to **John Haught**, evolutionary theory can seriously undercut the credibility of divine action. Daniel Dennett views evolution as a purely algorithmic process that leaves no room for God's action. Richard Dawkins argues that impersonal physical necessity drives genes to maximize opportunities for survival. Both conclude that Darwin has given atheism a solid foundation. Hence contemporary theology must include an apologetic dimension. At minimum, it should demonstrate that the scientific concepts involved in evolutionary theory—contingency, necessity, and the enormity of time—do not rule out the action of God. But theology should go beyond this to show that these concepts are open to metaphysical and theological grounding and that a careful understanding of God renders an evolving natural world more intelligible. Rather than a danger, Darwin offers theology a gift: the context for a doctrine of God as compassionate, suffering, and active in and fully related to the world.

This paper argues that such a theology is implied in the kenotic image of God's self-emptying, Christ-like love. Haught draws on Dietrich Bonhoeffer, Edward Schillebeeckx, and Jürgen Moltmann in stressing divine kenosis. An evolutionary theology extends this view backward to embrace the history of life on earth and forward to its eschatological completion. With Karl Rahner we are invited to accept the humility of God and to resist what Sallie McFague calls a "power of domination." Contrary to Dennett and Dawkins, the randomness of variations, the impersonality of natural selection, and the waste and suffering of evolution can be understood through the concept of a vulnerable God who renounces despotic force, who grounds evolution in divine love, and who participates in evolution to redeem nature.

But there is an additional problem here. Science is methodologically neutral, constrained to explain evolution without reference to the supernatural. Still, the assertions that science leaves no room for theology and that what physics depicts is the only reality lead beyond science to scientism and materialism, ideologies clearly in opposition to theology. Haught cites Stephen Jay Gould, Dennett, and Dawkins as conflating the science of evolution and the ideology of materialistic metaphysics. In response, theologians must employ a metaphysics of sufficient categorical breadth and philosophical depth to account for both the Christian experiences of God and evolutionary science, and one that counters materialism. It is Haught's conviction that some aspects are provided by Whiteheadian philosophy, with its emphasis on novelty and temporality as irreducible features of the world.

Faith's conviction that God's relationship to the world is one of complete self-giving can be elaborated, at least partially, through process theology's notion of the divine persuasive power which invites, though never forces, creation to engage in the process of becoming. Such emergent self-coherence in the evolving world is entirely consonant with the world's radical dependence on and intimacy with God. According to Haught, union with God actually differentiates the world from God rather than dissolving it into God. In the light of such a theology, rooted in the divine kenosis, we should not be surprised at nature's undirected evolutionary experimentation with multiple ways of adapting, or at the spontaneous creativity in natural process, or at the enormous spans of time involved in evolution. If God's incarnate love is expressed in persuasive and relational power, a world rendered complete and perfect in every detail by God's direct act would be metaphysically and theologically impossible. Such a world would not be truly distinct from God. It would be neither a truly graced universe, as is ours, nor meaningfully open to God's self-communication.

Kenotic process theology emphasizes God as the sole ground of the world's being. The sufferings and achievements of evolution take place within God's own experience and are graced by God's compassion. Such a theological stance, according to Haught, is not only consistent with, but ultimately explanatory of, the world seen in terms of evolutionary science—and in ways that go beyond the capabilities of materialism. Still, in light of theology's concerns for both creation and eschatology, Haught emphasizes a "metaphysics of the future" in which the fullness of being is found not in the past or present, but in what is yet to come. The ongoing creation of the universe and the evolutionary process are made possible by God's entering into the world from the realm of the future.

Haught realizes that science (or, more properly, scientism) is rooted in a "metaphysics of the past," but this is a view which he believes evolutionary theology will need to include even while surpassing it. To the objection that the future cannot 'cause' the present, he poses the metaphorical character of both theological and scientific language, and he invokes Paul Tillich's suggestions that we refer to God as "Ground," rather than cause, of being. Ultimately, biblical faith rules out unique mechanical causation from past events and is commensurate with process theology's insistence on the power of the future and on God as the ultimate source of all possibilities.

Haught returns then to his initial question: Does evolutionary theory leave room for theology? An affirmative response requires that there be an explanatory role for the idea of God in light of evolution which does not interfere with that of science. Haught provides this by pointing to three assumptions in the scientific explanation of life: the contingency of events, the laws of nature, and the irreversible, temporal character of the world. He believes theology's task is to provide an ultimate explanation and grounding for these assumptions. Theology does so by claiming that contingent events, such as genetic mutations, signal the inbreaking of the new creation, that necessity is an expression of God's faithfulness, and that the arrival of the divine *Novum* endows the world with its temporality.

In the first part of his paper, **Ian Barbour** describes the evolution of Darwinism over the past century. Charles Darwin actually shared many of the mechanistic assumptions of Newtonian science. By the early twentieth century, population genetics focused on statistical changes in the gene pool and the "modern synthesis" took a gradualist view of evolution. The discovery of the structure of DNA in 1953 led to the central dogma of molecular biology: information flows from DNA to protein. Recent theories have explored selection at a variety of levels including gene, organism, kin, group, and species, as well as punctuated equilibrium. Other biologists have noted that mutation and selection are not the only sources of novelty. While these new theories can be seen as extensions of Darwinism, a few scientists, such as Stuart Kauffman, claim they are moving beyond Darwinism by invoking principles of self-organization and holism.

Barbour then outlines four philosophical issues which characterize the interpretation of evolution. *Self-organization* is the expression of built-in potentialities and constraints in complex hierarchically-organized systems. This may help to account for the directionality of evolutionary history without denying the role of law and chance. *Indeterminacy* is a pervasive characteristic of the biological world. Unpredictability sometimes only reflects human ignorance, but in the interpretation of quantum theory, indeterminacy is a feature of the microscopic world and its effects can be amplified by non-linear biological systems. He also argues for *top-down causality* in which higher-level events impose boundary conditions on lower levels without violating lower-level laws and he places top-down causality within the broader framework of holism. He distinguishes between methodological, epistemological, and ontological reduction. *Communication of information* is another important concept in many fields of science, from the functioning of DNA to metabolic and immune systems and human language. In each case, a message is effective only in a context of interpretation and response.

According to Barbour, each of these has been used as a non-interventionist model of God's relation to the world in recent writings. If God is *the designer of a self-organizing process* as Paul Davies suggests, it would imply that God respects the world's integrity and human freedom. Theodicy is a more tractable problem if suffering and death are inescapable features of an evolutionary process for which God is not directly responsible. But do we end up with the absentee God of deism?

The *neo-Thomist* view of God as *primary cause* working through secondary causes as defended by Bill Stoeger tries to escape this conclusion, but Barbour thinks it undermines human freedom. Alternatively, God as providential *determiner of indeterminacies* could actualize one of the potentialities present in a quantum probability distribution. Selection of one of the co-existing potentialities would communicate information without energy input, since the energy of the alternative outcomes is identical. Does God then control all quantum indeterminacies—or only some of them? Barbour comments on the way these options have been discussed by George Ellis, Nancey Murphy, Robert Russell, and Thomas Tracy. God as *top-down cause* might represent divine action on "the world as a whole," as Arthur Peacocke maintains together with his "whole-part" models. But these are problematic according to Barbour since the universe does not have a spatial boundary, and the concept of "the-world-as-a-whole" is inconsistent with relativity theory. Grace Jentzen and Sallie McFague view the world as God's body but Barbour is concerned that this model breaks down when applied to the cosmos. God as *communicator of information* would act through the pattern of events in the world, in human thought, and in Christ's life as God's self-expression, but this model does not capture God's intention in creating loving and responsible people.

Process theology offers a fifth model of God's action in the world by providing a distinctive theme: the *interiority* of all integrated events viewed as moments of experience. Rudimentary forms of perception, memory, and response are present in lower organisms; sentience, purposiveness, and anticipation are found in vertebrates. But process authors maintain that consciousness occurs only at the highest levels of complex organisms. There is great diversity in the ways in which components are organized in complex systems, and therefore great differences in the types of experience that can occur.

The process model resembles but differs from each of the four models above. God as designer of self-organizing systems is a source of order, but the God of process thought is also a source of novelty. God acts in indeterminacies at the quantum level, but also within integrated entities at higher levels. God acts as top-down cause, not through the cosmic whole but within each integrated system which is part of a hierarchy of interconnected levels. Communication of information can occur through events at any level, not primarily through quantum events at the bottom or the cosmic whole at the top. God is persuasive, with power intermediate between the omnipotent God of classical theism and the absentee God of deism. God is present in the unfolding of every event, but God never exclusively determines the outcome. This is consistent with the theme of God's self-limitation in contemporary theology and with the feminist advocacy of power as empowerment. Process theology has much in common with the biblical understanding of the Holy Spirit as God's activity in the world. Barbour concludes by considering some objections to process thought concerning panexperientialism, God's power, the charge of being a "gaps" approach, and the abstract character of philosophical categories in the context of theology.

1.4.5 *Section Four: Biology, Ethics, and the Problem of Evil*

The opening papers in section four deal with the relation between biology and ethics. The first presents a phylogenetic model of biological and moral altruism (Cela-Conde and Marty). The second defends supervenience as a response to the challenge of reductionism (Murphy). The remaining two papers address the problem of sin and evil in the context of biology. The first focuses on the ethical and theological issues raised by germ-line genetic therapy (Peters). The last paper examines both human sinfulness and suffering in nature as it challenges our understanding of God as acting beneficently in nature and as it raises the question of theodicy (Tracy).

Camilo Cela-Conde and **Gisele Marty** focus on models of human evolution that account for the development of traits in individuals, including morphological traits such as large brains and functional traits such as speech. They also consider the development of collective traits in human populations, such as language, culture, and moral codes. These models raise important theological questions regarding divine action.

Most theological attempts to address human evolution treat the development of culture separately from, or in contrast to, the evolution of morphology, but these attempts run into serious problems since cultural evolution presupposes and builds on biological evolution. Interactionist models are thus needed. Such models must also address the question of when and how such traits as a complex brain and human language emerged during the past 2.5 million years. Some scientists hold for an almost instantaneous and isolated emergence of language, but since language does not fossilize, the conjecture is hard to test empirically. Others argue for a long, gradual, and early development of language going back to *Homo habilis,* and relate it to the slow development of the brain, an idea which is easier to test. Another question is how to differentiate the evolution of our species from other hominids and even from other primates. The theological task, in turn, is how to relate such an understanding of human evolution, especially of language and cognition, to divine action, taking into account the elements of continuity as well as discontinuity between humans and other hominids as well as between hominids in general and other primates.

Darwin was the first to speak of both a biological mechanism for moral behavior and a distinctive "moral sense" which he attributed uniquely to humans. He explained the evident diversity of moral codes in terms of adaptation to varying environments. But how does human moral sense (or "moral altruism") differ from the kind of biological altruism shared by so many species, and what are its genetic roots? This question leads the authors into a discussion of sociobiology as it has developed over the past two decades. In their view, it has focused on four key issues: the phenomena implied by human morality; the analogs to it at the animal level; the phylogenetic explanation of the emergence of these analogs; and the development of human morality within this framework. Though the debate has waned somewhat, Cela-Conde and Marty hope to show how it might now be reinvigorated.

Part of the challenge, in spite of the reductionism inherent in the debate, is the actual complexity of the phenomenon of human morality. A variety of approaches are being pursued. Some scientists point to the distinction between the capacity for, and the content of, moral thought. Others focus on group selection and kin selection models of altruism. Some argue for a strict separation between biological and moral altruism, while others stress their intimate connection. The authors note that, even if a strong connection is granted, reductionism can at least be partially avoided by appealing to the supervenience of moral language. Some sociobiologists have developed theories of reciprocal altruism, ultrasociality, and sociocultural fitness. Still the authors know of no model which includes all the elements required, from innate tendencies to empirical moral norms. Moreover, the complex cognitive processes implied in evaluating and making decisions suggest that the usual distinction between motive and criterion is inadequate.

Instead, in their model, Cela-Conde and Marty consider both the motive to act, the personal ethical criterion, and the set of collective values and norms. Individuals accumulate and actualize these values during the apprenticeship process, giving to the collective complex an evolutionary, changing character. They propose a phylogenetic argument that places biological and moral altruism as two successive stages in human evolution. Biological altruism is closely associated with the genetic code and belongs to the area of motivation; moral altruism is related to the personal ethical domain or the values of the group. Neither taken alone is able to explain the whole of human moral conduct. The combined development of cognitive capacity and moral behavior is sometimes called "co-evolution." They also draw on the cognitive sciences. Here, internal rewards to the individual may be available through religious rituals, acting before public crowds, integration into small communities, and so on. They conclude by exploring the idea of universal norms directing moral behavior as typified in the first stages of sociobiology, and the idea of universal tendencies to accept moral codes as found in later, more sophisticated sociological arguments. These results and their problems point, in turn, to the need for a more complete theory linking the biological substratum to moral conduct, the influence of social groups, and the role of emotions in maintaining moral behavior.

Nancey Murphy considers the role of supervenience in the relation of evolutionary biology and ethics. She argues that God acts in the moral as well as the biological sphere. She also challenges the attempt to reduce morality to biology and she affirms the claim that objective moral knowledge is dependent on a theological framework.

According to Murphy, the contemporary atomist-reductionist program can be traced to the rise of modern science with the transition from the Aristotelian hylomorphism to Galilean atomism. Following the success of Newtonian mechanics, atomism was extended to chemistry. In the nineteenth century, attempts were made to extend atomism to biology and, in our century, to the human sphere, including psychology and a general epistemological theory of the relations between the sciences. In this view, the world is seen as a hierarchy of levels of complexity with a corresponding hierarchy of sciences. Causation is strictly bottom-up from the lowest level to the top of the hierarchy. Murphy draws on Francisco Ayala's classification and lists six types of reductionism, including methodological reductionism which, by its success, lends credence to the other five. Bottom-up causality and the claim that the parts take priority over the whole embody what is, in fact, a metaphysical assumption. Modern thought also extended atomism to areas such as political philosophy and ethics through the writings of Thomas Hobbes and Immanuel Kant. Still, the determinism of the laws of physics has raised problems for human behavior. Materialists like Hobbes have simply accepted determinism, while dualists such as Descartes have sought to avoid it, but in its place have raised the mind-body problem, and the resulting separation of the natural and social sciences a century later.

Murphy concludes that a proper understanding of ethics and biology will require a thorough response to reductionism. The key will be nonreductive physicalism, originating in the work of Roy Wood Sellars, and now including the focus on emergent properties of hierarchical systems. Sellars, in emphasizing the significance of organizations and wholes, opposes Cartesian dualism, absolute idealism, and reductive materialism. Murphy then cites scientific evidence for non-reducibility: interactions with the environment portray an entity as part of a larger system and entail both top-down and whole-part analyses which complement the bottom-up approach and lead to language about emergence. The irreducibility of concepts entails the irreducibility of laws. But how are we to explain these problems with reducibility if our analysis, though multi-layered, still refers to one reality?

Here Murphy introduces the concept of supervenience in order to give a non-reducible account of morality and mental events. R. M. Hare introduced the term in 1952, and Donald Davidson developed it in 1970. It is now widely used in philosophy, though not without disagreement. Murphy's emphasis is on circumstance: identical behavior in different circumstances could constitute different moral judgments, and such supervenient properties are not necessarily reducible. Then she returns to the hierarchical model of the sciences with a branching between the human and physical sciences she and George Ellis suggested. The human sciences are incomplete without ethics. In this scheme, the moral supervenes on the biological if biological properties constitute moral properties under the appropriate circumstances. She draws on Alasdair MacIntyre and on Philip Kitcher's critique of E. O. Wilson to argue that claims to reduce ethics to biology are due to meta-ethical ignorance, challenging Richard Alexander and Michael Ruse in passing. Instead Murphy places ethics between the social sciences and theology. The former, rather than being value-free, now involves moral presuppositions and ethical questions. Ethics traditionally drew upon a concept of humankind's ultimate *telos*. The Enlightenment severed the ties to this tradition but kept the same moral prescriptions. Modern philosophers, failing to find an objective basis for moral discourse, reduced moral claims to empirical observations. All of this is due to our forgetting that the "ought" is only half of a moral truth; it is actually connected to a *telos* and a specific "is" statement. Murphy next discusses the supervenience relations that obtain as we move up the scale. The predicate of virtue supervenes on psychological characteristics, but only if there are nonreducible circumstances related to the ultimate goals of human existence.

Murphy concludes with the relation between ethics and biology. Here we extend the subvenience of virtues downward in the hierarchy. According to MacIntyre, virtues are acquired and not genetically determined traits. Still ethics will have much to learn from biology. Finally, all moral systems are dependent for their justification on beliefs about reality. Murphy dismisses claims like that of Jacques Monod regarding the meaningless and purposelessness in nature. Instead, we must reject reductionism by emphasizing the top-down connections from theology to ethics as well as the general inadequacies of reductionism.

The theory of evolution leads **Ted Peters** to emphasize that God's creation is not fixed but changing, and this may apply even to human nature. But should we seek to influence our own genetic future? Some denounce this as "playing God," especially when it comes to germ-line intervention, but Peters analyzes this term in light of a Christian theology of creation. He claims that God's creative activity gives the world a future, that humans are "created co-creators," as Hefner puts it, and that we should be open to improving the human genetic lot and thus to influencing our evolutionary future.

Peters suggests three meanings for the term, "playing God": learning about God's awesome secrets through science and technology; making life and death decisions in medical emergencies; and substituting ourselves for God. Concern for the third may lead to an attack on human pride in when we confuse knowledge for wisdom or ignore the problem of unforeseen consequences of germ-line intervention. A separate concern is that DNA is sacred and should be off-limits to humans. Here Peters cites Jeremy Rifkin, who appeals to naturalism or vitalism in defense of leaving nature alone. Still neither Christian theologians nor molecular biologists are likely to agree with Rifkin—though for different reasons.

Actually the term "playing God" raises the question of the relationship between God and creation in terms of both *creatio ex nihilo* and *creatio continua*. In Peters' view, giving the world a future is God's fundamental way of acting as creator. God creates new things including the new creation yet to come. The human is a "created co-creator": we are part of what God alone creates *ex nihilo*, yet we can have a special influence of the direction of what God continues to create. Indeed the meaning of *imago Dei* may well be creativity. Though the results of human creativity are deeply ambiguous ethically, we cannot not be creative. The ethical mandate concentrates instead on the purposes towards which we direct our work

Peters next turns to the Human Genome Project (HGP). The aim of the HGP is new knowledge and the betterment of human health, but the germ-line debate is over taking actions now that might potentially improve the health or relieve the suffering of people who do not yet exist. Should we instead only engage in somatic therapy or should we undertake germ-line therapy, and should the purposes of the latter include enhancement? Peters cites a number of church documents which call for caution or for limitations to somatic procedures; still others are open to germ-line therapy. At stake is the implicit association with eugenics.

By and large, religious ethical thinking is conservative, seeking to preserve the present human gene pool for the indefinite future. From this perspective, germ-line intervention triggers the sense of "playing God" both physically (we might degrade the biodiversity required for good health) and socially (we might contribute to stigma and discrimination). But Peters is critical of this perspective, because it assumes that the present state of affairs is adequate, and it ignores the correlation between conceiving of God as the creator and the human as created co-creator. Instead, we are called to envision a truly better future and press on towards it, though certainly with caution, prudence, and a clear concern for our *hubris*.

After recounting the summary of arguments for and against germ-line intervention stipulated by Eric Juengst, Peters turns to the position paper by the Council for Responsible Genetics (CRG). The strongest argument is that germ-line intervention will reinforce social discrimination. Peters endorses the CRG's concern: the definition of the ideal norm may be governed by economic and political advantage. He reaffirms human dignity regardless of genetic inheritance and the technical possibilities in genetic engineering, but he recognizes that the underlying reasons for prejudice and discrimination today are not germ-line intervention. A more serious challenge raised by the CRG concerns persons who do not yet exist: should they be given moral priority over present populations? Future generations might blame us for acting, or not acting, in terms of germ-line interventions, and, arguably, we are morally accountable to them. But the problem is complex, as Hardy Jones and John A. Robertson point out, since actual, generational differences would occur depending on whether or not we intervened. We clearly need an ethical framework that is grounded in God's will for the future, for the flourishing of all humanity, and for that which transcends particular concerns for contingent persons.

And so Peters asks the key question: would a future-oriented theology, with the human as co-creator, be more adequate than the CRG's proposal? His response is affirmative: a future-oriented theology would not give priority to existing persons; it is realistic about nature as inherently dynamic; and our task is to seek to discern God's purpose for the future and conform to it, while resisting the *status quo*. Rather than "playing God," in directing ourselves to that future we are being, in Peters' words, truly human.

Thomas Tracy addresses two of the central challenges that evolutionary theory poses for theology. First, how might we understand God as creatively at work in evolutionary history? Given the prominent role played by chance in evolution, Jacques Monod and others have contended that the meandering pathways of life cannot enact the purposes of God. Second, how can the affirmation of God's perfect goodness and creative power be reconciled with the ubiquitous struggle, suffering, and death that characterizes evolution? This, of course, is a form of the problem of evil.

The first question demands a theological interpretation of evolution. Tracy distinguishes his project both from natural theology, which attempts to argue from nature to God, and from any theological competition with biology, which attempts to show the inadequacy of naturalistic explanations in order to substitute theological ones. Instead, given the best current evolutionary theory, Tracy asks how might a theologian concerned with divine action and providence conceive of God's relation to the history of life?

In response, Tracy argues that there are several ways in which God might be understood to act in and through evolutionary processes. God acts universally to create and sustain all finite things. In so doing, God may choose to fix the course of events in the world by establishing deterministic natural laws. In this case every event in cosmic history could be regarded as an act of God. There are good reasons, both theological and scientific, to reject this universal determinism. From science it appears that indeterministic chance is built into the structures of nature, and that chance events at the quantum level can both constitute the stable properties of macroscopic entities and effect the course of macroscopic processes. Chaotic dynamics and evolutionary biology provide two key examples here. This in turn creates several fascinating possibilities for conceiving of divine action in the world. Perhaps a "hands-off" God leaves some features of the world's history up to chance. Or perhaps God chooses to act at some or all of these points of indeterminism. Then God could in this way initiate particular causal chains without intervening in the regular processes of nature. Tracy notes that there are conceptual puzzles raised by each of these ideas, such as whether God determines all or just some of these events—and thus the relative theological merits of each would need to be debated. But this variety of options for conceiving of divine action makes it clear that the first challenge can be met.

If we succeed in constructing a theological interpretation of evolution, we are immediately confronted by a compelling form of the problem of evil. How can belief in God's loving care for creation be reconciled not just with moral evil in the human sphere but also with the hardship, pain, suffering, and death that characterizes evolutionary processes, or what is called "natural evil"? According to Tracy, any morally sufficient response must identify the good for the sake of which evil is permitted, and it must explain the relation of evil to this good. One standard approach is to argue that God must permit some evils as a necessary condition for achieving various goods in creation. John Hick, for example, holds that the good of "soul making" requires that persons be free to develop their intellectual, moral, and spiritual capacities by acting in an environment that is lawful, impersonal, and at an "epistemic distance" from God. This entails both that we can do moral evil and that we (and other sentient beings) will suffer from natural evil. Clearly evolution can be accommodated within such a theodicy, though Tracy admits it is probably not required.

If we grant that the good cannot be achieved without permitting these evils, we may nonetheless object that the world contains far *more* of them than would be necessary to serve God's purposes. It is not difficult to think of evils that, as far as we can see, do not lead to any greater good and could readily have been prevented. Tracy's response is that God must permit evil that does not serve as the means to a greater good if it occurs as a necessary by-product of preserving moral freedom and the integrity of the natural order. Precisely because these "pointless" evils do not generate particular goods, they will appear to us to be unnecessary. However, evils of this type must be permitted by God, and it will be up to *us* to prevent or ameliorate them. From the fact that we cannot see a point to an evil, therefore, it does not follow that God should have prevented it. A world that includes the good of personal relationship with God must apparently include pointless evil.

But just how much pointless evil is really required? Does the world instead contain gratuitous evils? The problem here is, first of all, in assuming that we can calculate what could be considered the minimum amount of acceptable pointless evil, and thus that we could quantify and balance goods and evils. The deeper problem is in the assumption that the world really does include gratuitous evils. In fact we cannot even conclude that some evil is gratuitous merely because we cannot think of a reason for God's permitting it. Moreover we must recognize that we are in no position to see how each evil fits into the overall course of cosmic history, to comprehend all of the goods to which it may be relevant, or to recognize all of the consequences of eliminating it. In grappling with the reality of evil, we confront the limits of human comprehension and are forced to accept epistemic humility, as the Book of Job makes plain in God's speech from the whirlwind. We cannot expect to solve the problem of evil. Instead the central task for Christian faith in the face of evil is to proclaim and understand what God is doing to suffer with and to redeem creation.

MESSAGE TO THE VATICAN OBSERVATORY CONFERENCE ON EVOLUTIONARY AND MOLECULAR BIOLOGY[1]
28 June, 1996

Pope John Paul II

Ladies and Gentlemen,

1. I am pleased to greet you on the occasion of this fourth conference in the series devoted to dialogue between philosophy, theology and science. As you continue to consider God's action in the physical world, you turn now to the complex issue of the nature of life itself, seeking to arrive at a fuller understanding of the universe and man's place in it. Your dedication to this undertaking is in line with the Church's long tradition of intellectual commitment, as expressed for example by Saint Augustine: "*Intellectum valde ama*" (*Epist. 120*, 3, 16)—truly love the intellect, truly seek after understanding.

2. If scientific endeavor, philosophical inquiry and theological reflection are to bring genuine benefit to the human family, they must always be grounded in truth, the truth which "shines forth in all the works of the Creator and, in a special way, in man, created in the imaged and likeness of God" (Encyclical Letter *Veritatis splendor, Introduction*). This is the truth which "enlightens man's intellect and shapes his freedom" (ibid.). When related to this truth, advances in science and technology, splendid testimony of the human capacity for understanding and perseverance, spur men and women on to face the most decisive of struggles, those of the heart and of the moral conscience (cf. ibid., n. 1).

3. What you do as scientists, philosophers and theologians can contribute significantly to clarifying the vision of the human person as the focus of creation's extraordinary dynamism and the supreme object of divine intervention. Thus there is an intimate link between the development of scientific perspectives on divine action in the universe and the betterment of mankind. Those who work through the sciences, the arts, philosophy and theology in order to advance our understanding of what is true and beautiful are walking a path of discovery and service parallel and complementary to that followed by those who engage in the struggle to improve peoples' lives, fostering their genuine good and development. In the final analysis, *the true, the beautiful and the good are essentially one*.

4. From this point of view, I consider that this series of conferences, seeking to relate and unify the knowledge derived from many sources, offers an important contribution to that *exchange between religion and science* which I have made every effort to promote since the first days of my Pontificate. Grateful for the work you have already done in this field, I pray that you will continue to pursue with professional expertise this important inter-disciplinary dialogue. Upon you and your work I invoke the blessings of Almighty God.

---

[1] Originally printed in *L'Osservatore Romano*, 29 June, 1996. Reprinted with permission.

MESSAGE TO THE PONTIFICAL ACADEMY OF SCIENCES[1]
22 October, 1996

Pope John Paul II

To the Members of the Pontifical Academy of Sciences
taking part in the Plenary Assembly

With great pleasure I address cordial greetings to you, Mr. President, and to all of you who constitute the Pontifical Academy of Sciences, on the occasion of your plenary assembly. I offer my best wishes in particular to the new academicians, who have come to take part in your work for the first time. I would also like to remember the academicians who died during the past year, whom I commend to the Lord of life.

1. In celebrating the sixtieth anniversary of the Academy's refoundation, I would like to recall the intentions of my predecessor Pius XI, who wished to surround himself with a select group of scholars, relying on them to inform the Holy See in complete freedom about developments in scientific research, and thereby to assist him in his reflections.

He asked those whom he called the Church's *Senatus scientificus* to serve the truth. I again extend this same invitation to you today, certain that we will all be able to profit from the "fruitfulness of a trustful dialogue between the Church and science" (*Address to the Academy of Sciences*, n. 1, 28 October 1986; *L'Osservatore Romano* English edition, 24 November 1986, p. 22).

2. I am pleased with the first theme you have chosen, that of the origins of life and evolution, an essential subject which deeply interests the Church, since Revelation, for its part, contains teaching concerning the nature and origins of man. How do the conclusions reached by various scientific disciplines coincide with those contained in the message of Revelation? And if, at first sight, there are apparent contradictions, in what direction do we look for their solution? We know, in fact, that truth cannot contradict truth (cf. Leo XIII, Encyclical *Providentissimus Deus*). Moreover, to shed greater light on historical truth, your research on the Church's relations with science between the sixteenth and eighteenth centuries is of great importance.

During this plenary session, you are undertaking a "reflection on science at the dawn of the third millennium," starting with the identification of the principal problems created by the sciences and which affect humanity's future. With this step you point the way to solutions which will be beneficial to the whole human community. In the domain of inanimate and animate nature, the evolution of science and its applications gives rise to new questions. The better the Church's knowledge is of their essential aspects, the more she will understand their impact. Consequently, in accordance with her specific mission she will be able to offer criteria for discerning the moral conduct required of all human beings in view of their integral salvation.

---

[1] The official translation is reprinted here as it appeared in the English edition of the *L'Osservatore Romano*, 30 October, 1996, with two alterations noted below in footnotes 3 and 4 (see also *Origins*, CNS Documentary Service, vol. 26, no. 2, 14 November, 1996). Reprinted with permission. The spacing of the English text has been adjusted so that corresponding French and English paragraphs align with each other.

Pope John Paul II
[Official French Text[2]]

Aux Membres de l'Académie pontificale des Sciences
réunis en Assemblée plénière

C'est avec un grand plaisir que je vous adresse un cordial salut, à vous, Monsieur le Président, et à vous tous qui constituez l'Académie pontificale des Sciences, à l'occasion de votre Assemblée plénière. J'adresse en particulier mes vœux aux nouveaux Académiciens, venus prendre part à vos travaux pour la première fois. Je tiens aussi à évoquer les Académiciens décédés au cours de l'année écoulée, que je confie au Maître de la vie.

1. En célébrant le soixantième anniversaire de la refondation de l'Académie, il me plaît de rappeler les intentions de mon prédécesseur Pie XI, qui voulut s'entourer d'un groupe choisi de savants en attendant d'eux qu'ils informent le Saint-Siège en toute liberté sur les développements de la recherche scientifique et qu'ils l'aident ainsi dans ses réflexions.

À ceux qu'il aimait appeler le *Senatus scientificus* de l'Église, il demanda de servir la vérité. C'est la même invitation que je vous renouvelle aujourd'hui, avec la certitude que nous pourrons tous tirer profit de la «fécondité d'un dialogue confiant entre l'Église et la science» (*Discours à l'Académie des Sciences*, 28 octobre 1986, n. 1).

2. Je me réjouis de premier thème que vous avez choisi, celui de l'origine de la vie et de l'évolution, un thème essentiel qui intéresse vivement l'Église, puisque la Révélation contient, de son côté, des enseignements concernant la nature et les origines de l'homme. Comment les conclusions auxquelles aboutissent les diverses disciplines scientifiques et celles que sont contenues dans le message de la Révélation se rencontrent-elles? Et si, à première vue, il peut sembler que l'on se heurte à des oppositions, dans quelle direction chercher leur solution? Nous savons en effet que la vérité ne peut pas contredire la vérité (cf. Léon XIII, encyclique *Providentissimus Deus*). D'ailleurs, pour mieux éclairer la vérité historique, vos recherches sur les rapports de l'Église avec la science entre le XVIe et le XVIIIe siècle sont d'une grande importance.

Au cours de cette session plénière, vous menez une «réflexion sur la science à l'aube du troisième millénaire», en commençant par déterminer les principaux problèmes engendrés par les sciences, qui ont une incidence sur l'avenir de l'humanité. Par votre démarche, vous jalonnez les voies de solutions qui seront bénéfiques pour toute la communauté humaine. Dans le domaine de la nature inanimée, l'évolution de la science et de ses applications fait naître des interrogations nouvelles. L'Église pourra en saisir la portée d'autant mieux qu'elle en connaîtra les aspects essentiels. Ainsi, selon sa mission spécifique, ella pourra offrir des critères pour discerner les comportements moraux auxquels tout homme est appelé en vue de son salut intégral.

---

[2] Originally published in *L'Osservatore Romano*, 23 October, 1996. Reprinted with permission.

3. Before offering you several reflections that more specifically concern the subject of the origin of life and its evolution, I would like to remind you that the Magisterium of the Church has already made pronouncements on these matters within the framework of her own competence. I will cite here two interventions.

In his Encyclical *Humani generis* (1950), my predecessor Pius XII had already stated that there was no opposition between evolution and the doctrine of the faith about man and his vocation, on condition that one did not lose sight of several indisputable points (cf. *AAS* 42 [1950], pp. 575–76).

For my part, when I received those taking part in your Academy's plenary assembly on 31 October 1992, I had the opportunity, with regard to Galileo, to draw attention to the need of a rigorous hermeneutic for the correct interpretation of the inspired word. It is necessary to determine the proper sense of Scripture, while avoiding any unwarranted interpretations that make it say what it does not intend to say. In order to delineate the field of their own study, the exegete and the theologian must keep informed about the results achieved by the natural sciences (cf. *AAS* 85 [1993], pp. 764–72; Address to the Pontifical Biblical Commission, 23 April 1993, announcing the document on *The Interpretation of the Bible in the Church: AAS* 86 [1994] pp. 232–43).

4. Taking into account the state of scientific research at the time as well as of the requirements of theology, the Encyclical *Humani generis* considered the doctrine of "evolutionism" a serious hypothesis, worthy of investigation and in-depth study equal to that of the opposing hypothesis. Pius XII added two methodological conditions: that this opinion should not be adopted as though it were a certain, proven doctrine and as though one could totally prescind from Revelation with regard to the questions it raises. He also spelled out the condition on which this opinion would be compatible with the Christian faith, a point to which I will return.

Today, almost half a century after the publication of the Encyclical new knowledge has led us to realize that the theory of evolution is no longer a mere hypothesis.[3] It is indeed remarkable that this theory has been progressively accepted by researchers, following a series of discoveries in various fields of knowledge. The convergence, neither sought nor fabricated, of the results of work that was conducted independently is in itself a significant argument in favor of this theory.

What is the significance of such a theory? To address this question is to enter the field of epistemology. A theory is a metascientific elaboration, distinct from the results of observation but consistent with them. By means of it a series of independent data and facts can be related and interpreted in a unified explanation. A theory's validity depends on whether or not it can be verified; it is constantly tested against the facts; wherever it can no longer explain the latter, it shows its limitations and unsuitability. It must then be rethought.

Furthermore, while the formulation of a theory like that of evolution complies with the need for consistency with the observed data, it borrows certain notions from natural philosophy.

---

[3] *Editors' Note*: Here we follow George Coyne's translation of the French text originally printed in *L'Osservatore Romano*; see footnote #15, p. 14, in this volume.

3. Avant de vous proposer quelques réflexions plus spécialement sur le thème de l'origine de la vie et de l'évolution, je voudrais rappeler que le Magistère de l'Église a déjà été amené à se prononcer sur ces matières, dans le cadre de sa propre compétence. Je citerai ici deux interventions.

Dans son encyclique *Humani generis* (1950), mon prédécesseur Pie XII avait déjà affirmé qu'il n'y avait pas opposition entre l'évolution et la doctrine de la foi sur l'homme et sur sa vocation, à condition de ne pas perdre de vue quelques points fermes (cf. *AAS* 42 [1950], pp. 575–76).

Pour ma part, en recevant le 31 octobre 1992 les participants à l'Assemblée plénière de votre Académie, j'ai eu l'occasion, à propos de Galilée, d'attirer l'attention sur la nécessité, pour l'interprétation correcte de la parole inspirée, d'une herméneutique rigoureuse. Il convient de bien délimiter le sens propre de l'Écriture, en écartant des interprétations indues qui lui font dire ce qu'il n'est pas dans son intention de dire. Pour bien marquer le champ de leur objet propre, l'exégète et le théologien doivent se tenir informés des résultats auxquels conduisent les sciences de la nature (cf *AAS* 85 [1993], pp. 764–72; Discours à la Commission biblique pontificale, 23 avril 1993, annonçant le document sur *l'Interprétation de la Bible dans l'Église*: *AAS* 86 [1994], pp. 232–43).

4. Compte tenu de l'état des recherches scientifiques à l'époque et aussi des exigences propres de la théologie, l'encyclique *Humani generis* considérait la doctrine de l' "évolutionnisme" comme une hypothèse sérieuse, digne d'une investigation et d'une réflexion approfondies á l'égal de l'hypothèse opposée. Pie XII ajoutait deux conditions d'ordre méthodologique: qu'on n'adopte pas cette opinion comme s'il s'agissait d'une doctrine certaine et démontrée et comme si on pouvait faire totalement abstraction de la Révélation à propos des questions qu'elle soulève. Il énonçait également la condition á laquelle cette opinion était compatible avec la foi chrétienne, point sur lequel je reviendrai.

Aujourd'hui, près d'un demi-siècle après la parution de l'encyclique, de nouvelles connaissances conduisent à reconnaître dans la théorie de l'évolution plus qu'une hypothèse. Il est en effet remarquable que cette théorie se soit progressivement imposée à l'esprit des chercheurs, à la suite d'une série de découvertes faites dans diverses disciplines du savoir. La convergence, nullement recherchée ou provoquée, des résultats de travaux menés indépendamment les uns des autres, constitue par elle-même un argument significatif en faveur de cette théorie.

Quelle est la portée d'une semblable théorie? Aborder cette question, c'est entrer dans le champ de l'épistémologie. Une théorie est une élaboration métascientifique, distincte des résultats de l'observation mais qui leur est homogène. Grâce à elle, un ensemble de données et de faits indépendants entre eux peuvent être reliés et interprétés dans une explication unitive. La théorie prouve sa validité dans la mesure où elle suceptible d'être vérifiée; elle est constamment mesurée à l'étiage des faits; là où elle cesse de pouvoir rendre compte de ceux-ci, elle manifeste ses limites et son inadaptation. Elle doit alors être repensée.

En outre, l'élaboration d'une théorie comme celle de l'évolution, tout en obéissant à l'exigence d'homogénéite avec les données de l'observation, emprunte certaines notions à la philosophie de la nature.

And, to speak correctly,[4] rather than *the* theory of evolution, we should speak of *several* theories of evolution. On the one hand, this plurality has to do with the different explanations advanced for the mechanism of evolution, and on the other, with the various philosophies on which it is based. Hence the existence of materialist, reductionist, and spiritualist interpretations. What is to be decided here is the true role of philosophy and, beyond it, of theology.

5. The Church's Magisterium is directly concerned with the question of evolution, for it involves the conception of man: Revelation teaches us that he was created in the image and likeness of God (cf. Gen. 1:27–29). The conciliar Constitution *Gaudium et spes* has magnificently explained this doctrine, which is pivotal to Christian thought. It recalled that man is "the only creature on earth that God has wanted for its own sake" (n. 24). In other terms, the human individual cannot be subordinated as a pure means or a pure instrument, either to the species or to society; he has value *per se*. He is a person. With his intellect and his will, he is capable of forming a relationship of communion, solidarity and self-giving with his peers. St. Thomas observes that man's likeness to God resides especially in his speculative intellect, for his relationship with the object of his knowledge resembles God's relationship with what he has created (*Summa Theologica*, I-II, q.3, a.5 ad 1). But even more, man is called to enter into a relationship of knowledge and love with God himself, a relationship which will find its complete fulfilment beyond time, in eternity. All the depth and grandeur of this vocation are revealed to us in the mystery of the risen Christ (cf. *Gaudium et spes*, n. 22). It is by virtue of his spiritual soul that the whole person possesses such a dignity even in his body. Pius XII stressed this essential point: if the human body takes its origin from pre-existent living matter, the spiritual soul is immediately created by God ("*animas enim a Deo immediate creari catholica fides nos retinere iubet*"; Encyclical *Humani Generis*, *AAS* 42 [1950], p. 575).

Consequently, theories of evolution which, in accordance with the philosophies inspiring them, consider the mind as emerging from the forces of living matter, or as a mere epiphenomenon of this matter, are incompatible with the truth about man. Nor are they able to ground the dignity of the person.

6. With man, then, we find ourselves in the presence of an ontological difference, an ontological leap, one could say. However, does not the posing of such ontological discontinuity run counter to that physical continuity which seems to be the main thread of research into evolution in the field of physics and chemistry? Consideration of the method used in the various branches of knowledge makes it possible to reconcile two points of view which would seem irreconcilable. The sciences of observation describe and measure the multiple manifestations of life with increasing precision and correlate them with the time line. The moment of transition to the spiritual cannot be the object of this kind of observation, which nevertheless can discover at the experimental level a series of very valuable signs indicating what is specific to the human being. But the experience of metaphysical knowledge, of self-awareness and self-reflection, of moral conscience, freedom, or again, of aesthetic and religious experience, falls within the competence of philosophical

---

[4] *Editors' Note*: The *L'Osservatore Romano* gives the translation, "to tell the truth".

Et, à vrai dire, plus que de *la* théorie de l'évolution, il convient de parler *des* théories de l'évolution. Cette pluralité tient, d'une part, à la diversité des explications qui ont été proposées du mécanisme de l'évolution et, d'autre part, aux diverses philosophies auxquelles on se réfère. Il existe ainsi des lectures matérialistes et réductionnistes et des lectures spiritualistes. Le jugement ici est de la compétence propre de la philosophie et, au-delà, de la théologie.

5. Le Magistère de l'Église est directement intéressé par la question de l'évolution, car celle-ci touche la conception de l'homme, dont la Révélation nous apprend qu'il e été créé à l'image et à la ressemblance de Dieu (cf. *Gn* 1, 28–29). La Constitution conciliaire *Gaudium et spes* a magnifiquement exposé cette doctrine, qui est un des axes de la pensée chrétienne. Elle a rappelé que l'homme est «la seule créature sur terre que Dieu a voulue pour elle-même» (n. 24). En d'autres termes, l'individu humain ne saurait être subordonné comme un pur moyen ou un pur instrument ni à l'espèce ni à la société; il a valeur pour lui-même. Il est une personne. Par son intelligence et sa volonté, il est capable d'entrer en relation de communion, de solidarité et de don de soi avec son semblable. Saint Thomas observe que la ressemblance de l'homme avec Dieu réside spécialement dans son intelligence spéculative, car sa relation avec l'objet de sa connaissance ressemble à la relation que Dieu entretient avec son œuvre (*Somme théologique*, I–II, q.3, a.5, ad 1). Mais, plus encore, l'homme est appelé à entrer dans une relation de connaissance et d'amour avec Dieu lui-même, relation qui trouvera son plein épanouissement au-delà du temps, dans l'éternité. Dans le mystère du Christ ressuscité nous sont révélées toute la profondeur et toute la grandeur de cette vocation (cf. *Gaudium et spes*, n. 22). C'est en vertu de son âme spirituelle que la personne tout entière jusque dans son corps possède une telle dignité. Pie XII avait souligné ce point essentiel: se le corps humain tient son origine de la matière vivante qui lui préexiste, l'âme spirituelle est immédiatement créée par Dieu («*animas enim a Deo immediate creari catholica fides nos retinere iubet*» (Encycl. *Humani generis, AAS* 42 [1950], p. 575).

En conséquence, les théories de l'évolution qui, en fonction des philosophies qui les inspirent, considèrent l'esprit comme émergeant des forces de la matière vivante ou comme un simple épiphénomène de cette matière sont incompatibles avec la vérité de l'homme. Elles sont d'ailleurs incapables de fonder la dignité de la personne.

6. Avec l'homme, nous nous trouvons donc devant une différence d'ordre ontologique, devant un saut ontologique, pourrait-on dire. Mais poser une telle discontinuité ontologique, n'est-ce pas aller à l'encontre de cette continuité physique qui semble être comme le fil conducteur des recherches sur l'évolution, et ceci dès le plan de la physique et de la chimie? La considération de la méthode utilisée dans les divers ordres du savoir permet de mettre en accord deux points de vue qui sembleraient inconciliables. Les sciences de l'observation décrivent et mesurent avec toujours plus de précision les multiples manifestations de la vie et les inscrivent sur la ligne du temps. Le moment du passage au spirituel n'est pas objet d'une observation de ce type, qui peut néanmoins déceler, au niveau expérimental, une série de signes très précieux de la spécificité de l'être humain. Mais l'expérience du savoir métaphysique, de la conscience de soi et de sa réflexivité, celle de la conscience morale, celle de la liberté, ou encore l'expérience esthétique et

analysis and reflection, while theology brings out its ultimate meaning according to the Creator's plans.

7. In conclusion, I would like to call to mind a Gospel truth which can shed a higher light on the horizon of your research into the origins and unfolding of living matter. The Bible in fact bears an extraordinary message of life. It gives us a wise vision of life inasmuch as it describes the loftiest forms of existence. This vision guided me in the Encyclical which I dedicated to respect for human life, and which I called precisely *Evangelium vitae*.

It is significant that in St. John's Gospel life refers to the divine light which Christ communicates to us. We are called to enter into eternal life, that is to say, into the eternity of divine beatitude.

To warn us against the serious temptations threatening us, our Lord quotes the great saying of *Deuteronomy*: "Man shall not live by bread alone, but by every word that proceeds from the mouth of God" (8:3; cf. Matt. 4:4).

Even more, "life" is one of the most beautiful titles which the Bible attributes to God. He is the *living* God.

I cordially invoke an abundance of divine blessing upon you and upon all who are close to you.

From the Vatican, 22 October 1996.

[signed] Joannes Paulus II

religieuse, sont du ressort de l'analyse et de la réflexion philosophiques, alors que la théologie en dégage le sens ultime selon les desseins du Créateur.

7. En terminant, je voudrais évoquer une vérité évangélique susceptible d'apporter une lumière supérieure à l'horizon de vos recherches sur les origines et le déploiement de la matière vivante. La Bible, en effet, est porteuse d'un extraordinaire message de vie. Elle nous donne sur la vie, en tant qu'elle caractérise les formes les plus hautes de l'existence, une vision de sagesse. Cette vision m'a guidé dans l'encyclique que j'ai consacrée au respect de la vie humaine et que j'ai intitulée précisément *Evangelium vitæ*.

Il est significatif que, dans l'Évangile de saint Jean, la vie désigne la lumière divine que le Christ nous communique. Nous sommes appelés à entrer dans la vie éternelle, c'est-a-dire dans l'éternité de la béatitude divine.

Pour nous mettre en garde contre les tentations majeures qui nous guettent, notre Seigneur cite la grande parole du *Deutéronome*: «Ce n'est pas de pain seul que vivra l'homme, mais de toute parole qui sort de la bouche de Dieu» (8,3; cf. *Mt* 4,4).

Bien plus, la vie est un des plus beaux titres que la Bible ait reconnu à Dieu. Il est le *Dieu vivant*.

De grand cœur, j'invoque sur vous tous et sur ceux qui vous sont proches, l'abondance de Bénédictions divines.

Du Vatican, le 22 octobre 1996.

                                                        Joannes Paulus II

EVOLUTION AND THE HUMAN PERSON: THE POPE IN DIALOGUE[1]

George V. Coyne, S.J.

1 *Introduction*

The scope of this essay is much more limited than one might be led to believe from the rather ambitious-sounding title. And yet I do intend to offer some reflections on each of the topics enunciated in the title and on their nexus. In order to appreciate the recent message of John Paul II on evolution[2] one must see it against both the general backdrop of the science-faith relationship over the past four centuries (since the birth of modern science), and more specifically in light of the opening towards science generated under the current Papacy. An evaluation of the immediate circumstances in which the message was delivered is also important for an understanding of the message itself. I would like to do each of these in turn. Then I will identify the crux issue with which the Pope wrestles, namely, the consonance or dissonance between an evolutionary account of the origin of the human person and the classical religious view that God intervenes to create each person's individual soul. I will share in the Pope's wrestling on this, and I will argue that further dialogue may eventually resolve the issue.

2 *Four Centuries to Set the Large Context*

The message of John Paul II on evolution, received by the members of the Pontifical Academy of Sciences on 22 October 1996 during the Plenary Session of the Academy held in the shadow of St. Peter's Basilica, stirred a vast interest among both scientists and the wider public. This interest went well beyond the usual attention paid to papal statements. An attempt to answer why this was so will also help us, I believe, to appreciate the contents of the message. While the immediate circumstances in which the message occurred provide the principal reasons for the interest aroused, it requires, I believe, a return to three centuries ago to find a full explanation. The Pope himself, in fact, introduces his message in this vein when he asks:

> How do the conclusions reached by the various scientific disciplines coincide with those contained in the message of Revelation?... Moreover, to shed greater light on historical truth, your [the Pontifical Academy's] research on the Church's relations with science between the sixteenth and eighteenth centuries is of great importance.[3]

The relationship between religion and science has, in the course of three centuries, passed from one of conflict to one of compatible openness and dialogue. We might speak of the following four periods of history: the rise of modern atheism in the seventeenth and eighteenth centuries; anticlericalism in Europe in the nineteenth century; the awakening within the Church to modern science in the first six decades of the twentieth century; and the Church's view today. The Church's view of science in each of these periods can be characterized respectively as: Trojan horse, antagonist, enlightened teacher, and partner in dialogue.

---

[1] This essay originally appeared in *Science and Theology: The New Consonance*, ed. Ted Peters, (Boulder, Colo.: Westview Press, 1998), 153–61. Reprinted with permission.

[2] Pope John Paul II, "Message to the Pontifical Academy of Sciences," in this volume.

[3] Ibid., p. 2

2.1 *The Trojan Horse*

In his detailed study of the origins of modern atheism,[4] Michael Buckley concludes that it was paradoxically the attempt in the seventeenth and eighteenth centuries to establish a rational basis for religious belief that led to the corruption of religious belief. Religion yielded to the temptation to root its own existence in the rational certitudes characteristic of the natural sciences. This rationalist tendency found its apex in the enlistment of the new science by such figures as Isaac Newton and Rene Descartes. Although the Galileo case is typically recalled as the classical example of confrontation between science and religion, it is in the misappropriation of modern science to establish the foundations for religious belief that we find the roots of a much more profound confrontation. From these roots, in fact, sprung the divorce between science and religion in the form of modern atheism. Thus science served as a Trojan horse within religion.

2.2 *Antagonist*

As to the second movement in the dissonant symphony initiated by religion and science we turn to nineteenth century anticlericalism. The founding of the Vatican Observatory in 1891 by Pope Leo XIII is set very clearly in that climate of anticlericalism, and one of the principle motives that Leo XIII cites for the foundation is to combat such anticlericalism. His words show very clearly his view of the prevailing mistrust of many scientists for the Church:

> So that they might display their disdain and hatred for the mystical Spouse of Christ, who is the true light, those borne of darkness are accustomed to calumniate her to unlearned people and they call her the friend of obscurantism, one who nurtures ignorance, an enemy of science and progress...[5]

And so the Pope presents, in opposition to these accusations, a very strong, one might say even triumphalistic, view of what the Church wished to do in establishing the Observatory:

> ...in taking up this work we have become involved not only in helping to promote a very noble science, which more than any other human discipline, raises the spirit of mortals to the contemplation of heavenly events, but we have in the first place put before ourselves the plan...that everyone might see that the Church and its Pastors are not opposed to true and solid science, whether human or divine, but that they embrace it, encourage it, and promote it with the fullest possible dedication.[6]

The anticlerical climate of the nineteenth century made science appear to be an antagonist, while the Vatican in its own way was seeking to overcome the antagonism.

2.3 *Enlightened Teacher*

We now pass to a period of enlightenment, the awakening of the Church to science during the first six decades of the twentieth century, which is concretized in the person of Pope Pius XII, who was a man of a more than ordinary scientific culture and who even in his youth had become acquainted with astronomy through his association with astronomers at the Vatican Observatory. The Pope had an excellent

---

[4] Michael J. Buckley, S.J., *At the Origins of Modern Atheism* (New Haven, Conn.: Yale University Press, 1987).

[5] Pope Leo XIII, *Motu Proprio, Ut Mysticam*, published in Sabino Maffeo, S.J., *In the Service of Nine Popes, One Hundred Years of the Vatican Observatory* (Vatican City State: Vatican Observatory Publications, 1991, translated by G. V. Coyne, S.J.), 205.

[6] Ibid., 205.

knowledge of astronomy and he frequently discussed astronomical research with contemporary researchers. However, he was not immune from the rationalist tendency which I spoke about above; and his understanding of the then most recent scientific results concerning the Big Bang origins of the Universe led him to see in these results a rational support for the doctrinal understanding of creation derived from Scripture.

Treating science as an enlightened teacher risks repeating the invitation to the Trojan horse. A specific problem arose from the tendency of Pope Pius XII to identify the beginning state of the Big Bang cosmologies with God's act of creation. He stated, for instance, that

> contemporary science with one sweep back across the centuries has succeeded in bearing witness to the august instant of the primordial *Fiat Lux*, when along with matter there burst forth from nothing a sea of light and radiation... Thus, with that concreteness which is characteristic of physical proofs, modern science has confirmed the contingency of the Universe and also the well-founded deduction to the epoch when the world came forth from the hands of the Creator.[7]

Georges Lemaître, the father of the theory of the primeval atom which foreshadowed the theory of the Big Bang, had considerable difficulty with the Pope's view. Lemaître insisted that the Primeval Atom and Big Bang hypotheses should be judged solely as physical theories and that theological considerations should be kept completely separate.[8]

These contrasting views reached a climax when the time came for the preparation of an address which the Pope was to give to the Eighth General Assembly of the International Astronomical Union to be held in Rome in September 1952. Lemaître came to Rome to consult with the Cardinal Secretary of State concerning the address. The mission was apparently a success. In his discourse delivered on September 7, 1952,[9] Pius XII cited many specific instances of progress made in the astrophysical sciences during the previous half-century, yet he made no specific reference to scientific results from cosmology or the Big Bang. Never again did Pius XII attribute any philosophical, metaphysical, or religious implications to the theory of the Big Bang.

2.4 A Partner in Dialogue

Up until the recent papal discourse on evolution, which we shall discuss shortly, the principal sources for deriving the most recent view from Rome concerning the relationship of science and faith are essentially three messages of His Holiness John Paul II: 1) the discourse given to the Pontifical Academy of Sciences on 10 November 1979 to commemorate the centenary of the birth of Albert Einstein;[10] 2) the discourse given 28 October 1986 on the occasion of the fiftieth anniversary of the Pontifical Academy of Sciences;[11] 3) the message written on the occasion of the tricentennial of Newton's *Principia Mathematica* and published as an introduction

---

[7] Pope Pius XII, *Acta Apostolicae Sedis*, vol. 44, (Vatican City State: Tipografia Poliglotta Vaticana, 1952), 41-42.

[8] Georges Lemaître, "The Primeval Atom Hypothesis and the Problem of Clusters of Galaxies," in *La Structure et L'Evolution de L'Universe* (Bruxelles: XI Conseil de Physique Solay, 1958), 7.

[9] Pope Pius XII, *Acta Apostolicae Sedis*, 732.

[10] See *Discourses of the Popes from Pius XI to John Paul II to the Pontifical Academy of Sciences* (Vatican City State: Pontificia Academia Scientiarum, 1986), 151.

[11] Ibid.

to the proceedings of the meeting sponsored by the Vatican Observatory and the Center for Theology and the Natural Sciences in Berkeley to commemorate that same tricentennial.[12]

The public acceptance of the first two discourses has focused on statements made by Pope John Paul II concerning the Copernican-Ptolemaic controversy of the seventeenth century, especially the place of Galileo in those controversies. This has been an excessive emphasis, in my opinion. The Galileo affair is important, to be sure. However, if one reads the three papal documents which I have referred to above, it will be clear that there are many matters that are much more significant and forward-looking than a reinvestigation of the Galileo case. The newness in what John Paul II has said about the relationship consists in his having taken a position compellingly opposed to each of the three postures of Trojan horse, antagonist, or enlightened teacher. For instance, John Paul II clearly states that

> science develops best when its concepts and conclusions are integrated into the broader human culture and its concerns for ultimate meaning and value… Scientists… can come to appreciate for themselves that these discoveries cannot be a genuine substitute for knowledge of the truly ultimate. Science can purify religion from error and superstition; religion can purify science from idolatry and false absolutes. Each can draw the other into a wider world, a world in which both can flourish.[13]

Science and religion are distinct. Each has its own integrity. In dialogue, each can contribute positively to the welfare of the other.

The newest element in this new view from Rome is the expressed uncertainty as to where the dialogue between science and faith will lead. Whereas the awakening of the Church to modern science during the papacy of Pius XII resulted in a too facile appropriation of scientific results to bolster religious beliefs, Pope John II expresses the caution of the Church in defining its partnership in the dialogue: "Exactly what form that [the dialogue] will take must be left to the future."[14]

3 The Immediate Context for the Evolution Allocution

The message on evolution is in continuity with the partner-in-dialogue posture. While the encyclical of Pope Pius XII in 1950, *Humani Generis*, considered the doctrine of evolution a serious hypothesis, worthy of investigation and in-depth study equal to that of the opposing hypothesis, John Paul II states in his message: "Today almost half a century after the publication of the encyclical, new knowledge has led to the recognition that the theory of evolution is no longer a mere hypothesis."[15]

The sentences which follow this statement indicate that the "new knowledge" to which the Pope refers is for the most part scientific knowledge. He had, in fact, just stated that "the exegete and the theologian must keep informed about the results

---

[12] The message was first published in Robert J. Russell, William R. Stoeger, S.J. and George V. Coyne, S.J., eds., *Physics, Philosophy, and Theology: A Common Quest for Understanding* (Vatican City State: Vatican Observatory, 1988), pp. M3-M14. Comments on the papal message by a group of experts have been published in idem, eds., *John Paul II on Science and Religion: Reflections on the New View from Rome* (Vatican City State: Vatican Observatory, 1990).

[13] *Physics, Philosophy, and Theology*, M13.

[14] Ibid., M7.

[15] Pope John Paul II, "Message to the Pontifical Academy of Sciences," p. 14 (in this volume). The English translation of this sentence, printed on p. 7 of the Weekly Edition of *L'Osservatore Romano*, 30 October, 1996, is incorrect when it says, "… new knowledge has led to the recognition of more than one hypothesis in the theory of evolution."

achieved by the natural sciences." The context in which the message occurs strongly supports this. As the specific theme for its plenary session the Pontifical Academy of Sciences had chosen: *The Origin and Evolution of Life*, and it had assembled some of the most active researchers in the life sciences to discuss topics which ranged from "Molecular Phylogeny as a Key to Understanding the Origin of Cellular Life" to "The Search for Intelligent Life in the Universe" and "Life as a Cosmic Imperative"; from, that is, detailed molecular chemistry to sweeping analyses of life in the context of the evolving universe. Only months before the plenary session of the Academy the renowned journal *Science* published a research paper announcing the discovery that in the past there may have existed primitive life forms on the planet Mars. Furthermore within the previous two years a number of publications had appeared announcing the discovery of extra-solar planets. This ferment in scientific research not only made the plenary session theme very timely, but it also set the concrete scene for the papal message. Most of the scientific results cited were very tentative and very much disputed (as is true of almost all research at its beginning), but they were very exciting and provocative. Only three months after the plenary session the Pope would receive in private audience a group of scientists from Germany, Italy, and the United States who were responsible for the high-resolution observations being made by the satellite, *Galileo*, of the Jovian planets and their satellites. Within a few months of that audience NASA would announce the discovery of a huge ocean on Europa, a satellite of Jupiter.

These are the circumstances surrounding the papal message on evolution. Did they influence it? Normally the Pope receives the papal academicians at the time of their Plenary Session in a solemn, private audience, at times even in the presence of the College of Cardinals and the diplomatic corps. On this occasion he did not receive them at all, but rather sent his message to them.

Why? There can, of course, be many reasons for this, unknown and perhaps even unknowable to the historian. I would like to suggest, nonetheless, that a contributing factor to the nature of this message may be found precisely in the circumstances of the plenary session and the accompanying milieu of scientific research. A careful reading of the message is consistent with this suggestion. The Pope wished to recognize the great strides being made in our scientific knowledge of life and the implications that may result for a religious view of the human person; but at the same time he had to struggle with the tentative nature of those results and their consequences, especially with respect to revealed, religious truths. In other words an openness in dialogue appeared to be the most honest posture to take. Let us examine the message in this respect.

4 Science versus Revelation; Evolution versus Creation

In order to set the stage for dialogue the message distinguishes in traditional terms the various ways of knowing. The correct interpretation of observed, empirical, scientific data accumulated to date leads to a theory of evolution which is no longer a mere hypothesis among other hypotheses. It is an established scientific theory. But since philosophy and theology, in addition to the scientific analysis of the empirical facts, enter into the formulation of a theory, we do better to speak of several theories. And some of those theories are incompatible with revealed, religious truth. It is obvious that some theories are to be rejected outright: materialism, reductionism, spiritualism. But at this point the message embraces a true spirit of dialogue when it struggles with the two possibly incompatible stories about the origin of the human person: the evolutionist story and the creationist story. By "creationist" here I am not referring to the American evangelical or fundamentalist view that Genesis constitutes a scientific account of the world's origin. Rather, in the context of the papal discussion, "creationism" refers to the view that at conception God intervenes to create a new and unique soul for the individual person. It appears that no room exists within evolutionary theory to account for the divine creation of the human soul. This amounts to dissonance between science and religion, not consonance. This is the crux issue in Pope John Paul II's message.

The dialogue progresses in the following way: The Church holds certain revealed truths concerning the human person. Science has discovered certain facts about the origins of the human person. Any theory based upon those facts which contradicts revealed truths cannot be correct. Note the antecedent and primary role given to revealed truths in this dialogue; and yet note the struggle to remain open to a correct theory based upon the scientific facts. The dialogue proceeds, in anguish as it were, between these two poles. In the traditional manner of papal statements the main content of the teaching of previous Popes on the matter at hand is reevaluated. And so the teaching of Pius XII in *Humani Generis* is that, if the human body takes its origins from pre-existent living matter, the spiritual soul is immediately created by God. Is the dialogue resolved by embracing evolutionism as to the body and creationism as to the soul? Note that the word "soul" does not reappear in the remainder of the dialogue. Rather the message moves to speak of "spirit" and "the spiritual."

If we consider the revealed, religious truth about the human being, then we have an "ontological leap," an "ontological discontinuity" in the evolutionary chain at the emergence of the human being. Is this not irreconcilable, wonders the Pope, with the continuity in the evolutionary chain seen by science? An attempt to resolve this critical issue is given by stating that:

> The moment of transition to the spiritual cannot be the object of this kind of [scientific] observation, which nevertheless can discover at the experimental level a series of very valuable signs indicating what is specific to the human being.

The suggestion is being made, it appears, that the "ontological discontinuity" may be explained by an epistemological discontinuity. Is this adequate or must the dialogue continue? Is a creationist theory required to explain the origins of the spiritual dimension of the human being? Are we forced by revealed, religious truth to accept a dualistic view of the origins of the human person? Are we forced to be evolutionist with respect to the material dimension, and creationist with respect to the spiritual dimension? The message, I believe, when it speaks in the last paragraphs about the God of life, gives strong indications that the dialogue is still open with respect to these questions.

5 Consonance and Continuous Creation

I would like to use the inspiration of Pope John Paul II's address to suggest that reflections upon God's continuous creation may help to advance the dialogue with respect to the dualistic dilemma mentioned above. We might say that God creates through the process of evolution and that creation is continuous. Since we assume that ultimately there can be no contradiction between true science and revealed, religious truths, I propose that continuous creation is best understood in terms of the best scientific understanding of the emergence of the human being.

The concern for continuous creation arises in the following summary statement by the eminent evolutionary chemist, Christian de Duve, in his paper at the very Plenary Session of the Pontifical Academy of Sciences to which the papal message on evolution was directed:

> ...evolution, though dependent on chance events, proceeds under a number of inner and outer constraints that compel it to move in the direction of greater complexity if circumstances permit. Had these circumstances been different, evolution might have followed a different course in time. It might have produced organisms different from those we know, perhaps even thinking beings different than humans.[16]

Does such contingency in the emergence of the human being contradict religious truth? No; it appears to me that we might find consonance if theologians could develop a more profound understanding of God's continuous creation.

God in his infinite freedom continuously creates a world which reflects that freedom, a freedom that grows as the evolutionary process draws us toward greater and greater complexity. God lets the world be what it will be in its continuous evolution. God does not intervene arbitrarily, but rather allows, participates, loves. Is such thinking adequate to preserve the special character attributed by religious thought to the emergence of spirit, while avoiding a crude interventionist creationism? Only a protracted dialogue will tell.

---

[16] Christian de Duve, "Life as a Cosmic Imperative," *Pontifical Academy of Sciences*, October 1996; see also his book *Vital Dust: Life as a Cosmic Imperative* (New York: Basic Books, 1995).

I

SCIENTIFIC BACKGROUND

THE EVOLUTION OF LIFE: AN OVERVIEW

Francisco J. Ayala

1 *Introduction*

The great Russian-American geneticist and evolutionist Theodosius Dobzhansky wrote in 1973 that "Nothing in biology makes sense except in the light of evolution." The evolution of organisms, that is, their common descent with modification from simple ancestors that lived many million years ago, is at the core of genetics, biochemistry, neurobiology, physiology, ecology, and other biological disciplines, and makes sense of the emergence of new infectious diseases and other matters of public health. The evolution of organisms is universally accepted by biological scientists, while the mechanisms of evolution are still actively investigated and are the subject of debate among scientists.

The nineteenth-century English naturalist, Charles Darwin, argued that organisms come about by evolution, and he provided a scientific explanation, essentially correct but incomplete, of how evolution occurs and why it is that organisms have features—such as wings, eyes, and kidneys—clearly structured to serve specific functions. Natural selection was the fundamental concept in his explanation. Genetics, a science born in the twentieth century, revealed in detail how natural selection works and led to the development of the modern theory of evolution. Since the 1960s a related scientific discipline, molecular biology, has advanced enormously our knowledge of biological evolution and has made it possible to investigate detailed problems that seemed completely out of reach a few years earlier—for example, how similar the genes of humans, chimpanzees, and gorillas are (they differ in about 1 or 2 percent of the units that make up the genes.)

The diversity of living species is staggering. More than two million existing species of plants and animals have been named and described: many more remain to be discovered, at least ten million according to most estimates. What is impressive is not just the numbers but also the incredible heterogeneity in size, shape, and ways of life: from lowly bacteria, less than one thousandth of a millimeter in diameter, to the stately sequoias of California, rising 300 feet (100 meters) above the ground and weighing several thousand tons; from microorganisms living in the hot springs of Yellowstone National Park at temperatures near the boiling point of water—some like *Pyrolobus fumarii* are able to grow at more than 100° C (212° F)—to fungi and algae thriving on the ice masses of Antarctica and in saline pools at -23° C (-73° F); from the strange worm-like creatures discovered in dark ocean depths at thousands of feet below the surface to spiders and larkspur plants existing on Mt. Everest nearly 20,000 feet above sea level.

These variations on life are the outcome of the evolutionary process. All organisms are related by descent from common ancestors. Humans and other mammals are descended from shrew-like creatures that lived more than 150 million years ago; mammals, birds, reptiles, amphibians, and fishes share as ancestors small worm-like creatures that lived in the world's oceans 600 million years ago; plants and animals are derived from bacteria-like microorganisms that originated more than three billion years ago. Because of biological evolution, lineages of organisms change through time; diversity arises because lineages that descend from common ancestors diverge through the generations as they become adapted to different ways of life.

I intend to present in this article a brief summary of some central tenets of the theory of biological evolution. The process by which planets, stars, galaxies, and the universe form and change over time is a type of "evolution," but they evolve in a different sense, although both imply change over time. I will not discuss the evolution of the universe. The evolution of the hominids, the lineage that leads to our own species will be very briefly outlined at the end.

Contrary to popular opinion, neither the term nor the idea of biological evolution began with Charles Darwin and his foremost work *On the Origin of Species by Means of Natural Selection* (1859). The *Oxford English Dictionary* (1933) tells us that the word *evolution*, to unfold or open out, derives from the Latin *evolvere*, which applied to the "unrolling of a book." It first appeared in the English language in 1647 in a non-biological connection, and it became widely used in English for all sorts of progressions from simpler beginnings. Evolution was first used as a biological term in 1670 to describe the changes observed in the maturation of insects. However, it was not until the 1873 edition of *The Origin of Species* that Darwin first used the term. He had earlier used the expression "descent with modification," which is still a good brief definition of biological evolution.

A distinction must be drawn at the outset between the questions 1) *whether* and 2) *how* biological evolution happened. The first refers to the finding, now supported by an overwhelming body of evidence, that descent with modification has occurred during some 3.5 billion years of the Earth's history. The second refers to the theory explaining how those changes came about. The mechanisms accounting for these changes are still undergoing investigation; the currently favored theory is an extensively modified version of Darwinian natural selection.

2 Early Ideas About Evolution

Explanations for the origin of the world, humans, and other creatures are found in all human cultures. Traditional Judaism, Christianity, and Islam explain the origin of living beings and their adaptations to their environments—wings, gills, hands, flowers—as the handiwork of an omniscient God. The philosophers of ancient Greece had their own creation accounts. Anaximander proposed that animals could be transformed from one kind into another, and Empedocles speculated that they could be made of various combinations of preexisting parts. Closer to modern evolutionary ideas were the proposals of early Church Fathers like Gregory of Nazianzus and Augustine, who maintained that not all species of plants and animals were created as such by God; rather some had developed in historical times from creatures created earlier by God. Their motivation was not biological but religious. Some species must have come into existence only after the Noachian Flood because it would have been impossible to hold representatives of all species in a single vessel such as Noah's Ark.

Christian theologians of the Middle Ages did not directly explore the notion that organisms may change by natural processes, but the matter was, usually incidentally, considered as a possibility by many, including Albertus Magnus and his student Thomas Aquinas. Aquinas concluded, after detailed discussion, that the development of living creatures like maggots and flies from non-living matter like decaying meat was not incompatible with Christian faith or philosophy. But he left it to scientists to decide whether this actually happened.

In the eighteenth century, Pierre-Louis Moreau de Maupertuis proposed the spontaneous generation and extinction of organisms as part of his theory of origins, but he advanced no theory about the possible transformation of one species into another through knowable, natural causes. One of the greatest naturalists of the time, Georges-Louis Leclerc (Buffon) explicitly considered—and rejected—the possible descent of several distinct kinds of organisms from a common ancestor. However, he made the claim that organisms arise from organic molecules by spontaneous generation, so that there could be as many kinds of animals and plants as there are viable combinations of organic molecules.

Erasmus Darwin, grandfather of Charles Darwin, offered in his *Zoonomia or the Laws of Organic Life* some evolutionary speculations, but they were not systematically developed and had no real influence on subsequent theories. The Swedish botanist Carolus Linnaeus devised the hierarchical system of plant and animal classification that is still in use in a modernized form. Although he insisted on the fixity of species, his classification system eventually contributed much to the acceptance of the concept of common descent.

The great French naturalist Jean-Baptiste Lamarck held the view that living organisms represent a progression, with humans as the highest form. In his *Philosophical Zoology*, published in 1809, the year in which Charles Darwin was born, he proposed the first broad theory of evolution. Organisms evolve through eons of time from lower to higher forms, a process still going on and always culminating in human beings. As organisms become adapted to their environments through their habits, modifications occur. Use of an organ or structure reinforces it; disuse leads to obliteration. The characteristics acquired by use and disuse, according to this theory, would be inherited. This assumption, later called the inheritance of acquired characteristics, was thoroughly disproved in the twentieth century. The notion that the same organisms repeatedly evolve in a fixed sequence of transitions has also been disproved.

3 Darwin's Theory

Charles Darwin is appropriately considered the founder of the modern theory of evolution. The son and grandson of physicians, he enrolled as a medical student at the University of Edinburgh. After two years, however, he left to study at Cambridge University and prepare to become a clergyman. He was not an exceptional student, but he was deeply interested in natural history. On December 27, 1831, a few months after his graduation from Cambridge, he sailed as a naturalist aboard the HMS Beagle on a round-the-world trip that lasted until October 1836. Darwin was often able to disembark for extended trips ashore to collect specimens.

In Argentina he studied fossil bones from large extinct mammals. In the Galápagos Islands he observed numerous species of finches. These are among the events credited with stimulating Darwin's interest in how different species arise and become extinct. In 1859 he published *The Origin of Species*, a treatise providing extensive evidence for the evolution of organisms and proposing natural selection as the key process determining its course. He published many other books as well, notably *The Descent of Man and Selection in Relation to Sex* (1871), which provides an evolutionary account of human origins.

The origin of the Earth's living things, with their marvelous contrivances for adaptation, were generally attributed to the design of an omniscient God. In the nineteenth century, Christian theologians had argued that the presence of design, so evident in living beings, demonstrates the existence of a supreme Creator. The

British theologian William Paley in his *Natural Theology* (1802) used natural history, physiology, and other contemporary knowledge to elaborate this argument from design. If a person should find a watch, even in an uninhabited desert, Paley contended, the intricate design and harmony of its many parts would force him to conclude that it had been created by a skilled watchmaker. How much more intricate and perfect in design is the human eye, Paley went on, with its transparent lens, its retina placed at the precise distance for forming a distinct image, and its large nerve transmitting signals to the brain.

Natural selection was proposed by Darwin primarily to account for the adaptive organizations of living beings: it is a process that promotes or maintains adaptation and, thus, gives the appearance of purpose or design.[1] Evolutionary change through time and evolutionary diversification (multiplication of species) are not directly promoted by natural selection, but they often ensue as by-products of natural selection as it fosters adaptation to different environments. Darwin's theory of natural selection is summarized in the *Origin of Species* as follows:

> As more individuals are produced than can possibly survive, there must in every case be a struggle for existence, either one individual with another of the same species, or with the individuals of distinct species, or with the physical conditions of life... Can it, then, be thought improbable, seeing that variations useful to man have undoubtedly occurred, that other variations useful in some way to each being in the great and complex battle of life, should sometimes occur in the course of thousands of generations? If such do occur, can we doubt (remembering that many more individuals are born than can possibly survive) that individuals having any advantage, however slight, over others, would have the best chance of surviving and of procreating their kind? On the other hand, we may feel sure that any variation in the least degree injurious would be rigidly destroyed. This preservation of favorable variations and the rejection of injurious variations, I call Natural Selection.[2]

The publication of the *Origin of Species* produced considerable public excitement. Scientists, politicians, clergymen, and notables of all kinds read and discussed the book, defending or deriding Darwin's ideas. The most visible actor in the controversies immediately following publication was T. H. Huxley, knows as "Darwin's bulldog," who defended the theory of evolution with articulate and sometimes mordant words on public occasions as well as in numerous writings. Serious scientific controversies also arose, first in Britain and then on the Continent and in the United States.

One occasional participant in the discussion was the naturalist Alfred Russel Wallace, who had independently discovered natural selection and had sent a short manuscript to Darwin from the Malay archipelago. A contemporary of Darwin with considerable influence during the latter part of the nineteenth and early twentieth centuries was Herbert Spencer. He was a philosopher rather than a biologist, but he became an energetic proponent of evolutionary ideas, popularized a number of slogans, like "survival of the fittest" (which was taken up by Darwin in later editions of the *Origin*), and engaged in social and metaphysical speculations. His ideas considerably damaged proper understanding and acceptance of the theory of evolution by natural selection. Most pernicious was Spencer's crude extension of the notion of "struggle for existence" to human economic and social life that became known as social Darwinism.

---

[1] See my "Darwin's Devolution: Design Without Designer," in this volume.
[2] Charles Darwin, *Origin of Species* (New York: Avenel Books, 1979), 117, 130–31.

The most serious difficulty facing Darwin's evolutionary theory was the lack of an adequate theory of inheritance that would account for the preservation through the generations of the variations on which natural selection was supposed to act. Current theories of "blending inheritance" proposed that the characteristics of parents became averaged in the offspring. As Darwin became aware, "blending inheritance" could not account for the conservation of variations, because halving the differences among variant offspring would rapidly reduce the original variation to the average of the preexisting characteristics.

Mendelian genetics provided the missing link in Darwin's argument. About the time the *Origin of Species* was published, the Augustinian monk Gregor Mendel was performing a long series of experiments with peas in the garden of his monastery in Brünn (now Brno, Czech Republic). These experiments and the analysis of their results are an example of masterly scientific method. Mendel's theory accounts for biological inheritance through particulate factors (genes) inherited one from each parent, which do not mix or blend but segregate in the formation of the sex cells, or gametes. Mendel's discoveries, however, remained unknown to Darwin and, indeed, did not become generally known until 1900, when they were simultaneously rediscovered by a number of scientists on the Continent.

Darwinism, in the latter part of the nineteenth century, faced an alternative evolutionary theory known as neo-Lamarckism. This hypothesis shared with Lamarck's the importance of use and disuse in the development and obliteration of organs, and it added the notion that the environment acts directly on organic structures, which would explain their adaptation to their environments and ways of life. Adherents of this theory discarded natural selection as an explanation for adaptation to the environment. Prominent among the defenders of natural selection was the German biologist August Weismann, who in the 1880s published his germ-plasm theory. He distinguished two components in the make up an organism: the soma, which comprises most body parts and organs, and the germ-plasm, which contains the cells that give rise to the gametes and hence to progeny. The radical separation between germ and soma prompted Weismann to assert that inheritance of acquired characteristics was impossible, and it opened the way for his champion-ship of natural selection as the only major process that would account for biological evolution. The formulation of the evolutionary theory championed by Weismann and his followers toward the end of the nineteenth century became known as "neo-Darwinism."

4 *The Modern Theory of Evolution*

The rediscovery in 1900 of Mendel's theory of heredity ushered in an emphasis on the role of heredity in evolution. Hugo de Vries in the Netherlands proposed a new theory of evolution known as mutationism, which essentially did away with natural selection as a major evolutionary process. According to de Vries (joined by other geneticists such as William Bateson in England), there are two kinds of variation that take place in organisms. One is the "ordinary" variability observed among individuals of a species, which is of no lasting consequence in evolution because, according to de Vries, it could not "lead to a transgression of the species border even under conditions of the most stringent and continued selection." The other consists of the changes brought about by mutations, spontaneous alterations of genes that yield large modifications of the organism and gave rise to new species. Mutationism was opposed by many naturalists, and in particular by the so-called biometricians, led by Karl Pearson, who defended Darwinian natural selection as the major cause of evolution through the cumulative effects of small, continuous, individual variations.

Arguments between mutationists (also referred to at the time as Mendelians) and biometricians approached a resolution in the 1920s and '30s through the theoretical work of geneticists. They used mathematical arguments to show, first, that continuous variation (in such characteristics as size, number of eggs laid, and the like) could be explained by Mendel's laws; and second, that natural selection acting cumulatively on small variations could yield major evolutionary changes in form and function. Distinguished members of this group of theoretical geneticists were R. A. Fisher and J. B. S. Haldane in Britain and Sewall Wright in the United States. Their work provided a theoretical framework for the integration of genetics into Darwin's theory of natural selection. Yet their work had a limited impact on contemporary biologists because it was almost exclusively theoretical, formulated in mathematical language and with little empirical corroboration. A major breakthrough came in 1937 with the publication of *Genetics and the Origin of Species* by Theodosius Dobzhansky, a Russian-born American naturalist and experimental geneticist.

Dobzhansky advanced a reasonably comprehensive account of the evolutionary process in genetic terms, laced with experimental evidence supporting the theoretical argument. *Genetics and the Origin of Species* may be considered the most important landmark in the formulation of what came to be known as the synthetic theory of evolution, effectively combining Darwinian natural selection and Mendelian genetics. It had an enormous impact on naturalists and experimental biologists, who rapidly embraced the new understanding of the evolutionary process as one of genetic change in populations. Interest in evolutionary studies was greatly stimulated, and contributions to the theory soon began to follow, extending the synthesis of genetics and natural selection to a variety of biological fields. Other writers who importantly contributed to the formulation of the synthetic theory were the zoologists Ernst Mayr and Sir Julian Huxley, the paleontologist George G. Simpson, and the botanist George Ledyard Stebbins. By 1950 acceptance of Darwin's theory of evolution by natural selection was universal among biologists, and the synthetic theory had become widely adopted.

Since 1950, the most important line of investigation has been the application of molecular biology to evolutionary studies. In 1953 James Watson and Francis Crick discovered the structure of DNA (deoxyribonucleic acid), the hereditary material contained in the chromosomes of every cell's nucleus. The genetic information is contained within the sequence of components (nucleotides) that make up the long chainlike DNA molecules, very much in the same manner as semantic information is contained in the sequence of letters in an English text. This information determines the sequence of amino acids in the proteins, including the enzymes that carry out the organism's life processes. Comparisons of the amino-acid sequences of proteins in different species provides quantitatively precise measures of species divergence, a considerable improvement over the typically qualitative evaluations obtained by comparative anatomy and other evolutionary subdisciplines.

In 1968 the Japanese geneticist Motoo Kimura proposed the neutrality theory of molecular evolution, which assumes that at the level of DNA and protein sequence many changes are adaptively neutral and have little or no effect on the molecule's function. If the neutrality theory is correct, there should be a "molecular clock" of evolution; that is, the degree of divergence between species in amino acid or nucleotide sequence would provide a reliable estimate of the time since their divergence. This would make possible a reconstruction of evolutionary history that would reveal the order of branching of different lineages, such as those leading to humans, chimpanzees, and orangutans, as well as the time in the past when the lineages split from one another. During the 1970s and '80s it gradually became clear that the molecular clock is not exact; nevertheless, it has become a reliable source of evidence for reconstructing a history of evolution. In the 1990s, the techniques of DNA cloning and sequencing have provided new and more powerful means of investigating evolution at the molecular level.

Important discoveries in the earth sciences and ecology during the second half of the twentieth century have also greatly advanced our understanding of the theory of evolution. The science of plate tectonics has shown that the configuration and position of the continents and oceans are dynamic, rather than static, features of the Earth. Oceans grow and shrink, while continents break into fragments or coalesce into larger masses. The continents move across the Earth's surface at rates of a few centimeters a year, and over millions of years of geological history this profoundly alters the face of the Earth, causing major climatic changes along the way. These previously unsuspected massive modifications of the planet's environments have of necessity been reflected in the evolutionary history of life. Biogeography, the evolutionary study of plant and animal distribution, has been revolutionized by the knowledge, for example, that Africa and South America were part of a single landmass some 200 million years ago and that the Indian subcontinent was not connected with Asia until recent geologic times. The study of the interactions of organisms with their environments, known as the discipline of ecology, has evolved from descriptive studies—"natural history"—into a vigorous biological discipline with a strong mathematical component, both in the development of theoretical models and in the collection and analysis of quantitative data. Another active field of research in evolutionary biology is evolutionary ethology, the study of animal behavior. Sociobiology, the evolutionary study of social behavior, is perhaps the most active subfield of ethology and, because of its extension to human societies, the most controversial.

5 The Impact of Evolutionary Theory

Three different, though related, issues have been the main subjects of evolutionary investigations: 1) the fact of evolution; that is, that organisms are related by common descent with modification; 2) evolutionary history; that is, the details of when lineages split from one another and of the changes that occurred in each lineage; and 3) the mechanisms or processes by which evolutionary change occurs.

The fact of evolution is the most fundamental issue and the one established with utmost certainty. Darwin gathered much evidence in its support, but the evidence has accumulated continuously ever since, derived from all biological disciplines. As Pope John Paul II has noted, "It is indeed remarkable that this theory [of evolution] has been progressively accepted by researchers, following a series of discoveries in various fields of knowledge. The convergence, neither sought nor fabricated, of the results of work that was conducted independently is in itself a significant argument in favor of this theory."[3] Indeed, the evolutionary origin of organisms is today a scientific conclusion established with the kind of certainty attributable to such scientific concepts as the roundness of the Earth, the heliocentric

---

[3] Message to the Pontifical Academy of Sciences on 22 October 1996, reprinted in this volume.

motions of the planets, and the molecular composition of matter. This degree of certainty beyond reasonable doubt is what is implied when biologists say that evolution is a "fact"; the evolutionary origin of organisms is accepted by virtually every biologist.

The second and third issues go much beyond the general affirmation that organisms evolve. The theory of evolution seeks to ascertain the evolutionary relationships between particular organisms and the events of evolutionary history, as well as to explain how and why evolution takes place. These are matters of active scientific investigation. Many conclusions are well established; for example, that the chimpanzee and gorilla are more closely related to humans than is any of those three species to the baboon or other monkeys; or that natural selection explains the adaptive configuration of such features as the human eye and the wings of birds. Some other matters are less certain, others are conjectural, and still others—such as precisely when life originated on earth and the characteristics of the first living things—remain largely unresolved.

The theory of evolution has been seen by some people as incompatible with religious beliefs, particularly those of Christianity.[4] The first chapters of the book of Genesis describe God's creation of the world, the plants, the animals, and humans. A literal interpretation of Genesis seems incompatible with the gradual evolution of humans and other organisms by natural processes. Independently of the biblical narrative, the belief in the immortality of the soul and in humans as "created in the image of God" have appeared to some as contrary to the evolutionary origin of humans from nonhuman animals.[5]

Religiously motivated attacks started during Darwin's lifetime. In 1874 Charles Hodge, an American Protestant theologian, perceived Darwin's theory as "the most thoroughly naturalistic that can be imagined and far more atheistic than that of his predecessor Lamarck." He argued that the design of the human eye evinces that "it has been planned by the Creator, like the design of a watch evinces a watchmaker." He concluded that "the denial of design in nature is actually the denial of God." But, other Protestant theologians saw a solution to the difficulty in the idea that God operates through intermediate causes. The origin and motion of the planets could be explained by the law of gravity and other natural processes without denying God's creation and providence. Similarly, evolution could be seen as the natural process through which God brought living beings into existence and developed them according to his plan. Thus, A. H. Strong, the president of Rochester (N.Y.) Theological Seminary, wrote in his *Systematic Theology* (1885): "We grant the principle of evolution, but we regard it as only the method of divine intelligence." The brutish ancestry of human beings was not incompatible with their excelling status as a creature in the image of God.

More recently, biblical Fundamentalists have periodically gained considerable public and political influence in the United States. During the decade of the 1920s,

---

[4] An excellent collection of essays on the relations between Christianity and science is David C. Lindberg and Ronald N. Numbers, eds., *God and Nature: Historical Essays on the Encounter between Christianity and Science* (Berkeley and Los Angeles: University of California Press, 1986).

[5] The perception of conflict is far from universal. Pope John Paul II is cited below. On the matter of the immortality of the soul and, more generally, the uniqueness of humans as created "in the image of God," see Warren S. Brown, Nancey Murphy, and H. Newton Malony, eds., *Portraits of the Soul: Scientific and Theological Perspectives on Human Nature* (Minneapolis, Minn.: Fortress Press, 1998).

more than twenty state legislatures were influenced by them to debate anti-evolution laws, and four states—Arkansas, Mississippi, Oklahoma, and Tennessee—prohibited the teaching of evolution in their public schools. But, in 1968 the Supreme Court of the United States declared unconstitutional any law banning the teaching of evolution in public schools.

Arguments for and against Darwin's theory have come from both mainline Protestant and Catholic theologians. Gradually, well into the twentieth century, evolution by natural selection came to be accepted by the majority of Christian writers. Pope Pius XII in his encyclical *Humani Generis* (1950) acknowledged that biological evolution was compatible with the Christian faith, although he argued that God's intervention was necessary for the creation of the human soul. In 1981 Pope John Paul II stated in an address to the Pontifical Academy of Sciences:

> The Bible itself speaks to us of the origin of the universe and its make-up, not in order to provide us with a scientific treatise but in order to state the correct relationships of man with God and with the universe. Sacred scripture wishes simply to declare that the world was created by God, and in order to teach this truth it expresses itself in the terms of the cosmology in use at the time of the writer... Any other teaching about the origin and make-up of the universe is alien to the intentions of the Bible, which does not wish to teach how the heavens were made but how one goes to heaven.

The Pope's point was that it would be wrong to mistake the Bible for an elementary book of astronomy, geology, and biology. John Paul II returned in 1996 to the same topic. In the address to the Pontifical Academy of Sciences cited earlier, he stated that "the theory of evolution is no longer a mere hypothesis."

6 *The Evidence for Common Descent with Modification*

Evidence for relationship among organisms by common descent with modification has been obtained by paleontology, comparative anatomy, biogeography, embryology, biochemistry, molecular genetics, and other biological disciplines. The idea first emerged from observations of systematic changes in the succession of fossil remains found in a sequence of layered rocks. Such layers are now known to have a cumulative thickness of many scores of kilometers and to represent at least 3.5 billion years of geological time. The general sequence of fossils from bottom upward in layered rocks had been recognized before Darwin perceived that the observed progression of biological forms strongly implied common descent. The farther back into the past one looked, the less the fossils resembled recent forms, the more the various lineages merged, and the broader the implications of a common ancestry appeared.

Paleontology, however, was still a rudimentary science in Darwin's time, and large parts of the geological succession of stratified rocks were unknown or inadequately studied. Darwin, therefore, worried about the rarity of truly intermediate forms. Anti-evolutionists have then and now seized on this as a weakness in evolutionary theory. Although gaps in the paleontological record remain even now, many have been filled by the research of paleontologists since Darwin's time. Hundreds of thousands of fossil organisms found in well-dated rock sequences represent a succession of forms through time and manifest many evolutionary transitions. Microbial life of the simplest type (namely, prokaryotes, which are cells whose nuclear matter is not bounded by a nuclear membrane) was already in existence more than 3 billion years ago. The oldest evidence suggesting the existence of more complex organisms (namely, eukaryotic cells with a true nucleus) has been discovered in fossils that had been sealed in flinty rocks approximately 1.4 billion years old. More advanced forms like true algae, fungi, higher plants, and animals have been found only in younger geological strata. The following list presents the order in which progressively complex forms of life appeared:

Life Form	Millions of Years Since First Known Appearance (Approximate)
Microbial (prokaryotic cells)	3,500
Complex (eukaryotic cells)	1,400
First multicellular animals	670
Shell-bearing animals	540
Vertebrates (simple fishes)	490
Amphibians	350
Reptiles	310
Mammals	200
Nonhuman primates	60
Earliest apes	25
Australopithecine ancestors	5
Homo sapiens (modern humans)	0.15 (150,000 years)

Table 1. Order of Appearance of Life Forms on Earth

The sequence of observed forms and the fact that all except the first are constructed from the same basic cellular type strongly imply that all these major categories of life (including plants, algae, and fungi) have a common ancestry in the first eukaryotic cell. Moreover, there have been so many discoveries of intermediate forms between fish and amphibians, between amphibians and reptiles, between reptiles and mammals, and even along the primate line of descent from apes to humans, that it is often difficult to identify categorically when the transition occurs along the line from one to another particular genus or from one to another particular species. Nearly all fossils can be regarded as intermediates in some sense; they come between ancestral forms that preceded them and those that followed.

The fossil record thus provides compelling evidence of systematic change through time—of descent with modification. From this consistent body of evidence it can be predicted that no reversals will be found in future paleontological studies. That is, amphibians will not appear before fishes nor mammals before reptiles, and no complex life will occur in the geological record before the oldest eukaryotic cells. That prediction has been upheld by the evidence that has accumulated thus far: no reversals have been found.

Although some creationists have claimed that the entire geological record, with its orderly succession of fossils, is the product of a single universal flood that lasted a little longer than a year and covered the highest mountains to a depth of some seven meters a few thousand years ago, there is clear evidence in the form of intertidal and terrestrial deposits that at no recorded time in the past has the entire planet been under water. Moreover, a universal flood of sufficient magnitude to deposit the existing strata, which together are many scores of kilometers thick, would require a volume of water far greater than has ever existed on and in the earth, at least since the formation of the first known solid crust about 4 billion years ago. The belief that all this sediment with its fossils was deposited in an orderly sequence in a year's time defies all geological observations and physical principles concerning sedimentation rates and possible quantities of suspended solid matter. There were periods of unusually high rainfall, and extensive flooding of inhabited areas has occurred, but there is no scientific support for the hypothesis of a universal mountain-topping flood.

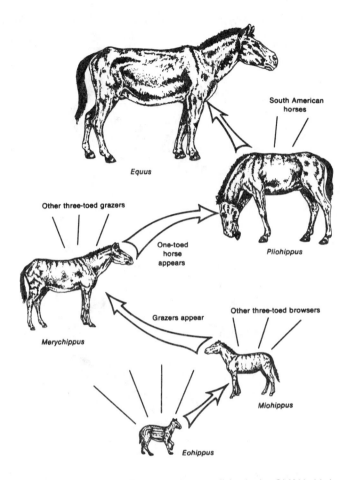

Evolution of the horse family. *Eohippus*, a browser living in the Old World about 50 million years ago, evolved into several forms. One of these (*Miohippus*) was a three-toed browser that evolved into several other browsers as well as into one form (*Merychippus*) that became a grazer. *Merychippus* was still three-toed and evolved into other three-toed grazers as well as into a one-toed horse (*Pliohippus*), which eventually gave rise to the modern horse (*Equus*) as well as to the South American horses.

Inferences about common descent derived from paleontology have been reinforced by comparative anatomy. The skeletons of humans, dogs, whales, and bats are strikingly similar, despite the different ways of life led by these animals and the diversity of environments in which they have flourished. The correspondence, bone by bone, can be observed in every part of the body, including the limbs: yet a person writes, a dog runs, a whale swims, and a bat flies with structures built of the same bones. Scientists calls such structures homologous and have concurred that they are best explained by common descent. Comparative anatomists investigate such homologies, not only in bone structure but also in other parts of the body as well, working out relationships from degrees of similarity. Their conclusions provide important inferences about the details of evolutionary history that can be tested by comparisons with the sequence of ancestral forms in the paleontological record.

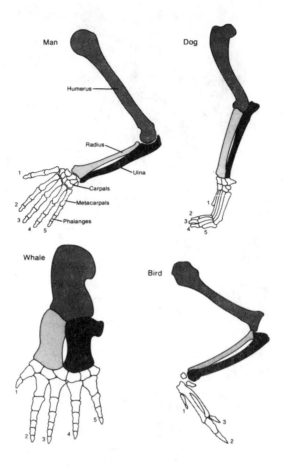

Bone composition of the forelimb of four vertebrates. The mammalian ear and jaw offer another example in which paleontology and comparative anatomy combine to show common ancestry through transitional stages. The lower jaws of mammals contain only one bone, whereas those of reptiles have several. The other bones in the reptile jaw are homologous with bones now found in the mammalian ear. What function could these bones have had during intermediate stages? Paleontologists have now discovered intermediate forms of mammal-like reptiles (*Therapsida*) with a double jaw joint—one composed of the bones that persist in mammalian jaws, the other consisting of bones that eventually became the hammer and anvil of the mammalian ear. Similar examples are numerous.

Biogeography also has contributed evidence for common descent. The diversity of life is stupendous. Approximately 250 thousand species of living plants, 100 thousand species of fungi, and 1.5 million species of animals and microorganisms have been described and named, each occupying its own peculiar ecological setting or niche, and the census is far from complete. Some species, such as human beings and our companion the dog, can live under a wide range of environmental conditions. Others are amazingly specialized. One species of the fungus *Laboulbenia* grows exclusively on the rear portion of the covering wings of a single species of beetle (*Aphaenops cronei*) found only in some caves of southern France. The larvae of the fly *Drosophila carcinophila* can develop only in specialized grooves beneath the flaps of the third pair of oral appendages of the land crab *Gecarcinus ruricola*, which is found only on certain Caribbean islands.

How can we make intelligible the colossal diversity of living beings and the existence of such extraordinary, seemingly whimsical creatures as *Laboulbenia*, *Drosophila carcinophila*, and others? Why are island groups like the Galápagos inhabited by forms similar to those on the nearest mainland but belonging to different species? Why is the indigenous life so different on different continents? Creationists contend that the curious facts of biogeography result from the occurrence of special creation events. A scientific hypothesis proposes that biological diversity results from an evolutionary process whereby the descendants of local or migrant predecessors became adapted to their diverse environments. A testable corollary of this hypothesis is that present forms and local fossils should show homologous attributes indicating how one is derived from the other. Also, there should be evidence that forms without an established local ancestry had migrated into the locality. Wherever such tests have been carried out, these conditions have been confirmed. A good example is provided by the mammalian populations of North and South America, where strikingly different endemic forms evolved in isolation until the emergence of the isthmus of Panama approximately 3 million years ago. Thereafter, the armadillo, porcupine, and opossum—mammals of South American origin—were able to migrate to North America along with many other species of plants and animals, while the placental mountain lion and other North American species made their way across the isthmus to the south.

The evidence that Darwin found for the influence of geographical distribution on the evolution of organisms has become stronger with advancing knowledge. For example, approximately two thousand species of flies belonging to the genus *Drosophila* are now found throughout the world. About one-quarter of them live only in Hawaii. More than a thousand species of snails and other land mollusks are also only found in Hawaii. The natural explanation for the occurrence of such great diversity among closely similar forms is that the differences resulted from adaptive

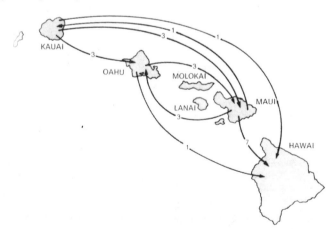

Minimum number of colonizations postulated to account for the evolution of the picture-winged *Drosophila* species of Hawaii. The arrows indicate the direction of migration; the numbers in each arrow indicate the minimum number of separate colonizations in the direction indicated.

colonization of isolated environments by animals with a common ancestry. The Hawaiian islands are far from, and were never attached to, any mainland or other islands, and they have had few colonizers. Organisms that reached these islands found many unoccupied ecological niches where they could then undergo separate evolutionary diversifications. No mammals other than one bat species lived on the Hawaiian islands when the first human settlers arrived; many other kinds of plants and animals were absent as well. The scientific explanation is that these kinds of organisms never reached the islands because of their great geographic isolation, while those that reached there multiplied in kind, because of the absence of related organisms that would compete for resources.

The vagaries of biogeography cannot be attributed to environmental peculiarities alone. The Hawaiian islands are no better than other Pacific islands for the survival of *Drosophila*, nor are they less hospitable than other parts of the world for many organisms not indigenous to them. For example, pigs and goats have multiplied in Hawaii after their introduction by humans. The general observation is that all sorts of organisms are absent from places well suited to their occupancy. Animals and plants vary from continent to continent and from island to island in a distribution pattern consistent with colonization and evolutionary change, rather than being simply responsive to the conditions of place.

Embryology, the study of biological development from the time of conception, is another source of independent evidence for common descent. Barnacles, for instance, are sedentary crustaceans with little apparent similarity to such other crustaceans as lobsters, shrimps, or copepods. Yet barnacles pass through a free-swimming larval stage, in which they look unmistakably like other crustacean larvae. The similarity of larval stages supports the conclusion that all crustaceans have homologous parts and a common ancestry. Human and other mammalian embryos pass through a stage during which they have unmistakable but useless grooves similar to gill slits found in fish—evidence that they and the other vertebrates shared remote ancestors that respired with the aid of gills.

Finally, the substantiation of common descent that emerges from all the foregoing lines of evidence is being validated and reinforced by the discoveries of modern biochemistry and molecular biology, a biological discipline that has emerged in the mid-twentieth century. This new discipline has unveiled the nature of hereditary material and the workings of organisms at the level of enzymes and other molecules. Molecular biology provides very detailed and convincing evidence for biological evolution.

7 *The Evidence from Molecular Biology*

The hereditary material, DNA, and the enzymes that govern all life processes hold information about an organism's ancestry. This information has made it possible to reconstruct evolutionary events that were previously unknown and to confirm and adjust the view of events that already were known. The precision with which events of evolution can be reconstructed is one reason the evidence from molecular biology is so compelling. Another reason is that molecular evolution has shown all living organisms, from bacteria to humans, to be related by descent from common ancestors.

The molecular components of organisms exhibit a remarkable uniformity in the nature of the components as well as in the ways in which they are assembled and used. In all bacteria, plants, animals, and humans, the DNA is made up of the same four component nucleotides, although many other nucleotides exist, and all of the various proteins are synthesized from different combinations and sequences of the same 20 amino acids, although several hundred other amino acids exist. The genetic "code" by which the information contained in the nuclear DNA is passed on to proteins is everywhere the same. Similar metabolic pathways are used by the most diverse organisms to produce energy and to make up the cell components.

This unity reveals the genetic continuity and common ancestry of all organisms. There is no other rational way to account for their molecular uniformity when numerous alternative structures are equally likely. The genetic code may serve as an example. Each particular sequence of three nucleotides in the nuclear DNA acts as a pattern, or code, for the production of exactly the same amino acid in all organisms. This is no more necessary than it is for a language to use a particular combination of letters to represent a particular reality. If it is found that many different combinations of letters, such as "planet," "tree," "woman," are used with identical meanings in a number of different books, one can be sure that the languages used in those books are of common origin.

Genes and proteins are long molecules that contain information in the sequence of their components in much the same way as sentences of the English language contain information in the sequence of their letters and words. The sequences that make up the genes are passed on from parents to offspring, identical except for occasional changes introduced by mutations. To illustrate, assume that two books are being compared; both books are 200 pages long and contain the same number of chapters. Closer examination reveals that the two books are identical page for page and word for word, except that an occasional word—say one in 100—is different. The two books cannot have been written independently; either one has been copied from the other or both have been copied, directly or indirectly, from the same original book. Similarly, if each nucleotide is represented by one letter, the complete sequence of nucleotides in the DNA of a higher organism would require several hundred books of hundreds of pages, with several thousand letters on each page. When the "pages" (or sequence of nucleotides) in these "books" (organisms) are examined one by one, the correspondence in the "letters" (nucleotides) gives unmistakable evidence of common origin.

Two arguments attest to evolution. Using the alphabet analogy, the first argument says that languages that use the same dictionary—the same genetic code and the same 20 amino acids—cannot be of independent origin. The second argument, concerning similarity in the sequence of nucleotides in the DNA or the sequence of amino acids in the proteins, says that books with very similar texts cannot be of independent origin.

The evidence of evolution revealed by molecular biology goes one step further. The degree of similarity in the sequence of nucleotides or of amino acids can be precisely quantified. For example, cytochrome-c (a protein molecule) of humans and chimpanzees consists of the same 104 amino acids in exactly the same order, but differs from that of rhesus monkeys by one amino acid, that of horses by 11 additional amino acids, and that of tuna by 21 additional amino acids. The degree of similarity reflects the recency of common ancestry. Thus, the inferences from comparative anatomy and other disciplines concerning evolutionary history can be tested in molecular studies of DNA and proteins by examining their sequences of nucleotides and amino acids.

The authority of this kind of test is overwhelming; each of the thousands of genes and thousands of proteins contained in an organism provides an independent test of that organism's evolutionary history. Not all possible tests have been performed, but many hundreds have been done, and not one has given evidence contrary to evolution. There is probably no other notion in any field of science that has been as extensively tested and as thoroughly corroborated as the evolutionary origin of living organisms. There is no reason to doubt the evolutionary theory of the origin of organisms any more than to doubt the heliocentric theory of the orbit of the planets around the sun.

8 *The Genetic Basis of Evolution*

The central argument of Darwin's theory of evolution starts from the existence of hereditary variation. Experience with animal and plant breeding demonstrates that variations can be developed that are "useful to man." So, reasoned Darwin, variations must occur in nature that are favorable or useful in some way to the organism itself in the struggle for existence. Favorable variations are ones that increase chances for survival and procreation. Those advantageous variations are preserved and multiplied from generation to generation at the expense of less advantageous ones. This is the process known as natural selection. The outcome of the process is an organism that is well adapted to its environment, and evolution often occurs as a consequence.

Biological evolution is the process of change and diversification of living things over time, and it affects all aspects of their lives—morphology, physiology, behavior, and ecology. Underlying these changes are changes in the hereditary materials. Hence, in genetic terms, evolution consists of changes in the organism's hereditary makeup.

Natural selection, then, can be defined as the differential reproduction of alternative hereditary variants, determined by the fact that some variants increase the likelihood that the organisms having them will survive and reproduce more successfully than will organisms carrying alternative variants. Selection may be due to differences in survival, in fertility, in rate of development, in mating success, or in any other aspect of the life cycle. All of these differences can be incorporated under the term "differential reproduction" because all result in natural selection to the extent that they affect the number of progeny an organism leaves.

Darwin explained that competition for limited resources results in the survival of the most effective competitors. But natural selection may occur not only as a result of competition but also as an affect of some aspect of the physical environment, such as inclement weather. Moreover, natural selection would occur even if all the members of a population died at the same age, if some of them produced more offspring than others. Natural selection is quantified by a measure called Darwinian fitness, or relative fitness. Fitness in this sense is the relative probability that a hereditary characteristic will be reproduced; that is, the degree of fitness is a measure of the reproductive efficiency of the characteristic.

Evolution can be seen as a two-step process. First, hereditary variation takes place; second, selection occurs of those genetic variants that will be passed on most effectively to the following generations. Hereditary variation also entails two mechanisms: the spontaneous mutation of one variant to another, and the sexual process that recombines those variants to form a multitude of variations.

The "gene pool" of a species is the sum total of all of the genes and combinations of genes that occur in all organisms of the same species. The necessity of hereditary variation for evolutionary change to occur can be understood in terms of the gene pool. Assume, for instance, that at the gene locus that codes for the human *MN* blood groups, there is no variation; only the *M* form exists in all individuals. Evolution of the *MN* blood groups cannot take place in such a population, since the allelic frequencies have no opportunity to change from generation to generation. On the other hand, in populations in which both forms *M* and *N* are present, evolutionary change is possible.

The more genetic variation that exists in a population, the greater the opportunity for evolution to occur. As the number of genes that are variable increases and as the number of forms of each gene becomes greater, the likelihood that some forms will change in frequency at the expense of their alternates grows. The British geneticist R. A. Fisher mathematically demonstrated a direct correlation between the amount of genetic variation in a population and the rate of evolutionary change by natural selection. This demonstration is embodied in his fundamental theorem of natural selection: "The rate of increase in fitness of any organism at any time is equal to its genetic variance in fitness at that time." This theorem has been confirmed experimentally.

Given that a population's potential for evolving is determined by its genetic variation, evolutionists are interested in discovering the extent of such variation in natural populations. Techniques for determining genetic variation have been used to investigate numerous species of plants and animals. Typically, insects and other invertebrates are more varied genetically than mammals and other vertebrates; and plants bred by outcrossing exhibit more variation than those bred by self-pollination. But the amount of genetic variation is in any case astounding.

Consider as an example humans, whose level of variation is about the same as that of other mammals. At the level of proteins, the human heterozygosity (the measure of genetic variation) value is stated as $H = 0.067$, which means that individuals are heterozygous (has two different gene forms) at 6.7 percent of their genes. It is not known how many genes there are in humans, but estimates range from 30,000 to 100,000. Assuming the lower estimate, a person would be heterozygous at $30,000 \times 0.067 = 2,010$ genes. This implies a typical human individual has the potential to produce $2^{2,010}$, or approximately 10^{605} (1 with 605 zeros following), different kinds of sex cells (eggs or sperm). But that number is vastly much larger than the estimated number of atoms in the visible universe, 10^{76}, which is trivial by comparison. This calculation becomes yet more dramatic if it is made at the level of the DNA. Among the three billion nucleotides (the letters in the DNA) that we inherit from each one of our parents, more than one per thousand (that is, more than three million) are different between the paternal and maternal. By random recombination they have the potential to produce more than $10^{750,000}$ (1 with 750,000 zeros following) different sex cells; and the set that can be produced by each person is different from the set of every other person.

The same can be said of all organisms that reproduce sexually; every individual represents a unique genetic configuration that will never be repeated. This enormous reservoir of genetic variation in natural populations provides virtually unlimited opportunities for evolutionary change in response to the environmental constraints and the needs of the organisms, beyond the new variations that arise every generation by the process of mutation, which I shall now discuss.

9 *The Origin of Genetic Variation*

All living things have evolved from primitive living forms that lived about 3,500,000,000 years ago. At present there are more than two million known species, which are widely diverse in size, shape, and ways of life, as well as in the DNA sequences that contain their genetic information. What has produced the pervasive genetic variation within natural populations and the genetic differences among species?

Heredity is not a perfectly conservative process; otherwise, evolution could not have taken place. The information encoded in the nucleotide sequence of DNA is, as a rule, faithfully reproduced during replication, so that each replication results in two DNA molecules that are identical to each other and to the parent molecule. But, occasionally "mistakes," or mutations, occur in the DNA molecule during replication, so that daughter cells differ from the parent cells in at least one of the letters in the DNA sequence. A mutation first appears on a single cell of an organism, but it is passed on to all cells descended from the first. Mutations can be classified into two categories: gene, or point, mutations, which affect only a few letters (nucleotides) within a gene; and chromosomal mutations which either change the number of chromosomes or change the number or arrangement of genes on a chromosome (chromosomes are the elongated structures that store the DNA of each cell).

Gene mutations can occur spontaneously; that is, without being intentionally caused by humans. They can also be artificially induced by ultraviolet light, X rays, and other high-frequency radiations, as well as by exposure to certain mutagenic chemicals, such as mustard gas. The consequences of gene mutations may range from negligible to lethal. Some have a small or undetectable effect on the organism's ability to survive and reproduce, because no essential biological functions are altered. But when the active site of an enzyme or some other essential function is affected, the impact may be severe.

Newly arisen mutations are more likely to be harmful than beneficial to their carriers, because mutations are random events with respect to adaptation; that is, their occurrence is independent of any possible consequences. Harmful mutations are eliminated or kept in check by natural selection. Occasionally, however, a new mutation may increase the organism's adaptation. The probability of such an event's happening is greater when organisms colonize a new territory or when environmental changes confront a population with new challenges. In these cases, the established adaptation of a population is less than optimal, and there is greater opportunity for new mutations to be better adaptive. This is so because the consequences of mutations depend on the environment. Increased melanin pigmentation may be advantageous to inhabitants of tropical Africa, where dark skin protects them from the Sun's ultraviolet radiation; but it is not beneficial in Scandinavia, where the intensity of sunlight is low and light skin facilitates the synthesis of vitamin D.

Mutation rates are low, but new mutants appear continuously in nature, because there are many individuals in every species and many genes in every individual. The process of mutation provides each generation with many new genetic variations. More important yet is the storage of variation, arisen by past mutations, that is present in each organism, as calculated earlier for humans. Thus, it is not surprising to see that when new environmental challenges arise, species are able to adapt to them. More than 200 insect and rodent species, for example, have developed resistance to the pesticide DDT in parts of the world where spraying has been intense. Although the insects had never before encountered this synthetic compound, they adapted to it rapidly by means of mutations that allowed them to survive in its presence. Similarly, many species of moths and butterflies in industrialized regions have shown an increase in the frequency of individuals with dark wings in response to environmental pollution, an adaptation known as industrial melanism. The examples can be multiplied at will.

10 *Dynamics of Genetic Change*

The genetic variation present in natural populations of organisms is sorted out in new ways in each generation by the process of sexual reproduction. But heredity by itself does not change gene frequencies. This principle is formally stated by the Hardy-Weinberg law, an algebraic equation that describes the genetic equilibrium in a population.[6]

The Hardy-Weinberg law assumes that gene frequencies remain constant from generation to generation—that there is no gene mutation or natural selection and that populations are very large. But these assumptions are not correct; indeed, if they were, evolution could not occur. Why, then, is the Hardy-Weinberg law significant if its assumptions do not hold true in nature? The answer is that the Hardy-Weinberg law plays a role in evolutionary studies similar to that of Newton's first law of motion in mechanics. Newton's first law says that a body not acted upon by a net external force remains at rest or maintains a constant velocity. In fact, there are always external forces acting upon physical objects (gravity, for example), but the first law provides the starting point for the application of other laws. Similarly, organisms are subject to mutation, selection, and other processes that change gene frequencies, and the effects of these processes are calculated by using the Hardy-Weinberg law as the starting point. There are four processes of gene frequency change: mutation, migration, drift, and natural selection.

The allelic variations that make evolution possible are generated by the process of mutation; but new mutations change gene frequencies very slowly, since mutation rates are low. Moreover, gene mutations are reversible. Changes in gene frequencies due to mutation occur, therefore, very slowly. In any case, allelic frequencies usually are not determined by mutation alone, because some alleles are favored over others by natural selection. The equilibrium frequencies are then decided by the interaction between mutation and selection, with selection usually having the greater consequence.

Migration, or gene flow, takes place when individuals migrate from one population to another and interbreed with its members. The genetic make-up of populations changes locally whenever different populations intermingle. In general, the greater the difference in allele frequencies between the resident and the migrant individuals, and the larger the number of migrants, the greater effect the migrants have in changing the genetic constitution of the resident population.

Gene frequencies can also change from one generation to another by a process of pure chance known as genetic drift. This occurs because populations are finite in

---

[6] If there are two alleles, A and a, at a gene locus, three genotypes are possible, AA, Aa, and aa. If the frequencies of the two alleles are p and q, respectively, the equilibrium frequencies of the three genotypes are given by $(p + q)^2 = p^2 + 2pq + q^2$, for AA, Aa, and aa, respectively. The genotype equilibrium frequencies for any number of alleles are derived in the same way.

number, and thus the frequency of a gene may change in the following generation by accidents of sampling, just as it is possible to get more or less than 50 "heads" in 100 throws of a coin simply by chance. The magnitude of the change in gene frequency due to genetic drift is inversely related to the size of the population; the larger the number of reproducing individuals, the smaller the effects of genetic drift. The effects of genetic drift in changing gene frequencies from one generation to the next are quite small in most natural populations, which generally consist of thousands of reproducing individuals. The effects over many generations are more important. Genetic drift can have important evolutionary consequences when a new population becomes established by only a few individuals, as in the colonization of islands and lakes. This is one reason why species in neighboring islands, such as those in the Hawaiian archipelago, are often more heterogeneous than species in comparable continental areas adjacent to one another.

11 *The Process of Natural Selection*

The phrase "natural selection" was used by Darwin to refer to any reproductive bias favoring some hereditary variants over others. He proposed that natural selection promotes the adaptation of organisms to the environments in which they live because the organisms carrying such useful variants would leave more descendants than those lacking them. The modern concept of natural selection derives directly from Darwin's but is defined precisely in mathematical terms as a statistical bias favoring some genetic variants over their alternates (recall that the measure to quantify natural selection is called "fitness"). Hereditary variants, favorable or not to the organisms, arise by mutation. Unfavorable ones are eventually eliminated by natural selection; their carriers leave no descendants or leave fewer than those carrying alternative variants. Favorable mutations accumulate over the generations. The process continues indefinitely because the environments that organisms live in are forever changing. Environments change both physically—in their climate, physical configuration, and so on—and biologically, because the predators, parasites, and competitors with which an organism interacts are themselves evolving.

If mutation, migration, and drift were the only processes of evolutionary change, the organization of living things would gradually disintegrate, because these processes are random with respect to adaptation. These three processes change gene frequencies without regard for the consequences that such changes may have in the ability of the organisms to survive and reproduce. The effects of such processes alone would be analogous to those of a mechanic who changed parts in a motorcar engine at random, with no regard for the role of the parts in the engine. Natural selection keeps the disorganizing effects of mutation and other processes in check because it multiplies beneficial mutations and eliminates harmful ones. But natural selection accounts not only for the preservation and improvement of the organization of living beings but also for their diversity. In different localities or in different circumstances, natural selection favors different traits, precisely those that make the organisms well adapted to their particular circumstances and ways of life.

The effects of natural selection can be studied by measuring the ensuing changes in gene frequencies; but they can also be explored by examining changes on the observable characteristics—or phenotypes—of individuals in a population. Distribution scales of phenotypic traits such as height, weight, number of progeny, or longevity typically show greater numbers of individuals with intermediate values and fewer and fewer toward the extremes (the so-called normal distribution). When individuals with intermediate phenotypes are favored and extreme phenotypes are selected against, the selection is said to be stabilizing. The range and distribution of phenotypes then remains approximately the same from one generation to another. Stabilizing selection is very common. The individuals that survive and reproduce more successfully are those that have intermediate phenotypic values. Mortality among newborn infants, for example, is highest when they are either very small or very large; infants of intermediate size have a greater chance of surviving.

But the distribution of phenotypes in a population sometimes changes systematically in a particular direction. The physical and biological aspects of the environment are continuously changing, and over long periods of time the changes may be substantial. The climate and even the configuration of the land or waters vary incessantly. Changes also take place in the biotic conditions; that is, in the other organisms present, whether predators, prey, parasites, or competitors. Genetic changes occur as a consequence, because the genotypic fitnesses may be shifted so that different sets of variants are favored. The opportunity for directional selection also arises when organisms colonize new environments where the conditions are different from those of their original habitat. The process of directional selection often takes place in spurts. The replacement of one genetic constitution for another changes the genotypic fitnesses of genes for other traits, which in turn stimulates additional changes, and so on in a cascade of consequences.

The nearly universal success of artificial selection and the rapid response of natural populations to new environmental challenges are evidence that existing variation provides the necessary materials for directional selection, as Darwin already explained. More generally, human actions have been an important stimulus to this type of selection. Humans transform the environments of many organisms, which rapidly respond to the new environmental challenges through directional selection. Well-known instances are the many cases of insect resistance to pesticides, synthetic substances not present in the natural environment. Whenever a new insecticide is first applied to control a pest, the results are encouraging because a small amount of the insecticide is sufficient to bring the pest organism under control. As time passes, however, the amount required to achieve a certain level of control must be increased again and again until finally it becomes ineffective or economically impractical. This occurs because organisms become resistant to the pesticide through directional selection. The resistance of the housefly, *Musca domestica*, to DDT was first reported in 1947. Resistance to one or more pesticides has now been recorded in more than 100 species of insects.

Sustained directional selection leads to major changes in morphology and ways of life over geologic time. Evolutionary changes that persist in a more or less continuous fashion over long periods of time are known as evolutionary trends. Directional evolutionary changes increased the cranial capacity of the human lineage from the small brain of *Australopithecus*, human ancestors of four million years ago, which weighed somewhat less than one pound, to a brain three and a half times as large in modern humans, *Homo sapiens*. The evolution of the horse family from more than 50 million years ago to modern times is another of the many well-studied examples of directional selection.

Sometimes, two or more divergent traits in an environment may be favored simultaneously, which is called diversifying selection. No natural environment is homogeneous; rather, the environment of any plant or animal population is a mosaic consisting of more or less dissimilar sub-environments. There is heterogeneity with respect to climate, food resources, and living space. Also, the heterogeneity may be temporal, with change occurring over time, as well as spatial, with dissimilarity found in different areas. Species cope with environmental heterogeneity in diverse ways. One strategy is the selection of a generalist genotype that is well adapted to all of the sub-environments encountered by the species. Another strategy is genetic polymorphism, the selection of a diversified gene pool that yields different genetic make-ups, each adapted to a specific sub-environment.

One important factor in reproduction is mutual attraction between the sexes. The males and females of many animal species are fairly similar in size and shape except for the sexual organs and secondary sexual characteristics such as the breasts of female mammals. There are, however, species in which the sexes exhibit striking dimorphism. Particularly in birds and mammals, the males are often larger and stronger, more brightly colored, or endowed with conspicuous adornments. But bright colors make animals more visible to predators; for example, the long plumage of male peacocks and birds of paradise and the enormous antlers of aged male deer are cumbersome loads in the best of cases. Darwin knew that natural selection could not be expected to favor the evolution of disadvantageous traits, and he was able to offer a solution to this problem. He proposed that such traits arise by "sexual selection," which "depends not on a struggle for existence in relation to other organic beings or to external conditions, but on a struggle between the individuals of one sex, generally the males, for the possession of the other sex." Thus, the colored plumage of the males in some bird species makes them more attractive to their females, which more than compensates for their increased visibility to potential predators. Sexual selection is a topic of intensive research at present.

The apparent altruistic behavior of many animals is, like some manifestations of sexual selection, a trait that at first seems incompatible with the theory of natural selection. Altruism is a form of behavior that benefits other individuals at the expense of the one that performs the action; the fitness of the altruist is diminished by its behavior, whereas individuals that act selfishly benefit from it at no cost to themselves. Accordingly, it might be expected that natural selection would foster the development of selfish behavior and eliminate altruism. This conclusion is not so compelling when it is noticed that the beneficiaries of altruistic behavior are usually relatives. They all carry the same genes, including the genes that promote altruistic behavior. Altruism may evolve by kin selection, which is simply a type of natural selection in which relatives are taken into consideration when evaluating an individual's fitness.

Kin selection is explained as follows. Natural selection favors genes that increase the reproductive success of their carriers, but it is not necessary that all individuals with a given genetic make-up have higher reproductive success. It suffices that carriers of the genotype reproduce more successfully on the average than those possessing alternative genotypes. A parent shares half of its genes with each progeny, so a gene that promotes parental altruism is favored by selection if the behavior's cost to the parent is less than half of its average benefits to the progeny. Such a gene will be more likely to increase in frequency through the generations than an alternative gene that does not promote parental care. The parent spends some energy caring for the progeny because it increases the reproductive success of the parent's genes. But kin selection extends beyond the relationship between parents and their offspring. It facilitates the development of altruistic behavior when the energy invested, or the risk incurred, by an individual is compensated in excess by the benefits ensuing to relatives.

In many species of primates (as well as in other animals), altruism also occurs among unrelated individuals when the behavior is reciprocal and the altruist's costs are smaller than the benefits to the recipient. This reciprocal altruism is found, for example, in the mutual grooming of chimpanzees as they clean each other of lice and other pests. Another example appears in flocks of birds that post sentinels to warn of danger. A crow sitting in a tree watching for predators, while the rest of the flock forages, incurs a small loss by not feeding, but this is well compensated by the protection it receives when it itself forages and others of the flock stand guard.

12 *The Origin of Species*

Darwin sought to explain the splendid diversity of the living world: thousands of organisms of the most diverse kinds, from lowly worms to spectacular tropical birds, from yeasts and molds to oaks and orchids. His *Origin of Species* is a sustained argument showing that the diversity of organisms and their characteristics can be explained as the result of natural processes. As Darwin noted, different species may come about as the result of gradual adaptation to environments that are continuously changing in time and differ from place to place. Natural selection favors different characteristics in different situations.

In everyday experience we identify different kinds of organisms by their appearance. Everyone knows that people belong to the human species and are different from cats and dogs, which in turn are different from each other. There are differences among people, as well as among cats and dogs; but individuals of the same species are considerably more similar among themselves than they are to individuals of other species. But there is more to it than that; a bulldog, a terrier, and a golden retriever are very different in appearance, but they are all dogs because they can interbreed. People can also interbreed with one another, and so can cats, but people cannot interbreed with dogs or cats, nor these with each other. It is, then, clear that although species are usually identified by appearance, there is something basic, of great biological significance, behind similarity of appearance; namely, that individuals of a species are able to interbreed with one another but not with members of other species. This is expressed in the following definition: *species are groups of interbreeding natural populations that are reproductively isolated from other such groups.*[7]

The ability to interbreed is of great evolutionary importance, because it determines that species are independent evolutionary units. Genetic changes originate in single individuals; they can spread by natural selection to all members of the species but not to individuals of other species. Thus, individuals of a species share a common gene pool that is not shared by individuals of other species, because they are reproductively isolated.

Although the criterion for deciding whether individuals belong to the same species is clear, there may be ambiguity in practice for two reasons. One is lack of knowledge; it may not be known for certain whether individuals living in different sites belong to the same species, because it is not known whether they can naturally interbreed. The other reason for ambiguity is rooted in the nature of evolution as a

---

[7] Bacteria and blue-green algae do not reproduce sexually, but by fission. Organisms that lack sexual reproduction are classified into different species according to criteria such as external characteristics and morphology, chemical and physiological properties, and genetic constitution. The definition of species given above applies only to organisms able to interbreed.

gradual process. Two geographically separate populations that at one time were members of the same species later may have diverged into two different species. Since the process is gradual there is not a particular point at which it is possible to say that the two populations have become two different species. It is an interesting curiosity that some anti-evolutionists have referred to the existence of species intermediates as evidence against evolution; quite the contrary, such intermediates are precisely expected.

A similar kind of ambiguity arises with respect to organisms living at different times. There is no way to test whether or not today's humans could interbreed with those who lived thousands of years ago. It seems reasonable that living people, or living cats, would be able to interbreed with people, or cats, exactly like those that lived a few generations earlier. But what about the ancestors removed by one thousand or one million generations? The ancestors of modern humans that lived 500 thousand years ago (about 20 thousand generations) are classified in the species *Homo erectus*, whereas present-day humans are classified in a different species, *Homo sapiens*, because those ancestors were quite different from us in appearance and thus it seems reasonable to conclude that interbreeding could not have occurred with modern-like humans. But there is not an exact time at which *Homo erectus* became *Homo sapiens*. It would not be appropriate to classify remote human ancestors and modern humans in the same species just because the changes from one generation to the next are small. It is useful to distinguish between the two groups by means of different species names, just as it is useful to give different names to childhood, adolescence and adulthood, even though there is not one moment when an individual passes from one to the other. Biologists distinguish species in organisms that lived at different times by means of a commonsense rule. If two organisms differ from each other about as much as two living individuals belonging to two different species differ, they will be classified in separate species and given different names.

The biological properties of organisms that prevent interbreeding are called reproductive isolating mechanisms (RIMs). Oaks on different islands, minnows in different rivers, or squirrels in different mountain ranges cannot interbreed because they are physically separated, but not necessarily because they are biologically incompatible. Geographic separation, therefore, is not an RIM, since it is not a biological property of organisms. There are two general categories of RIMs: prezygotic (those that take effect before fertilization) and postzygotic (those that take effect after). Prezygotic RIMs prevent the formation of hybrids between members of different populations through ecological, temporal, ethological (or behavioral), mechanical, and gametic isolation. Postzygotic RIMs reduce the viability or fertility of hybrids or their progeny. As the descendants of one species become gradually divergent and eventually evolve into different species, RIMs appear and often accumulate in mutual reinforcement. Species that diverged quite some time ago are typically kept from interbreeding by several RIMs.

13 *A Model of Speciation*

Since species are groups of populations reproductively isolated form one another, asking about the origin of species is equivalent to asking how reproductive isolation arises between populations. Two theories have been advanced to answer this question. One theory considers isolation as an accidental by-product of genetic divergence. Populations that become genetically less and less alike (as a consequence, for example, of adaptation to different environments) may eventually be unable to interbreed because their gene pools are disharmonious. The other theory regards isolation as a product of natural selection. Whenever hybrid individuals are less fit than non-hybrids, natural selection will directly promote the development of RIMs. This occurs because genetic variants interfering with hybridization have greater fitness than those favoring hybridization, given that the latter are often present in poorly fit hybrids.

Scientists have shown that these two theories of the origin of reproductive isolation are not mutually exclusive. Reproductive isolation may indeed come about incidentally to genetic divergence between separated populations. Consider, for example, the evolution of many species of plants and animals in the Hawaiian archipelago. The ancestors of these species arrived in the Hawaiian Islands several million years ago. There they evolved as they became adapted to the environmental conditions and colonizing opportunities found on the islands. Reproductive isolation between the populations evolving in Hawaii and the continental populations was never directly promoted by natural selection because their geographic remoteness forestalled any opportunities for hybridizing. Nevertheless, reproductive isolation became complete in many cases as a result of gradual genetic divergence over thousands of generations. Frequently, however, the course of speciation involves the processes postulated by both theories; reproductive isolation starts as a by-product of gradual evolutionary divergence but is completed by natural selection directly promoting the evolution of prezygotic RIMs.

The two sets of processes identified by the two speciation theories may be seen, therefore, as two different stages in the splitting of one evolutionary lineage into two species. The process can start only when gene flow is somehow interrupted between two populations. Interruption may be due to geographic separation, or it may be initiated by some genetic change that affects some but not other individuals living in the same territory. Absence of gene flow makes it possible for the two populations to become genetically differentiated as a consequence of adapting to diverse local conditions (and also as a consequence of genetic drift, particularly in small populations, such as occur in new colonizations). Two isolated groups are likely to become more and more different as time goes on. Eventually some incipient reproductive isolation may take effect because the two gene pools have become sufficiently different. These circumstances may persist for so long that the populations become completely differentiated into separate species, as in the example of the Hawaiian Islands. It happens quite commonly, however, in both animals and plants, that opportunities for hybridization arise between two populations that are becoming genetically differentiated. Two outcomes are possible. One is that the hybrids manifest little or no reduction of fitness, so that gene exchange between the two populations proceeds freely, eventually leading to their integration into a single gene pool. The second possible outcome is that reduction of fitness in the hybrids is sufficiently large for natural selection to favor the emergence of prezygotic RIMs preventing the formation of hybrids altogether. This situation may be identified as the second stage in the speciation process.

14 *Adaptive Radiation*

As indicated earlier, the geographic separation of populations derived from common ancestors may continue long enough so that the populations become completely differentiated species before coming together again. As the separated populations continue evolving independently, RIMs develop and morphological differences may arise. The second stage of speciation—with natural selection directly stimulating the evolution of RIMs—never comes about in such situations, because reproductive isolation takes place simply as a consequence of the continued separate evolution of the populations.

This form of speciation is particularly apparent when colonizers reach geographically remote areas, such as islands, where they find few or no competitors and have an opportunity to diverge as they become adapted to the new environment. Sometimes a multiplicity of new environments becomes available to the colonizers, giving rise to several different lineages and species. This process of rapid divergence of multiple species from a single ancestral lineage is called adaptive radiation.

Examples of speciation by adaptive radiation in archipelagos removed from the mainland have already been mentioned. The Galápagos Islands are about 600 miles off the west coast of South America. When Darwin arrived there in 1835, he discovered many species not found anywhere else in the world—for example, 14 species of finch (known as Darwin's finches). These passerine birds have adapted to a diversity of habitats and diets, some feeding mostly on plants, others exclusively on insects. The various shapes of their bills are clearly adapted to probing, grasping, biting, or crushing—the diverse ways in which these different Galápagos species obtain their food. The explanation for such diversity (which is not found in finches from the continental mainland) is that the ancestor of Galápagos finches arrived in the islands before other kinds of birds and encountered an abundance of unoccupied ecological opportunities. The finches underwent adaptive radiation, evolving a variety of species with ways of life capable of exploiting niches that in continental faunas are exploited by different kinds of birds.

Striking examples of adaptive radiation occur in the Hawaiian Islands. The archipelago consists of several volcanic islands, ranging from less than one million to more than ten million years in age, far away from any continent or other large islands. An astounding number of plant and animal species of certain kinds exist in the islands while many other kinds are lacking. Among the species that have evolved in the islands, there are about two dozen species (about one-third of them now extinct) of honeycreepers, birds of the family *Drepanididae*, all derived from a single immigrant form. In fact, all but one of Hawaii's 71 native bird species are endemic; that is, they have evolved there and are found nowhere else. More than 90 percent of the native species of the hundreds of flowering plants, land mollusks, and insects in Hawaii are also endemic, as are two-thirds of the 168 species of ferns. About one-third of the world's total number of known species of *Drosophila* flies (more than 500) are native Hawaiian species. The species of *Drosophila* in Hawaii have diverged by adaptive radiation from one or a few colonizers which encountered an assortment of ecological opportunities that in other lands are occupied by different groups of flies or insects.

15 *Quantum Speciation*

The first stage of speciation is sometimes achieved in a short period of time. Rapid modes of speciation are known by a variety of names, such as "quantum," "rapid," and "saltational" speciation, all suggesting the short time involved. An important form of quantum speciation is polyploidy, which occurs by the multiplication of entire sets of chromosomes. A typical (diploid) organism carries in the nucleus of each cell two sets of chromosomes, one inherited from each parent; a polyploid organism has several sets of chromosomes. Many cultivated plants are polyploid: bananas have three sets of chromosomes, potatoes have four, bread wheat has six, some strawberries have eight. All major groups of plants have natural polyploid species, but they are most common among flowering plants (angiosperms) of which about 47 percent are polyploids.

In animals, polyploidy is relatively rare because it disrupts the balance between chromosomes involved in the determination of sex. But polyploid species are found in hermaphroditic animals (individuals having both male and female organs), which include snails and earthworms, as well as in forms with parthenogenetic females (which produce viable progeny without fertilization), such as some beetles, sow bugs, goldfish, and salamanders.

16 Reconstruction of Evolutionary History

It is possible to look at two sides of evolution: one, called "anagenesis," refers to changes that occur within a lineage; the other, called "cladogenesis," refers to the split of a lineage into two or more separate lineages. Anagenetic evolution has, over the last four million years, more than tripled the size of the human brain; in the lineage of the horse, it has reduced the number of toes from four to one. Cladogenetic evolution has produced the extraordinary diversity of the living world, with more than two million species of animals, plants, fungi, and microorganisms.

The evolution of all living organisms, or of a subset of them, can be represented as a tree, with branches that divide into two or more as time progresses. Such "trees" are called phylogenies. Their branches represent evolving lineages, some of which eventually die out while others persist down to the present time. Evolutionists are interested in the history of life and hence in the topology, or configuration, of evolution's trees. They also want to know the anagenetic changes along lineages and the timing of important events.

Tree relationships are ascertained by means of several complementary sources of evidence. First, there is the fossil record, which provides definitive evidence of relationships among some groups of organisms, but is far from complete and often seriously deficient. Second, there is comparative anatomy, the comparative study of living forms, as well as the related disciplines of comparative embryology, cytology, ethology, biogeography, and others. In recent years the comparative study of informational macromolecules—proteins and nucleic acids—has become a powerful tool for the study of evolution's history (see section 18 below). We saw earlier how the results from these disciplines demonstrate that evolution has occurred. Advanced methods have now been developed to reconstruct evolution's history.

These methods make it possible to identify when the correspondence of features in different organisms is due to inheritance from a common ancestor, which is called "homology." The forelimbs of humans, whales, dogs, and bats are homologous. The skeletons of these limbs are all constructed of bones arranged according to the same pattern because they derive from an ancestor with similarly arranged forelimbs (see Figure 2). Correspondence of features due to similarity of function but not related to common descent is termed "analogy." The wings of birds and of flies are analogous. Their wings are not modified versions of a structure present in a common ancestor but rather have developed independently as adaptations to a common function, flying.

Homology can be recognized not only between different organisms but also between repetitive structures of the same organism. This has been called serial homology. There is serial homology, for example, between the arms and legs of humans, among the seven cervical vertebrae of mammals, and among the branches or leaves of a tree. The jointed appendages of arthropods are elaborate examples of serial homology. Crayfish have 19 pairs of appendages, all built according to the same basic pattern but serving diverse functions—sensing, chewing, food handling, walking, mating, egg carrying, and swimming. Serial homologies are not useful in reconstructing the phylogenetic relationships of organisms, but they are an important dimension of the evolutionary process.

Relationships in some sense akin to those between serial homologs exist at the molecular level between genes and proteins derived from ancestral gene duplications. The genes coding for the various hemoglobin chains are an example. About 500 million years ago a chromosome segment carrying the gene coding for hemoglobin became duplicated, so that the genes in the different segments thereafter evolved in somewhat different ways, one eventually giving rise to the modern gene coding for alpha hemoglobin, the other for beta hemoglobin. The beta hemoglobin gene became duplicated again about 200 million years ago, giving rise to the gamma (fetal) hemoglobin. The alpha, beta, and gamma hemoglobin genes are homologous; similarities in their DNA sequences occur because they are modified descendants of a single ancestral sequence.

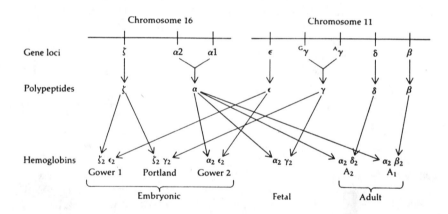

Position of several genes for hemoglobin in the human chromosomes 11 and 16. These genes are active ("turned on") at different stages of human life. Their transcripts (polypeptides) combine in pairs (one from chromosome 16 and one from chromosome 11) to produce the hemoglobin molecules essential for respiration.

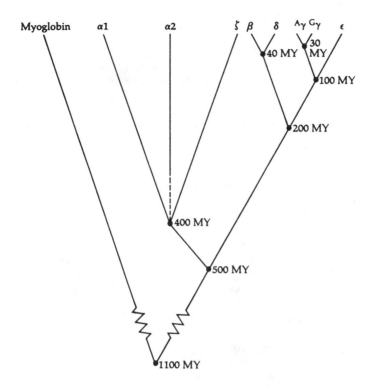

Evolutionary history of the hemoglobin genes, which share a common ancestor with the gene for myoglobin, which is a protein active in the muscle of most animals.

17 *Gradual and Punctuational Evolution*

Morphological evolution is by and large a gradual process, as shown by the fossil record. Major evolutionary changes are usually due to a building up over the ages of relatively small changes. But the fossil record is discontinuous. Fossil strata are separated by sharp boundaries; accumulation of fossils within a geologic deposit (stratum) is fairly constant over time, but the transition from one stratum to another may involve gaps of tens of thousands of years. Different species, characterized by small but discontinuous morphological changes, typically appear at the boundaries between strata, whereas the fossils within a stratum exhibit little morphological variation. That is not to say that the transition from one stratum to another always involves sudden changes in morphology; on the contrary, fossil forms often persist virtually unchanged through several geologic strata, each representing millions of years.

Paleontologists attributed the apparent morphological discontinuities of the fossil record to the discontinuity of the sediments; that is, to the substantial time gaps encompassed in the boundaries between strata. The assumption is that, if the fossil deposits were more continuous, they would show a more gradual transition of form. Even so, morphological evolution would not always keep progressing gradually, because some forms, at least, remain unchanged for extremely long times. Examples are the lineages known as "living fossils": the lamp shell *Lingula*, a genus of brachiopod that appears to have remained essentially unchanged since the Ordovician Period, some 450 million years ago; or the tuatara (*Sphenodon punctatus*), a reptile that has shown little morphological evolution since the early Mesozoic Period some 200 million years ago.

According to some paleontologists, however, the frequent discontinuities of the fossil record are not artifacts created by gaps in the record, but rather reflect the true nature of morphological evolution, which happens in sudden bursts associated with the formation of new species. The lack of morphological evolution, or stasis, of lineages such as *Lingula* and *Sphenodon* is in turn due to lack of speciation within those lineages. The proposition that morphological evolution is jerky, with most morphological change occurring during the brief speciation events and virtually no change during the subsequent existence of the species, is known as the "punctuated equilibrium" model of morphological evolution.

The question whether morphological evolution in the fossil record is predominantly punctuational or gradual is a subject of active investigation and debate. The imperfection of the record makes it unlikely that the issue will be settled in the foreseeable future. Intensive study of a favorable and abundant set of fossils may be expected to substantiate punctuated or gradual evolution in particular cases. But the argument is not about whether only one or the other pattern ever occurs; it is about their relative frequency. Some paleontologists argue that morphological evolution is in most cases gradual and only rarely jerky, whereas others think the opposite is true. Much of the problem is that gradualness or jerkiness is in the eye of the beholder.

Consider the evolution of rib strength (the ratio of rib height to rib width) within a lineage of fossil brachiopods of the genus *Eocelia*. An abundant sample of fossils from the Silurian Period in Wales has been analyzed, with the results shown in the figure. One possible interpretation of the data is that rib strength changed little or not at all from 415 to 413 million years ago; rapid change ensued for the next one million years, with virtual absence of change from 412 to 407 million years ago; another short burst of change occurred around 406 million years ago, followed by a final period of stasis. On the other hand, the record shown in the figure may be interpreted as not particularly punctuated but rather as a gradual process, with the rate of change somewhat greater at particular times.

Evolution of rib strength in the brachiopod *Eocoelia* (opposite). The time scale in the graph ranges from about 405 to 415 million years ago; rib strength is indicated on the horizontal scale. It remained rather constant until near 413 million years ago, when it changed rapidly for a while, and then continued to change more gradually until 405 million years ago. This example of punctuated equilibrium shows the alternation of periods during which a trait (in this case, rib strength) changes gradually or rapidly with periods when little or no change occurs ("stasis"). Stasis is manifest between 410.5 (collection 3) and 407.7 (collection 9) Mya; gradual change occurs between 413.3 (collection 7) and 412.3 (collection 15) Mya. Notice also that, although the average rib strength decreases in the period represented (from 416 to 406 Mya, bottom to top), there is not a consistent trend towards decreasing rib strength. This somewhat erratic pattern of morphological change is the rule rather than the exception in the fossil record. Often, trends (as in the horses represented on p. 31, showing gradual increase in size and reduction in the number of toes) come about because only some species are considered at each stage in the lineage of descent or because many species have become extinct. also, the more details are examined, the less smooth the trend appears. (After A. M. Ziegler)

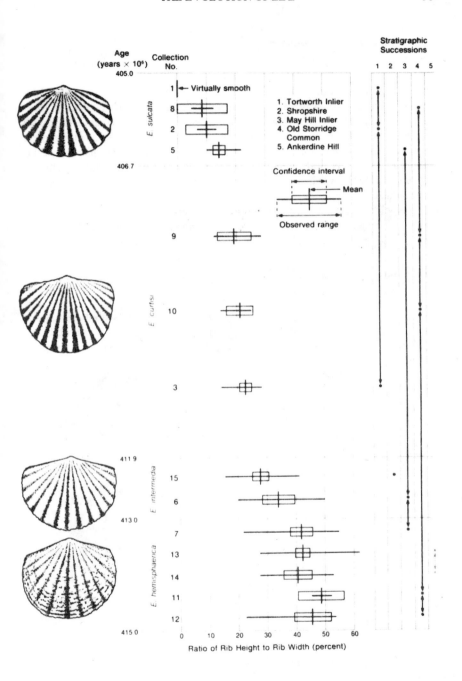

Stratigraphic Successions

1. Tortworth Inlier
2. Shropshire
3. May Hill Inlier
4. Old Storridge Common
5. Ankerdine Hill

Ratio of Rib Height to Rib Width (percent)

18 *DNA and Protein Evolution*

The advances of molecular biology have made possible the comparative study of proteins and the nucleic acid DNA, which is the repository of hereditary (evolutionary and developmental) information. The relationship of proteins to the DNA is so immediate that they closely reflect the hereditary information. This reflection is not perfect, because the genetic code is redundant and, consequently, some differences in the DNA do not yield differences in the proteins. Moreover, it is not complete, because a large fraction of the DNA (about 90 percent in many organisms) does not code for proteins. Nevertheless, proteins are so closely related to the information contained in the DNA that they, as well as the nucleic acids, are called informational macromolecules.

Nucleic acids and proteins are linear molecules made up of sequences of units—nucleotides in the case of nucleic acids, amino acids in the case of proteins—

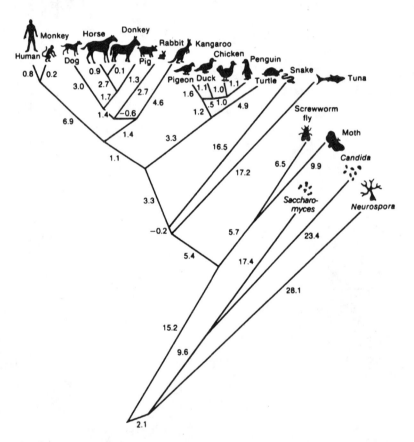

Evolutionary tree with twenty species derived from the composition of the protein cytochrome-c. The numbers are estimates of the changes in this protein that have occurred during evolution. First developed in the 1960s, this tree is one of the earliest reconstructions of evolution's history using data from molecular evolution. It agrees well with the pattern of phylogenetic relationships worked out by classical techniques of comparative morphology and from the fossil record. (After W. Fitch and E. Margoliash)

which retain considerable amounts of evolutionary information. Comparing two macromolecules establishes the number of their units that are different. Because evolution usually occurs by changing one unit at a time, the number of differences is an indication of the recency of common ancestry. Changes in evolutionary rates may create difficulties, but macromolecular studies have two notable advantages over comparative anatomy and other classical disciplines. One is that the information is more readily quantifiable. The number of units that are different is precisely established when the sequence of units is known for a given macromolecule in different organisms. The other advantage is that comparisons can be made even between very different sorts of organisms. There is very little that comparative anatomy can say when organisms as diverse as yeasts, pine trees, and human beings are compared; but there are homologous DNA and protein molecules that can be compared in all three.

Informational macromolecules provide information not only about the topology of evolutionary history (that is, the configuration of evolutionary trees), but also about the amount of genetic change that has occurred in any given branch. It might seem at first that determining the number of changes in a branch would be impossible for proteins and nucleic acids, because it would require comparison of molecules from organisms that lived in the past with those from living organisms. But this determination can actually be made using elaborate methods developed by scientists who investigate the evolution of DNA and proteins (see overleaf).

19 *The Molecular Clock of Evolution*

One conspicuous attribute of molecular evolution is that differences between homologous molecules can readily be quantified and expressed as, for example, proportions of nucleotides or amino acids that have changed. Rates of evolutionary change can, therefore, be more precisely established with respect to DNA or proteins than with respect to morphological traits. Studies of molecular evolution rates have led to the proposition that macromolecules evolve as fairly accurate clocks. The first observations came in the 1960s when it was noted that the numbers of amino-acid differences between homologous proteins of any two given species seemed to be nearly proportional to the time of their divergence from a common ancestor.

If the rate of evolution of a protein or gene were approximately the same in the evolutionary lineages leading to different species, proteins and DNA sequences would provide a molecular clock of evolution. The sequences could then be used not only to reconstruct the evolutionary tree (that is, the configuration of the branches), but also the time when the various branching events occurred. Consider, for example, the tree shown on p. 52. If the substitution of nucleotides in the gene coding for cytochrome-c occurred at a constant rate through time, one could determine the time elapsed along any branch of the tree simply by examining the number of nucleotide substitutions along that branch. One would need only to calibrate the clock by reference to an outside source, such as the fossil record, that would give us the actual geologic time elapsed in at least one specific lineage.

The molecular evolutionary clock is not a metronomic clock, like a watch or other timepiece that measures time exactly, but a stochastic clock like radioactive decay. In a stochastic clock, the probability of a certain amount of change is constant, although some variation occurs in the actual amount of change. Over fairly long periods of time, a stochastic clock is quite accurate. The enormous potential of the

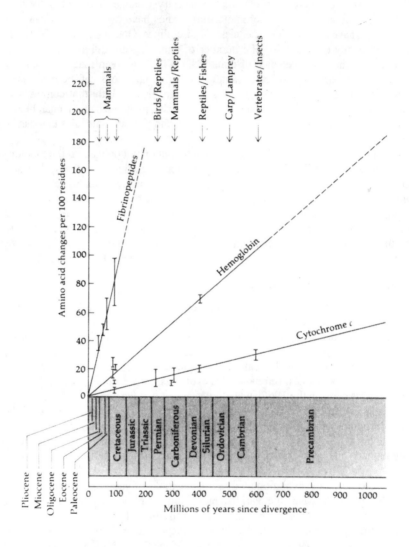

Rate of molecular evolution in three different proteins. Each gene and protein evolves at a distinct rate. Several can be used in combination in order to elucidate with precision a particular question. (After R. Dickerson)

molecular evolutionary clock lies in the fact that each gene or protein is a separate clock. Each clock "ticks" at a different rate—the rate of evolution characteristic of a particular gene or protein—but each of the thousands and thousands of genes or proteins provides an independent measure of the same evolutionary events.

Evolutionists have found that the amount of variation observed in the evolution of DNA and proteins is greater than is expected from a stochastic clock; in other words, the clock is inaccurate. The discrepancies in evolutionary rates along different lineages are not excessively large, however. It turns out that it is possible to time phylogenetic events with as much accuracy as may be desired; but more genes or proteins (about two to four times as many) must be examined than would be required if the clock were stochastically accurate. The average rates obtained for several DNA sequences or proteins taken together become a fairly precise clock, particularly when many species are investigated. This conclusion is illustrated in the figure below, which plots the cumulative number of changes in seven proteins against the paleontological dates of divergence of 17 species of mammals. The overall rate of substitution is fairly uniform, although some primates (shown by the dots at the lower left of the figure) evolved at a slower rate than the average for the rest of the species.

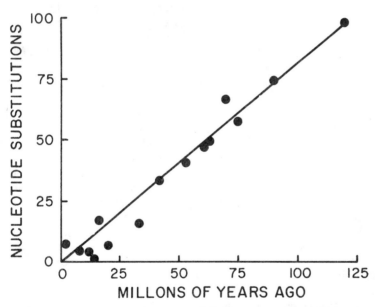

Cumulative change ("nucleotide substitutions") for seven proteins in 17 species of mammals. The solid line has been drawn from the origin to the outermost point, and corresponds to the average rate of evolution for all species. (After W. Fitch)

20 *Human Evolution*[8]

Many fossils representing recent ancestors of our species have been discovered, particularly in the last two decades. A very brief summary of what is known goes as

---

[8] For a more detailed discussion, see Camilo Cela-Conde's survey of hominid evolution in this volume.

follows. Our closest biological relatives are the great apes and, among them, the chimpanzees, who are more related to us than they are to the gorillas, and much more than to the orangutans. The hominid lineage diverged from the chimpanzee lineage 5–7 million years ago (Mya) and evolved exclusively in the African continent until the emergence of *Homo erectus*, somewhat before 1.8 Mya. The first known hominid, *Ardipithecus ramidus*, lived 4.4 Mya, but it is not certain whether it was in the direct line of descent to modern humans, *Homo sapiens*, or even bipedal. The recently described *Australopithecus anamensis*, dated 3.9–4.2 Mya, was bipedal and has been placed in the line of descent to *Australopithecus afarensis*, *Homo habilis*, *H. erectus*, and *H. sapiens*. Other hominids, not in the direct line of descent to modern humans, are *Australopithecus africanus*, *Paranthropus aethiopicus*, *P. boisei*, and *P. robustus*, who lived in Africa at various times between 3 and 1 Mya, a period when three or four hominid species lived contemporaneously in the African continent.

Shortly after its emergence in tropical or subtropical Eastern Africa, *H. erectus* spread to other continents. Fossil remains of *H. erectus* are known from Africa, Indonesia (Java), China, the Middle East, and Europe. *H. erectus* fossils from Java have been dated 1.81±0.04 and 1.66±0.04 Mya, and from Georgia between 1.6 and 1.8 Mya. Anatomically distinctive *H. erectus* fossils have been found in Spain, deposited earlier than 780 thousand years ago, the oldest in southern Europe.

The transition from *H. erectus* to *H. sapiens* occurred around 400 thousand years ago, although this date is not well determined owing to uncertainty as to whether some fossils are *erectus* or "archaic" forms of *sapiens*. *H. erectus* persisted for some time in Asia, until 250 thousand years ago in China and perhaps until 100 thousand years ago in Java, and thus was coetaneous with early members of its descendant species, *H. sapiens*. Fossil remains of Neandertal hominids (*Homo neanderthalensis*) appeared in Europe around 200 thousand years ago and persisted until thirty or forty thousand years ago. The Neandertals had, like *H. sapiens*, large brains. Until recently, they were thought to be ancestral to anatomically modern humans, but now we know that modern humans appeared at least 100 thousand years ago, much before the disappearance of the Neandertals. Moreover, in caves in the Middle East, fossils of modern humans have been found dated 120–100 thousand years ago, as well as Neandertals dated at 60 and 70 thousand years ago, followed again by modern humans dated at 40 thousand years ago. It is unclear whether the two forms repeatedly replaced one another by migration from other regions, or whether they coexisted in some areas. Recent genetic evidence indicates that interbreeding between *sapiens* and *neanderthalensis* never occurred.

There is considerable controversy about the origin of modern humans. Some anthropologists argue that the transition from *H. erectus* to archaic *H. sapiens* and later to anatomically modern humans occurred consonantly in various parts of the Old World. Proponents of this "multiregional model" emphasize fossil evidence showing regional continuity in the transition from *H. erectus* to archaic and then modern *H. sapiens*. In order to account for the transition from one to another species (something which cannot happen independently in several places), they postulate that genetic exchange occurred from time to time between populations, so that the species evolved as a single gene pool, even though geographic differentiation occurred and persisted, just as geographically differentiated populations exist in other animal species as well as in living humans. This explanation depends on the occurrence of persistent migrations and interbreeding between populations from different continents, of which no direct evidence exists. Moreover, it is difficult to reconcile the multiregional model with the contemporary existence of different species or forms in different regions, such as the persistence of *H. erectus* in China and Java for more than one hundred thousand years after the emergence of *H. sapiens*.

Other scientists argue instead that modern humans first arose in Africa or in the Middle East somewhat prior to 100 thousand years ago and from there spread throughout the world, replacing elsewhere the preexisting populations of *H. erectus* or archaic *H. sapiens*. The African (or Middle East) origin of modern humans is supported by a wealth of recent genetic evidence and is, therefore, favored by many evolutionists.

THE HOMINID EVOLUTIONARY JOURNEY: A SUMMARY

Camilo J. Cela-Conde

1 *Introduction*

Thirty years ago, C. Loring Brace suggested a model of human evolution that has been a great success.[1] Brace's model traces a straight and simple line that covers our entire family and leads directly from our common ancestors, shared with great apes, to current human beings.[2] It is very common to find this kind of straightforward model behind any argument on the theological implications of human emergence (such as God's action, original sin, and so forth). On the one hand, we have an ape-like species as the common ancestor of human beings and great apes, and on the other hand, we have the human species—the task of the theory of evolution being to explain how one and the other are related. Human evolution is just one step in the journey which leads us from that ape-like species to our own.

Is this simple and clear idea correct enough to deal with evolutionary (or theological) problems? I will try to give a negative answer, showing that no straight line can be drawn from our ancestors to the modern human species. A very complex model is needed—and some conspicuous holes will remain.

I should first note that I am not being entirely fair to Brace. Before reaching the Modern Stage, Brace's straight-line model had, it is true, several stages that may be identified with the Australopithecine Stage, the Pithecanthropine Stage and the Neandertal Stage. All these hominids belong to a sole family, *Hominidae*, although it is possible, of course, to identify different genera within it. Brace's model was characterized by its great parsimony when identifying different genera and species of the hominid family. Brace suggested, for instance, that all the different forms of *Australopithecus* found in South Africa, being normally classified as different species and even in different genera, "are very much closer in size than is commonly reported."[3]

---

[1] C. L. Brace, *The Stages of Human Evolution*, 3rd ed., (Englewood Cliffs, N.J.: Prentice-Hall, 1965/1988).

[2] According to traditional taxonomy (G. G. Simpson, "A New Classification of Mammals," *Bulletin of the American Museum of Natural History*, 85 (1931): 1–350), the *Hominidae* family consists of *Australopithecus* and *Homo*, and together with the *Pongidae* (chimpanzees, gorillas and orang-outans) it forms part of the superfamily *Hominoidea*. Biomolecular studies have negated the idea that hominoid evolution may be understood by grouping together, on the one hand, the higher apes, and on the other, human beings, given that gorillas, chimpanzees and humans are closer to each other than any of them is to the orang-outans; M. Goodman, "Man's Place in the Phylogeny of the Primates as Reflected in Serum Proteins," in *Classification and Human Evolution*, S. L. Washburn, ed., (Chicago, Ill.: Aldine, 1963), 204–34. This is why authors associated with the cladistic school distinguish two subgroups within the *Hominidae* family: *Ponginae* (orang-outans) and *Homininae* (gorillas, chimpanzees, and human beings); C. P. Groves, *A Theory of Human and Primate Evolution* (Oxford: Clarendon Press, 1989). As Tobias says, the molecular biologists are correct, but, given the lack of consensus on a new classification, it is prudent to continue with the old one; P. V. Tobias, "The Environmental Background of Hominid Emergence and the Appearance of the Genus *Homo*," *Human Evolution*, 6 (1991): 129–42. In this paper, *Hominidae* is used in the traditional sense (see Table 1 on p. 67).

[3] To sustain this, Brace compares an *Australopithecus africanus* skull (Sts 5) from

We must recognize that Brace's book was written some years before many of the current Pliocene hominid remains were found.[4] However, the most important aspects of his book are not the details belonging to different fossils, but the idea of the human evolutionary journey as a line which leads directly, with no ramifications, from early hominids to modern human beings. This idea was still held in the third edition of his book (1988), that is to say, at the time when many specimens of the different species of East Africa hominids had already been found. Why was this so? Perhaps because the straight-line idea agrees with a very widely held and popular image of what the evolutionary process is. Pre-hominids became Australopithecines; in turn, these became Pithecanthropines; and through the Neandertal stage, these latest hominids finished up as modern humans. In addition, the journey through successive stages had a common protagonist, bipedalism, and a shared environment, the savanna. Bipedalism is, therefore, seen as the hominid's opportunity to compete with other animals that intended to colonize the open spaces of the savannas. Brace emphasizes the use of tools by the early Australopithecines, even if those tools were made of perishable materials. "Although we can never 'know' this, we can guess that the wielding of a pointed stick was the crucial element which led to the change in selective forces that produced a tool-dependent biped in the first place."[5] If we take into account the success of this identification of bipedalism and tool using with the appearance of the first hominid, Brace's model happens to be a very attractive one. It not only offers an ecological explanation of hominid appearance, but also gives us a strong theory of what a hominid is.

2 What is a hominid?

Technically speaking, the task of taxonomy requires that any distinct group must have some exclusive—autapomorphic—traits (one, at least) which distinguish the members of that group from the members of different groups. To ask what a hominid is entails asking what its autapomorphic traits are. This question has been answered in several ways. As there are some very conspicuous traits which clearly distinguish humans from great apes, such as a large brain, an articulated language, and a large coefficient of encephalization, it is not surprising that these traits have been used to characterize the hominid family.[6] The idea of a large brain as the main trait of the evolutionary journey to humankind gave theoretical support to the Piltdown Man. Not all paleontologists agreed, of course, with the reconstruction of the Piltdown skull when this specimen was shown, and some of them clearly discarded the existence of any such being like the one discovered by Dawson (and anatomically reconstructed by Smith Woodward in 1912).[7] But it was not until 1953 that Le Gros Clark pronounced the specimen a fake. Why did it take so long to discredit it? It is not easy to give just one answer. Nevertheless, it seems clear that assigning the

---

Sterkfontein (South Africa) to the jaws of a *Paranthropus robustus* (SK 23) from Swartranks (South Africa, also).

[4] Regarding Pliocene hominids, only South Africa's and Olduvai's (Tanzania) specimens were known.

[5] Brace, *The Stages of Human Evolution*, 85.

[6] It is important not to mistake "hominid," that is, a member of the family Hominidae, with "human being," that is, a member of the species *Homo sapiens*.

[7] The story is very well known: Piltdown Man was the faked combination of a modern human skull and a orang-outan jaw. An accurate dissertation of Piltdown Man's episode is J. Reader, *Missing Links. The Hunt for Earliest Man* (London: Collins, 1981), chap. 3.

status of earliest hominid to the Piltdown fossil was fully compatible with the idea of a very developed brain as being the autapomorphic trait of the *Hominidae* family.

The Taung child fossil, an *Australopithecus africanus* specimen found in South Africa,[8] gave a different idea of what a hominid is. Instead of proposing large skulls as the autapomorphic trait (Taung had a very small one, even when taking into account that it was a child), the attention was focused on the bipedalism/culture combination. Walking in an upright position and using stone tools were both supposed to be the distinctive traits which allowed *A. africanus* to colonize the new savanna biota. With the general cooling process of the Earth during the Pliocene and early Pleistocene, rain forests were replaced by ever-increasing areas of open areas that could offer a new environment for a different primate, which *A. africanus* happened to be. Becoming more and more carnivorous—first scavenging and later hunting—the hominid felt an evolutionary urge to develop better implements, more accurate means of communication, and superior brains; these developments mutually contributed something to the other, that is, they constituted a feedback process.

The existence of implements can explain how an early hominid could survive in the open savanna. However, culture—the ability to make tools—as an autapomorphic trait of the *Hominidae* family, was shown to be an incorrect assumption when the East Africa finds started to give a different picture. Some of the tool specimens found at East Africa sites were much older than any belonging to the first known cultural traditions. Oldowan culture from Olduvai (Tanzania) was dated at 1.7 million years (Myr), KBS of Koobi Fora (Kenya) is even older (1.7–2.5 Myr); and there are, perhaps, implements as old as 2.6 Myr in Hadar (Ethiopia). The most widely held opinion attributes to a relatively young species, *Homo habilis*, the distinction of being the earliest maker of stone tools. *Homo habilis* was considered not older than 2 Myr in the 1960s.[9] However, much older Pliocene hominids, such as *Australopithecus afarensis*,[10] had come to light by that time, or a little bit later.

We have to propose, therefore, a different trait as the autapomorphic characteristic to identify the Hominidae family. Neither large brains nor culture are traits which could have existed in early hominid times. What can be said about a different trait already mentioned, namely, that of walking in an upright position?

Because bipedalism and savanna adaptation were conditions so intrinsically combined in Dart's original idea (see footnote 8), these traits also remained strongly linked even when culture disappeared from the scheme. Bipedalism was interpreted then as the main system which allowed hominids to adapt themselves to open fields.[11] Hominids with no cultural tradition, but endowed with bipedalism, could act both as scavengers and gatherers, roaming farther and farther, and consequently

---

[8] R. Dart, "*Australopithecus africanus*: The Man-Ape of South Africa," *Nature*, 115 (1925): 195–99.

[9] L. S. B. Leakey, P. V. Tobias, and J. R. Napier, "A New Species of the Genus *Homo* from Olduvai," *Nature*, 202 (1964): 7–9.

[10] D. Johanson, T. White, and Y. Coppens, "A New Species of the Genus Australopithecus (Primates: Hominidae) from the Pliocene of Eastern Africa," *Kirtlandia*, 28 (1978): 1–14; D. Johanson and M. Taieb, "Plio-Pleistocene Hominid Discoveries in Hadar, Ethiopia," *Nature*, 260 (1976): 293–97.

[11] B. Latimer, "Locomotor Adaptations in *Australopithecus afarensis*: The Issue of Arboreality," in *Origine(s) de la bipédie chez les hominidés*, Y. Coppens and B. Senut eds., (Paris: Editions du CNRS, 1991), 169–76, denies the importance of anatomical evidence, maintaining the idea that the "directional vector" of natural selection gives the possibility of arboreal locomotion as a residual and secondary option for *A. afarensis*.

colonizing very extensive areas. Even if the earliest Australopithecines had no culture, their bipedal character seemed to be well established. *All* hominids, found either in South Africa or in East Africa, seemed to be bipedal. Nevertheless, there was a problem lurking in this identification of bipedalism as the autapomorphic trait of hominids. If any early Australopithecine fossil morphology suggested a lack of bipedalism as a functional, well-developed trait, and projected, consequently, some doubts on the bipedal model, this suggestion was quickly rejected. As an alternative, non-fully bipedal morphological traits were attributed to sexual dimorphism. Lovejoy held that *Australopithecus afarensis* females were less bipedal because of the early work division—when males went out to scavenge/gather in the open spaces to feed their families, females stayed near the trees or within the most shielded forest areas in order to care for the children.[12]

Once again, not all scholars agree with the idea of all hominids as wholly functionally bipedal.[13] Since 1965, Phillip Tobias has forcefully discussed this idea, distinguishing the erectness of the trunk from the bipedal gait.[14] However, there seems to exist a proof of great importance for bipedal locomotion at a very early date—the Laetoli footprints. The Laetoli (Tanzania) site includes, in addition to many valuable fossils, tracks left there by three bipedal hominids that had walked over a layer of volcanic ash 3.5 Myr before the present (B.P.).[15] But even this rare find does not have the same meaning for everyone. Some scholars take it as definite proof of very early developed full bipedalism.[16] Others see in the same footprints traits that are close to pongid's feet.[17]

One fact in particular was of great importance in casting doubt on the bipedalism/savanna identification. Careful studies of the Pliocene climatic episodes have shown that the extension of open savannas is a relatively recent phenomenon, not older than 2.5 Myr.[18] Later on, I will come back to this point. In the meantime it seems clear that *Australopithecus afarensis*, which dates from at least 3 Myr, could be viewed as hominids not adapted to the savanna's open conditions.

Were these hominids fully bipedal? Once the link between savanna and bipedalism disappears, the fossils may be seen in a different light. Evidence of their incomplete bipedalism becomes increasingly importance. Perhaps the earliest hominid locomotive behavior was something between complete bipedalism and ape-

---

[12] C. O. Lovejoy, "Biomechanical Perspectives on the Lower Limb of Early Hominids," in *Primate Functional Morphology and Evolution*, R.H. Tuttle, ed., (The Hague: Mouton, 1975), 291–326.

[13] See C. J. Cela-Conde, "¿Comienza Africa en los Pirineos?" *Investigación y ciencia*, 232 (1996): 32–34; idem, "Bipedal/Savanna/Cladogeny Model. Can it Still Be Held?" *History and Philosophy of the Life Sciences*, 18 (1996): 213–24.

[14] For instance, see P. V. Tobias, "Man the tottering biped: the evolution of his erect posture," in *Proprioception, Posture and Emotion*, D. Garlick, ed., (Sydney: Committee in Postgraduate Medical Education, Univ. of NSW, 1982), 1–13.

[15] M. D. Leakey and R. L. Hay, "Pliocene Footprints in the Laetoli Beds at Laetoli, Northern Tanzania," *Nature*, 278 (1979): 317–23.

[16] M. H. Day and E. H. Wickens, "Laetoli Pliocene Hominid Footprints and Bipedalism," *Nature*, 286 (1980): 385–87; T. D. White, "Evolutionary Implications of Pliocene Hominid Footprints," *Science*, 208 (1980): 175–76.

[17] Y. Deloison, "El pie de los primeros homínidos," *Mundo científico*, 164 (1996): 20–22.

[18] M. L. Prentice and G. H. Denton, "The Deep-Sea Oxygen Isotope Record: The Global Ice Sheet System and Hominid Evolution," in *Evolutionary History of the "Robust" Australopithecines*, F. E. Grine, ed., (New York, N.Y.: Aldine de Gruyter, 1988), 383–403.

like locomotion. All early forms can be seen nowadays from this point of view: *Australopithecus africanus*,[19] *Australopithecus afarensis*[20] and *Ardipithecus ramidus*.[21] In the light of this we must once again ask the main question, what is a hominid?

It appears that we have finished with our initial list of autapomorphic traits. No trait already examined (large brain, culture, bipedalism) can fulfill that condition. Do we have any more traits to propose? Or, on the contrary, must we hold that there is no trait shared by *all* hominids? It seems clear that the idea of *Hominidae* as an evolutionary group (a family, as we have seen, but the actual category of the group does not matter) obliges us to define the autapomorphic traits which separate this group from another. If this is not possible, we will have no criterion by which to decide whether new fossil finds belong to *Hominidae* or not. This is not just a hypothetical question. Some very old fossils recently found, such as, *Ardipithecus ramidus*,[22] can be interpreted either as hominid or pongid, as their age and morphology appears to be very close to the great ape/hominid divergence.[23] How then can we decide which is the correct place for them?

Before answering this question, I would like to sketch the present-day picture of hominids by giving a list of the different species that are considered as belonging to the hominid family. Comparing all of them may be a good way to conclude what kind of characteristics they all share and, consequently, what, if any, are the autapomorphic traits that distinguish *Hominidae*. To clarify the ecological phenomena underlying hominid evolution I will refer to five different evolutionary episodes: 1) the appearance of early hominids; 2) the 2.5 Myr cladogeny; 3) the *Homo erectus*' out-of-Africa journey; 4) the Neandertal problem; and finally, 5) the appearance of modern human beings.

3 Evolution of the Hominidae family

3.1 The early hominids

In recent years, a considerable amount of very early hominid fossils have been found. Leaving to one side the *Australopithecus afarensis* from Laetoli and Hadar, well

---

[19] R. J. Clarke and P. V. Tobias, "Sterkfontain Member 2 Foot Bones of the Oldest South African Hominid," *Science*, 269 (1995): 521–24.

[20] B. Senut, *Contribution à l'étude de l'humérus et de ses articulations chez les Hominidés du Plio-Pleistocène*, thèse doctorat 3ème cycle, (Paris: Université Pierre et Marie Curie, 1978); R. L. Susman, J. T. Stern Jr., and W. L. Jungers, "Arboreality and Bipedality on the Hadar Hominids," *Folia Primatologica*, 43 (1984): 113–56; R. L. Susman, J. T. Stern, and W. L. Jungers, "Locomotor Adaptations in the Hadar Hominids," in *Ancestors*, Delson, ed., 184–92; B. Senut and C. Tardieu, "Functional Aspects of Plio-Pleistocene Hominid Limb Bones: Implications for Taxonomy and Phylogeny," in *Ancestors*, Delson, ed., 193–201; W. L. Jungers, "Relative Joint Size and Hominoid Locomotor Adaptations with Implications for the Evolution of Hominid Bipedalism," *Journal of Human Evolution*, 17 (1988): 247–65; M. D. Rose, "The Process of Bipedalization in Hominids," in *Origine(s) de la bipédie chez les hominidés*, Coppens and Senut, eds., 137–48.

[21] W. H. Kimbel, D. C. Johanson, and Y. Rak, "The First Skull and Other New Discoveries of *Australopithecus afarensis* at Hadar, Ethiopia," *Nature*, 368 (1994): 449–51.

[22] T. D. White, G. Suwa, and B. Asfaw, "*Australopithecus ramidus*, a New Species of Early Hominid from Aramis, Ethiopia," *Nature*, 371 (1994): 306–12; idem, "*Australopithecus ramidus*, a New Species of Early Hominid from Aramis, Ethiopia." *Nature*, 375 (1995): 88.

[23] P. Andrews, "Ecological Apes and Ancestors," *Nature*, 376 (1995): 555–56, for instance, holds that *ramidus* is not a hominid; see below.

documented as far back as the late 1970s, we have at least two very old members of our family, namely, *Australopithecus anamensis*, a 4.0 Myr-old species from Kanapoi and Allia Bay (Kenya),[24] and *ramidus*, a 4.4 Myr-old species from Aramis, Ethiopia.[25] The latter seems to be so archaic that its discoverers classified it first as an *Australopithecus*, but later they proposed a new genus, *Ardipithecus*.[26] The first Australopithecine discovered west of the Rift Valley, in northern Chad, a 3.0–3.5 Myr-old mandible (KT 12/H1), was classified as belonging to *Australopithecus afarensis*,[27] but a new species, *Australopithecus bahrelghazali*, has been lately proposed with this mandible as holotype.[28]

Keeping in mind that molecular genetic evidence refers to a date of 5–6 Myr for the divergence between great apes and hominids,[29] *A. ramidus* is quite close to that date. Therefore, we can ask whether this specimen belongs to the hominid family, or whether it is perhaps the last fossil form in existence before divergence occurred. A third possibility may be suggested: perhaps *ramidus* is even a chimpanzee ancestor. This is not such a bizarre option, for scholars such as Peter Andrews consider that the dental characteristics of *ramidus* are "more of what you'd expect from a fossil chimp"; some other characteristics, such as the postcrania, are chimpanzee-like too.[30] The thin dental enamel of *ramidus* is a real problem because all hominids have thick enamel. Even 10 Myr-old hominoid ancestors have relatively thick enamel, since thinner tooth enamel is just a derived characteristic of chimpanzees. The doubts as to how to classify *Ardipithecus ramidus* are the same as those that could be expressed for the new species recently found by Meave Leakey and collaborators in Allia Bay and Kanapoi (Kenya).[31] *Australopithecus anamensis*, at 4 Myr-old, is an intermediate form between *Ardipithecus anamensis* and *Australopithecus afarensis*, but its being a hominid is not completely taken for granted. Kanapoi specimens are a "remarkable combination of advanced *Homo*-like features of the postcrania... with primitive jaws and teeth similar to those of Miocene apes."[32]

However, it is also possible to find scholars that place *ramidus* and *anamensis* deep inside the hominid family.[33] Enamel thickness is, from White's point of view, a widely variable characteristic depending on diet, which could have changed with time.[34] In brief, no concrete conclusion has yet been reached—though it must be said that in paleoanthropology there are very few definitive conclusions.

---

[24] M. G. Leakey, C. S. Feibel, I. McDougall, and A. Walker, "New Four-Million-Year-Old Hominid Species from Kanapoi and Allia Bay, Kenya," *Nature*, 376 (1995): 565–72.

[25] White et al., "*Australopithecus ramidus*," *Nature*, no. 371.

[26] White et al., "*Australopithecus ramidus*," *Nature*, no. 375.

[27] M. Brunet, A. Beauvilain, Y. Coppens, E. Heintz, A. H. E. Moutaye, and D. Pilbeam, "The first australopithecine 2,500 kilometers west of the Rift Valley (Chad)," *Nature*, 378 (1995): 273–75.

[28] Idem, "*Australopithecus bahrelghazali*, une nouvelle espèce d'Hominidé ancien de la région de Koro Toro (Tchad)," *C. R. Acad. Sci. Paris*, 322, série IIa (1996): 907–13.

[29] V. Sarich, "A Molecular Approach to the Question of Human Origins," in *Background for Man*, V. Sarich and P. Dolhinow, eds., (Boston, Mass.: Little, Brown and Co., 1971), 60–81. See p. 65 for the "late divergence hypothesis."

[30] Andrews, "Ecological Apes and Ancestors." See also J. Fischman, "Putting Our Oldest Ancestors in Their Proper Place," *Science*, 265 (1994): 2011–12.

[31] M. G. Leakey et al.,"New Four-Million-Year-Old Hominid Species."

[32] Andrews, "Ecological Apes and Ancestors."

[33] White et al., "*Australopithecus ramidus*," *Nature*, 371; M. G. Leakey et al.,"New Four-Million-Year-Old Hominid Species."

[34] Fischman, "Putting Our Oldest Ancestors in Their Proper Place."

The different ways of viewing the *A. ramidus* and *A. anamensis* characteristics help us to better understand the enormity of the problem. If we hold that some very old species like *A. anamensis*, *A. afarensis* and (perhaps) *A. ramidus* belong to the hominid family, but have neither culture nor large brains, why are we supposing that they are hominids? Why can we not classify them as pongids, as the last specimen of the common ancestors shared by great apes and human beings? Peter Andrews resolves the argument with an elegant bias: even if the phylogenetic condition of early hominids is not clear, we can consider them, ecologically speaking, still as apes.[35] This idea takes into account (and explains) the presence of both non-bipedal traits and ancestral dental characteristics. But Andrews' proposal does not give an answer to the main question. If, as Meave Leakey holds, hominids radiated 4 Myr ago into several species—one of them eventually leading to *Homo*[36]—why are we talking of "hominids" related to all these different and very old species? Why has the threshold which separates hominids from the rest of primates been crossed? Whether we discard complete bipedalism as a trait shared by all those forms (and not by pongids), as well as the *ad hoc*-like answer of "not full, but transitional bipedalism,"[37] we still need to identify what constitutes the early hominids' autapomorphic trait.

Paleontologists like to relate any hypothesis to actual fossil specimens rather than to speculations about non-fossilized traits (such as some kind of behavior). Nevertheless, a special kind of behavior is the guideline normally mentioned which allows us to identify a hominid—a being which acts according to a non-specialized behavior (feeding behavior, mainly) to adapt to an environment. As a matter of fact, this seems to be a negative identification (any highly specialized specimen is *not* a hominid). Luckily, there is a trait which is related to fossils and gives a positive identification of a hominid too. This trait is enamel thickness/size of teeth, considered always from a relative point of view. There is a tendency toward increased thickness of the enamel and decreased size of incisors and canines. This tendency gives us evidence of the very wide (that is, non specialized) range of food in the hominid lineage diet. However, these tendencies started in the Miocene at least, that is, a long time before that the early hominids appeared. In the 1960s a Miocene specimen, the so-called *Ramapithecus*, was considered to be the first member of our family because of its thick enamel and relatively small teeth.[38] This idea was absolutely rejected by molecular geneticists, as the immunological studies and the molecular clock gave a relationship which was too close among gorillas, chimpanzees and human beings, ruling out their supposed divergence as far back as 8–10 Myr.[39] The "late divergence hypothesis" proposed by molecular genetics is

---

[35] Andrews, "Ecological Apes and Ancestors."

[36] E. Culotta, "New Hominid Crowds the Field," *Science*, 269 (1995): 918.

[37] The transition stage to bipedalism is an important trait shared by all forms which are not fully bipedal. Although I do not want to give the impression of paying scant attention to this, it seems clear that this kind of transitional trait must be accompanied by other traits to justify the appearance of something as important in evolutionary terms as a new family.

[38] Elway L. Simons, "The Phyletic Position of *Ramapithecus*," *Postilla* 57 (1961): 1–9.

[39] M. Goodman, "Evolution of the Immunologic Species Specificity of Human Serum Proteins," *Human Biology*, 34 (1962): 104–50; idem, "Man's Place in the Phylogeny of the Primates as Reflected in Serum Proteins"; Sarich, "A Molecular Approach to the Question of Human Origins," in *Background for Man*, Sarich and Dolhinow, eds.; V. Sarich and A. C. Wilson, "Rates of Albumin Evolution in Primates," *Proceedings of the National Academy of Sciences*, 58 (1967): 142–48.

now widely accepted, and *Ramapithecus* does not exist anymore as a separate genus (the finds are considered female specimens of *Sivapithecus*).[40]

Consequently, even if relatively small teeth with thick enamel are normally used to identify a hominid, this assumption must be specified. As White says, we must take into account different characteristics—such as the shape of the teeth—in addition to enamel thickness in order to classify specimens.[41] And these are traits that refer to non-specialized behavior as well. Since early times, hominids have fed themselves on a varied diet, and gradually developed several traits (such as bipedalism) to improve their adaptive strategy. Later, in the 2.5 Myr cladogeny age—as we shall see—some of the hominids (the robust ones) developed a more specialized behavior, but this is just a relative assumption. Even robust Australopithecines had thick enamel. To be sure about what kind of different diets early hominids had, one needs very sophisticated means of proof, such as micrographies of the enamel surfaces.[42]

I have recently pointed out[43] that it is difficult with the criteria normally used, to place all the specimens found in the hominid family—from early *Australopithecus* to late *Homo sapiens*. Thus, we may need to consider excluding *Australopithecus* from the hominids, as Dart and Broom suggested when they found the first South Africa specimen. The idea of an intermediate Pongidae/Hominidae family was then proposed. However, Le Gros Clark placed *Australopithecus* so definitively among the members of our family that any attempt to alter this classification almost seems wishful thinking nowadays.[44] However, as I mentioned before, Peter Andrews recently emphasized that both *Ardipithecus* and early *Australopithecus* should be considered as "ecological apes," in spite of what their morphology may tell us.[45] Perhaps Simons' classification of the order Primates, separated into two different subfamilies, Australopithecine ("ape-men") and Homininae ("humans"), is still the best solution.

---

[40] Not all scholars agree on the close relationship between gorillas, chimpanzees, and human beings proposed by molecular genetics. Schwartz, for instance, considers that the orang-outan is a closer relative to human beings than any other great ape; J. H. Schwartz, "Hominoid Evolution: A Review and a Reassessment," *Nature*, 308 (1984): 501–5; idem, "The Evolutionary Relationships of Man and Orang-outans," *Nature*, 308 (1984): 501–5.

[41] The use of enamel thickness as the single characteristic to classify hominid specimens leads us to a "slippery substrate." See Fischman, "Putting Our Oldest Ancestors in Their Proper Place."

[42] R. F. Kay and F. E. Grine, "Tooth Morphology, Wear and Diet in *Australopithecus* and *Paranthropus* from Southern Africa," in *Evolutionary History of the 'Robust' Australopithecines*, Grine, ed., 427–47.

[43] Symposium on "La evolución humana. Homenaje al Profesor Phillip V. Tobias." Palma de Mallorca (Spain), May 7–8, 1996. See my "¿Qué es un homínido?" in *Senderos de la evolución humana: Estudios en homenaje a Phillip Tobias*, Camilo J. Cela-Conde, Raul Gutiérrez Lombardo, and Jorge Martínez Contreras, eds. (*Ludus Vitalis, Journal of Philosophy of Life Sciences*, special number 1, 1997).

[44] W. L. G. Clark, *The Fossil Evidence for Human Evolution: An Introduction to the Study of Paleo-anthropology* (Chicago, Ill.: Chicago University Press, 1955).

[45] Andrews, "Ecological Apes and Ancestors."

Superfamily	Family	Subfamily
Hominoidea (apes and humans)	Oreopithecidae (oreopithecids)	
	Hylobatidae (lesser apes)	Pliopithecine (fossil small apes)
		Hylobatinae (gibbons)
	Pongidae (great apes)	Dryopithecinae (fossil large apes)
		Ponginae (modern great apes)
	Hominidae (hominids)	Australopithecine (ape-men)
		Homininae (humans)

Table 1. Formal Classification of the order of Primates (Superfamily Hominoidea)[46]

In my opinion, only the convergence of morphological, functional, and behavioral traits (such as language, or the early development of altruistic links inside family groups) may be able to give a fair portrait of what a hominid is. However, some of these aspects are only known to us in a speculative way. We have no fossil proof of such behavior. On the other hand, these speculations give us a rich field of discussion for the purposes of this conference.

3.2 The 2.5 Myr cladogeny

A cladistic episode is an evolutionary phenomenon which implies the appearance through evolution of two species coming from a former species, which disappears at the same time. Even if not all evolutionary episodes can be interpreted as cladogenies—a species can evolve through anageny, converting itself into a different one, rather than into two—some of the most important evolutionary episodes imply cladogeny. One of them is of great importance in explaining how hominid evolution took place.

Coinciding with the cooling phenomenon already mentioned, about 2.5 Myr B.P., open savannas offered an obvious alternative to the former tropical forest environment. The hominids of that time seem to have responded to this selective pressure in two different ways. First, they developed morphological traits, such as heavy mandibles and skulls, suitable to the savanna habitat. Second, they adapted themselves to the new conditions through their ability to make flint implements, using these tools as a means of obtaining food. These alternative journeys lead us to the important and very well known division of the *Hominidae* family—the "robust" and "gracile" specimens.[47]

If we accept that the *Hominidae* family members—which would belong to several species with distinct morphological traits—developed at the same time but with different adaptive strategies, Brace's model of hominid evolution as a single and simple line collapses. Since 1959, hominid remains with significant differences in their morphological traits have been found in the Rift Valley, thus giving a very different picture of what hominid evolution was. For instance, we have 3.5 Myr-old

---

[46] Elway L. Simons, *Primate Evolution: An Introduction to Man's Place in Nature* (New York, N.Y.: Macmillan, 1972), slightly modified by R. D. Martin, *Primate Origins and Evolution: A Phylogenetic Reconstruction* (London: Chapman and Hall, 1980).

[47] Not all authors agree with this interpretation. For instance, R. L. Susman, "Hand of *Paranthropus Robustus* from Member 1, Swartkrans: Fossil Evidence for Tool Behavior," *Science*, 240 (1988): 781–84, holds that *Australopithecus robustus*, a robust form, may have been responsible for the Swartranks (South Africa) implements.

forms with very small skulls,[48] 2.5 Myr-old forms with heavy skull crests (such as pongids have),[49] 2 Myr-old forms with quite large skulls and no crests at all,[50] and 1.5 Myr-old robust forms with skull crests.[51] All these new fossils, found in East Africa, increase the Pliocene hominid fossil register. On the other hand, they also make the classification task harder. To put these finds, including the South Africa fossils, all together into an evolutionary sequence seems a weighty task. Or perhaps it is an easy one, considering the plethora of evolutionary trees that exist. As a brief summary, I include below a figure showing three hypotheses of the evolutionary relationships among *Australopithecus afarensis*, *Australopithecus africanus*, *Paranthropus aethiopicus*, *Paranthropus boisei*, *Paranthropus robustus*, and the human lineage.

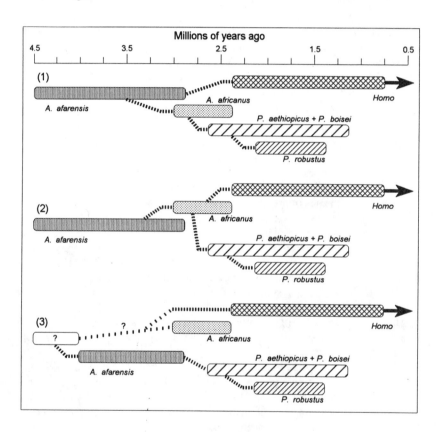

Figure 1. Three Hypotheses of Australopithecine Evolution[52]

[48] Johanson et al., "A New Species of the Genus Australopithecus."

[49] A. Walker, R. E. Leakey, J. M. Harris, and F. H. Brown, "2.5 Myr *Australopithecus boisei* from West of Lake Turkana, Kenya," *Nature*, 322 (1986): 517–22.

[50] R. E. F. Leakey, "Evidence for an Advanced Plio-Pleistocene Hominid from East Rudolf, Kenya," *Nature*, 242 (1973): 447–50.

[51] For instance, R. Leakey, "New Hominid Remains and Early Artifacts from Northern Kenya," *Nature*, 226 (1970): 223–24.

[52] Figure adapted from B. Wood, "Evolution of Australopithecines," in S. Jones, R. Martin and D. Pilbeam, eds., *The Cambridge Encyclopedia of Human Evolution*

Even this scenario, with just three different trees, is conservative. Groves[53] accepts the following hominid Pliocene genera:

1. Genus unnamed (which includes the famous AL 288-1, "Lucy" from Hadar, Ethiopia, and also *Homo aethiopicus* and *Australopithecus africanus aethiopicus*).

2. Genus *Paranthropus* (which includes several robust forms, namely, *Paranthropus robustus* and *Paranthropus crassidens* of South Africa, and also *Paranthropus boisei* and *Paranthropus walkeri* from East Africa).

3. Genus *Australopithecus* (including *Australopithecus africanus* and also *Australopithecus afarensis* found in Laetoli, but excluding "Lucy," perhaps the best known example of *afarensis* that any paleontologist could give, which Groves classifies as belonging to the unnamed genus listed as 1 above).

4. Genus *Homo* (*Homo sp.*[54] from Hadar, that is, the AL-333 series; *Homo rudolfensis*, including some specimens from Koobi Fora, Kenya,—such as ER 1470, normally attributed to *Homo habilis*, and, obviously, *Homo habilis* himself, with special reference to the very well known fossils of Olduvai, Tanzania, such as OH7). *Homo ergaster*, from Koobi Fora, Kenya, may be included as one more species of the Pliocene hominids, genus *Homo* as well.

Even for the scholar who agrees with the cladistic system of classification, it is not easy to make an evolutionary tree which includes all these different kinds of Pliocene hominids. If we take into account that, since the publication of Groves' book, more fossils (like *Australopithecus ramidus*) have been found, which are older and very different from those already known at that time, the idea of a direct and simple line of evolution from early hominids to modern humans is obviously untenable. Sometimes it seems even impossible, using the almost endless list of specimens, to arrive at any rational idea of how the evolutionary process of hominids took place. However, I will try to provide some clues later, that, in my opinion, may help to paint a picture of the general panorama of hominid evolution.

3.3 *Out of Africa: The* Homo erectus *dilemma*

I have mentioned the presence of very different hominid forms during the Pliocene, and their different ways of adaptation. This variety of forms decreases dramatically during the Middle Pleistocene, even if, in those times, hominids radiated from Africa to many different places throughout the ancient world (Asia and Europe).

It is common practice to classify all Middle Pleistocene hominids as belonging to a single species, that of *Homo erectus*, even if there are doubts regarding its evolutionary characteristics. Since African and Asian forms have very different morphologies, the origin of these differences must be explained. Some scholars see *H. erectus* as a species that gradually evolved from its first appearance (about 1.8 Myr ago) to the times when it gives way to *Homo sapiens*.[55] Others hold that *H. erectus* underwent very few changes during most of its life as a species and then, in a very short space of time, evolved into *Homo sapiens*.[56] But this is not the only alternative we can consider. The difference between African and Asian Middle

---

(Cambridge: Cambridge University Press, 1992), 231–40.

[53] Groves, *A Theory of Human and Primate Evolution*.

[54] This abbreviation is commonly used to designate a tentatively proposed species.

[55] M. H. Wolpoff, "Evolution in *Homo erectus*: The Question of Stasis," *Paleobiology*, 10 (1984): 389–406; J. E. Cronin, N. T. Boaz, C. B. Stringer, and Y. Rak, "Tempo and Mode in Hominid Evolution," *Nature*, 292 (1981): 113–22.

[56] G. P. Rightmire, "The Tempo of Change in the Evolution of Mid-Pleistocene *Homo*," in *Ancestors*, Delson, ed., 255–64.

Pleistocene hominids is strong enough for some scholars to distinguish between *Homo erectus* (Asian forms) and *Homo ergaster* (East African forms).[57]

The radiation of either *Homo erectus* or *Homo ergaster*, or both, colonizing wide territories outside Africa led hominids to colder climates, that is, climates where a tropical primate cannot easily adapt itself. The Acheulean culture is assumed to be responsible for the presence of *H. erectus* in such places as continental China and Europe. Acheulean culture distinguishes itself from Oldowan implements in several ways, including sharper edged bifacial handaxes with symmetric forms and the use of fire. An evolution of mental abilities is normally related to that kind of cultural presence, such as the different axis of symmetry and the controlled curves of bifacial handaxes, which indicate that a preconceived form was held in the *H. erectus* mind.[58] The use of fire, even if it was ignited spontaneously, supports the presence of improved mental abilities as well, although it is not easy to explain how an animal could get rid of its instinctive fear of fire and start to use it as a means of obtaining heat, defense, or perhaps even tastier meals.

Even if the Acheulean culture is associated with every form of both *Homo ergaster* or *Homo erectus*, local differences exist too. Asian forms, for instance, do not seem to have made bifacial handaxes, something that can be explained by means of an early hominid migration. If Asian *H. erectus* ancestors arrived there before having developed more sophisticated Acheulean tools, the absence of bifacial axes is not strange. Nevertheless, the problem is to decide what the dates of the *H. erectus* presence in Asia are and if they conform with the dating techniques. Since the Rift Valley in East Africa had plenty of volcanic activity during the Pliocene and Middle Pleistocene epochs, it is possible to use K/Ar (potassium/argon) dating, a very accurate technique. Unfortunately, K/Ar dating cannot be used in Chinese *erectus* sites, due to the absence of volcanic materials. Java, the other important area in Asia having *erectus* remains, had volcanic activity too, but the main problem there is that of attributing the fossil finds to their site origin, as many of them were collected by natives, who later showed where the remains were found. This circumstance leads us to a very dubious compromise, as dating fossils depends intrinsically on assigning them to the exact location where they were found. Guessing the remains' origin, Swisher gave to Java *H. erectus* a date of 1.8 Myr,[59] but Javanese site dating is a very debatable matter.[60]

On the other hand, very old *Homo* finds from sites on both sides of Europe—Dmanisi (Georgia, East Europe) and Atapuerca (Spain, West Europe)—have been reported in recent years. Before the appearance of these finds, *Homo*'s presence in Europe was assumed to be relatively recent, not more than half a million years ago.

---

[57] C. Stringer, "The Definition of *Homo erectus* and the Existence of the Species in Africa and Europe," in *The Early Evolution of Man*, P. Andrews and J. Frazen, eds., (Frankfurt, Senckenberg Museum, 1984); P. Andrews, "An Alternative Interpretation of the Characters Used to Define *Homo erectus*," in *The Early Evolution of Man*, Andrews and Frazen, eds.

[58] J. A. J. Gowlett, "Early Human Mental Abilities," in *The Cambridge Encyclopedia of Human Evolution*, S. Jones, R. Martin and D. Pilbeam eds., (Cambridge: Cambridge University Press, 1992), 341–45.

[59] C. C. Swisher III, G. H. Curtis, T. Jacob, A. G. Getty, and A. S. Widiasmoro, "Age of the Earliest Known Hominids in Java, Indonesia," *Science*, 263 (1994): 1118–21.

[60] See, for instance, G. P. Rightmire, *The Evolution of Homo Erectus: Comparative Anatomical Studies of an Extinct Human Species* (Cambridge: Cambridge University Press, 1990).

Fossils around that date, or younger, have been found in Central Europe (Mauer, Germany; Vérstesszöllös, Hungary; Montmaurin, France), giving credence to the idea of a European transition from *H. erectus* to *H. sapiens*. Nowadays, the Dmanisi site has a date of 1.8 Myr,[61] and TD-6 site of Atapuerca more than 0.7 Myr.[62] The date of the Dmanisi fossil (a mandible) is suspect, but the Atapuerca dates are well established.[63] The primitive aspect of their culture gives rise to the possibility of a very old presence of *Homo* in Europe. Another site, Orce (Granada, Spain), with partial and much discussed finds but very primitive tools, could drive the West European *Homo* dates as far back as 1.2 Myr, but more accurate field work is needed to obtain acceptable dating and classification of Orce's hominid fossils.[64]

3.4 Homo sapiens: *The Neandertal Problem*

The appearance of our own species is an episode that, as we have seen, must be related to many different former species in a "bush-like" phylogenetic tree (the metaphor comes from Stephen Jay Gould). We are already very far from the straight line proposed by Brace and perhaps more confused than cognizant of the evolutionary relations among these different hominid lineages. *Homo sapiens'* story will add even more doubts, but, on the other hand, these new doubts could perhaps be of great interest for the purposes of this conference. At least, some biblical names do appear very often in the context of arguments about modern human origins, such as "Mitochondrial Eve," the "Garden of Eden" and "Noah's Ark." However, I would like to start with an earlier problem—that of Neandertal man.

If we use *Homo sapiens neanderthalensis* as a scientific name, it refers to individuals belonging to the same species as our own, but to a different subspecies. The evolutionary implications of a classification like this are not clear enough. In maintaining the idea of the same species, we are admitting that a population of Neandertal, *Homo sapiens neanderthalensis*, could have mixed with a different population of *sapiens*, giving place, perhaps, to morphologically modern human beings (subsequently abbreviated to MMHB).[65] In other words, we could be, genetically speaking, something like grandchildren of *Homo sapiens neanderthalensis*. Later we will come back to this. Let us first ask, what kind of being was a Neandertal?

Some participants in this conference have dealt with the question of how God's action can be seen within a biological context, entertaining the possibility of the existence of other planets where consciousness and mentality could have emerged. In this paper, I have been proposing the idea of a gradual and slow evolution of mental processes, which can be identified in members of other species. Nevertheless,

[61] L. Gabunia and A. Vekua, "A Plio-Pleistocene Hominid from Dmanisi, East Georgia, Caucasus," *Nature*, 373 (1995): 509–12.

[62] E. Carbonell, J. M. Bermúdez de Castro, J. L. Arsuaga, J. C. Díez, A. Rosas, G. Cuenca-Bescós, R. Sala, M. Mosquera, and X. P. Rodríguez, "Lower Pleistocene Hominids and Artifacts from Atapuerca-TD6 (Spain)," *Science*, 269 (1995): 830–32.

[63] J. M. Parés and A. Pérez-González, "Paleomagnetic Age for Hominid Fossils at Atapuerca Archaeological Site, Spain," *Science*, 269 (1995): 830–32.

[64] See C. J. Cela-Conde, "Choosing between Two Conflicting Scientific Hypotheses: The Orce Dilemma," in *Human Evolution* (in press).

[65] Phillip Tobias dislikes the expression "morphologically" (or "anatomically") modern human beings, finding it redundant and useless. However, it is often used to identify forms with a modern aspect that can be classified as *Homo sapiens sapiens*, to distinguish them from archaic forms and Neandertal forms.

it might be argued that a *true* emergence of human consciousness has nothing in common with low-level cognitive abilities. The human mental state includes some kinds of characteristics that we can identify only with our own species. I cannot give any opinion on the possibility of the existence of similar mental abilities on a different and remote planet. However, I will offer some evidence of the emergence of very similar characteristics, apart from ourselves, on our own planet. I am talking, obviously, of *Homo neanderthalensis*.

Even for someone not skilled in anthropology, it is very easy to give a list of the main traits which are particular to our human condition: aesthetic sense, abstract thought, creative language, moral sense, and so forth. If we guess at the presence of that kind of phenomena in fossils of, say, thirty to thirty-five thousand years ago, we could be accused of contributing to a fallacy, as none of those characteristics fossilize. But nobody can deny that Aurignacian culture, with its abundance of ornamental objects, magnificent portraits of animals, and delicately hand-carved items—such as the La Madelaine engraved in fossilized ivory—were made by beings with similar feelings and thoughts as ourselves. Ancestors of Aurignacian times are classified as Cro-Magnon and are assumed to be the first true modern human beings. Cro-Magnon people had religious beliefs as well. They buried their dead in a special position (the "crouch burial"), leaving offerings in their graves.

Neandertals were not, morphologically speaking, Cro-Magnons in any way. They also had a different culture—the Mousterian one. However, if we have detected the presence of very highly developed mental abilities in Cro-Magnons, as seen in the Aurignacian objects, the same can also be said, on a close examination, of Neandertal beings and their culture. Almost all those characteristics previously mentioned to identify MMHB—aesthetic sense, moral sense, abstract thought— which we can observe in the Aurignacian culture, are present in the Mousterian culture as well, though at a lesser level. Only a fully developed language would be lacking, in Lieberman's opinion, owing to the anatomy of the Neandertal larynx.[66]

According to some authors, Neandertals were, as a matter of fact, "intentionally buried"[67] and their graves give "evidence not merely of the mechanical act of burial but also of ritual behavior."[68] The finds of Neandertal skeletons, such as at Teshik-Tash (Uzbekistan), encircled by pairs of horns from wild goats, gave rise to interpretations galore of Neandertal spiritualism related to the Cave Bears' cult.[69] However, other authors think that neither ritual burial, nor religious cult, was developed by Neandertals.[70]

Much easier to interpret are some Neandertal attitudes toward physical incapacities. The "Old Man" from La-Chapelle-aux-Saints lived to an advanced age (30 years old), having lost most of his teeth (molars) before dying. By the time of his

---

[66] P. Lieberman, "Human Language and Human Uniqueness," *Language & Communication*, 14 (1994): 87–95. See the discussion about the appearance of language in my second essay in this volume.

[67] A. Bouyssonie, J. Bouyssonie, and L. Bardon, L., "Découverte d'un squelette humain mousterian à la bouffia de la Chapelle-aux-Saints (Corrèze)," *L'Anthropologie*, 19 (1908): 513–19.

[68] Erik Trinkaus and Pat Shipman, *The Neandertals: changing the image of mankind* (New York, N.Y.: Knopf, 1993), 255, quoting Alberto Blanc's change of mind on the explanation of Neandertal burial rituals.

[69] Trinkaus and Shipman, *The Neandertals*, 258.

[70] C. Stringer and C. Gamble, *In Search of Neandertals: Solving the Puzzle of Human Origins* (London: Thames and Hudson, 1993).

death, he was suffering from several illnesses, such as arthritis of the lower neck, back and shoulders, a broken rib (already healed), and a badly deteriorated left hip.[71] Even if the first interpretation of the Old Man skull (Marcellin Boule's 1911–13 well-known monography) characterized Neandertals as individuals close to apes, it is impossible to explain how the La Chapelle specimen could have reached such a relatively old age without the help of his relatives.

Shanidar I, a well preserved burial site of a Neandertal man, aged between thirty and forty-five at death, gives one more similar example. This specimen was probably blind in the left eye and had severe atrophies on his right side (arm, hand, foot and leg)[72] due to some traumatic injury suffered years after his birth, so he would by no means have survived without being well taken care of. His survival is "a testament of Neandertal compassion and humanity."[73]

The Neandertal's high level of humanity could easily be explained if *Homo neanderthalensis* were actually a direct ancestor of MMHB.[74] The "Neandertal problem" relates, in fact, to this question—what is the evolutionary relationship between *H. neanderthalensis* and *H. sapiens*?

Authorities in the Neandertal field such as Trinkaus hold that with regard to the anatomical differences existing between them, we can no longer consider MMHB simply as descendants of the Neandertals. But, modern humans could be the result of genetic changes between different populations, the Neandertals included.[75] Genetic continuity between Neandertals and MMHB is regarded as a documented certainty.[76] However, there seems to be a general consensus that Neandertal and MMHB differ from one another in many conspicuous features of the skull and postcranial skeleton, with no intermediate morphology.[77] This absence of intermediate morphology speaks in favor of assigning Neandertal and MMHB to different species, something suggested by Stringer[78] (and recently by Zollikofer et al. through a computer reconstruction of the Neandertal skull which had the honor of being on the cover of *Nature*[79]). Nevertheless, we have no direct proof of what kind of evolutionary relationship could link archaic and modern humans, unless we consider the presence of *H. neanderthalensis* and MMHB in some Near Eastern sites. In this case, the coexistence of both at the same time would make a direct evolutionary ancestor/descent relationship between them impossible.

---

[71] Trinkaus and Shipman, *The Neandertals*, 191.

[72] E. Trinkaus, *The Shanidar Neandertals* (New York, N.Y.: Academic Press, 1983).

[73] Trinkaus and Shipman, *The Neandertals*, 341.

[74] Ales Hrdlicka, the founder of the American Journal of Physical Anthropology, was one of the first (among very few) defenders of Neandertals as ancestors of modern humans. A good history of the scientific arguments on Neandertals is that of Trinkaus and Shipman, *The Neandertals*.

[75] E. Trinkaus, and F. H. Smith, "The Fate of Neandertals," in *Ancestors*, Delson, ed., 325–33.; E. Trinkaus, "Les néandertalians," *La Recherche*, 180 (1986): 1040–47.

[76] G. A. Clark and J. M. Lindly, "The Case for Continuity: Observations on the Biocultural Transition in Europe and Western Asia," in *The Human Revolution: Behavioral and Biological Perspectives on the Origins of Modern Humans*, P. Mellars and C. Stringer, eds., (Princeton, N.J.: Princeton University Press, 1989), 626–76.

[77] G. Bräuer, "The Evolution of Modern Humans: A Comparison of the African and non-African Evidence," in *The Human Revolution*, Mellars and Stringer, eds., 123–54.

[78] C. B. Stringer, "¿Está en Africa nuestro origen?" *Investigación y ciencia*, 173 (1991): 66–73.

[79] C. P. E. Zollikofer, M. S. Ponce de León, R. D. Martin, and P. Stucki, "Neandertal Computer Skulls," *Nature*, 375 (1995): 283–85.

3.5 Homo sapiens: *The Appearance of Morphologically Modern Humans*

The main argument against a direct link between Neandertal and modern humans lies in the very early presence of the latter. It is true that Aurignacian culture is nearly thirty thousand years old, but human fossils with a very modern appearance exist from at least fifty thousand years before that date. Thermoluminescence and electronic spin resonance have been used to give datings around eighty to one hundred thousand years B.P. to the Near East MMHB remains,[80] while Neandertals first appeared about the beginning of the last Interglacial event (some seventy thousand years B.P.)[81] in Europe and moved to the Near East later.

Israeli sites show the coexistence of Neandertals and MMHB, each living quite close to the other's caves. Tabun cave (Mount Carmel, Israel) has provided enough evidence to reconstruct the chrono-cultural sequence of Mousterian culture shared, at that time, by Neandertals and MMHB. By the time Neandertals disappeared, Aurignacian culture had come into its own, and Near East sites (Sefunim cave, Mount Carmel; Boker Tachtit, Negev; K'sar Akil, Lebanon) give some evidence of the cultural transition too—industry that is Mousterian in typology but Upper Paleolithic in technology.[82] Nevertheless, as Clark and Lindly hold, it seems very difficult (even impossible) to relate both phenomena—the morphological and technological transitions.[83] Therefore, the question of evolutionary relationships existing between Neandertals and MMHB must be confined to the morphological and genetic fields.

Two opposite models (Multiregional Transition and Mitochondrial Eve) try to give an account of MMHB's first presence. To understand the existing differences between them, we need to clarify a prior question. The genetical change across an evolutionary transition within a given region depends on:

1. an *in situ* transition with neither gene flow nor population replacement, or
2. a gene flow, with no population replacement, or
3. a population replacement, with a minimal amount of gene flow.[84]

Thorne and Wolpoff give a very clear idea of these different possibilities with their metaphor of a pool where different individuals are splashing about. These individuals keep their individual characteristics, but the waves made by them can be understood to be the equivalent of genes flowing from one side of the pool to the other. Multiregional Transition holds that different individuals appear in different

---

[80] H. Valladas, J. L. Reyss, J. L. Joron, G. Valladas, O. Bar-Yosef, and B. Vandermeersch, "Thermoluminescence Dating of Mousterian 'Proto-Cro-Magnon' Remains from Israel and the Origin of Modern Man," *Nature*, 331 (1988): 614–16.

[81] Bräuer, "The Evolution of Modern Humans."

[82] Clark and Lindly, "The Case for Continuity," in *The Human Revolution*, Mellars and Stringer, eds.

[83] As Bräuer reminds us, no human remains associated with the early Aurignacian have yet been discovered; Bräuer, "The Evolution of Modern Humans."

[84] F. H. Smith, "Upper Pleistocene Hominid Evolution in South-Central Europe: A Review of the Evidence and an Analysis of Trends," *Current Anthropology*, 23 (1982): 667–86. The three alternatives are theoretical extremes. If we hold that Neandertals and MMHB belong to the same species, it is clear that some amounts of *in situ* genetic transition, gene flow, and population replacement may be found together. The question is, which was the most important one? Anyway, I am defending neither the Neandertal-MMHB continuity, nor the two-species theory; I am simply using Smith's summary to clarify the different possibilities of genetic change within a given population.

corners of the pool and they then exchange waves. Mitochondrial Eve (mtEve), on the contrary, obliges us to consider the arrival of a new swimmer, who jumps into the pool with such physical force that all the other individuals drown.[85]

From the MtEve Theory point of view, all mitochondrial DNA (mtDNA) existing in modern human populations is inherited from a single common ancestor who lived in Africa some 200,000 years ago.[86] This ancestral population later emigrated from Africa and replaced the former *Homo* populations existing throughout the ancient world (Africa, Asia and Europe).[87] MtDNA has some characteristics shared with family names, which we can use to better explain the hypothesis. Mitochondria are cytoplasmic organelles, that is, cellular structures existing outside the nucleus which have their own "genome," a single circular strand of DNA, probably coming from the origin of eukaryotic cells.[88] Forming part of the cytoplasms, and not of the nuclei, only the mitochondria of the mother's ovule will pass to the new being, so mtDNA is inherited strictly through the maternal line with neither segregation, nor recombination. Consequently, it is possible to trace the genealogy through maternal ancestry. Something very similar occurs with family names, although now it is paternal ancestry that counts (in Western societies). Going back in time we can trace, generation after generation, the origin of our name. Family names suffer many cultural changes (I may have changed my name to adapt it to a different language, for instance), but this is not the case with mtDNA. If any difference appears, it must be attributed to a mutation.[89] Comparing the differences in mtDNA existing in 35 mtDNA types found in 200 people living in four distinct geographic regions and minimizing (through a computer program) the total number of mutations needed to account for these differences, Johnson et al. obtained a clear-cut separation between African and non-African populations.[90] Two branches of

---

[85] A. G. Thorne and M. H. Wolpoff, "Evolución multirregional de los humanos," *Investigación y ciencia*, 189 (1992): 14–20.

[86] M. Stoneking, K. Bhatia, and A. C. Wilson, "Rate of Sequence Divergence Estimated from Restriction Maps of Mitochondrial DNAs from Papua New Guinea," *Cold Spring Harbor Symposia in Quantitative Biology*, 51 (1986): 433–39; R. L. Cann, M. Stoneking, and A. C. Wilson, "Mitochondrial DNA and Human Evolution," *Nature*, 325 (1987): 31–36.

[87] Before the mtDNA Eve proposal, Howells called the theory of a common ancestral population for MMHB "Noah's Ark"; W. W. Howells, "Explaining Modern Man: Evolutionists versus Migrationists," *Journal of Human Evolution*, 5 (1976): 477–96. Combining African origin, the descent from a single Eve, and the subsequent spread of modern peoples out of Africa, Wolpoff likes to call the model "Garden of Eden"; M. H. Wolpoff, "Multiregional Evolution: The Fossil Alternative to Eden," in *The Human Revolution*, Mellars and Stringer, eds., 62–108. As one can see, biblical references are not absent in human paleontology.

[88] This could occur through a symbiotic process among different organisms (such as prokaryotic cells and bacteria) that initiate a new and shared evolutionary journey. Bacterial DNA may have remained as a residual part in the mitochondria.

[89] MtDNA has a much higher mutation rate than nuclear DNA.

[90] M. J. Johnson, D. C. Wallace, S. D. Ferris, M. C. Ratazzi, and L. L. Cavalli-Sforza, "Radiation of Human Mitochondria DNA Types analyzed by Restriction Endonuclease Cleavage Patterns," *Journal of Molecular Evolution*, 19 (1983): 255–71. Other studies have been done that vary the number of mtDNA types and the individuals' origin. The phenomenon has been explained many times. For instance, see M. Stoneking and R. L. Cann, "African Origin of Human Mitochondrial DNA," in *The Human Revolution*, Mellars and Stringer, eds., 17–30. I like the Cavalli-Sforza comparison of gene flow and language evolution; L. L. Cavalli-Sforza, A. Piazza, P. Menozzi, and J. Mountain, "Reconstruction of Human Evolution: Bringing Together Genetic, Archaeological, and Linguistic Data,"

mtDNA types were found: the first one comprising ten African types, and the second one comprising all the other types, but some African types as well. The conclusion points to a common African ancestor.

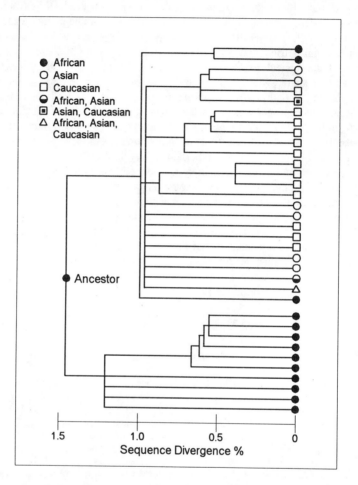

Figure 2. Genealogy relating 35 mtDNA types found by M. Johnson et al. in their survey of 200 mtDNAs from four geographic regions[91]

Contrary to what it seems to be, this assumption does not support the existence of just *one* ancestral grandmother shared by all humankind. It is easy to fall into such a pitfall, but it is still an error even if nearly everyone takes for granted the existence of a single African Eve. If we come back to the example of family names, it is very possible to find a small remote village in which everybody shares the same family

Proceedings of the National Academy of Sciences U.S.A., 85 (1988): 6002–6. A very good criticism is that of F. J. Ayala, "The Myth of Eve: Molecular Biology and Human Origins," *Science*, 270 (1995): 1930–36.

[91] Figure adapted from Stoneking and Cann, "African Origin of Human Mitochonrial DNA," in *The Human Revolution*, Mellars and Stringers, eds., 17–30.

name. This does not mean that people living in the village have only one ancestor—it just means that all other family names have disappeared. How is this so? If we consider one particular name, perhaps all the individuals sharing this name in the X number of previous generations were females.

What the MtEve Model assumes is that current human beings come from a small population which probably lived in Africa. Dating this earliest population depends on the average of sequence divergence rates. With a rate for human mtDNA evolution of 2% to 4% per million years and assuming it to be constant, the divergence accumulated among the individuals examined may have appeared within a range of 142,500–285,000 years.[92] Roughly 200,000 years B.P. is the most common dating attributed to the ancestral African population.[93]

There are no fossil proofs of the existence of this ancestral population. The earliest MMHB remains are those of South Africa, which date to 30–40,000 years B.P. (more precise estimates being problematic),[94] and those of the Near East, which date close to 100,000 years B.P. in the oldest case.[95] Putting aside the exact accuracy of the dates, this situation speaks in favor of the earlier appearance of MMHB in some region situated between South Africa and the Near East. But the lack of fossil remains is a strong argument against the African Eve model.

Paleontologists like to construct their theories from tangible evidence such as fossils. It is not strange, then, that the Multiregional Transition Model is very popular. As Clark and Lindly say,[96] the specialized literature overwhelmingly supports the theory of an *in situ* transition between archaic and modern humans with no population replacement and very little gene flow, both in the Near East and in Europe. Perhaps Milford Wolpoff[97] is the best known reference, but other highly regarded scholars, such as Trinkaus, F. H. Smith, as well as Clark and Lindly, agree with the Multiregional Transition Model. Much criticism has come from paleontologists due to the low confidence in sequence divergence rate estimates and to ambiguous uses of the rate of divergence *along* a lineage and *between* lineages. Nevertheless, this question only applies to the estimated dating of the ancestral population. Wolpoff adds a different argument against mtDNA Eve. If different lineage survivorship is the cause of mtDNA variation (as in the case of family names), stochastic lineage extinctions would insist on tracing all the mitochondrial lines to a single female, although—as Wolpoff points out—the ancestral population would have had multiple nuclear DNA descent, possibly from most of the founders. This criticism lies in the confusion about the meaning of the MtEve theory which I mentioned before. "Noah," "Garden of Eden," and "Africa Eve" models do not compel us to consider all MMHB as descending from a single female. They just hold that a particular female was among the ancestors of all humankind. From the point

---

[92] Stoneking and Cann, "African Origin of Human Mitochondrial DNA," in *The Human Revolution*, Mellars and Stringer, eds.

[93] R. L. Cann et al.,"Mitochondrial DNA and Human Evolution."

[94] R. G. Klein, "Biological and Behavioral Perspectives on Modern Human Origins in Southern Africa," in *The Human Revolution*, Mellars and Stringer, eds., 529–46.

[95] For instance, O. Bar-Yosef, "The Date of the South-West Asian Neandertals," in *L'Homme de Néandertal. Vol. 3: L'Anatomie*, M. Otte, ed., (Liège: Etudes et Recherches Archéologiques de l'Université de Liège, 1988), 31–38.

[96] Clark and Lindly, "The Case for Continuity," in *The Human Revolution*, Mellars and Stringer, eds.

[97] For instance, Wolpoff, "Multiregional Evolution," in *The Human Revolution*, Mellars and Stringer, eds.

of view of the MtEve Model, we may have DNA inside our present genome that comes from most of the founders' population. This circumstance even supports the MtEve model, as we are relating the whole of humankind to a single ancestral population. More interesting is Rouhani's comment on the disadvantage of the mtDNA model for exploring patterns of human migrations, as its uniparental mode of inheritance only reflects the migratory pattern of females, which may bias the results in the case of a highly asymmetrical migratory pattern.[98]

Francisco Ayala has recently published a clear and well documented paper on the MtEve hypothesis which explains, much better than I have, the possible molecular techniques used in searching for MMHB origins.[99] Given the range of molecular evidence that cannot be ignored, it must be emphasized that, apart from mtDNA Eve, some other related genetic-based models, such as Y chromosome DNA polymorphism,[100] Alpha and Beta-Globin gene cluster polymorphisms[101] and MHC polymorphism,[102] all provide support for the out-of-Africa spread of modern humans.

---

[98] S. Rouhani, "Molecular Genetics and the Pattern of Human Evolution: Plausible and Implausible Models," in *The Human Revolution*, Mellars and Stringer, eds., 47–61.

[99] See Ayala, "The Myth of Eve."

[100] G. Lucotte, "Evidence for the Paternal Ancestry of Modern Humans: Evidence from a Y-Chromosome Specific Sequence Polymorphic DNA Probe," in *The Human Revolution*, Mellars and Stringer, eds., 39–46.

[101] J. S. Wainscoat, A. V. S. Hill, S. L. Thein, J. Flint, J. C. Chapman, D. J. Weatherall, J. B. Clegg, and D. R. Higgs, "Geographic Distribution of Alpha- and Beta-Globin Gene Cluster Polymorphisms," in *The Human Revolution*, Mellars and Stringer, eds., 31–38.

[102] J. Klein, N. Takahata, and F. J. Ayala, "MHC Polymorphism and Human Origins," *Scientific American*, 269 (1993): 46–51.

THE PHENOMENON OF THE EUKARYOTIC CELL

Julian Chela-Flores

1 *From Cosmic Evolution to Chemical Evolution*

1.1 *Introduction*

Long before the advent of science and philosophy, theologians were raising questions that we still cannot answer fully. Some of the deepest questions that have persistently remained with us since biblical times are: What is the origin of the universe? What is it made of? What is its ultimate destiny? How did life in general, and humans in particular, originate? Are we alone in the cosmos? The first philosophical attempts to answer these questions were made a few centuries before our own era by the ancient Greeks, notably by Thales of Miletus (who was active during the time of the Prophet Jeremiah and the destruction of Jerusalem in 587 B.C., which had been founded by King David almost half a millennium earlier). A significant contribution to the birth of science was made by another ancient Greek, Democritus of Abdera, one of the founders of atomism (who was active during the time of Ezra's reform, which, by the promulgation of the by-then-ancient Mosaic Law in 430 B.C., gave Judaism its distinctive character).

I feel that there is no better way to organize the information available to scientists than to reconsider those questions originally asked in antiquity, in order to appreciate, in an orderly manner, the considerable progress that has been achieved up to the present in our understanding of cosmological models and the appearance of intelligent life on Earth. In spite of the impressive progress of science, many fundamental questions remain unanswered. Therefore, before beginning our task it may be prudent to recall that Cosmology suggests that there is sufficient time available for science to progress to a stage in which further, and possibly better attempts than are now possible, will be made in order to search for answers of some of the deepest questions (mentioned above) which humans have been asking themselves from time immemorial.

Up to the present partial answers have often led to controversy between scientists, philosophers, and theologians. We should be constantly aware of the limited scope of scientific method, a point that has been stressed by Bertrand Russell. Recognizing a limit to the applicability of present day science, he noted that, "almost all the questions of most interest to speculative minds are such as science cannot answer."[1]

1.2 *Cosmological Models*

Our first topic is suggested by the Book of Genesis ("In the beginning God created the heavens and the earth" Gen. 1:1). To approach this problem from the point of view of science, a preliminary step is to grasp the significance of the scale of time involved. For this purpose we must return to the first instants of cosmic expansion. Our sketch must be brief, as we already have available the previous review of

---

[1] Bertrand Russell, *History of Western Philosophy and its Connection with Political and Social Circumstances from the Earliest Times to the Present Day* (London: Routledge, 1991), 13.

George Coyne at the 4th Trieste Conference as well as several previous conferences in this series.[2]

The American scientist Edwin Hubble discovered in 1929 that the velocity of recession of a certain galaxy under observation is proportional to its distance, which is normally expressed in megaparsecs (1 Mpc = 10^6 pc; 1 pc = 3.26 light years). The constant of proportionality, which is known as the Hubble constant H_0 is, consequently, given by the ratio of the speed of recession of the galaxy to its distance; H_0 represents quantitatively the current rate of expansion of the universe. The measurements of H_0 at present[3] lie, in one case (Mould and Freedman) in the range of 55–61 km s^{-1} Mpc^{-1} and, in a second case (Sandage, Tammann, and Saha) in the range of 68 to 100 km s^{-1} Mpc^{-1}. The uncertainty in H_0 decreases as systematic errors are eliminated. The two results mentioned above, together with other current values, suggest a value of H_0 in the range 65–82 km s^{-1} Mpc^{-1} .

The implication of, for instance, H_0 = 70 km s^{-1} Mpc^{-1} in the standard cosmological model (cf. the Friedmann model below) of an expanding universe implies an age of 9–14 billion years, depending on the particular assumption we may adopt for the matter density present in the universe. With life on Earth extending back to some 4 billion years; the chemical evolution scenario faces no particular difficulties with the above values of H_0.

The classical cosmological solution assumes isotropy and homogeneity in the equations of the theory of gravitation of Albert Einstein, known as General Relativity (GR).[4] Cosmological models may be discussed in terms of a single function R of time t. This function may be referred to, quite appropriately, as a "scale factor" or, sometimes in reference to the particular solution of the GR field equations in a geometric background, the expression "radius of the universe" may be preferred for the function R. As the universal expansion sets in, R is found to increase in a model that assumes homogeneity in the distribution of matter (the "substratum"), as well as isotropy of space.

The functional dependence of R, as a function of time t is a smooth increasing function for a specific choice of two free parameters which have a deep meaning in the GR theory of gravitation, namely, the curvature of space and the cosmological constant. The functional behavior of the scale factor R was found by the Russian mathematician Alexander Friedmann in 1922. This solution is also attributed to Howard Percy Robertson and A.G. Walker for their work done in the 1930s. Such a (standard) model is referred to as the Friedmann model.

In fact, R is inversely proportional to the substratum temperature T. Hence, since R is also found to increase with time t, T decreases; this model implies, therefore, that as t tends to zero (the "zero" of time) the value of the temperature T

---

[2] G. Coyne, S.J., "Cosmology: The Universe in Evolution," in *Chemical Evolution: Physics of the Origin and Evolution of Life*, J. Chela-Flores, and F. Raulin, eds. (Dordrecht, The Netherlands: Kluwer Academic Publishers, 1996), 35-49; Robert J. Russell, William R. Stoeger, and George V. Coyne, eds., *Physics, Philosophy and Theology: A Common Quest for Understanding* (Vatican City State: Vatican Observatory, 1988); Robert J. Russell, Nancey C. Murphy, and Chris J. Isham, eds., *Quantum Cosmology and the Laws of Nature: Scientific Perspectives on Divine Action* (Vatican City State; Berkeley, Calif.: Vatican Observatory Publications; Center for Theology and the Natural Sciences, 1993).

[3] R. C. Kennicutt, Jr., "An old galaxy in a young universe," *Nature* 381 (1996): 555-56.

[4] H. Bondi, *Cosmology*, 2nd ed. (Cambridge: Cambridge University Press, 1968).

is large. (The temperature goes to infinity as T tends to zero.) In other words, the Friedmann solution suggests that there was a "hot" initial condition.

As the function R represents a scale of the universe (in the sense we have just explained), the expression "Big Bang," due to Sir Fred Hoyle, has been adopted for the Friedmann model. The almost universal acceptance of Big Bang cosmology is due to its experimental support. As time t increases the universe cools down to a certain temperature, which at present is close to $3°$ K. The work performed during the 1960s by the engineers of the Bell Telephone Company Arno Penzias and Robert Wilson was rewarded with a Nobel Prize in Physics in 1978. They provided solid evidence for the "$T = 3°$ K" radiation, which may be confidently considered to be a relic from the Big Bang.

Andre Linde envisions a vast cosmos in a model in which the current universal expansion is a bubble in an infinitely old "superuniverse." In other words, in this model the universe is assumed to be a bubble amongst bubbles, which are eternally appearing and breeding new universes. The scale parameter R evolves as a function of time t in such a cosmology, but as in the earlier model of Alan Guth, the Linde model differs from the Friedmann solution in the first instants of cosmic evolution. A word of warning: depending on the way the word 'standard' is used, both the Linde and Guth models can be considered standard Big Bang models. Indeed, Guth proposed the original inflationary model which solved many of the problems inherent in the preliminary Big Bang models. In fact, the inflationary universe, as we are realizing, is a class of Big Bang models that incorporates a finite period of accelerated expansion during its early development. The inflation process releases a large amount of energy retrieved from the vacuum of spacetime. It should be pointed out that Linde also assumed inflation.

1.3 *The Birth of the Studies of the Origin of Life*

Science today aims at inserting biological evolution into the context of cosmic evolution. We will mainly attempt to convey the idea of an ongoing transformation in origin-of-life studies. The subject first began to take shape as a scientific discipline in the early 1920s when an organist chemist, Alexander Oparin applied the scientific method of conjecture and experiments to the origin of the first cell,[5] thereby allowing scientific enquiry to shed valuable new light on a subject that has traditionally been the focus of philosophy and theology. Recent progress in understanding our own origins began about three decades ago, triggered by the successful retrieval of some key biomolecules in experiments which attempted to simulate prebiotic conditions. Some of the main experiments of the 1950s and early 60s were done by organic chemists including Melvin Calvin, Stanley Miller, Sidney Fox, Juan Oro, Cyril Ponnamperuma and their coworkers.

These efforts have led to a view of the cosmos as a matrix in which organic matter is inexorably self-organized by the laws of physics and chemistry into what we recognize as living organisms. However, in this context it should be stressed that chemical evolution experiments have been unable to reproduce the complete pathway from inanimate to living matter. In particular, the prebiotic synthesis of all the RNA bases is not clear; and cytosine, for instance, may not be prebiotic—it may even have been imported from space. This aspect of chemical evolution has been

---

[5] C. Ponnamperuma "The origin of the cell from Oparin to the present day," in *Chemical Evolution: The Structure and Model of the First Cell*, C. Ponnamperuma and J. Chela-Flores, eds. (Dordrecht, The Netherlands: Kluwer Academic Publishers, 1995), 3–9.

repeatedly pointed out by Robert Shapiro.[6] Thus, at present the physical and chemical bases of life that have been delineated remain a strong possibility, but further research is still needed.

1.4 *The Growth of the Studies of the Origin of Life*

The still ongoing second large step in the development of origin-of-life studies consists in taking the subject from the laboratories of the "card-carrying" organist chemists, where Oparin had introduced it, into the domain of the life, earth, and space scientists. In fact, the subjects of exobiology and bioastronomy have come of age due to the space missions of the last 30 years (Mariner, Apollo, Viking, and particularly the most recent one, Galileo, which is currently sending valuable information on the Jovian system). Many further missions are being planned for the rest of this decade as well as the beginning of the next century

New areas of research have come into existence as a consequence of this change of emphasis, such as *planetary protection,* whose aim is to prepare ourselves in advance for the retrieval of samples from Mars and Europa in the missions that may be scheduled, in the case of Mars, as early as the year 2003.

The experiments and sample retrieval from Europa are in the process of discussion. Indeed, a NASA and National Science Foundation (NSF) meeting to discuss such matters was organized in 1996 (*Europa Ocean Conference,* San Juan Institute, San Juan Capistrano, California). Planetary Protection has been pioneered by Donald DeVincenzi at NASA-Ames, John Rummel at the Marine Biological Laboratory, and Margaret Race at the SETI Institute.

A second discipline that has arisen from the exploration of the Solar System is *comparative planetology,*[7] which is needed in the formulation and development of devices which in principle will allow us to search the Solar System for relics of the earliest stages of the evolution of life.

The problem of the search for extraterrestrial life may be approached in order of increasing complexity. Alexandra MacDermott, from the University of Cambridge, has led in the search for extraterrestrial homochirality, SETH. (Homochirality refers to the single-handedness observed in the main biomolecules, namely, amino acids, nucleic acids and phospholipids.) Secondly, we have argued in favor of the search for extraterrestrial single-celled organisms of a nucleated type. (For a discussion of several types of microorganisms wee section 2.3 below, in which the word of Greek origin 'eukaryote' shall be assigned to such nucleated cells. The search for extraterrestrial eukaryotes shall be abbreviated as SETE.)

Finally, at the other extreme, we have the search for extraterrestrial intelligence (SETI), a time-honored subject that was pioneered by Frank Drake in the early 1960s.[8] The eventual success of SETI may depend on the adoption of truly significant enterprises, such as the extension of SETI to the far side of the Moon in the 100 km diameter crater Saha, as advocated by Jean Heidmann from the Observatory of Paris and Claudio Maccone from the G. Colombo Center for

---

[6] R. Shapiro, *Prebiotic synthesis of the RNA bases: a critical analysis.* Summary presented at the ISSOL Conference, *Book of Abstracts* (Orleans, July 1996), 39.

[7] B. Murray, M. C. Malin, and R. Greely, *Earthlike Planets. Surfaces of Mercury, Venus, Earth, Moon, Mars* (San Francisco: W. H. Freeman& Co., 1981), 317.

[8] Frank Drake, "Progress in searches for extraterrestrial intelligent radio signals," in *Chemical Evolution: Physics of the Origin and Evolution of Life,* Chela-Flores and Raulin, eds., 335–42; F. Drake and D. Sobel, *Is there anyone out there? The scientific search for extraterrestrial intelligence* (New York: Delacorte Press, 1992).

Astrodynamics in Turin. None of the extremely ambitious engineering projects underlying this proposal seem beyond present technological capabilities, such as the construction of a 340 km road linking laboratories in Mare Smythii (at the equatorial level and at the Moon's limb) and crater Saha (103° E, 2° S).[9] In fact, the only limitation foreseen is the will of our society to maintain continued progress by providing adequate budgets to scientists and engineers involved in the main programs of bioastronomy and exobiology during the coming decade.

To continue the overview of the origin-of-life studies, we must first return to cosmological models.

1.5 *The Origin of the Elements*

In less than one million years after the beginning of the general expansion of the universe, the temperature T was already sufficiently low for electrons and protons to be able to form hydrogen atoms. Up to that moment these elementary particles were too energetic to allow atoms to be formed. Once "recombination" of electrons and protons was possible, due to falling temperatures, thermal motion was no longer able to prevent the coulomb interaction from forming atoms. This is the "moment of decoupling" of matter and radiation. At this stage of universal expansion the force of gravity was able to induce the hydrogen gas to coalesce into stars and galaxies. Gravity itself has induced the coalesced nuclei of stars to initiate thermonuclear reactions that created all the heavier elements. At the supernova stage these elements are expelled for later new cycles of condensation into stars and planets. The Earth arose in this manner some 4.6 billion years ago.

It is well known that stars evolve as nuclear reactions convert mass to energy. In fact, stars such as our Sun follow a well known pathway (the *main sequence)* along the *Hertzprung-Russell* (HR) diagram, which was introduced independently by Ejnar Hertzprung and Henry Russell in the early part of this century. Many nearby stars exhibit a certain regularity (such stars are called *main sequence* stars) when we plot luminosity (the total energy of visual light radiated by the star per second) versus its surface temperature or, alternatively, its spectroscopic type. We may ask, how do stars move on the HR plot as hydrogen is burned? Extensive calculations show that main sequence stars are funneled into the upper right hand of the HR diagram, where we find stars of radii that may be 10 to 100 times the Sun radius. Such *giant stars* are characterized by high luminosity and red color, and hence are called "red giants."

It is instructive to consider a sample of stars, emphasizing some that are now known to be much more similar to our own Solar System,[10] as they have Jupiter-like planets; such a property militates in favor of life having been provided a variety of favorable environments in our own galaxy.

Stellar evolution puts a significant constraint on the future of life on Earth. The radius of the Earth's orbit is small (its eccentricity may be neglected), and the radius of the Sun is bound to increase as it evolves along the main sequence in the HR diagram, thus eventually ending life on Earth. The current estimate is that life may

---

[9] J. Heidmann, "SETI from the Moon. A case for a XXIst Century SETI—Dedicated Lunar Farside Crater," in *Chemical Evolution: Physics of the Origin and Evolution of Life*, Chela-Flores and Raulin, eds., 343–53.

[10] M. Mayor, D. Queloz, S. Udry, and J.-L. Halwachs, "From brown dwarfs to planets," in *Astronomical and Biochemical Origins and the Search for Life in the Universe*, C. B. Cosmovici, S. Bowyer and D. Werthimer, eds. (Bologna: Editrice Compositore, 1997), 313–30.

continue on Earth for another 4–5 billion years until the Earth's orbit is no longer in a habitable zone. The implication of such duration of geologic time is indeed profound for our species, as we shall suggest in the next section.

1.6 *Implications of Stellar and Biological Evolution*

Homo sapiens has evolved, according to the standard view of paleoanthropology, in less than a few million years since the last common ancestor of the hominids, who must have lived at the Miocene/Pliocene transition, just some 5 million years ago. Such fast evolutionary tempo has been effective only within a small fraction of 1% of the geologic time that is still available for natural selection to continue the process of evolution of all living species, including *H. sapiens*. (Such time is measured in billions of years).

Science is at present unable to anticipate the future of *H. sapiens*; but this has not deterred fiction writers from envisaging a model for a future speciation event into the imaginary *eloi* and *morlock* species of the *Homo* genus.[11]

What is clear at present, from a combined consideration of cosmic, stellar, biological, and cultural evolution, is that the contemporary stage of human development will inevitably continue to evolve. This is due to the mechanism of natural selection, which is expected to continue for at least four billion years to play its role on all living organisms on Earth. (Once again, this would be true as long as the Earth remains within the habitable zone of the Solar System.)

Although the human brain is considerably larger than that of the great apes, it will still be subject to evolutionary pressures for over a thousand times as long as the period in which natural selection gave rise to the primitive brain of the earliest hominids (the australopithecines, who lived in the Lower Pliocene Epoch) with cranial capacity from under 400 cc, to over 300% of that capacity in the most evolved early humans of the Pleistocene. This broad scenario for the possible future effects of natural selection on the *Homo* genus gives us an optimistic outlook on the future of life on Earth.

2 *From Chemical Evolution to the Origin of Single Cells*

2.1 *Chemical Evolution in the Cosmos*

Given the importance of the origin and evolution of biomolecules, we feel that it is appropriate to begin with comments on the origin of some of the most important biomolecules. We have to recall that there has been considerable progress in our understanding of the interstellar organic molecules. There are also many indications that the origin of life on Earth may not exclude a strong component of extraterrestrial inventories of the precursor molecules that gave rise to the major biomolecules. About 98% of all matter in the universe is made of hydrogen and helium. The other five biogenic elements C, N, O, S, and P make up about 1% of the cosmic matter. The abundance of biogenic elements would suggest that the major part of the molecules in the universe would be organic (that is, compounds of carbon atoms); in fact, out of over a hundred molecules that have been detected, either by microwave or infrared spectroscopy, 75% are organic.[12]

Once again, we should underline that chemical evolution experiments fare well in comparison with the observation of the interstellar medium. Some of the identified

---

[11] H. G. Wells, *The Time Machine*, abridged ed. (London: Phoenix, 1996).

[12] J. Oro, "Chemical synthesis of lipids and the origin of life," in *Chemical Evolution: The Structure and Model of the First Cell*, Ponnamperuma and Chela-Flores, eds., 135–47.

molecular species detected by means of radioastronomy are precisely the same as those shown in the laboratory to be precursor biomolecules.

2.2 *From "Precursor-biomolecules" to Biomolecules*

In our studies of the origin of life we may encounter the major biomolecules: amino acids, nucleic acids, polysaccharides and lipids. Altogether, about twelve interstellar molecules have been shown to be precursors of the main biomolecules. Our insights into the origin of life are based on the formation of compounds of H, N, O, and P. The evolution of organic compounds in the interstellar medium has been exhaustively studied. Interstellar gas, dust, ices, and grains are the precursors of the Solar System, after a primordial condensation process. Biomolecular precursors have been detected in the interstelllar medium, such as molecular hydrogen, water (the dominant ice component in dense clouds), and carbon monoxide, amongst others. Research in the field of what has been called "*cosmochemistry*" began in the 1950s and extends into the present, but will undoubtedly continue its robust progress in the future. Comprehensive reviews are given in the Trieste Series on Chemical Evolution.[13]

2.3 *Modern Taxonomy and Terminology of Single-celled Organisms*

The focus of our attention in what follows will be a group of microorganisms that are either single-celled or multicellular; their genetic material is enclosed inside a membrane different from the cell membrane. For this reason they are called eukaryotes, from the Greek words '*karyon*' (which means nucleus) and '*eu*'(which means true).

Taxonomically the totality of such organisms is said to form the domain *Eukarya*, which contains kingdoms, such as the well-known kingdom *Animalia*. (The most recent taxon introduced in the scientific literature at the highest level is called a "domain.")

On the other hand, there are unicellular organisms referred to as "prokaryotes" that, as their name implies, normally lack a nuclear membrane around their genetic material. Generally they are smaller than eukaryotes. Well-known examples are bacteria. However, not all prokaryotes are bacteria! There is a whole domain that encompasses a group of single-celled organisms which are neither bacteria nor eukaryotes. They are generally referred to as *archaebacteria*.[14] In other words, all prokaryotes are encompassed in two domains: Bacteria and Archaea. They may be adapted to extreme conditions of temperature (up to just over $100°$ C), in which case they may be called *thermophiles*. They may also be adapted to extreme acidic conditions, in which case they may be referred to as being *acidophilic*. The alternative expressions of *extremophiles,* or *hyperthermophiles* are sometimes used for these organisms to distinguish different degrees of adaptability to such extreme ranges of conditions. As we said above, all archaebacteria are said to form the third domain of life, namely the domain Archaea, which is divided into several kingdoms.

---

[13] C. Ponnamperuma and J. Chela-Flores, eds., *Chemical Evolution Series: Origin of Life* (Hampton, Virginia: A. Deepak Publishing, 1993); idem, eds., *Chemical Evolution: The Structure and Model of the First Cell*; J. Chela-Flores, M. Chadha, A. Negron-Mendoza, and T. Oshima, eds., *Chemical Evolution: Self-Organization of the Macromolecules of Life* (Hampton, Virginia: A. Deepak Publishing, 1995); Chela-Flores and Raulin, eds., *Chemical Evolution: Physics of the Origin and Evolution of Life*.

[14] O. Kandler, "Cell wall biochemistry in Archaea and its phylogenetic implications," in *Chemical Evolution: The Structure and Model of the First Cell*, Ponnamperuma and Chela-Flores, eds., 165–69.

2.4 *Evolution of Prokaryotic Cells in the Precambrian*

The main steps of chemical evolution on Earth should have taken place from 4.6–3.9 billion years ago, the preliminary interval of geologic time which is known as the Hadean Subera. It should be noticed, however, that impacts by large asteroids on the early Earth do not necessarily exclude the possibility that the period of chemical evolution may have been considerably shorter. Indeed, it should not be ruled out that the Earth may have been continuously habitable by non-photosynthetic ecosystems from a very remote date (approximately 4.44 billion years ago).[15] The content and the ratios of the two long-lived isotopes of reduced organic carbon in some of the earliest sediments (retrieved from the Isua peninsula, Greenland, some 3.8 billion yeras ago), may convey a signal of biological carbon fixation.[16] This reinforces the expectation that chemical evolution may have occurred in a brief fraction of the Hadean Subera, in spite of the considerable destructive potential of large asteroid impacts which took place during the same geologic interval in all the terrestrial planets, the so-called *heavy bombardment period*.

An early date for the origin of life may be considered by interpreting the molecular analysis of ribosomal RNA (rRNA) in a variety of organisms in terms of the molecular clock hypothesis.[17] This work in molecular biology suggests that life emerged early in Earth's history, even before the end of the heavy bombardment period. In those remote times life may have already colonized extreme habitats, which allowed at least two prokaryotic species to survive the large impacts that were capable of nearly boiling over an ocean. In subsequent suberas of the Archean (3.9–2.5 billion years ago) life, as we know it, was present. This is well represented by fossils of the domain Bacteria, which is well documented by many species of cyanobacteria.[18]

To sum up, we may not exclude from the geochemical data earlier dates, in the Lower Archean, for the first prokaryotic microflora, although some critical considerations from the point of view of geochronology still do not rule out the possible origin of life immediately after the end of the Hadean Subera.[19] However, the earliest prokaryotic fossils that have not been subjected to regional metamorphism are those dating back to 3.5 billion years ago.[20] Eukaryotes—nucleated cells—arose in the next eon, which is called the Proterozoic (2.5–0.6 billion years ago). This eon saw the progression from a largely anoxic atmosphere to a weakly oxic one about 2 billion years ago. The increment of atmospheric oxygen was a key geophysical factor which allowed for the formation of the eukaryotic cells.

---

[15] N. H. Sleep, K. J. Zahnle, J. F. Kasting, and H. J. Morowitz, "Annihilation of ecosystems by large asteroid impacts on the early Earth," *Nature* 342 (1989): 139–42.

[16] M. Schidlowski, "Early terrestrial life: Problems of the oldest record," in *Chemical Evolution: Self-Organization of the Macromolecules of Life*, Chela-Flores et al., eds., 65–80.

[17] M. Gogarten-Boekels, E. Hilario and J. P. Gogarten, "The effects of heavy meteorite bombardment on the early evolution—The emergence of the three domains of life," *Origin and Evolution of the Biosphere* 25 (1995): 251–64.

[18] J. W. Schopf, "Microfossils of the Early Archean Apex Chert: New Evidence of the Antiquity of Life," *Science* 260 (1983): 640–46.

[19] S. Moorbath, "Age of the oldest rocks with biogenic components," in *Chemical Evolution: The Structure and Model of the First Cell*, Ponnamperuma and Chela-Flores, eds., 85–94.

[20] J. W. Schopf, ed., *Earth's earliest biosphere its origin and evolution.* (Princeton: Princeton University Press, 1983).

3 *From the Age of the Prokaryotes to Eukaryogenesis*

3.1 *Role of O_2 and Fe in the First Appearance of Eukaryotes (Eukaryogenesis)*

Several lines of research suggest the absence of current values of O_2 for a major part of the Earth history. There are some arguments, nevertheless, that militate in favor of Archean atmospheres with values of the partial pressure of atmospheric oxygen O_2 about 10^{-12} of present atmospheric level (PAL).

Some rocks, called "igneous," have originated by solidification from a molten condition as they poured out from volcanos. Shale is a sedimentary rock resulting from the wastage of pre-existing rocks. It has played a role in our understanding of biological evolution. The onset of atmospheric oxygen is demonstrated by the presence in the geologic record of red shale colored by ferric oxide. The age of such "red beds" is estimated to be about 2 billion years. At that time oxygen levels may have reached 1–2% PAL, sufficient for the development of a moderate ozone protection for the microorganisms of the Proterozoic from ultraviolet (UV) radiation. In fact, UV radiation is able to split the O_2 molecule into the unstable O-atom, which, in turn, reacts with O_2 to produce ozone O_3, which is known to be an efficient filter for the UV radiation.

The paleontological record suggests that the origin of the eukaryotic cell occurred earlier than 1.5 billion years ago. Some algae may even date from 2.1 billion years ago,[21] a period comparable to the onset of red beds. This is still rather late, compared to the earliest available prokaryotic fossils of some 3.5 billion years ago.[22]

However, if we keep in mind certain affinities between eukaryotes and archaebacteria (such as homologous factors in protein synthesis), we may argue that archaebacteria and the stem group of eukaryotes may have diverged at about the same time.[23] This conjecture, combined with the lightest carbon isotope ratios from organic matter, implies that bacteria capable of oxidizing methane CH_4 ("methylotrops") may have been using methane produced by archaebacteria that were able to produce it as a byproduct of their metabolism. From the age of such fossils a tentative date of 2.7 billion years ago, or earlier, may even be assigned to eukaryogenesis.

There are Archean rock formations (which may be found up to 2 billion years ago) that are significant in the evolution of life. These are laminated compounds of dioxide of silicon (silica) and iron. In reference to their laminated structure they are called "banded iron formations" (BIFs). The period in which the BIFs were laid out ended some 1.8 billion years ago. In the anoxic atmosphere of the Archean, iron compounds could have been dispersed over the continental crust. They could have absorbed some oxygen, thereby protecting photosynthesizers that could not tolerate oxygen. Such microorganisms in turn produced oxygen that combined with their environment to produce iron oxide (for example, hematite Fe_2O_3), which makes up the BIFs.

---

[21] T.-M. Han and B. Runnegar, "Megascopic eukaryotic algae from the 2.1-billion-year-old Negaunee iron-formation, Michigan," *Science* 257 (1992): 232–35.

[22] Schopf, "Microfossils of the Early Archean Apex Chert: New Evidence of the Antiquity of Life."

[23] B. Runnegar, "Origin and Diversification of the Metazoa," in *The Proterozoic Biosphere*, J. W. Schopf, and C. Klein, eds. (Cambridge: Cambridge University Press, 1992), 485.

In strata dating prior to 2.3 billion years ago it has been observed that there is an abundance of the easily oxidized mineral form of uranium (IV) oxide (urininite, for example the well-known variety *pitchblende)*. This supports the conclusion that we had to wait until about 2 billion years ago for a substantial presence of free O_2. Once the eukaryotes enter the fossil record, its organization into multicellular organisms followed in a relatively short period (in a geological time scale).

Metazoans are hypothesized to have arisen as part of a major eukaryotic radiation in the Riphean Period of the of the Late Proterozoic, approximately 0.8–1 billion years ago,[24] when the level of atmospheric O_2 had reached 4–8% PAL. There is some evidence in the Late Proterozoic during the Vendian Period, for the existence of early diploblastic grades (Ediacaran faunas). These organisms were early metazoans with two germ layers, such as the modern coelenterates (jellyfishes, corals, and sea anemones).

Later on, when the level of atmospheric O_2 had reached values in excess of 10% PAL, these grades were overtaken in numbers by triploblastic phyla as the level of atmospheric O_2 had reached 40% PAL. These organisms are called the Cambrian faunas, that is, Early Paleozoic faunas (500 million years ago), which were mainly metazoans with three germ layers. They constitute at present the greater majority of multicellular organisms, including *H. sapiens*.

We may obtain further insights from paleontology: acceleration in the evolutionary tempo is observed after the onset of eukaryogenesis, as it is clearly demonstrated by the microfossils of algae from the Late Proterozoic[25] and by the macrofossils of the Early Phanerozoic (Cambrian Period of the Paleozoic Era).[26] Such evolutionary changes to the simple prokaryotes within the first billion years of atmospheric oxygen gave rise to eukaryotes, metazoans, and metaphytes.

3.2 Identification of Eukaryotes Whose Morphology Is Not Radically Different from Prokaryotes

Primitive eukaryotic organisms have been studied in detail and many difficulties are inherent in the eventual design of an assay to identify them in an extraterrestrial environment, a question which begins to be important in view of the forthcoming space missions. In the following paragraphs we discuss some properties of the organisms, either alive or fossilized, that may be difficult to distinguish in a robotic mission.

1. One such taxon is the family Cyanidiophyceae. These organisms are rhodophytes, commonly known as red algae.[27] Joseph Seckbach from the Hebrew University of Jerusalem has argued eloquently that these acido-thermophilic algae may constitute a bridge between cyanobacteria and red algae.[28] In particular,

---

[24] A. H. Knoll, "Proterozoic and Early Cambrian protists: Evidence for accelerating evolutionary tempo," *Proc. Natl. Acad. Sci.* 91 (1994): 6743–50.

[25] Ibid.

[26] S. Conway-Morris, "The fossil record and the early evolution of the Metazoa," *Nature* 361 (1993): 219–25.

[27] Joseph Seckbach, "The natural history of *Cyanidium* (Geitler, 1933): past and present perspectives," in *Evolutionary pathways and enigmatic algae:* Cyanidium caldarium *(Rhodophyta) and related cells*, J. Seckbach, ed. (Dortrecht, The Netherlands: Kluwer Academic Publishers, 1994), 99–112.

[28] J. Seckbach, "The first eukaryotic cells—Acid hot-spring algae," in *Chemical Evolution: The Structure and Model of the First Cell*, Ponnamperuma and Chela-Flores, eds., 335–45.

Cyanidium caldarium has a primitive eukaryotic cellular structure. Other remarkable properties of *C. caldarium* is that it may live at temperatures of up to 57° C and shows better rates of growth and photosynthesis when cultured in an "atmosphere" of pure CO_2.[29]

2. Silicification experiments of microorganisms have led to the identification of one artefact, which can cause confusion in the identification of the transition organisms.Frances Westall and coworkers at the University of Bologna have observed an artificial nucleus formed during the process of fossilization. If such a phenomenon is preserved in the natural fossil record, then it can lead to a confusion with the eukaryotes.[30]

3. Prokaryotic symbionts of the ciliated protozoan *Euplotidium itoi* has ultrastructure which is more complicated than the majority of prokaryotes. Strictly from the point of view of planning future SETE experiments, the most striking feature discovered by Giovanna Rosati and coworkers at the University of Pisa is the presence of "basket tubules," consisting of the protein tubulin, which is normally present only in eukaryotes.[31]

4 From Eukaryogenesis to the Appearance of Intelligent Life on Earth

4.1 Modern Taxonomy Emphasizes Single-celled Organisms

In evolutionary terms, we have chosen not to emphasize complex multicellular organisms. Instead, we have shifted our attention to the single-celled nucleated organism (eukaryote), whose evolution is known to have led to intelligence, at least on Earth. This point of view is forced upon us by the present taxonomical classification of organisms into domains, which stresses single-celled organisms. Previously, an older taxonomical classification, with kingdoms as the highest taxon, highlighted multicellular organisms, incorrectly in our view.

The older approach was due to biologists' lack of understanding of molecular biology, which only made its first appearance in the early 1950s. Biology has been able to provide us with sufficient insights into the cell constituents to permit the wide acceptance of a comprehensive taxonomy, which places Bacteria, Archea, and Eukarya as the highest groups (taxa) of organisms. We can paraphrase Sir Julian Huxley's comments in the introduction of Pierre Teilhard de Chardin's *The Phenomenon of Man*,[32] by remarking that there is an evident inexorable increase towards greater complexity in the transition from Bacteria to Eukarya. To borrow the phrase of Christian De Duve,[33] we may say that the laws of physics and chemistry imply an "imperative" appearance of life during cosmic evolution, a view which is not in contradiction with the relevant remarks of Sir Peter Medawar in his comment on Teilhard's work.[34]

---

[29] J. Seckback, "On the fine structure of the acidophilic hot-spring alga *Cyanidium caldarium*: a taxonomic approach," *Microbios* 5 (1972): 133–42.

[30] F. Westall, L. Boni, and E. Guerzoni, "The experimental silicification of microorganisms," *Paleontology* 38 (1995): 495–528.

[31] G. Rosati, P. Lenzi, and U. Franco, "'Epixenosomes': peculiar epibionts of the protozoon ciliate *Euplotidium itoi*: Do their cytoplasmic tubules consist of tubulin?" *Micron* 24 (1993): 465–71.

[32] Pierre Teilhard de Chardin, *The Phenomenon of Man* (New York: Collins, 1954).

[33] Christian De Duve, *Vital dust: Life as a cosmic imperative* (New York: Basic Books, 1995).

[34] Peter Medawar, *The strange case of the spotted mice and other classic essays on*

4.2 *The Phenomenon of the Eukaryotic Cell*

We are going one step beyond De Duve. We defend the hypothesis that not only is life a natural consequence of the laws of physics and chemistry, but once the living process has started, then the cellular plans, or blueprints, are also of universal validity: the lowest cellular blueprint (prokaryotic) will lead to the more complex cellular blueprint (eukaryotic). This is a testable hypothesis. Within a decade or two a new generation of space missions may be operational. Some are currently in their planning stages, such as the hydrobot/cryobot, which is aiming to reach the Jovian satellite Europa in the second decade of next century.[35] We shall present the rationalization behind this effort in section 6 below.

Closely related to the above hypothesis (the proposed universality of eukaryo-genesis), is a question concerning the different positions which are possible regarding extraterrestrial life. Is it reasonable to search for Earth-like organisms, such as a eukaryote, or should we be looking for something totally different? We will discuss in turn some of the arguments involved.

First, the more widely accepted belief on the nature of the origin of life is that it evolved according to the principles of deterministic chaos. Evolutionary developments of this type never run again through the same path of events.[36]

Second, the possibility for similar evolutionary pathways on different planets of the Solar System has been defended recently.[37] Indeed, even if some authors may consider this to be a remote possibility, Paul Davies bases his work on the increasing acceptance that catastrophic impacts have played an important role in shaping the history of terrestrial life. Thus there may be some common evolutionary pathways between the microorganisms on Earth and those that may have developed on Mars during its "clement period" (roughly equivalent to the early Archean in Earth stratigraphy). The means of transport may have been the displacement of substantial quantities of planetary surface due to large asteroid impacts on Mars.

Third, even in spite of the second possibility raised above, many researchers still see no reason to assume that the development of extraterrestrial life forms followed the same evolutionary pathway to eukaryotic cells, as it is known to have occurred on Earth. Moreover, a widespread point of view is that it would seem reasonable to assume that our ignorance concerning the origin of terrestrial life does

---

science (Oxford: Oxford University Press, 1996).

[35] S. Trowell, J. Wild, Joan Hovrath, Jack Jones, Elizabeth Johnson, and James Cutts, "Through the Europan Ice: Advanced Lander Mission Options," in *Europa Ocean Conference*, (Book of Abstracts), Capistrano Conference No. 5 (San Juan Capistrano, Calif.: San Juan Capistrano Research Institute [31872 Camino Capistrano, 92675, USA. Telephone: (714) 240 2010; Fax: (714) 240 0482], 12–14 November 1996), 76; Joan Horvath, Frank Carsey, James Cutts, Jack Jones, Elizabeth Johnson, Bridget Landry, Lonne Lane, Gindi Lynch, Julian Chela-Flores, Tzyy-Wenand Jeng, and Albert Bradley, "Searching for ice and ocean biogenic activity on Europa and Earth," in *Instruments, Methods and Missions for Investigation of Extraterrestrial Microorganisms*: Proceedings of Optical Science, Engineering, and Instrumentation SD 97 Symposium (Bellingham, Wash.: The International Society for Optical Engineering, in press).

[36] S. A. Kauffman, *The origins of order: Self-Organization and Selection in Evolution* (Oxford: Oxford University Press, 1993).

[37] P. C. W. Davies, "Evolution of Hydrothermal Ecosystems on Earth (and Mars?)," in *Proceedings of the CIBA Foundation Symposium*, no. 202 (London, 30 January–1 February, 1996).

not justify the assumption that any extraterrestrial life form has to be based on just the same genetic principles that are known to us.

In sharp contrast to the position denying that common genetic principles may underlie the outcome of the origin of life elsewhere, De Duve has pointed out[38] a fourth point of view in the discussion of the various possible ways of approaching the question of the nature of extraterrestrial life. We may conclude that we all agree that the final outcome of life evolving in a different environment would not be the same as the Earth biota. But De Duve has broken new ground when he raised the question as to how different would the outcome be of the origin of life elsewhere? This has led to a clear distinction that there is no reason for the details of the phylogenetic tree to be reproduced elsewhere (except for the possibility of biogenic exchange in the Solar System discussed above.)[39] The tree of life constituted by Earth's biota may be unique to Earth.

On the other hand, there is plenty of room for the development of differently shaped evolutionary trees in an extraterrestrial environment, where life may have taken hold. But, "certain directions may carry such decisive selective advantages as to have a high probability of occurring elsewhere as well."[40]

4.3 The Phenomenon of Multicellularity

The current view on terrestrial evolution spanning from the first appearance of the eukaryotic cell to the evolution of metazoan organisms is as follows: in the Proterozoic Eon, a geologic period which extends from 2.5 to 0.545 billion years ago, we come to the end of the Precambrian that had seen the first appearance of the eukaryotic cell. The Proterozoic was followed by the current eon, the Phanerozoic, which has seen the spread of multicellular organisms throughout the Earth on its oceans and continents. We should comment on some aspects of the eras in this current eon.

The Paleozoic Era saw the first appearance of fish (510 million years ago) and some other vertebrates.

The Mesozoic Era began after the massive extinctions at the end of the Permian Period, 250 million years ago. This era saw the first appearance of mammals (205 million years ago). Current evolutionary thinking sees the first appearance of humans, a question of central interest to both theologians and philosophers, as a natural consequence of the series of events which began in the Mesozoic and culminated in the Cenozoic Era, particularly in the Upper Miocene some 5 million years ago, when according to anthropological research, our ancestors diverged from the other primates.

The Quaternary Era saw the arrival of the first traces of intelligence, as we understand it, in our ancestors. A more evident demonstration of the appearance of intelligent life on Earth than the habilines' tools, or the ceremonial burials, had to wait till the Magdalenian "culture" (in the archaeologic classification of southwestern Europe and north Africa this group of human beings flourished from about 20,000 to 11,000 years ago. The Magdalenians left some fine works of primitive art as, for instance the 20,000 year old paintings on the walls of the cave discovered in December 1994 by Jean-Marie Chauvet in southeastern France. Indeed, the birth of art in the Magdalenians' caves is one of the most striking additions to the output of

---

[38] De Duve, *Vital dust*.

[39] Davies, "Evolution of Hydrothermal Ecosystems on Earth (and Mars?)."

[40] De Duve, *Vital dust*.

humans that entitle us to refer to the groups that produced these fine works as *cultures* rather than *industries*, a term which is reserved for groups of humans that produced characteristic tools rather than works of art.

The purpose of this sketch has been to bring into focus the different branches of science, including the social sciences, which have contributed to a huge canvas depicting a great deal of knowledge that has been put together to provide partial answers to the very questions that humans have recorded since the most ancient times, particularly the Israelites over two thousand years ago when the Old Testament of the Bible was written down. Those disciplines include, among others: anthropology, paleontology, prehistoric research, geochronology, biogeology,and geochemistry.

5 *Extraterrestrial Sources for Life's Origin: the Evidence from Two Meteorites*

5.1 *The Case for a SETE Program*

The problem of eukaryogenesis occupies a central position in a wide range of problems which concern the origin, evolution, and distribution of life in the universe, ranging from chemical evolution to "exobiology." A program for the search for extraterrestrial eukaryotes (SETE) is relevant to SETI, as the only form of intelligent life that we are familiar with is based on multicellular organisms of eukaryotic cells (Domain Eukarya). Rather than employing the empirical SETI approach, exploring the consequences of the laws of physics and chemistry may give us some insights into the question of whether or not we are alone in the cosmos. This has been attempted by De Duve.[41] This concept of the constraint on chance militates against the older criterion of Monod[42] who, during his lifetime, had less information than we have accumulated now on the mechanisms that are behind the origin and evolution of life on Earth.

Indeed, chance (rate of mutations) and necessity (natural selection) do not imply that life elsewhere in the cosmos is unlikely. Thus, the prospect for detecting life in the newly discovered planets remains a possibility by detecting life-supporting volatiles, as strongly advocated in the past by Christopher McKay. The rationale behind this approach to the search for life is that we now know that a substantial fraction of the volatiles on Earth were deposited by the in-fall of comets in the past. It seems plausible that such a phenomenon may not have been restricted just to Earth. We should add that many researchers now feel that searching for life is equivalent to searching for water, since as we shall argue in section 6, water on a silicate body (either on a planet or on a satellite), near a star or even far from a star, but near a Jupiter-like planet, seem to be sufficient to engender life, as tectonic activity in the silicate body is triggered by the vicinity of the large planet or star. This is related to the general consensus that neither oxygen nor photosynthesis are necessary to trigger off the simplest (prokaryotic) blueprint for a microorganism.[43]

The possibility of an exogenous source of earthly volatiles was suggested by John Oro in the early 1960s. Current spectroscopic analysis of the nuclei of comets, such as the Hale-Bopp, have supported this point of view, since it has been observed

---

[41] Ibid.

[42] Jacques Monod, *Chance and Necessity* (New York: Knopf, 1971).

[43] J. Delaney, J. Baross, M. Lilley, and D. Kelly, "Hydrothermal systems and life in our solar system," in *Europa Ocean Conference* (Book of Abstracts); J. Chela-Flores, "Testing for evolutionary trends of Europan biota," in *Instruments, Methods and Missions for Investigation of Extraterrestrial Microorganisms.*

that most of the precursors of the prebiotic molecules are present in comets. An alternative means of probing for extraterrestrial life is by directly detecting their radio messages, as persistently maintained by Frank Drake. Preliminary steps in the search for intelligent life on new planets have already been taken.[44]

5.2 The Murchison Meteorite: the Inventory of Earthly Amino Acids May Have an Extraterrestrial Component

It has been clear for some time now that the extraterrestrial origin of life should not be ruled out. First, a significant illustration of the plausibility of an extraterrestrial origin of the precursors of the biomolecules is based on the meteorite which fell in the town of Murchison, Australia on September 21, 1969. The laboratory of Cyril Ponnamperuma was able to obtain a piece of it. At that time they were preparing for the first analysis of the lunar rocks. They were able to get the first conclusive evidence of extraterrestrial amino acids.[45] The results obtained by Ponnamperuma's group have demonstrated the universality of the formation of some organic compounds which are essential for life today.

However, the above sketch of the chemical evolution of the origin of life is incomplete. This fact is illustrated, for instance, by the question of which gases were present in the early Earth atmosphere, a topic which is not settled.[46] Besides, carbon dioxide must have been sufficiently abundant to have prevented the Archean Earth from freezing under the Sun at a lower level in the main sequence of the Hertzprung-Russell diagram (some 30% less luminous at the time of the origin of life). This is referred to as the "faint young Sun paradox." Through a greenhouse effect there would have been the appropriate temperatures for producing the microflora that we know must have been in existence some 3.5 billion years ago. Also in need of clarification is the relative importance of the precursors of life that were brought to Earth by comets, meteorites, and micrometeorites, compared with the inventories that were part of the Earth as it formed out of our own planetary nebula.

5.3 The Allan Hills Meteorite: Have There Been Prokaryotes on Mars?

A second important meteorite for exobiology was retrieved in 1984 by a NSF mission from the wastelands of Antarctica in a field of ice called the Allan Hills. It weighs about two kilograms. A preliminary analysis of this meteorite (ALH84001) was reported during the 1995 Trieste Conference on Chemical Evolution;[47] subsequently a fuller report was made known to the public independently.[48] This work has shown the presence of an important biomarker, polycyclic aromatic hydrocarbons (PAHs), which on Earth are abundant as fossil molecules in ancient sedimentary rocks, and as components of petroleum.

----

[44] F. Biraud, J. Heidmann, J. C. Tarter, and S. Airieau, "A search for artificial signals from the newly discovered planetary systems," in *Astronomical and Biochemical Origins and the Search for Life in the Universe*, Cosmovici, Bowyer and Werthimer, eds., 689–92.

[45] Ponnamperuma "The origin of the cell from Oparin to the present day," in *Chemical Evolution: The Structure and Model of the First Cell*, Ponnamperuma and Chela-Flores, eds.

[46] J. F. Kasting, "Earth's Early Atmosphere," *Science* 259 (1993): 920–26.

[47] C. P. McKay, "Oxygen and the rapid evolution of life on Mars," in *Chemical Evolution: Physics of the Origin and Evolution of Life*, Chela-Flores and Raulin, eds., 177–84.

[48] D. S. McKay, E. K. Gibson Jr., K. L. Thomas-Keprta, H. Vali, C. S. Romanek, S. J. Clemett, X. D. F. Chiller, C. R. Maechling, and R. N. Zare, "Search for past life on Mars: Possible Relic Biogenic Activity in Martian Meteorite ALH84001," *Science* 273 (1996): 924–30.

The presence of PAHs in the ALH84001 meteorite is compatible with the existence of past life on Mars. This result, which requires confirmation, underlines the importance in the future of formulating comprehensive questions (as suggested in this section) regarding life on Mars to include *the degree of biological evolution* covering the complete range known to us, from simple bacteria to the eukaryotes.

It should be added that since the first report of preliminary evidence of life in the Allan Hills meteorite, some scientists have found chemical traces in a separate meteorite that seem to be consistent with the presence of life on Mars. They have found organic matter in two meteorites—the Allan Hills one and in a second meteorite (named 79001) that crashed to Earth 600,000 years ago.[49] The rock itself was sedimented some 150 million years ago. Residues and chemicals in the rock that could only be formed by living organisms.

The meteorite had been considered earlier in the context of whether life existed on Mars. Experiments on parts of 79001 may be significant as the meteorite had become sealed in a glass-like substance before it came to Earth and was, to a certain extent, insulated from the Earth's organic matter. The main result is that amounts of organic material were found (up to 1,000 parts per million).

To be fair we should point out that there is still some skepticism about the fact that these trace elements found in Martian meteorites are sufficient evidence to say that it represents proof that there was life on Mars.

6 Habitability of Europa by Eukaryotic Life

6.1 In Which Solar System Habitats Can Extremophiles Survive?

It is useful to know in which atmospheres extremophiles can survive, as we now know of several Solar-System planets and satellites that have atmospheres, not all similar to our own. This consideration may help the reader narrow down the possible Solar System sites in which to focus our attention in our search for "clement" habitats. For example, four large planetary satellites in the outer Solar System are known to have atmospheres: Europa and Io (Jupiter), Titan (Saturn) and Triton (Neptune).[50] Mars currently has a CO_2 dominated atmosphere. The possibility of extending the biosphere deep into the silicate crust of Mars has some implications.

This question seems pertinent to exobiology: we cannot exclude the possibility that organisms, which have been found to live deep in the silicate crust of the Earth, may been deposited with the original sediment, and survived over geologic time. McKay has considered this question in some detail from the point of view of geophysics.[51] From all of these options we feel that the most promising candidate is Europa. The reasons shall be considered below.

6.2 Favorable Sites in the Solar System for Extremophiles and Other Microorganisms

In the present "Golden Age" of the search for life in the Solar System, there are two promising sites to be investigated.

---

[49] G. D. McDonald and J. L. Bada, "A search for endogenous amino acids in the Martian meteorite EETA79001," *Geochem. Cosmochem. Acta* 59 (1995): 1179–84.

[50] D. T. Hall, D. F. Strobel, P. D. Feldman, M. A. McGrath, and H. A. Weaver, "Detection of an oxygen atmosphere on Jupiter's moon Europa," *Nature* 373 (1995): 677–79.

[51] McKay, "Oxygen and the rapid evolution of life on Mars," in *Chemical Evolution: Physics of the Origin and Evolution of Life*, Chela-Flores and Raulin, eds.

1. The first site is Mars, but the possibility of finding life elsewhere in the Solar System is not restricted to planetary habitats.[52] As this possibility has been exhaustively discussed elsewhere, we shall focus attention on an attractive alternative.

2. The second site is Europa, the Jovian satellite. The existence of internal oceans in the Jovian satellites goes back to 1980 when Gerald Feinberg and Robert Shapiro speculated on this possibility in the specific case of the satellite Ganymede; more recently, Shapiro has further refined the Feinberg-Shapiro hypothesis by suggesting the presence of hydrothermal vents at the bottom of the ocean in Europa, in analogy with the same phenomenon on Earth.[53] In fact, for these reasons Europa is a candidate for a SETE program, which we have proposed.

A large proportion of the spectroscopically detectable material on the surface of Europa is water.[54] According to the results on the 1979 Voyager fly-by, Europa showed a cracked icy crust suggesting that underneath there might be an ocean of water. Also, since Europa must be subject to strong Jovian tides, one would expect the ocean to be heated by the proximity to the planet. For these reasons it was estimated by Oro, Squares, Reynolds and Mills that the temperature underneath the icy crust should be $4°$ C.[55] The preliminary results of the photographs of Europa retrieved from the Galileo Mission add some plausibility to the older conjectures regarding the Europa ocean.

6.3 Can Thermophilic Archaebacteria Evolve in the Europa's Ocean?

From the similarity of the processes that gave rise to the solid bodies of the Solar System, we may expect that hot springs may lie at the bottom of Europa's ocean. The main thesis supporting this expectation is that, as Jupiter's primordial nebula must have contained many organic compounds, then possibly, organisms similar to thermophilic archaebacteria could evolve at the bottom of Europa's ocean.

The argument of Oro et al. correctly pointed out the most important requirements for the maintenance of life in Europa; their work is in agreement with the above list (water, an energy source, and organic compounds). However, the analogous Earth ecosystems considered by them excluded eukaryotes. The case mentioned above of the cyanidiophycean algae is a warning that we should keep an open mind while discussing the possible degree of evolution of Europan biota, or indeed of living organisms elsewhere.

---

[52] J. Chela-Flores, "A Search for Extraterrestrial Eukaryotes: Biological and Planetary Science Aspects," in *Astronomical and Biochemical Origins and the Search for Life in the Universe*, Cosmovici, Bowyer and Werthimer, eds.; McKay, "Oxygen and the rapid evolution of life on Mars," in *Chemical Evolution: Physics of the Origin and Evolution of Life*, Chela-Flores and Raulin, eds.; McKay et al., "Search for past life on Mars: Possible Relic Biogenic Activity in Martian Meteorite ALH84001."

[53] R. Shapiro, *L'origine de la vie* (Paris: Flammarion, 1994), 402–4; G. Feinberg, and R. Shapiro, *Life Beyond Earth: The Intelligent Earthling's Guide to Life in the Universe* (New York: William Morrow and Co., 1980), 328–32.

[54] J. Oro, "Cosmic evolution, life and man," in *Chemical Evolution: Physics of the Origin and Evolution of Life*, Chela-Flores and Raulin, eds., 3–19; J. Oro, S. W. Squyres, R. T. Reynolds, and T. M. Mills, "Europa: Prospects for an ocean and exobiological implications," in *Exobiology in Solar System Exploration*, G. C. Carle, D. E. Schwartz, and J. L. Huntington, eds. *Symposium Proceedings from the Exobiology Program of NASA's Division of Life Sciences, SP 512* (Moffet Field, Calif.: NASA Ames Research Center, 1992), 103–25.

[55] Oro, "Cosmic evolution, life and man," in *Chemical Evolution: Physics of the Origin and Evolution of Life*, Chela-Flores and Raulin, eds.

We may add that presently the divergence into the three domains, arising from the evolution of the "*progenote*" (the earliest ancestor of all living organisms), is not well understood. Indeed, plate tectonics has obliterated fossils of early organisms from the Earth's crust, which is the only record currently available, with the possible exception of the SNC meteorites. (The initials are taken from the names of sub-groups: shergottites, nakhlites, and chassignites.) These are a small group of basaltic meteorites from relatively young (120–1300 million years ago) lava flows on Mars.

6.4 A Case for Evolutionary Trends in the Europan Biota

Several additional factors arising from current experience with eukaryotes may help to clarify the case for not excluding these microorganisms from those microorganisms to be sought in new Solar System environments.

1. A critical step in the diversification of single-celled organisms may have been the loss of the ability to manufacture a cell-wall biomolecule (peptidoglycan). Without the constraint that this biomolecule imposes on cell shape, both Archaea and Eukarya have been able to diversify beyond the Domain Bacteria.

2. In spite of the fact that eukaryotes and archaebacteria have both lost the peptidoglycan, we can nevertheless distinguish between them, as all eukaryotes have ester lipids in their membranes, whereas archaebacteria have ether lipids.

3. Earth-bound eukaryotes are not extremophiles, but their diversification may share a common thread with archaebacteria. Eukaryotes, in spite of not being able to exploit fully all the extreme environments, may invade, to a certain extent, environments normally at the disposal of the extremophiles.

We have remarked that the identification of primitive eukaryotes is not straightforward; one additional difficulty is that the simplest criterion, namely the search for a membrane-bounded sets of chromosomes, does not help to identify unambiguously eukaryotes, as there are prokaryotic organisms that do have a membrane-bounded nucleoid.[56]

Finally, if we are granted the possibility that archaebacteria may be present at the bottom of the Europan Ocean (section 6.3), then there really is a case for testing the degree of evolution of such biota.[57]

Because they are so foreign to our everyday experience and also far removed by catastrophes that may exterminate other ecosystems, the hot-spring environment has been assumed to be a sort of refuge against evolutionary pressures. If these ecosystems were true refuges against evolution, no extinctions would be expected. Hence, no organisms that once dwelt in them would be expected to be completely missing from contemporary hydrothermal vents.

It has also been observed by *in situ* submarine investigation that hot-spring communities of animals are remarkably similar throughout the world. This "refuge" concept has prevailed even among exobiologists that have speculated on the possible origin of life at the bottom of the Europan putative ocean.

Previous authors have only considered that archaebacteria may have also evolved in the bethnic regions heavily dependent on bacterial chemosynthesis. However, scientists at the Urals branch of the Russian Academy[58] have identified

---

[56] J. A. Fuerst and R. J. Webb, "Membrane-bounded nucleoid in the eubacterium *Gemmata oscuriglobus," Proc. Natl. Acad. Sci.* 88 (1991): 8184–88.

[57] Chela-Flores, "Testing for evolutionary trends of Europan biota," in *Instruments, Methods and Missions for Investigation of Extraterrestrial Microorganisms.*

[58] C. T. S. Little, R. J. Herrington, V. V. Maslennikov, N. J. Morris, and V. V. Zaykov, "Silurian hydrothermal-vent community from the southern Urals, Russia," *Nature* 385

fossils from the earliest hydrothermal-vent community found so far, dating from the late Silurian Period, over 400 million years ago. This particular community has its own case of species extinction, as fossils have been identified and correspond to lamp shells (inarticulate brachiopods) and snail-like organisms (monoplacophorans). The implication of this discovery is that since neither the above-mentioned brachiopods nor the mollusks inhabit contemporary hydrothermal vent communities, these environments have known their own cases of species extinction. This weakens considerably the refuge hypothesis, namely, that any environment on Earth may serve as a refuge against the inevitable consequences of natural selection.

We cannot maintain with certainty today that the analogous conditions that may exist in Europa will induce the appearance of archaebacteria-like organisms. This is purely an observational question. However, given the common origin of all the Solar System, the favorable conditions in Europa may have been conducive to the first steps in evolution, including the first appearance of archaebacteria.

7 Conclusions

We should underline that from the experience that we have gathered on Earth[59] we are aware of at least two important groups of abundantly distributed biomarkers which have been characterized from oils and sediments (the fossil triterpanes and steranes), whose parental materials are found *almost exclusively in eukaryotes*. Therefore, if the degree of biological evolution in Mars has gone as far as eukaryogenesis (or taken the first steps towards multicellularity), there are means available for detecting this important aspect of exobiology in future space missions (namely, the search for the parental materials for steranes: tetracyclic steroids).

To sum up, the presence of archaebacteria-like organisms as the most likely biota in the Europa ocean is plausible.[60] Based on the fact that hydrothermal vents should not be considered refuges against evolution[61] (section 6.4), we have argued that evolution should have occurred in Europa. This has led to the attempt to identify parameters that may be considered to be indicators of the degree of evolution of putative Europan biota, both at the ice surface as well as in the putative ocean.

These conjectures may be put to the test in the not too distant future as we are now in the process of planning possible missions in which these ideas may be tested. Indeed, the present work has been stimulated by the possibility of designing for the early part of next century an advanced lander mission that may melt through the ice layer above the Europan ocean in order to deploy a tethered submersible,[62] which may allow a direct test for life and its evolutionary status.

(1997): 146–48.

[59] M. Chadha, "Role of transient and stable molecules in chemical evolution," in *Chemical Evolution: Physics of the Origin and Evolution of Life*, Chela-Flores and Raulin, eds., 107–22.

[60] Oro, "Cosmic evolution, life and man," in *Chemical Evolution: Physics of the Origin and Evolution of Life*, Chela-Flores and Raulin, eds.; Delaney et al., "Hydrothermal systems and life in our solar system," in *Europa Ocean Conference* (Book of Abstracts).

[61] Little et al., "Silurian hydrothermal-vent community from the southern Urals, Russia."

[62] Horvath et al., "Searching for ice and ocean biogenic activity on Europa and Earth," in *Instruments, Methods and Missions for Investigation of Extraterrestrial Microorganisms*, Trowell et al., "Through the Europan Ice: Advanced Lander Mission Options" in *Europa Ocean Conference* (Book of Abstracts); Chela-Flores, "Testing for evolutionary trends of Europan biota," in *Instruments, Methods and Missions for Investigation of Extraterrestrial Microorganisms*.

Acknowledgment. The author would like to thank Mr. Kirk Wegter-McNelly for a critical reading of the manuscript. He would also like to thank the editors of the present volume for a number of suggestions that led to the improvement of the text.

II

EVOLUTION AND DIVINE ACTION

DARWIN'S DEVOLUTION: DESIGN WITHOUT DESIGNER

Francisco J. Ayala

It is also frequently asked what our belief must be about the form and shape of heaven according to Sacred Scripture. Many scholars engage in lengthy discussions on these matters... Such subjects are of no profit for those who seek beatitude, and, what is worse, they take up very precious time that ought to be given to what is spiritually beneficial. What concern is it of mine whether heaven is like a sphere and the earth is enclosed by it and suspended in the middle of the universe?... In the matter of the shape of heaven the sacred writers... did not wish to teach men these facts that would be of no avail for their salvation.

As for the other small creatures... there was present from the beginning in all living bodies a natural power, and, I might say, there were interwoven with these bodies the seminal principles of animals later to appear... each according to its kind and with its special properties.

St. Augustine, The Literal Meaning of Genesis[1]

1 *Summary of the Argument*

I advance three propositions. The first is that Darwin's most significant intellectual contribution is that he brought the origin and diversity of organisms into the realm of science. The Copernican revolution consisted in a commitment to the postulate that the universe is governed by natural laws that account for natural phenomena. Darwin completed the Copernican revolution by extending that commitment to the living world.

The second proposition is that natural selection is a creative process that can account for the appearance of genuine novelty. How natural selection creates is shown with a simple example and clarified with two analogies, artistic creation and the "typing monkeys," with which it shares important similarities and differences. The creative power of natural selection arises from a distinctive interaction between chance and necessity, or between random and deterministic processes.

The third proposition is that teleological explanations are necessary in order to give a full account of the attributes of living organisms, whereas they are neither necessary nor appropriate in the explanation of natural inanimate phenomena. I give a definition of teleology and clarify the matter by distinguishing between internal and external teleology, and between bounded and unbounded teleology. The human eye, so obviously constituted for seeing but resulting from a natural process, is an example of internal (or natural) teleology. A knife has external (or artificial) teleology, because it has been purposefully designed by an external agent. The development of an egg into a chicken is an example of bounded (or necessary) teleology, whereas the evolutionary origin of the mammals is a case of unbounded (or contingent) teleology, because there was nothing in the make-up of the first living cells that necessitated the eventual appearance of mammals.

---

[1] Book 2, Chap. 9; Book 3, Chap. 14.

I conclude that Darwin's theory of evolution and explanation of design does not include or exclude considerations of divine action in the world any more than astronomy, geology, physics, or chemistry do.

2 *The Darwinian Revolution*

The publication in 1859 of *The Origin of Species* by Charles Darwin ushered in a new era in the intellectual history of humankind. Darwin is deservedly given credit for the theory of biological evolution: he accumulated evidence demonstrating that organisms evolve and discovered the process, natural selection, by which they evolve. But the import of Darwin's achievement is that it completed the Copernican revolution initiated three centuries earlier, and thereby radically changed our conception of the universe and the place of human beings in it.

The discoveries of Copernicus, Kepler, Galileo, and Newton in the sixteenth and seventeenth centuries, had gradually ushered in the notion that the workings of the universe could be explained by human reason. It was shown that Earth is not the center of the universe, but a small planet rotating around an average star; that the universe is immense in space and in time; and that the motions of the planets around the sun can be explained by the same simple laws that account for the motion of physical objects on our planet. These and other discoveries greatly expanded human knowledge, but the intellectual revolution these scientists brought about was more fundamental: a commitment to the postulate that the universe obeys immanent laws that account for natural phenomena. The workings of the universe were brought into the realm of science: explanation through natural laws. Physical phenomena could be accounted for whenever the causes were adequately known.

Darwin completed the Copernican revolution by drawing out for biology the notion of nature as a lawful system of matter in motion. The adaptations and diversity of organisms, the origin of novel and highly organized forms, even the origin of humankind itself could now be explained by an orderly process of change governed by natural laws.

The origin of organisms and their marvelous adaptations were, however, either left unexplained or attributed to the design of an omniscient Creator. God had created the birds and bees, the fish and corals, the trees in the forest, and best of all, humans. God had given us eyes so that we might see, and He had provided fish with gills to breathe in water. Philosophers and theologians argued that the functional design of organisms manifests the existence of an all-wise Creator. Wherever there is design, there is a designer; the existence of a watch evinces the existence of a watchmaker.

A well-known example is the English theologian William Paley. His *Natural Theology* (1802) elaborated the argument-from-design as forceful demonstration of the existence of the Creator, drawing a comparison between the obvious design of a watch and the apparent design of a human eye. A few decades later, the *Bridgewater Treatises*, published between 1833 and 1840, were written by eminent scientists and philosophers to set forth "the Power, Wisdom, and Goodness of God as manifested in the Creation." The structure and mechanisms of the human hand were, for example, cited as incontrovertible evidence that the hand had been designed by the same omniscient Power that had created the world.

The advances of physical science had thus driven our conception of the universe to a split-personality state of affairs, which persisted well into the mid-nineteenth century. Scientific explanations, derived from natural laws, dominated the world of non-living matter, on Earth as well as in the heavens. Supernatural explanations, depending on the unfathomable deeds of the Creator, accounted for the origin and configuration of living creatures—the most diversified, complex, and interesting realities of the world. It was Darwin's genius to resolve this conceptual schizophrenia.

3 Darwin's Discovery: Design without Designer

The conundrum faced by Darwin can hardly be overestimated. The strength of the argument-from-design to demonstrate the role of the Creator is easily set forth. Wherever there is function or design we look for its author. A knife is made for cutting and a clock is made to tell time; their functional designs have been contrived by a knifemaker and a watchmaker. The exquisite design of Leonardo da Vinci's *Mona Lisa* proclaims that it was created by a gifted artist following a preconceived purpose. Similarly, the structures, organs, and behaviors of living beings are directly organized to serve certain functions. The functional design of organisms and their features would therefore seem to argue for the existence of a designer. It was Darwin's greatest accomplishment to show that the directive organization of living beings can be explained as the result of a natural process, natural selection, without any need to resort to a Creator or other external agent. The origin and adaptation of organisms in their profusion and wondrous variations were thus brought into the realm of science.

Darwin accepted that organisms are "designed" for certain purposes, that is, they are functionally organized. Organisms are adapted to certain ways of life and their parts are adapted to perform certain functions. Fish are adapted to live in water, kidneys are designed to regulate the composition of blood, the human hand is made for grasping. But Darwin went on to provide a natural explanation of the design. He thereby brought the seemingly purposeful aspects of living beings into the realm of science.

Darwin's revolutionary achievement is that he extended the Copernican revolution to the world of living things. The origin and adaptive nature of organisms could now be explained, like the phenomena of the inanimate world, as the result of natural laws manifested in natural processes. Darwin's theory encountered opposition in some religious circles, not so much because he proposed the evolutionary origin of living things (which had been proposed before, and accepted even by Christian theologians), but because the causal mechanism, natural selection, excluded God as the explanation for the obvious design of organisms.[2] The Roman

---

[2] See David C. Lindberg and Ronald L. Numbers, eds., *God and Nature* (Berkeley, Calif.: University of California Press, 1986), chaps. 13–16. Charles Hodge (1793–1878), an influential Protestant theologian, published in 1874 *What is Darwinism?*, one of the most articulate attacks against evolutionism. However, a principle of solution was seen by other Protestant theologians in the notion that God operates through intermediate causes, such as the origin and motion of the planets and evolution. Thus, A. H. Strong, president of Rochester Theological Seminary, wrote in 1885 in his *Systematic Theology*: "We grant the principle of evolution, but we regard it as only the method of divine intelligence." Strong drew an analogy with Christ's miraculous conversion of water into wine: "The wine in the miracle was not water because water had been used in the making of it, nor is man a brute because the brute has made some contributions to its creation."

Arguments for and against Darwin's theory came from Catholic theologians as well. Gradually, well into the twentieth century, evolution by natural selection came to be accepted by the enlightened majority of Christian writers. Pius XII accepted in his encyclical *Humani*

Catholic Church's opposition to Galileo in the seventeenth century had been similarly motivated not only by the apparent contradiction between the heliocentric theory and a literal interpretation of the Bible, but also by the unseemly attempt to comprehend in the workings of the universe, the "mind of God." The configuration of the universe was no longer perceived as the result of God's Design, but simply as the outcome of immanent, blind, processes.

There were, however, many theologians, philosophers, and scientists who saw no contradiction then nor see it now between the evolution of species and Christian faith. Some see evolution as the "method of divine intelligence," in the words of the nineteenth-century theologian A. H. Strong. Others, like the American contemporary of Darwin, Henry Ward Beecher (1818–87), made evolution the cornerstone of their theology. These two traditions have persisted to the present. Pope John Paul II has recently (October 1996, reprinted in this volume) stated that "the theory of evolution is no longer a mere hypothesis. It is... accepted by researchers, following a series of discoveries in various fields of knowledge." The views of "process" theologians, who perceive evolutionary dynamics as a pervasive element of a Christian view of the world, are well represented in this volume.

4 Natural Selection as a Directive Process

The central argument of the theory of natural selection is summarized by Darwin in *The Origin of Species* in terms of natural selection, the fact that individuals who inherit useful variations have greater chances of survival and reproduction.

Darwin's argument addresses the problem of explaining the adaptive character of organisms. Darwin argues that adaptive variations ("variations useful in some way to each being") occasionally appear, and that these are likely to increase the reproductive chances of their carriers. Over the generations favorable variations will be preserved while injurious ones will be eliminated. In one place, Darwin adds: "I can see no limit to this power [natural selection] in slowly and beautifully *adapting* each form to the most complex relations of life." Natural selection was proposed by Darwin primarily to account for the adaptive organization, or "design," of living beings; it is a process that promotes or maintains adaptation. Evolutionary change through time and evolutionary diversification (multiplication of species) are not directly promoted by natural selection (hence, the so-called "evolutionary stasis," the numerous examples of organisms with morphology that has changed little, if at all, for millions of years, as pointed out by the proponents of the theory of punctuated equilibrium). But change and diversification often ensue as by-products of natural selection fostering adaptation.

Darwin formulated natural selection primarily as differential survival. The modern understanding of the principle of natural selection is formulated in genetic and statistical terms as differential reproduction. Natural selection implies that some

Generis (1950) that biological evolution was compatible with the Christian faith, although he argued that God's intervention was necessary for the creation of the human soul. In 1981, Pope John Paul II addressed the Pontifical Academy of Sciences: "The Bible itself speaks to us of the origin of the universe and its make-up, not in order to provide us with a scientific treatise but in order to state the correct relationship of man with God and with the universe." The Pope's argument is that it is wrong to mistake the Bible for an elementary book of astronomy, geology, and biology. The argument goes clearly against the biblical literalism of Fundamentalists and shares with most Protestant theologians a view of Christian belief that is not incompatible with evolution and, more generally, with science.

genes and genetic combinations are transmitted to the following generations on the average more frequently than their alternates. Such genetic units will become more common in every subsequent generation and their alternates less common. Natural selection is a statistical bias in the relative rate of reproduction of alternative genetic units.

Natural selection has been compared to a sieve which retains the rarely arising useful genes and lets go the more frequently arising harmful mutants. Natural selection acts in that way, but it is much more than a purely negative process, for it is able to generate novelty by increasing the probability of otherwise extremely improbable genetic combinations. Natural selection is thus creative in a way. It does not "create" the entities upon which it operates, but it produces adaptive genetic combinations which would not have existed otherwise.

The creative role of natural selection must not be understood in the sense of the "absolute" creation that traditional Christian theology predicates of the divine act by which the universe was brought into being *ex nihilo*. Natural selection may rather be compared to a painter which creates a picture by mixing and distributing pigments in various ways over the canvas. The canvas and the pigments are not created by the artist but the painting is. It is conceivable that a random combination of the pigments might result in the orderly whole which is the final work of art. But the probability of Leonardo's *Mona Lisa* resulting from a random combination of pigments, or St. Peter's Basilica resulting from a random association of marble, bricks, and other materials, is infinitely small. In the same way, the combination of genetic units which carries the hereditary information responsible for the formation of the vertebrate eye could never have been produced by a random process like mutation—not even if we allow for the three billion years plus during which life has existed on earth. The complicated anatomy of the eye, like the exact functioning of the kidney, is the result of a non-random process—natural selection.

Critics have sometimes alleged as evidence against Darwin's theory of evolution examples showing that random processes cannot yield meaningful, organized outcomes. It is thus pointed out that a series of monkeys randomly striking letters on a typewriter would never write *The Origin of Species*, even if we allow for millions of years and many generations of monkeys pounding at typewriters.

This criticism would be valid if evolution depended only on random processes. But natural selection is a non-random process that promotes adaptation by selecting combinations that "make sense," that is, combinations that are useful to the organisms. The analogy of the monkeys would be more appropriate if a process existed by which, first, meaningful words would be chosen every time they appeared on the typewriter; and then we would also have typewriters with previously selected words rather than just letters in the keys, and again there would be a process to select meaningful sentences every time they appeared in this second typewriter. If every time words such as "the," "origin," "species," and so on, appeared in the first kind of typewriter, they each became a key in the second kind of typewriter, meaningful sentences would occasionally be produced in this second typewriter. If such sentences became incorporated into keys of a third type of typewriter, in which meaningful paragraphs were selected whenever they appeared, it is clear that pages and even chapters "making sense" would eventually be produced.

We need not carry the analogy too far, since the analogy is not fully satisfactory, but the point is clear. Evolution is not the outcome of purely random processes, but rather there is a "selecting" process, which picks up adaptive combinations because these reproduce more effectively and thus become established in populations. These adaptive combinations constitute, in turn, new levels of organization upon which the mutation (random) plus selection (non-random or directional) process again operates.

The manner in which natural selection can generate novelty in the form of accumulated hereditary information may be illustrated by the following example. Some strains of the colon bacterium, *Escherichia coli*, in order to be able to reproduce in a culture medium, require that a certain substance, the amino acid histidine, be provided in the medium. When a few such bacteria are added to a cubic centimeter of liquid culture medium, they multiply rapidly and produce between two and three billion bacteria in a few hours. Spontaneous mutations to streptomycin resistance occur in normal (that is, sensitive) bacteria at rates of the order of one in one hundred million (1×10^{-8}) cells. In our bacterial culture we expect between twenty and thirty bacteria to be resistant to streptomycin due to spontaneous mutation. If a proper concentration of the antibiotic is added to the culture, only the resistant cells survive. The twenty or thirty surviving bacteria will start reproducing, however, and allowing a few hours for the necessary number of cell divisions, several billion bacteria are produced, all resistant to streptomycin. Among cells requiring histidine as a growth factor, spontaneous mutants able to reproduce in the absence of histidine arise at rates of about four in one hundred million (4×10^{-8}) bacteria. The streptomycin resistant cells may now be transferred to a culture with streptomycin but with no histidine. Most of them will not be able to reproduce, but about a hundred will start reproducing until the available medium is saturated.

Natural selection has produced in two steps bacterial cells resistant to streptomycin and not requiring histidine for growth. The probability of the two mutational events happening in the same bacterium is of about four in ten million billion ($1 \times 10^{-8} \times 4 \times 10^{-8} = 4 \times 10^{-16}$) cells. An event of such low probability is unlikely to occur even in a large laboratory culture of bacterial cells. With natural selection, cells having both properties are the common result. It may be noticed that the process described in this example is indeed *natural* selection, in contrast with the *artificial* selection that occurs when an animal breeder selects, for example, the cows that produce the most milk in order to breed the next generation. In the case of bacterial selection, human intervention is restricted to changing the environments in which bacteria breed, rather than selecting the bacteria that breed best in each particular case. In nature, environments change endlessly from place to place and from one time to the next. This persistent environmental variation prompts evolution by natural selection, as variations favored in one environment become replaced by those favored in the next.

As illustrated by the bacterial example, natural selection produces combinations of genes that would otherwise be highly improbable because natural selection proceeds stepwise. The vertebrate eye did not appear suddenly in all its present perfection. Its formation requires the appropriate integration of many genetic units, and thus the eye could not have resulted from random processes alone. The ancestors of today's vertebrates had for more than half a billion years some kind of organs sensitive to light. Perception of light, and later vision, were important for these organisms' survival and reproductive success. Accordingly, natural selection favored genes and gene combinations increasing the functional efficiency of the eye. Such genetic units gradually accumulated, eventually leading to the highly complex and efficient vertebrate eye. Natural selection can account for the rise and spread of genetic constitutions, and therefore of types of organisms, that would never have existed under the uncontrolled action of random mutation. In this sense, natural selection is a creative process, although it does not create the raw materials—the genes—upon which it acts.[3]

5 Chance and Necessity

There is an important respect in which an artist makes a poor analogy for natural selection. A painter usually has a preconception of what he wants to paint and will consciously modify the painting so that it represents what he wants. Natural selection has no foresight, nor does it operate according to some preconceived plan. Rather it is a purely natural process resulting from the interacting properties of physico-chemical and biological entities. Natural selection is simply a consequence of the differential multiplication of living beings. It has some appearance of purposefulness because it is conditioned by the environment: which organisms reproduce more effectively depends on what variations they possess that are useful in the environment where the organisms live. But natural selection does not anticipate the environments of the future; drastic environmental changes may be insuperable to organisms that were previously thriving.

The team of typing monkeys is also a bad analogy for evolution by natural selection, because it assumes that there is "somebody" who selects letter combinations and word combinations that make sense. In evolution there is no one selecting adaptive combinations. These select themselves because they multiply more effectively than less adaptive ones.

There is a sense in which the analogy of the typing monkeys is better than the analogy of the artist, at least if we assume that no particular statement was to be obtained from the monkeys' typing endeavors, but just any statements making sense. Natural selection does not strive to produce predetermined kinds of organisms, but only organisms that are adapted to their present environments. Which characteristics will be selected depends on which variations happen to be present at a given time in a given place. This in turn depends on the random process of mutation, as well as on the previous history of the organisms (in other words, on the genetic make-up they have as a consequence of their previous evolution). Natural selection is an "opportunistic" process. The variables determining in what direction it will go are the environment, the preexisting constitution of the organisms, and the randomly arising mutations.

Thus, adaptation to a given environment may occur in a variety of different ways. An example may be taken from the adaptations of plant life to desert climate. The fundamental adaptation is to the condition of dryness, which involves the danger of desiccation. During a major part of the year, sometimes for several years in

---

[3] A common objection posed to the account I have sketched of how natural selection gives rise to otherwise improbable features, is that some postulated transitions, for example from a leg to a wing, cannot be adaptive. The answer to this kind of objection is well known to evolutionists. For example, there are rodents, primates, and other living animals that exhibit modified legs used for both running and gliding. The fossil record famously includes the reptile *Archaeopterix* and many other intermediates showing limbs incipiently transformed into wings endowed with feathers. One challenging transition involves the bones that make up the lower jaw of reptiles but have evolved into bones now found in the mammalian ear. What possible function could a bone have, either in the mandible or in the ear, during the intermediate stages? However, two transitional forms of therapsids (mammal-like reptiles) are known from the fossil record with a double jaw joint—one joint consisting of the bones that persist in the mammalian jaw, the other composed of the quadrate and articular bones, which eventually became the hammer and anvil of the mammalian ear.

succession, there is no rain. Plants have accomplished the urgent necessity of saving water in different ways. Cacti have transformed their leaves into spines, having made their stems into barrels containing a reserve of water; photosynthesis is performed in the surface of the stem instead of in the leaves. Other plants have no leaves during the dry season, but after it rains they burst into leaves and flowers and produce seeds. Ephemeral plants germinate from seeds, grow, flower, and produce seeds—all within the space of the few weeks while rainwater is available; the rest of the year the seeds lie quiescent in the soil.

The opportunistic character of natural selection is also well evidenced by the phenomenon of adaptive radiation. The evolution of *Drosophila* flies in Hawaii is a relatively recent adaptive radiation. There are about 1,500 *Drosophila* species in the world. Approximately 500 of them have evolved in the Hawaiian archipelago, although this has a small area, about one twenty-fifth the size of California. Moreover, the morphological, ecological, and behavioral diversity of Hawaiian *Drosophila* exceeds that of *Drosophila* in the rest of the world.

Why should have such "explosive" evolution have occurred in Hawaii? The overabundance of drosophila flies there contrasts with the absence of many other insects. The ancestors of Hawaiian drosophila reached the archipelago before other groups of insects did, and thus they found a multitude of unexploited opportunities for living. They responded by a rapid adaptive radiation; although they are all probably derived from a single colonizing species, they adapted to the diversity of opportunities available in diverse places or at different times by developing appropriate adaptations, which range broadly from one to another species.

The process of natural selection can explain the adaptive organization of organisms; as well as their diversity and evolution as a consequence of their adaptation to the multifarious and ever changing conditions of life. The fossil record shows that life has evolved in a haphazard fashion. The radiations, expansions, relays of one form by another, occasional but irregular trends, and the ever present extinctions, are best explained by natural selection of organisms subject to the vagaries of genetic mutation and environmental challenge. The scientific account of these events does not necessitate recourse to a preordained plan, whether imprinted from without by an omniscient and all-powerful designer, or resulting from some immanent force driving the process towards definite outcomes. Biological evolution differs from a painting or an artifact in that it is not the outcome of a design preconceived by an artist or artisan.

Natural selection accounts for the "design" of organisms, because adaptive variations tend to increase the probability of survival and reproduction of their carriers at the expense of maladaptive, or less adaptive, variations. The arguments of Aquinas or Paley against the incredible improbability of chance accounts of the origin of organisms are well taken as far as they go. But neither these scholars, nor any other authors before Darwin, were able to discern that there is a natural process (namely, natural selection) that is not random but rather is oriented and able to generate order or "create." The traits that organisms acquire in their evolutionary histories are not fortuitous but determined by their functional utility to the organisms.

Chance is, nevertheless, an integral part of the evolutionary process. The mutations that yield the hereditary variations available to natural selection arise at random, independently of whether they are beneficial or harmful to their carriers. But this random process (as well as others that come to play in the great theater of life) is counteracted by natural selection, which preserves what is useful and eliminates the harmful. Without mutation, evolution could not happen because there would be no variations that could be differentially conveyed from one to another generation. But without natural selection, the mutation process would yield disorganization and extinction because most mutations are disadvantageous. Mutation and selection have jointly driven the marvelous process that, starting from microscopic organisms, has spurted orchids, birds, and humans.

The theory of evolution manifests chance and necessity jointly intertwined in the stuff of life; randomness and determinism are interlocked in a natural process that has spurted the most complex, diverse, and beautiful entities in the universe: the organisms that populate the earth, including humans who think and love, are endowed with free will and creative powers, and are able to analyze the process of evolution itself that brought them into existence. This is Darwin's fundamental discovery, that there is a process that is creative though not conscious. And this is the conceptual revolution that Darwin completed: that everything in nature, including the origin of living organisms, can be accounted for as the result of natural processes governed by natural laws. This is nothing if not a fundamental vision that has forever changed how we perceive ourselves and our place in the universe.

6 Teleology and Teleological Explanations

Explanation by design, or teleology, is "the use of design, purpose, or utility as an explanation of any natural phenomenon" (*Webster's Third New International Dictionary*, 1966). An object or a behavior is said to be teleological when it gives evidence of design or appears to be directed toward certain ends. For example, the behavior of human beings is often teleological. A person who buys an airplane ticket, reads a book, or cultivates the earth is trying to achieve a certain end: getting to a given city, acquiring knowledge, or getting food. Objects and machines made by people also are usually teleological: a knife is made for cutting, a clock is made for telling time, a thermostat is made to regulate temperature. Similarly features of organisms are teleological as well: a bird's wings are *for* flying, eyes are *for* seeing, kidneys are constituted *for* regulating the composition of the blood. The features of organisms that may be said to be teleological are those that can be identified as adaptations, whether they are structures like a wing or a hand, or organs like a kidney, or behaviors like the courtship displays of a peacock. Adaptations are features of organisms that have come about by natural selection because they serve certain functions and thus increase the reproductive success of their carriers.

Inanimate objects and processes (other than those created by people) are not teleological in the sense just explained because we gain no additional scientific understanding by perceiving them as directed toward specific ends or for serving certain purposes. The configuration of a sodium chloride molecule (common salt) depends on the structure of sodium and chlorine, but it makes no sense to say that this structure is made up so as to serve a certain purpose, such as tasting salty. Similarly, the shape of a mountain is the result of certain geological processes, but it did not come about so as to serve a certain purpose, such as providing slopes suitable for skiing. The motion of the earth around the sun results from the laws of gravity, but it does not exist in order that the seasons may occur. We may use sodium chloride as food, a mountain for skiing, and take advantage of the seasons, but the use that we make of these objects or phenomena is not the reason why they came into existence or why they have certain configurations. On the other hand, a knife and a car exist and have particular configurations precisely in order to serve the purposes of cutting and transportation. Similarly, the wings of birds came about precisely because they permitted flying, which was reproductively advantageous. The mating display of peacocks came about because it increased the chances of mating and thus of leaving progeny.

The previous comments point out the essential characteristics of teleological phenomena, which may be encompassed in the following definition: *teleological explanations account for the **existence** of a certain feature in a system by demonstrating the feature's contribution to a specific property or state of the system*. Teleological explanations require that the feature or behavior contribute to the persistence of a certain state or property of the system: wings serve for flying; the sharpness of a knife serves for cutting. Moreover—and this is the essential component of the concept—this contribution must be the reason why the feature or behavior exists at all: the reason why wings came to be is because they serve for flying; the reason why a knife is sharp is that it is intended for cutting.

The configuration of a molecule of sodium chloride contributes to its property of tasting salty and therefore to its use as food, not vice versa; the potential use of sodium chloride as food is not the reason why it has a particular molecular configuration or tastes salty. The motion of the earth around the sun is the reason why seasons exist; the existence of the seasons is not the reason why the earth moves about the sun. On the other hand, the sharpness of a knife can be explained teleologically because the knife has been created precisely to serve the purpose of cutting. Motorcars and their particular configurations exist because they serve transportation, and thus can be explained teleologically. Many features and behaviors of organisms meet the requirements of teleological explanation.[4] The human hand, birds' wings, the structure and behavior of kidneys, and the mating displays of peacocks are examples already given.[5]

---

[4] Not all features of a car contribute to efficient transportation—some features are added for aesthetic or other reasons. But as long as a feature is added because it exhibits certain properties—such as appeal to the aesthetic preferences of potential customers—it may be explained teleologically. Nevertheless, there may be features in a car, a knife, or any other man-made object that need not be explained teleologically. That knives have handles may be explained teleologically, but the fact that a particular handle is made of pine rather than oak might simply be due to the availability of material. Similarly, not all features of organisms have teleological explanations.

[5] In general, as pointed out above, those features and behaviors that are considered adaptations are explained teleologically. This is simply because adaptations are features that come about by natural selection. Among alternative genetic variants that may arise by mutation or recombination, the ones that become established in a population are those that contribute more to the reproductive success of their carriers. "Fitness" is the measure used by evolutionists to quantify reproductive success. But reproductive success is usually mediated by some function or property. Wings and hands acquired their present configuration through long-term accumulation of genetic variants adaptive to their carriers. How natural selection yields adaptive features may be explained with examples where the adaptation arises as a consequence of a single gene mutation. One example is the presence of normal hemoglobin rather than hemoglobin S in humans. One amino acid substitution in the beta chain results in human hemoglobin molecules less efficient for oxygen transport. The general occurrence in human populations of normal rather than S hemoglobin is explained teleologically by the contribution of hemoglobin to effective oxygen transport and thus to reproductive success. A second example, the difference between peppered-gray moths and melanic moths is also due to one or only a few genes. The replacement of gray moths by melanics in polluted regions is explained teleologically by the fact that in such regions melanism decreases the probability that a moth be eaten by a bird. The predominance of peppered forms in non-polluted regions is similarly explained.

It is useful to distinguish different kinds of design or teleological phenomena. Actions or objects are *purposeful* when the end-state or goal is consciously intended by an agent. Thus, a man mowing his lawn is acting teleologically in the purposeful sense; a lion hunting deer and a bird building a nest have at least the appearance of purposeful behavior. Objects resulting from purposeful behavior exhibit *artificial* (or *external*) teleology. A knife, a table, a car, and a thermostat are examples of systems exhibiting artificial teleology: their teleological features were consciously intended by some agent.

Systems with teleological features that are not due to the purposeful action of an agent but result from some natural process exhibit *natural* (or *internal*) teleology. The wings of birds have a natural teleology; they serve an end, flying, but their configuration is not due to the conscious design of any agent. We may distinguish two kinds of natural teleology: *bounded*, or *determinate* or *necessary*, and *unbounded* or *indeterminate* or *contingent*.

Bounded natural teleology exists when specific end-state is reached in spite of environmental fluctuations. The development of an egg into a chicken is an example of bounded natural teleological process. The regulation of body temperature in a mammal is another example. In general, the homeostatic processes of organisms are instances of bounded natural teleology.[6]

Unbounded design or contingent teleology occurs when the end-state is not specifically predetermined, but rather is the result of selection of one from among several available alternatives. The adaptations of organisms are designed, or teleological, in this indeterminate sense. The wings of birds call for teleological explanation: the genetic constitutions responsible for their configuration came about because wings serve to fly and flying contributes to the reproductive success of birds. But there was nothing in the constitution of the remote ancestors of birds that would necessitate the appearance of wings in their descendants. Wings came about as the consequence of a long sequence of events, where at each stage the most advantageous alternative was selected among those that happened to be available; but which

---

Not all features of organisms need to be explained teleologically, since not all come about as a direct result of natural selection. Some features may become established by random genetic drift, by chance association with adaptive traits, or in general by processes other than natural selection. Proponents of the neutrality theory of protein evolution argue that many alternative protein variants are adaptively equivalent. Most evolutionists would admit that at least in certain cases the selective differences between alternative amino acids at a certain site in a protein must be virtually nil, particularly when population size is very small. The presence in a population of one amino acid sequence rather than another, adaptively equivalent to the first, would not then be explained teleologically. Needless to say, in such cases there would be amino acid sequences that would *not* be adaptive. The presence of an adaptive protein rather than a nonadaptive one would be explained teleologically; but the presence of one protein rather than another among those adaptively equivalent would not require a teleological explanation.

[6] Two types of homeostasis are usually distinguished—physiological and developmental—although intermediate conditions exist. Physiological homeostatic reactions enable organisms to maintain certain physiological steady state in spite of environmental fluctuations. The regulation of the concentration of salt in blood by the kidneys, or the hypertrophy of muscle owing to strenuous use, are examples of physiological homeostasis. Developmental homeostasis refers to the regulation of the different paths that an organism may follow in the progression from fertilized egg to adult. The process can be influenced by the environment in various ways, but the characteristics of the adult individual, at least within a certain range, are largely predetermined in the fertilized egg.

alternatives were available at any one time depended, at least in part, on chance events.[7]

Teleological explanations are fully compatible with (efficient) causal explanations.[8] It is possible, at least in principle, to give a causal account of the various physical and chemical processes in the development of an egg into a chicken, or of the physicochemical, neural, and muscular interactions involved in the functioning of the eye. (I use the "in principle" clause to imply that any component of the process can be elucidated as a causal process if it is investigated in sufficient detail and in depth; but not all steps in almost any developmental process have been so investigated, with the possible exception of the flatworm *Caenorhabditis elegans*. The development of *Drosophila* fruit flies has also become known in much detail,

----

[7] In spite of the role played by stochastic events in the phylogenetic history of birds, it would be mistaken to say that wings are not teleological features. As pointed out earlier, there are differences between the teleology of an organism's adaptations and the non-teleological potential uses of natural inanimate objects. A mountain may have features appropriate for skiing, but those features did not come about so as to provide skiing slopes. On the other hand, the wings of birds came about precisely because they serve for flying. The explanatory reason for the existence of wings and their configuration is the end they serve—flying—which in turn contributes to the reproductive success of birds. If wings did not serve an adaptive function they would have never come about, and would gradually disappear over the generations.

The indeterminate character of the outcome of natural selection over time is due to a variety of non-deterministic factors. The outcome of natural selection depends, first, on what alternative genetic variants happen to be available at any one time. This in turn depends on the stochastic processes of mutation and recombination, and also on the past history of any given population. (What new genes may arise by mutation and what new genetic constitutions may arise by recombination depend on what genes happen to be present—which depends on previous history.) The outcome of natural selection depends also on the conditions of the physical and biotic environment. Which alternatives among available genetic variants may be favored by selection depends on the particular set of environmental conditions to which a population is exposed.

It is important, for historical reasons, to reiterate that the process of evolution by natural selection is not teleological in the purposeful sense. The natural theologians of the nineteenth century erroneously claimed that the directive organization of living beings evinces the existence of a divine designer. The adaptations of organisms can be explained as the result of natural processes without recourse to consciously intended end-products. There is purposeful activity in the world, at least in humans; but the existence and particular structures of organisms, including humans, need not be explained as the result of purposeful behavior.

Some scientists and philosophers who held that evolution is a natural process erred, nevertheless, in seeing evolution as a determinate, or bounded, process. Lamarck (1809) thought that evolutionary change necessarily proceeded along determined paths from simpler to more complex organisms. Similarly, the evolutionary philosophies of Bergson (1907), Teilhard de Chardin (1959), and the theories of *nomogenesis* (Berg 1926), *aristogenesis* (Osborn 1934), *orthogenesis*, and the like are erroneous because they all claim that evolutionary change necessarily proceeds along determined paths. These theories mistakenly take embryological development as the model of evolutionary change, regarding the teleology of evolution as determinate. Although there are teleologically determinate processes in the living world, like embryological development and physiological homeostasis, the evolutionary origin of living beings is teleological only in the indeterminate sense. Natural selection does not in any way direct evolution toward any particular kind of organism or toward any particular properties.

[8] See my "Teleological Explanations in Evolutionary Biology," *Philosophy of Science*, 37 (1970): 1–15; and "The Distinctness of Biology," in *Laws of Nature: Essays on the Philosophical, Scientific, and Historical Dimensions*, Friedel Weinert, ed., (Berlin and New York: Walter de Gruyter, 1995), 268–85.

even if not yet completely.) It is also possible in principle to describe the causal processes by which one genetic variant eventually becomes established in a population by natural selection. But these causal explanations do not make it unnecessary to provide teleological explanations where appropriate. Both teleological and causal explanations are called for in such cases.

Paley's claim that the design of living beings evinces the existence of a divine designer was shown to be erroneous by Darwin's discovery of the process of natural selection, just as the pre-Copernican explanation for the motions of celestial bodies (and the argument for the existence of God based on the unmoved mover) was shown to be erroneous by the discoveries of Copernicus, Galileo, and Newton. There is no more reason to consider anti-Christian Darwin's theory of evolution and explanation of design than to consider anti-Christian Newton's laws of motion. Divine action in the universe must be sought in ways other than those that postulate it as the means to account for gaps in the scientific account of the workings of the universe.

The Copernican and Darwinian revolutions have jointly brought all natural objects and processes within the realm of scientific investigation. Is there any important missing link in the scientific account of natural phenomena? I believe there is, namely, the origin of the universe. The creation or origin of the universe involves a transition from nothing into being. But a transition can only be scientifically investigated if we have some knowledge about the states or entities on both sides of the boundary. Nothingness, however, is not a subject for scientific investigation or understanding. Therefore, as far as science is concerned, the origin of the universe will remain forever a mystery.

7 Coda: Science as a Way of Knowing

Science is a wondrously successful way of knowing. Science seeks explanations of the natural world by formulating hypotheses that are subject to the possibility of empirical falsification or corroboration. A scientific hypothesis is tested by ascertaining whether or not predictions about the world of experience derived as logical consequences from the hypothesis agree with what is actually observed.[9]

---

[9] Testing a scientific hypothesis involves at least four different activities. First, the hypothesis must be examined for internal consistency. An hypothesis that is self-contradictory or not logically well-formed in some other way should be rejected.

Second, the logical structure of the hypothesis must be examined to ascertain whether it has explanatory value, i.e., whether it makes the observed phenomena intelligible in some sense, whether it provides an understanding of why the phenomena do in fact occur as observed. An hypothesis that is purely tautological should be rejected because it has no explanatory value. A scientific hypothesis identifies the conditions, processes, or mechanisms that account for the phenomena it purports to explain. Thus, hypotheses establish general relationships between certain conditions and their consequences or between certain causes and their effects. For example, the motions of the planets around the sun are explained as a consequence of gravity, and respiration as an effect of red blood cells that carry oxygen from the lungs to various parts of the body.

Third, the hypothesis must be examined for its consistency with hypotheses and theories commonly accepted in the particular field of science, or to see whether it represents any advance with respect to well-established alternative hypotheses. Lack of consistency with other theories is not always ground for rejection of an hypothesis, although it will often be. Some of the greatest scientific advances occur precisely when it is shown that a widely-held and well supported hypothesis is replaced by a new one that accounts for the same phenomena that were explained by the preexisting hypothesis, as well as other phenomena it could not

Science as a mode of inquiry into the nature of the universe has been successful and of great consequence. Witness the proliferation of science departments in universities and other research institutions, the enormous budgets that the body politic and the private sector willingly commit to scientific research, and its economic impact. The Office of Management and the Budget of the U.S. government has estimated that fifty percent of all economic growth in the United States since the Second World War can directly be attributed to scientific knowledge and technical advances. Indeed, the technology derived from scientific knowledge pervades our lives: the high-rise buildings of our cities, thruways and long span-bridges, rockets that bring us to the moon, telephones that provide instant communication across continents, computers that perform complex calculations in millionths of a second, vaccines and drugs that keep bacterial parasites at bay, gene therapies that replace

---

account for. One example is the replacement of Newtonian mechanics by the theory of relativity, which rejects the conservation of matter and the simultaneity of events that occur at a distance—two fundamental tenets of Newton's theory.

Examples of this kind are pervasive in rapidly advancing disciplines, such as molecular biology at present. The so-called "central dogma" holds that molecular information flows only in one direction, from DNA to RNA to protein. The DNA contains the genetic information that determines what the organism is, but that information has to be expressed in enzymes (a particular class of proteins) that guide all chemical processes in cells. The information contained in the DNA molecules is conveyed to proteins by means of intermediate molecules, called messenger RNA. David Baltimore and Howard Temin were awarded the Nobel Prize for discovering that information could flow in the opposite direction, from RNA to DNA, by means of the enzyme reverse transcriptase. They showed that some viruses, as they infect cells, are able to copy their RNA into DNA, which then becomes integrated into the DNA of the infected cell, where it is used as if it were the cell's own DNA.

Other examples are the following. Until very recently, it was universally thought that only the proteins known as enzymes could mediate (technically "catalyze") the chemical reactions in cells. However, in 1989 Thomas Cech and Sidney Altman received the Nobel Prize for showing that certain RNA molecules act as enzymes and catalyze their own reactions. One more example concerns the so-called "co-linearity" between DNA and protein. It was generally thought that the sequence of nucleotides in the DNA of a gene is expressed consecutively in the sequence of amino acids in the protein. This conception was shaken by the discovery that genes come in pieces, separated by intervening DNA segments that do not carry genetic information; Richard Roberts and Philip Sharp received the 1993 Nobel Prize for this discovery.

The fourth and most distinctive test is the one I have identified, which consists of putting on trial an empirically scientific hypothesis by ascertaining whether or not predictions about the world of experience derived as logical consequences from the hypothesis agree with what is actually observed. This is the critical element that distinguishes the empirical sciences from other forms of knowledge: the requirement that scientific hypotheses be empirically falsifiable. Scientific hypotheses cannot be consistent with all possible states of affairs in the empirical world. An hypothesis is scientific only if it is consistent with some but not with other possible states of affairs not yet observed in the world, so that it may be subject to the possibility of falsification by observation. The predictions derived from a scientific hypothesis must be sufficiently precise that they limit the range of possible observations with which they are compatible. If the results of an empirical test agree with the predictions derived from an hypothesis, the hypothesis is said to be provisionally corroborated; otherwise it is falsified.

The requirement that a scientific hypothesis be falsifiable has been called by Karl Popper the *criterion of demarcation* of the empirical sciences because it sets apart the empirical sciences from other forms of knowledge. An hypothesis that is not subject to the possibility of empirical falsification does not belong in the realm of science. See my "On the Scientific Method, Its Practice and Pitfalls," *Hist. Phil. Life Sci.*, 16 (1994): 205–40.

DNA in defective cells. All these remarkable achievements bear witness to the validity of the scientific knowledge from which they originated.

Scientific knowledge is also remarkable in the way it emerges by way of consensus and agreement among scientists, and in the way new knowledge builds upon past accomplishment rather than starting anew with each generation or each new practitioner. Surely scientists disagree with each other on many matters; but these are issues not yet settled, and the points of disagreement generally do not bring into question previous knowledge. Modern scientists do not challenge that atoms exist, or that there is a universe with a myriad stars, or that heredity is encased in the DNA.

Science is a way of knowing, but it is not the only way. Knowledge also derives from other sources, such as common sense, artistic and religious experience, and philosophical reflection. In *The Myth of Sisyphus*, the great French writer Albert Camus asserted that we learn more about ourselves and the world from a relaxed evening's contemplation of the starry heavens and the scents of grass than from science's reductionistic ways.[10] One needs not endorse such a contrasting claim in order to uphold the validity of the knowledge acquired by non-scientific modes of inquiry. This can be simply established by pointing out that science, in the modern sense used here, dawned in the sixteenth century, but humankind had for centuries built cities and roads, brought forth political institutions and sophisticated codes of law, advanced profound philosophies and value systems, and created magnificent plastic art, as well as music and literature. We thus learn about ourselves and about the world in which we live and we also benefit from products of this non-scientific knowledge. The crops we harvest and the animals we husband emerged millennia before science's dawn from practices set down by farmers in the Middle East, Andean sierras, and Mayan plateaus.

It is not my intention in this essay's final section to belabor the extraordinary fruits of non-scientific modes of inquiry. But I have set forth the view that nothing in the world of nature escapes the scientific mode of knowledge, and that we owe this universality to Darwin's revolution. Here I wish simply to state something that is obvious, but becomes at times clouded by the hubris of some scientists. Successful as it is, and universally encompassing as its subject is, a scientific view of the world is hopelessly incomplete. There are matters of value and meaning that are outside science's scope. Even when we have a satisfying scientific understanding of a natural

---

[10] This point has been made with sensuous prose by John Updike in his recent *Toward the End of Time* (1995): "It makes no sense: all those blazing suns, red and swollen or white and shrunken or yellow like our moderate own, blue and new or black and collapsed, madly spinning neutron stars or else all-swallowing black holes denser yet, not to mention planets and cinder-like planetoids and picturesque clouds of glowing gas and dark matter hypothetical or real and titanic streaming soups of neutrinos, could scarcely be expected to converge exactly upon a singularity smaller, by many orders of magnitude, than a pinhead. The Weyl curvature, in other words, was very very very near zero at the Big Bang, but will be much larger at the Big Crunch. But, I ignorantly wonder, how does time's arrow know this, in our trifling immediate vicinity? What keeps it from spinning about like the arrow of a compass, jumping broken cups back on the table intact and restoring me, if not to a childhood self, to the suburban buck I was when still married." The mystery of aging and ultimate personal demise receive, in Updike's view, but little help from considerations of the immensity and endurance of the physicist's universe. I sometimes make the same point with a witticism that I once learned from a friend: "In the matter of the value and meaning of the universe, science has all the answers, except the interesting ones."

object or process, we are still missing matters that may well be thought by many to be of equal or greater import. Scientific knowledge may enrich aesthetic and moral perceptions, and illuminate the significance of life and the world, but these are matters outside science's realm.

On April 28, 1937, early in the Spanish Civil War, Nazi airplanes bombed the small Basque town of Guernica, the first time that a civilian population had been determinedly destroyed from the air. The Spanish painter Pablo Picasso had recently been commissioned by the Spanish Republican Government to paint a large composition for the Spanish pavilion at the Paris World Exhibition of 1937. In a frenzy of manic energy, the enraged Picasso sketched in two days and fully outlined in ten more days his famous *Guernica*, an immense painting of 25 feet, 8 inches by 11 feet, 6 inches. Suppose that I now would describe the images represented in the painting, their size and position, as well as the pigments used and the quality of the canvas. This description would be of interest, but it would hardly be satisfying if I had completely omitted aesthetic analysis and considerations of meaning, the dramatic message of man's inhumanity to man conveyed by the outstretched figure of the mother pulling her killed baby, bellowing faces, the wounded horse or the satanic image of the bull.

Let *Guernica* be a metaphor of the point I wish to make. Scientific knowledge, like the description of size, materials, and geometry of *Guernica*, is satisfying and useful. But once science has had its say, there remains much about reality that is of interest, questions of value and meaning that are forever beyond science's scope.

EVALUATING THE TELEOLOGICAL ARGUMENT FOR DIVINE ACTION

Wesley J. Wildman

1 *Introduction*

1.1 *Divine Action and Evolutionary Biology*

There are many ways to conceptualize divine action in nature and history, ranging from attribution to God of natural-law suspending miracles or natural-law conforming activity, to virtual identification of the laws and processes of nature with the initiating creative act of God or with the divine nature itself. It must be recognized from the outset that some of these conceptions cannot possibly profit from insights drawn from the natural sciences, including evolutionary biology. One example is Rudolf Bultmann's assertion that divine action occurs only in the realm of human existence and leaves no traces in history and nature; this depends upon a dualism of being or language. Another is John Locke's reliance on the miraculous as a mode of special divine action. To the extent that miraculous and various forms of dualistic theories of divine action are defensible—and I think they are if the right approach is taken—a theory of divine action that is independent of considerations from the natural sciences, including evolutionary biology, is still feasible. Theories of divine action that take the natural sciences to have something crucial to offer, however, have much better chances of achieving the virtues of specificity and plausibility.

If we accept this, then we will be inclined to try to establish substantive connections between theories of divine action and all kinds of scientific theories, including evolutionary biology. One type of connection begins with the appearance of purposes or ends in nature and attempts to construe this as evidence of the reality of divine action by means of the argument that such apparent ends indicate genuine teleology in natural objects and processes, and that this teleology (in any of a number of possible forms) is the mode of God's action. I shall call this argument "the teleological argument for divine action." The English divine William Paley appealed to the teleological argument for divine action when he drew his famous analogy between a watch and the wondrous structures and processes of nature: both demand a designer.[1] Likewise, Alfred Russel Wallace, co-discoverer with Charles Darwin of the principle of natural selection, found the complexity of some features of biology so amazing that he invoked an active designer God to explain it.[2] This peculiarly aggressive form of the teleological argument for divine action (the so-called design argument) is comparatively rare in our day because evolutionary biology has made impressive advances in explaining how complex organs and biological systems developed from simpler forms. That has made it exceedingly difficult to attempt to move from the *products* of biological evolution to divine action by means of the argument that the beauty and functionality of those products is so wonderful as to demand a divine mind whose intention they are; or from the *process* of biological

---

[1] William Paley, *Natural Theology—of Evidences of the Existence and Attributes of the Deity Collected from the Appearances of Nature*, 2nd ed. (Oxford: J. Vincent, 1828).

[2] This is so according to Michael Shermer, *Why People Believe Weird Things* (New York: W. H. Freeman, 1997), who cites an article of Wallace in *Quarterly Review* (April, 1869).

evolution to divine action by means of the argument that the evolutionary process requires occasional divine moderation, adjustment, acceleration, or specific directing to account for the forms of life that exist. The theory of evolution is increasingly well justified in asserting that wonderful forms of life result from the evolutionary process regardless of what any mind intends, and that this process is automatic, in need of no occasional, special adjustments.[3] The argument from design has been thoroughly undermined as a result.

The teleological argument for divine action, however, has more modest, more viable forms. One is driven by the question of the significance and possible "ultimate purpose" of the evolutionary trajectory that has produced human life.[4] Another finds a congenial starting point in one of the intuitions guiding neo-Darwinian evolutionary theory, namely, that increases in biological complexity probably occur at different speeds (albeit virtually always gradually) within the evolutionary process,[5] suggesting the possibility of higher-level laws of complexity,[6] and leading to the question of whether teleological categories are needed for the adequate description of the conditions for the possibility of punctuation in evolutionary equilibrium. Yet another seeks to move from the *laws and capacities of nature*—the conditions of the possibility of biological evolution—to the reality of divine action by means of an argument that nature is purposefully designed by God to have the laws and capacities it has, in which case divine design of nature is the primordial divine act.[7] There are other motivations for exploring the teleological argument for divine action, but what has been said is enough to show that this type of connection between evolutionary biology and divine action might be well worth examining closely. The special virtue of the teleological argument for divine action is its promise of *relevant, detailed support for the reality of divine action*. Other advantages of centralizing the category of teleology when examining the relation between divine action and evolutionary biology will become evident later.

1.2 *The Argument of this Paper and its Significance*

The argument of this paper leads to my provisional conclusion that no relevant, detailed, supportive relation between evolutionary biology and the reality of divine action is possible using this approach. This is a negative result as far as the teleological argument for divine action is concerned, but it does not imply that evolutionary biology bluntly rebuts the claim that God acts in nature and history.

---

[3] For marvelous descriptions of many particular case studies, see Richard Dawkins, *The Blind Watchmaker* (New York, London: W. W. Norton, 1986), and even more impressively, idem, *Climbing Mount Improbable* (New York, London: W. W. Norton, 1996).

[4] See, for example, Pierre Teilhard de Chardin, *The Phenomenon of Man* (London: Collins, New York: Harper, 1959; tr. from 1955 French ed.), and *Man's Place in Nature* (London: Collins, New York: Harper, 1966; tr. from 1956 French ed.).

[5] See, for example, Stephen Jay Gould, *Wonderful Life: The Burgess Shale and the Nature of History* (New York: W. W. Norton, 1989), and Niles Eldredge, *Macroevolutionary Dynamics: Species, Niches, and Adaptive Peaks* (New York: McGraw-Hill, 1989), and idem, *Time Frames: The Rethinking of Darwinian Evolution and the Theory of Punctuated Equilibria* (New York: Simon and Schuster, 1985).

[6] This topic is explored perhaps most vigorously by Stuart Kauffman. See *The Origins of Order: Self-organization and Selection in Evolution* (Oxford and New York: Oxford University, 1993), and idem, *At Home in the Universe: The Search for the Laws of Self-Organization and Complexity* (New York and Oxford: Oxford University, 1995).

[7] See, for example, Arthur Peacocke, *Theology for a Scientific Age: Being and Becoming—Natural, Divine and Human*, 2nd ed. (Philadelphia, Pa.: Fortress, 1993).

Rather, evolutionary biology is one of many considerations that can influence theories of divine action without having much evidentiary effect either way.

This argument will merely confirm what many theorists of divine action seem already to hold, but it may challenge the assumptions of some. For example, those who think that an argument can be constructed leading from apparent purposes in nature to the activity of God will need to refute the conclusion of the present argument. Similarly, those who think that evolution—particularly its portrayal of ends in nature as epiphenomenal side-effects of the evolutionary process—destroys affirmations of divine action will need to grapple with the argument of this paper.

It might seem, therefore, that I am preaching to the converted, and that the argument is only problematic for those whose views can be depreciated in many other ways besides that taken here. Making the argument has other benefits, however. Most importantly, it exhibits in detail a small part of the diversity of possible connections between evolutionary biology and theories of divine action, and drives home the scope of the metaphysical ambiguity that attends every step of the movement from one to the other.[8] This means that the argument may be of value even to those who would be inclined at the outset to agree with its conclusion.

The argument from apparent ends in nature to the affirmation of the reality of divine action—the teleological argument for divine action—has three logically distinguishable stages. The first (section 3) must conclude that apparent ends in nature are indications of genuinely teleological capacities of natural objects and processes. This necessarily involves grappling with the problem of reductionism, and with the evolutionary critique of teleological terminology. The second (section 4) must situate the affirmation of the genuinely teleological capacities of natural objects and processes in a wider metaphysical context that is rich enough to refer to fundamental teleological principles, because it is only through metaphysical generalization that particular teleological capacities can be connected with God, who is assumed to be the ontological ground of such capacities, or at least metaphysically connected with them.[9] The third (section 5) must show that these fundamental teleological principles support particular theories of divine action. Any argument from apparent ends in nature to the reality of divine action includes these three stages.

Some preliminaries are also needed. To that end, I address some basic philosophical concerns about the definition and application of "having an end," and propose a schema that draws attention to key features of a number of views of teleology in the evolutionary process (section 2).

Each stage of the evolutionary argument for divine action is negotiable only with complex and subtle argument. I will show that each of the intermediate conclusions required for the overall argument to work cannot be secured without recourse to presuppositions that have far more influence on the conclusions than do

---

[8] I am using "metaphysics" throughout this essay roughly in the sense of Charles Hartshorne, for whom it encompasses propositions true in every possible world. "Cosmology" can be distinguished from this as involving propositions characterizing the actual world of our experience; it is less abstract than metaphysics. Ontology concerns the meaning and value of being itself and includes inventory questions about what exists.

[9] This is not the place to defend the possibility of such reflection. Suffice to say that I do not suppose that Kant's strictures on metaphysics can be set aside lightly. On the contrary, the pragmatism of C. S. Peirce offers a way around them while taking them with proper seriousness. My interpretation of the task of inquiry, and my general indebtedness to pragmatism (not, however, to Richard Rorty's neo-pragmatism), is laid out briefly in "Similarities and Differences in the Practice of Science and Theology," *CTNS Bulletin* 14.4 (Fall, 1994).

considerations from evolutionary biology. That is, there is no chain of sound implications from appearances of ends in nature, to the reality of the teleological capacities of natural objects and processes, to the identification of fundamental teleological principles in a wider metaphysical theory, to particular theories of divine action; each proposition is crucially under-determined by the previous one, and other premises are required to make the implications valid. This paper will examine what some of those additional premises might be. It will turn out that, for every such premise that facilitates the movement of the teleological argument for divine action to its next stage, there are many equally plausible premises that lead not in the direction of divine action but in other directions altogether.

In concluding this introduction, I want to make two further remarks. First, with regard to limitations, because this examination of teleology will concentrate on the place of teleology in biological evolution, *I forgo the chance to state or criticize cumulative arguments for a fundamentally teleological universe*—and this is unquestionably where many of the debates in the theological literature focus their attention. The narrowing of focus is needed, however, and it does not interfere with my more limited goal of assessing the teleological argument for divine action.[10]

Second, with regard to my motivation, this essay seeks to do partial justice to the many criticisms of the very idea of divine action in history and nature. These depreciations range from the denial of the reality of God and the God-affirming denial that "divine action" is a meaningful phrase, to the rejection of nature and history as metaphysically significant categories, as a result of the contention (typical of much Indian and Buddhist philosophy) that ultimate reality lies deeply beneath its misleading natural, historical appearance. Centralizing the category of teleology helps because it is possible to specify its meaning for a wide variety of metaphysical and religious traditions; the idea of divine action cannot be generalized to the same degree. After conclusions about the conceptual relations between teleology and biological evolution are drawn, the possibility will then exist of relating these conclusions to other concepts, such as divine action—or, for that matter, the Indian philosophical concepts of *samsara* and *maya*, though I will not be pursuing this.[11] The teleological argument for divine action follows this procedure precisely.

2 *Speaking of Teleology*

Teleological categories have been generally out of favor in the West for some time, so it is necessary to clear some terminological ground.

2.1 *The Meaning of "Having an End"*

The ancient Greek philosopher, Aristotle (384–322 B.C.E.),[12] contended that the essence (and so the behavior) of a thing is understood when four questions about it

---

[10] For an example of such an ambitious undertaking, see William R. Stoeger's paper in this volume.

[11] This two staged approach to the problem of teleology and divine action has been adopted before to good effect, notably and influentially as the distinguishing principle for the two books constituting Paul Janet, *Final Causes*, tr. from the 2nd French ed. by William Affleck (New York: Charles Scribner's Sons, 1892; 1st French ed., 1876).

[12] The following translations of Aristotle's works are referred to or quoted in what follows: *Physics (Physica)*, tr. R. P. Hardie and R. K. Gaye; *Metaphysics (Metaphysica)*, tr. by W. D. Ross; *On the Parts of Animals (De partibus animalium)*, tr. William Ogle; *On the Gait of Animals (De incessu animalium)*, tr. A. S. L. Farquharson; and *On the Generation of Animals (De generatione animalium)*, tr. Arthur Platt.

can be answered: What is it made of? What are its essential attributes? What brought it into being? What is its purpose? (*Physics* II.3, 194^b.16–195^a.2; *Metaphysics* V.2, 1013^a.24–1013^b.2)[13] These questions correspond to what scholastic philosophers aptly called the material, formal, efficient and final causes.[14] The fourth of Aristotle's questions is answered by identifying the "end" of a thing. But how was this conceived?

Aristotle implicitly defined an end when he spoke of the cause of a thing "in the sense of end or 'that for the sake of which' a thing is done. For example, health is the cause of walking about" (*Physics* II.3, 194^b.33). Thus, an end "causes" its means by virtue of the fact that the means (as cause) are capable of securing that end (as effect). Now, the end, since it lies in the future relative to the means, cannot obviously be their cause, though the *idea* of the end can certainly be the cause of the means. Thus, we arrive at a definition: *An end, E, is the cause of means, M, insofar as E is the foreseen effect of M.* This is a common definition of "having an end," and fits Aristotle's view rather well. It captures the meaning of "end" through being explicit about how it is that ends cause.

The usual way of allowing for literal application of teleological categories is through the concept of *intending*: since human beings and some other animals *intend*, their behavior is genuinely purposeful and causal. In the context of intentional agents, therefore, since "foreseeing effects" can be spoken of literally, the definition of end just given is uncontroversial. Extending this definition to cover some cases of habitual, preconscious, unconscious, goal-directed, and even some acquired and instinctive behavior poses comparatively few problems. Outside the realm of intending and its physiological derivatives, however, making sense of "having an end" is far more difficult. Aristotle accepted human beings as free agents and allowed human intending to be the basis of many kinds of events that are for the sake of something, such as habitual and what we would call unconscious behaviors. Contemporary philosophy will go that far with Aristotle, but rarely much further. In particular, Aristotle's attribution of ends to inanimate natural processes is genuinely difficult to justify.

Aristotle was fully aware of the problems with this more ambitious usage of "end." In the context of a discussion of the various kinds of processes that have ends,[15] Aristotle dealt with the problem of assigning ends to spontaneous natural processes—an important consideration in the context of evolutionary biology—by distinguishing between intelligent and natural ends. When an end is consciously entertained or habitually assumed by a moral agent, it can be spoken of as an intelligent end; other ends are natural. What intelligence does for the one by way of foreseeing, nature does for the other in a kind of natural, teleological analogue of foreseeing (*Physics* II.6, 198^a.1–12). The definition of end given above will work for every sphere of nature if this natural, teleological foreseeing is legitimate.

Philosophers who have affirmed ends in this more ambitious way have also tried to offer compelling arguments for their interpretations. Alfred North White-head's argument turned on a sophisticated theory of causation that had the attractive virtue of claiming to solve the freedom-determinism problem. Aristotle's argument

---

[13] References are in the form book.chapter, page^{column}.line of the Berlin Greek text.

[14] Aristotle himself used only nouns or nominal phrases to designate the four causes (e.g. *to telos*); the adjectival forms are later Latin creations.

[15] For details of the classification, see *Physics* II.5, and especially the discussion of spontaneous and chance processes in *Physics* II.6, 197^b.18–21.

flowed from a grand teleological vision of reality in which nature itself is a vast teleological organism and each object and process has natural ends fitted to the actualization of its natural potential—a view notable for its explanatory and ethical power. In these two cases and all others of which I am aware, the reality of natural ends is affirmed as a consequence of a wider metaphysical theory that commends itself based on numerous considerations apart from the question of the reality of natural ends.

2.2 A Criterion for "Having an End"

If we are to use phrases such as "the reality of natural ends"—and evaluating the teleological argument for divine action demands such talk—there needs to be a criterion for "endedness" in natural objects and processes that does not beg the question about the reality of ends in nature. Forcing the definition of "having an end" to serve as a criterion for detecting apparently-ended natural objects and processes does not meet this condition. A widely held criterion for "endedness" that does meet this condition is as follows: *a natural process or object can be said to be ended (to have an end) if it exhibits a tendency toward some endpoint that persists through changing circumstances.*[16]

This criterion is especially apt for designating processes that might be called "closed-ended," in which the process really does have a particular endpoint. It is less useful for the situations of most interest in biology, the "open-ended" processes, because an open-ended process potentially yields very different outcomes, and so there can be no definitely known "endpoint."[17] Open-ended processes, however, do have the predictable appearance of closed-ended processes in their stable regimes (typically when they are close to thermodynamic equilibrium). In practice, therefore, it is possible to apply the criterion for endedness even when we are dealing with open-ended processes of some kinds. We just need to remember to allow for the possibility that the final outcome of an apparently ended process may not be known in advance, even when a proximate endpoint is known, because of: 1) the complexity of the process; 2) the ability of the environment to alter available end states of the process; or, 3) the role that chance factors play in the transition of a system between relatively stable regimes of behavior. Our criterion might not capture all open-ended processes, therefore, but it does include all of the processes with a relatively stable

---

[16] Something akin to this is defended in R. B. Braithwaite, *Scientific Explanation* (Cambridge: Cambridge University, 1953), and in many other writers. Dawkins, who uses the term "designoid" for "apparently designed," introduces statistical measures that reflect human intuitions about what is designed and what is not designed. This approach seems useful also for furnishing an approach to apparent endedness. See Dawkins, *Climbing Mount Improbable*, chapter 1.

[17] This appears to be the reason for Francisco J. Ayala's approach to the problem. He begins with a vague and general criterion: "An object or a behavior is said to be teleological or telic when it gives evidence of design or appears to be directed toward certain ends." He then partially overcomes the vagueness of this definition by distinguishing between artificial (external) teleology, due to deliberate purposefulness, and natural (internal) teleology, when no deliberate purposefulness is involved; he makes a distinction within the category of natural teleology between determinate teleology (what I am calling closed-endedness) and indeterminate teleology (open-endedness). The vagueness of the initial definition is understandable in view of what it must cover. See Theodosius Dobzhansky et al., eds., *Evolution* (W. H. Freeman, 1977), 497; reprinted as "Teleological Explanations" in Michael Ruse, ed., *Philosophy of Biology* (New York: Macmillan, London: Collier Macmillan, 1989). See also Ayala's contribution to this volume.

appearance, whether part of a larger open-ended process or not. In view of what we need it for, this is sufficient.

With this criterion in place, we have selected out a class of nominally ended natural objects and processes that is even richer than Aristotle's class of events and processes that are for the sake of something. It is important to note that this class is stratified, as in Table I.[18]

Realm of Nature	Characteristics
Self-conscious animals (human beings)	Conscious, deliberate (e.g. strategizing)
	Habitual (e.g. walking, talking)
	Preconscious (e.g. subliminal perception)
	Unconscious (e.g. projecting desires)
Other higher animals	Goal-directed (e.g. seeking food)
All animals	Acquired (e.g. learning skills)
	Instinctive (e.g. self-protection, mating)
Human-made objects	Feedback-guided, goal-seeking (e.g. thermostat, metal detector)
Biological organisms	Appropriately systemic (e.g. operation of organs, body parts)
Everything	Functional (e.g. anything in its functional aspect)

Table 1. Hierarchical class of events, objects, and processes with nominal ends

The items at the top of the table are better placed to win assent from contemporary thinkers to the thesis that teleological categories are *necessary* for adequate explanations, while those lower in the table are less well placed. Unsurprisingly, it is Aristotle's intelligent ends that are at the top of the table (especially conscious and habitual behavior).

For each of the objects and processes falling under one of the categories in this table, it is possible—and this is the point of the criterion for endedness—to ask: Is the appearance of endedness in this instance due to real ends in nature, or is it merely a misleading epiphenomenon of complex natural processes without ends? If the epiphenomenal explanation is to be preferred in every case, then this constitutes a strong argument for eliminating the more loaded usages of teleological language from all descriptions and explanations of nature, though of course speaking of ends and purposes may still serve a useful heuristic function. If in some cases the explanation for apparent ends is that they are real, then teleological categories will be needed for adequate explanations of the processes in question, and some mediating theory of causality and teleology will be needed also.

2.3 Dangers and Virtues of Teleology

If the teleological argument for divine action is to move even a step forward, then it is necessary first to deflect a fundamental objection to teleology. To that end, let us venture a brief evaluation of Aristotle's teleological vision so as to illumine the modern suspicion of teleological categories.

According to Aristotle, everything has a natural, in-built purpose, a purpose fitted to its nature (the ambiguity of the English word "nature" reflects Aristotle's viewpoint). This purpose is expressed in the form nature gives to each thing, which makes the purposes of things immanent within the things themselves. For example, the oak tree has a life principle that explains both its development from an acorn, and

---

[18] Edwin Levy presents a hierarchy that is a subset of this one in "Networks and Teleology," 159–186, in Mohan Matthen and Bernard Linsky, eds., *Philosophy and Biology*, Canadian Journal of Philosophy, Supplementary Volume 14 (Calgary: University of Calgary, 1988).

its shape, color, and acorn-producing capacity. This life principle cannot be abstracted from the oak, as if the purpose of the acorn-to-oak growth process were in the mind of some other being, or the tree's death were the result of the withdrawal of the life principle from the tree's essence. By extension, therefore, nature can be likened to a vast, integrated, purposive organism, with ends fitted to each thing for the optimal fulfillment of that thing's potential: "nature is a cause, a cause that operates for a purpose" (*Physics* II.8, 199^b.32). The final teleological principle of this great organism resides with a perfectly unified, fully actualized prime mover that transcends the world, and toward which the world is drawn.[19]

This magnificent vision of reality was the basis for much of Aristotle's philosophical achievement, from his ethics to his studies of plants and animals. In practice, however, his own answers to the four questions that were intended to guide the investigation of nature (the four "causes") were of limited use because he failed to maintain a balance among them, emphasizing final causes and muting efficient causes.

Aristotle's studies of plants and animals,[20] for example, while taxonomically brilliant, were occasionally contaminated with implausible explanations of behavior in terms of supposed natural purposes. Now, it must be admitted that Aristotle had generally excellent success in interpreting the parts and motion of animals with the aid of such telic assumptions as: "Nature makes the organs for the function, and not the function for the organs" (*On the Parts of Animals*, IV.12, 694^b.13); "Nature never fails nor does anything in vain so far as is possible in each case" (*On the Generation of Animals*, V.8, 788^b.22); "Nature creates nothing without a purpose, but always the best possible in each kind of living creature by reference to its essential constitution" (*On the Gait of Animals*, 2, 704^b.15), and "Nature never makes anything that is superfluous" (*On the Parts of Animals*, IV.11, 691^b.4). The very fact that this kind of interpretation is ever successful is testimony to the ubiquity of *apparent* purpose in nature. However, Aristotle was so enamored with his guiding presuppositions about nature's purposes that he described mollusks as a "mutilated class" owing to their odd means of locomotion; he called the seal and bat "quadrupeds but misshapen" (*On the Gait of Animals*, 19, 714^b.10–15); and he explained the small amount of blood in the chameleon by means of its timid nature (inferred from frequent color changes), and the principle that "fear is a refrigeration, and results from deficiency of natural heat and scantiness of blood" (*On the Parts of Animals*, IV.11, 692^a.25).

---

[19] This makes the prime mover something like the life principle of the entire cosmos, which might seem inconsistent with Aristotle's rejection of life principles in living beings. It is his view nonetheless. This tension is closely related to a complex corner of Aristotle interpretation having to do with his distinction between active and passive reason. Aristotle's need to find in human beings something akin to Plato's indestructible soul is the basis for attributing a mixture of active and passive reason to them. Active reason suggests a life principle that requires no body and it is active reason that is generalized and perfected in Aristotle's concept of God. Nonetheless, Aristotle insists in relation to all beings apart from God that their soul is their principle of unity and not a mystical life principle separable from their constitution as formed matter.

[20] The works on zoology include, in addition to those mentioned above, *On the Motion of Animals (De motu animalium)*, tr. A. S. L. Farquharson; *History of Animals (Historia animalium)*, tr. D'Arcy Wentworth; and the so-called *Short Physical Treatises (Parva naturalia)*.

Likewise, in his ethics,[21] Aristotle was insufficiently suspicious of his readings of the natural purposes of certain types of people. For example, Aristotle held that women should be treated with honor fitting to their place as the helpers of men; this was (we might say) the Golden Mean between Plato's admission of them to the ruling class and the common treatment of women as virtual slaves. This view of the place of women was determined by Aristotle's view of their natural purpose, which flowed from his interpretation of their essential nature. He assumed, on the basis of experience, and admitting a few "contrary-to-nature" exceptions, that women have a partially ineffective reasoning faculty. Slaves have no reason at all, and so need to be ruled outright, according to Aristotle, but the kind of partially irrational soul possessed by women determines that their natural purpose and thus their social place is to be the helpers of, and ruled by, men, who have fully functional faculties of reason, and can regulate the irrational tendencies of women.[22] Though Aristotle's view, in his context, was relatively generous toward women—though not to slaves, whom he regarded as "living tools" and "living possessions" (*Politics* I.4, $1253^b.23–1254^a.17$)—it is evident that his analyses of purposes are too often indistinguishable from conservative rationalizations of social practices he found desirable.

The more dubious aspects of Aristotle's use of final causes were amplified in much subsequent philosophy, some of which was not characterized by a steadying critical instinct to the extent that Aristotle's was. Thus, it is not surprising that modern Western thinkers have frequently been quite aggressive in banishing consideration of purposes from most natural, and even many ethical, inquiries. This anti-teleological posture has secured many desirable results, including protection of scientific research and social policy from the negative effects of unchecked speculation, and increased efficiency of the powerful process of scientific discovery and theorizing. The main reason for the decline of interest in teleology, however, is that analyses based on efficient causation proved to be far more specific and fruitful than teleological analyses. Instead of resting content with the statement that the final, internal purpose of an acorn is to grow into an oak, for example, the dynamism of natural change is now explained primarily through efficient causes: the acorn's genetic capacities decisively constrain the chemical processes of growth made possible by the causal interactions between acorn and environment. That is an explanation that fosters further detailed development and leads out into testable consequences, so it is far better suited to scientific theorizing.

This abandonment of the explanatory contribution of final causes in favor of the greener pastures of efficient causes also has a significant disadvantage. It obscures some important perspectives that the teleological approach keeps in the forefront, such as the question of the *ultimate* basis for the amazing capacities of acorns. For this reason, final causes have never vanished into the realm of philosophical curiosities. There have always been thinkers willing to argue forcefully that ultimately satisfying explanations of nature cannot be achieved in isolation from the category of purpose, that ethics is untenable without final causes, or that God's action in the world is impossible to discern if teleological categories are not admitted into metaphysical explanations, if not physical ones. Moreover—and for my

---

[21] See especially *Nicomachean Ethics (Ethica Nicomachea)*, tr. W. D. Ross; and *Politics (Politica)*, tr. Benjamin Jowett.

[22] See *Politics* I, esp. I.12–13, $1259^a.37–1260^b.25$; and *Nicomachean Ethics* V.11, $1138^b.5–9$, VIII.11, $1161^a.10–1161^b.10$.

purposes this is crucial—ends in nature *seem* to be everywhere, and denying their reality on the basis of an efficient-causal reduction carried out only part-way in theoretical detail and the rest of the way in the imagination is probably hasty, and is certainly difficult to justify. Thus, there is no way preemptively to block the teleological argument for divine action by invoking the achievements of modernity against Aristotelian natural science and ethics.

2.4 *A Schema for Views of Teleology in the Evolutionary Process*

There have been many systematic interpretations of teleology, especially from West Asian philosophical traditions, but also from South and East Asia. Most influential in the West, without question, has been Aristotle's vision of nature as a vast teleological organism. Most notable during the last two centuries is Hegel's theory of the ever more profoundly reflexive, and logically determined, self-realization of *Geist* in history. Important in the last half of the twentieth century has been Whitehead's cosmology, which offers an elaborately worked out and fundamentally teleological doctrine of causality in which the basic entities (occasions) of reality become actual through resolving prehended, antecedent influences under the sway of an initial aim fitted to the character of each occasion. As to South Asian philosophy, the basic concepts shared by many Indian philosophical schools, both Hindu and Buddhist, also yield the possibility of the literal application of teleological categories, though in quite a different way. In this case, generally speaking, nature is for, and only for, the sake of the liberation of souls; indeed, liberation consists in attaining the discrimination required fully to grasp (to put it in Hindu terms) that human consciousness is fundamentally different from, and actually more real than, nature. Another instance of understanding reality in teleological categories is the East Asian conception of the natural and social world as fundamentally a flowing together of events in harmony—originally and always at least potentially—with some larger cosmic pattern that is usually described in terms of a heaven or principle.

There are other teleological visions of the world that, like these, have been developed in great detail over many centuries by philosophical traditions whose achievements are comparable in grandeur. All of these theories remain useful for rendering teleological categories literally applicable to natural objects and processes. Recent years have seen newer theories that are typically less philosophically developed but peculiarly well placed to deal with current understandings of nature from biology, chaos and complexity, and self-organization. In fact, the creation of these contemporary views of teleology in the evolutionary process has been inspired as much by evolutionary biology and the natural sciences generally as by the need to extend long-standing philosophical traditions, and so they are of special interest for my purposes. This is not the place to offer a survey of such views, however, because the book within which this article stands already contains a number of them. It is enough to note that they have been marked in recent years by a number of controversies that are relevant (positively or negatively) to the question of divine action. A review of four of these controversies permits a schematization of part of the range of theoretically possible positions on the issue of teleology, evolution, and divine action.

The First Dispute: Teleology or No Teleology? To take a position on this issue is to answer the question posed in the first stage of the teleological argument for divine action about the possibility of finding a place for teleological categories in furnishing an ultimately satisfying account of apparent ends in nature. Obviously enough, apparent ends can be taken to be *only apparent*, leading to the denial that there is a fundamental, teleological principle at work in nature. This is the view of Richard Dawkins, who expounds Darwin's theory precisely to show that ends in nature are only apparent. Picking up (as it were) William Paley's analogy of the watch, mentioned above, Dawkins' thesis is as follows:

> The analogy between the telescope and the eye, between watch and living organism, is false. All appearances to the contrary, the only watchmaker in nature is the blind forces of physics, albeit deployed in a very special way. A true watchmaker has foresight: he designs his cogs and springs, and plans their interconnections, with a future purpose in his mind's eye. Natural selection, the blind, unconscious, automatic process which Darwin discovered, and which we now know is the explanation for the existence and apparently purposeful form of all life, has no purpose in mind. It has no mind and no mind's eye. It does not plan for the future. It has no vision, no foresight, no sight at all.[23]

Dawkins freely admits that nature appears to be full of ends.[24] It is this apparent design he intends to explain, and to explain away *as apparent*, without impugning its beauty and complexity. But the entire argument is directed toward the conclusion that the explanatory reduction achieved by Darwin and later theorists can be extended to an ontological reduction.[25] A more strident or colorful statement of this case can scarcely be imagined.

The opposite position is that at least some objects and process in nature appear to have ends because they really do have them in some metaphysically profound sense. This view has been affirmed in a variety of ways, corresponding to various strategies for locating the teleological grounding for apparent ends in natural objects and processes, as we shall see.

Between these two opposed views lie intermediate possibilities that take a yes-and-no position in relation to the first dispute. An intriguing instance of this is based on—perhaps it is an imaginative extension of—Jacques Monod's account of molecular and evolutionary biology in *Chance and Necessity*.[26] Monod analyzes the appearance of ends in nature specifically in the realm of living beings. Living beings, he argues, have two distinguishing primary characteristics: teleonomy (being endowed with apparent purposes or projects), and reproductive invariance (the ability invariantly to pass information expressed in the structure of a living being

---

[23] Dawkins, *The Blind Watchmaker*, 5.

[24] *The Blind Watchmaker*, chapter 2 is an extended appreciation of the apparently designed character of so much in nature; this is also a prominent theme throughout his *Climbing Mount Improbable*.

[25] In *The Blind Watchmaker*, Dawkins' approach begins from an idiosyncratic definition of biological complexity, which functions as a criterion for the class of objects and processes with apparent ends. After dismissing a few problematic alternatives, Dawkins defines a complex object as "statistically improbable in a direction that is not specified with hindsight" (15). He is quite prepared to work with an alternative definition for the sake of argument, however, as the crux of his argument lies elsewhere. Note that he includes human-made objects as "honorary living things" (1–2, 10). He considers that this class of objects and processes will be explained when an account of it is provided that is consistent with, and relies on nothing other than, the basic laws of physics. Chapter three is devoted to spelling out the special way that the laws of physics are deployed in evolutionary theory.

[26] Jacques Monod, *Chance and Necessity: An Essay on the Natural Philosophy of Modern Biology*, tr. from the French by Austryn Wainhouse (New York, Alfred A. Knopf, 1971). Monod advances a form of existentialist polemic against all manner of vitalisms and animisms, superstitions, and self-deceptive metaphysics, in the name of a materialist "ethic of knowledge."

through reproduction to other living beings). The interlocking of these two characteristics is what makes possible the increase of complexity through invariant reproduction in spite of the second law of thermodynamics:

> [I]nvariance is bought at not one penny above its thermodynamic price, thanks to the perfection of the teleonomic apparatus which, grudging of calories, in its infinitely complex task attains a level of efficiency rarely approached by man-made machines. This apparatus is entirely logical, wonderfully rational, and perfectly adapted to its purpose: to preserve and reproduce the structural norm. And it achieves this, not by departing from physical laws, but by exploiting them to the exclusive advantage of its personal idiosyncrasy. (20–21)

This interlocking of teleonomy and invariance is not only wonderful, Monod argues, but also in flagrant contradiction with what he calls the "cornerstone" of the scientific method: nature's objectivity. While objectivity, on Monod's reading, requires "the *systematic* denial that 'true' knowledge can be got at by interpreting phenomena in terms of final causes" (21), it also calls for the frank recognition that living organisms realize and pursue purposes in their structure and activity (22).

This is the epistemological contradiction that biology sets out to resolve. And resolve it biology does, with its answer that invariance is logically and physically *prior* to teleonomy. In a beautiful sentence, Monod describes this solution as:

> the Darwinian idea that the initial appearance, evolution, and steady refinement of ever more intensely teleonomic structures are due to perturbations occurring in a structure *which already possesses the property of invariance*—hence is capable of preserving the effects of chance and thereby submitting them to the play of natural selection. (23–24)

This is the key to Monod's argument that the explanatory reduction of apparent ends in nature can be—must be—extended to an ontological reduction. Invariance only is ontologically primary; teleonomy is entirely derivative. In the logic of the case, this is the only conclusion possible if biology is to be epistemologically coherent (24).

Monod does not fail to draw out the philosophical significance of this viewpoint. As he insists, it is spectacularly opposed to all other answers to the question about the strangeness of living beings. These answers Monod classifies into two groups: the vitalist identification of a teleological principle that operates in the sphere of living beings, and the animist affirmation of a universal teleological principle. Both the vitalist and animist views are well represented in religious, philosophical, political, and even scientific ideologies. Moreover, both assume the opposite answer to the one defended by biology, namely, that invariance is a manifestation of a fundamental teleological principle (24). Why do so many powerful ideologies find themselves at odds with biology? According to Monod, "All religions, nearly all philosophies, and even a part of science testify to the unwearying, heroic effort of mankind desperately denying its own contingency" (44).

This, then, is the entry point to Monod's urging that the choice be taken—it cannot be scientifically or politically compelled—to embrace an ethic of knowledge. The ethic of knowledge, contrary to the ethic of vitalism or animism, distinguishes rigidly between value judgments and statements of knowledge. But it needs to be *adopted*, and this requires a subjective commitment to the fundamental value of the objectivity of knowledge. This is a crucial decision, because modern societies

> owe their material wherewithal to this fundamental ethic upon which knowledge is based, and their moral weakness to those value systems, devastated by knowledge, to which they still try to refer. The contradiction is deadly. It is what is digging the pit we see opening under our feet. The ethic of knowledge that created the modern world

is the only ethic compatible with it, the only one capable, once understood and accepted, of guiding its evolution. (177)

More importantly, for my purposes, the existential commitment to the ethic of knowledge requires the rejection of fundamental teleological principles. This takes courage, in Monod's view, because we are culturally and (he thinks) probably genetically predisposed to desire security. The vision of ourselves as the products of teleonomic structures, grounded on reproductive invariance, and functioning essentially as amplifiers of random noise, brings us face to face with our contingency. Of course, it is the structure of natural laws and not chance that accounts for the emergence of complexity and the generally upward driving character of evolution. Nevertheless, natural selection "operates *upon* the products of chance and can feed nowhere else" (118–119). And just as chance bespeaks the contingency of our origins and development, so in the encounter with it, "man knows at last that he is alone in the universe's unfeeling immensity, out of which he emerged only by chance. His destiny is nowhere spelled out, nor is his duty. The kingdom above or the darkness below: it is for him to choose" (180).

While Monod clearly states and argues for the ontological reduction of teleonomy to reproductive invariance, he also speaks of chance as the feeding ground of natural selection, and as the great revealer of the contingency of all life, especially human life. This indicates that, in Monod's view, what might perhaps be called an "anti-teleological" principle is at work in nature. This principle of anarchy is necessary for all complex systems, and is potentially fruitful; yet it is also infinitely threatening, driving entropic dispersal of energy, promising the ultimate destruction of all order, and kept in check only by the capacity of natural selection to make creative use of it. In fact, to put the point so as to make its irony more evident, nature is utterly dependent on chance for its ability to stimulate adaptations in nature capable of deflecting the threat of chance. Therefore, while Monod clearly denies fundamental teleological principles, his viewpoint is pregnant with suggestions that nature is something like a battle between teleonomic and chaotic tendencies; more precisely, it is ultimately an inexplicable, symbiotic dualism between a disruptive, anti-teleological principle (chance) and a constructive, ordering principle (natural laws). This is what makes his view an intriguing middle position in relation to the first dispute.

The Second Dispute: Teleology Permits Achievement of Specific Goals? Another dispute bears on whether the fundamental teleological principle affirmed in any case is so constituted as to be amenable to the realization of specific purposes. To convey what is meant here requires first saying what is not meant. To use language introduced earlier, teleological processes can be open-ended, in which case they have a trajectory without a specific goal; or closed-ended, in which case they do have a specific goal. Though the question of whether there are any closed-ended complex systems is debated by some—and there are others who argue that evolution itself is closed-ended—these are peripheral views. All thinkers in the mainstream in this respect hold that complex systems are open-ended. The dispute is not about open-endedness versus closed-endedness.

So, what is meant by this second dispute? It is possible to ask whether a fundamental teleological principle allows for the possibility that one specific goal out of the possible ends of an open-ended teleological process could somehow be achieved. This is easiest to conceive when an intentional agent (such as some tricky supernatural entity, perhaps) is thought to be the ultimate ground of, or to know how to manipulate, the fundamental teleological principle. In that case, would the supposed fundamental teleological principle permit this agent to bring about an intended goal? The point would need to be generalized from the case of an intentional agent to make sense of the views of the historical-evolutionary process expressed in, for example, Hegel's logic, Rahner's Christology, or Teilhard's Omega Point, but this can be done.

We must note that a fundamental teleological principle could be confined to the laws of nature, in deistic fashion, with the result that there is in this case no possibility of the principle or its divine wielder selecting out a particular end for realization in a teleological process.[27] Alternatively, a fundamental teleological principle could embrace both law and chance, thus making conceivable a process of top-down causation or whole-part constraint whereby God might elect to manipulate complex systems so as to select out for realization one particular end from among those permitted by natural laws.[28] This would be a form of teleological realization that is consistent with natural laws in the sense that it does not involve breaking or suspending them.

The Third Dispute: Internal Relations or Complexity? Most thinkers involved in evolutionary biology hold that high-level characteristics of living systems are due to the complexity of arrangement of component parts. On this view, for example, the biological feature of the human brain called consciousness is only the highest level property of a hierarchy of large scale characteristics of the brain, including in the middle reaches its structure and function, and at the lower levels its texture, color, weight, and size. On the other hand, some thinkers hold that emergence due to complexity of arrangement is inadequate as an explanation of the products of biological evolution. Rather, complex organisms must be interpreted as communities of fundamental elements, each of which has the character it does only in relation to the other constituents of the organism. This is the doctrine of internal relations, and it promotes the contention that properties of a living organism do not emerge

---

[27] Paul Davies interprets what I am calling the fundamental teleological principle in this way. See *The Cosmic Blueprint: New Discoveries in Nature's Creative Ability to Order the Universe* (New York: Simon and Schuster, 1988), and *The Mind of God: The Scientific Basis for a Rational World* (New York: Simon and Schuster, 1992). Also see Davies' essay in this volume.

[28] This approach is taken by many theologians. Arthur Peacocke in *Theology for a Scientific Age* tends not to use explicitly teleological categories, but he does take this approach to divine action. See also Peacocke's essay in this volume. Process philosophers and theologians are interesting on this point. Some would allow that the fundamental teleological principle could select out a particular end for realization in an open-ended teleological process, as appears to be the case in Marjorie Hewitt Suchocki's Christology, for example, though not everywhere in her writing. See *God, Christ, Church* (New York: Crossroad, 1989), especially Part III, "God as Presence." Other phases of that book appear to be in sharp tension with the tendency to require specific outcomes of open-ended process that appears at times in connection with the Christology, especially as regards the perfection of Jesus' response to the (divine) initial aim. Others (such as Alfred North Whitehead, Charles Hartshorne, and Charles Birch) would reject this possibility. John Cobb usually tilts decisively in the latter direction, but on rare occasions, perhaps anxious to find continuities with traditional Christian teaching, he seems to lean in the former direction. See, for example, the view of Jesus Christ espoused in John Cobb, *Christ in a Pluralistic Age* (Philadelphia: Westminster, 1975), which I would say requires certain specific outcomes to be effected in open-ended teleological processes. This is a complex case, however, and cannot be argued here. Process philosophy does, however, allow that nature is open to *persuasion* toward specific, possible outcomes at every point.

inexplicably from thin air, but from incipient possibilities already present in its constituent elements.[29] On one view, the doctrine of internal relations is superfluous, a philosophical enthusiasm; while on the other it is necessary to make sense of self-organization, and is even an unacknowledged implication of the emergence-due-to-complexity-of-arrangement view.[30]

The Fourth Dispute: Ground of Teleology—Laws, Chance, or Basic Constituents? When a fundamental teleological principle is affirmed, it is natural to inquire as to how it is expressed in nature. Perhaps it is expressed only in the laws of nature. Perhaps it is expressed also in anarchic chance or—which probably amounts to the same thing in view of the sensitivity of complex systems close to bifurcations—in boundary conditions. Or perhaps the fundamental teleological principle is also expressed in the basic constituents of nature, which we might expect to be the case for some forms of panpsychism or dipolar metaphysics. It is certainly the case for those views affirming the doctrine of internal relations.

There is an important correlation between positions taken on the first three disputes and positions taken on the fourth. This correlation appears in the following table in the similarity between the pairs of columns marked *A, B,* and *C,* where "Y" and "N" denote "Yes" and "No" respectively, "N/A" denotes "not applicable," and "Disputes" refers to the disputes described in this section. Four hypothetical positions are assigned Roman numerals in the first column; I have already mentioned examples of each.

View	Key disputes about teleology			How teleology is expressed in nature		
	Dispute 1: Teleological categories are non-reducible? [A]	*Dispute 2: Teleology permits specific goals?* [B]	*Dispute 3: Internal relations needed?* [C]	*Dispute 4: In the laws of nature?* [A]	*Dispute 4: In chance or boundary conditions?* [B]	*Dispute 4: In the basic constituents of nature?* [C]
I	N	N/A	N	N	N	N
II	Y	N	N	Y	N	N
III	Y	Y	N	Y	Y	N
IV	Y	Y	Y	Y	Y	Y

Table 2. Four types of views on teleology in biological evolution

The great virtue of this schema is that it highlights some of the decisions that need to be settled in the three stages of the teleological argument for divine action. In so doing, it illumines the complexity of that argument and the difficulty of prosecuting it—especially its second stage—without heavy reliance on highly contentious premises. The column for "Dispute 1" corresponds to one aspect of the first stage of the teleological argument for divine action, from apparent ends in nature to affirmation of the teleological capacities of natural objects and processes. The column for "Dispute 2" corresponds to one of the factors influencing the third stage, which moves from a metaphysical theory affirming a fundamental teleological

---

[29] Here again, process philosophers make an interesting contribution. See, for example, the affirmation of the doctrine of internal relations in Charles Birch and John B. Cobb, Jr., *The Liberation of Life: From the Cell to the Community* (Cambridge and New York: Cambridge University, 1981), and Birch, *A Purpose for Everything: Religion in a Postmodern Worldview* (Kensington: New South Wales University Press; Mystic: Twenty-Third Publications, 1990).

[30] This contrast is most evident when Davies' view is compared to that of Birch and Cobb.

principle to an interpretation of divine action. The columns for "Dispute 3" and "Dispute 4" correspond to some of the aspects of the second stage—though the correlation between the pairs of columns marked A, B, and C means that the whole diagram is needed for understanding the character of the second stage.

It must be pointed out immediately that, although these four types of views cover some interesting waters, the range of options is much wider when the metaphysical net is cast deeper into the richness of Western or wider to East and South Asian philosophy. Even so, this schematization offers one way to conceptualize part of the range of views that may be taken (with varying degrees of justification) on the question of teleology in the evolutionary process.

3 The First Stage: Teleology and Nature

From physical cosmology's cosmic anthropic principle to zoological morphology, from the status of natural laws to the analysis of tool-wielding animals, from the interpretation of literature to the ascribing of responsibility in legal traditions, the appearance of ends in nature is ubiquitous. The first stage of the teleological argument for divine action begins with this observation and attempts to establish that real purposes give rise to at least some of these appearances. But this raises the question: How can we tell whether ends are merely apparent or real? More generally: Do teleological categories have some advantages in spite of the objections to them in contemporary science?[31]

3.1 The Evolutionary Objection to Real Ends in Nature

The debate over the reducibility of natural ends has classic status in Western philosophy. It is evident, for example, in Aristotle's critiques of his predecessors, especially Democritus:

> Democritus, however, neglecting the final cause, reduces to necessity all the operations of Nature. Now they are necessary, it is true, but yet they are for a final cause and for the sake of what is best in each case.... [T]o say that necessity is the only cause is much as if we should think that the water has been drawn off from a dropsical patient on account of the lancet, not on account of health, for the sake of which the lancet made the incision. (*On the Generation of Animals* V.8, 789^b.3–6, 11–15)

Leaving aside Aristotle's questionable agreement with Democritus on the necessity of nature's processes, but following his main point, this debate can be expressed briefly in the form of a single question: Does the usefulness of efficient-causal explanations of apparent ends in nature justify the conclusion that apparent ends are *only* apparent?

Evolutionary biology has produced the strongest possible reason for answering this question affirmatively, thereby threatening to bring the teleological argument for divine action to a grinding halt before it has completed its first step. The evolutionary objection to real ends in nature in its philosophical form is actually ancient in origins. Aristotle himself, drawing on the thought of Democritus, stated and attempted to refute it:

> [W]hy should not nature work, not for the sake of anything, nor because it is better so, but just as the sky rains, not in order to make the corn grow, but of necessity? ... if a man's crop is spoiled on the threshing-floor, the rain did not fall for the sake of

---

[31] It is because of this bias that Richard Feynman's demonstration that classical mechanics can be based entirely on least action principles (which are teleological in a certain sense) is so striking.

this—in order that the crop might be spoiled—but that result just followed. Why then should it not be the same with the parts in nature, e.g. that our teeth should come up of necessity—the front teeth sharp, fitted for tearing, the molars broad and useful for grinding down the food—since they did not arise for this end, but it was merely a coincident result; and so with all other parts in which we suppose that there is a purpose? Wherever then all the parts came about just what they would have been if they had come to be for an end, such things survived, being organized spontaneously in a fitting way; whereas those which grew otherwise perished and continue to perish, as Empedocles says his 'man-faced ox progeny' did. (Physics II.8, 198^b.17–32)

This is a remarkable passage, partly because it mentions ideas such as fitness for survival and spontaneous organization, which eerily anticipate contemporary discussions, and partly because of its sensitivity to the logical possibility that the *appearance* of natural ends may not be due to the existence of ends in nature. But Aristotle's multi-pronged attack on this beautiful statement of the evolutionary objection is ineffective (*Physics* II.8, 198^b.34–199^b.32), so I will not take space to criticize his replies.

What is the logical force of this ancient objection after more than a century of development of the theory of biological evolution? It is clear that evolutionary theory imparts tremendous momentum to the evolutionary objection: whereas Democritus was simply speculating, Darwin and others adduced powerful evidence that those speculations were right on target. But I do not think the evolutionary objection is any more *logically* forceful because of evolutionary biology. To see this, consider the two-fold logical point of the evolutionary objection, in either its ancient or modern form.

The most forceful argument flowing from the evolutionary objection is not that evolutionary biology furnishes proof that Aristotle's teleology is *mistaken*—after all, metaphysical speculation can render almost any hypothesis secure from threat—but only that it is *arbitrary*, a charge fierce enough to worry a philosopher. If the evolutionary theory of Darwin (or Darwin's successors, or Democritus—it makes no difference) is correct, real ends in nature are superfluous: explanations in terms of ordinary efficient causes can account for all apparently ended natural objects and processes. In this way, evolutionary biology supposedly removes all the good reasons in support of grand teleological visions, leaving their assertion in any form—from Aristotle to Whitehead—merely an imposition of philosophical taste.

Thus, the evolutionary objection undermines arguments for real ends in nature without directly attacking the teleological hypothesis itself. To develop a direct attack—again, in Democritus' time or our own—it is necessary to invoke an Ockhamistic minimalism that seeks to keep the metaphysical shelves as free as possible of amusing but pointless trinkets such as real ends that do not explain anything. This brings to the evolutionary objection a more metaphysical cast, as follows: theoretical explanations based on efficient causation account fully for *apparent* ends in nature, and human (and perhaps other animal) intending lies at the basis of everything with *real* purpose in nature and history. Therefore, there is no need to clutter the shelves with a second level of pseudo-explanations in the form of ends in natural objects and processes. Keep things simple, and the apparent ends in most of nature are justifiably concluded to be epiphenomenal appearances of a complex and wonderful biological process. The ground of that process as a whole is a separate question that may call for a teleonomic answer with regard to the fundamental laws of nature—that is, one that ascribes inherently telic characteristics to those laws—but it does not change anything about the conclusion just reached concerning the ends of most objects and processes of nature being epiphenomenal.

It is common to see rhetorical flourishes in which this argument overextends itself—perhaps by hiding its reliance on Ockham's razor, by ignoring the final caveat about the need for an explanation of the laws of nature themselves, or by trying to treat even conscious human purposes as epiphenomenal.[32] When its conceptual forcefulness is not squandered in these ways, however, the evolutionary objection is genuinely impressive. It forces the first stage of the teleological argument for divine action—and indeed any assertion of real ends in nature—to overcome the reasonable principle of metaphysical minimalism and the blunt charge of philosophical arbitrariness. Is this possible?

3.2 Evaluating the Evolutionary Objection

The two points at which the evolutionary objection is vulnerable are its heavy reliance on a principle of metaphysical minimalism, and its sweeping claim that all ends in nature outside of purposes associated with the act of intending can be exhaustively explained by means of efficient causes and without reference to final causes.

First, there are plenty of good ethical and theoretical arguments for metaphysical minimalism, but both ethics and the theory of inquiry demand that a balance be achieved among all relevant considerations. It is conceivable, then, that real ends could sneak back in through being necessary for theoretical consistency, even though efficient causes exhaustively explain the appearance of ends. That is the essence of the reply to the evolutionary objection that teleological philosophers from Aristotle to Whitehead offer. For example, Aristotle's failure to rebut the evolutionary objection in detail makes little difference, because he relies most heavily on his constructive metaphysical theory to establish that ends in nature are more than merely apparent.

This heavy reliance on a more general metaphysical theory is typical in this area. As I said before, though in absence of a definite argument to support my claim, affirmations of real ends in nature can be made *only indirectly* by means of arguments for a large-scale metaphysical theory that implies real ends in nature. We simply cannot read through apparent ends to real ends, as Paley famously contended we could. So, then, we are able to state a necessary condition for the success of the first stage of the teleological argument for divine action: it requires that a metaphysical scheme postulating real natural ends can be shown to be superior to its competitors. And in this battle, the principle of metaphysical minimalism is but one of many criteria for superiority that must be collectively evaluated. Of course, such a metaphysical scheme must also be consistent with some interpretation of divine action, but that is a mere detail at this stage.

Second, and more pointedly, how we are to decide that evolutionary explanations based on efficient causes do indeed exhaustively account for the appearance of ends in the biological sphere, so that we may justifiably conclude that explanations based on final causes are superfluous? This is a much more perplexing question than it may seem at first glance, and the perplexity has both scientific and philosophical wings. On the scientific side, recent attempts within some branches of neo-Darwinian theory to theorize about and perhaps to identify higher-level laws of self-organization and complexity suggest that the biological data itself may not admit of

---

[32] For a similar critique, see George Ellis in this volume.

exhaustive accounts in terms of efficient causes. But this enterprise is still in its infancy, and thus too difficult to evaluate. On the philosophical side, where debate on this question has been extensive, the considerations are too many to evaluate in passing. I will therefore mention just two issues; both are representative of the wider discussion.

On the one hand, at the most basic level, the theory of efficient causation faces many famous problems, some of vagueness and others of consistency. With regard to vagueness, if the theory of causation is to be spelled out in any detail, the door is opened to inherently teleological accounts such as Whitehead's, and then real ends in nature come flooding back in with enough conceptual integrity to overcome the objections of Ockhamistic minimalism. With regard to consistency, quantum mechanics and quantum cosmology seem to demand an atemporal theory of causation, which throws the common sense account of efficient causation—with its fundamentally temporal cast—into profound doubt. The need for reconstruction invites many visions of causation, including possibly some for which teleological categories are basic.[33]

On the other hand, the very task of showing that evolutionary theory exhaustively accounts for apparent ends by means of efficient causation is challenging. The efficient-causal story in any instance is more complex than we can now, or possibly ever, manage in detail. Some process philosophers and other thinkers leap into this gap and predict that the efficient-causal account will always remain incomplete on its own terms because teleological categories are essential even for an adequate empirical analysis of nature (this is the third dispute, above). Other thinkers, including some other process philosophers, see no gap at all but simply assume that achieving completeness of the efficient-causal account on its own terms is a task limited only by time, energy, money, and other practical considerations. I have little confidence in the intuition of the former group and, based on the ever-increasing detail of efficient-causal accounts of episodes in evolutionary biology, I am inclined to throw my lot in with the latter group.

The weaknesses of the evolutionary objection, it seems, are thoroughly metaphysical in character. If so, then empirical tests will be unable ever to demonstrate that teleological categories are indispensable for adequate efficient-causal accounts within evolutionary biology. Does this, then, mean victory for the evolutionary objection to the first stage of the teleological argument for divine action? Has this objection demonstrated that the use of teleological categories is philosophically arbitrary, allowing the clean use of Ockham's razor to cut away all teleological speculations?

No. The evolutionary objection is much more ambitious than merely establishing that efficient-causal accounts can do the explaining without help from teleological categories, as I have shown. It seeks to justify the use of the criterion of Ockhamistic minimalism to block the use of teleological categories, and showing philosophical arbitrariness in the use of those categories is the means for achieving that end. The first stage of the teleological argument for divine action has a strike against it because teleological categories are superfluous in empirical explanations, but it has not yet struck out. To avoid striking out, it is necessary to engage the philosophical questions associated with judgments of arbitrariness—and that is precisely what the second stage of the argument seeks to do. The burden of proof has shifted, though: the evolutionary objection has seized the initiative and the

---

[33] See, for example, the contribution of Robert Russell to this volume.

teleological argument now must show good cause why anyone ought to think that teleological categories might have some virtue.

To that end, consider a simple example. A genetically-based capacity to regulate blood composition more effectively conferred on animals possessing it a survival advantage that could be transferred to at least some offspring. Random variations, sometimes in competitive environments, then led both to the development of extremely efficient waste-processing organs, such as kidneys, and to the misleading appearance that kidneys are *for the sake of* waste processing, that waste processing is their natural *end*. This is a good point, of course, but—dare Aristotle's reply to Democritus be invoked here?—kidneys *are* for the sake of waste processing. What precisely is wrong with the teleological language here: "for the sake of"? How is it ruled out by furnishing a detailed story of the origin and development of the kidney?

This problem can be cast into a helpful light by noticing another misleading appearance of kidneys, namely, that they look *designed* for the sake of waste processing, in the sense of being the result in their current form of a specific intention. This really is a misleading appearance, because it suggests some other story at the level of efficient causation than actually applies. The history of the design argument (in William Paley's form, for example) bears this out: to the extent that it made any suggestions about efficient causation, it has collapsed, and has only been able to reestablish itself at the level of the laws of nature, removed from the realm of efficient causation to the realm of the condition for the possibility of the operation of efficient causes, in which context "design" is a thoroughly abstract notion. Saying that the kidney is *for the sake of* waste processing, however, says nothing about efficient causes. It is much easier to push the appearance of design off the playing field of efficient causation, therefore, than it is to provide an exhaustive explanatory reduction of apparent ends in terms of efficient causes.

If nothing else, this shows that Aristotle thought a lot harder and more clearly about causation than is sometimes assumed. This is essentially his reply, after all. But more needs to be said, and to take the discussion further, it is necessary to ask about the nature of those ends on account of which we say "kidneys are for the sake of waste processing." Two answers present themselves, and the distinction between these two is of the utmost importance for the teleological argument for divine action.

On the first level, "being for the sake of" may be a functional way of speaking about the properties of the thing in question in some larger context. For example, waste processing is a property that is only functional in the context of a living body, and "being for the sake of" expresses that context silently. To see this, imagine a change of context, which for me brings up memories of having to eat steak and kidney pie as a child. In that case, kidneys are for the sake of eating. The examples can be multiplied. The signification of "for the sake of" shifts with the context in which the kidney is considered. Now, if this was all there was to be said about the ends, then ends in nature could be admitted without interfering with efficient-causal explanations, and the richer structure of a teleological metaphysics really would be superfluous.

On the second level, however, one context may have priority over the others in the sense that it is the *natural* context for thinking about the natural end of kidneys. This is, of course, a way to say that the functional analysis just given may not exhaust what of significance can be said about the end of kidneys. Indeed, it is the story furnished at the level of efficient causation about the development and function of kidneys that determines the natural context for assessing the natural end of kidneys. In that context, asserting that "kidneys are for the sake of waste processing" has a more fundamental status than the statement "kidneys are for the sake of eating" has in *any* context. It is the fundamental status of the *natural* end that so impressed Aristotle; it has always driven, and will continue to drive, teleologically minded thinkers to try to speak of natural ends as a way of capturing what is important in nature, even if such ends have no part in functional-empirical accounts of evolutionary biology.

This is a subtle point, so let me be as clear as I can. We know roughly how kidneys developed the capacities and functions that they have. We can tell this story of origins and development in some detail without recourse to categories of purpose. We can show how this process gives kidneys the appearance of having been designed, even though no self-conscious, intentional designer needs to be invoked for the empirical story to work—and this exclusion of an intentional designer in no way inhibits our sense of wonder about kidneys. We can also show how this process gives kidneys a purpose relative to their function in the animal bodies that have them. But we can't treat the appearance of purpose in the same way as we do the appearance of design. A better analogy is this: we can speak of qualia without affecting neurobiological accounts of brain function one way or another, so the decision about whether to speak of qualia must turn on other issues. Likewise, our speaking of purposes (or not) has no effect on the causal story of biological evolution, so other reasons must decide whether to use teleological categories. Just as there is a reason to speak of qualia (they just *seem* so indispensable for saying what is important about a person even though we know they are biologically produced), so there is a reason to speak of purposes (they just *seem* so indispensable for saying what is important about nature even though we know they are biologically produced). The question is, therefore, whether the reasons for speaking of real natural ends are good enough to outweigh the contention of the evolutionary objection that their use is philosophically arbitrary.

3.3 A Place for Teleology

With this, then, we come to the crux of the debate about natural ends (other than purposes associated with acts of intending). The first stage of the teleological argument for divine action cannot be negotiated without a philosophical judgment about the value of using teleological categories. They are arbitrary in that they offer no help to empirical accounts of nature, but they are useful for expressing what is important and natural about natural processes. Weighing all such considerations together is the only alternative. Naturalness is an aesthetic category like beauty, however, so "accounting for naturalness" is not a task to which efficient-causal explanations are well-suited. The same goes for accounting for value, importance, and the like. If teleological categories help us deal with such matters, then it is genuinely difficult to remove the need for teleological categories in any *complete* explanation of biological evolution (notwithstanding the completeness of the efficient-causal account on its own terms).

At this stage, with teleology reappearing, it is vital to remember that the evolutionary objection has had an effect on this debate. For example, due to its influence, any claim that teleological categories are necessary for efficient-causal explanations of apparent ends is desperately weak. But teleological categories can no more be kept from the task of "accounting for naturalness" than can metaphysics in general be kept from the human imagination. Kant thought of these as understandable but misleading impulses, but I see no sound reason decisively to ban either,

Kant to the contrary notwithstanding. Teleology may only appear as teleonomy, at the level of the laws of nature, but appear it ought.

So, while admitting that this is a complex judgment involving balancing competing considerations, I conclude that there is a place for teleological categories in accounting for apparent ends in nature. But exactly what place is this? This question brings us to the first crossroads of the teleological argument for divine action, with two more to come later. The way teleological categories are actually wielded varies. Some philosophers, theologians, and scientists would be inclined to find real ends underlying apparent ends by virtue of the laws of nature (for example, Davies). Some would make use of a philosophical strategy hinging on supervenience, whereby multiple independent descriptions of the same process can each be true on its own level (for example Murphy).[34] Some (such as myself) are inclined to resort to teleology to engage the topics of value and importance in nature. And, as I have mentioned, there are even a few (including some extremists in the process philosophy camp) who contend that teleological categories are needed even to produce adequate efficient-causal accounts of apparent ends in nature. I have argued only that teleological categories cannot be entirely ruled out of comprehensive explanations for apparent ends in nature, and I have suggested that I find the causal-gap prediction of the last option breathtaking but implausible. To that I will add only that the other options seem compatible, and that every option, even the supervenience strategy, requires contextualization in a wider metaphysical theory to achieve intelligibility.[35]

4 The Second Stage: Teleology and Metaphysics

The second stage of the teleological argument for divine action attempts to situate the affirmation of the reality of natural ends in a broader theory that is capable of presenting real natural ends as instances of a more general fundamental teleological principle. This metaphysical context is the bridge between real ends in nature and a theory of divine action, and must be compatible with both. It is clear that real ends in nature can be metaphysically contextualized in a variety of ways. The question for evaluation here is whether the second stage of the teleological argument for divine action can successfully move from real ends in nature to only those metaphysical theories that are amenable to divine action (in some sense), avoiding all otherwise adequate metaphysical theories that are antagonistic to divine action. The answer to

---

[34] See Nancey Murphy's essay in this volume for a definition and discussion of supervenience (primarily with regard to ethics). See also William Stoeger's use of this concept in his essay for this volume.

[35] Nancey Murphy denies this; see her essay in this volume. Murphy adopts the supervenience strategy in order to argue for the feasibility of higher-order language about ethics and theology, yet feels no need to explain how those higher order languages relate in detail to other levels of discourse about the world, for which metaphysics is indispensable. This freedom from the worries of metaphysics is held to be a desirable state of affairs to which we are propelled by Wittgenstein's later philosophy. By contrast, I take this attempt at maintaining higher-order worlds of discourse while bypassing questions of metaphysical and all manner of intellectual connections to other language games to be strategically futile (it fails to secure the long-term stability of ethical and theological discourse) and philosophically wrong-headed (it is mistaken in its assumption of substantial independence between such language games and presupposes an inadequate theory of inquiry). A partial argument for the operating theory of inquiry from which these critiques may be inferred is in my "Similarities and Differences in the Practice of Science and Theology."

this question is negative, I shall argue, notwithstanding the fact that the science-religion literature at the present time exhibits views with a strong correlation between being friendly to teleology and being friendly to divine action. This, therefore, is the second crossroads at which a wealth of metaphysical choices obstructs the clear lines of inference needed by the teleological argument for divine action.

4.1 Counterexamples: Teleology without Divine Action

The obvious place to begin is with arguments that the second stage of the teleological argument for divine action cannot succeed. For this, it is necessary to find examples clearly illustrating that real ends in nature can be contextualized in metaphysical systems that are both antagonistic to divine action and otherwise adequate, or at least comparable in adequacy to metaphysical systems within which divine action can be imagined. There are a number of such counterexamples, and I shall mention several from a variety of philosophical traditions here.

First, and most obviously, Aristotle's teleological metaphysics posits real natural ends but is antagonistic toward divine action in all of the usual senses. On the one hand, by holding that the universe is everlasting, Aristotle tried to block the specter of creation, which he seems to have thought robbed the God-world relation of its aesthetically pleasing, necessary character. On the other hand, God's role as the ground of the giant teleological organism that is the universe was understood by Aristotle as thoroughly automatic, which is to say, precisely the opposite of deliberate. To be the prime mover in Aristotle's understanding does not imply that God undertakes any specific actions. On the contrary, just as motion has to be understood as change in accordance with the fulfillment of the nature of a thing, so a thing's nature or purpose cannot be understood unless there is a principle of order in which all natures participate. God is this principle, for Aristotle. God neither begins a chain of efficient causes as an efficient cause, nor interferes with it, nor creates the universe in which this dynamism of change occurs.

Perhaps it might be argued that Aristotle's God does act in the sense of being creative; after all, God does at times seem to be thought of by Aristotle along the lines of the creative part of human rationality. Furthermore, this is how God acts in Whitehead's teleology. But Aristotle's God neither persuades nor reconciles the actuality of the world in the consequent nature, as Whitehead's God does. And the analogy for God of the active human intellect goes nowhere when such creative characteristics are not affirmed. Whitehead's God *does* act, even though not through creation as such, nor through the expression of particular, specific intentions (which has not stopped process *theologians* from affirming the divine expression of such intentions necessary to speak of special events within "salvation history"). But Aristotle's God does not act, because it is fully actualized with no potential. It is not creative, but rather the serenely all-present principle of nature.

This view of Aristotle's was arguably also present in Plato, in a related way. By the time of the middle Platonists, however, it had already weakened because the forms came to be identified with the ideas of God, thus making God more closely analogized by the active intellect of human creativity. Slowly and unsurprisingly after the middle Platonists, the concept of creation became firmly established in Western and especially Christian philosophy—*creatio ex nihilo*, no less—and then, no matter what else is said about God, God at least acts in creating determinate reality. After the time of Aristotle, therefore, his view is hard to find in the West, even though conceiving nature as a vast teleological organism cannot easily be argued to be less persuasive than thinking of it as the result of an act of divine creation. Indeed, the former view has the advantage of a less stringent form of the problem of theodicy. Outside of the West, views similar to Aristotle's are found in Chinese philosophy, both ancient and modern. In this case, the concept of *li*, in the sense of principle, is central. It is what is expressed in the nature of individual objects and processes, and in their coming together to make an orderly world. Yet it is usually not considered as an active principle, but rather as a changeless one. Divine action makes little sense on this view, too, yet it is one of the greatest and stablest metaphysical systems the world has known.

A second type of teleology-without-divine-action viewpoint is widespread in South Asian philosophy, but requires a shift in the focus of teleology. Mādhyamaka Buddhist philosophy—say, in the thought of Bhāvaviveka—offers an example. Here we have a metaphysics without God, and so without divine action. Yet nature has a definite purpose, albeit one that dissipates into nothing as properly discriminating human beings see the world for what it really is. What is that purpose? Nature is dependently co-arising with human consciousness; the suffering, frustration, and weary repetition of nature reflect our own delusion. We achieve liberation when we attain the discrimination needed to end all attachment to conventional reality, including our own being. Western philosophers are quick to ask why our attachment results in so interesting and public a delusion. Buddhist philosophy, and South Asian philosophy in general, has been relatively weak in answering this question, but for a good reason. To appeal to a famous image from the Buddhist fire sermon, when a house is burning down around you, the only important thing to do is to escape; studying the intricate patterns on the wallpaper on the way out is absurdly, dangerously beside the point. Somehow, our deluded state creates impressions of things with apparent reality, including ourselves, and it is neither possible nor ultimately interesting or important to know why it is so. However, the suffering ubiquitous in this dependently co-arising world is the great clue to its ultimate unreality, and so to its ultimate purpose: to help us wake up, and flee the flames.

Here, then, we have a teleological principle for nature as a whole and for individual instances of suffering (including most ordinary events and processes in one respect), but one that says little explicitly about real ends in natural objects and processes. However, this teleological principle is embedded in a truly powerful metaphysical perspective with an enormously sophisticated history of development. Conceivably, the development of this view could lead to the answering of wallpaper-type questions about evolutionary biology, in which case it could be rendered a more fully fleshed-out counterexample to the second stage of the teleological argument for divine action. Though Buddhist philosophy typically has not been interested in such explanatory possibilities, it may have to become more engaged in them if it is effectively to engage the natural science of the modern world. Indeed, signs of growing interest are already evident as Buddhism becomes better established in the West.

A third type of counterexample derives from certain mystical theologies in theistic traditions. These theologies have two characteristics: the affirmation of nature as a telic process within the life of God (in a sense) and the denial that talk of divine action makes religious or philosophical sense. God on this view is infinitely hard to describe: every symbolic characterization of God is needed in the path by which the soul ascends to unity with the divine and yet each fails decisively and must be contradicted and refused on that same path. This embrace of contradictions and frank acknowledgment of the failure of human cognition are neither needed nor desirable for the comprehension of much in nature and human life, but they are essential in approaching divine realities. Thus, this view is not irrational but rather supremely rational through clearly recognizing the limits of human wisdom at precisely the point where reason's self-deception can have the most harmful effects. This view can adopt a highly teleological analysis of nature along any number of lines and yet typically will speak of divine action and divine intentions only as a first-order approximation to a deeper mystery; better approximations leave intentional and personal categories for God behind. Here again, then, we have a viewpoint that can be highly sympathetic to fundamental teleological principles in various forms and yet can be profoundly uninterested in talk of divine action. This view has made its presence known more recently in the radical theologies of the twentieth century. The blending of atheism and religious sensibility in these theologies is profound, in my judgment, and truly expressive of the richness of Western theological insight. Moreover, in all such cases, the category of divine action is ultimately inapplicable.

It is interesting to me that there are so few examples in the West of philosophical systems that are teleological in character and yet unsympathetic to all three classes of divine action: creation, creativity, and the expression of specific divine intentions. The perspective of mystical theology has usually been marginalized in the history of Western theology, and Aristotle is ancient. We might be inclined to suppose, by sheer weight of Western habit since the invention of the concept of creation, that the second stage of the teleological argument for divine action is successful in moving from real ends in nature to theories of fundamental teleological principles that are only ever amenable, and never antagonistic, to divine action in some form. Even a rudimentary knowledge of South and East Asian philosophy will save us from this mistake. But just one counterexample is sufficient to block the second stage of the teleological argument for divine action, and the ancient Western example of Aristotle fills the bill, providing it is a basically viable philosophical view. To assess this crucial caveat, some evaluation of these teleological views is in order.

4.2 Evaluating the Counterexamples

To give the teleological argument for divine action its due, let us continue by noting how few are the options for providing a metaphysical framework for real ends in nature without introducing a conception of a God who acts: there are just two. On the one hand, we may decline to furnish a fundamental teleological principle as an explanation for real natural ends. That is, we could try to affirm real ends in nature while denying that any fundamental teleological principle is expressed therein, which prevents real natural ends from ever receiving a metaphysical contextualization that might be relevant to divine action. This amounts to denying that real natural ends are coordinated with each other, much as human intentional ends are frequently uncoordinated with each other (as human societies demonstrate). The key philosophical move here parallels the pluralistic rejection of monism. This view threatens to be philosophically unstable, however, because it is probably simpler to drop real ends and dispense with teleology in nature altogether. That is, this view is likely to trip over the criterion of philosophical adequacy I have been calling Ockhamistic minimalism.

On the other hand, we might admit a fundamental teleological principle—called God by some, *li* by others, and strategically unnamed by yet others—and conceive this principle so as to block any further move toward divine action. This fundamental teleological principle would be impersonal, without specific intentions, neither creative nor a creator, so it could not meaningfully be said to act. All of the candidates for counterexamples in the previous section are of this kind. Yet in all of these cases the second stage of the teleological argument for divine action retains a fighting chance. Even a small philosophical nudge—by the questions about there being something rather than nothing, or about the public character of nature— threatens to push such a fundamental teleological principle into a conception of God that creates or is creative. While this threat can be effectively rebuffed, I think the history of Western philosophy shows how difficult it is (at least for *that* philosophical tradition) to resist the impulse of such questions toward positing a God that acts as a creator or as creative.

This accounts for the fundamental attractiveness of the teleological argument for divine action: if only we can show that there are real ends in nature (supposedly the hard part), then it is only a short hop to the reality of divine action (supposedly the easy leg of the journey). Well, the second stage of the argument does not live up to this promise, but it is interesting to see how close it gets. It gets close enough, in fact, that another question presents itself: What more, if anything, can be done to make the second stage of the teleological argument for divine action successful? It would be necessary to solve the problem of the conflict induced by a plurality of (at least superficially) adequate metaphysical contextualizations of real ends in nature.

It may be that some would assert that these alternative teleological visions are fundamentally inadequate, perhaps just because they are not amenable to divine action, or for other reasons. That certainly takes courage, at least at this early stage of the process of systematic comparative metaphysics. This is not the place to establish the relative adequacy of a rainbow of metaphysical views. But it is appropriate to insist here that such well-attested, long-standing views of reality cannot be dismissed cavalierly. Indeed, at least with regard to majestic world-views such as those I have mentioned, the presumption of adequacy must be granted until a preponderance of evidence to the contrary is established. Nor will it do to satisfy oneself with identifying a weakness merely in one respect, for all metaphysical systems have weaknesses, and evaluating overall superiority must comprehend questions of balance and emphasis. The task of comparative metaphysics is genuinely difficult. Reality seems susceptible of description by multiple, conflicting, adequate metaphysical schemes, and a non-arbitrary approach finds soundly-argued decisions among such theories infuriatingly difficult to construct. This is the famous problem of metaphysical ambiguity.

4.3 *Metaphysical Ambiguity*

This problem of metaphysical ambiguity has been the chief source of despair over metaphysics in the West from ancient times. The Sophists cited it as evidence of the intellectual corruption of Socrates, Kant of his Leibnizian heritage, Kierkegaard of Hegel, Ayer and Wittgenstein of the entire metaphysical tradition. Every philosophical tradition worldwide shows signs of skepticism induced by the specter of metaphysical ambiguity. Now, we ought to recall that the second stage of the teleological argument for divine action does not need to justify the one true metaphysics but only the more modest result of merely excluding teleological views antagonistic toward divine action (though the third stage needs more). Perhaps, after all, the problem of metaphysical ambiguity can be overcome to the extent needed through an ongoing process of diligent comparison and analysis in relation to carefully examined and constantly revised criteria for theoretical adequacy. Unfortunately, it is obvious that the problem of metaphysical ambiguity is *very far* from being overcome, even to this modest extent. Moreover, we appear to lack even some crucial tools for accomplishing it, such as a tradition of systematic inquiry into categories used in cross-cultural, metaphysical comparisons.[36]

We must ask, then, exactly how bad is the problem? The dimensions of the problem of metaphysical ambiguity can be estimated from the side of metaphysics in the following way. Philosophical contextualization of real ends in nature by means of a fundamental teleological principle does not require the idea of a God that acts, nor even the idea of God, as I have pointed out already. However, it is usually Western traditions that have been explicitly interested in teleology, so the idea of God has appeared frequently in teleologically concerned metaphysics. If the idea of God does show up, it may not be (and often has not been) a deistic or theistic idea of God. And if deism or theism is implied in the teleological metaphysics, it may or may not be one of the traditional ideas of God familiar to the major Western theistic religions: Judaism, Christianity, or Islam.

To be a little more specific at the level of this sprawling metaphysical wildness that is closest to the sphere in which divine action can be conceived, there is important variation even in traditionally recognized forms of theism, both within and among the three major Abrahamic traditions. One debate that appears within all three is that over whether or not God is ontologically fundamental. In Christianity, it is usually debate over omnipotence and creation that signals the presence of this question, with process and classical theism taking opposed views on both doctrines. In Judaism it appears in legal debates over the ontological primacy of the law, and in metaphysical and ethical debates surrounding Kabbalah (Jewish mysticism) over whether God needs "salvation" through human cooperation. In Islam it shows up in some of the ethical and legal debates between the competing medieval Mu`tazilite and Ash`arite schools, as when they argued that God forbade killing because it is bad, and that killing is bad because God forbade it, respectively. All three traditions, therefore, have ways to think of God either as subject to fundamental teleological principles, or as their absolute ground—and kenotic theories of creation try to have both at once. Thus, it appears that, even when systematic metaphysical accounts of fundamental teleological principles include some form of theism, multiple ways of envisaging the relation between God and teleology are still possible.

4.4 Metaphysical Ambiguity and Evolutionary Biology

In spite of this staggering diversity, these views of teleology in the evolutionary process do have common features. Three of these common characteristics become evident from the point of view of evolutionary biology. In fact, these shared characteristics apply even to philosophical contextualizations of teleology that reject teleological categories. There are metaphysical ways of understanding teleology that do not have all of these characteristics, and so stand outside of the diverse mainstream I seek to characterize here, but they seem to be relatively rare and theoretically fragile. These common features suggest a somewhat skeptical conclusion about the usefulness of biological evolution for resolving debates about teleology in the short term.

---

[36] An attempt to develop such a tradition of inquiry out of fragmentary, extant efforts has been funded for 1995–6 and subsequent years by the National Endowment for the Humanities in conjunction with some private foundations. The Principle Investigator for the three year project is Robert C. Neville, and the Co-Investigators are Peter Berger and John Berthrong.

1. *Current knowledge of biological evolution is consistent with all of these views of the place of teleology in the evolutionary process.* The obvious upshot is that none of these views can be rejected on the grounds of simple inconsistency with the contemporary account of biological evolution. Of course, process philosophers sometimes argue that the doctrine of internal relations is indispensable for any satisfactory account of the emergence of life and consciousness. This subtle debate seems to be unresolvable on the basis of biological considerations alone; certainly the doctrine of internal relations is not a *popular* position, in view of the fact that virtually all biologists appear to believe that emergent properties such as life and consciousness can be explained on the basis of complex, stratified organization. Granted, on this view, the mystery of life as such persists. However, it is not demystified in the doctrine of internal relations, but only shifted to another level of discussion—the level of the ultimate constituents of nature and the theory of causation. This may well be the right level on which to conduct the debate. However, the debate at any level is sufficiently complex that there is scant justification for the expulsion of views that hold to a doctrine of emergence based simply on complex organization.

2. *All of these views of the place of teleology in the evolutionary process are neutral toward all short-term controversies in biological evolution.* These short-term controversies include, with regard to biological evolution, whether or not gaps in the biochemical story about the origin of life can be filled; whether or not apparent variations in the rate of variation and selection can be explained; and whether higher level laws or tendencies of complex systems can be identified. Further debates likely to be short-term in nature pertain to evolutionary psychology: whether or not law-like relations can be identified between gene-perpetuation interests and social practices; and whether or not it is possible to specify the senses in which human freedom mediates between gene-perpetuation and other, possibly competing, interests. Mainstream views are indifferent to the outcome of such inquiries. At worst, there might be a failure of the new paradigm to turn up solutions to some of these problems, and that may threaten the progressive status of the research program it defines. That, however, would not be a short-term crisis. It must be admitted, of course, that this is a rather curtailed list of short-term puzzles—there are hundreds of major research problems that can reasonably be expected to find solutions in the relative near term in the ordinary course of scientific advance. But I am aware of no short-term controversies whose resolution could justify the exclusion of any of the large class of mainstream views.

3. *Each of these views of the place of teleology in the evolutionary process is vulnerable (if at all) only to long-term metaphysical controversies that are unlikely to be affected by biological evolution.* The class of long-term disputes includes many debates that may not be resolvable in principle, and others that may not be resolvable in practice. Long-term controversies include, with regard to biological evolution, whether or not compelling evidence can be adduced for the irreducibility of teleological categories to the description of what is essential in complex biological systems; whether or not God acts undetectably to influence evolutionary development; and whether or not any given form of life can be demonstrably and exhaustively accounted for in detail in terms of specific chemical processes and evolutionary principles. Another long-term debate, pertaining to evolutionary psychology, is whether or not culture, ethics and religion can be exhaustively explained in terms of gene-perpetuation interests, or other principles connected to genetic heritage.

It is important not to be too presumptuous about what might or might not eventually fall under the auspices of the scientific method. While the long-term problems just mentioned are presently at least as much philosophical as biological in character, it is possible that some of them might one day be considered more completely a part of evolutionary biology and biochemistry than they are now. In any event, the point is that all of the debates in which views of teleology in biological evolution have something at stake lie in the class of long-term-disputes. Mainstream views that reject fundamental teleological principles (for example, Dawkins and Monod) have the most to lose, since they could potentially stumble on an unfavorable result in every one of the long-term disputes mentioned. I take such vulnerability to be a sign of profound intelligibility, for the intelligibility of a hypothesis partly involves being able to indicate clearly what counts as evidence against it. However, such vulnerability by itself is not necessarily a reliable indicator of truth.

5 The Third Stage: Teleology and Divine Action

It remains now to consider the third stage of the teleological argument for divine action—and after what has been said, this is relatively simple. Here again, for the third time, the specter of metaphysical options interferes with the easy inferences that would make the teleological argument for divine action simpler than it is.

5.1 The Connection between Teleology and Divine Action

Some dimensions of the question of divine action are not highlighted when teleology is the source of illumination, but there are compensating advantages. Among these is the fact that, because apparent endedness is a highly effective category for expressing what is interesting about nature, it is useful as a principle for organizing conceptions of divine action. So, then, what possibilities for divine action are suggested by this discussion of the place of teleology in biological evolution?[37]

Let us begin by noting that, if the apparent ends of objects and processes are *only apparent*, then the rough and ready conclusion—certainly the one that we are entitled to assume Dawkins would draw—is that there is no possibility of divine action. Strictly speaking, I suppose it is conceivable that God might act without leaving apparently teleological traces, but the metaphysical and theological viability of such a view is low, as it would shut creation, all patterns in nature, and all stories in history out of the domain of divine interest, leaving miscellaneous, unintelligible (to us) interference as the sole mode of divine action. Similarly, if Monod's view is correct, then traditional deism and theism are highly misleading accounts of ultimate reality. The more natural metaphysical contextualization of his view (Monod does not propose this himself) is the dualist one of a primal battle between principles of order and anarchy, such as was and is still found in Zoroastrianism, except that these two principles must be symbiotically related. If this symbiosis itself is named God (rather than the more obvious Nature), then we are speaking of a kind of pantheism in which divine action is synonymous with "event," which renders this God

---

[37] Owen Thomas distinguishes six ways to parse the question "How does God act?" in *God's Activity in the World: The Contemporary Problem* (Chico: Scholar's Press, 1983): By what means? In what way or manner? To what effect? With what meaning or purpose? To what extent? On analogy with what? (234–236). While these six questions considerably enlarge the ordinary sense of the original query, they also helpfully draw attention to the fact that divine action probably cannot be discussed thoroughly without suggesting answers to all or most parts of this six-fold battery of questions. The following discussion focuses chiefly only on the first two questions, and so stops short of complete thoroughness.

profoundly morally ambivalent and evacuates divine action of specific meaning. Against these anti-teleological views is ranged an array of metaphysical contextualizations of fundamental teleological principles, many of which are not amenable to divine action. I will not revisit these views here. It is enough to see that the teleological argument can break down when trying to speak of divine action even in the context of emphatically teleological metaphysics.

Now, moving by these open metaphysical options, let us suppose for the sake of argument that the second stage of the teleological argument for divine action has been successful and that we begin the third stage from within the ambit of the usual deistic, theistic, or panentheistic world-views that allow us to speak in recognizable ways about divine action. In this case, the locus in nature of the fundamental teleological principle—the fourth dispute discussed earlier—will be the key insight into the possible modes of divine action. So let us reflect on the relations between the locus of teleology in nature and divine action.

When the locus in nature of this fundamental teleological principle is natural laws only, then divine action cannot include the expression of specific divine intentions in the context of an ongoing providential relationship with that creation because this requires the fundamental teleological principle also to be expressed in chance (or boundary conditions), as discussed earlier. Nor can divine action presuppose teleological characteristics within the constituents of nature. That leaves two modes of divine action, both bearing on creation, and both expressed in the laws of nature: the universal determination of natural possibilities and the ontological grounding of nature.

The locus in nature of this fundamental teleological principle might include chance, understood as a general category including the influence on complex systems of their boundary conditions. If so, then divine intentions (or analogues thereof) can conceivably be expressed either directly—there are a number of proposals for such mechanisms—or less specifically in the striving for general ideals of harmony, complexity, and intensification of value in history and nature (as in Whitehead's version of process philosophy).

Finally, when the locus in nature of this fundamental teleological principle also includes the constituents of nature, two other ideas of divine action come into play. On the one hand, process philosophy stipulates a rich theory of causality that posits specifically teleological characteristics in the fundamental constituents of nature. In this case, divine action consists in the performance of the necessary regulative tasks associated with that theory of causation: offering initial aims to concrescent actual occasions from out of a primordial envisagement of possibilities, and reconciling the actuality of the world in the maximally harmonized consequent nature. On the other hand, it is possible to conceive of God as furnishing the material conditions for the possibility of the emergence of complex, self-organizing systems through creation. These conditions would be realized in the constituents of matter itself, but the mode of divine action would be creation rather than creativity, the latter being the category to which process philosophy appeals in explaining the emergence of complex and novel forms of self-organization. This view of divine action is implied whenever complexity and self-organization require the constituents of nature to have particular capacities in addition to the constraints on their interaction stipulated by the laws of nature. An example of such a view is the philosophy of Robert Neville, whose theory of causality is similar to Whitehead's process philosophy, but affirms the metaphysical theory of creation *ex nihilo*, and denies that God furnishes initial aims to actual occasions.[38]

5.2 A Schema for Further Discussion

These six modes of divine action and their relationships to the loci in nature of fundamental teleological principles are represented in the following table. Note that all three classes of divine action appear here. Creation appears in modes 1, 2, and 5; creativity shows up in modes 3 and 6; and the expression of specific divine intentions is covered in mode 4, which can be specified in a number of different ways.

Locus in nature of teleological principle	Mode of divine action
Laws of nature	1. Creation as universal determination of natural possibilities (e.g. *creatio ex nihilo*, Paul Davies)
	2. Creation as ongoing ontological grounding of nature (e.g. divine faithfulness)
	3. Creativity as striving for harmony, complexity, intensification of value (e.g. *creatio continua*)[39]
	4. Expression of specific divine intentions via: • top-down causation or whole-part constraint (e.g. Arthur Peacocke) • manipulating boundary conditions of chaotic systems (e.g. John Polkinghorne) • chaotic amplification of quantum field actualizations (e.g. Robert Russell) • lawfully widening the canalizing of complex processes (given feedback mechanism from environment to operation of natural laws) • means associated with atemporal theories of causation[40] • and perhaps other means as well...
Fundamental constituents of nature	5. Creation as furnishing the material conditions for the possibility of the emergence of complex, self-organizing systems (e.g. Robert Neville, but not process philosophy, which denies creation)
	6. Creativity as expressed in a theory of causation that assigns a necessary regulative role to God (e.g. Whitehead, Birch, and Cobb)

Table 3. Correlation between teleological loci in nature and modes of divine action

---

[38] See Robert Cummings Neville, *Creativity and God: A Challenge to Process Theology* (New York: Seabury, 1980).

[39] This use of *creatio continua* is problematic on some views of causality. It is, of course, quite natural in the context of process philosophy. On some other views, however, the teleological capacities of natural laws as currently understood are *by themselves* sufficient for fostering trajectories toward complexity, which implies that divine action would not be needed for the maintenance of processes of complexification, except in the most basic sense that God, on this view, is the ultimate ground of all natural processes (this is mode 2). This narrows the meaning of *creatio continua* as it applies in these cases, accordingly. It also illustrates the intimate connection between the meaning of *creatio continua* and theories about causality and the fundamental constituents of nature.

[40] For example, it has been proposed by Troy Catterson in conversation with me that superspace versions of quantum cosmology, in conjunction with an interpretation of the Heisenberg uncertainty principle that applies to the relation of space and time, allow for the possibility of understanding natural-law-conforming action of a non-temporal deity in temporal nature. See his "No Time for Time: Trans-Temporal Creation of a Time-Bound Realm," *Journal of Faith and Science Exchange*, 1998.

It is clear from this table that there are many possibilities for envisaging modes of divine action, even *after* the philosophical contextualization of real ends in nature is specified so as to be compatible with one or more types of divine action. If the locus in nature of teleology is limited to the laws of nature, then there are fewer options. If it extends into the processes and constituents of nature, however, the possibilities multiply rapidly. Deciding among them depends not upon teleological considerations but upon other philosophical issues including such tough problems as causality and time.

Note, too, how the contrast between essentially deistic proposals (Davies) and more traditional theistic proposals (Peacocke, Russell) appears here. Their difference, while genuine, is not so conceptually large as is often thought. In fact, both affirm a fundamental teleological principle expressed at least in the laws of nature, which is an agreement of considerable proportions in view of the fact that it involves assent to the irreducibility of teleology, and dramatic, almost identical narrowings of the breadth of metaphysical possibilities. The move from deism to theism is then accomplished by a relatively minor enlargement of the locus of teleology so as to include chance (or boundary conditions). This suggests that the move to theism from deism is hard to block from the deistic side without arbitrariness.

6 Conclusions

The main conclusion to be stated here is that the teleological argument for divine action is not very teleological. That is, there is no sound chain of implications from analysis of apparent ends in nature, to judgments about the ontological irreducibility of those apparent ends, to estimations of the locus in nature of fundamental teleological principles, and then to specification of the modes of divine action. In fact, the implications run more smoothly in the reverse direction. In the order stated, the chain breaks down at each link, at least when biological evolution remains the sphere of discussion. Additional premises are needed to move from apparent ends in nature to the affirmation of real ends, from there to metaphysical theories affirming a fundamental teleological principle consistent with divine action, and from any such teleological theory to the reality of divine action in particular modes. None of these missing premises is furnished by biological evolution, and I have tried to spell out what some of them might be at each stage. Because the additional premises needed to make the teleological argument for divine action valid characteristically have little specifically to do with teleology, we need to conclude that the argument does not depend as much on its starting point of the ubiquity of apparent ends in nature as the way it is stated promises.

Discussions about divine action in connection with biological evolution must not casually assume that these missing premises are unproblematic. In particular, it would be easy to fall into a kind of blinkered or perhaps ideological ignorance of alternative, profound teleological visions that are antagonistic toward divine action and that are as well supported by biological evolution as any that permit us to speak of divine action. I have adverted to a number in this essay. In discussions of divine action in the context of teleology in nature, and especially in evolutionary biology, therefore, let us hesitate to conflate articulation of theories about divine action with justification of those theories. The gap between what is possible by way of divine action and what can be justified is rather large.[41]

The failure of the teleological argument for divine action, especially the second stage, has been traced to the problem of metaphysical ambiguity, and I argued that the specter of metaphysical ambiguity is largely immune from considerations drawn from contemporary biology. Therefore, just as it is unwise to expect to be able to decide among competing views of teleology in biological evolution based on any short-term considerations from biology, so it is over-hasty to expect to narrow the range of possibilities for metaphysical construals of divine action on the basis of considerations drawn from contemporary biology.

A second, subsidiary conclusion is, I hope, a sound conjecture. It is related more to teleology than to the teleological argument for divine action itself, but is strongly suggested by the various pieces of argumentation presented here. It is this: the case for affirming a fundamental teleological principle is far stronger than that for rejecting it, given the premise that the cosmos (note: *not* ultimate reality) is meaningful rather than absurd. This premise seems not infrequently denied by biologists and philosophers of biology who dare to treat such questions. And philosophically it is notorious for being impossible to justify except in obviously self-referential ways. But, if it is granted, then it is genuinely difficult to maintain, as many popular writers in evolutionary biology do, that the universe has no overarching teleological sweep. How do they do this, then?

It seems to me that all attempts to avoid postulating a fundamental teleological principle require either an arbitrary arresting of inquiry, or the assumption of an absurd cosmos. Dawkins and Monod, who have been mentioned a number of times by now, make interesting case studies in this regard. In Dawkins' case, in both *The Blind Watchmaker* and *Climbing Mount Improbable*, curiosity is inexplicably terminated, at least by my standards. His assumption of a self-explanatory and ontologically self-sufficient universe ought to be examined, if only to understand why his other apparent assumption that the universe is a meaningful and wonderful place is to be believed. That is, Dawkins denies *both* cosmic absurdity *and* a teleological sweep to the cosmos, and the conceptual strain that results forces the contrived truncation of his inquiry. Future books may correct this, perhaps by extending his intriguing images of the biosphere as a distributed supercomputer, of the cosmos as dancing, or of evolution as an enchanted loom, "weaving a massive database of ancestral wisdom."[42] These are images whose precise articulation and evaluation demands metaphysical categories and arguments that Dawkins so far seems unwilling to engage.

By contrast, it is possible to construe Monod as abandoning inquiry *not* arbitrarily, but in recognition of cosmic absurdity, in the context of which humans are simply confronted with the choice to create proximate meanings or not. That is, Monod denies a teleological sweep to the cosmos, but courageously and consistently

---

[41] The scope of this essay prevents me from arguing to the more adventurous conclusion that the gap between what is possible by way of divine action and what can be justified is large, *no matter what the context of discussion*. That is, this fundamental kind of metaphysical ambiguity can be found not only in relation to evolutionary biology, but also in relation to everything from cosmology to religious experience, from history to mysticism, from sociology to hermeneutics. This is perhaps equivalent to a thesis as to the limitations of human rationality.

[42] Dawkins, *Climbing Mount Improbable*, 326.

pays the intellectual price by explicitly surrendering the hypothesis of cosmic intelligibility. This marks out a genuine intellectual possibility, albeit a paradoxical one, for it admits the possibility of inquiry (it does not affirm *ultimate* absurdity, which is extreme skepticism, but only *cosmic* absurdity) while characterizing it as an anomalous activity that peters out into deferential silence—existentially in our experience through ubiquitous limitations and contradictions, and in history and nature through the eventual vanishing of all life.

The cosmic absurdity view is affirmed systematically in Vedanta philosophy (notably and with important differences in Śaṅkara and Rāmānuja). It is also expounded in Mādhyamika Buddhism, especially in the thought of Nāgārjuna and Bhāvaviveka, who denied that any fundamental metaphysical principle (teleological or not) can be identified without distorting more than illumining. Many other thinkers and sub-traditions of Buddhism affirm more or less the same position, as do various strands of apophatic mysticism in the West. These views tend to use the conjunction of apparently contradictory statements as a form of reference to "states of affairs" essentially beyond categorial experience, and so beyond discussion. This form of reference (the so-called "Middle Way," which is the meaning of "Madhyamaka") is similar to that used in the Copenhagen interpretation of quantum mechanics, and corresponds functionally to Monod's deferential arresting of inquiry.

The choice to move beyond the cosmic absurdity thesis in any direction, including by means of affirming a fundamental teleological principle, cannot be coerced, for neither the circumstances of life nor argument can force the abandonment of the vision of cosmic absurdity. However, it is a choice that can still be entertained. My conjectural conclusion amounts to the contention that *every* move beyond the cosmic absurdity thesis involves positing a fundamental teleological principle. Put differently, this conclusion fails only if there is a way simultaneously to affirm the overall meaningfulness of the cosmos (against cosmic absurdity) and yet to deny a fundamental teleological sweep to that cosmos—and, based on the absence of actual examples in addition to the other reasons I have given, I think there is no such possibility.

Of course, this is *not* to say that a meaningful cosmos necessitates a God; that is a separate case, and (as I have already said) there are many ways of speaking of fundamental teleological principles that do not advert to divine action, or even to divinity. Nor is it to say that the cosmos *must* be meaningful; in fact the case for cosmic absurdity in the idiosyncratic sense in which I have used the term is, I would say, every bit as strong as the case for its rejection. But that, too, is a separate case. The final conclusion does, however, significantly narrow the metaphysical choices available to those who affirm that biological evolution suggests a meaningful (as against an absurd) cosmos: to say this is implicitly to be committed to a fundamental teleological principle of some kind. And that is an awkward conclusion for a number of writers on the philosophical significance of evolutionary biology.

TELEOLOGY WITHOUT TELEOLOGY:
PURPOSE THROUGH EMERGENT COMPLEXITY

Paul Davies

1 *Introduction*

In this essay I will offer a modified version of the standard uniformitarian view of divine action.[1] According to the standard view, God acts not only by creating the universe "at the beginning" together with its laws, something which even deism asserts. God also acts uniformly by sustaining the universe moment by moment, since the existence of the universe is contingent on God's continuous creative activity. In addition, the unfolding of the potentialities embedded in these laws through the processes of nature can be considered God's action as well. The modification I am proposing is that, in selecting the laws of nature, God chooses *very specific* laws with very remarkable properties. In particular, the laws allow not only for chance events in the routine sense but for the genuine emergence of complexity in nature, an emergence that requires these laws but which goes far beyond a mere unfolding of their consequences.

Thus my view of divine action depends on a much more fecund understanding of nature than the one usually given by scientists. As I see nature, the development of complexity is not just an outworking of the laws of nature; it also depends on the kind of radical chance permitted by, and yet transcending the determination of, these very laws.[2] This has striking scientific implications: it means that the full gamut of natural creativity cannot be explained if one is limited in biology to neo-Darwinism and in physics to the fundamental laws of relativity and quantum mechanics. Still these laws are not only valid but essential, for they give rise to genuine openness in nature. The intrinsic creativity is entirely the result of natural processes which are infused with inherent powers of self-organization based on, though not reducible to, these very specific laws of nature. Thus, the emergence of complexity in no way warrants appeal to special divine action in particular events, particularly when this in turn entails an interventionist view of divine agency. (My openness, however, to newly developing "non-interventionist" views is noted below.) In essence, though nature's complexity gives every appearance of intentional design and purpose, it is entirely the result of natural processes. In effect it renders a teleology without providing for one, leading to my theme of "teleology without teleology."

In this paper I start with a brief discussion of three models of divine action: interventionism, non-interventionism, and uniformitarianism. I then introduce the

---

[1] In this essay I am working with a modified version of the typology developed in the "Introduction" to *Chaos and Complexity: Scientific Perspectives on Divine Action*, Robert John Russell, Nancey Murphy, and Arthur R. Peacocke, eds., (Vatican City State: Vatican Observatory Press, and Berkeley: The Center for Theology and the Natural Sciences, 1995), 9–13, hereafter *CAC*.

[2] These ideas are a development of previous arguments I gave in "The Intelligibility of Nature," in *Quantum Cosmology and the Laws of Nature: Scientific Perspectives on Divine Action*, Robert John Russell, Nancey Murphy, and C. J. Isham, eds., (Vatican City State: Vatican Observatory Press, and Berkeley: The Center for Theology and the Natural Sciences, 1993, 1996), hereafter *QCLN*. See also Paul Davies, *The Mind of God: The Scientific Basis for a Rational World* (New York: Simon & Schuster, 1992).

modified uniformitarian approach, modeled after the game of chess and elaborated through an extended discussion of complexity in nature. Finally, I respond to Darwinian-style attempts to explain away what I consider truly creative events in nature as mere unfoldings of underlying natural laws, mentioning briefly both the anthropic principle and the eternal creation of the universe by God. I include specific suggestions about how my ideas can be tested scientifically in terms of assessing the complexifying trends in nature and the search for extraterrestrial life in the universe.

2 How Can God Act In the World?

According to theological tradition, God not only creates and sustains the world but acts to bring about special events in nature and history. But what does it mean to say that God acts in nature? Can such acts be reconciled with the scientific picture of nature subject to lawlike principles? We may distinguish three ways in which God might be said to act in nature:

2.1 Interventionist Divine Action

Perhaps God's special actions break the ordinary flow of physical processes and entail a violation of the laws of nature. Conservative Christians typically approach divine action in this "interventionist" way.

According to one form of interventionism, God acts like a physical force in the world. God moves atoms and other microscopic objects about, but to do so, God must violate the physical laws studied by science. Thus if a particle would naturally follow trajectory X, then as a result of God's intervention it contravenes the laws of physics and follows trajectory Y. If God intervenes at the atomic level, we may call this "bottom-up" action, since the effects at the atomic level may percolate up to the macroscopic realm where they result in events such as those recorded as miracles in the Bible. Of course, once the principle of divine intervention is accepted, we may also entertain the possibility of God acting directly on a macroscopic scale.

The problem with this approach is that the idea of a god as *a force* pushing and pulling matter alongside the other forces of nature is decidedly unappealing to scientists—and to many theologians, since it reduces god to an aspect of nature. In its place, one may prescind from any physical representation or explanation of divine action and assert the purely abstract claim that God acts per se. Even then, if such action entails violating or suspending the laws of nature, the claim is still deeply anti-scientific. Moreover it is subject to internal theological contradiction, since the laws of nature being violated are themselves a reflection of God's ubiquitous and orderly action moment by moment in sustaining the world in all its physical regularity. This approach suggests the image of the cosmic magician, who creates a flawed universe and prods it whimsically from time to time to correct the errors. In spite of its intellectual problems and its apparent derivation from pagan notions of deity, this version of God's action is still the one favored by many religious people.

2.2 Non-interventionist Divine Action

According to a new approach being developed by several scholars, God effectively acts in special events in nature, making a difference in what actually comes to pass, but without interrupting these processes or violating the laws of nature. Instead, it is precisely these laws which make a "non-interventionist" action possible.[3]

---

[3] See the articles by Nancey Murphy, George Ellis, Tom Tracy, and Bob Russell in this volume and in *CAC*.

One option is to locate the effects of God's acts at the atomic level where quantum physics pertains. Here the concept of a unique, causally closed trajectory becomes problematic. Instead, quantum indeterminism permits a range of possible trajectories consistent with the laws of physics. The future evolution of a system is not fully determined by the physical forces at work; instead, nature is ontologically indeterministic.[4] Thus indeterminism, in turn, provides a possible means for God to act without violating the laws of physics.

One objection is that this sounds like a variation on the traditional God-of-the-gaps theme. But this objection can be easily met, since in this case the gaps are not explanatory gaps due to human ignorance, but physical gaps inherent in the structure of nature itself. A second, stronger objection that if God were to act repeatedly in this way on a specific physical system (e.g. on a particular atom) then the spirit, though not necessarily the letter, of the statistical laws of quantum physics would be violated, thus running into some of the objections associated with the interventionist option (section 2.1, above). Nevertheless, it is easy to imagine that God may achieve his/her purposes with ease by merely "loading the quantum dice" slightly across the universe. Such a stratagem would almost certainly go unnoticed. More elegantly, God could act entirely within a "fair dice" universe by introducing a degree of determinism within the range of indeterminism allowed by quantum physics. That is, within the range of quantum alternatives offered to a set of physical systems, God could "pick the choices" in each case in such a way as to respect the statistical laws, but nevertheless to determine the outcome (perhaps only to some level of probability). There is a spectrum of intervention between occasional and minor acts of "selection," and the extreme case of God's will totally "soaking up" the residual indeterminism left by the quantum laws.

Quantum indeterminism is not the only possibility for grounding the apparent behavioral openness of physical systems. No real finite physical system is physically closed, if for no other reason than that external gravitational disturbances cannot be shielded. Furthermore, many physical systems are "chaotic," which means that their behavior is exceedingly sensitive to minute external influences, including gravity.[5] Now epistemic unpredictability does not necessarily entail ontological indeterminism, as Wildman and Russell have carefully argued.[6] Moreover, we still use deterministic equations to describe chaotic systems in many domains, suggesting that they may indeed be deterministic systems. Still, unless one assumes that the universe as a whole is closed and deterministic,[7] it may in fact turn out that they are

---

[4] Ontological indeterminism is only one of several defensible interpretations of quantum physics. It is the one being adopted here, but none of these interpretations has greater predictive power than the others, leaving the choice open at this point.

[5] To take a simple example, suppose a box of gas subject to Newtonian laws were in fact closed save for the influence of a single electron located at the edge of the observable universe, ten billion light years away. Furthermore, let us suppose that the electron interacted with the molecules in the box not electrically, but via the much weaker inverse square law gravitational force. Now ask the question: After how many collisions with its neighbors would the predicted and actual trajectories of a given molecule diverge to the point that continued prediction is completely compromised? The answer turns out to be about thirty, which for air molecules at room temperature would take less than a millisecond!

[6] Wesley J. Wildman and Robert John Russell, "Chaos: A Mathematical Introduction with Philosophical Reflections," in *CAC*.

[7] The question of whether the universe as a whole is closed is a subtle one, depending on the definition of universe, but if we take "universe" to mean the region of space within our

ontologically open. Thus, in my opinion it is unwise to discuss physical systems as if they were closed and deterministic, even when neglecting quantum effects.[8]

The fact that physical systems may well be "open to the future" implies that there is room for such systems to be affected by God. Indeed, a common analogy for such divine action is the kind of "top-down" causality presupposed in many discussions of human agency. The mind-body interaction may include both bottom-up "nudging" of specific key quantum events as well as various forms of top-down causation.[9] The precise mechanism whereby downward causation might operate remains mysterious—we do not yet understand how minds and brains relate to one another! But in considering God's action in the world the fact is often overlooked that *our* minds can act in the world (perhaps through downward causation), so there would seem to be no logical impediment to God also acting in a similar manner.[10]

2.3 Uniform Divine Action

Finally, I should like to mention a type of divine action that is in a sense weaker than interventionism or non-interventionism, and yet in some ways more impressive. This is the traditional liberal Protestant and Anglo-Catholic view of God's action "in, through, and under" the laws of nature, to use Arthur Peacocke's helpful phrase.[11] It too takes on various modalities:

In its more deistic modes, this approach views God as playing the role of designer or grand architect by initially selecting a suitable set of laws from among the infinity of possibilities. These laws are chosen because of the inherent self-organizing and self-complexifying properties they confer to matter when it is subject to them.[12] This is the model of the universe I sought to develop in my book *The Cosmic Blueprint*.

In its more uniformitarian view of divine agency, this approach adds to God's initial choice an emphasis on God's continuing role of creating the universe afresh in each moment, though without in any way bringing about particular events which nature "on its own" would not have produced.

---

particle horizon at any given time, then this region certainly does not constitute a closed system, because unknowable physical influences can intrude across that horizon.

[8] John Polkinghorne makes a similar argument about chaos and complexity; see, for example, "The Metaphysics of Divine Action," in *CAC*; see also his *The Faith of a Physicist: Reflections of a Bottom-Up Thinker*, Gifford Lectures 1993–94 (Princeton, N.J.: Princeton University Press, 1994), 26. William Alston makes a similar argument about the lack of a sufficient warrant for determinism in general; see, for example, "Divine Action, Human Freedom, and the Laws of Nature," in *QCLN*, and *Divine Nature and Human Language: Essays in Philosophical Theology* (Ithaca, N.Y.: Cornell University Press, 1989).

[9] See, for example, Paul Davies, *The Cosmic Blueprint: New Discoveries in Nature's Creative Ability to Order the Universe* (New York: Touchstone, 1988); Arthur Peacocke, *Creation and the World of Science* (Oxford: Oxford University Press, 1979).

[10] For detailed discussion see the forthcoming publication resulting from the CTNS/Vatican Observatory conference on the neurosciences and cognitive sciences held in June, 1998.

[11] See Peacocke, *Creation and the World of Science*.

[12] For further discussions of design arguments, see G. F. R. Ellis, "The Theology of the Anthropic Principle," and N. Murphy, "Evidence of Design in the Fine-Tuning of the Universe," both in *QCLN*. See also George Ellis, *Before the Beginning: Cosmology Explained* (London: Boyars/Bowerdean, 1993); Nancey Murphy and George F. R. Ellis, *On the Moral Nature of the Universe: Theology, Cosmology, and Ethics* (Minneapolis, Minn.: Fortress Press, 1996).

It is the latter view which I wish to explore in more depth in the following sections of this paper, though in a new and sharply modified form, and I also wish to remain open to a combined approach which takes seriously non-interventionism (section 2.2, above) as well.

3 Modified Uniformitarianism: God as Chess Player

T. H. Huxley likened nature to the game of chess, with the pieces representing physical systems, the rules of chess the laws of nature, and the game itself the evolution of the universe.[13] It is a useful analogy. Suppose you were given a checkerboard and pieces and asked to invent a game. If you didn't give a lot of thought to the problem, the chances are that the game you invented would be either boringly repetitive or a total chaotic shambles.[14] In contrast, the rules of chess have been carefully selected to ensure a rich and interesting variety of play. More importantly, the end of any given game of chess is not determined by these alone rules, *but also by the specific sequence of moves* taken by each player and thus open to human whim. The rules serve to constrain, canalize and encourage certain patterns of behavior, but they do not fix it in advance. The game thus becomes an exquisite mix of order and unpredictability, which is why it is so fascinating.

We may exploit the chess analogy and suggest that God, on the one hand, acts by selecting from the set of all possible laws of nature those laws that encourage or facilitate rich and interesting patterns of behavior, and these laws are inherently statistical. On the other hand, the details of the actual evolution of the universe are left open to the "whims" of the players (including chance operating at the quantum or chaos level, the actions of human minds, etc.). I will call this proposed mode of divine action "modified uniformitarianism." I believe it has a number of appealing features.

First, God need never suspend, manipulate, bend or violate his/her own laws since their statistical character allows for the action of divine—and perhaps human—agency. There are no interventions, save for the miracle of existence itself. Second, God does not exercise an overbearing influence on the evolution of the universe, thus reducing it to a pointless charade. There is room for human freedom, and room for even inanimate systems to explore unforseen pathways into the future. A third advantage of this approach is that it enables one to discuss a concept of design in nature that is impervious to a Darwinian-style rebuttal. So long as it is agreed that the universe as it exists is not necessary—that it could have been otherwise—then clearly the actual universe has selected (or had selected for it) a particular set of laws from the (probably infinite) list of all possible laws. In my experience, almost all scientists, including hard-nosed atheists, concede this point. Thus the contingent nature of the world inevitably begs the questions: Why those laws? and, Is there anything special or peculiar about the actual set of laws selected?

The answer to the latter question seems to be a definitive yes. Much has been written about the various "anthropic" coincidences, the astonishing intelligibility of

---

[13] See Peter Coveney & Roger Highfield, *Frontiers of Complexity* (New York: Faber & Faber, 1995), 89.

[14] Generally, totally rule-bound games, like the Tower of Hanoi, are tedious and pointless, soon losing their appeal, while games of *pure* chance, like the card game snap, also have limited appeal. The best games are those that combine elements of chance and choice, for example chess, monopoly, and whist. Wesley Wildman has suggested to me that the game Yahtzee provides the most appealing analogy to nature in its mix of order and openness.

nature, its beauty and harmony, and so on.[15] Here I wish to dwell on the fact that the laws encourage the universe to behave creatively. This property of the laws makes them look to me as though they are the product of an ingenious—even loving—designer. Before such a conclusion can be drawn, however, we must consider the alternative possibility: whether the remarkable nature of the laws of the universe might only *appear* to be designed, but are in fact the product of pure chance in the "blind watchmaker" sense, to use Dawkins' apt phrase.[16] After all, Darwinism in the biological realm mimics design where there is none. Contrary to creationists and proponents of so-called "intelligent design,"[17] biological complexity evolved by natural selection acting on a large ensemble of randomly varying systems. So might a kind of cosmic Darwinism account in a similar way for the distinctive nature of the laws of physics? Might the suggestion of an element of intentional input in the laws be an illusion covering over the fact that there is none?

I believe the answer is no, since I assume that there is but one universe with its one set of laws. Unless multiple universes are allowed, the concept of Darwinian selection is simply inapplicable. Of course, some people have tried to argue that the hypothesis of a single universe is wrong: there is an ensemble of universes, and we have evolved in one compatible with our requirements as biological organisms.[18] In some versions of inflationary big bang cosmology, the visible universe occupies only one domain in a vast, possibly infinite, set of domains of the whole universe. Proponents of quantum gravity depict our universe as part of a mega-universe characterized by "eternal chaotic inflation" to use the phrase of Andrei Linde.[19] In all these examples, the laws and constants of nature may vary considerably among different domains.[20] Lee Smolin has even gone so far as to suggest a sort of cosmic Darwinism whereby cosmic natural selection operates on baby universes in a way analogous to biological natural selection.[21]

There are, however, many objections to these ideas: 1) First a familiar objection: Invoking an infinity of unseen (and perhaps unseeable) universes just to explain the one we do see is the antithesis of Occam's razor, and is fundamentally unscientific. In any case, it is scarcely more plausible than a single unseen God.[22] Now for two more novel objections: 2) It is hard to see how the appearance of law and rationality can emerge from total randomness. Anthropic selection cannot explain regularities in nature, the failure of which would not be life threatening. Yet we know of many such regularities. For example, the law of conservation of electric charge is known to great precision. Slight and rapid violations of this law in atoms would have only

---

[15] John Barrow and Frank Tipler, *The Anthropic Cosmological Principle* (Oxford: Oxford University Press, 1986); P. C. W. Davies, *The Accidental Universe* (Cambridge: Cambridge University Press, 1982).

[16] Richard Dawkins, *The Blind Watchmaker* (London: W. W. Norton & Company, 1986, 1987).

[17] Michael J. Behe, *Darwin's Black Box: The Biochemical Challenge to Evolution* (New York: Free Press, 1996).

[18] For a helpful overview, see John Leslie, *Universes* (London: Routledge, 1989).

[19] Andrei Linde, *Particle Physics and Inflationary Cosmology* (New York: Harwood Academic Publishers, 1990).

[20] For a helpful discussion and extensive references, see Willem B. Drees, *Beyond the Big Bang: Quantum Cosmologies and God* (La Salle, Ill.: Open Court, 1990), 48–51.

[21] Lee Smolin, *The Life of the Cosmos* (Oxford: Oxford University Press, 1997).

[22] In fact I suspect that a single unseen God and an infinity of unseen universes are ontologically equivalent for many versions of the multiple-universes theory.

trivial consequences for chemistry and life. Why don't such violations occur? 3) An ensemble of universes, each lawlike in differing ways, would still require an explanation for the origin of "law" as such, even if one could explain how, given such lawfulness, the actual laws of this universe obtained.

A key concept in the model of the divine selection of laws is that the laws themselves are in a certain sense timeless and eternal. To appropriate the ancient wisdom of Augustine for our purposes here,[23] God does not act *within* a pre-existing and endless time, picking a suitable set of laws at some moment in the past and then making a universe to try them out on. Instead God acts to create all that is, including space, time, and the laws of nature, from the mode of God's eternity, and thus these laws are in this sense eternal, too. Indeed, one of the purposes in choosing these laws is that they permit the universe—including space and time—to originate spontaneously "from nothing" in a lawlike manner, without the need for a further, special divine act.[24]

Thus God as the eternal selector is, in this function at least, outside time altogether. Still "creation" is not a once-and-for-all act at the big bang, but ongoing and inherent in nature itself. This is true since the continued existence of nature in time depends on God's will, and since nature, in turn, is highly creative in ways that go beyond the usual view of the genetic evolution of complexity. These insights have led me to the modified view of uniformitarian divine action which I will present next.

4 Modified Uniformitarianism and the Creative Cosmos

Among the infinite variety of possible laws will be a subset—some assert a very small subset—that permits what we would now call self-organizing complexity to emerge in the universe. One example that has been much discussed is the complexity associated with life and consciousness. The existence of these phenomena imposes rather stringent restrictions on the values of the fundamental constants of nature and on the cosmological initial conditions. If the laws of physics and the structure of the universe were not rather similar to the actual state of affairs then it is unlikely that life and consciousness, at least as we presently understand them, could exist.

It is likely that many complex systems (snowflakes, turbulent eddies, sand piles) are highly sensitive in their specific details to the form of the laws of physics. The source of most natural complexity can be traced to the existence of nonlinearity in the controlling forces. Such nonlinearity depends in turn upon the form of the underlying laws of physics (e.g., the specific Lagrangian), and on such quantitative

---

[23] Augustine, *Confessions*, trans. F. J. Speed (London: Sheed & Ward, 1960), Book 11.

[24] The meaning of "nothing" here is highly problematic, of course. I take the concept to mean nothing existed out of which our present, visible universe arose, with the exception of the eternal laws and constants of nature. If so, then the emergence of this universe may be an entirely "natural" event not requiring further special action by God. Of course, God would still have had to create the laws and choose the constants of nature. Alternatively, if our universe arose out of an eternally inflating mega-universe, as suggested by quantum cosmology, one would need still to explain why the mega-universe, with its fields and laws, exist. In any case, the point is that God creates from eternity, not at a moment in time. For further discussion see Paul Davies, *The Mind of God,* chap. 2. Also see Robert John Russell, "Finite Creation without a Beginning," in *QCLN*, especially 315–18, and articles by C. J. Isham and W. R. Stoeger in the same volume. See also C. J. Isham, "Creation of the Universe as a Quantum Process," in *Physics, Philosophy and Theology: A Common Quest for Understanding*, Robert J. Russell, William R. Stoeger and George V. Coyne, eds., (Vatican City State: Vatican Observatory, 1988).

details as the values of coupling constants, relative particle masses, etc. Of course, the very complexity of the subject of complexity precludes any general conclusions at the present state of our knowledge. All one can really say at present is that if one were able to "twiddle the knobs" and alter some of these fundamental quantities then the nature of most complex systems would vary greatly; indeed, the very existence of these systems may well be compromised. This is a subject area ripe for scientific investigation, and the results would have an important bearing on the whole question of divine selection and purpose. The narrower the range of lawlike possibilities that permit rich complexity, the more contrived, ingenious, and purposeful the universe will appear. Some scientists expect that there are quasi-universal principles of organization and complexity that will describe similar features in systems as disparate as embryos and spin glasses. These "organizing principles" would not usurp the fundamental laws of physics, but complement them. They would not be reducible to, or derivable from, the laws of physics, but neither would they be some sort of mystical or vitalistic addition to them. Instead, the organizing principles I have in mind would arise from the logical and mathematical structure inherent in all forms of complexity. At least that is the hope.

Central to this entire philosophy is that the emergence of organized complexity is lawlike, spontaneous and natural, and not the result of divine tinkering or vitalistic supervision. In other words, life and consciousness *emerge* as part of the natural outworking of the laws of physics. Thus from the uniformitarianist perspective taken in this paper, I am proposing that God "initially"[25] selects the laws, which then take care of the universe, both its coming-into-being at the big bang and its subsequent creative evolution, without the need for direct supernatural intervention. By selecting judiciously, God is able to bestow a rich creativity on the cosmos, because the actual laws of the universe have a remarkable ability to canalize, encourage, and facilitate the evolution of matter and energy along pathways leading to greater organizational complexity. Indeed, I have suggested that there may be a strict mathematical sense in which the laws optimize organizational complexity—what Freeman Dyson has termed "the principle of maximum diversity." Again, this is a subject ripe for scientific investigation: one may be able to prove that the familiar laws of physics form an optimal set.[26] If such ideas are correct, then in a certain scientific sense, we may well live in the best of all possible worlds.

Still it is important to realize that the particular form of the uniformitarianist argument being proposed here goes far beyond its usual "garden-variety" form of theism which often tends towards deism. My point is that in choosing these particular laws God also chose not to determine the universe in detail but instead to give a vital, co-creative role to nature herself. Remember, the selection of the laws, and even the cosmological initial conditions, does not serve to determine the fate of the universe in detail, since the laws that have in fact been selected contain an element of indeterminism. Instead, the role of chance is a two-edged sword. In most uniformi-

---

[25] By "initially" I do not mean in the temporal sense, but in the sense of being logically prior in the explanatory scheme.

[26] Let me give a specific example: Standard quantum mechanics assigns probabilities to components of the wave function according to a well-known calculus. One might consider a world in which such assignments were made according to an alternative prescription. I conjecture that the actual calculus that pertains to the real universe is such as to optimize some information theoretic quantity, such as the mutual information between components of entangled physical systems.

tarianist approaches to divine action, chance could be viewed as neglect or abandonment by God—the world left vulnerable to the vagaries of happenstance. In my modified approach, however, God's choice of chance bestows an openness—a freedom—upon nature that is crucial for its impressive creativity, for without chance the genuinely new could not come into existence and the world would be reduced to a pre-programed machine.

The exquisite and crucial feature of the actual arrangement of the statistical laws of physics—the specific amalgam of chance and necessity that pertains to the actual universe—is that chance is not mere anarchy, as it could so easily be if the laws were simply chosen entirely at random. Instead, chance and lawlike necessity conspire *at the basic physical level* felicitously to produce (incredibly!) emergent lawlike behavior at the higher levels of complexity.

This point cannot be overstressed. Many scientists have the misconception that complex order (e.g. mental activity associated with brains, universality in chaos, the ubiquity of fractal structures) is really no surprise because nature is ordered at the fundamental level of the basic laws of physics, and that this underlying order somehow reappears in complex systems. This argument has even been used to "explain" how it is that human beings are able to create the mathematics needed to describe the lawfulness of nature. In my view this is a totally erroneous connection.[27] The regularities observed in complex systems, which are often quasi-universal (e.g. Feigenbaum's numbers in chaos theory), are *emergent* phenomena, not pale manifestations of the underlying laws of physics. To be sure, if the laws were different, these regularities would be different (probably, I have argued, nonexistent), but the regularities are not derivable from, or reducible to, those underlying laws, for they depend in a crucial way on the openness and indeterminism of the complex systems involved. Thus the laws of physics produce order—the order of simplicity—at the micro-, reductionistic level, while the felicitous interplay of chance and necessity leads to the emergence of *a different sort of order*—the order of complexity—at the macro-, holistic level. It is this *specific* view of chance, which acts as a prerequisite for emergence and thus bequeaths the universe the conditions for the possibility of creativity, that underlies my modified uniformitarian view of divine action.

The fact that there has emerged an entirely new form of order—the order of complexity—in the organization of matter and energy on the macro-level, is part of the striking ingenuity of the laws of physics. How much easier it would be for an omnipotent deity to "cobble together" the complex systems "along the way" by crude manipulative intervention (as in the first approach, above), but how much less impressive! To select a set of laws that, through their subtle interplay, bring about a *natural* creativity of an orderly and organizational form, a spontaneous self-organizing potency that is not anarchic but hierarchical and constructive, which is ordered and yet open, determinate in its general trend and yet undecided in the specifics, is altogether more wonderful and cause for celebration!

In the earlier divine teleological schemes of pre-Darwinian Christianity, God directly selected a final outcome (e.g. the existence of "Man") and simply engineered the end product by supernatural manipulation. By contrast, the concept I am discussing is "teleology without teleology." God selects very special laws that guarantee a trend towards greater richness, diversity, and complexity through

---

[27] See Paul Davies, *The Mind of God,* especially chap. 6. See also Jack Cohen and Ian Stewart, *The Collapse of Chaos* (New York: Viking, 1994).

spontaneous self-organization, but the final outcome in all its details is open and left to chance. The creativity of nature mimics pre-Darwinian teleology, but does not require the violation or suspension of physical laws. Nature behaves *as if* it has specific preordained goals—it exhibits purpose-like qualities—while in fact it is, at least to a limited extent, open to the future.

How can we test these ideas? The clearest example to begin with is the emergence of life, consciousness and culture, and—the crowning achievement—intellectual schemes such as mathematics and science that capture (accurately describe, resonate with) the very laws upon which this magnificent edifice is constructed, completing the loop of existence that links the highest organizational level (mind) back to the lowest level (particles and fields of matter). How many of these features are due to chance and how many to necessity? I contend that the general trend of matter→mind→culture is written into the laws of nature at a fundamental level; that is, it is part of the natural outworking of ingeniously selected laws. But I also contend that the specific details (the human form, our mental make-up, the character of our culture) depend on the myriad accidents of evolution. If this argument is correct, we would expect the same universal laws to work themselves out along a similar trend elsewhere in the universe, to bring forth life, consciousness, and culture in other planetary systems too.

The acid test of my thesis is therefore whether or not we are alone in the vastness of the universe. If we are, then contrary to my hypothesis it suggests that life on Earth is either the product of a supernatural interventionist act in a universe of mind-numbing over-provision (given the 10^{20} or so stars within the observable universe), or else a hugely improbable but purely accidental series of events of staggering irrelevance. My hope and expectation is that we are not alone, and that life on Earth, including the emergence of mind, will be seen as a natural consequence of the outworking of universal laws. That is why I attach such importance to the search for extraterrestrial life.[28]

Finally I should say that even the modification to uniform divine action that I am proposing here will be regarded by many as too impoverished and remote a concept of God. There is, however, the option to combine my proposal with non-interventionism and consider that God may be more immediately involved in the process of evolutionary change when the laws of nature are themselves compatible—and by God's choice!—with non-interventionist divine agency! I'll leave it to my colleagues developing the non-interventionist approach to pursue this combined option.[29]

5 *How Does This Fit With Darwinism?*

I will be very interested in the reaction of biologists to the concept of "teleology without teleology." I expect the reaction might be favorable for at least two reasons.

First, biologists have already incorporated self-organization and emergent complexity as part of the neo-Darwinian account of biological evolution in a variety of ways. For example, in the self-organization of a protein, genes carry information about the linear sequence ("primary structure") of amino acids in the protein and play a "supervisory" role in its formation. The secondary, tertiary, and quaternary structures of the protein, however, all having to do with three-dimensional

---

[28] See Paul Davies, *Are We Alone?* (London: Penguin, 1995) and *The Fifth Miracle: the search for the origin of life* (New York: Simon & Schuster, 1998).

[29] See papers by George Ellis, Nancey Murphy, Bob Russell, and Tom Tracy in this and previous volumes of the series.

organization and functionality, are excellent examples of spontaneous self-organization. The developing embryo provides another remarkable example of these various kinds of self-organization.[30]

Secondly, there seems to be clear evidence of complexifying trends in biological phenomena. For example, the discovery, description, and interpretation of "trends" has been the stock-in-trade of paleontology for two hundred years. Another example is the so-called encephalization quotient, which compares brain size to body weight.[31]

Though diverting us from my central point, I feel obliged to add a word of caution here. Language about "trends" raises a highly contended issue: although I am *not* suggesting this, for some it seems appropriate to predicate of such trends the terms "directionality" and (particularly) "progress." Here we enter into semantic and, more importantly, philosophical questions, and the discussion of them among philosophically-inclined evolutionists has been extensive and at times rabid.

Some biologists find even the suggestion of trends unacceptable since they seem to imply "progress." Any such talk of "trends" or "progress," they argue, is the misguided result of anthropocentrism, of our chauvinistically placing value on our own species and regarding humans to be at the top of an evolutionary ladder.[32] Instead, evolution is blind; it is a random walk through the space of possibilities. Because life necessarily began as (relatively) simple, and diffused into the boundless space of possibilities as a result of random variation, there is the illusion of progress, of the trend-like advance of complexity. In fact, no such trend exists. Stephen Jay

---

[30] In general, self-organization is used rather loosely to describe any phenomenon in which complex organization emerges; still, physicists and biologists tend to use it in somewhat different ways. Physics provides scores of examples of what can be called spontaneous self-organization, processes which involve just general physical laws and random variables. The famous Bénard instability in fluid mechanics involving the formation of hexagonal convection cells provides a classic example of spontaneous self-organization, a curious mix of chance and necessity. Although general mathematical principles legislate that the cells shall be hexagons, the exact size, shape and location of the hexagons is decided by random fluctuations at the microscopic level. (Actually there are additional problems here. Suppose we assume that the exact pattern of the convection cells is determined by an information-bearing ingredient that could say: "A hexagon edge here please!" How could this be achieved by an ingredient within the fluid *that is distributed homogeneously through the fluid* in its initial uniform state?)

The situation in biology is more complex. Consider the problem of cell development. Although this is one of the most active areas of research in biology, it seems clear that the formation of cell structure results from a three-fold combination: 1) random factors, such as chemical diffusion and spontaneous symmetry breaking; 2) the laws of nature provided by physics and chemistry; and 3) deterministic factors such as the genetic information contained in the DNA. Thus the appearance of complex organization in a genetically programed system would seem to have a somewhat different quality from the spontaneous appearance of organization in, say, convection cells, the latter merely involving random factors operating together with general laws. Let me call the former phenomenon "supervised self-organization" and the latter spontaneous or "unsupervised" self-organization. My question to my biologist colleagues, then, is which biological features can be accounted for in terms of genetically supervised processes, and which are of the spontaneous, unsupervised variety?

[31] Once nature discovered the value of brainpower, it seems that the encephalization quotient accelerated. The growth of brain capacity among hominids has been especially fast. See the papers by Camilo Cela-Conde in this volume.

[32] In his earlier writings, Richard Dawkins forcefully argued that there is not only no evidence from the fossil record for "progress" or "directionality" in evolution, but also no evidence of "trends." Dawkins, *The Blind Watchmaker,* 181.

Gould has forcefully argued against any evidence for trends.[33] Richard Dawkins has been the subject of Gould's excoriating attacks[34] for seeming to propound progress, advance in complexity, directionality, and trend in *Climbing Mount Improbable*.[35]

Others, however, take a more moderate view of the notion of progress. Francisco J. Ayala, for example, has written a number of scholarly papers on "progress in evolution" beginning in 1974.[36] Yet he, too, is wary of overstating the case, as suggested by his recent criticisms of Michael Ruse. In Ayala's view, Ruse, in his recent book, *Monad to Man*,[37] attributes too much progress to the views of virtually all eminent evolutionists from Darwin to the present, including J. Maynard Smith and himself.[38]

The controversies not withstanding, if at least some biologists find language about self-organization and trends in evolution acceptable and even routine, it seems to me that these concepts provide evidence in favor of what I am calling "teleology without teleology." Again, I am using this phrase to point to the perspective in which self-organization and trends, though suggestive in a limited sense of teleology, are natural phenomena entirely amenable to scientific analysis. Moving into the context of theology, I suspect that this term will open up the possibility for theologians to speak in a more subtle and scientifically congenial way about divine purposes being achieved in evolution—that is, in a way entirely consistent with science. Of course this proposal would be an anathema if it led to the interventionist guiding hand of the divine. However, as I have been at pains to point out, that need not be so. Instead, in my view there is no miracle, no crudely-direct teleology, no supernatural tinkering; only the outworking of peculiarly creative and felicitous laws chosen, in my view, by God for these very purposes. And although the general trend of this process is basic to the laws, the actual details of evolution are left to the vagaries of chance. Moreover, the theological interpretation of this phenomenon in terms of divine purposes would not be obligatory on anyone. A scientist (or a theologian) could still shrug it aside with the remark: "That is just the way the world is. It's amazing, but I'll simply accept it as a brute fact." However, to others, it might be strong circumstantial evidence which is entirely consistent with physics and biology and which is in favor of a modified uniformitarian view of God who not only presides over a universe of subtle ingenuity and creativity, but of a universe in which the emergence of life and consciousness were part of God's blueprint at the most fundamental level.

Acknowledgment. My thanks to Robert Russell for extensively revising an earlier draft of this essay and to Francisco Ayala for helpful comments on the final version.

---

[33] See Stephen Jay Gould, "On Replacing the Idea of Progress with an Operational Notion of Directionality," in *Evolutionary Progress*, Matthew H. Nitecki, ed., (Chicago: University of Chicago Press, 1988).

[34] Stephen Jay Gould, "Self-Help for a Hedgehog Stuck on a Molehill," *Evolution* 51 (1997): 1020–23.

[35] Richard Dawkins, *Climbing Mount Improbable* (New York: Norton, 1966).

[36] See for example, Francisco J. Ayala, "Reduction in Biology: A Recent Challenge," in *Evolution at a Crossroads: The New Biology and the New Philosophy of Science*, David J. Depew and Bruce H. Weber, eds., (Cambridge: MIT Press, 1985), 65–79; idem, "Can 'Progress' be Defined as a Biological Concept?" in *Evolutionary Progress*, 75–96.

[37] Michael Ruse, *Monad to Man: The Concept of Progress in Evolutionary Biology* (Cambridge: Harvard University Press, 1996).

[38] Francisco J. Ayala, "Review of *Monad to Man*, by Michael Ruse," *Science* 275 (24 January, 1997): 495–96.

THE IMMANENT DIRECTIONALITY OF THE EVOLUTIONARY PROCESS, AND ITS RELATIONSHIP TO TELEOLOGY

William R. Stoeger, S.J.

1 *Introduction*

When confronted with the wonderful variety, intricate interrelatedness and hierarchical levels of complexity of our world, some scientists and philosophers conclude that there must be an overarching plan or directionality to it all. Are there holistic laws of nature—not just universal regularities and relationships like gravity, but principles of behavior which govern each entity in virtue of its role as a part of the larger whole, the universe itself—which embody that directionality more globally than do the four fundamental forces, and endow the universe and its components with a definite goal and purpose? Does time's arrow already in some way adequately embody a global directionality, or is something more required? Careful work of many scientific and philosophical researchers comes up empty-handed in the quest for any such controlling directionality or holistic regularities beyond those already given by physics, chemistry, and biology. From a global, synthetic perspective such an immanent directionality—many would say "immanent teleology," but that word is very ambiguous and unclear—*seems* essential. Without it there seems to be no sufficient reason for the observable universe to be globally ordered as it is, with the same laws and regularities holding throughout, and at the same time to manifest the coherent tendency towards complexification and life that it does. But from a local, analytic viewpoint there *seems* to be nothing more needed than what is already given by the laws of physics and chemistry, and what follow from, or is presupposed by, them, together with such key mechanisms as natural selection, symbiosis, and molecular drive at the biotic and prebiotic levels. Methodological reductionism triumphs. Each time a concerted effort to discern a vital force, a whole-part or top-down controlling factor, or a nomogenetic influence within evolutionary history and process is mounted, the results are negative. Though the mechanisms effecting evolutionary development lead to the emergence of new and more complex systems, they seem peculiarly blind and insensitive to future improvements and evolutionary advances. Chance events and randomness enter the evolutionary process in an essential and irreducible way. As Stephen J. Gould points out, if we were to return to any stage of evolutionary history and then run evolution forward from that point, such chance events would prevent the system from repeating the subsequent stages achieved in the first run.[1] What we would end up with would most probably be much different from what we now have.

This apparent lack of foresight and long-range "goal-awareness" and "purposiveness" in the laws of nature as we know them is philosophically counterintuitive, and *seems* to prevent us from establishing a clear connection between God's creative action in the world and what the natural sciences reveal to us in great detail about the dynamics, the history, and the underlying structure of reality. The aim of this paper is to demonstrate that this apparent disjunction or lack of connection is illusory, and that as a matter of fact the account of the emergence of structure, complexity, life,

---

[1] Stephen Jay Gould, *Wonderful Life* (New York: W. W. Norton and Co., 1989), 48, 50–52.

and consciousness in our universe given by the natural sciences can be brought into harmony with an adequate understanding of God's creative action in the world, *without* demanding or postulating global or holistic regularities, directionalities or teleological mechanisms or embodiments beyond the types of such regularities and mechanisms which are discovered and described at the level of the sciences. This does not mean that we know the laws of nature adequately or comprehensively as they really function. We do not. What it does mean is that we do not need, and will probably not find, global or teleologically comprehensive regularities or connections at the level of the sciences other than those they are already revealing to us, regularities and connections which would serve to determine or channel the later eventualities of the evolutionary process relative to any earlier stage towards a goal, or the fulfilment of its purpose. The regularities, processes, and relationships within physical and biological reality that we have not yet understood and modeled will not constitute such holistic laws nor possess the teleological reach some yearn for. Such holistic and teleologically comprehensive principles *may* be accessible to philosophical or theological investigation—they may function only at a metaphysical level. But, if so, they should have important manifestations in, and links with, the regularities and processes of nature as the natural sciences describe them.

We are really interested in three related questions. The first is: To what extent is there a directionality—and, further, even a teleology—immanent in the evolutionary dynamics of material reality, as modeled by physics, chemistry, and biology? The second is: How should we interpret at the level of metaphysics and theology what the sciences reveal in this regard concerning the directionality and uncertainty of the overall evolutionary process? The third question is complementary and more foundational: What are the key elements of a philosophy of reality, of creation, of being and action, of time, space, and matter, which together are adequate to what the sciences reveal and which at the same time are adequate to the relationships, structures, and grounds of reality which philosophical and theological analysis reveal? I shall be concerned primarily with first two of these questions, and leave a careful consideration of the third to future work. However, in the course of this paper I shall also indirectly indicate some of the elements necessary for constructing an adequate answer to that further question.

Other writers have also been moving towards answers to these questions. Bernd-Olaf Küppers and Paul Davies in their essays, research papers, and books have dealt with the first and second issues.[2] Paul Davies, Ian Barbour, Bob Russell, Wim Drees, Wesley Wildman, and Tom Tracy have discussed issues related to the second question.[3] And Barbour, Wildman, Charles Birch, John Haught, Philip Hefner, Colin Gunton, Arthur Peacocke, and John Polkinghorne have treated many issues related to the third.[4]

---

[2] Bernd-Olaf Küppers, "Understanding Complexity," in *Chaos and Complexity: Scientific Perspectives on Divine Action*, Robert John Russell, Nancey Murphy, and Arthur R. Peacocke, eds. (Vatican City State/Berkeley: Vatican Observatory Publications and The Center for Theology and the Natural Sciences, 1995), 93–105; Paul Davies in this volume.

[3] Paul Davies, Ian Barbour, Robert John Russell, Willem B. Drees, and Wesley J. Wildman in this volume; Thomas F. Tracy, "Particular Providence and the God of the Gaps," in *Chaos and Complexity*, Russell, Murphy, and Peacocke, eds., 289–334; idem, "Evolution, Divine Action and the Problem of Evil," in this volume.

[4] Ian G. Barbour, Wesley J. Wildman, John Haught, Philip Hefner, and Charles Birch in this volume; Colin Gunton, *The One, the Three and the Many* (Cambridge: Cambridge University Press, 1993); Arthur R. Peacocke, *God and the New Biology* (London: Dent,

Before giving a scientific account of the directionality of evolution, I shall briefly indicate the motivations I have for writing this paper, some of my metaphysical and epistemological presuppositions, and why it is necessary to summarize the technical details in our discussion of evolutionary directionality and God's action in the world.

I have three main reasons for crafting this paper. The first is to demonstrate that there really is a directionality manifest at the level of the sciences in the evolutionary process considered as a whole. This is very important to affirm, since so many scientists themselves seem to deny this. That is one reason why I include so much scientific detail in what follows, particularly from evolutionary biology and biophysics. The second reason is to investigate the connection between the directionality evident at the level of the sciences and various notions of teleology. To what extent does this directionality imply an immanent goal or purpose in the universe and in its evolutionary processes? The third motivation is to relate the notions of directionality and the teleology it implies to divine action, determining what constraints they put on our articulation of divine creative action.

With regard to my metaphysical presuppositions, I can not adequately articulate them all. Furthermore, I am in process of re-examining them. First of all, I am a critical realist. As such, I presuppose that the regularities and inter-relationships we describe in our scientific theories have some adequate cause or basis in what actually obtains in reality. I accept and presuppose the principle of sufficient reason as applied to contingent reality, and I also see no reason for not extending it to deal with ultimate questions. Along with many others who have contributed to this volume, I believe constitutive relationships are key to understanding reality. However, I also hold that at any level relationships must be between definite terms. Thus, I hold generally that at every level there is a substantiality to entities which allows them to enter into relationship at that level. This substantiality is given by constitutive relationships at more basic levels, and ultimately by an entity's particular constitutive relationship with God.

Although I hold that God does act in the world in both the universal creative sense and in a special sense as revealed to us by Godself, my emphasis in this paper is on God's universal creative action. In developing an adequate account of this, I am convinced that we must take what the sciences discover about the origin, development, and structure of reality—about reality itself—very seriously. Even though the account they give is provisional and limited, general features of physical, biological, and psychological reality and their proximate origins are convincingly substantiated in a coherent way by the methods and applications of the natural sciences themselves. Given their compelling explanatory force (Tom Tracy makes a similar point in his essay in this volume), these features must partially constrain how we conceive and articulate God's creative action in the world, and to a lesser extent, God's special action. The alternatives are either to consider the knowledge the sciences give us as illusory or superficial, or to consider that God's creative action in the world is not immanent in the regularities, processes, and inter-relationships we discover and try to describe. Either one of these alternatives is possible, but leads to consequences which are in severe conflict with philosophical and theological

1986); idem, *Theology for a Scientific Age* (Minneapolis, Minn.: Fortress Press, 1993); John Polkinghorne, "The Metaphysics of Divine Action," in *Chaos and Complexity*, Russell, Murphy, and Peacocke, eds., 147–56.

commitments we consider to be fundamental—philosophical and theological commitments we have a great deal of justification for making.

It is fair to state baldly that I hold that God's creative action is immanent in the processes revealed by the sciences—a statement with which everyone at this conference agrees. Why then the scientific details? We want to say more about God's action than that. If God's action is immanent in these processes, and in the evolution of the universe as a whole, then the details must tell us something about God's action—they must constrain how we articulate the concrete manifestations of God's creative action. The fact that chance is an important constituent, for instance, uncomfortable as it maybe from some points of view, must tell us something about God's involvement and activity in physical, chemical, and biological systems. And the fact that there is a perceptible complementary directionality—which is not rigidly deterministic, and somewhat but not completely open—must tell us something, too.

In particular, we need to recognize that, if there is no directionality manifest in the evolution of the universe, and of life, as the sciences describe them, then there is no directionality obvious at the level of features of reality which are subsumable under general laws and descriptions pertaining to material reality (only, perhaps, at the level of particularities). If that is true, then there is no end-directedness evident at these levels either. But, if there is no evident end-directedness, then, even if we believe that God is active in the world, God's intentions and purposiveness are not immanent or reflected in the world at this basic level (though, perhaps, other aspects of God's action may still be—God's power, for example. Anything depending on God purposiveness, for example, God's wisdom, could not be immanent at the level of general laws and regularities). As a consequence, there would be a real puzzle about what God's immanence in the world means concretely. Conversely, if there is a directionality and therefore some type of end-directness manifest in nature at the level of the sciences, then this already gives some concrete expression to the asserted immanence of God's universal creative action, intention, and purposiveness. Then, we can say that what the sciences deal with is at least to some extent in contact with what is significant from a philosophical and theological point of view. Further, and perhaps most importantly, there may then be no need to appeal to other hidden forces to complement what is given by the sciences, whereas otherwise we might be tempted to invoke such things if we discerned that there is no directionality given by the evolutionary process itself.

2 The Scientific Account of Directionality—Cosmology and Astronomy

The scientific account of the directionality of evolution begins really in cosmology with the Big Bang, moves through astronomy with the origin of structure, the stars and the elements, and enters a rather new phase with the origin of life on Earth. A second transition occurs with the origin of human consciousness. Here I want to discuss the cosmological, prebiotic, and biotic phases—focusing on the last—and leaving the directionality of evolution as revealed in the origin and development of human consciousness, which constructs culture, to another time.

In the very first trillionths of a second after the Big Bang, the fundamental laws of physics—those of gravitation, electromagnetism, and the strong and weak nuclear interactions—were determined. Together with them an arrow of time was established, either by the direction of increasing entropy or by the direction determined by the relationship between efficient causes and their effects. Long before stars and galaxies, and even before there was any neutral hydrogen, much less helium or the other elements, these fundamental regularities, which underlie, enable, and constrain all of material reality and its development, were fixed. Here I shall not delve into the complex and delicate issues involved in how those fundamental interactions and the arrow of time may have emerged. That has been discussed elsewhere by many people.

Under the interactions allowed and determined by those forces, remarkable structures emerged on both the macroscopic and the microscopic levels. It is likely that long before that occurred, however, there was an extremely early epoch of exponential expansion of the universe—inflation. After inflation, under the influence of gravity, small overdensities in the almost homogeneous cosmic fluid, which probably originated and were frozen into the cosmic plasma towards the end of the inflationary epoch, gradually grew as the universe expanded and cooled until they began to collapse and fragment to form the clusters of galaxies, galaxies, clusters of stars, and stars which now characterize our universe. Much earlier, in a series of crucial phase transitions as the universe rapidly cooled from unimaginably high temperatures—long before we had these structures—the neutrons, protons, electrons, neutrinos, and the other fundamental particles emerged together with the fundamental interactions which dictate their relationships and interactions with one another. More basically, those forces reflect the way those particles themselves are constructed—for instance, how neutrons and protons are constituted of three quarks bound together by the strong color force, which is essentially the strong nuclear force. Once the temperature of the universe—and of the homogeneous cosmic plasma—fell below about 4,000° K, much of the matter (only hydrogen, helium, and a trace of lithium at that time) became neutral, and decoupled from the radiation bathing it. It is only after this occurred that overdense seeds which had been sown at inflation could begin to grow to form galaxies, stars, and giant agglomerations of those. In all of this, the role of gravity is crucial and essential.

We need to stop at this stage now and ask what determined the diversification and consequent complexification which occurred in the universe up to this point. What determined this direction—that there should be the fundamental particles that we now have, and the simple atoms of hydrogen, helium, and lithium on the microscopic level? And on the macroscopic level what determined that we have such things as galaxies and stars? To what extent were these developments inevitable—or coded into the laws of nature—from the very beginning? And to what extent was the outcome of this early but important stage the result of unforeseen accidents? These are difficult but important questions to consider if we want to determine whether or not—and to what extent—there is a primordial directionality inherent in physical reality which in some sense guides or determines its evolution, and the emergence of new and more complex entities within it. That such new and more complex entities have continually emerged over the course of time is beyond doubt. Whether that was inevitable given conditions at earlier stages—that is, whether such a direction was in some sense predetermined, or predisposed given the regularities and interactions, and the conditions at earlier stages—is not beyond doubt.

In this relatively simple case—compared with the enormous complications which are introduced later in the biological phase—the answers to these questions are relatively clear-cut. As long as the universe is in a classical (nonquantum-cosmological) regime, then, *given* the conditions at any stage and the interactions according to which the systems and entities—particles, fields, perturbations—behave, evolve, and relate, the general situation at a subsequent stage is determined. In affirming this, I am not espousing a rigid and thorough determinism—that every detail of future configurations is determined. There are significant elements of indeterminism at the quantum level, obviously, and at least of unpredictability at the classical level due to chaotic structures. What I am stressing here is a flexible determinism relative to a given set of conditions and to given laws of nature as we have imperfectly formulated them. Given such conditions and laws, the system will evolve in a way yielding configurations within a certain definite narrow range of possibilities—a collapsing gas cloud into a cluster of stars, or a given distribution of chemical species into a definite final distribution of molecular products. Thus, in a definite sense the series of outcomes—the direction of evolution—*is* coded into both the regularities of nature and the conditions at earlier times. Obviously both contribute crucially to what eventuates at later stages of development. The laws and relationships of nature by themselves are inadequate to determine what occurs. Initial conditions must be given, but they—without the laws of physics and chemistry—are also woefully insufficient. Thus, within this classical descriptive framework, whatever or whoever establishes the physics and whatever or whoever sets the initial conditions encodes the evolutionary development we have charted up to the formation of galaxies and stars. This is covering very old ground! But it is worth revisiting in this context in order to orient ourselves with regard to these issues of the underlying directionality of evolution, and the level of its teleological content, according to the models of reality the natural sciences have given us.

As we move from a given stage to successive ones in this unfolding of physical reality, there is a certain inevitability. Genuine surprises are possible only because we are ignorant of some aspects of the processes dominating the system or of the conditions which obtain. As long as the universe is expanding and cooling at the rate it is, and as long as we have a spectrum of perturbations to work with, the formation of structure will occur. The instabilities which are generated define local direction-ality. Each instability automatically (as given by the laws of physics and chemistry) tends to growth or to decay and has a determined direction of resolution, which may be the formation of new structure or environment, or the relaxation into a former state. The environment in each locale, then, may be constantly changing, and the laws of physics and chemistry work on that environment in well-defined and determined ways to effect the emergence of a new environment containing at certain key transition points new systems and entities, like galaxies, stars, clusters of galaxies, neutral hydrogen and helium atoms, molecular hydrogen, etc.

Of course, we can ask about the quantum-cosmological stage which preceded this, and during which there was evidently no deterministic link between earlier and subsequent quantum states of the system. In this regime, apparently, there is no directionality—and certainly no definite and inevitable outcome from a particular stage. From this extremely early perspective, the establishment of the conditions which led to galaxies and stars, and perhaps even to the physical laws and particles we have now, was accidental. There was a certain quantum probability of a given set of conditions emerging from the quantum regime which would lead to something interesting, but that probability may have been very small. How do we interpret this?

Certainly, from the point of view of the quantum configurations and quantum cosmological regularities which dominated before the universe exited from the Planck era, the emergence of our region as it is now was indeterminate. It would depend on which of the quantum-statistical possibilities was actually realized. And which one was "selected" was apparently not determined by anything—that the universe exited from the Planck era, and later on from the inflationary epoch, was determined by the conditions and the laws of physics operative at that time, and that one particular set of classical initial conditions was selected from all the quantum-statistical possibilities was determined by the laws of quantum cosmology, but *which one* was not. At least that is the standard answer that physics can give at present. It is important to realize that these puzzles are intimately related with the problem of measurement and the collapse of the wave function in standard quantum mechanics and with the problem of quantum decoherence and a host of other issues in quantum cosmology, all of which remain to be adequately resolved. The primary point I wish to make here is that, however that happened, once we are outside the quantum-dominated realm and outside the inflationary epoch, the conditions and regularities obtaining at any stage are, from what we well know, directed towards—are set for—the formation of the structures in the universe we now have.

Once we have stars, we have the proper conditions for significantly enriching the microscopic world with a marvelous array of heavier elements—carbon, phosphorous, oxygen, iron, copper, nitrogen, all the way to uranium. Stars are the factories in which all the heavy elements are synthesized—either in their internal fusion processing of material, or in their rapid nucleosynthesis of it in supernova explosions, which also serve to scatter the newly formed heavy elements far and wide. Here we see again how the emergence of different and novel conditions, here macroscopic ones—the existence of stars, a new type of environment—provides the essential, and in this case sufficient, requirements for the manufacture of further novel entities—in this case microscopic ones. In fact, given stars together with the laws of nuclear physics and the constants of nature, the fabrication of elements heavier than lithium is inevitable. Once those heavier elements are scattered throughout the galaxies by the explosions of dying stars (supernovae), the interstellar material is now different than it was before—there are again new conditions, a much different type of material, upon which the laws of physics and chemistry can work. New and different macroscopic objects and microscopic objects become not only possible, but inevitable—new types of stars, Jupiter-like planets, and rock planets, asteroids, comets, crystals, dust, and a wondrous array of molecules. With these new objects pervading the universe, the conditions and environment has again evolved to something different, leading in turn to further new and exciting possibilities.

I have repeated this well-known story in this context to emphasize the order and especially the directedness inherent in physical reality—in the laws, regularities, and evolving conditions as they function together to constitute the processes and relationships which emerge at each stage of cosmic history. From the point of view of science, and perhaps even of philosophy, we cannot say where that directedness—these strong, persistent, and dynamic orientations towards diversity and complexity—originates, but we do know that it is embodied and actively functioning within physical reality, because we find it there. It is part of the order which characterizes our universe as a whole.

At this stage of our discussion it helps to stress two rather obvious philosophical points. The first is that this directedness and overall orientation resides in the totality of ordered and coordinated processes, systems, and entities, and not in any one interaction, relationship, or condition. They all contribute. And it is relatively clear how the interactions and processes we have studied—gravity, electromagnetism, and the nuclear interactions—operate on given conditions to constitute this directedness and orientation towards complexity. The physics of nonlinear, nonequilibrium systems, and chaos add important details—illuminating how certain types of self-organization occur—for instance of vortices, such as the giant red spot on Jupiter, of Saturn's rings, and of collective effects in plasmas. These all are manifestations of the intricate and extremely rich mutual interactions and influences with which the totality is endowed. This totality indicates a fundamental togetherness in space, so to speak. This togetherness is strongly reflected in the physical origin of reality and its fundamental interactions. Space and time and all that they "contain" were originally "together" and completely unified in the Planck epoch, immediately after the Big Bang. All reality was in the form of a unified quantum cosmological state, and all the basic laws of nature coincided in a single "superforce." Diversification and complexification began later and only gradually unfolded.

The second point is that the laws of nature indicate as well as a fundamental "togetherness" in time.[5] Past, present, and future are not separate realities. What is present depends radically on what happened in the past—obviously! And what will occur in the future depends on what happens in the past and present. Furthermore, what has occurred in the past, and what is happening and obtains in the present (for instance, a certain unstable condition—the very notion of an instability implies a future outcome) points to and is directed towards future possibilities—and, in many cases, relative certainties. It is very possible that the Earth will be struck by a large meteorite within the next thousand years. It is virtually certain that summer will come this year, and that a magnitude 7+ earthquake will occur somewhere in the world within the next six months. In some sense, the past is preserved and continues to exist in present and future realities. And the future is anticipated as either certain or very possible in present conditions and the processes and relationships which operate within them. This togetherness is fundamentally what space and time, and the laws of nature operating within them, do for us—at a very basic level—bringing disparate entities into relationship. And the fundamental interactions provide a dynamics to that togetherness, enriching it, and moving it forward to realize new possibilities and new entities situated in new environments, as we have already seen.

From this analysis, there are several other important points which come into view—points which have been made by Küppers and Mikhail Volkenstein.[6] The first is that methodological reductionism is an extremely powerful tool, especially when complemented by a synthetic appreciation and perspective aided by computer simulations, the physics of complex systems, nonlinear dynamics and nonequilibrium statistical mechanics, etc. At the level of physical and chemical reality all can in principle be explained using the sciences. There is no need for introducing radically new additions to our description of physical reality, hidden forces, or divine interventions. Thus, from what we have seen to this point—combining our analytic powers and our synthetic prowess—the directedness of cosmic evolution does not demand any other embodiment than the regularities and relationships describable by physics and chemistry in conjunction with the conditions and systems upon which they operate at any given stage.[7]

---

[5] William R. Stoeger, "Faith Reflects on the Evolving Universe: Divine Action, the Trinity and Time," in *Finding God in All Things*, Michael J. Himes and Stephen J. Pope, eds. (New York: Crossroad Publishing Co., 1996), 162–82, and references therein.

[6] Küppers, *Information and the Origin of Life* (Cambridge: MIT Press, 1990); idem, "Understanding Complexity"; Mikhail V. Volkenstein, *Physical Approaches to Biological Evolution* (New York: Springer-Verlag, 1994).

[7] William Pollard, *Chance and Providence: God's Action in a World Governed by Scientific Law* (New York: Scribner, 1958); Robert J. Russell, "Special Providence and Genetic Mutation," in this volume; George F. R. Ellis, "Ordinary and Extraordinary Divine Action: The Nexus of Interaction," in *Chaos and Complexity*, Russell, Murphy, and Peacocke, eds., 359–95; Nancey Murphy, "Divine Action in the Natural Order: Buridan's Ass

As we have already implied, the synthetic dimension is just as important as the analytic. It is critical to realize that certain phenomena can only be explained as the result of the nonlinear interaction of several forces—thus in these cases superposition does not work. Oftentimes the properties of a given entity—or perhaps the given entity itself—are very different when it is in relationship with another entity, or an integral part of system, instead of being free and independent. Neutrons which are bound in the nucleus, for instance, do not decay—free neutrons do, with a half-life of ten minutes or so. Free electrons interact with photons very strongly via Thompson scattering, to provide rather significant opacity at high temperature. Bound electrons interact with photons much less strongly, and differently. The characteristics of chemical compounds cannot be easily deduced from the characteristics of the elements making them up—unless the complex relationships among the components are carefully taken into account—including not just primary but also secondary and tertiary structure.

A second important point is that the phenomena of emergence and top-down causality are already evident at these levels. So is the related characteristic that "the whole is greater than the sum of its parts." (This is somewhat related to what we were just discussing above in terms of the crucial synthetic contribution.) These pervasive features of evolving reality are *not*, as both Küppers and Volkenstein emphasize, special or peculiar to biological systems.[8] And they do not as such require any other force to explain them. They are often the result of the regularities, processes, and relationships inherent in material reality acting upon entities and systems under different conditions—different boundary or initial conditions, different temperatures, different chemical potentials, etc. At this level at least they require no new physics or chemistry, as long as we take the synthetic dimension into account.

3 The Scientific Account of Directionality—Chemistry and Geophysics

We can easily see that the same assertions that we have made concerning the inherent directionality of evolution during the formation of galaxies and stars and the consequent synthesis of the elements apply during the next phase, the formation of planets, asteroids, comets, and cool gas clouds. These become the environments— the laboratories—within which new and more complex molecules are synthesized— water, ammonia, oxides, nitrides, simple and even more complex organic molecules such as formaldehyde, methane, and the building blocks for life, such as sugars, the amino acids, and the nucleic acids. The key idea here again is obvious—given certain conditions of temperature, radiation, concentration of certain elements and molecules, equilibrium or nonequilibrium, etc., the processes leading to the fabrication of new molecules will occur in cool dark clouds or on planets, asteroids, and comets. Similarly for these natural laboratories themselves—given proper

---

and Schrödinger's Cat," in *Chaos and Complexity*, Russell, Murphy, and Peacocke, eds., 325–57; Thomas F. Tracy, "Particular Providence and the God of the Gaps," in *Chaos and Complexity*, Russell, Murphy, and Peacocke, eds.; and idem, "Evolution, Divine Action and the Problem of Evil," in this volume, discuss the possibility that God may directly act in the window provided by quantum indeterminacy without violating the laws of nature to guide and direct evolution and other processes in nature. While this remains a possibility, there is no need to rely upon such a mode of divine action. The directedness evident in the universe and in physical and biological evolutionary development seems to be sufficiently accounted for by what the sciences give us—at that level.

[8] Küppers, *Information and the Origin of Life*; idem, "Understanding Complexity"; Volkenstein, *Physical Approaches to Biological Evolution*.

conditions in the stellar nebular disks around young stars, including their enrichment with heavier elements from previous generations of stars—planets, asteroids, meteorites, and gas clouds of various sorts will form from the dust and gas in the vicinity. In each case these new objects and new molecules radically modify the existing environment and conditions, or constitute completely new environments which in turn fulfil the conditions for the development of further new systems and environments, some of which may have even more immediate potential for evolutionary advance. And so on.

In fact these planets, asteroids, and gas clouds begin to evolve in their own very peculiar ways, depending on the ever-changing conditions which characterize them—or characterize various regions on them. As planets cool, contract, shift in their structure, develop atmospheres, are subjected to storms, electrical discharges, and showers of meteorites and cometary debris, they will change in certain— sometimes radical—ways, as will the conditions of temperature, chemical composition, atmosphere, etc. on their surfaces. As certain types of conditions develop, what were only remote possibilities become near certainties, and what were possibilities are bypassed and excluded from realization. New regularities and relationships emerge among the new entities and systems—regularities and relationships which are thoroughly consistent with the four fundamental physical forces and rely upon them, but which characterize the relationships and inherent behavior of the new entities and systems themselves. None of what happens to any particular system or object is absolutely inevitable. Unforeseen events and occurrences from outside the immediate environment of a given entity or system can intrude. A large meteoroid can collide with the planet, a nearby star can become a supernova bathing everything in powerful radiation, the whole system can collide with another system suddenly altering all the conditions. However, given certain conditions at any stage, as we have already said, and barring alterations of those conditions from such "outside" and infrequent catastrophic influences, the evolution of these systems is, in overall character, determined by those conditions, by the processes, and by the component entities which constitute them.

Thus, there is directionality coded into physical reality as it is constituted by the systems and entities, by the conditions, and by the processes, regularities, and relationships which obtain at any given time. There is more to this than merely a pattern of regularities that we observe. That pattern must have some sufficient cause in the structure of nature itself. With Paul Davies,[9] I locate this cause in the laws of nature, which in the weak sense are our models of how these regularities and inter-relationships operate and interconnect.[10] As Davies says, and as I emphasize in other terms, the directedness and propensities we observe in nature and in the evolutionary process can be derived from, or explained, by the laws of nature. In the strong sense, they are the inherent inter-relationships within nature as they actually are, and not necessarily as we imperfectly describe them.[11]

---

[9] Paul Davies, "Teleology Without Teleology: Purpose Through Emergent Complexity," in this volume.

[10] See William R. Stoeger, "Contemporary Physics and the Ontological Status of the Laws of Nature," in *Quantum Cosmology and the Laws of Nature: Scientific Perspectives on Divine Action*, Robert John Russell, Nancey Murphy and C. J. Isham, eds., (Vatican City State/Berkeley: Vatican Observatory Publications and The Center for Theology and the Natural Sciences, 1993), 209–34.

[11] Ibid.

How about unforeseen catastrophes and disruptive events? In the deepest sense they also should be included. The presence of a relatively distant asteroid which undergoes a series of orbit modifications in interaction with its surroundings so that it ends up on a collision course with a planet whose evolution we are monitoring is, at least remotely, part of the environment of that planet. The catastrophic collision which ensues is only unforeseen from the point of view of those who did not take surrounding objects and their evolving relationship to the planet into adequate consideration. What we refer to as chance, or contingent, events do not disrupt the directionality of evolution. They contribute strongly to it. As I have stressed above, directionality is coded into the concrete totality of physical reality, and that includes these "unforeseen" accidents or chance events—whether they be of ultimately classical or quantum provenance. The fact that they cannot always be predicted with certainty does not disqualify their contributions to directionality. They are somewhat like macroscopic mutations of the system. They can be incredibly destructive of what has already been attained. But they can also function as stimuli to novelty, suddenly altering conditions so that large previously unexplored regions of parameter space can be probed and realized. Just because we cannot always predict their occurrence or their long-term effects does not mean that they are not contributors to realizing the intrinsic possibilities of the whole system and therefore to embodying its directionality (see section 6 below).

However, in light of the possible amplification of quantum effects and the uncertain deterministic status of the physical structures which are modeled by deterministic chaos, it is not clear that what a given part of a complex system will yield in the distant future in terms of novelty and enhanced complexity is determined by the conditions now. Nevertheless, even here the possibilities for it are maintained, and the general patterns of the occurrence of novelty and increasing organization are apparent. This in itself indicates the embodiment of directionality and tendency in the concrete totality. Directionality, or even teleological content, does not require rigid determinism relative to any moment, but rather a manifold of possibilities together with the capability of gradually actualizing some of those possibilities, which are more and more complex.

At this point we begin to consider those molecules which are the immediate precursors of life on Earth. We are, of course, interested in the fabrication of amino acids and nucleic acids, the basic constituents of proteins and DNA/RNA, respectively. How did the rich variety of proteins and DNA sequences coding for them arise? Could this have happened just as a result of the chemical and physical regularities and relationships operating under certain constraints and conditions?

4 *The Scientific Account of Directionality—Biology*

The answer to the preceding question is, of course, "Yes!" Though we do not know how this happened, we do know, as Eigen, Schuster, and Küppers have emphasized, that it could have happened rather easily, by the operation of hierarchies of catalytic hypercycles consisting of mutating DNA or RNA sequences coevolving with the proteins for which they code.[12] We shall not go into the details, which have been

---

[12] Manfred Eigen and P. Schuster, *The Hypercycle—A Principle of Natural Self-Organization* (New York: Springer-Verlag, 1978); Manfred Eigen, *Steps Towards Life: A Perspective on Evolution* (Oxford: Oxford University Press, 1992); Küppers, *The Molecular Theory of Evolution* (New York: Springer-Verlag, 1983); idem, *Information and the Origin of Life*; idem, "Understanding Complexity."

extensively discussed elsewhere, but merely give a rough sketch of the basic ideas before going on to describe briefly directionality in biological evolution.

As John Maynard Smith and Eörs Szathmáry stress, along with others, the two principal characteristics of living organisms, metabolism and replication (heredity), have a mutual interdependence.[13] How could this have arisen? Did metabolism arise first, and then replication? Or vice-versa? Or did they develop together? Smith and Szathmáry present some of the alternative scenarios, and Eigen, Schuster, and Küppers discuss the coevolutionary possibilities.[14] Clifford Matthews has espoused a scheme which envisions abiotic (non-replicative) metabolism as primary.[15]

Before summarizing the status of our knowledge of the origins of biological molecules, we need to review the fundamentals. The first is that, as is very well known, the operation of natural selection is absolutely crucial at all phases of biological, and even in the earlier phases of pre-biological, evolutionary development—in order for self-organizing systems to adapt to given environmental conditions, and thus at the same time accumulate and preserve further information enabling the gradual improvement of emerging systems. And as Eigen stresses they have no choice but to do that—if they are going to continue to thrive.[16] What is needed for the processes of natural selection to operate? Obviously, there must be systems which 1) can replicate—this almost always involves the formation of templates—and interact with their environment, 2) are susceptible to mutations, changes in the structures of their templates which alter their interaction with their environment, and 3) possess metabolism, by which the replicating systems are made and maintained. The most difficult requirement to fulfil here is the fabrication of the polymers which are capable of template replication.

Thus, second, a range of conditions must be fulfilled for the gradual emergence of such hereditary replicating metabolic systems. The most fundamental and obvious one is the earlier abiotic formation of the various molecules which constitute them and enable them to replicate and function, the sugars, amino acids, nucleotides, or other molecules capable of base pairing, and the lipids needed for membrane formation—so that in subsequent evolutionary stages the unit of natural selection will be more complex protocells or cells rather than the replicating molecule itself.[17] For such a rich biochemical environment to form, and later on for the production of enzymes to effect and catalyze the proto-genetic chemical manipulations, many complex processes operating cooperatively are necessary—including autocatalytic cycles of nucleic acid sequences and proteins, subject to mutations.[18]

---

[13] Eigen, *Steps Towards Life*; John Maynard Smith and Eörs Szathmáry, *Major Transitions in Evolution* (San Francisco: W. H. Freeman and Co., 1995); Volkenstein, *Physical Approaches to Biological Evolution*; Küppers, *The Molecular Theory of Evolution*; idem, *Information and the Origin of Life*; idem, "Understanding Complexity."

[14] Smith and Szathmáry, *Major Transitions in Evolution*; Eigen and Schuster, *The Hypercycle*; Küppers, *The Molecular Theory of Evolution*.

[15] Clifford Matthews, "Cosmic Metabolism: The Origin of Macromolecules," in *Bioastronomy—the Next Steps*, G. Marx, ed. (The Netherlands: Kluwer Academic Publishers, 1988), 167–78; idem, "Origin of Life: Polymers before Monomers?" in *Environmental Evolution*, Lynn Margulis and Lorraine Olendzenski, eds. (Cambridge: MIT Press, 1992), 29–38; idem, "Dark Matter in the Solar System: Hydrogen Cyanide Polymers," in *Origins of Life and Evolution of the Biosphere*, 21 (1992): 421–34.

[16] Eigen, *Steps Towards Life*.

[17] Smith and Szathmáry, *Major Transitions in Evolution*, 27.

[18] Ibid., 33–34.

Third, as Küppers has pointed out, there are clearly three separate major epochs in the evolution of life.[19] There is, first of all, chemical evolution, during which *non-instructed* synthesis of the pre-biotic molecules we have mentioned above is accomplished. Second, there is the phase of molecular self-organization, during which there is the *instructed* synthesis of nucleic acids and proteins, along with their organization into self-reproducing genetic systems. Finally, there is biological evolution, in which complex multi-cellular systems developed from primitive genetic systems.

The status of acceptable and adequate scenarios for the formation of the necessary prebiotic molecular components for self-replicators and for their evolution into more complex genetic entities is briefly and well discussed by John Maynard Smith and Eörs Szathmáry.[20] They indicate that possible mechanisms for the formation of large numbers of amino acids, sugars, and even nucleic acid bases have been suggested for reasonable conditions on the primitive Earth. But there are some of the required components whose origins are more difficult to account for—particularly ribose and the pyramidines—and some necessary processes which are difficult to fit into the overall scenarios. It seems to me very likely, however, that further work will solve these problems. As many researchers are fond of mentioning, it is unlikely that we shall ever be able to reconstruct exactly how life originated on Earth. What people are concentrating on is showing how life *could have* originated. We seem well on the way to doing that in terms of the laws of physics and chemistry, the sequence of conditions which developed on the earth at very early times, and the chance events which occurred within that context. There is no indication whatsoever that something more than is studied or presupposed by physics and chemistry is needed in accounting for life. This does not mean that higher-level organization is predictable from lower-level entities and their behavior. Often it is not. But it does mean that, at the level of the sciences, it is totally explainable in terms of them.

There are further scenarios which seem adequate to explain the synthesis of the first replicators from this chemically diverse environment,[21] as well as the second (molecular self-organization) stage of evolution. Again, this is perfectly explicable on the basis of the known laws of physics and chemistry. The key elements in this transition from non-living to living organisms systems are: 1) the laws of chemical reaction kinetics; 2) molecular systems which are far from equilibrium (that is, there is an injection of energy or material); 3) non-linearity—the synthetic reactions are either autocatalytic or co-catalytic; 4) a random source of mutations, which can induce continuous novelty into the species represented in the system; 5) enzymes like the polymerases that can control mutability and integrate and stabilize favorable mutations into the system; 6) the origin and stabilization of the genetic code, by which the information in the nucleic acids is translated into functional proteins; and finally 7) systems which participate in co-operative (symbiotic) development, or co-evolution, thus providing a principle of selection and stabilization.[22] A leading

---

[19] Küppers, *The Molecular Theory of Evolution*; idem, *Information and the Origin of Life*.

[20] Smith and Szathmáry, *Major Transitions in Evolution*, 20–37.

[21] Ibid., 41–78; Ernest Nagel, "Teleology Revisited," chap. in *Teleology Revisited and Other Essays in the Philosophy and History of Science* (New York: Columbia University Press, 1979), 275–341.

[22] Küppers, *The Molecular Theory of Evolution*; William R. Stoeger, "Key Developments in Physics Challenging Philosophy and Theology," in *Religion and Science: History, Method, Dialogue*, W. Mark Richardson and Wesley J. Wildman, eds. (New York

proposal for this is Eigen and Schuster's "catalytic hypercycle."[23] All the suggested scenarios for the origin of life presuppose incorporate these fundamental elements.

Turning to biological evolution itself, let us recapitulate the key ideas needed in our more philosophical analysis of directionality, and later teleology, below. As we have already indicated, the role of natural selection and one of its preconditions, genetic mutation, is central. It is one of the key processes, if not the key process, which drives evolution, namely, the emergence of new organisms and the gradual modification of old ones as a result of their differential fitness, or ability to survive and propagate as populations, relative to the changing environmental conditions. Several clarifications are important here for our future discussions of directionality.

The first involves the relationship of natural selection to the environment and to the systems or entities upon which it works. Natural selection is severely constrained and conditioned by both of these factors. Obviously it can only select among the types of entities or systems which already exist. In itself it is not a process for producing new entities or systems—that is accomplished by a whole range of other processes determined by the laws and regularities of physics and chemistry affecting the behavior and interactions of the chemical species which are available at a given time, including random mutations and DNA recombination which lead to new DNA and RNA sequences. Natural selection sifts out the less successful or less fit (in terms of their reproductive capabilities) species or types and enhances those in the population which are more fit—relative to the environment—among those types which *already* exist. Thus it is a mechanism for preserving and propagating those species of molecules or organisms which are best adapted to the environmental conditions at a given time and in a given place. As new organisms or molecules form, more complex ones, say, it continues to operate on those. What is worth noting here is that the results of natural selection at one stage determine the range of new organisms via other natural processes at the next stage. In this sense evolution is always building on the types of organisms successful at earlier times. Some of these will have eventually disappeared and left no descendants, of course. But some of them will have descendant species far into the future—as long as life exists on earth. And the species existing at any given time will all have long lines of ancestors reaching all the way back to the beginning of life. Furthermore, the characteristics of organisms present at any given stage will always strongly depend on the characteristics of those at earlier stages. We have four limbs because all of our ancestral species going way back have four appendages. Natural selection is strongly constrained by what it has to work with![24]

Natural selection is also strongly constrained and conditioned by the environment, simply because it chooses those organisms for population enhancement which are optimally adapted to a range of *particular* environmental conditions. In fact, the organisms, or species of molecules, present in any given environment can only exist if a very narrow range of environmental conditions is maintained. Raising or lowering the temperature, changing the chemical composition of the medium however slightly, or increasing the surrounding radiation, will severely affect them. If conditions were different at any stage of evolutionary development, the range of organisms on which natural selection would act would be different (because different types of organisms in the past would have been selected), and the ones it would

---

and London: Routledge, 1996), 183–200.

[23] Eigen and Schuster, *The Hypercycle*; Küppers, *The Molecular Theory of Evolution.*
[24] Volkenstein, *Physical Approaches to Biological Evolution.*

select would also be different (even supposing those from which it can select are the same), since it is selecting them precisely on the basis of their viability and suitability in rather different conditions. It is also obvious that natural selection can strongly alter environmental conditions at a subsequent stage. Selecting a new organism for enhancement will alter both the range of organisms in subsequent selection and their ambience. The new organisms will affect the environment differently than the old ones, what they prefer to eat, how they live, what their waste products are, and how they interact with their context in other ways. The subsequent action of natural selection will be relative to these new conditions.

The second key point or perspective involved here, which casts important light on what is occurring, is that provided by the concept of information. As Manfred Eigen says, "Evolution as a whole is the steady generation of information—information that is written down in the genes of living organisms."[25] Once we have replicating molecules like DNA and RNA we have a chemical mechanism for preserving and transmitting vast quantities of information—and thus for gradually accumulating the extremely detailed programs necessary for the generation, growth, and operation of the most complex organisms. The results of the myriads of successful steps nature has taken, by law, by chance, and by the continual conspiracy of law and chance, in fashioning a given type of organism and enabling it to function and reproduce in a given range of environmental conditions are preserved in reproducible and implementable form (because of the genetic code linking DNA and RNA to the proteins which perform the necessary organic functions) in the genome or genotype of the organism. Its evolutionary history is contained therein, and is rendered effective as this present organism develops and lives, and eventually passes that information on to its progeny. Küppers has dealt at length with this aspect of molecular evolutionary biology.[26]

Though natural selection plays a critical role in sifting the information that is preserved and passed on in any population of organisms, the information contains much, much more than is attributable to natural selection alone. A population of organisms must exist for natural selection to work on. The genotype contains information about every structure and every process, including timing, levels of compartmentalization and morphogenesis, which is necessary for initiating and maintaining the generation and growth of an organism, given the proper environmental conditions and resources.

Furthermore, information is radically context-dependent, as all information is, relative to each pair of levels between which it is communicated.[27] Its meaning and its functional expression is always relative to the whole organism—the organism itself establishes a proximate context for understanding what the information means and how it is to be used (what it is for)—and to the conditions obtaining at each stage of its development and its life.

This raises the issue of boundary conditions. The information coded in the DNA is extremely sensitive to environmental conditions for its preservation and its functional expression—that is, it is context-dependent. But it also specifies, and under proper environmental conditions actualizes, the biological boundary conditions, which at each stage are essential for the full realization of the information the genotype contains. In another sense, the DNA sequence itself constitutes the

---

[25] Eigen, *Steps Towards Life*, 17.
[26] Küppers, *Information and the Origin of Life*.
[27] Ibid.

biological boundary conditions for the population of organisms—the extremely complex and intricately related constraints according to which the laws of nature must function within this limited hierarchy of complex and interrelated systems.[28]

In addition, in the generation and use of biological information, feedback loops are prevalent. There are innumerable feedback mechanisms enabling the control of development, and insuring homeostasis. But, more important for our considerations, in the very generation of this information over tens and hundreds of millions of years, feedback is occurring via natural selection and symbiosis to match the evolutionary development of populations of organisms to the environmental conditions, leading to viable novelty at each stage. Successful functionality for which a genotype codes is reinforced and preserved for future generations by natural selection. Weak or dysfunctional genotypes are weeded out and disappear. Again the success or failure of a complex of functionalities is relative to the environmental conditions with which the organism is confronted; that success or failure is transferred back to the survival or the extinction of the genotype responsible for it. As the title of one of Eigen's chapters has it, "Evolution means the optimization of functional efficiency."[29] This implies continuing overarching feedback and control of a very real sort.

The genotype, which contains all this information is looked upon by some as a"blueprint" for the organism, or the "computer program" according to which it is produced and functions. David Cole (unpublished comments) prefers "computer program," since it, like the genotype, does not prescribe a single unique outcome. There is some truth in these analogies. Volkenstein suggests, however, that it is really neither.[30] It is more like a "recipe"—it specifies the necessary ingredients and the stage, place, and way at which they are to be introduced, but the actual shape and structure is determined by the hundreds and sometime thousands of proteins acting at just the right moments and with the right strengths.[31] The recipe is at least one stage removed from a plan or blueprint. We cannot recover the genome by analyzing the structure of the organism; nor can we predict the organism's structure by knowing the genome. Besides being too complex, the genome sets in motion processes of self-organization which the genome itself controls but does not describe.[32]

Thus, the susceptibility to the surrounding conditions which characterizes all physical and chemical processes—expressing concretely the layered inter-relationality of reality—finds much fuller and more sophisticated realization in biological systems. They enhance and diversify in an amazing variety of ways, and at many different levels, the relationships between simultaneously existing and functioning parts and entities. They also deepen and complexify the relationships of parts and entities to one another in time—the links from those in the present to those in the past via heredity, and from those in the present to those in the future in terms of new proximate possibilities and the modification of present conditions for future evolutionary development. A good example of this would be the establishment of an oxygen atmosphere on the earth by cyanobacteria, which opened a wonderful niche for the development of oxygen-breathing organisms.

Moving from our rather extensive discussion of the informational aspects of evolution, the third key point is that there are evidently a number of other extremely

---

[28] Ibid., 157–58.
[29] Eigen, *Steps Towards Life*, 22.
[30] Volkenstein, *Physical Approaches to Biological Evolution*, 77–78, 249–50.
[31] Ibid., 78.
[32] Ibid.

important types of processes in biological evolution at the molecular and at the organismic levels which complement natural selection and the role of chance point mutations in the evolution of the genotype. Among the more important of these are symbiosis, which seems pervasive, molecular drive (gene conversion, transposition, slippage, transfer, etc.),[33] the role of neutral point mutations, redundancy of genes and functional stability, recombination in sexual reproduction, and compartmentaliza tion on different levels. As Charles Birch has pointed out (unpublished comments), all of these processes are themselves the result of natural selection. However, my point is that once they are in place they function in important ways along with natural selection. Furthermore, they do not arise solely as the result of natural selection—but from natural selection acting upon other biological processes and regularities. In particular, these mechanisms are crucial for the self-organization and stabilization of biomatter into even more complex entities, which then become subject to natural selection, instead of their parts or subsystems, which were formally the direct objects of natural selection.[34] They are also crucial for the rapid generation of novelty. Even at the level of very primitive replicators there are processes which involve "quasi-species" centered on a wild-type evolving by natural selection relatively rapidly over mountainous regions of the selection-value landscape.[35] Where a given genome evolving alone would take an enormous amount of time to reach a configuration of optimal fitness in a given environment, groups of genomes interacting with one another and with the environment evolve quickly as a group towards regions of maximal fitness (the mountainous regions). A whole class of genotypes is being selected for, and there are obviously cooperative phenomena involved in this.

The fourth key point is that all these processes which are involved in evolutionary development are non-linear and must take place far from equilibrium, that is, in an environment nourished by a continuous flux of matter and energy. In such conditions surprising, highly ordered forms—often referred to as "dissipative structures"—can develop. They begin as fluctuations which grow and stabilize—instead of damp, as they would in an equilibrium regime.[36]

The fifth key point is that in biological systems in general, there are multiple redundancies and a great deal of systemic "play" throughout. They can tolerate a wide range of alteration or injury without essential incapacity. This is one reason why these systems, and the many levels of component subsystems of which they are constituted, have been successful. These features enable them to achieve and maintain homeostasis and a range of flexibility, and to evolve significantly without loosing functionality and viability. Much of this "play" in biological systems is due to the rather great degeneracy of genotype relative to phenotype—that is, there are many genotypes which code for the same phenotype. This, in turn, is in large part due to the degeneracy of the primary structure of proteins relative to their tertiary and quaternary structure (a range of different sequences of amino acids may fold into functionally equivalent protein molecules).

The final point is that, though chance events are frequent and important in biological evolution, rendering its actual course indeterminate or unpredictable in

---

[33] Ibid., 237ff.

[34] Smith and Szathmáry, *Major Transitions in Evolution*, 6–7.

[35] Eigen, *Steps Towards Life*; Küppers, *The Molecular Theory of Evolution*.

[36] Arthur R. Peacocke, "Chance and Law in Irreversible Thermodynamics, Theoretical Biology, and Theology," in *Chaos and Complexity*, Russell, Murphy, and Peacocke, eds., 134–35, reprinted from *Philosophy in Science*, 4 (1990): 145–80.

exact outcome from any particular stage, these events and their short- and long-term effects—whether they be of point mutations at the level of molecular DNA, or the impact of a meteorite—are always within a context of regularities, constraints, and possibilities. They themselves are always among those possibilities—and occur with certain frequencies—even though their time and occurrence cannot be predicted. And, more importantly, when they do occur, the modifications they effect are always measured and understandable within the laws of nature. Thus, to refer to them as "pure chance," or to assert blithely that evolution proceeds by purely chance events, is much less than a precise description of this source of unpredictability in biological evolution. Though its actual course is indeterminate, its general course towards complexity, self-organization, and even the emergence of self-replicating molecules and systems, given the hierarchies of global and local conditions which are given, can be interpreted as inevitable in the universe in which we live.[37]

5 Self-Organization and Boolean Networks

In this section I want to summarize briefly the tentative but potentially very important finding of Stuart Kauffman and others on the self-organizational capabilities of Boolean networks, which quite possibly provide adequate models for the self-organizational capability of complex systems in general, including those in biology. An example is the way a large system of different genes interrelate and interact as a whole in their evolutionary development while still being able to guide effectively the development of the organism to which they belong.[38] Kauffman's principal point is that natural selection must be complemented by the self-organizing capabilities of matter at its different levels of interaction. Natural selection by itself cannot explain what has happened. This emphasis is in basic agreement with what we have been saying above, based on the work of Eigen, Volkenstein, Maynard Smith, and Küppers. As Francisco Ayala and David Cole have pointed out (unpublished comments), this assertion is very controversial. Some biologists maintain that these mechanisms, though possibly important in prebiotic evolutionary development, no longer function effectively once natural selection "cuts in." Davies (private communication) and many other physicists insist that they continue to remain important even after natural selection begins to function.

Kauffman details his findings regarding what he calls NK networks, systems of N components each of which is directly connected to K other components. An example would be N = 500 light bulbs linked together in an electrical network in such a way that each one is directly connected to K = 10 others. This means that whether or not a given light bulb is on or off at a given moment will be directly determined by whether each of the ten bulbs to which it is directly connected is on or off, along with a rule expressed in the wiring which expresses how many directly related light bulbs must be on for a given bulb to be on. Kauffman points out certain very surprising regularities which characterize the behavior of these systems, particularly the way certain classes of these systems—those with relatively very low K—self-organize extremely quickly, and zero in on developing within a very, very small part of their phase space. This seems always to happen when such systems are right on the boundary between subcriticality and supracriticality, between overly ordered boring behavior and chaotic behavior. Systems in this wonderful borderland

---

[37] Ibid., 136, quoting Eigen.
[38] Stuart A. Kauffman, *The Origin of Order* (Oxford: Oxford University Press, 1993); idem, *At Home in the Universe* (Oxford: Oxford University Press, 1995).

demonstrate an ability to diversify their behavior and their organization in very rich combinations which they are then able to preserve and build upon (by finding attractors)—for instance by establishing autocatalytic cycles among their components. Natural selection is then able to act on these varieties of subsystems. Systems which are too far from the boundary either settle into a frozen, static order (if their K is too low) or into chaotic behavior, in which all of parameter space is traversed by the system, and the states which are found quickly evanesce and disappear in the chaos. Thus, there seems to be emerging a set of "laws" which describe the behavior of these systems.

Connected with this is the character of the fitness landscapes within the phase spaces (possibility space or sequence space) of these systems. High fitness regions, as we have seen, are mountainous; low fitness regions are lowlands. A frozen system has a phase space which is random, in which it gets quickly and irretrievably trapped on one or two peaks. A very chaotic system has a very smooth phase space and roams almost at will throughout all its regions. Borderline systems tend to have very correlated phase spaces with significant mountainous regions—not just isolated mountain peaks of high fitness towering above the plains, but extensive areas covered with mountains, signifying many high fitness configurations closely related to one another. A system right on the boundary between boring and chaotic behavior strongly tends to evolve in these mountainous regions and can relatively quickly find not only the local maxima, but also the region where the global maximum is located, for a given set of conditions. It may not reach the global maximum for a long time— if ever—but it moves into that region with comparative alacrity, just as a quasi-species of genotypes is able to do (see previous section). Furthermore, as the system is evolving, the fitness landscape is also evolving. Each little change—for instance, the extension of a species—causes the landscape to change. Of course, what is desired for the growth of complexity is the situation not only where small changes do not alter the landscape too much—otherwise there is catastrophe for many of the entities in the system—but also where there is enough change, so that new possibilities can be explored without undoing the key accomplishments of the past. In fact, what happens, as seen in evolution itself, is that the stable complex systems which have already emerged become the foundation upon which further complex configurations are explored and discovered.

The significance of this work is, I believe, that it highlights the importance of the emerging interrelationships which the basic laws of nature at different levels constitute between the entities—the subsystems. These interrelationships, as long as they are not too strong or overwhelming—as long as they respect the integrity of the relata—provide the glue, so to speak, which allows further self-organization and complexification to occur and to stabilize. Nature evolves as a constantly diversifying unity, which at the same time respects—to a greater or lesser extent—the individuality and autonomy of the components, and yet uses those components as the basis for higher level organization.

The regularities Kauffman has found he considers as new laws of nature. And they very well may be. However, their status at present is not certain or clear. I tend to believe that they may be the first discoveries of what may become the laws of organization for complex systems. But, if that is the case, then, though they are important, they still do not provide the overarching holistic laws of nature, or the long-range teleological embodiment of purpose and goal in natural process for which some people yearn. Again, as I have indicated above I think that quest is misguided, and that such laws are not needed, even from a theological point of view, and will not be discovered at the level of the sciences.

There is a striking basic coherence between Kauffman's findings and the key aspects of evolution, which I have described above, and those of many other philosophers and theologians, for example process thinkers and others, like Colin Gunton, who have emphasized the essential importance of the interrelationships within reality—that things are constituted by their relationship to one another, and that there is a unity to all things, but a richly differentiated and evolving unity.[39] What we have been exploring in this paper, in part, are some of the concrete ways in which those evolving relationships within this unity are realized, function, and evolve.

6 *Teleological Reflections on the Scientific Account of Directionality*

We have seen that there is very rich directionality inherent in nature and in natural processes at every level. With more complex systems this directionality becomes ever more focused and specific. In what sense does this immanent directionality constitute a teleology? Here, of course, we are principally concerned with "long-range teleology." When we speak of the teleology inherent in evolution and in nature, what do we mean—and what do we not mean? Do we mean that there is a predetermined, precise goal or target configuration towards which the evolutionary process ineluctably proceeds? Does this directionality need to imply conscious purpose in order for it to embody a teleology? Or is it enough that definite ends result de facto, without any seeking or intending of those ends as goals? The immanent directionality in the overall evolutionary character of the universe, and of life on our planet, certainly embodies much more than the fact that definite ends result. Or, finally, is end-directed or goal-seeking behavior, without evident conscious purpose, adequate for attributing an immanent teleology to the processes of evolution? Answering these questions requires great care with respect to the different levels of investigation and discourse we are spanning. The last alternative is the one I shall end up accepting at the level of scientific and philosophical knowledge.

First, we must emphasize that we are primarily interested in determining what can be said about the purposiveness and finality of the dynamic complex of processes which constitute evolution in all its phases—cosmic, prebiotic, biological, cultural—at the level of the natural sciences, and then at the level of philosophy and of theology. Does the directionality we have identified and described constitute such a purposiveness and finality, and of what type is it? We are not specifically concerned with the purposiveness or goal-directed character of any particular development or transition—though that is an important and often controversial question in its own right—but rather of the universal phenomenon of evolution itself. The implications for our description of God's action in the world are obvious.

Does having a teleology or purposiveness inherent in nature require a definite predetermined goal or target? The way teleology is often discussed seems to presuppose a rigidly defined terminus. This would require some type of blueprint of the final configuration to be operatively present or at least available to the manifold of evolutionary processes and laws of nature so that it could somehow (*how* that information about the goal is transferred to the process, even if it were present, is a key question) guide them towards that target. Obviously, there would have to be feedback mechanisms which would keep the overall congeries of dynamic processes

---

[39] Gunton, *The One, the Three and the Many*.

moving toward the construction of the configuration represented in that blueprint or exemplar. Although there is certainly something like this occurring in biological systems themselves on a local level, there does not seem to be anything of this sort going on in the evolutionary process itself at the global level. (As we have already seen, even the genome of an organism is not a blueprint in the strict sense.) It is not at all evident that such a blueprint exists, or is effectively coded into the laws of nature. Neither are there long-range supervisory feedback mechanisms which control the gradual approach to a final configuration, nor evident cosmological principles which determine the destiny of components in light of the destiny or structure of the whole. This is not what the directionality, much less the teleology, of nature involves.

To have a teleology operative in a system does not, then, necessarily involve having a blueprint for the final product. What is essential is that the system move towards realizing its proximate and more remote possibilities in an ordered way. (What is also essential for a teleology, strictly speaking, is that these ends, or range of possible ends, are the very reason for the processes of the system to be as they are. From a scientific point of view this cannot usually be determined, but it cannot be ruled out, either. It often can be affirmed from a philosophical or a theological perspective, however.) Even when the exact outcome in evolution is not targeted from a distance, some gradually focused range of outcomes is being targeted depending on the complex conditions it spawns and encounters—to which it responds—on its evolutionary path. This is what some people have referred to as an "open teleonomic system."[40] There is what might be described as a developing nested set of directionalities which gradually emerge with ever greater specificity in certain locales within the overall evolutionary manifold. The manifold itself is a nested set of developing submanifolds, a hierarchy of macroscopic structures, descending in length scale but ascending in the possible degree of complexity and organization—universe, cluster of galaxies, galaxy, cluster of stars, star, planet, continent, locale, community of objects or organisms, object or organism, etc. The cosmic manifold, as we have already seen, manifests some very basic directionality in terms of time, expansion, and rate of global temperature decrease. At any given point in the universe's evolution we can say that it will continue to expand and cool for tens or hundreds of billions of years. This cooling sets very crucial and important conditions on the laws of physics and chemistry—and later biology—which dominate, and therefore, as we have seen, on the evolutionary processes which occur—for example, the formation of stars, and with them planets. Here, as we have already indicated, the role of gravity is essential. As this substructure develops, a more focused local directionality within a given substructure—planet or star—develops.

The evolving conditions around and within each local evolutionary submanifold establish a particular directionality there, which becomes more particular, more focused as its ecology stabilizes and evolves. The vast range of possibilities

---

[40] Ernst Mayr, "The Multiple Meanings of Teleological," chap. in *Towards a New Philosophy of Biology* (Cambridge: Harvard University Press, 1988), 38–66; idem, "How Biology Differs from the Physical Sciences," in *Evolution at a Crossroads*, David J. Depew and Bruce H. Weber, eds. (Cambridge: MIT Press, second printing 1986), 43–63; Richard T. O'Grady and Daniel R. Brooks, "Teleology and Biology," in *Entropy, Information, and Evolution*, Bruce H. Weber, David J. Depew, and James D. Smith, eds. (Cambridge: MIT Press, 2nd printing 1990), 285–316; John H. Campbell, "Evolution as Nonequilibrium Thermodynamics: Halfway There?" in *Entropy, Information and Evolution*, Weber, Depew, and Smith, eds., 275–84.

accessible to the entire universe are maintained for extremely long periods of time, but most of them are very remote relative to its basic structure and kinematic characteristics. Within each subsystem—especially those of stellar or planetary size—some of those remote possibilities are either eliminated or rendered more focused and proximate by the conditions developing there. For example, the conditions around a variable star, or one which undergoes periodic episodes of violent accretion with consequent bursts of X-radiation, will not preserve the possibilities for the development of complex systems, which require a very stable and temperate environment over long periods of time. Conditions on planets belonging to a stable, nonvariable star, like our own, as long as those planets are rich in minerals and water and neither too close nor too far from the star—so that the temperatures are moderate and never extreme—maintain rich possibilities for future evolutionary development, rendering some possibilities ever more proximate, and others virtually certain. The nested directionalities on that planet become ever more focused or well defined towards greater complexity and intricacy—towards life itself. Of course, as we have already mentioned, accidents and relatively unforeseen events occur—like a catastrophic collision with a meteor or asteroid. That may alter the conditions on the planet radically, causing sudden defocusing of possibilities and directionality, rendering some possibilities more remote and uncertain, or eliminating outright a broad range of potential future configurations. Thus, the overall directionality of nature is gradually discovered and realized in different ways in the different subsystems and subsubsystems which emerge within the cosmic manifold.

In appreciating what is continually occurring in these situations, we need to balance our analytical perspective with a synthetic one, as we stressed much earlier. What in fact is being given at each moment in the evolutionary process of some system or manifold, whether it be of a star, a planet, a region on the planet, or a compartment within a larger system, is an ecology—a rich interrelated network of conditions, processes, and entities. That ecology is not completely isolated—there is a flow of influences from the outside—but its mutually beneficial and intimate internal interrelationships are normally much stronger than those with the exterior, and it possesses boundaries partially separating and isolating it from its larger environment, and controlling the influences from the outside. What is concretely given is the whole interacting system or ecological complex as it and its components are mutually constituted and affected by their structural and functional interrelationships, by their own internal characteristics, and by the processes and conditions which obtain at any given time. Thus, the individual events that constitute this evolutionary development are all interrelated—those that take place more or less at the same time are related by being influenced by the same environment, conditions, processes and history, and those that are separated in time, depend on one another as present and future depend on the past, and as past and present move towards the future—particularly when feedback mechanisms are at work.

Another way of describing the directionality of nature is as weakly interacting evolving chains of realized possibilities. The realization of any given possibility presupposes the prior realization of other possibilities, which are the stepping stones to those involving greater complexity or organization. The actualizations of possibilities are not independent. They build on one another as physical and chemical processes operate on the initial and boundary conditions that constrain them at each stage. What was only a remote possibility relative to a previous stage of evolution later becomes a very strong possibility, or even a virtual certainty, relative to a more advanced stage. Some chains of actualized possibilities are very short—as the conditions of possibility for greater complexification are eliminated. Other chains are very long—like that of the Earth—since the prior and later conditions of possibility for more highly organized organisms and behaviors have been progressively fulfilled. Chains continue to lengthen—preserving the achievements of the past and storing information in their rich variety of genomes about how to adapt successfully to the conditions which dominate in the environment. Barring destabilization or disruption, such chains often create the conditions for their further growth towards more complex and self-organized entities and organisms.

Thus, the directionality and teleology hidden in the universe explores avenues of possibility and actualization in billions of experiments running concurrently, building in different ways upon whatever happens according to the conditions and entities which emerge at each stage in a given avenue or chain. Oftentimes a particular avenue reaches a certain level and stabilizes with no further possibilities open to it. Sometimes, under special conditions, the evolutionary process achieves a level—for instance the level of biological molecules—which establishes a firm basis for moving far beyond what has been achieved in most other situations. With time nature explores and actualizes many different avenues of possibility in this way, the concrete emergence of one new stage preparing the way for the emergence of more advanced stages, if the conditions are right.

The lack of definite pre-planned or pre-designed sets of entities does not render this any less directionality, or even any less teleology. There is a direction and an orientation towards the realization of a large number of possibilities overall. That is clear enough. The fact that the channeling and specification only occurs with long-range unpredictability and in widely varying degrees and achieves the highest levels of organization only in a certain few randomly selected locations certainly does not disqualify this from being directionality, nor from being genuine embodied teleology of a specific type (see below). It is just that this is not *our* usual way of conceiving them in our well and meticulously orchestrated, but very limited, activities. In fact, it is a very rich notion of directionality and teleology, which gives freedom and autonomy to the laws and processes of nature and encourages them to explore and realize the full range of proximate and more remote potentialities of the universe. Though we cannot say so from the data of science, this may very well be why the "laws of nature" have the characteristics they have.

But do these nested directionalities we have detected in the evolutionary process indicate *consciously* directed intention or purpose? From the point of view of the natural sciences, they do *not*—nor do they rule it out. There is nothing about the directionality we discern in evolution which inclines us to suspect that anything analogous or comparable to the conscious intentionality with which we are familiar is at work. Of course, what is implied here is that our experience of consciously intended processes and products provides the patterns necessary for recognizing consciously intended processes and products in general, in particular those of divine provenance. Signs of divine purposiveness within creation may be much different and much more subtle than those we associate with human intentionality, and are probably not susceptible to detection by the sciences. Thus, there *may be* such a divine intention or purpose driving evolution. However, the evidence for it is not embodied in concrete evolutionary history and products in an obvious or unambiguous way, susceptible to scientific analysis. The self-organizational properties of matter on so many different levels and the pervasive action of natural selection in the prebiotic and the biotic phases, along with the seemingly irreducible roles chance and contingency play throughout the evolutionary process, serve to mask and obscure any possible indications of conscious purpose or intentionality detectable by scientific methods. This is partially due to the limitations of the sciences themselves.

If we turn to philosophical analysis of the models and conclusions provided by evolutionary science, we also seem unable to reach such a strong conclusion—certainly not on the basis of arguments considered satisfactory by most philosophers. However, there may in fact be valid philosophical approaches that would yield one. The situation is unclear. The obstacle is that any philosophical treatment rests on metaphysical presuppositions, which are usually very difficult to justify. Personally, I believe that we can construct an argument for a primordial initiating cause on the basis of both the existence and the order of the universe. From there it is probably possible to argue to the conclusion that this order has been consciously intended in some definite sense. But this two-step move would have to be pursued very cautiously—particularly in articulating and justifying the metaphysical underpinnings we would need, for example the principle of sufficient reason, applied to ultimate questions.

When we move finally to the level of revelation and theology, we find that, within the Christian traditions certainly, we inevitably must conclude to a conscious divine purpose and intentionality in creation, and therefore by extension to the overall process of evolution, despite our inability to do so on purely scientific and philosophical grounds. Thus, at this level we become aware that God is somehow working within the immanent dynamisms and interlocking directionalities of the evolutionary process—despite and even through its autonomy, contingency, inner freedom, and apparent blindness. This conscious divine purposiveness is only unambiguously manifest in God's revelation of God's action and intention to us.

Though we cannot scientifically or philosophically support the assertion that evolution is intentionally or consciously directed towards a definite, or even general goal, can we assert that it is directed towards a definite goal or range of goals? Can the directionality we have uncovered in evolution be translated into an end-directed or goal-directed formulation of teleology, and not just a minimalist end-resulting one? We have already seen from our account and analysis of the full gamut of scientific evidence that it definitely can.

We turn now to consider the philosophical evidence for this less exalted teleological option. Before doing this, however, we should refer to Mayr's threefold classification of teleological concepts.[41] He reserves the term "teleological" for processes which involve a consciously intended purpose or goal. Programed goal-directness, as in organisms, whose development and functions are governed by the "program" encoded in their DNA, Mayr calls "teleonomic." Finally, automatic attainment of an end, as in the action of gravity or electromagnetism, he classifies as "teleomatic." Along with several others (for example, Nagel, O'Grady, and Brooks),[42] I am not convinced that the programed nature of certain types of goal-directedness adds anything substantially more teleological to Mayr's category "teleomatic." I rather see a more basic definite distinction between processes which are end-resulting and those which are end-directed,[43] neither of which is necessarily consciously goal-intending, that is, "teleological" in Mayr's sense. Further, we might also attach the general teleological category "goal-seeking" to the distinctive case,

[41] Mayr, "The Multiple Meanings of Teleological"; idem, "How Biology Differs from the Physical Sciences."

[42] Nagel, "Teleology Revisited"; O'Grady and Brooks, "Teleology and Biology."

[43] See O'Grady and Brooks, "Teleology and Biology."

discussed by Nagel, in which a system or process actively seeks a particular goal or configuration in such a way that the relevant variables or parameters, such as the power expended by a steam engine and the rotational speed of its turbine, are orthogonal to one another—that is, not connected to one another by a law of nature (instead, in this case, the governor establishes the connection and implements the feedback between the two important parameters).[44]

This last teleological category includes the functional end-seeking category which is so prominent in biology—evolutionary biology in particular—as described by Ayala in his paper.[45] The existence of the eye is explained by its function—it is for seeing. That is why it evolved. Although functional teleology contributes to the overall end-directedness of evolution at the level of the individual processes which constitute it, it is not directly related to the general issue upon which we are focusing here—the universal end-directedness of the evolutionary complex of processes itself.

What are the key differences between these teleological categories, which are not necessarily consciously determined: "end-resulting," "end-directed," and "goal-seeking"? From what we have already seen, the key difference between the last and the first two is the tightness, or the inevitability, of the connection between the process and the end or function it attains. If it is specified by a law of nature—if it is inevitable—then, by our definition, it is not goal-seeking. The principal difference between "end-resulting" and "end-directed" is simply that in the former the end state results but there is no necessary orientation towards it—the process is not directed towards any specific or general range of ends—whereas in an end-directed process there is. In Mayr's classification, end-resulting processes are not even teleomatic. From what we have discussed so far, we see that the laws of nature, both as we know them and as they function in themselves, are intrinsically directed towards a certain range of ends, even though of themselves they do not specify a unique outcome.

One of the consequences of there being laws of nature at all—regularities characterizing physical reality—is precisely that there are processes which are oriented towards certain general ends! If there were no end-directed, or end-seeking, behavior in physical reality, there would be no regularities, functions, or structures about which we could formulate laws of nature. Whether these ends are automatically attained or not, relative to the laws of nature as we understand them, has no bearing on this general point. Both automatic and programed end-directed processes fulfil this requirement. Without regularities or inter-relationships all processes would be end-resulting, probably, in the sense that definite events would occur—things would happen. But those outcomes would in no way be determined by, and could not be foreseen from a knowledge of, the structures, processes, and conditions obtaining at any particular time. From the perspective of the contemporary natural sciences, we formulate the laws of nature and carry out our investigations and analyses in terms of efficient causes, often ignoring final causes or ends. However, it becomes clear from this discussion that any ordering of efficient causes and their effects implicitly acknowledges and presupposes that the efficient causes and the processes which embody them are directed towards the realization of certain specific types of ends. Efficient causes always have certain specifiable *effects*.

How about end-resulting, but not necessarily end-directed, processes? This is an important case for our immediate considerations, because many scientists and philosophers consider the whole process of evolution to fall into this category. To the

---

[44] Nagel, "Teleology Revisited."

[45] Francisco J. Ayala, "Darwin's Devolution: Design without Designer," in this volume.

extent that a process is end-resulting, but not end-directed, we say that it is undetermined or indeterministic. Its outcome is a matter of chance, either classical chance, the intersection in an event of two or more apparently independent causal chains, or ontological, quantum mechanical, indeterminacy.

It is helpful to realize, as we have already begun to see, that in speaking of evolution as a process, we are really assigning a simple designation to an incredibly complex and interrelated set of processes. Each component process has certain particular ends—some very specific, and some more general. Many, perhaps even most, are deterministic, at least within certain well-defined limits. Given the conditions at any stage and location, the laws of physics, chemistry, and biology will determine the general, and in some cases even the specific, character of subsequent stages there and nearby. What we often refer to as chance, in the classical sense, which prevents us from predicting specific details of generally determined outcomes, is merely our lack of precise, comprehensive knowledge of all the conditions. As we have already seen, quantum effects can introduce an essential indeterminacy. However, except at the very beginning—at the universe's exit from the Planck era, or before that—quantum indeterminacy is contained within very narrow and well-defined limits, and does not undo the determination of a general range of ends in the evolutionary process. It seems that many writers in this area are quick to conclude that, because no specific form is uniquely determined by a given process, the process itself is oriented towards no general class of end products. This is patently incorrect. Given that the regularities and interrelationships of nature are operative, we can always identify a class of ends towards which the process under consideration is directed, within which the specific outcome of a process will inevitably fall.

There are two further points which shed light on the finality of the evolutionary process, and indeed on the finality of any process. The first is the importance of identifying the end, or range of ends, to which a given process is directed, given the regularities it embodies and the conditions and context within which it operates. A density perturbation in a gas cloud will grow, resulting in its collapse, fragmentation, and formation of stars and clusters of stars. If I observe an embryo developing inside a female monkey in a normal way, I know that it is going to result in a baby monkey. These examples lead to my second point: The end to which a process in directed and the degree of its specificity is relative to the conditions obtaining at any given stage.

Now that we have specified the type of teleology expressed in the directionality evident in evolution as a whole, we need to return to examine the possibility of conceiving the laws of nature as they actually function as constituting overarching holistic principles which embody specific long-range goal-directedness—for instance, life or human consciousness. We can begin with some very simple and almost trivial observations. The first is that, as Bob Russell points out (unpublished comments), overarching holistic principles are much more than the universal laws, such as gravitation, with which we are familiar. If they exist on the cosmic level, they would constrain all entities in the universe to behave in well-defined ways in virtue of their being parts related in specific ways to a well-defined whole. Then, if this whole—the universe, for instance—has a discernable purpose or end, the parts making up the universe would of necessity participate in this purpose or end. Though it is conceivable that the laws of nature as they function are the result or expression of such cosmic whole-part relationships, neither the existence nor the possible character of such relationships is at all evident at the level of the sciences. Furthermore, the observable universe, and *a fortiori* the universe of all things which exist, is not an entity whose essential characteristics as a whole and whose relationships with either its components or its ground are well defined and accessible to either the natural sciences or philosophy, except in the laws of nature as we know them. Thus, further, it is not possible to specify from the natural sciences and philosophy alone the ultimate end or goal of the universe, and to elaborate that end or goal at the level of the entities which compose it. This is possible, however, at the level of revelation and our reflection upon it—theology.

The second trivial observation in this regard is that, as the universe evolves, it breaks into hierarchies of hierarchies of relatively independently evolving objects, as we have already discussed. Once our galaxy had formed, it evolved relatively independently of other galaxies in the universe—as long as the conditions for its continued evolution are fulfilled, and barring disruption by neighboring galaxies. Our solar-planetary system behaves similarly. Thus, throughout the universe at any given time after the formation of structure, we have billions and billions of separate evolutionary experiments being carried out on different levels. At each hierarchical level we can, in principle, compare the different experiments and ask what is common and what is different in each case, giving answers to these questions in terms of astronomy, geology, chemistry, biochemistry, biology, etc.

In many of these venues at any given hierarchical level, we shall find that evolutionary development has stalled at a certain point and has eventually been terminated without further advance—for example, by the variability of a star in a planetary system, by a supernova, or by the collision of a planet with a large asteroid. Thus, if we envision the possibility of discovering overarching, holistic laws of nature which embody a specific end-directedness of the overall evolutionary process, then these same holistic laws must also explain why in so many instances—perhaps in all except one—that intended goal is *not* attainable.

If there are holistic laws—or if our laws of nature are subtly holistic—there are avenues available for incorporating this teleologically unsettling disappointment. A very promising one which incorporates our description of the teleology of the universe in terms of nested directionalities would be on the basis of a statistical ensemble, which has often been employed in a similar way with possible universes. The idea is that we have an enormous multitude of relatively separate individual evolutionary experiments in progress, as indeed we have, all relying on the same basic laws of nature at the level of physics and chemistry. They differ only in how conditions emerge at intermediate levels of cosmic evolution—say, after stars and planets form. Advancing to the highest levels of complexity and function demands the negotiation of millions of intermediate stages beyond those established in the formation of galaxies, stars, planets, elements, and molecules. In a very small percentage of these experiments these more advanced possibilities will be realized— life, consciousness, culture. Within the ensemble it is virtually certain, as long as the conditions of possibility remain open, that these will be achieved at least once, if not more often. It is likely, furthermore, that, if they are realized more than once, they will probably be realized somewhat differently in each case.

This approach has much to recommend it, obviously. It is basically consonant, as I have indicated, with the character of the directionality we have detected in the evolutionary process, and it relies heavily on what the sciences give us. It does not appeal to principles or forces which are unsupported by scientific evidence, or unjustified by careful philosophical analysis. We do have hierarchies of hierarchies of entities which develop over time. At the level of each hierarchy certain laws of nature are operative, and imply certain classes of end states which are possible within that context, and within the eventual sub-hierarchies of each given hierarchy.

The directionality or end-directedness of reality at that level is further focused and specified in the development of successive sub-hierarchies of organization. In many the emerging conditions will render further advance impossible. But, in some advancing complexification and self-organization will continue to the highest level of possibility, depending on the layers of conditions and structures which emerge and are stabilized. We have, in effect, not only many independent evolutionary experiments, but many different experiments in the evolution of teleology.

In the latter half of this discussion I have tried to stay close to what the sciences give us, relying on somewhat transparent and uncontroversial observations. I have also engaged in some low-level philosophical reflection on these scientifically based observations. Some of my philosophical presuppositions need further development and examination. However, further metaphysical treatment of teleology and the character of God's action in the universe, it seems to me, must take careful account of the general points I have made, and of the distinctions upon which I have insisted.

When we move to the theological level, we also need to take the preliminary conclusions we have reached above very seriously. As we mentioned towards the beginning of this concluding section, the account we have proposed is consistent with God's conscious intentions and expressed motivations as given in revelation. These are not evident in what the natural sciences tell us about evolutionary history and development. Nor are they contradicted by it. How we understand them and how they have been, and are being, realized, is, however, partially constrained by what we know of that history and development, and of the processes upon which it has relied. For instance, unless we do considerable violence to our notions of God's relationship with creation, we must admit that the laws of nature which we have come to know, however imperfectly, constitute one of the key ways in which God is active in the world. Much of what we then do theologically will depend on how we model and articulate divine conscious intending and acting toward created reality. With regard to the chance events which are crucial to evolutionary development, for example, are they limitations of God's intentional action within creation, or are they rather one of the ways of bringing about what God intends? Are they subject to divine intention, or are they something God "puts up with" in order that higher ends be realized (autonomy, freedom, richness of variety, surprise)? We can ask the same questions of the apparently free conscious intended activity of human beings. There the issues become even more acute, relative to the problem of moral evil and the harm that we are capable of and often do visit on creation and upon one another. How one resolves both of these theological issues will depend very strongly on how one resolves the problem of temporality and divine action. From the point of view of a completely atemporal God, for example, chance presents no problem.[46] But, obviously, if God is completely insulated from time, even more serious problems arise.[47] It is at this point that contact is made between teleological issues and a number of other key problems related to divine action.

Acknowledgment. My special thanks to all at the conference for their help and encouragement, particularly Bob Russell, Nancey Murphy, David Cole, Ian Barbour, Paul Davies, and George Ellis.

---

[46] See Ernan McMullin, "Evolutionary Contingency and Cosmic Purpose," in *Finding God in All Things*, Himes and Pope, eds., 140–61.

[47] Stoeger, "Faith Reflects on the Evolving Universe," in *Finding God in All Things*, Himes and Pope, eds., 162–82.

SPECIAL PROVIDENCE AND GENETIC MUTATION:
A NEW DEFENSE OF THEISTIC EVOLUTION[1]

Robert John Russell

The phenomenon of *gene mutation* is the only one so far known in these sciences which produces gross macroscopic effects but seems to depend directly on changes in individual molecules which in turn are governed by the Heisenberg indeterminacy principle.[2]

William Pollard

γενηθήτω τὸ θέλημά σου
("let it come about the will of thee," Matthew 6:10b)

1 *Introduction*

For well over a century, a variety of Christian theologians have found ways to accommodate or even to integrate the Darwinian theory of biological evolution into systematic and philosophical theology. Antagonism to evolution has also characterized this period, in part as a defense against those voices of atheism which have sought to use Darwinian evolution to attack Christianity. Today the majority of scholars who take seriously the mutually constructive interaction between theology and science have found evolution compatible with the core conviction that the God of the Bible is the creator of the universe and life within it. Evolution, in short, is God's way of creating life. God is both the absolute, transcendent source of the universe and the continuing, immanent creator of biological complexity. God gives the universe its existence at every moment *ex nihilo* and is the ultimate source of nature's causal efficacy, faithfully maintaining its regularities which we describe as the laws of nature. God provides the world with rich potentialities built into nature from the beginning, including the combination of law and chance which characterize physical and biological processes. God also acts in, with, under, and through these processes as immanent creator, bringing about the order, beauty, complexity, and wonder of life in what can either be called God's general providence (appropriate to a more traditional, static conception of nature) or continuous creation (*creatio continua*, emphasizing the dynamic character of the universe). This broad set of views is frequently called "theistic evolution;" it has representatives across the conservative/liberal spectrum, including Roman Catholics, Anglicans, Protestants, and Anabaptists.

In this paper I will start with theistic evolution and attempt to press the case for divine action further. Given the neo-Darwinian theory of evolution, can we also think of God as acting with specific intentions in particular biological events? In other words, can we think in terms of special providence, when the scene is nature and not

---

[1] This paper is a development of ideas previously published in "Cosmology from Alpha to Omega" *Zygon* 29.4 (1994): 557–77; "Theistic Evolution: Does God Really Act in Nature?" *Center for Theology and the Natural Sciences Bulletin*, 15.1 (Winter 1995): 19–32; "Does the 'God Who Acts' Really Act?: New approaches to divine action in light of science," *Theology Today* 54.1 (April 1997): 43–65.

[2] William G. Pollard, *Chance and Providence: God's Action in a World Governed by Scientific Law* (London: Faber and Faber, 1958), 56, my italics.

just God's special action, or "mighty works," in personal life and history? And can we do so without being forced to argue that God's special action constitutes an intervention into these processes and a violation of the laws of nature which God has established and which God maintains? To many theologians, the connection between special providence and intervention has seemed unavoidable, leaving them with a forced option. Liberals usually restrict language about God's action to our subjective response to what is really only God's uniform, general action. This strategy may avoid interventionism but, in my opinion, it reduces the theological meaning of God's action to a uniform, single enactment at best, often drifting to a kind of deism. Conservatives tend to argue for special providence as the particular and objective acts of God in history and nature attested to by faith. They thus offer a robust account of divine action but at the price of viewing God's action as interventionist. Moreover both options must still face the question of theodicy: what do we say about the pain, disease, suffering, and death which pervade the long history of life on earth?

Meanwhile critics of Christianity, such as Richard Dawkins, Carl Sagan, Daniel Dennett and Jacques Monod, have claimed that since chance pervades biological evolution and is built into its cornerstones—variation and natural selection—any claim about God's action in evolution is unintelligible. Liberal defenders of theistic evolution typically respond by emphasizing that God creates through the combination of chance and law; these reflect God's intentions in creating the universe as it is in the first place. But this strategy, unfortunately, does not really settle the matter. Chance in biology, from cell to organism to population and the environment, usually stands for our ignorance of what are in fact underlying, though exceedingly complex, deterministic processes. If chance is only epistemic ignorance and nature is really a closed causal system, then to claim that God acts through chance in evolution and not just "at the beginning" leads to an impasse. Either it means nothing, since nature does what it does entirely by law, or it forces liberals to the option they have tried verbally to reject but must logically affirm: that God intervenes in nature, breaking the web of nature, causing gaps in its closed causal order, and then acting in these gaps.[3] Conservative defenders of theistic evolution have been more willing to speak in terms of occasional intervention, but this option threatens to undercut the conversation with science and minimize their credibility in a scientifically informed culture; at worst, it can move them in the direction of scientific creationism.[4]

---

[3] Process theologians argue that an actual occasion comes to be as a result of God's action as subjective lure, together with the causal efficacy of the past and the event's intrinsic novelty. Still, if the outcome—the actual occasion—is in fact fully determined just by the past, as the deterministic equations of classical science together portend, it seems hard to understand how either divine lure or intrinsic novelty can be said to have been truly effective. If, however, the scene is quantum physics and with it, irreducible chance, it might be appropriate to argue that where science suggests that the past is insufficient to determine the outcome, metaphysics may now speak of novelty and lure both as operative in each actual quantum occasion.

[4] Unfortunately, when Christians attack Darwinian science and seek to replace it with "creation science" or "intelligent design," they play directly into the hands of the atheist, since they implicitly agree that it is Darwinism, and not its atheistic interpretation, which must be attacked. In doing so they ignore the fact that theistic evolution offers the real attack on atheism by successfully giving a Christian interpretation to science—thus undermining the very assumption they share with atheists, namely, that a Darwinian account of biological evolution is inherently atheistic. Not only does this abandon vast realms of biology to the atheistic camp, it implicitly undercuts the integrity of those Christians who faithfully pursue research in mainstream biology, as well as the vast number of Christians who, while not being

The purpose of this paper is to move us beyond these options to a new approach: a non-interventionist understanding of special providence. This approach is only possible theologically if nature, at least on some level, can be interpreted philosophically as ontologically indeterministic in light of contemporary science. This would mean that nature, according to science, is not an entirely closed causal system. Instead, the laws which science discover at least at one level would suggest that nature throughout that level is open: there are what could be called "natural gaps"[5] in the causal regularities of nature, they occur everywhere, and they are simply part of the way nature is constituted.

My claim is that chance in evolution, at the level of quantum mechanics underlying genetic mutation, is a sign not of epistemic ignorance but of ontological indeterminism. If this claim is sustained, we can view nature theologically as genuinely open to objective special providence without being forced into interventionism. I will refer to this as a "non-interventionist view of objective special providence."[6]

With this approach, God can be understood theologically as acting purposefully within the ongoing processes of biological evolution without disrupting them or violating the laws of nature. God's special action results in specific, objective consequences in nature, consequences which would not have resulted without God's special action. Yet, because of the irreducibly statistical character of quantum physics, these results would be entirely consistent with the laws of science, and because of the (ex hypothesis) indeterminism of these processes, God's special action would not entail their disruption. Essentially what science describes without reference to God is precisely what God, working invisibly in, with, and through the processes of nature, is accomplishing. Moreover, although these results may originate directly at the quantum level underlying genetics, they could lead eventually to indirect effects on populations and species. This insight is crucial if we are to make good on our claim that this non-interventionist approach is centered on *special* and *objective* providence. Finally, regarding the problem of theodicy, my suggestion is that we place the topic of biological evolution within a broader theology of both creation and redemption instead of focusing narrowly on creation alone, for God is not the source of pain and death but its redeemer.

I want to emphasize the importance of this approach. It offers a synthesis of the strengths of both liberal and conservative approaches by combining special

---

biologists as such, accept it and give it a Christian interpretation. How much better it would be if those promoting "creation science" and "intelligent design" would attack atheism instead of evolutionary biology!

[5] I hope that this new term conveys the idea of gaps as an ordinary part of nature found everywhere at the quantum level. It is basically equivalent to what Thomas Tracy calls "causal gaps." See sections 3.3 and 4.2 below.

[6] This term was first introduced in *Chaos and Complexity* together with a working typology on divine action. Both emerged through numerous conversations with Nancey Murphy and Tom Tracy, in which we sought to express a growing consensus at previous CTNS/VO conferences about how to characterize types of views regarding divine action. See *Chaos and Complexity: Scientific Perspectives on Divine Action*, Robert John Russell, Nancey Murphy and Arthur Peacocke, eds. (Vatican City State: Vatican Observatory and Berkeley, Calif.: The Center for Theology and the Natural Sciences, 1995), hereafter *CAC*. In a recent review essay of *CAC*, Owen Thomas makes calls for a more nuanced discussion of concepts like intervention, God of the gaps, and so on. I agree with him, and hope that my remarks here move us towards this goal. See Owen Thomas, "Chaos, Complexity and God: A Review Essay," in *Theology Today*, 54.1 (April, 1997): 66–76. See specifically pp. 74–75.

providence and non-interventionism. It undercuts the atheistic claim that evolution makes divine action impossible as well as the assumption that to defend divine action we must attack evolution. It may thereby lead to a new period of creative discussions of evolution and Christian faith.

The paper is divided into four sections. In section 2, the claim is elaborated as an hypothesis expressed in five steps, a comment on methodology is offered, and arguments in support of the hypothesis are given. Section 3 reviews the history of the hypothesis, including early sources, critical voices, and recent constructive developments. Section 4 addresses three caveats to the argument: Is it an explanation of how God acts? Is it a "gaps" argument? Is a "bottom-up" approach to divine action warranted and does it exclude other approaches? Finally, section 5 turns to the questions of divine purpose in evolution and the challenge of theodicy, concluding with two promising issues for further research.

2 The Hypothesis and Its Warrants

2.1 Special Providence and Genetic Mutation: A Theological Hypothesis

The theological hypothesis to be explored in this paper can be expressed in five steps:

1. The basic perspective of the hypothesis is theistic evolution: *Grounded in. Christian faith and life, systematic theology speaks of the Triune God as Creator: the absolute source and sustainer of an intelligible and contingent universe (*creatio ex nihilo*), and the continuing Creator who, together with nature, brings about what science describes in a neo-Darwinian framework as the biological evolution of life on earth (*creatio continua*). Thus the 3.8 billion years of biological evolution on earth is God's way of creating life (theistic evolution).*

2. Theistic evolution is explicitly expanded to include providential divine action: *God is not only the ultimate cause of all existence, God is also the source of its meaning and ultimate purpose. Thus God not only creates but guides and directs the evolution of life towards the fulfilling of God's overall, eschatological, purposes.*

3. Providence includes both general and special providence, and both are taken here to be objective: *God's action is here to be understood not only in terms of general providence, that is, not only in terms of God's providing evolution as a whole with an overall goal and purpose. It is also understood in terms of special providence: God's special action having specific and objective consequences for evolution. These consequences would not otherwise have occurred within God's general providence alone, though they can only be recognized as due to God's action through faith. Note: here I am assuming that God's special action, though special and objective, is mediated through the natural processes with which God works. Thus God's action is neither unilateral or unmediated, nor is it entirely reducible to a natural process. Finally, I am assuming that God may act directly in particular ways at one level in nature, that this action may have indirect consequences at another level, and that these indirect consequences may be identified as acts of divine objective special providence.*[7]

---

[7] By *mediated*, I mean 1) that God acts with and through the existing processes of nature, and 2) that God's action assumes and takes up their histories and what has been accomplished thereby. I am drawing the distinction between *direct* and *indirect* acts from the context of the philosophy of action as it has developed around the problem of human agency. By a "direct act" I mean an act which an agent accomplishes without having to a perform any

4. Now in a crucial move responding to the problem of rendering our theological program intelligible in light of science, we require that objective special providence be understood as non-interventionist: *God does not act by violating or suspending the stream of natural processes but by acting within them. God does not act in a way which violates or suspends the known laws of nature but rather in a way which is intelligible from a theological perspective in light of them (though without such a theological perspective we cannot recognize the effects of God's actions as, in fact, the effects of God's action). Our account of divine action does not rely on a gap in our current scientific knowledge but on the positive content of that knowledge. Because of that knowledge, our account of divine action does not require God to create gaps in an otherwise closed causal order but relies on the intrinsically open character of natural processes.*

5. With these requirements in mind, we present the core hypothesis: The non-interventionist effects of God's special action occur directly at the level of, and are

---

prior act to bring it about. By an "indirect act" I mean an act which an agent eventually accomplishes by setting into motion a sequence of events stemming from a direct act which the agent actually performs. So, for example, when I turn on a light switch, the indirect act of my finger moving the switch is the result of a sequence of biological events in my body stemming, originally, from a direct act by which I initiate this sequence of events, presumably through a form of "top-down" causality between my mind and my brain. I will use the distinction between direct and indirect acts analogously for divine action, recognizing the severe apophatic limitations on any such analogy. In the context of this paper, I am considering the possibility that when we view events at the level of the phenotype as acts of God they are actually the indirect acts of God. According to the idea being explored here, God acts directly at the level of the genotype, and the sequence of events so initiated may result in an effect in the phenotype.

From this it follows that *all indirect divine acts are mediated*. A direct divine act may be mediated, but it is possible to conceive of a direct divine act which is *immediate* or *unmediated*. In this case, God performs such an act unilaterally without taking up the existing material or natural processes. If this were to include violating or suspending the laws of nature, it would be roughly equivalent to what I mean by *interventionism*. I prefer to leave open the question here whether an immediate direct divine act is necessarily interventionist.

I want to emphasize that the theological context of this discussion of divine acts is special providence. These acts presuppose the existence of the universe as a whole, in each part, and at every moment, as well as the laws of nature, and thus the doctrine of providence presupposes the doctrine of creation *ex nihilo*. This means that God's acts of creation *ex nihilo* should be considered as direct: each event in the universe, and not just the first (if there were one, such as at t=0), depends on God's direct act as its Creator. It also means that each direct act of creation is an immediate divine act: God creates each event *ex nihilo*, not in, with, and through the events which precede it. So, whereas in the context of providence one can refer to a mediated direct act, in the context of creation one can not.

The present focus on special providence in nature would need extensive refinement if it were to be extended to include a theology of miracles, including the nature miracles, the healing miracles, and the central three-fold miracle of the Incarnation, Resurrection, and Ascension of Christ. I hope to develop a fuller treatment of the relation between providence and miracle, and their relation to science, in future research.

Note: I am indebted to the discussion of divine action found in Thomas Tracy in "Particular Providence and the God of the Gaps,"in *CAC*, particularly pp. 294–96, and in private conversations. I am using the term "direct act" as roughly equivalent to the term "basic act" as it is frequently found in Tracy and elsewhere. The distinction between direct and indirect acts and Tracy's distinction between basic and "instrumental" acts may also be roughly equivalent. On the other hand, Tracy occasionally uses the term "instrumental" to mean "mediated; " clearly this is not equivalent to how I wish to use the term "indirect."

mediated by, those genetic variations in which quantum processes play a significant role in biological evolution. *Admittedly it is possible, as others argue and although I doubt it, that the effects of God's creative and providential action may occur directly and in a non-interventionist manner at many, perhaps even every, level of biological complexity when and as these levels evolve, and that if these direct effects occur there they may in turn initiate sequences of subsequent, indirect effects both via top-down and whole-part patterns of interaction between and within these levels.[8] Nevertheless, the current hypothesis is that if divine action within evolution is to be both direct and non-interventionist, the most likely locus of these effects, and perhaps (contrary to the arguments of others noted above) the **only** such locus, is the level of genetic mutation. It is specifically at this level that I claim such direct effects may arise in a non-interventionist way, that is, in a way entirely consistent with the known laws of physics, chemistry, and molecular biology. Moreover, if the effects of God's action do indeed occur directly at the level of genetic variation as we claim they do, they may have bottom-up conse- quences which indirectly affect the course of evolution.[9] However, for this claim to hold, we must show that quantum processes may be interpreted philosophically in terms of ontological indeterminism and that quantum processes are relevant scientifically to genetic variation.*

2.2 Comment on Methodology

This is a project in constructive theology, with special attention to a theology of nature. This hypothesis should be taken not as a form of natural theology, nor one of physico-theology, and most certainly not an argument from design. Instead it is part of a general constructive trinitarian theology pursued in the tradition of *fides quaerens intellectum*, whose warrant and justification lies elsewhere and which incorporates the results of science and the concerns for nature into its broader framework mediated by philosophy. Although I believe God's special action is intelligible in the context of, and coherent with, our scientific view of the world because of the indeterministic character of the natural processes as suggested by a plausible philosophical interpretation of quantum physics, such special action cannot be discovered by the natural sciences as such nor can it be based on them. Where faith will posit it, science will see only random events described, as far as they can be by science, by the theories of physics, chemistry, biology, ecology, and so on. The positive grounds for an alleged divine action are theological, not scientific. This hypothesis is not drawn from science even though it aims to be consistent with science.

Science would not be expected to include anything explicitly about God's action in nature as part of its scientific explanation of the world. Theology, however, in *its* explanation of the world, can and should include both. This is as it should be for the mutual integrity of, and distinction between, the two fields of inquiry, and for the order of containment entailed by emergence views of epistemology which requires

---

[8] Usually "top-down causality" refers to relations between a higher and a lower level, as in mind/brain, or organ/cell, while "whole-part causality" or constraint refers to relations between boundary conditions and internal states at the same level, such as vortices produced in a liquid when heated in a pot, or quantum eigenstates in a square potential well.

[9] "Bottom-up causality" usually refers to effects occurring at a higher level due to processes at a lower level, such as temperature and pressure of a gas arising from the kinetic energy of its atoms and their exchange of momentum with the container.

that theology include and be constrained by, while transcending, science in its mode of explanation.

2.3 Arguments in Support of the Hypothesis

Point (1) above is the standard move in theistic evolution which is presupposed here without further discussion. In support of (2), (3), and (4), this section begins with discussions of the importance of a theology of objective special providence, the problem of interventionism, and the possibility of a non-interventionist approach. To address (5) I then offer support for an indeterministic interpretation of quantum mechanics, for the importance of genetic mutations in biological evolution, and for the role of quantum mechanics in these mutations. I also need to explain why I have chosen a bottom-up account of divine action here, and how this account should be placed within a larger framework that includes both top-down and whole-part causality, but I will defer these questions to section 4.3.

2.3.1 Why Is a "Non-interventionist View of Objective Special Providence" Important Theologically?

Background to the problem of divine action: historical perspectives. The notion of God's acting in the world is central to the biblical witness. From the call of Abraham and the Exodus from Egypt to the birth, ministry, death, and raising of Jesus and the founding of the church at Pentecost, God is represented as making new things happen.[10] Through these "mighty acts," God creates and saves. From the Patristic period through the Protestant Reformation, faith in God the creator was articulated through two distinct but interwoven doctrines: creation and providence. The doctrine of creation asserts that the ultimate source and absolute ground of the universe is God. Without God, the universe would not exist, nor would it exist as "universe."[11] Creation theology, in turn, has often included three related but distinct claims: 1) the universe had a beginning; 2) the universe depends absolutely and at every moment on God for its sheer existence; and 3) the universe is the locus of God's continuing activity as Creator. The first two have traditionally been grouped in terms of *creatio ex nihilo*, and the third in terms of *creatio continua*.

The doctrine of providence[12] presupposes a doctrine of creation, but adds significantly to it. While creation stresses that God is the cause of all existence, providence stresses that God is the cause of the meaning and purpose of all that is. God not only creates but guides and directs the universe towards the fulfilling of God's purposes. These purposes are mostly hidden to us, though they may be partially seen after the fact in the course of natural and historical events. The way God achieves them is hidden, too. Only in the eschatological future will God's action throughout the history of the universe be fully revealed and our faith in it confirmed. General providence refers to God's universal action in guiding all events; special providence refers to God's particular acts in specific moments, whether they be found in personal life or in history.

---

[10] See, e.g., Gen. 45:5; Job 38:22–39:30; Ps. 148:8–10; Is. 26:12; Phil. 2:12–13; 1 Cor. 12:6; 2 Cor. 3:5.

[11] This is a delicate point since it is not shared with process theology, and I would prefer to include process theology in the typology.

[12] For a helpful introduction, see Michael J. Langford, *Providence* (London: SCM Press Ltd., 1981); see also Julian N. Hartt, "Creation and Providence," in *Christian Theology: An Introduction to Its Traditions and Tasks*, 2nd Edition, Peter C. Hodgson and Robert H. King, eds. (Philadelphia, Pa.: Fortress Press, 1985).

The rise of modern science in the seventeenth century and Enlightenment philosophy in the eighteenth, however, led many to reject the traditional view of providence. Newtonian mechanics depicted a causally closed universe with little, if any, room for God's *special* action in specific events. A century later, Pierre Simon Laplace combined the determinism of Newton's equations with epistemological reductionism (the properties and behavior of the whole are reducible to those of the parts) and metaphysical reductionism (the whole is simply composed of its parts), thus portraying nature as a closed, impersonal mechanism. This in turn led to the concept of interventionism: if God were really to act in specific events in nature, God would apparently have to break the remorseless lock-step of natural cause and effect by intervening in the sequence and violating the laws of nature in the process. The eighteenth century was also shaped by David Hume's attack on the arguments for God as first cause and as designer, and by Immanuel Kant's critical philosophy, with its attempt to relocate religion from our knowing (the activity of pure reason) to our sense of moral obligation (the activity of practical reason).

Protestant theology in the nineteenth century produced a variety of responses to these movements.[13] We may very roughly group these into the "liberal" response, which involved a fundamental questioning not only of theological content and structure but even of its method, and the "conservative" response, which upheld traditional formulations and tended to reject "modernity." Thus, conservatives either rejected evolution as a whole or gave it a limited acceptance with the proviso that the objective acts of special providence constitute divine interventions in nature. For Continental liberals, the century's most influential theologian was Friedrich Schleiermacher. He sought to relocate religion from its objective grounds in Scripture or nature to its subjective grounds in personal piety, specifically the feeling of absolute dependence. Schleiermacher collapsed the distinction between creation and providence and reduced the concept of miracle to "... the religious name for [an otherwise ordinary] event."[14] Anglo-Catholic liberals in Britain and America tended to accommodate or even integrate Darwinism into Christianity without interventionism, since they viewed God as immanent in nature.

Aspects of the situation in the twentieth century. During the first half of the twentieth century, Protestant theology was largely shaped by the "neo-orthodoxy" of Karl Barth, who returned theology to its biblical roots and thematized God as the "wholly other." Recognizing that liberal theology tended eventually to dead end in a Feuerbach or a Freud, Barth and his followers held fast to the objective action of God as depicted biblically in creating and redeeming the world. In the 1940s and 1950s, the "God who acts" became a hallmark of the ensuing "biblical theology" movement.[15] These approaches seemed to offer a *tertium quid* between liberal and conservative theologies, but by the 1960s many recognized their seemingly overwhelming internal problems. In 1961 Langdon Gilkey forcefully argued that neo-orthodoxy is an unhappy composite of conservative language, with its

---

[13] See Claude Welch, *Protestant thought in the Nineteenth Century, Volume 2, 1870–1914* (New Haven: Yale University Press, 1985); chap. 6, "Evolution and Theology: Détente or Evasion?" provides a detailed study of the reactions to Darwin.

[14] Friedrich Schleiermacher, *On Religion: Speeches to its Cultured Despisers* (New York: Harper Torchbook, 1958), 88. For a related discussion, see *The Christian Faith* (Edinburgh: T&T Clark, 1968), #47.3/183.

[15] See for example G. Ernest Wright, *God Who Acts: Biblical Theology as Recital* (London: SCM Press, 1952); Bernard Anderson, *Understanding the Old Testament* (Englewood Cliffs: Prentice-Hall, Inc., 1957).

objectivist/interventionist view of divine action, and the (Continental) liberal interpretation of nature as a closed causal system and its subjectivist view of God's action. The result, according to Gilkey, is equivocation. "[T]he Bible is a book descriptive not of the acts of God but of Hebrew religion.... [It] is a book of the acts Hebrews believed God might have done and the words he [sic] might have said had he done and said them—but of course we recognize he did not."[16]

And so we find ourselves back once again to a key theological problem facing contemporary theology: Should special providence be understood entirely as our subjective response to God's uniform and undifferentiated action, or can it include an objective dimension of divine agency which grounds our response to special events? The question is whether we are right to call certain events "special" because we are actually responding to God's distinctive action in, together with, and through them. This claim can be made even more sharply: Is it the case that, had God not acted in a special way in this particular event, the event would *not* have occurred in precisely the way it did? Theologians such as Rudolf Bultmann, Gordon Kaufman, and Maurice Wiles represent the subjectivist interpretation which denies the claim, while others such as Charles Hodge, Donald Bloesch, and Millard Erickson take the objectivist view and affirm it.[17]

For those who take a subjective view, special providence tends to be absorbed into general providence and the latter is usually blended in with creation to give a single undifferentiated view of divine action. For those who hold for objective special providence, our response to specific events or experiences is based on God's special action in these events and experiences. To quote once again the inimitable Gilkey, "for those of faith [an act of God] must be objectively or ontologically different from other events. Otherwise, there is no mighty act, but only our belief in it.... Only an ontology of events specifying what God's relation to ordinary events is like, and thus what his relation to special events might be, could fill the now empty analogy of mighty acts..."[18]

A key to the problem: The presumed link between objective special providence and intervention. Since the Enlightenment, the idea of objective special providence seemed to entail divine intervention: for God to act in particular events, God must intervene in nature, violating or at least suspending the laws of nature. As Nancey Murphy has convincingly argued,[19] the cause of this perceived linkage between

---

[16] Langdon B. Gilkey, "Cosmology, Ontology, and the Travail of Biblical Language," *The Journal of Religion*, 41 (1961): 194–205, esp. 198.

[17] Rudolf Bultmann, *Theology of the New Testament*, Complete in One Volume, translated by Kendrick Grobel (New York, N.Y.: Charles Scribner's Sons, 1951/1955); idem, *Kerygma and Myth*, ed. by H. W. Bartsch, vol. 1, 197–99; idem, *Jesus Christ and Mythology* (New York, N.Y.: Charles Scribner's Sons, 1958), particularly chap. V; Gordon D. Kaufman, *Systematic Theology: A Historicist Perspective* (New York, N.Y.: Charles Scribner's Sons, 1968); idem, "On the Meaning of 'Act of God'," in *God the Problem* (Cambridge: Harvard University Press, 1972); Maurice Wiles, *God's Action in the World: The Bampton Lectures for 1986* (London: SCM Press: 1986); idem, "Religious Authority and Divine Action," *Religious Studies*, 7 (1971): 1–12, reprinted in *God's Activity in the World*, Owen Thomas, ed., (Chico, Calif.: Scholars Press), 181–94; Charles Hodge, *Systematic Theology*, 3 vols. (New York, N.Y.: Scribner's Sons, 1891); Donald G. Bloesch, *Holy Scripture: Revelation, Inspiration and Interpretation* (Downers Grove, Ill.: InterVarsity Press, 1994); Millard Erickson, *Christian Theology*, one-volume ed. (Grand Rapids, Mich.: Baker, 1983).

[18] Gilkey, "Cosmology, Ontology, and the Travail of Biblical Language," reprinted in *God's Activity in the World*, Thomas, ed., 37.

[19] Nancey Murphy, *Beyond Liberalism and Fundamentalism* (Valley Forge, Pa.: Trinity

objective special providence and interventionism was the combination of mechanistic physics and reductionistic philosophy. If the physical world is a causally closed,
deterministic system, and if the behavior of the world as a whole is ultimately
reducible to that of its physical parts, the action of a free agent—whether human or
divine—must entail a violation of natural processes. Though disagreeing on
practically everything else, most liberal, neo-orthodox, and conservative theologians
took for granted the link between objective special providence and intervention.
Thus, by and large the choice has been either to affirm objective special providence
at the cost of an interventionist and, in some extreme cases, an anti-scientific
theology, or abandon objective special providence at the cost of a scientifically
irrelevant and, in many cases, privatized and tame theology. Seen in this light, a third
option is crucial.

   Breaking the link: a non-interventionist view of objective special divine action.
It seems clear that any purported "third option" will require an intelligible concept
of objectively special providence which does *not* entail divine intervention. Such a
concept could serve as a genuine *tertium quid* to conservative and liberal notions of
special providence, combining strengths borrowed from each.[20] In specific, we will
seek to speak about special divine acts in which God acts objectively in an unusual
and particularly meaningful way in, with, and through these events which thus serve
to mediate God's action. We will seek to do so without being forced into the
additional claim that God must intervene in, or at least suspend, the laws of nature
(which are themselves the result of God's general providence/continuous creation).
I refer to this type of divine action as a "non-interventionist view of objective special
providence."[21]

2.3.2 *Quantum Physics and Ontological Indeterminism*

In classical physics, nature is a closed causal system described by deterministic
equations. Thus in classical physics—and this carried over into the other classical
sciences including evolutionary biology—the notion of chance is purely epistemic.
From dice being cast, chromosomes moving in a thermal gradient, or the random
selection of mates, we refer to an event happening by chance when we simply do not
know all the underlying deterministic factors, though we believe they are there in
principle. Chance can also stand for a more complicated incidence involving the

---

Press International, 1996), chap. 3. I am indebted to her analysis here, and for her references
to the conservative sources cited above. See also Peacocke, *Theology for a Scientific Age:
Being and Becoming—Natural, Divine, and Human,* enlarged edition (Minneapolis, Minn.:
Fortress Press, 1993), 139–140.

   [20] I leave as an open question whether the events and processes of nature described by
physics and biology are best understood: 1) merely in terms of efficient causality and divine
action, or 2) as process theologians would insist, they are also shaped by intrinsic novelty and
creativity. *Both* positions are consistent with the argument being studied here.

   [21] In "Chaos, Complexity and God: A Review Essay," Thomas makes the important
point—with which I agree—that many authors who turn to a non-interventionist approach in
theology and science treat interventionism pejoratively. This is unfortunate since it confuses
the issue and undercuts a potentially useful concept. My concern here for a non-interventionist
approach is not meant to disparage interventionism *per se*, only its unnecessary application.
For key issues such as the Incarnation, the Resurrection of Christ and Pentecost, an
interventionist approach might be justifiable and necessary after suitable nuancing, since when
the domain of God's action is eschatological the "laws of nature" (i.e., God's faithful action)
themselves will be different and "intervention" may cease to be a useful concept. Here we
might refer to God's action as unmediated and unilateral.

juxtaposition of two apparently unrelated causal trajectories, such as the coincidences leading up to a car crash or the random relation between a given genetic mutation and the adaptivity of its phenotype.[22] In general, classical chance is represented mathematically by Boltzmann statistics (the familiar "bell-curve") and pervades not only physics but biology as well, from Mendel's laws to the Hardy-Weinberg theorem.[23]

Quantum physics severely challenges this understanding of chance.[24] Instead, quantum chance, in an important sense, is formative of the basic features of the classical world, features taken for granted but never really explained by classical physics or biology. One reason for this is that the statistics employed by quantum mechanics come in two varieties and both are strikingly different from classical statistics.[25] Particles such as protons and electrons obey Fermi-Dirac statistics and thus the Pauli exclusion principle. These lead to the impenetrability of matter and its "space-filling" character. They also produce chemical properties such as valency and color. Particles such as photons and gravitons obey Bose-Einstein statistics and carry the fundamental forces in nature, such as electromagnetism and gravity. In this sense quantum statistics, with its two distinct forms and their differences from everyday chance, underlies and accounts for the fundamental properties of our everyday world.[26]

But is quantum chance a sign of ontological indeterminism, or is it, too, just a product of our ignorance of still further hidden causes? From its beginnings in 1900 up to the present, physicists and philosophers of science have wrestled with these and other philosophical implications of quantum mechanics, exploring a variety of interpretations that represent both deterministic and indeterministic alternatives.[27]

---

[22] This is of course the famous example of chance which led Jacques Monod to argue against belief in God's action in the world. See Jacques Monod, *Chance and Necessity* (New York, N.Y.: Alfred A. Knopf, 1971). I will offer a response to Monod in section 5.1 below.

[23] The salient point here is that we could in principle give a completely deterministic description; we chose a statistical description out of convenience. Chaos theory is not an exception. Recent studies show that chaotic systems, though often unpredictable in principle, are nevertheless describable by deterministic equations, and thus they support a deterministic view of nature philosophically. Their theological importance is discussed in *CAC*.

[24] For an introduction for science undergraduates, see P. C. W. Davies, *Quantum Mechanics* (London: Routledge & Kegan Paul, 1984). Technical sources include Eugen Merzbacher, *Quantum Mechanics* (New York, N.Y.: John Wiley & Sons, 1961); Kurt Gottfried, *Quantum Mechanics* (New York, N.Y.: W. A. Benjamin, 1966); and the classic work, P. A. M. Dirac, *The Principles of Quantum Mechanics* (Oxford: Clarendon Press, 1958, revised fourth edition). Recent works for the general reader include J. C. Polkinghorne, *The Quantum World* (Princeton, N.J.: Princeton Scientific Library, 1989).

[25] Technically, the following results are part of *relativistic* quantum mechanics.

[26] See the discussion in R. J. Russell, "Quantum Physics in Philosophical and Theological Perspective," in *Physics, Philosophy and Theology: A Common Quest for Understanding*, Robert J. Russell, William R. Stoeger, and George V. Coyne, eds. (Vatican City State: Vatican Observatory, 1988), hereafter *PPT*.

[27] Quantum physics can be interpreted *philosophically* as posing fundamental limitations on epistemology (Bohr's "Copenhagen" interpretation),or as indicative of ontological determinism (Einstein, Bohm), or of ontological indeterminism (Heisenberg). Other interpretations include many-worlds (Everett), idealism/the role of consciousness (Von Neumann, Wigner, Wheeler), non-standard logic (Gribb), observer-free formulations such as the decoherent histories approach (Griffiths, Omnès, Gell-Mann, and Hartle), and so on. See Ian G. Barbour, *Issues in Science and Religion* (New York, N.Y.: Prentice-Hall, 1966), chap. 10; Max Jammer, *The Philosophy of Quantum Mechanics: The Interpretations of Quantum*

To complicate matters even further, during the past three decades reflections on Bell's theorem have underscored the non-local and non-separable character of quantum phenomena, making each of these earlier interpretations even more problematic.[28]

While the debate is not settled, there are strong arguments supporting Heisenberg's view that quantum chance points to a fundamental *ontological indeterminacy* in nature.[29] From this perspective, the use of statistics in quantum mechanics is not a mere convenience to avoid a more detailed causal description. Instead, quantum statistics is all we can have, for *there is no underlying, fully deterministic natural process.* For example, consider a sample of uranium (^{238}U) in which a specific atom suddenly decays into thorium (^{234}Th) by emitting an α-particle. We can calculate the probability of the event to occur, but we cannot explain why this particular uranium decayed when it did and why its neighbors did not. All the atoms in the sample are absolutely identical, and the decay event is independent of any physical or chemical conditions imposed on them.[30] Similar quantum mechanical

---

Mechanics in Historical Perspective (New York, N.Y.: John Wiley & Sons, 1974); R. J. Russell, "Quantum Physics in Philosophical and Theological Perspective," in *PPT*; Sheldon Goldstein, "Quantum Theory without Observers," *Physics Today* (March and April, 1998). For an accessible account, see Nick Herbert, *Quantum Reality: Beyond the New Physics* (Garden City: Anchor Books, 1987). For a more specific focus on the "measurement problem," see footnotes 37 and 39 below.

[28] See for example Michael Redhead, *Incompleteness, Nonlocality, and Realism: A Prolegomenon to the Philosophy of Quantum Mechanics* (Oxford: Clarendon Press, 1987); James T. Cushing & Ernan McMullin, eds., *Philosophical Consequences of Quantum Theory: Reflections on Bell's Theorem* (Notre Dame: University of Notre Dame Press, 1989); Chris J. Isham, *Lectures on Quantum Theory: Mathematical and Structural Foundations* (London: Imperial College Press, 1995).

[29] Heisenberg's ontological interpretation of quantum physics has a number of current supporters. C. J. Isham writes:

> The most common meaning attached to probability in classical physics is an epistemic one.... However, unless hidden variables are posited, the situation in quantum theory is very different.... In particular, there are *no* underlying microstates of whose precise values we are ignorant. If taken seriously, such a view of the probabilistic structure in quantum theory entails a radical departure from the philosophical position of classical physics... (*Lectures on Quantum Theory*, 131–32).

According to Paul Davies,

> Prior to quantum theory, physics was ultimately *deterministic.* . . . The quantum factor... implies that we can never know in advance what is going to happen.... We shall see that this indeterminism is a universal feature of the micro-world (*Quantum Mechanics,* 4).

Ian Barbour writes:

> ...[A]*lternative potentialities* exist for individual agents. We urged, in accordance with critical realism, that the Heisenberg Principle is an indication of objective indeterminacy in nature rather than the subjective uncertainty of human ignorance (*Issues,* 315–16; see also his *Religion in an Age of Science* (New York: HarperCollins, 1990), 123).

For earlier sources and references on the ontological interpretation of indeterminacy, see H. Margenau, "Reality in quantum mechanics," *Phil. Science,* 16 (1949): 287–302; W. Heisenberg, *Philosophic Problems of Nuclear Science* (Greenwich: Fawcett Publications, 1952); K. Popper, *Quantum Theory and the Schism in Physics* (London: Hutchinson, 1956).

[30] Bohm's elegant formulation accounts for this process deterministically. However some of the deterministic elements in his approach do not have clear classical analogues, and the metaphysics he offers to explain them move us even further from the classical view of the world. Thus even if one opted for Bohm's approach it would hardly constitute a return to a

descriptions apply to the making and breaking of covalent bonds in molecules, to the tunneling of electrons in transistors, to the absorption and emission of photons by atomic electrons, and so on. In each case, the total set of natural conditions affecting the process, and thus the total possible set of conditions which science can discover and describe through its equations, are *necessary but insufficient* to determine the precise outcome of the process. The future is ontologically open, influenced but *under*-determined by the factors of nature acting in the present. In, through, and beyond the causal conditions we can describe scientifically, things just happen.

Following Heisenberg's suggestion,[31] we can characterize a quantum system in terms of potentialities and actualities. The system starts off in a superposition of "coexistent potentialities": a variety of distinct states are simultaneously possible for the system, but none of them are fully actual. Suddenly one of them becomes realized or actual at a specific moment in time, though we cannot attribute the process of actualization to interaction with other processes. The uranium atom moment by moment comes to be what it has been, uranium, and then it comes to be something else, something which it could always have been, thorium.

Returning to our theme of divine action, we can see the importance of Heisenberg's interpretation of quantum mechanics.[32] If his interpretation is correct, we can view nature theologically as genuinely open to God's participation in the bringing to actuality of each state of nature in time. Where science employs quantum mechanics and philosophy points to ontological indeterminism, faith sees God acting with nature to create the future. This is neither a disruption of the natural process nor a violation of the laws of physics. Instead, it is God fulfilling what nature offers, providentially bringing to be the future which God promises for all creation, acting specifically in all events, moment by moment.[33]

Two more aspects of quantum physics particularly relevant to this paper are related to what is often called the "measurement problem." Granted that quantum processes give rise to the classical world and its ordinary, bulk properties due to the kind of statistics which governs quantum processes. Does this mean that the specific effects of *every* quantum event get "averaged out" by the sheer number of such events? This claim, if true, could serve as a reason against using quantum physics in a discussion of God's action.[34] Actually, however, specific quantum processes *can* have an irreversible effect on the classical world, and they can do so in a way that is entirely consistent with quantum mechanics. A particularly vivid example is vision, in which a single photon absorbed by one's retina can produce a mental

---

classical view of causality and chance.

[31] Werner Heisenberg, *Physics and Philosophy: The Revolution in Modern Science* (New York, N.Y.: Harper & Row, 1958), 185.

[32] Clearly an alternative interpretation of quantum physics may come to prevail in the future, but it will be instructive to discover what insights we may gain regarding divine action in evolution by provisionally adopting Heisenberg's interpretation. For an extended discussion of the value of proceeding this way, see section 3.2 below.

[33] Process philosophy can suggest here that the openness of these quantum states includes the efficacy of inherent novelty or creativity, along with God's lure and the necessary conditions of the causal past. Thus quantum mechanics can fit nicely with a process metaphysics, though it also fits nicely with other metaphysical systems. Which system is most appropriate for quantum physics remains, in my opinion, undecided but critically important.

[34] For an early example of the subtleties involved here, see Barbour, *Issues*, 308, and compare with p. 309, including footnote #4. For a more recent example, see John Polkinghorne, "The Quantum World," in *PPT*, 334.

impression![35] A somewhat *artificial* but now infamous example is the "Schröd-inger's Cat" thought experiment.[36] I cite it both because it underscores the fact that a specific quantum event can lead to important effects in the ordinary world and because I take the genotype-phenotype relation as its *natural* and ubiquitous "biological instantiation," as I'll argue below. In fact, whenever we make a measurement of a quantum process, we are registering the effect in the lab due ultimately to a process taking place submicroscopically.[37] The second aspect is the ubiquity of what we mean by measurement. What is crucial to recognize, however, is that we need *not* restrict the term 'measurement' to a laboratory experiment. Instead such events occur *constantly* in the universe[38] whenever elementary particles interact irreversibly with molecules, gases, solids, and plasmas.[39]

Let me summarize what we have seen so far. 1) Quantum processes can be interpreted philosophically in terms of ontological indeterminism, and this in turn opens the possibility of a non-interventionist account of divine action. 2) Quantum

---

[35] See my "Quantum Physics in Philosophical and Theological Perspective" in *PPT*, endnote 2, p. 369.

[36] Here the life of a cat hangs in the balance over a single radioactive decay event. If the event does not occur, the cat is spared; if it does, it triggers a Geiger counter whose voltage spike causes lethal gas to be released into a chamber holding the cat. For a readable account and a helpful discussion of the underlying philosophical issues, see John Gribbin, *In Search of Schrödinger's Cat* (New York, N.Y.: Bantam, 1984); idem, *Schrödinger's Kittens and the Search for Reality* (Boston: Little, Brown and Co., 1995). I dislike this story for obvious reasons, but it has become too famous to easily "sanitize."

[37] An obvious example of a measurement is the detection of a charged particle by a Geiger counter. Here a charged particle passes through a cylindrical ionization chamber filled with gas and containing a wire stretched along its center axis. The chamber walls are negatively charged, the wire charged positively, and a counter measures their relative voltage. The passage of the charged particle triggers a cascade of ionized gas which eventually produces a voltage spike on the counter.

[38] Examples include: dust motes jostled by gas molecules (Brownian motion); light illuminating a surface and heating it (blackbody radiation) or freeing electrons from it (photoelectric effect); or fission and fusion processes.

[39] Yet another aspect of the measurement problem involves the so-called "collapse of the wave function." When we represent elementary particles as interacting with each other, the wave function which describes them evolves continuously in time. Yet when we consider them as interacting with a classical system—such as the charged particle and the "macroscopic" Geiger counter—the wave function describing the quantum system changes instantaneously and discontinuously, a change often referred to as "the collapse of the wave function." But if the classical system—here, the Geiger counter—is itself composed of fundamental particles, why doesn't the wave function for the entire system—particle plus Geiger counter—evolve continuously? Alternatively, when and where does the wave function of the entire system collapse. The options range all the way from the first interaction between the elementary particle and a gas molecule to the conscious observation of the Geiger counter. Any careful attempt to answer this question leads us directly into complex and unsettled issues in the philosophical foundations of quantum mechanics, and they are made even more complicated when Bell's theorem and its implications about non-locality and non-separability are considered. The theoretical issues lying behind the measurement process thus form one of the most subtle and most controverted topics in all of quantum physics. A full treatment would take far more space than is possible here. Still, for the purposes of this paper, the point I wish to emphasize is that *whatever position one takes about the measurement problem, the phenomenon it refers to is ubiquitous in nature and is not restricted to the performing of measurements in the lab*. This point will arise again in discussing the positions of Polkinghorne and of Murphy below.

statistics are dramatically different from classical statistics. They are foundational for the ordinary properties of the everyday world and they allow for individual quantum events to trigger irreversible and significant effects in that world.[40] In doing so they offer us a clue to how things in general come to be as they are *as well as* how things in particular happen within the general environment. This in turn opens up the possibility both for non-interventionist general divine action, which indirectly results in creating and sustaining the world, and for non-interventionist special divine action, which can indirectly result in special events in the world.

But why is this relevant specifically to divine action in evolution? This leads to our next section.

2.3.3 *What Role Does Genetic Mutation Play in Biological Evolution?*

According to contemporary biological science, life on earth extends back approximately 3.5–3.9 billion years. All living species have evolved from extremely simple and small organisms whose origin is barely understood. According to the neo-Darwinian theory of evolution,[41] the vast biological complexity we see in the fossil record along with the 2 million species we now know to exist can be explained in terms of two fundamental principles: variation and natural selection. It starts with the fact that variations occur in the hereditary material which alter the chance for survival and procreation of organisms carrying that material. Those variations which happen to be favorable to the survival of the organism tend to spread throughout the species from generation to generation while less favorable or harmful variations tend to decrease. This process of natural selection, or "the differential reproduction of alternative hereditary variants," results in the species being increasingly adapted to its environment.[42] Hereditary variation, in turn, involves both spontaneous mutations which change one variant to another and sexual reproduction, during which these variations are recombined in countless ways. With the discovery of the molecular structure of DNA by Watson and Crick in 1956, we know that hereditary information is carried by the sequence of nucleotides whose groupings as genes form the DNA molecule. Mutations in DNA during replication can involve either a substitution, an insertion, or a deletion of one or a few pairs of nucleotides in the DNA. For its consequences to be hereditary, the mutation must have occurred in the germ-line of an organism, after which it can be passed on to its progeny.[43]

Mutations can occur spontaneously in DNA, or they can be induced by ultraviolet light, X-rays, or exposure to mutagenic chemicals, often as the result of human activity. Because genetic mutations occur at random with respect to the environment of the organism, their consequences for progeny are more likely to be

---

[40] Ultimately, one can ask the question why there is a classical world at all and not just a world described by quantum mechanics.

[41] See Francisco Ayala, "The Evolution of Life: An Overview," in this volume. Also see Neil A. Campbell, *Biology*, 2nd ed. (Redwood City, Calif.: Benjamin Cummings Publishing Company, 1987/1990). Dawkins, Wilson, Gould, Mayr, and others have sought to expand neo-Darwinism, while others such as Kauffman, Wicken, Goodwin, Ho, and Saunders suggest we move outside the neo-Darwinian paradigm. For a helpful discussion and references, see the paper by Ian G. Barbour in this volume. I will leave a theological response to these suggestions to another essay.

[42] Ayala, "The Evolution of Life," in this volume, p. 36.

[43] "Chance" in evolution thus involves *both* genetic variation and the juxtaposition of these variations with complex changes in the environment of the species; in this volume, see Peacocke, p. 361ff and Ayala, p. 108. For the problems this second kind of chance raises to divine action, see section 5.1 below.

neutral or harmful. Occasionally, however, they increase the adaptive fit of an organism to its environment, and the mutation is passed on through successive reproduction. "Natural selection keeps the disorganizing effects of mutations and other processes in check because it multiplies beneficial mutations and eliminates harmful ones."[44] Although the rate of variation in a population can occur due to a variety of factors including mutations, genetic drift, gene flow, sexual reproduction, and non-random mating, variation per se is ultimately due to genetic mutation. According to W. F. Bodmer and L. L. Cavalli-Sforza, "mutation in its broadest sense is, by definition, the origin of new hereditary types. It is the ultimate origin of all genetic variation; without it there would be no genetic differences, and so no evolution." Masatoshi Nei claims that "mutation is the driving force of evolution at the molecular level. I have also extended this view to the level of phenotypic evolution and speciation.... I have challenged the prevailing view that a population of organisms contains virtually all sorts of variation and that the only force necessary for a particular character to evolve is natural selection." Francisco J. Ayala argues that "...the science of genetics has provided an understanding of the processes of gene mutation and duplication by which new hereditary variations appear.... Gene mutation and duplication [are] the ultimate sources of all genetic variability."[45]

2.3.4 *What Role Does Quantum Physics Play in Genetic Variation?*

In this paper I am adopting the *theological* view that God's special action can be considered as objective and non-interventionist if the quantum events underlying genetic mutations are given an indeterminist interpretation philosophically. If it can be shown scientifically that quantum mechanics plays a role in genetic mutations, then by extension it can be claimed theologically that God's action in genetic mutations is a form of objectively special, non-interventionist divine action. Moreover, since genetics plays a key role in biological evolution, we can argue by inference that God's action plays a key role in biological evolution, and our hypothesis is warranted.

Thus it is of central importance to this paper that we discuss the precise role of quantum mechanics in certain types of genetic mutation. In specific, we must ask: 1) to what extent is variation the result of classical processes such as hydrodynamics, thermodynamics, statistical mechanics, chaotic dynamics, chemistry, geology, ecology, and so on, with their presupposition of classical chance; and 2) to what extent is variation the result of quantum physics acting at the atomic and subatomic levels, principally in genetic mutations? I believe a reasonable answer to these questions is the following:

Classical sources. Sources of variation in organisms which probably have an entirely classical explanation include: chemical mutagens, mechanical/physical mutagens (including physical impacts), and chromosome segregation. Sources of variation in species include genetic drift, gene flow; and non-random mating.

Quantum sources. Sources of variation in organisms which may include a quantum process, or at least involve a semi-classical (classical/quantum) process,

---

[44] Ibid., 40.

[45] W. F. Bodmer and L. L. Cavalli-Sforza, *Genetics, Evolution and Man* (San Francisco, Calif.: W. H. Freeman and Company, 1976), 139; Masatoshi Nei, *Molecular Evolutionary Genetics* (New York, N.Y.: Columbia University Press, 1987), 431; Francisco J. Ayala, "The Theory of Evolution: Recent Successes and Challenges," in *Evolution and Creation*, Ernan McMullin, ed. (Notre Dame, Ind.: University of Notre Dame Press, 1985), 78, 82. See also Ayala in this volume, pp. 34–37.

include: point mutations, including base-pair substitutions, insertions, deletions; spontaneous mutations, including errors during DNA replication, repair, recombination; radiative physical mutagens (including X-rays and ultraviolet light); and crossing over.

There are, however, some very interesting and as yet unsettled questions here. Further scientific research is required for us to gain a clearer understanding of the relative importance of quantum processes and classical processes in variation, including such specific topics as chromosome number mutation (aneuploidy and polyploidy), chromosome structure mutation (including deletion, duplication, inversion, and reciprocal translocation), transposons, and DNA mutagenesis. Some of the outstanding questions yet to be explored include:

1. To what extent do point mutations arise from the interaction of a single quantum of radiation and a single proton in a hydrogen bond in a specific base?

2. How important are quantum effects in the phenomenon of crossing over?

3. To what extent does cooperativity, being a semi-classical effect extending over several base pairs, minimize the quantum aspects of genetic variation?

4. Do discrete changes in phenotype, from more interwoven macroscopic changes (for example, single versus multiple insect wings and gross macroscopic changes such as the proverbial "wings to arms" change), to changes traceable to a point mutation of a single base-pair of DNA, result in turn from a single quantum interaction, or from a series of separate base-pair mutations, each the result of a single quantum interaction?

5. What is the relative importance of monogenetic effects compared with the polygenetic effects and/or effects of the entire genome in phenotypic expression? Here "polygenetic" includes gene-gene interactions either within the same chromosome or between different chromosomes, the role of the physical structure of chromosomes, the effects of ploidy or segmental rearrangements, and one could add the interactions between clusters of adjacent genes, and so on.

6. To what extent are the linkages between genes and expression so non-linear that the possibility of an analysis of the genetic basis of expression is seriously impaired?

7. Which are the most crucial factors leading from a point-mutation or a series of point-mutations to significant changes in the population? Clearly the amplification of the mutation to the level of the phenotype, the survival and reproduction of the phenotype, and its adaptive advantage in the population, are complex and pivotal factors. However, two other very specific factors seem particularly important. 1) The mutation must occur in the germ-line. Somatic mutations will not affect progeny, and cannot be amplified by population increase through reproduction. 2) Amplification requires the faithful replication of the DNA mutation, producing billions of copies of the original mutation. Further research may help clarify the relative importance of these specific factors to the overall process of genotype-phenotype amplification.[46]

8. To what extent do environmental factors, from such quantum factors as radiation to such classical factors as chemical and physical effects at the sub-cellular level and ecological factors at the level of populations contribute to evolution?

Thus, it is hoped that further scientific research will clarify the roles of classical and quantum sources of genetic variation and shed further light on the central

---

[46] I am indebted here, and elsewhere, to very fruitful comments and corrections from David Cole.

theological hypothesis of this paper regarding a non-interventionist, objective interpretation of special divine action in the evolution of life.

3 Theological Sources, Criticisms, and Recent Developments

3.1 Early Sources

We begin our account of theological sources with the writings of Karl Heim and William Pollard. In 1953 Heim suggested that, since Laplacian determinism had been overturned by quantum indeterminism, God could now be thought of as acting at the quantum level.[47] In a paraphrase of Matthew 10:29, Heim wrote: "No quantum-jump happens without your Father in heaven." In Heim's opinion, this led to the further claim that the world of ordinary experience is entirely determined by God's action at the quantum level. "All events, however great, we now know to be the cumulation of decisions which occur in the infinitesimal realm."[48] He did not, however, make the connection between quantum events and genetic mutations as did physicist and Episcopalian priest William Pollard, and in this paper I will *not* follow Heim's preference for divine omnideterminism.

In 1958 Pollard advanced a more complex form of the argument.[49] To the scientist, quantum processes are entirely random; to the Christian, God can be seen as choosing the outcome from among the quantum mechanically allowed options.[50] Pollard then added two key reservations: 1) this view of God's action does not imply that God acts *as* a natural force; and 2) it is not a form of natural theology, since belief in divine action is based on theological, not scientific, grounds. "Science, for all its wonderful achievements, can of itself see nothing of God..."[51] With these

---

[47] Karl Heim, *The Transformation of the Scientific World* (London: SCM Press, 1953). For a lucid discussion of Heim and others, see John Y. Fenton, "Random Events and the Act of God," *Journal of the AAR*, 25 (March, 1967): 50–57.

[48] Heim, Ibid., 156.

[49] William G. Pollard, *Chance and Providence: God's Action in a World Governed by Scientific Law* (London: Faber and Faber, 1958).

[50] In his 1956 Bampton Lectures, E. L. Mascall made claims similar to those Pollard was to make in 1958. Mascall first focused on classical physics and its inherent determinism. The theist will account for physical events as due to finite agents acting through secondary causes. God, as primary cause, will be seen as acting by creating and conserving these finite agents and their causal efficacy. Thus the account on the physical level will be complete, while a metaphysical account will include reference to God's action as well as secondary causes.

If now we abandon the classical standpoint and adopt that of quantum physics, we cannot give a complete account, even on the physical level, simply in terms of finite causes. The degree of autonomy with which God has endowed the finite agents is sufficient to specify the relative frequency or probability with which specified types of event occur, but nothing more... The situation in fact is as if... God has reserved to himself the final decision as to whether a specified event occurs of not.

Referring to the example of a click in a Geiger-counter, Mascall concludes that a click "may be due solely to the primary causality of God." Mascall thus anticipates several of Pollard's claims, though he does not extend the argument to evolution—somewhat surprisingly to me, given his discussion of genetic mutations and Schrödinger's ideas in the same lectures. My reservation with Mascall's position is that I see God's activity in quantum events as mediated by nature (both materially and nomologically); I would not see quantum events as due solely to primary causality. I am grateful to Kirk Wegter-McNelly for calling my attention to Mascall's ideas. E. L. Mascall, *Christian Theology and Natural Science: Some Questions in their Relations*, The Bampton Lectures, 1956 (London: Longmans, 1956), 200–1.

[51] Pollard, *Chance and Providence*, 86.

reservations in mind, we turn to Pollard's central idea of linking quantum indeterminism with genetic mutation. "[T]he phenomenon of gene mutation is the only one so far known in these sciences which produces gross macroscopic effects but seems to depend directly on changes in individual molecules which in turn are governed by the Heisenberg indeterminacy principle."[52]

Next Pollard made the connection between chance in science and divine causality. First he pointed out that chance is not a cause: "...chance as such simply cannot be the cause or reason for anything happening." Instead it stands for our ignorance of hidden, but real, causes. Thus, since science knows it cannot discover them, it makes sense to posit God as providing the ultimate cause for particular natural events. Finally, Pollard countered the critics of religion who claimed that chance undermines providence; instead, "...those secular writers who feel that they have demolished the biblical view of creation and evolution as soon as they have established the statistical character of the phenomena involved, have unwittingly done the one thing necessary to sustain that view."[53] Though I strongly support Pollard's advance over Heim's formulation of the thesis, I do *not* support their advocacy of divine omnideterminism.

These arguments resurfaced two decades later in the writings of Mary Hesse and Donald M. MacKay. Chance, as MacKay put it, "...stands for the *absence* of an assignable cause."[54] Thus what appears to us as a random event may be taken as the action of a supremely sovereign God. Hesse too defended a theistic interpretation of evolution:

> Just because chance is necessary to the evolutionary history that Monod accepts, there must be irreducibly random outcomes that scientific theory cannot explain. It follows that this theory cannot refute a theistic hypothesis according to which God is active to direct the course of evolution at points that look random from the purely scientific point of view.[55]

But Hesse went further than MacKay and Pollard, recognizing that, in order to launch a robust response to Monod, this view would have to be housed within a systematic framework, including:

> ...theology, ethics and a theory of knowledge that can be shown to be more intimately related to each other and to what we know of the facts than Monod and other scientific humanists have been able to show in the case of their own systems.[56]

Though I strongly agree with Hesse here, she did not cite the possible role of quantum physics in genetics, nor did she refer to Pollard.

3.2 *Critical Voices*

As early as 1966, Ian G. Barbour, though appreciative of much of Pollard's work, raised serious questions about the kind of total divine sovereignty Pollard supports.[57] He also stressed the idea of God as acting through both order and novelty, and the extension of the domain of God's action to all levels of reality. I agree with

---

[52] Ibid., 56.

[53] Ibid., 92, 97.

[54] Donald M. MacKay, *Science, Chance and Providence* (Oxford: Oxford University Press, 1978), 33.

[55] Mary Hesse, "On the Alleged Incompatibility between Christianity and Science," in *Man and Nature*, Hugh Montefiore, ed. (London: Collins, 1975), 121–31; see p. 128 for the quotation.

[56] Ibid., 130.

[57] Barbour, *Issues in Science and Religion*, 428–30.

Barbour's insightful criticisms here.[58] In his 1979 Bampton Lectures Arthur Peacocke launched a major challenge to Monod's atheistic evolution, with its attendant claim that the universe is purposeless. Instead, evolution provides the matrix of God's continuous and immanent creative action; working in, under, and through it, God brings forth the full potentialities of nature. Here and elsewhere, Peacocke—and Barbour as well—has made an invaluable contribution to theistic evolution.

In passing, Peacocke supported Pollard's claim that quantum phenomena represent the *only* domain of ontological indeterminacy, at least below the level of consciousness in animals: "Apart from events at the level of the fundamental subatomic particles, the [statistical character of the laws of science] represent[s] simply our ignorance of all the factors contributing to the situation—they do not imply any lack of causality in the situation itself..."[59] Still, Peacocke dismissed this view as allowing one arbitrarily "to pick and chose" which chance events are to be credited to providence. He then shifted to the thermodynamics of living organisms where, as his own agreement with Pollard implies, the determinism suggested by science makes non-interventionist bottom-up divine action very problematic.[60]

Although Peacocke has published widely on these issues in recent years,[61] Pollard's view has not reappeared. In a recent essay, Peacocke explained why he rejects the relevance of quantum indeterminacy to the problem of divine action. "[The] inherent unpredictability [of quantum events] represents a limitation of the knowledge even an omniscient God could have of the values of these variables and so of the future trajectory... of the system."[62] Unfortunately I fear this comment misconstrues the issue at hand in two ways. First, recall the Heisenberg interpretation of "coexistent potentialities"; roughly speaking, a quantum system is, ontologically, a simultaneous variety of distinct but merely potential states. Suddenly one of them becomes actual at a specific moment in time. This interpretation now lends itself to our theological claim: Before God acts, the quantum system is in a superposition of potential states. But when God *acts*, the effect of that action, together with that specific superposition, decides which quantum outcome becomes actual. Now the system is in a definite state, and it can lead to specific results in the future. Second, God does not (fore)know these future results by predicting them from the present. Instead, God knows the future in its own present. Thus, God acts in the present to indirectly bring about a future which God chooses eternally.[63]

---

[58] His comments, written over thirty years ago now, have stood as a significant challenge to this view of divine action, to which I hope we can finally give an adequate response here.

[59] Peacocke, *Creation and the World of Science* (Oxford: Clarendon, 1979), 95–96.

[60] In its place Peacocke has increasingly turned to top-down and whole-part approaches, emphasizing that God acts on the whole of reality and that these actions can eventually bring about specific events which we claim for special providence.

[61] See for example *God and the New Biology* (San Francisco, Calif.: Harper & Row, 1986), 62, for a reference to Monod but no reference to Pollard. See also *Theology for a Scientific Age: Being and Becoming—Natural, Divine, and Human*, enlarged edition (Minneapolis, Minn.: Fortress Press, 1993), 117–18; there is no reference to Pollard.

[62] Arthur Peacocke, "God's Interaction with the World: The Implications of Deterministic 'Chaos' and of Interconnected and Interdependent Complexity," in *CAC*, 279. In the mid 1980s, Bartholomew provided a careful analysis with special attention to Peacocke's views, but he too downplayed the role of quantum physics, claiming that "from the theological point of view, it matters little whether or not all chance can be expressed in deterministic terms." See *God of Chance*, (London: SCM Press, 1984), 68.

[63] See section 5.1 for an extended discussion.

In writings dating from 1986, physicist and Anglican priest John Polkinghorne has reflected on the possibilities offered by quantum physics for our understanding of divine action. In *One World* he cited Pollard's argument, but asserted it had "... an air of contrivance..."[64] In 1988, he again rejected Pollard's argument: "The idea of such a hole-and-corner deity, fiddling around at the rickety roots of the cosmos, has not commended itself to many."[65] A year later, however, Polkinghorne gave the idea a more objective response, citing two problems: 1) "...[it] founder[s] on the propensity for randomness to generate regularity, for order to arise from chaos," and 2) "[If] the everyday certainties of the world of Newtonian mechanics arise from out of their fitful quantum substrate... [it is] unlikely [that they, by themselves] provide a sufficient basis for human or divine freedom"[66] Regarding point (1), I believe Pollard clearly saw that chance is not a form of causality but the lack of it. As for the point (2), one can claim that quantum indeterminism contributes something important to the problem of human or divine action without claiming reductively that it contributes all that is necessary to it, even if Pollard may have done so.

Finally, in 1991, Polkinghorne gave three reasons why the appeal to quantum physics raises real complications for the problem of divine action.[67] 1) The first focuses on chaos in classical science, where small changes in the initial conditions are amplified rapidly into large change as the system develops in time. Some have speculated that quantum physics may be the ultimate source of these initial changes. The problem is that the equations of quantum physics are not chaotic, unlike their classical counterparts; but if this is so, how can chaos arise at the classical level out of the underlying non-chaotic quantum processes? 2) Quantum physics is subject to competing interpretations, including deterministic ones. This should caution us from drawing metaphysical conclusions too quickly from quantum physics. 3) Finally, there is the quantum measurement problem: How can a piece of apparatus yield exact measurements on a quantum system if it too is composed of elementary particles obeying the indeterminacy principle?

Polkinghorne returned to these issues in 1995, referring in his footnotes to Pollard's early work. Here he actually sets to rest his second concern: though other interpretations are possible, ontological indeterminism "... is a strategy consciously or unconsciously endorsed by the great majority of physicists."[68] But Polkinghorne reiterates points (1) and (3), drawing out their implications in more detail. The indeterminacies in quantum behavior only arise in "those particular events which qualify, by the irreversible registration of their effects in the macro-world, to be described as measurements. In between measurements, the continuous determinism of the Schrödinger equation applies." Because measurements only occur occasionally, this approach would limit God's action to "occasions of measurement" and suggest an "episodic account of providential agency."[69] For these and other reasons, Polkinghorne turned from quantum physics to chaos theory as an indication that new

---

[64] John Polkinghorne, *One World: The Interaction of Science and Theology* (London: SPCK, 1986), 72.

[65] John Polkinghorne, *Science and Creation: The Search for Understanding* (London: SPCK, 1988), 58; idem, "The Quantum World," in *PPT*, 339–40.

[66] John Polkinghorne, *Science and Providence: God's Interaction with the World* (Boston: New Science Library, 1989), 27–28.

[67] John Polkinghorne, *Reason and Reality: The Relationship between Science and Religion* (Philadelphia: Trinity Press International, 1991), 40–41.

[68] John Polkinghorne, "The Metaphysics of Divine Action," in *CAC*, 147–56.

[69] Ibid., 152–53.

holistic laws, carrying an indeterministic interpretation, will be found at a more complex level of nature in the future, and that they will be consistent with special divine action.

Polkinghorne's reasons for not using quantum physics in the context of divine action are important and certainly deserve extensive discussion. Here I can only offer a preliminary response. With respect to (1) the connection I wish to make is between quantum mechanics and genetic mutation (which is an interdisciplinary issue of a phenomenological nature), not between quantum mechanics and chaos theory (which is an intra-physics problem of a theoretical nature). I'll defer my response to (2) to section 4.2 below. On point (3), it is certainly true that the measurement problem is connected with tremendously complex issues in the philosophical foundations of quantum mechanics, but from this it does not follow that the measurement problem makes God's actions "episodic." Polkinghorne may be correct in asserting that indeterminacies in quantum behavior only arise when an irreversible registration of their effects occurs in the macro-world. My point, however, is that these events are not limited to physical measurements in the lab, as I am sure Polkinghorne would agree. Instead, they occur constantly throughout the universe (see section 2.3.2). Thus, if one holds that these events provide one domain (among others) for the non-interventionist effects of divine action, then it suggests a God who is *acting providentially everywhere and at all times* in and through all of nature—a God whose agency is hardly "episodic."

3.3 Recent Constructive Developments

Beginning in 1993, several scholars have developed the theme initiated by Heim and Pollard, pointing to quantum indeterminacy as providing grounds for an understanding of special divine action from a *non-interventionist* perspective. Here I will focus specifically on the writings of Thomas Tracy and Nancey Murphy.

Thomas Tracy[70] is committed to a biblical understanding of God's action in the world. He rejects a "gaps" view of divine action in which we use "God" to explain things we simply do not (yet) understand. Moreover, Tracy rejects this kind of gaps argument for *theological* reasons, citing Dietrich Bonhoeffer: "[It is wrong] to use God as a stop-gap for the incompleteness of our knowledge... [Instead] we are to find God in what we know, not in what we don't know."[71] But suppose what we know from science suggests that there are "causal gaps"in nature?[72] Granted, the message of classical physics is that there are no such gaps; nature is a closed causal system. Theologians such as Rudolf Bultmann then built on the assertion that nature is causally closed to conclude that there is no room for divine action short of intervention.[73] Now, however, science has changed, and with it our belief in a closed

[70] Thomas F. Tracy, "Particular Providence and the God of the Gaps," in *CAC*, 289–324.

[71] Dietrich Bonhoeffer, *Letters and Papers from Prison*, Eberhard Bethge, ed., enlarged ed. (New York, N.Y.: MacMillan, 1979), 311.

[72] Tracy defines "causal gaps" as "breaks in the order of what is commonly called 'event causation'... which occur when events are not uniquely determined by their antecedents. Where causal gaps occur, later events cannot be deduced from a description of their antecedents and deterministic laws of nature"; "Particular Providence and the God of the Gaps," in *CAC*, 290.

[73] Rudolf Bultmann, *Jesus Christ and Mythology* (New York: Scribner, 1958), 65. See also Gordon Kaufman, "On the Meaning of 'Act of God'," in *God the Problem* (Cambridge, Mass.: Harvard University Press, 1972). John Macquarrie, *Principles of Christian Theology*, 2nd ed. (New York: Scribner, 1977).

causal world. Instead, if we accept an indeterministic interpretation of quantum mechanics,[74] nature is ontologically open, suffused with causal gaps. Thus quantum mechanics may be relevant to a theology of non-interventionist special providence.

To strengthen the case, Tracy puts conditions on the kind of causal gaps in the processes of nature that would be relevant to theology.[75] First, causal gaps must be part of the macroscopic world's regular, ordered structure in order that God's action through them will play an "ongoing and pervasive role in contributing to the direction of events." Secondly, causal gaps should make a difference in that order. According to Tracy, who cites my previous work here, this is precisely what quantum physics provides.[76] On the one hand, the overwhelming majority of quantum events do indeed give rise to the ordinary, ordered structure called the classical world. On the other hand, individual quantum events, though falling within the statistical distributions of quantum theory, occasionally have "significant macroscopic effects over and above contributing to the stable properties and lawful relations of macroscopic entities." Here Tracy cites the "Schrödinger's cat" thought experiment as well as the connection between quantum events acting as a source of genetic mutation and its effects on evolution.[77]

In sum, quantum mechanics allows us to think of special divine action as the "providential determination of otherwise undetermined events." Moreover, though pervasive in its effects on the world's structure, God's action will remain hidden within that structure. God's action will take the form of realizing one of several potentials in the quantum system, not of manipulating subatomic particles as a quasi-physical force. Meanwhile we are still free to think in terms of divine action at other levels of nature, including God's interaction with humanity and God's primary act of creating and sustaining the world. Tracy closes with a key question. Does God act to determine all quantum events or merely some? Pollard took the first option since it was consistent with his commitment to divine sovereignty, but Tracy is hesitant about this for theological reasons. Instead he finds it helpful to explore the latter view briefly while keeping both options open for further thought.[78]

Nancey Murphy gives one of the most sustained and balanced arguments to date for divine action in light of quantum mechanics, appropriating many of the previous positions while developing her own, creative argument.[79] According to Murphy, a theologically acceptable understanding of divine action must be consistent with the claim that God creates, sustains, cooperates with, and governs all things. It must include special divine action without exacerbating the problem of evil. Though it must be consistent with science, it might offer a new metaphysical view of matter and causality which takes into account the non-reducible hierarchy of complexity through either "bottom-up" or "top-down" causation.[80] Moreover, divine action is mediated: God never acts (apart from creation *ex nihilo*) except through cooperation

---

[74] Tracy of course is well aware of the fact that there are competing interpretations of quantum indeterminism, and this fact, he writes, should "keep us cautious." "Particular Providence and the God of the Gaps," in *CAC*, 315ff.

[75] Ibid. 316–18.

[76] Here Tracy cites an earlier version of this paper, "Theistic Evolution: Does God Really Act in Nature?" *CTNS Bulletin*, 15.1 (Winter 1995): 19–32; see also, "Quantum Physics in Philosophical and Theological Perspective," in *PPT*, esp. 345.

[77] Tracy, "Particular Providence and the God of the Gaps," in *CAC*, 318, footnote #64.

[78] Ibid., 320–22; see also Tracy's paper in this volume, section 2.3.2.

[79] Nancey Murphy, "Divine Action in the Natural Sciences," in *CAC*, 324–57.

[80] Ibid., 329–38.

with created agents.[81] This means that "God's governance at the quantum level consists in activating or actualizing one or another of the quantum entity's innate powers at particular instants." It also means that "these events are not possible without God's action."[82]

Murphy then turns to the relation between God's action at the quantum level and its macroscopic effects via bottom-up causality. She claims that the ordinary macroscopic properties of matter "…are consequences of the regularities at the lowest level, and are *indirect* though intended consequences of God's direct acts at the quantum level."[83] Thus God acts as sustainer of the macro-level by means of God's action at the quantum level. But does God act in *every* quantum event to bring it to actualization, or only rarely? Unlike Tracy, who wants to keep the options open, Murphy claims that every quantum event involves a direct divine act. One reason is that "God must not be made a competitor with processes that on other occasions are sufficient in and of themselves to bring about a given effect." Another is to avoid any sense of God's action as intermittent or occasional; instead, God's action is "a necessary but not sufficient condition for every (post-creation) event."[84]

Murphy sees her proposal as coming closest to, but actually being preferable to, that of Pollard since she believes hers allows for human freedom by insisting on both top-down and bottom-up causality and by stressing God's under-determination of the outcome of events. Similarly, she believes that her proposal minimizes God's responsibility for evil, thereby softening the problem of theodicy.

In her closing section Murphy turns to several "unanswered questions." She cites Polkinghorne's reason for not turning to quantum mechanics in connection with special divine action: namely, that macroscopic events stemming from specific quantum events are restricted to those describable as measurements, and that this leads to an "episodic" account of divine action. In response she cites my claim that "the general character of the entire macroscopic world" is in some sense a result of quantum mechanics, and thus the universe as a whole is the realm of divine action.

> "I have been assuming Russell's position throughout this paper. Yet even if Russell is correct… does the fact that God is affecting the whole of reality… *in a general sense* by means of operation in the quantum range allow for the sort of special or extraordinary divine acts that I claim Christians need to account for? Or would such special acts be limited to the few sorts of instances that Polkinghorne envisions?"[85]

Murphy leaves the matter unsettled. I would like to close by reflecting briefly on four issues raised by Tracy and Murphy.

1. I've responded in part to Murphy's closing question in my earlier reply to Polkinghorne's charge (section 3.2). I think that indeterminacies in quantum behavior arise in a much more pervasive way than the term "measurement" suggests. Instead, they arise *constantly*, everywhere and at all times, in every part of the universe. If so, this claim can increase the theological intelligibility of our faith in general providence of the Triune God who is *everywhere and at all times* at work in and through all of nature. Moreover, quantum events can lead to a significant

---

[81] "To say that each sub-atomic event is solely an act of God would be a version of occasionalism, with all [its] attendant theological difficulties…" These include: exacerbating the problem of evil, verging on pantheism, and conflicting with the belief that God gives the world an independent existence. Ibid., 340.

[82] Ibid., 342.

[83] Ibid., 346. The italics are Murphy's.

[84] Ibid., 343.

[85] Ibid., 357.

difference in the bulk state of a macroscopic object or process. If so, this claim can provide at least one non-interventionist way for thinking about God's special providence in nature. All we need to add is the insight that the relation between particular quantum events and their indirect macroscopic effects is realized not only in the context of physics but even more prominently and pivotally in the context of evolutionary biology. In particular, the phenotypic expression of a genetic mutation is the biological analog of amplifying a quantum effect into a macroscopic result in physics. God has indeed created a universe in which God's special providence can come about in the long stretches of evolutionary biology without intervention.

2. I agree with Tracy and Murphy that the insistence on mediated direct divine action helps to counter two charges: occasionalism[86] and divine omnideterminism. In my view, though, it is still unclear whether mediation alone settles the question of how we *enact* free will. If quantum events are due entirely to direct divine action mediated by physical causality, how do we think about conscious agents enacting mental states somatically without incurring the charge of over-determinism?[87]

3. Finally, we turn to the closely related issue: does God act directly in some or in all quantum events? Tracy explores the option that God acts in some but not all quantum events. This option seems to violate the principle of sufficient reason, since some quantum events would occur without sufficient prior conditions, constraints, or causes. On the other hand it underscores the "special" character of "special providence": God's direct acts in key quantum events are special not only because their indirect outcome is special, but also because God normally does not act in other quantum events beyond creating them and sustaining them in being. Moreover, Tracy's approach provides a fruitful basis for thinking of God's occasional, special action in terms of self-limitation: God *could* act together with nature to co-determine all quantum events, but God *abstains* from such action in most cases.

Murphy's option—God acting in all quantum events—is supported by the principle of sufficient reason. However, her option may not entirely avoid the problem of *divine* omnipotence that beset Pollard's view. Similarly her option might undercut our ability to *exercise* free will *somatically* as mentioned above. However, Murphy is careful throughout her paper to make the crucial claim that God restricts God's action so as to respect the integrity of creatures at every level of complexity. If this claim can be sustained by further analysis, then these problems could dissolve.

I would suggest one alternative view. We may think of God as acting in all quantum events in the course of biological evolution until the appearance of organisms capable of even primitive levels of consciousness. From then on, God may continue to work in terms of the quantum-genetic domain, but God may then abstain from acting in those quantum events underlying bodily dispositions, thereby allowing the developing levels of consciousness to act out their intentions somatically. This approach combines aspects of Murphy's and Tracy's approaches, it includes the idea of divine self-limitation, and it gives to them all a temporal character. God bequeaths us not only the capacity for mental experience via God's special action in evolution

---

[86] This view denies the causal interaction between events in nature. Instead sequences of events in time provide the "occasion" for God, as the sole cause, to produce the expected effects in the world.

[87] Whiteheadian philosophy offers an interesting three-fold alternative here: every actual occasion includes an irreducible element of intrinsic novelty along with physical causality and divine lure. Perhaps this third element provides the basis for understanding how our conscious decisions can be enacted somatically.

and the resulting rise of the central nervous system, but God also bequeaths to us the capacity for free will and the capacity to enact our choices by providing at least one domain of genuine indeterminacy in terms of our somatic dispositions.

4. I believe the problem of theodicy is stunningly exacerbated by all the proposals, including my own, that God acts at the level of genetics—certainly much more so than either Tracy or Murphy allow. The development of an adequate theological response is an important goal for future theological research. I will offer a brief suggestion regarding a possible direction for this research in section 5 below.

4 Three Remaining Caveats

There are a number of subsidiary issues raised here, including some immediate reactions to, and possible misinterpretations of, the position put forward here. I attempt to do deal with them in the form of caveats drawn in part from arguments which can be found already in the sources cited above, and in part from new suggestions I wish to bring to the conversation.

4.1 Should My Proposal Be Read as an Explanation of How God Acts?

No. I am not in the least suggesting this as an explanation of *how* God acts,[88] nor even an argument *that* God acts. It is only a claim that *if* one believes for theological reasons that God acts in nature, quantum physics provides a clue as to one possible location or domain of action. More precisely, it locates the domain in which that action—however mysterious it truly is in itself—may have an *effect* on the course of nature, namely in the domain of the gene.

4.2 Isn't This a Kind of "Gaps" Argument?[89]

There are two kinds of gaps arguments which I do not believe correctly describe this proposal.

Type I: Epistemic gaps in science as a token of our ignorance. Many gaps in our current understanding of nature will eventually be filled by new discoveries or changing paradigms in science. The argument is that we ought not stake our theological ground on transitory scientific puzzles.

I agree with this concern. Epistemic gaps as such in science *will* be filled by science, and theology should *not* be evoked as offering the type of explanation which science could eventually provide. My proposal, however, is *not* about epistemic gaps. Instead it is a claim that depends on the contents of what is *known* by one branch of science, namely quantum mechanics, and on a reasonable interpretation of quantum mechanics, namely ontological indeterminism. According to this view of quantum mechanics, what we *know* is that we cannot explain why a specific quantum mechanical outcome occurs, only its probability of occurring.

Type II: Ontological gaps in nature caused by occasional divine intervention. The concern here is that God occasionally intervenes in nature, causing gaps in the

---

[88] For example, it is not meant as a solution to the problem of "double agency" or to what Austin Farrer referred to, when addressing the God-world relation, as the "causal joint."

[89] I am using terms in a slightly different way than Tracy does. According to Tracy, there are two kinds of "explanatory gaps" In-practice-gaps are holes in our current theories which we expect to fill in the future. In-principle-gaps are holes in our current theories which our theories suggest can never be filled. The latter may, but need not, arise from another kind of gap, namely "causal gaps," or holes in the flow of natural processes. Quantum mechanics may be interpreted as pointing to such causal gaps, as Heisenberg first suggested. See Tracy, "Particular Providence and the God of the Gaps," in *CAC*, 290–92.

otherwise seamless natural processes by violating, or at least suspending, these processes in an act which breaks the very laws of nature which God previously created and constantly maintains. Such a claim depicts God as, in effect, against God! It also tends to suggest that God is normally absent from the web of natural processes, acting only in the gaps God causes.

Again, I agree with these concerns; we *should* avoid an interventionist argument. *This* proposal if fact does so because it does not view the gaps in nature as disruptions of what would otherwise have been a closed, causal process. Instead it assumes that 1) the laws and processes of nature are neither violated nor suspended by God's action, but maintained by God, 2) it is precisely the laws which God maintains that tell us that there already are *ontological* or *causal* gaps throughout nature. Moreover, 3) since these gaps are ubiquitous in nature (as my response to Polkinghorne's restriction on "measurement" suggests), the God who acts through them is the God who is universally present in and through that action, not an occasional and normally absent actor.[90]

Still this proposal could be considered a "gaps" argument of a more subtle variety, since it appeals to quantum mechanics, and like any theory, quantum mechanics may one day be replaced. Why then take its metaphysical implications as seriously as I do here in deploying a theology of divine action if the specter of "historical relativism" looms large?

Actually this is a concrete example of a very general issue: How should the historical relativity that inevitably surrounds *any* scientific theory affect the philosophical and theological discussions of that theory? It is, in fact, a crucial methodological issue lying at the heart of *any* conversation about "theology and science." A decision regarding it is required of every scholar in the field. I will try to describe mine here, though all too briefly.

One option would be to disregard theories which are at the frontier of science, and instead stick with proven theories. I don't agree with such an overly cautious approach for two reasons. 1) The theories which we know are "proven" are the ones which have been the most clearly falsified![91] We know precisely in which domains classical physics applies for all practical purposes, namely in the limits $h \to 0$ and $c \to$ infinity. But classical physics is in principle false; as a fundamental perspective, its view of nature and its explanations of the world are wrong. 2) Unfortunately, however, it is with this classical view of nature that the theology of previous centuries, and much of contemporary theology, has in general operated. It has contributed to the divisions between conservatives and liberals over issues such as divine agency and special providence, and it was a major fact underlying what Gilkey called the "travail" of neo-orthodoxy and biblical theology. In addition, both the

---

[90] Indeed, regardless of the issue of quantum mechanics, God is already present and acting ubiquitously in nature in and through all the laws of nature and as the source of nature and the laws of nature. In making this case I am implicitly adopting a trinitarian doctrine of God here, framed within the broad outlines of panentheism. Such a God is seen as supremely active in all reality, as its absolute origin (first person), the form and wisdom which structure and guide its processes (second person), and the power which sanctifies and empowers it towards completion (third person). Thus God does not intervene because God is already universally present in nature. In the famous phrase of Jürgen Moltmann, if we start with a trinitarian conception of God's relation to the world we avoid what monotheism otherwise renders us, namely a "worldless God and a godless world." We know we've fallen into that trap when we use language such as "God's interaction with the world."

[91] This is Charles Misner's argument about the museum of discarded, "true" theories.

atheistic challenge to theistic evolution and the religiously motivated attack on evolution have almost without exception ignored the quantum mechanical aspects of genetic variation and presupposed classical science and a mechanistic, deterministic metaphysics. Thus their arguments, too, are fundamentally flawed. So sticking only with proven theories is out.

Another option is to "pick and choose" among frontier issues, but omit quantum physics. After all, along with the possibility of historical relativism is that of *metaphysical under-determinism:* quantum physics can, after all, be given a deterministic interpretation as Bohm did, and this fact, when combined with its historical relativity, makes its appropriation seem doubly risky.[92] My response is two-fold. First, the determinism suggested by Bohm's approach is definitely not classical determinism but a highly non-local view in which the whole of nature determines each part to be as it is.[93] Even if we were convinced of his approach we would not fall back into classical metaphysics. Perhaps after a theology of divine action in light of the metaphysical implications of Bohm's views has not yet been worked out in detail we will be in a better position to evaluate this alternative. Second, why stop with quantum mechanics? *Any* scientific theory is open to competing metaphysical interpretations; indeed, metaphysics is *always* under-determined by science. So this concern about quantum mechanics applies, in principle, to *any* metaphysical interpretation of *any* scientific theory. Indeed, it is an issue not only for a theistic but even for a naturalist or an atheistic interpretation of science!

My approach, instead, is to engage in the conversation with quantum mechanics, as with any scientific theory, in full realization of the tentativeness of the project—but to engage in it, nevertheless. This is warranted for three reasons. 1) We are doing constructive theology, *not natural theology let alone physico-theology*. Hence a change in science or in its philosophical interpretation would at most challenge the constructive proposal at hand, but not the overall viability of a theology of divine action in nature, whose warrant and sources lie elsewhere in Scripture, tradition, reason and experience. 2) As the experimental violation of Bell's Theorem shows us, any future theory will have to deal with some aspects of current quantum phenomena, and it is these general features and their metaphysical implications which are our focus here. 3) We should welcome the specificity of this approach and the vulnerability it produces to problems like these, for by illuminating the actual implications of a concrete example of a non-interventionist approach to objective special divine action it enhances the strengths as well as reveals the limitations of that approach, and this in turn leads to further insight and research.

4.3 *Is a "Bottom-up" Approach to Divine Action Warranted and Does it Exclude Other Approaches?*

Does my focus on God's action in the domain of quantum mechanics and genetic mutations limit or restrict our understanding of God's action to a "bottom-up" approach to divine action?

---

[92] John Polkinghorne stresses this concern frequently as a reason not to use quantum physics for an indeterministic metaphysics within a theory of divine action, citing David Bohm for his a realist and determinist account of quantum phenomena.

[93] Bohm first introduces a divergenceless field, the quantum potential, which "guides" elementary particles in terms of their environment, and then reformulates the underlying ontology in terms of what he calls the "implicate order."

My response is that we should *not* see the present focus as a *general* limitation or restriction of divine action to "bottom-up" causality alone. Instead, I see the present argument as located within a much broader context, namely a theology of divine action in personal experience and human history, because that is primarily where we, as persons of faith, encounter the living God as God reveals Godself to us. Here we clearly need to consider a variety of models, including both "top-down," "whole-part," and "bottom-up" causes and constraints, and their roles within both embodiment and non-embodiment models of agency. Moreover, we will eventually need to work out the relations between these models in detail by integrating them into a consistent and coherent, adequate and applicable metaphysical framework.

The question here, though, is why and how God might be thought of as acting within the evolutionary processes via a form of bottom-up causality. Granted that God is the creator of the universe *per se*; without God there would be no universe, nor would the universe exist moment by moment. Granted that God maintains the efficacy of nature, whose regularities, which we call the laws of nature, manifest God's faithfulness and rational intelligibility as Creator. Granted that these laws have just the right statistical ingredients to allow for the production of "order out of chaos" as part of God's creative actions. Granted that in some situations, such as our personal encounter through faith with God, we might introduce top-down language about God's action. Granted all this, can we adequately understand God's action within the evolutionary processes out of which we arose as expressing God's intention in ways that go beyond that of maintaining the existence of these processes and allowing their built in "potentialities" to work themselves out over time? And can such an understanding of God's action be rendered in an intelligible way if we restrict ourselves to top-down causality and whole-part constraint *alone*?

I believe it cannot. Top-down causality is helpful when considering the action of conscious and self-conscious creatures with some capacity to respond to God's self-communication as gracious love.[94] But it is hard to see what constitutes the "top" through which God acts in a top-down way when conscious, let alone self-conscious, creatures capable of mind/brain interactions, have not yet evolved. Remember, we are trying to understand the evolution of organisms over a period of nearly four billion years, ranging from the simplest primitive forms to the present vastly rich profusion of life.

In fact, a top-down approach would be incompatible with our choice of a non-interventionist strategy for divine action. If God acts at the "top" level of complexity at a given stage in evolutionary history, that level of complexity must be ontologically open, that is, it must be described by laws which can be interpreted in terms of metaphysical indeterminism.[95] Yet, until the evolution of organisms capable of even primitive mentality, the "top" levels would presumably have been within the domain of the "classical" sciences reflecting the ontological determinism of Newtonian physics. Hence special divine action would be unintelligible *without* intervention

---

[94] Process theology may claim to respond to this question nicely, since all actual occasions are understood as capable of responding to God's subjective lure, but it too runs into problems. Process philosophy distinguishes between societies of actual occasions organized as mere aggregates which cannot respond to God, such as a rock or a molecule, and societies which can, such as an animals. So the problem remains: How can God "lure" a DNA *molecule* (which is an *aggregate, not a society,* of occasions) to undergo a genetic mutation?

[95] George Ellis makes a related claim in "Ordinary and Extraordinary Divine Action: The Nexus of Interaction," in *CAC*, 359–95.

during those early stages of evolutionary biology prior to and leading to the development of a central nervous system. But if we do not include this early period within the scope of our discussion of special providence, then we once again risk limiting God's special action: it can only occur after a sufficient degree of biological complexity has been achieved, but it cannot be effective within the processes by which that degree of complexity is achieved. For both these reasons, then, the top-down strategy seems stymied.

Perhaps we should try whole-part constraint arguments instead. The challenge here is to find phenomena in pre-animate biology which display *genuinely* holistic characteristics. The ecological web is often cited as a candidate, due to its inherent complexity and its seemingly endless openness to external factors, but in my opinion it fails to be genuinely holistic in principle because of the underlying determinism of the processes involved, no matter how complex or distant. Unless one returns to the quantum level, where holism and indeterminism are displayed ubiquitously, I see little hope that God's action within the early stages of evolution can be can be described in non-interventionist ways using whole-part constraint arguments.[96]

Thus on critical reflection, and contrary to the hopes of most previous attempts at theistic evolution, it seems unlikely that top-down or whole-part approaches are of much value when we are seeking to interpret evolution at the pre-cognitive and even pre-animate era in terms of special divine action.

5 *Divine Purpose and Theodicy: Two Promising Issues for Further Research*

5.1 *Chance and the Challenge to Purposeful Divine Action*

"Chance" in evolution raises at least two distinct challenges to divine action: 1) to the possibility of divine action as such, and 2) to the possibility of God achieving a *future* purpose by acting *in the present*. So far I have focused on the first challenge; here I will turn, briefly, to the second. The challenge arises because "chance" in evolution, as Jacques Monod notably stressed, involves more than genetic variation. In addition we must consider a variety of domains which involve random factors: the multiple paths leading to the expression of genotype in phenotype, processes influencing the survival and reproduction of progeny, changes in the ecosystem in which natural selection plays out. Finally, we must consider the juxtaposition of these two streams: the stream of (ontological) chance events at the molecular level and the second stream of (epistemic) chance events at the environmental level, where natural selection occurs. Since these streams are uncorrelated, *their* juxtaposition is random. It is in *this* sense in particular that evolution is often called "blind." How, then, can God anticipate the eventual consequences of God's action at the quantum level of genes given all these varying factors?

To respond we must first locate these questions within the perennial issue of the relation of time and eternity. In specific, I would start with the claim that *God can*

---

[96] The universe has other important holistic characteristics, but its status as a unitive object of scientific study—the meaning of speaking of the universe "as a whole"—is problematic. Moreover in both cases (the ecological web, the universe) we are again at the "classical" level where divine action would be limited, presumably, to intervention. Quantum gravity/quantum cosmology may be an important exception. See Robert J. Russell, Nancey C. Murphy, and Chris J. Isham, eds., *Quantum Cosmology and the Laws of Nature: Scientific Perspectives on Divine Action* (Vatican City State; Berkeley, Calif.: Vatican Observatory Publications; Center for Theology and the Natural Sciences, 1993), hereafter *QCLN*.

know what are for us the future consequences of God's actions in our present. In the spirit of the non-interventionist approach of this paper, I would add: *God having such knowledge must not entail a violation or suspension of the laws of nature.* Next, this claim assumes that God does not *foresee* our future from our present or *foreknow* our future by calculating the outcome from our present.[97] Instead, God as eternal *sees* and *knows* the future in its own present time and determinate state. Such a view was given a highly nuanced formulation in classical theism, and it has been reshaped in important new ways during the twentieth century by a number of Roman Catholic and Protestant theologians.[98] The basic point, however, is that God's knowledge of what is for us the indeterminate future is God's eternal knowledge of an event in what is its own present, determinate state. Thus theologically, God can have knowledge of the future consequences of God's actions in the present.

But the commitment to non-interventionism raises an additional issue here: How are we to think about the ontological status of the future which God is to have knowledge of in light of special relativity? There are two alternative interpretations of special relativity.[99] According to one view, the flow of time is real. However, this view undercuts the reality of the future from the point of view of the present, and thus seems to undermine our claim for God's knowledge of it. According to the other view, the flow of time is an illusion (the so-called block universe view). But this view undercuts the purported indeterminism of the future from the point of view of the present, and with it, some would say, free will. My response would be to argue that both alternatives fail to do justice to the actual implications of special relativity. Instead, an ontology of events and a relational definition of time can allow for the reality of time's flow consistent with special relativity. Weaving this model eventually into the contemporary trinitarian understanding of eternity promises to add further credibility to the claim that God can indeed act purposefully in the context of biological evolution.

5.2 Theodicy and the Need for a Theology of Redemption

If God is intimately at work at the level of the gene, is not God also responsible for the disease, pain, suffering, and death brought about by these genes? Why does God allow the overwhelming majority of genetic mutations to end in failure of the organism? This is, of course, the perennial problem of theodicy—Why does a just, good, and powerful God allow real evil?—but now with the domain extended beyond the human world to include all of life on earth.[100]

My first response is that to stress that pain, suffering, disease, death and extinction are facts which any theological interpretation of evolution must deal with; it is not unique to this approach. An all-too-frequent response is to remove God from

---

[97] This is one of the assumptions that I believe underlie Peacocke's reasons for rejecting quantum indeterminism as helpful for the problem of divine action. See p. 210 above.

[98] Here I am drawing on an understanding of the relation of time and eternity generally found, though with important differences, in the writings of Karl Barth, Karl Rahner, Wolfhart Pannenberg, Jürgen Moltmann, Ted Peters, and others. The essential claim here is that eternity is not a truncated, point-like *nunc* and thus the opposite of time, but rather the endlessly rich source of time.

[99] For a recent discussion see C. J. Isham and J. C. Polkinghorne, "The Debate over the Block Universe," in *QCLN*.

[100] Indeed, if evidence of even primitive life is ever found on other worlds—and current discussion of the "Martian meteor" raises this possibility—the domain could be extended almost without limit. See the papers by Paul Davies and Julian Cela-Flores in this volume.

the detailed history of nature. Instead God created the universe with certain potentialities, and the history of life is the mere unfolding of the these potentialities, at least until humanity comes along to respond to God's personal revelation. Pain and suffering are seen strictly as the result of human sin, whose consequences ravage humanity and an innocent environment. Disease and death are simply the natural prerequisites for the evolution of life. But restricting God's action in this way does not resolve the problem of theodicy or the question of the origin of sin. It simply raises these problems to the level of cosmology: Why did God choose to create this universe with this particular set of laws of nature and their unfolding consequences? Could not God have produced a universe in which life evolved without death, pleasure without pain, joy without sorrow, free will without moral failure?[101] Moreover, a world thus stripped of God's special providence and tender, constant attention seems a much more troubling one to me —since its suffering would be *both* real and beyond God's care—than a world in which God is genuinely, even if inscrutably, at work, caring for every sparrow that falls. By keeping God at a distance from the suffering of nature, we thereby render that suffering all the more pointless, its outcome all the more hopeless.

A more fruitful response begins with the insight that God created this universe with the evolution of moral agents in mind. In such a universe suffering, disease, and death are in some way coupled with the conditions for genuine freedom and moral development. A variety of promising responses are being pursued to this question in the current literature. For example, those offered by Ian Barbour, George Ellis, Philip Hefner, David Ray Griffin, Arthur Peacocke, John Polkinghorne, Bill Stoeger, and Thomas Tracy, while differing in important ways, mostly tend to stress the (either voluntary or metaphysical) limitation on God's action required by genuine creaturely freedom—a theme I suggested above (section 3.3)—with a corresponding emphasis on the suffering of God with nature.[102] More recently, Nancey Murphy and George Ellis have drawn on the Radical Reformation/Anabaptist tradition to develop a "kenotic ethic" closely connected with a "fine-tuning" response to the Anthropic Principle.[103] But can these responses account for the magnitude of suffering in nature across billions of years and the essential role of death in the evolution of biological complexity and, in at least one species, moral agency?[104]

I can only suggest ways in which my own response to theodicy seeks to incorporate these themes and ground them in the central biblical insight that the God who creates is the same God who redeems,[105] particularly as it has been developed

---

[101] One could try to answer these questions, of course, by a "many-worlds" strategy, a "these laws are the only possible ones" strategy, or a "best of all possible laws" strategy, and so on. My point is simply that removing God from the detailed history of nature does not automatically eliminate the challenge of theodicy.

[102] See, for example, the papers by Barbour, Peacocke, and Tracy in this volume.

[103] Nancey Murphy and George F. R. Ellis, *On the Moral Nature of the Universe: Theology, Cosmology, and Ethics* (Minneapolis, Minn.: Fortress Press, 1996), esp. chap. 10, section 4.

[104] See, for example, Holmes Rolston, III, "Does Nature need to be Redeemed?" *Zygon*, 29.2 (June, 1994): 205–29; see also Tracy in this volume and Ellis in *CAC*.

[105] In Christian theology, the link between creation and redemption is made very explicit: the Word made flesh, the Word through whom all things were made, is also the same Word whose self-emptying kenosis in revealed in the Incarnation and through the Cross. The Logos and Wisdom of creation is the slain Lamb who atones for all human sin. The Alpha Logos,

in twentieth century trinitarian theologians such as Karl Barth, Denis Edwards, Catherine John Haught, Mowry LaCugna, Elizabeth Johnson, Jürgen Moltmann, Wolfhart Pannenberg, Ted Peters, and Karl Rahner.[106] Redemption, in turn, takes as its cornerstone the Cross of Christ, in which the power of God is revealed through suffering as transformative love. That such love will ultimately—eschatologically— overcome the evils of this world is a wager Christians make in the wake of often overwhelmingly contradictory evidence, as the history of this century's human anguish bespeaks. Now as we extend the scope of history to include the billion-year drama of life on earth, this means that evolution must somehow provide the occasion, albeit hidden, not only for *creatio continua* but even more profoundly for healing, cruciform grace to continuously *redeem* creation and guide it along with humanity into the new creation. In a formal sense, what I am proposing is that the doctrine of creation, and with it providence, offers a necessary but, in the final analysis, inadequate framework for responding to the problem of theodicy. Instead, theodicy and the implicit problem of evil can only be addressed (though never "resolved") within the framework of a theology of redemption and new creation. In fact, I would venture a further step. The long sweep of evolution may not only suggest an unfinished and continuing divine creation but even more radically a creation whose theological status as "good" may be fully realized only in the eschatological future.[107] If these ideas are at all sound, they underscore how their further development is not only a daunting task but an imperative one if we are to answer not just the critics of Christian faith, but also our own internal criticisms over the inadequacy of generic theistic evolution.

Acknowledgment. I wish to express my immense gratitude for the extraordinarily careful reading of an early draft by A. Durwood Foster. I also wish to thank Ian Barbour, Dave Cole, Wim Drees, Nancey Murphy, Mark Richardson, Tom Tracy, and Kirk Wegter-McNelly for very incisive comments on recent drafts.

---

whose "trace" is the laws of physics, is the mothering healer of the suffering and death of all that lives and the eschatological hope of the new creation, the Omega of the universe.

[106] For a helpful perspective on the many voices, see Ted Peters, *God as Trinity: Relationality and Temporality in Divine Life* (Louisville, Ky.: Westminster/John Knox Press, 1993); see also Edwards and Haught in this volume.

[107] Pannenberg stresses this point at the close of his systematics. See Wolfhart Pannenberg, *Systematic Theology*, vol. 3 (Grand Rapids, Mich.: William B. Eerdmans Publishing Company, 1998), chap. 15, esp. 645–46. See also Karl Barth, *Church Dogmatics*, vol. 3.2, *The Doctrine of Creation*, G. W. Bromiley and T. F. Torrance, eds. (Edinburgh: T. & T. Clark, 1958), esp. 385–414. Peters makes the point indirectly by setting it within a broader holism in which the term "creation" includes the whole history of the universe. See Ted Peters, *God, the World's Future: Systematic Theology for a Postmodern Era* (Minneapolis, Minn.: Fortress Press, 1992), 134–39, 307–9; idem, ed., *Cosmos as Creation: Theology and Science in Consonance*, (Nashville, Tenn.: Abingdon Press, 1989), 96–97.

NEO-DARWINISM, SELF-ORGANIZATION,
AND DIVINE ACTION IN EVOLUTION

Charles Birch

I cannot think that the world is a result of chance; and yet I cannot look at each separate thing as the result of Design.

Charles Darwin[1]

The universe is a creative advance into novelty. The alternative to this doctrine is a static morphological universe.

Alfred North Whitehead[2]

Tom to the mother of creation "I heard that you were always making new beasts out of old." "So people fancy" she said "I make things make themselves."

Charles Kingsley[3]

1 *Introduction*

Four years after Charles Darwin published *The Origin of Species* the Church of England vicar and novelist Charles Kingsley wrote for his children the evolutionary fairytale *The Water Babies*. Kingsley was convinced that the Darwinian theory of evolution was the context within which it was possible to find the working of "a living, immanent, ever-working God."[4] The concept of God as immanent in creation was understood by Kingsley as a creation in which God made "things make themselves." Self-determination, for him, was consistent with divine action. Indeed, divine creativity for him was of such a nature that it could only work through entities that had their own degree of creativity and self-determination. I, too, shall argue that self-determination is consistent with Darwinian natural selection and with a view of divine creativity as understood especially in process thought.

Neo-Darwinism and self-organization, as understood by biologists, are mechanistic operations in the evolutionary process. That is to say, the procedure of the biologist is to investigate living organisms *as if* they are machines such as cars.[5]

---

[1] Francis Darwin, ed., *The Life and Letters of Charles Darwin* (London: John Murray, 1888), 353.

[2] Alfred North Whitehead, *Process and Reality*, corrected ed., David Ray Griffin and D. W. Sherbourne, eds. (New York: Free Press, 1978), 222.

[3] Charles Kingsley, *The Water Babies* (London: Hodder & Stoughton 1930, original edition 1863), 248.

[4] Charles E. Raven, *Science, Religion and Christian Theology*, vol.1 (Cambridge: Cambridge University Press, 1953), 177.

[5] Charles Birch, *On Purpose* (Kensington: New South Wales University Press, 1990), also published as *A Purpose for Everything* (Mystic, Conn.: Twenty-third Publications, 1990), 57–61; idem, *Regaining Compassion: for humanity and nature* (St. Louis, Miss.: Chalice Press, 1993), 58–59; idem, *Feelings* (Kensington: New South Wales University Press, 1995), 7; Charles Birch and John B .Cobb, Jr. *The Liberation of Life: from the cell to the community* (Cambridge: Cambridge University Press, 1981), 68–75.

While the modern understanding of evolution renders highly improbable the concept of an interventionist view of God, as I shall indicate, it is consistent with a process view of God as ever active in all the entities of creation and in their evolution. But first it is necessary to understand what biologists mean by neo-Darwinism and self-organization. Since these scientific concepts are understood by most biologists as related to completely mechanistic processes, they have led many biologists and others to a thoroughgoing materialistic world-view. This is because not only are organisms investigated as if they are machines, but also the conclusion is drawn by these biologists that they are machines. Methodological mechanism leads them to metaphysical mechanism.

I shall argue that these same facts of nature are better understood within the framework of a completely different world-view variously called process thought, panexperientialism and panpsychism. I want to put the science of evolution in context, to show its borders not as sharply defended frontiers but as complex interweavings with other ways of thinking. I do not see the various discourses of science, philosophy, and theology as clearly distinct. They flow into one another. The warrant for this point of view goes beyond the scope of this paper. However, I trust that enough is argued to indicate its plausibility. The world of thought divides knowledge into separate factions of science, religion, and philosophy. But there is only one reality. We need to break down the artificial walls, or at least try to make holes in the walls through which we might see a bit more of the whole. I see this as a specifically Christian responsibility in a divided world which we are called upon to heal. Ours is a universe, not a multiverse, despite the disciplinolatry of learning today.

I accept neo-Darwinism and self-organization as facts of biology. However, I find they make more sense in the context of a non-materialistic and in particular a process view of nature. It makes no sense to talk about a philosophical framework for evolution and self-organization without briefly explaining first what biologists understand by these concepts. I do not attempt to present the whole argument for process thought but to indicate those elements of it particularly relevant to evolution. Had I been writing about physics no doubt I would have emphasized other elements. I give enough references hopefully to indicate where this partial account leads.

While most of my teachers in biology have been thoroughgoing mechanists in both meanings of the term, I am grateful that one of my first teachers in biology and one of my last helped me to understand evolution within the framework of process thought. The first was Wilfred Agar, professor of zoology at the University of Melbourne, whose book *A Contribution to the Theory of the Living Organism* was first published in 1943.[6] The latest of my teachers in biology, who was a process thinker, was Sewall Wright, professor of zoology at the University of Chicago, who was one of the founders of neo-Darwinism.[7]

The basic reasons I was led to an alternative framework to the sheer materialism of most of my colleagues in biology have to do with my particular understanding of the following philosophical concepts. These are: the distinction between object and subject, the concept of reductionism and its opposite which is the interpretation of

---

[6] Wilfred. E. Agar, *A Contribution to the Theory of the Living Organism* (Melbourne: Melbourne University Press, 1943).

[7] Sewall Wright, "Gene and Organism," *American Naturalist* 87: 5–18; idem, "Biology and the Philosophy of Science," in W. L. Reese and E. Freeman, *Process and Divinity* (La Salle, Ill.: Open Court, 1964), 101–25.

the lower in terms of the higher, and the concept of emergence. Hence my discussion of neo-Darwinism and self-organization are followed by sections on these subjects as each is informed by process thought. Only in the context of these subjects can my discussion of process thought in evolution make any sense at all.

2 Neo-Darwinism

Neo-Darwinism is the dominant biological theory accounting for the order of the living world. There is, however, a second approach to the understanding of order in the living world, namely the concept of self-organization. This is the proposition that many of life's building blocks have their own ordered properties that account for some of the order in living organisms.

Neo-Darwinism, which is Darwinism in the light of modern genetics, is a thoroughly mechanistic explanation of evolution. It emphasizes three propositions: chance genetic variation, struggle for existence, and natural selection. It is from the pool of genetic variability derived from mutation of genes and the recombination of genes in sexual reproduction that *natural selection* operates to change the type in *the struggle for existence*. Natural selection is not a force so much as a *necessary outcome* of genetic variation and the struggle for existence. Hence the title of Jacques Monod's book *Chance and Necessity* and his statement that, "drawn from the realm of pure chance, the accident enters into that of necessity, of the most implacable certainties."[8]

A simple and accurate definition of natural selection is that it is the differential survival and reproduction of individuals in a population. It is an ordering process because it moves the genetic constitution of the population in the direction of greater adaptiveness. There is much discussion among biologists as to what is selected: some argue for the gene as the unit of selection (the selfish gene); some argue that it is the genotype that is selected; others argue that it is the population of genotypes; still others argue that the phenotype—the individual as molded by the interaction of the genotype with its environment—is the unit of selection. Whatever the unit of selection, most biologists agree that natural selection is basic to any understanding of evolution.

An aspect of genetic variation that was a problem for evolution for a long time was that most mutations are deleterious to the organism and are lost almost as soon as they are created. However, it can be shown that, given enough time, sufficient favorable mutations occur to lead to greater adaptiveness of the population. Because of the length of time involved, it is only in special circumstances that we can witness the creation of new species except in plants, where it is not an unusual phenomenon. There is no problem in our witnessing microevolution within species of microorganisms, plants, and animals. Such changes have been extensively studied, especially in microorganisms and insects.

The process of evolution by natural selection of chance variations has been much misunderstood, leading, for example, to Sir Fred Hoyle's howler that no complex protein could be assembled from its constituent molecules of amino acids by natural selection any more than a Boeing 747 could be assembled by the random collection of parts in a junk yard. This is a variant of the older metaphor of the million monkeys hitting at random on a million typewriters to produce one of

---

[8] Jacques Monod, *Chance and Necessity: an essay on the natural philosophy of modern biology* (New York: Alfred Knopf, 1971), 118.

Shakespeare's plays. Another variant is that evolution by natural selection of chance variations is as incredible as the production of a great picture by a blind painter sprinkling a canvas at random with a brush dipped in unseen colors.

The analogy is misleading for several reasons. The analogy of the blind painter has to be changed as follows. Not one, but a million, nay billions, of blind painters each sprinkle a few splashes of colors on millions of canvases. Of these, only the few that show the first feeble suggestion of a meaningful picture are preserved. The rest are destroyed. The selected rudimentary pictures are reproduced a million fold and again millions of blind painters add a few random touches of paint to them. Again the best are selected and reproduced, and so on millions of times corresponding to the number of generations that have elapsed since life appeared on Earth. This metaphor comes a bit closer to reality because it provides for reproduction and selection, step by step over many generations, not just in one fell swoop. Yet even this analogy misses many of the subtleties of the genetic process of natural selection as now understood.

Another misinterpretation is to compare natural selection to a sieve which retains the beneficial mutations and lets the deleterious ones through to be lost. The analogy is a poor one because it leaves out of account sexual recombination of genes and the interaction between genes. The usefulness or harmfulness of a genetic variant often depends upon the other genes in whose company it finds itself. A variant that is beneficial in one company can be detrimental in another. Dobzhansky has said that for the sieve analogy to have any validity at all one would have to imagine an extraordinary sort of sieve. It would have to be designed so as to retain or to discard a particle, not on account of its size alone, but in consideration of its many other qualities and those of other particles present in the sieve.[9]

The concept of what is known as the genetic assimilation of environmental effects as discussed by Conrad Waddington is another example of the subtle workings of natural selection that cannot be understood in terms of a sieve analogy.[10] Genetic assimilation is the selection of those variants that have the genetic propensity to react to environmental stress. The thesis is, for example, that not all ostriches in their evolution responded to friction on the ground of the underparts of the body by forming calluses. Those that had the genetic propensity to do so developed calluses that were protective. Selection resulted in the further accumulation of genes that would result in calluses forming independently of the environmental effect. So offspring of ostriches eventually were born with calluses on the appropriate parts of their bodies. Waddington demonstrated genetic assimilation in fruit flies in laboratory experiments. This is not a form of Lamarckian inheritance of acquired characters. It is rather a thoroughly Darwinian process.

An example from natural populations of the genetic assimilation of acquired behavior is the evolution of a walnut race of codlin moth. The codlin moth is a pest of apples. It was not found in walnuts in the United States before 1918. A few years later walnuts became consistently infested. What apparently happened was that a few moths changed their habits and were attracted to walnuts. There they laid their eggs. Their caterpillars produced moths that themselves were conditioned to lay their eggs on walnuts. The moth is conditioned to choose the smell of the food on which it is

---

[9] Theodosius Dobzhansky, *The Biology of Ultimate Concern* (New York: New American Library, 1967), 42.

[10] Conrad. H. Waddington, *The Strategy of the Genes: a discussion of some aspects of theoretical biology* (London: Allen and Unwin, 1957).

raised. The walnut is a different environment from the apple, so selection for the walnut is different from selection for the apple, and a new race of walnut codlin moths evolved. There are other examples of "purposive behavior" leading to evolutionary change.[11]

Not only do organisms select their environments but they also modify their environments which, in turn, influence natural selection. For example, soil inhabited by earthworms is different from soil without earthworms. A changed environment is one in which selection will be different. Levins and Lewontin have analyzed these somewhat neglected interactions between organism and environment and the pathways by which the interaction leads to evolutionary change.[12]

The modern study of evolution by natural selection, both in the laboratory and in natural populations, demonstrates, beyond doubt, that the natural selection of random genetic variation is a major ordering influence in evolution. It is a mechanical process. It is possible to look at the major evolutionary transformations, brought about by natural selection of chance variations, in the evolutionary sequence of plants and animals, from first life to modern forms, in terms of a relatively small number of major adaptive transitions involving quite complex processes that include mutation and selection.[13]

Much of the explosion of scholarship centered on Darwinian evolution in the last twenty years has been well summarized by Depew and Weber.[14] They include an account of what is known as the neutral theory of evolution, which some have regarded as an alternative to natural selection. It is not. The theory does not deny the role of natural selection in determining the course of adaptive evolution. It may eventually come to be regarded as a supplement to Darwinism. It is the theoretical proposition that most genetic variation is selectively neutral and that evolutionary change is dependent upon other sorts of causes such as genetic drift and gene flow.

Much disputation has also taken place in recent years as to whether evolution occurs by gradual change or by major jumps (punctuated evolution). I come down on the side of those such as Maynard Smith who are persuaded that evolution leaps as well as creeps. It is not a case of one or the other. Darwin thought this to be the case as did indeed G. G. Simpson, the paleontologist whose primary intention was to show that the fossil record supports both gradual change as well as changes through large steps, and that both are consistent with Darwinian natural selection.

The only significant contestant to Darwin's views in his lifetime was Lamarck's theory of evolution by the inheritance of acquired characteristics. Despite the fact that people from time to time reassert the claim that Lamarckian inheritance occurs, there are as yet no reasons to accept it as a possible mechanism that is important in evolution.

Christians have been slow to come to terms with two critical insights of Darwinism: the role of chance (in the production of variation), and the struggle for existence. This is despite the fact that most non-fundamentalist theologies have accepted these as facts of life. In a world completely determined by divine design,

---

[11] Birch and Cobb, *The Liberation of Life*, 56–59.

[12] Richard Levins and Richard Lewontin, *The Dialectical Biologist* (Cambridge, Mass.: Harvard University Press, 1985), 57.

[13] John Maynard Smith and Eörs Szathmáry, *The Major Transitions of Evolution* (Oxford: W. H. Freeman, 1995).

[14] David J. Depew and Bruce H .Weber, *Darwinism Evolving* (Cambridge Mass.: MIT Press, 1995).

as in Paley's natural theology, there is no room for chance. Nor is there a place for struggle when all creatures are perfectly adapted to fulfil their role in a benign nature.

In the struggle for existence many are called but few are chosen. It is a struggle against a hostile environment, which includes predators, storms, and anything that influences the chance to survive and reproduce. It also includes a role for cooperation where that enhances the chance to survive and reproduce. In reflecting upon cruelty in the struggle for existence, Whitehead said "life is robbery. It is at this point that with life morals become acute. The robber requires justification."[15] The reality of struggle, suffering, and chance raises the question how they can be reconciled with a God of love. Darwin lost his faith with the help of a wasp. "I cannot persuade myself that a beneficent and omnipotent God would have designedly created the Ichneomonidae with the express intention of their feeding within the living bodies of caterpillars, or that a cat should play with mice."[16] In so far as theology comes to terms with the role of chance and the facts of struggle and suffering, it will have to be a less deterministic and a more creative theology, as I suggest in the last part of this paper.

3 Self-organization in evolution

Physicists and cosmologists regard self-organization as the source of order in cosmic evolution up to and including the origin of life. Some physicists, notable among them Paul Davies, have argued that the onus is now on biologists to demonstrate its importance, not only in the origin of life but in the subsequent evolution of life.[17] Some have, and it should now be regarded as an addition to neo-Darwinism. As far as I know there is no reference in either Darwin's writings nor of those of the founders of neo-Darwinism to self-organization. Kauffman has remarked: "Darwin did not know the power of self-organization. Indeed, we hardly glimpse that power ourselves. Such self-organization, from the origin of life to its coherent dynamics, must play an essential role in this history of life...But Darwin was also correct. Natural selection is always acting."[18] As understood by biologists, self-organization is a strictly mechanistic process. Since examples in biology are not widely known, even among biologists, I give a number of them.

The spores of slime molds on germination produce amoeba-like cells. These amoebae immediately disperse as though mutually repelled from each other. Provided they have sufficient food in the soil in which they live, which consists of bacteria, they divide like amoebae by simple fission. When food becomes scarce the amoeba-like cells tend to distribute themselves uniformly and no longer repel one another. They then aggregate at a number of centers to form at each center a creature that slithers like a slug. It may reach a diameter of twenty-five centimeters or more. From this apparently undifferentiated mass a stalk grows upwards, at the top of which a fruiting body is formed that develops spores. The fruiting body bursts to distribute the spores and so the strange life-cycle continues.

---

[15] Whitehead, *Process and Reality*, 105.

[16] Quoted in Birch, *On Purpose*, 40.

[17] See Davies' paper in this volume.

[18] Stuart A. Kauffman, "What is Life: was Schrodinger right?" in M. P. Murphy and A. J. O'Neill, *What is Life: the next fifty years: speculations on the future of biology*, (Cambridge: Cambridge University Press, 1995), 111.

For many years biologists believed that the aggregation process was coordinated by specialized cells known as pacemaker cells. According to this theory each pacemaker cell sent out a chemical signal, "telling" the other cells to gather around it, resulting in a cluster. In 1970 Keller and Segal proposed an alternative model,[19] which was later substantiated by Hagan and Cohen.[20] The understanding now is that a chemical substance, which is called acrasin, is secreted by the amoeba-like cells when they run out of food. The cells then move up a gradient of acrasin resulting in their aggregation to form the slug-like creature. When an amoeba finds itself in a gradient of acrasin produced by other amoebae it not only moves up the gradient but it itself begins to produce acrasin and so amplifies the message. There is also evidence that concentration waves of a chemical substance govern the production of the stalk and fruiting body. At that stage the effect of the concentration waves is to activate specific genes whose message is to produce fruiting bodies.[21]

Resnick simulated the aggregation of slime mold cells in a computer model.[22] Each "creature" in his model is given the characteristic corresponding to the emission of a chemical substance while also following the gradient of this chemical substance. The chemical is given a finite life corresponding to its evaporation. With this decentralized strategy the "creatures" aggregate into clusters. Many different species of slime mold may live in the same place. How is it then that the cells of the different species do not get mixed up in aggregation? The answer is that different species secrete different acrasins.

The aggregation of slime mold cells and the production of the fruiting body are examples of self-organization as contrasted with centralized organization. Resnick defines self-organization as "patterns determined not by some centralized authority, but by local interactions about decentralized components."[23] Charles Taylor similarly describes self-organization as "local interactions at one level of organization that give rise to structure at a higher level."[24] Self-organization is thus an ordering process; it leads from a less ordered state to a more ordered state.

The significance of self-organization in evolution has tended to be ignored until quite recently. This is probably due to the extraordinary success of molecular biology in pinpointing the role of DNA and RNA in centralized organization and the emphasis that evolution is change in the organizing molecules DNA and RNA.

A simple virus, such as the tobacco mosaic virus, can form by self-assembly in which all participating proteins are organized into a complexly structured virus. Such viruses spontaneously reassemble following dissociation of their parts. More complex viruses cannot do this.[25] Local interactions among amino acids give rise to the complexly folded protein molecule. Local interactions among nucleotides give

---

[19] E. F. Keller and L. Segal, "Initiation of slime mold: aggregation viewed as an instability," *Journal of Theoretical Biology* 26 (1970): 399–415.

[20] P. Hagan and M. Cohen, "Diffusion-induced Morphogenesis in *Dictyostelium*," *Journal of Theoretical Biology* 37 (1981): 881–909.

[21] John Tyler Bonner, "Chemical Signals of Social Amoebae," *Scientific American* 248. 4 (1983): 106–12; Susannah Eliott & Keith L. Williams, "Modelling people using cellular slime moulds," *Australian Natural History* 23.8 (1991): 608–16.

[22] Mitchel Resnick, "Learning about Life," in Christopher Langton, ed., *Artificial Life: an overview* (Cambridge, Mass.: MIT Press, 1995), 229–41.

[23] Ibid., 229.

[24] Personal communication.

[25] W. B. Wood, "Virus Assembly and its Genetic Control," in F. E. Yates., ed., *Self-Organizing Systems: the emergence of order* (New York: Plenum, 1987).

rise to complexly structured transfer RNAs. In both examples the complex molecule can be dissociated from a complex three dimensional state to a simpler linear state to be followed by spontaneous reassembly. If they did not reassociate it would not be correct to refer to their original assembly as self-organization. They are not always able to do this without guidance from the higher level of organization—hence the charmingly named chaperone proteins that guide the three dimensional folding of proteins.

Spontaneous self-organization is usually invoked in the evolution of the prebiotic world, in which a variety of atoms assemble into organic molecules which lead to further complex assemblies and eventually to cells.[26] It took a billion years or so for the first cells to form on Earth and about 3 billion years more for these to evolve into multicellular organisms. The first confirmed biochemical example of self-organization is said to be the glycolytic cycle within cells. This energy-producing cycle involves a feedback between the molecules adenosine triphosphate and adenosine diphosphate. Many others are now known to exist.[27]

When motile cells of the bacterium *Escherichia coli* are grown from a single point on certain substrates they form stable symmetrical patterns. These patterns can be explained by chemical attractants the bacteria secrete. The chemical attractants form gradients along which the bacteria are attracted to move and come to rest to form patterns on the substrate. What is self-organized is the spatial pattern of the bacteria one to another.[28]

In a way that is quite similar, termites construct their complex nests in an orderly fashion that, at least in part, is guided by chemical gradients. Termites are among the master architects of the animal world. When a termite deposits a lump of earth on the base of the forthcoming nest it deposits at the same time a chemical that attracts other termites to the same place to deposit their lumps of earth there and so form a pillar. That is but one aspect of the building of a complexly structured nest by a large number of individual termites cooperating together. Each termite colony has a queen.[29] But, as in ant colonies, the queen does not "tell" the termites what to do. The queen is more like a mother than a ruling queen. There is no one in charge of a master plan. Rather, each termite carries out a relatively simple task. Termites are practically blind, so they must interact with each other and the world around them primarily through the senses of touch and smell. From local interactions among thousands of termites impressive structures emerge.

Resnick has made computer models of some of the steps in this direction.[30] As he points out, the construction of an entire termite nest would be a monumental project. Instead he proceeded with a simple model to program termites to collect wood chips and put them in piles. At the start of the program wood chips were scattered randomly throughout the termites' world. The challenge was to make the termites organize the wood chips into a few orderly piles. He made each individual termite obey the following rules:

---

[26] Harold J. Morowitz, "A Hardware View of Biological Organization," in Yates, *Self-organizing systems*.

[27] Peter Coveney and Roger Highfield, *Frontiers of Complexity: the search for order in a chaotic world* (London: Faber & Faber, 1995).

[28] Elena O. Burdene and Howard C. Berg, "Dynamics of formation of symmetry patterns by chemotactic bacteria," *Nature* 376 (1995): 49–53.

[29] Ilya Prigogine and Isabelle N. Stengers, *Order out of Chaos: man's new dialogue with nature* (New York: Bantam), 186.

[30] Resnick, "Learning about Life."

1. If you are not carrying anything and you bump into a wood chip, pick it up.
2. if you are carrying a wood chip and you bump into another wood chip, put down the wood chip you are carrying.

The program worked. At first the termites gathered the wood chips into hundreds of small piles, but gradually the number of piles declined while the number of wood chips in surviving piles increased.

AntFarm is a computer program that simulates the foraging strategies of ants.[31] Many ants obeying simple rules produce the complex foraging behavior. Self-organizing ants require just four rules to be followed:

1. If you find food take it to the nest and mark the trail with a chemical substance called a pheromone.
2. If you cross the trail and have no food, follow the trail to food.
3. If you return to the nest, deposit the food and wander back along the trail.
4. If the above three rules do not apply, wander at random.

Another striking example of self-organization in animal behavior is the swarm-raid of the army ant.[32] A raid consists of a dense phalanx of up to 200,000 workers that march relentlessly across the forest floor. The phalanx can be up to twenty meters wide. It leaves in its wake not only a trail of carnage but also a series of connected columns along which the victorious army ant workers run with their booty. These columns all lead to the principal trail of the raid, which links the swarm front to the temporary bivouac. It is inconceivable that a tiny individual within the 200 meter long raid has any knowledge of the plan of the raid as a whole. The structure of the swarm can be achieved by simple, self-organizing interactions among the raiding ants. Franks devised computer simulations in which moving ants lay down chemical trails which organize the movement patterns of other individuals throughout the developing raid. His simulations mimic real raids with some precision. His models show how the collective behavior of the swarm can be achieved with no central coordination, but instead through the communication between foragers by the laying down of, and reaction to, chemical trails.

Christopher Langton gives a vivid description of the life of an ant colony as follows:

> There is no one ant that is calling the shots, picking from among all the other ants which one is going to get to do its thing. Rather, each ant has a very restricted set of behaviors, but all the ants are executing their behaviors all the time, mediated by the behaviors of the ants they interact with and the state of the local environment. When one takes these behaviors on aggregate, the whole collection of ants exhibits a behavior, at the level of the colony itself, which is close to being intelligent. But it is not because there is an intelligent individual telling all the others what to do. A collective pattern, a dynamic pattern, takes over the population, endowing the whole with modes of behavior far beyond the simple sum of the behaviors of its constituent individuals.[33]

Resnick emphasizes that in trying to make sense of decentralized systems and self-organizing phenomena, the idea of levels is critically important. "Interactions among objects at one level give rise to new types of objects at another level. Interactions

---

[31] Coveney and Highfield, *Frontiers of Complexity*, 250–51.
[32] Nigel R. Franks, "Army Ants: a collective intelligence," *Scientific American* 77: 139–45.
[33] Christopher G. Langton, "A Dynamical Pattern," in John B. Brockman, ed., *The Third Culture: beyond the scientific revolution* (New York: Simon & Schuster, 1995), 350.

among slime-mold cells give rise to slime-mold clusters. Interactions among ants give rise to foraging trails. Interactions among cars give rise to traffic jams. Interactions among birds give rise to flocks."[34]

What has self-organization got to do with evolution? In the first place evolution is the creation of ordered systems from less ordered ones. Most models of evolution put the emphasis on centralized agencies such as genes determining the order. Self-organization demonstrates that complex order is also achieved when individual entities have quite simple behaviors that together lead to a complex order without any centralized directing. Natural selection selects those behaviors that are adaptive. In the self-organizing process that leads to a higher level of order the individuals, be they cells, ants, or termites, have goal-like behavior. The goal of the termite is not to build a termite nest but to move a piece of wood if it meets one. In the computer model the modeler gives the termite this goal. In real life no doubt chemical attractants are involved, but there is nothing to preclude the proposition that there is in addition something analogous to purpose in the "mind" of the termite. I return to this in a later section.

In the second place self-organization may help to elucidate one of the most complex ordering processes that have evolved, namely development of the egg to the adult. In the example given earlier of the bacterium *Escherichia coli* Budene and Berg point out that the formation of spatial patterns from a mass of initially identical cells is one of the central problems of developmental biology.[35] This example illustrates that the motile response of cells to diffusible substances can have surprising consequences, so surprising indeed that the journal *Nature,* in which the paper was published, put a large colored picture of one of the patterns on its front cover. The critical question of developmental biology is one of ordering: how an egg turns into an adult, how a single cell becomes a multitude of many different cells that form muscles, nerves and so on, of the different body plans that characterize the major groups of animals.[36] Muscle cells are different from nerve cells not because they contain different genes but because different genes are switched on in different environments. How the appropriate genes are switched on in the appropriate places remains a problem.

As Maynard Smith points out, one possibility involves chemical gradients. In the fruit fly the first difference between the front and back end of the egg is caused by the cells of the mother's ovary, external to the egg, that release at the anterior end a specific chemical which then diffuses backwards, giving rise to a chemical gradient of concentration. This in turn causes different genes to be switched on in different places. A single gradient cannot set up a whole pattern, but a succession of such processes might. Are we seeing in morphogenesis similar sorts of ordering being produced by chemical gradients as seems to be the case with slime molds and nest-building in termites?

The examples I have given start with quite particular facts in biology which self-organizing behavior helps to explain. However, there is also a much more theoretical approach to self-organization which begins with computer models and then looks for

---

[34] Resnick, "Learning about Life," 238.
[35] Budene and Berg, "Dynamics of Formation of Symmetry Patterns by Chemotactic Bacteria."
[36] John Maynard Smith, "Life at the edge of chaos?" *New York Review of Books* March 2 (1995): 28–30.

facts. This approach is characterized by Kauffman.[37] The behavior of his models turns out to be what are called, in the mathematical sense, chaotic. The proponents of this approach claim that in their models complex systems give rise to organizations that can explain evolution without natural selection or at the very least can explain how the organizations between which natural selection chooses come to exist in the first place. Kauffman argues that chaotic systems can give rise to structures "for free," instead of each detail of structure having to be forged independently by natural selection. *The system leaps spontaneously into a state of greater organized complexity.* This is what Kauffman calls self-organization. The devotees of this approach have been criticized by geneticists as practicing "fact-free" science in contrast to ordinary "earth-bound" genetics.[38] It is not surprising that self-organization, in Kauffman's meaning of the term, creates this sort of response among evolutionists. The founders of Neo-Darwinism were, without exception, naturalists as was Darwin himself. By contrast the proponents of self-organization, in Kauffman's meaning, are not naturalists. They are interested in complex systems that could give rise to some form of order. The question for the future is whether Kauffman's models will be heuristically valuable for biology. Self-organization is part of the program of so-called complexity theory at the Santa Fe Institute for the Study of Complex Systems. As Lewontin remarks "It [complexity theory] proposes that sufficiently large systems of parts with enough interactions will generate totally new, but simple, 'laws of organization' that will explain, among other things, us."[39]

Kauffman's concept of self-organization is more readily accepted in physics and chemistry where real examples can be identified. One such is the laser, where disordered atoms cooperate to emit an extremely pure beam of light. Another example is the self-organization that gives rise to the myriad symmetrical shapes of snowflakes. Probably no two snowflakes are alike. Snowflakes are formed by crystals of ice that generally have a hexagonal pattern, often beautifully intricate. The shape and size of the crystals depends mainly on temperature and the amount of water vapor available as they develop. At temperatures above -40 degrees Celsius ice crystals form around minute particles of dust that float in the air. At lower temperatures, crystals form directly from water vapor. If the air is humid the crystals grow rapidly and develop branches and clump together to form snowflakes. In colder and drier air the particles remain small and compact. The snowflake illustrates the production of a complex order by self-organization, the details of which depend upon the environment.

I need to emphasize that all the examples I have given of self-organization are explained by scientists in strictly mechanistic terms, complex though those mechanisms may be. As I have indicated, many of the examples can be replicated on a computer. But this does not necessarily imply that the behaving entities are in all respects machines. This is the subject of the next section.

---

[37] Stuart A. Kauffman, *The Origins of Order: self organization and selection in evolution* (Oxford: Oxford University Press, 1993).

[38] Gabriel A. Dover, "On the edge," *Nature* 365: 704–6; Smith, "Life at the edge of chaos?"

[39] Richard Lewontin, "The last of the nasties," *New York Review of Books* 43.4 (1996): 26.

4 *Organisms as subjects*

The modern theory of the mechanism of evolution discussed in the previous sections explains a lot. However, it has little, if anything, directly to say about freedom, choice, purpose, self-determination and the possibility of divine action in the living world. Darwin's theory did nothing to disprove such concepts. But it did destroy the only argument by which many people thought the existence of God, for example, could possibly be established. For those of us who believe in freedom as a reality in nature and in purposive processes of living organisms, these issues become important. The question then becomes, how can we find a role for freedom, self-determination, purpose, and divine action within the context of the modern mechanistic theory of evolution?

The quotation from Darwin's contemporary Charles Kingsley that heads this paper proposes that God makes things that make themselves. They are not made by a *deus ex machina* of those deists who proposed that God made living things as a watchmaker might make a watch. It is not a case of watchmaker, blueprint and then watch. Darwinism rules that out as a credible proposition. There has to be room for chance (variation) and the sorting out process of natural selection in which few survive and most go to the wall. Darwinism very quickly sounded the death knell for the deism of William Paley's natural theology. Darwinism showed that this particular concept of divine action was not necessary. But it did not rule out the possible validity of other concepts of divine action as Kingsley was quick to point out.

However, Darwinism did not go far enough in its reform. In the dominant interpretation of Darwinian evolution, individual living organisms are treated like machines, devoid of self-determination or spontaneity in any sense and so subject only to external forces. They are objects, not subjects, just as classical physics regarded the so-called fundamental particles to be objects pushed around by external forces as billiard balls are pushed around on a table. So Darwinism readily became a thoroughgoing materialistic world-view. Darwin became the Newton of biology.

The evolution of mind and consciousness has remained an enigma for Darwinism. No strict Darwinian has ever given a convincing reason why mind and consciousness evolved. Of course consciousness has survival value, but that is not the point. A robot that is programed to retreat from danger and to enable itself to be provided with all necessary resources would have survival value without being conscious. It does not have to be conscious to survive, provided it is well adapted to its environment. So why are not all organisms perfectly adapted robots? The difficult question, which has never been answered, is which functions, if any, can be subserved *only* by consciousness? That is the question to which biology has not as yet given a satisfactory answer.[40]

There is another way of looking at this problem. It is that we have gotten into this dilemma about mind and consciousness because we think of mentality and subjectivity as elements that do not enter the evolutionary sequence until quite high up. There is an alternative, and that is that from its beginnings evolution is the evolution of subjects.

Whitehead anticipated with great insight this view long ago when he wrote about biological evolution:

A thoroughgoing evolutionary philosophy is inconsistent with materialism. The aboriginal stuff, or material, from which a materialistic philosophy starts is incapable of evolution. The material is in itself the ultimate substance. Evolution, on the

---

[40] Birch, *On Purpose*, 28ff; idem, *Feelings*.

materialistic theory, is reduced to the role of being another word for the description of the changes of the external relations between portions of matter. There is nothing to evolve, because one set of external relations is as good as any other set of external relations. There can merely be change, purposeless and unprogressive... The doctrine thus cries aloud for a conception of organism as fundamental to nature.[41]

The principle is that machines cannot evolve. They can only have rearrangement of parts. Whitehead enunciated more clearly than anyone how creative evolution of living organisms cannot be understood if the elements composing them are conceived as individual entities that maintain exactly their identity throughout all the changes and interactions, as is the case with the parts of a machine. Whitehead called his theory of nature the philosophy of organism or the philosophy of organic mechanism.[42] Alternative names today are panexperientialism, panpsychism and process thought.

The problem can be put another way. Cosmic and biological evolution are not simply the evolution of objects that are reorganized by changes in their external relations. It is the evolution of subjects. A subject is something that has some degree of self-determination. Only subjects can exhibit creativity. In the above quotation Whitehead proposes that evolution is not simply change in external relations of individual entities but *change in the internal relations of subjects*.

A subject is what it is by virtue of its relations with other entities. This is its internal relations as contrasted to its external relations. External relations, that push or pull, do not affect the nature of the things related. An internal relation is different. It is part of the entity that is related. An internal relation determines the nature of, even the existence of an entity. An internal relation is a taking account of the environment without being totally determined by it. We may best understand what is meant by internal relations by recalling those that we consciously experience in our own lives. I am what I am partly by virtue of the internal experiences of friendship and life's other adventures.

The analysis begins in process thought (the philosophy of organism) with the postulation of two elements in the internal relations of a creative entity, be that entity a human person or a proton. One is that the entity has internal relations with its immediate past which we could call memory (which Whitehead called its physical feelings). The second is that the entity has the aim of constituting its present occasion both for immediate "satisfaction" and for the sake of the anticipated possible future. Possibilities are real (final) causes just as the past is a real cause, as are also mechanical (efficient) causes. The evolutionary history of life suggests that there is an ever-present urge which can be interpreted as purposive. It can be seen also as an aim to greater richness of experience or "higher modes of subjective satisfaction."[43] Whitehead sees this as "a three-fold urge: (i) to live, (ii) to live well, (iii) to live better. In fact the art of life is first to be alive, secondly to be alive in a satisfactory way, and thirdly to acquire an increase in satisfaction."[44] Whitehead was writing about life in general. Value and the aim at value are present in every

---

[41] A. N. Whitehead, *Science and the Modern World* (Cambridge: Cambridge University Press, 1933), 134–5.

[42] Ibid., 99.

[43] Thomas E. Hosinski, *Stubborn Fact and Creative Advance: an introduction to the metaphysics of Alfred North Whitehead* (Lanham, Maryland: Rowman & Littlefield, 1993), 83.

[44] A. N. Whitehead, *The Function of Reason* (Boston: Beacon, 1929, reprinted 1958), 8.

individual. This is deemed by many as being an "unscientific" understanding of evolution. This is probably because biologists often speak of evolution as if mere survival were the only value at stake in the history of life. But as Whitehead pointed out, if survival is the only value how is it that life appeared at all?

> In fact life itself is comparatively deficient in survival value. The art of persistence is to be dead. Only inorganic things persist for great lengths of time. A rock survives for eight hundred million years; whereas the limit for a tree is about a thousand years, for a man or an elephant about fifty or one hundred years... It may be possible to explain "the origin of species" by the doctrine of the struggle for existence among such organisms. But certainly this struggle throws no light whatever upon the emergence of such a general type of complex organism, with faint survival power.[45]

In process thought the subjective elements that constitute internal relations of a creative entity apply all down the line from living persons to protons. All are subjects. There is no sense in which the atoms and molecules of classical physics could be called creative subjects. Nor is there any sense in which the classical notion of the gene could be said to be creative, since there is no sense in which they were considered to have any degree of self-determination. But neither atoms nor genes are any longer regarded as pellets of matter (like beads on a string, as we used to say of genes). Geneticists no longer teach "particulate genetics" as I was taught as an undergraduate. Genes are not like particles at all. What a gene (DNA) does depends upon neighboring genes, on the same and on different chromosomes, and on other aspects of its environment in the cell. DNA makes nothing by itself, not even more DNA. Its actions also seem to involve a choice, at least in some circumstances.[46] The phenomenon of self-organization is just what we would expect if individual entities are what they are by virtue of their internal relations. Anticipation has its subjective element of purpose and choice. All individual entities, as defined in this paper, are characterized by self-organization. They are not like puppets controlled by a central puppeteer. Pattern is generated by the behavior together of individual entities. However mechanical the macroprocesses of self-organization, discussed earlier in this paper, maybe they are based on microprocesses of self-organization that are not mechanistic.

How are we to begin to think in terms of subjects and not just objects? Whitehead takes human experience as the event or process that points best to the nature of all individual entities from protons to people. This is the radical theme of process thought which "sees human experience as a high level exemplification of reality in general."[47] Leonardo da Vinci had this insight long ago when he said "Each man is an image of the world." This is the principle of interpreting the lower in terms of the higher. Reductionism is the principle of interpreting the higher in terms of the lower. There is a place for both, which is discussed below.

The philosophy of organism is also known as the philosophy of panexperiential- ism because internal relations are relations of experience. The presence of mentality in some form, however attenuated, in all individual entities from protons to people, overcomes the problem of the evolution of mentality from no mentality, of freedom from no freedom, of subjectivity from no subjectivity and the problem of what is left for God to do in a seemingly mechanistic world.

---

[45] Ibid., 4–5.

[46] Birch and Cobb, *The Liberation of Life*, 81–82.

[47] John B .Cobb and David R. Griffin, *Process Theology: an introductory exposition* (Philadelphia: Westminster, 1976), 13.

Two contemporary thinkers, not in the tradition of process thought so far as I know, have independently supported the proposition that experience, in some form, is an aspect of nature right down the line to the so-called fundamental particles. David Chalmers points out how a remarkable number of phenomena have turned out to be explicable wholly in terms of entities simpler than themselves, but that this is not universal. Occasionally in physics it happens that an entity has to be taken as fundamental. Fundamental entities are not explained in terms of something simpler. They are taken as basic. In the nineteenth century it turned out that electromagnetic processes could not be explained in terms of the wholly mechanical processes of classical physics. So Maxwell and others introduced electromagnetic charge and electromagnetic force as new fundamental components of physical theory. Other features that physical theory takes as fundamental include mass and space-time. Chalmers' proposition is that an adequate theory of consciousness should take experience as fundamental alongside mass, charge, and space-time.[48] Of course taking experience as fundamental does not tell us why there is experience in the first place. But nothing in physics tells us why there is matter in the first place.

On lines similar to Chalmers' argument Strawson and Stapp each independently start with the proposition that experience is a real aspect of our world, yet nothing in current physics covers the fact that entities experience. Therefore there must be physical properties of which physics is so far ignorant. They ask, What kind of physical theory might capture the nature of mental properties? Their answer is, One that includes experience as fundamental.[49]

The view of mentality and experience as existing in some form all down the line from humans to protons is at variance with the widely accepted doctrine of the emergence of mind. This is the proposition that mind and consciousness arose from no-mind at a late stage in biological evolution. The next section therefore deals with this proposition.

5 *Emergence*

The concept of emergence in biology has to do with the origin of organs. For example, no animal had feathers until birds appeared. Feathers are said to have emerged from things which were not feathers, namely the scales of reptiles. The five-toed limb is said to have emerged from fish that had no such limbs but fins. Analogously some emergentists say mentality emerged from something that had no mentality. This is the orthodoxy of evolutionary thinking today. There is *no problem* about the doctrine of emergence, except for this last step: the possibility that subjectivity, mentality and consciousness, arose from a world which was completely devoid of such in any form. This last step contains a serious flaw which is philosophically known as a category mistake.[50] Feathers, bones, eyes, ears, limbs, and the rest of organs of animals are externalistic properties knowable to sensory

---

[48] David J. Chalmers, "Facing up to the problem of consciousness," *Journal of Consciousness Studies* 2 (1995): 200–19; idem, *The Conscious Mind* (New York: Oxford University Press, 1996).

[49] Galen Strawson, *Mental Reality* (Cambridge Mass.: MIT Press, 1995); Henry Stapp, "The Hard Problem: a quantum approach," *Journal of Consciousness Studies* 3 (1996): 194–210.

[50] David R. Griffin, ed., *The Reenchantment of Science: postmodern proposals* (Albany, N.Y.: State University of New York Press, 1988), 19, 147, 151; idem, *Unsnarling the World-knot: consciousness, freedom and the mind-body problem* (Berkeley, Calif.: University of California Press, forthcoming).

experience from an external point of view, something known as a third person point of view. But experience itself does not belong in this category. It is not something that is observed through the eyes, ears, or hands of another organism. It is something that can only be known from an internal perspective. So it can only be described in first person language. As Griffin says,

> To put experience itself in the same class as those properties that are objects of experience is a *category mistake* of the most egregious kind.[51]

Thomas Nagel makes the same point when he argues that it is unintelligible to speak of the emergence of experience, which is something *for* itself, out of things that are purely physical.[52] It is difficult (arguably impossible) to understand how something with an inside, something that is for itself, could emerge out of things that have no inside, that are nothing for themselves. Because this seems *prima facie* impossible, the burden of proof is on those who think that it is possible.

In chemistry the concept of emergence has to do with properties that have emerged in particular combinations of atoms such as saltiness or wetness. This is a valid use of the concept of emergence. Some supporters of the idea that mentality emerged from no mentality draw the analogy between the emergence of properties such as saltiness or wetness. But this is again to commit a category mistake. It is quite logical to speak of saltiness and wetness as emerging, but it is incorrect to regard these as valid analogies to the proposition of mentality emerging from no-mentality. Saltiness is a property of sodium chloride. It is not a property of either sodium alone (which is a metal) nor of chlorine alone (which is a gas), which together constitute sodium chloride. Wetness is a property of hydrogen and oxygen combined in the ratio H_2O. It is not a property of either of the gases hydrogen and oxygen alone. So it is argued that mentality and experience have arisen out of a particular configuration of nerve cells, none of which by themselves have mentality or experience. But to compare mentality with saltiness and wetness is to commit the category mistake. Saltiness and wetness are properties of things as they appear to us from without. But a human experience is itself not a property of things as they appear to us from without. It is what we are, in and for ourselves.

A corollary of the proposition that mentality emerged from non-mentality implies there was a stage in biological evolution when mentality first emerged and the first sentient creatures came into existence. Where then is the line to be drawn between the sentient and the non-sentient? Descartes drew it between the human soul and the rest of nature. But drawing an absolute line anywhere, be it between humans and all other organisms, between fish and frogs, or between living and non-living matter, is completely arbitrary. There is no clear line between a living cell and a virus. Are then the cell and the virus sentient while their DNA and RNA molecules are not? My conclusion is that the concept of emergence, while relevant to some problems in biology and chemistry, is not a solution to the evolution of subjectivity and self-determination.

6 *The lower-to-higher approach*

The concept of emergence is closely tied to the concept of reductionism, which is the interpretation of higher levels of organization in terms of lower levels. Much of the

---

[51] Griffin, *The Reenchantment of Science*, 147.
[52] Thomas Nagel, *Mortal Questions* (Cambridge: Cambridge University Press, 1979), 202.

interpretation of science is reductionistic. Reductionists, for example, try to explain the properties of complex wholes such as living organisms or molecules in terms of the units of which they are composed, or of aspects of human behavior in terms of biochemistry. There is a valid form of reductionism. Different sciences investigate the world at different levels. Thus at the level of the living organism much can be understood in terms of the biochemistry of cells that constitute the organism. Biochemistry throws light on biology. Physics throws light on chemistry. The naive reductionist believes that biological phenomena such as evolution will only be fully understood through molecular biology and that any other approach is a waste of time. Molecular biology becomes the theophany of reductionism in biology.

If complex things, such as living organisms, can be broken down into their component parts, how is it that the whole has properties the components do not have? It is evident that the properties of the whole are not found in the parts, except as they are organized in that whole. It is for this reason that the reductionist program is deficient. One response has been to say that the whole is more than the sum of its parts. There is an element of truth in this statement, but it does not go far enough.

It is not just that the whole is more than the sum of its parts. It is that parts become qualitatively different by being parts of a whole. Lewontin, Rose, and Kamin discuss the different levels on which atoms are assembled to make molecules, molecules to make cells and so on: "as one moves up a level the properties of each larger whole are given not merely by the units of which it is composed but of the organizing relations between them... these organizing relationships mean that the properties of matter relevant at one level are just inapplicable at other levels."[53]

Carbon atoms possess a number of properties, but they manifest different ones in different environments. Carbon atoms in a diamond (which consists entirely of carbon atoms) reveal different properties from carbon atoms in graphite (which also consists entirely of carbon atoms). But what could give carbon these different properties? Lewontin, Rose and Kamin, in the quotation above, say it is in the "organizing relations" between the parts. Bernd-Olaf Küppers says, "the whole determines the behavior of the parts," which he calls "downward causation."[54] The self-organizing relations involve physical relations that are different between carbon atoms in diamond and those in graphite. But these self-organizing relations may also involve internal relations in the sense already discussed.

The most fundamental answer to this question of the relation of the whole to the parts is in terms of the doctrine of internal relations. As Cobb has said, the most fundamental basis for rejecting reductionism as adequate to explain the physical world is the doctrine of internal relations.[55] Hartshorne made the same point when he said "materialism overlooks... the internal relations involved in individuals."[56] The examples above illustrate the principle of process thought that an individual entity is what it is by virtue of its internal relations. It reveals different qualities in

---

[53] Richard Lewontin, Steven Rose, and Leon J. Kamin, *Not in our Genes: biology, ideology and human nature* (New York: Pantheon, 1984), 278.

[54] Bernd-Olaf Küppers, "Understanding Complexity," in Robert John Russell, Nancey Murphy and Arthur R. Peacocke, eds., *Chaos and Complexity: scientific perspectives on divine action* (Vatican City State: Vatican Observatory Publications, 1995), 94.

[55] John B. Cobb, "Overcoming Reductionism," in John B. Cobb and Franklin I. Gamwell, *Existence and actuality: conversations with Charles Hartshorne* (Chicago: University of Chicago Press, 1984), 151.

[56] Charles Hartshorne, *The Philosophy and Psychology of Sensation* (Chicago: University of Chicago Press, 1934), 20.

different environments. This concept is to be contrasted with the concept of external relations which do not influence the nature of the entities so related. For example, the bricks that go into the construction of an office block remain the same if that office block is torn down and these same bricks are assembled into a cathedral. From a process perspective a brick is not an individual entity. It is an aggregate of individual entities. One brick is not influenced by another brick or rock or anything else next to it in the building. Not so for *individual entities* such as protons or *compound entities* such as atoms in a molecule or molecules in a cell. These are postulated to act and feel (internally relate) as one. Bricks, tables, and computers do not have this property. They are called *aggregates*. There is a group of living organisms that do not readily fall into the category of individual entities (or their compounds) or the category of aggregates. Such are plants and sponges. They would appear not to have unified experiences. Whitehead called them living democracies.

Whitehead argued that process thought, with its doctrine of internal relations, is mainly devoted to the task of making clear the notion of "being present in another entity."[57] He went on to explain that this is not the crude notion that one entity is simply added to another. This is a central aspect of creative evolution and self-organization, which a mechanical analysis is unable to deal with.

The doctrine of internal relations on the understanding of the nature of the physical world is radical. It points to the limitations of reductionism, and it destroys the notion of individual substance and substitutes that of event or relation.

7 The higher-to-lower as opposed to the lower-to-higher approach

The proposition of this section is that instead of putting all our eggs into the reductionist basket of analysis, the lower to higher approach, we should make room also for the higher to lower approach. As Freeman Dyson has said "If we try to squeeze science into a single scientific viewpoint such as reductionism, we are like Procrustes chopping off the feet of his guests when they do not fit into his bed."[58]

We gain a truer perspective of the whole evolutionary process if we study it from its most recent results rather than only from its beginnings. A universe that can produce human beings is different from a universe that is unable to produce human beings. Our inner human experience of life is the aspect of nature which we know most directly. The higher to lower approach leads naturally to a panexperiential view of nature. Far more individual entities are experiential, that is to say they are subjects, than most of us recognize. But not all experience is conscious. The proposition is that atoms and molecules, for example, take account of their environment internally, which is analogous to conscious experience at a higher level of organization. In process thought the term experience is applied to all levels of entities from protons to people whether conscious or not. Consciousness is a high level experience.

The doctrine of internal relations in panexperientialism has important implications for the scientific enterprise. That enterprise has, for the most part, been committed to the reductionist approach. Science *until now* has studied things as machines that have no internal relations. Yet there is no reason why the whole world of inner experience should not be included in the domain of a "re-enchanted" science. Brain physiologist Roger Sperry, who regarded himself as an exception

---

[57] Whitehead, *Process and Reality*, 50.
[58] Freeman Dyson, "The Scientist as Rebel," *New York Review of Books* 42.9 (1995): 33.

among brain physiologists, has said just this, recognizing its implications that science can then speak of causation from higher to lower, exerted by events of inner experience. Feelings, in whatever form, are recognized as causes.[59] Sperry argued that they should therefore enter into the scientific analysis of events of the brain. The developmental biologist Waddington, who was much influenced by process thought, criticized those of his colleagues who argued that metaphysical considerations should have no impact on the direction in which science advances. On the contrary, argued Waddington, they should and do have a profound influence. "I am quite sure that many of the two hundred or so experimental papers I have produced have been definitely affected by consciously held metaphysical beliefs, both in the types of problems I set myself and the manner in which I tried to solve them."[60]

In contrast to Waddington's view, another distinguished biologist, Sewall Wright, who, like Waddington, was also much influenced by process thought, wrote "science is a limited adventure, concerned with the external and statistical aspect of events and incapable of dealing with the unique creative aspect of each individual event."[61] Nevertheless he was willing as a biologist to write about the gene as organism in the Whiteheadian sense.[62] The line is not easily drawn, for science today has not the sharply defined frontiers it was given in earlier periods.

In the perspective of process thought, biological evolution is seen not just as involving mechanical changes, say to the heart as a pump, but internal changes whereby the experience or internal relations become richer in a human being as compared to a mosquito. "Creativity," said Whitehead, "is the principle of novelty."[63] We may then ask what is creatively novel about evolutionary change? There is novelty all along the route in the sense that human experience is novel compared to the experience of a dinosaur. A world of dinosaurs without humans is a different world from one that contains humans. But human experience has a continuity in origin from the feelings that constituted the being of the first mammals, the reptiles from which they evolved, and all individual entities prior to them in the evolutionary sequence going back, if you will, to the physicist's initially featureless universe— hence Whitehead's proposition that the cosmic evolution of the universe "is a creative advance into novelty."[64] Something is achieved from beginning to end. I call that 'creative advance' or 'novelty' which I think is less ambiguous than the term 'progress.'

8 Divine action in evolution

In the first two sections of this paper I gave an account of the processes of evolution as understood by the majority of biologists today. The processes of mutation, natural selection, and self-organization are usually understood within a mechanistic and reductionist framework in which the higher levels of organization are understood in terms of the lower levels. In process thought these same data are understood within

---

[59] David R. Griffin, "The Restless Universe: a postmodern vision.," in K. J. Carlson , ed., *The Restless Earth* (San Francisco, Calif.: Harper & Row, 1990), 85.

[60] C. H. Waddington, "The practical consequences of metaphysical beliefs on a biologist's work: an autobiographical note," in C. H. Waddington, ed., *Towards a Theoretical Biology. Vol. 2: Sketches.* (Edinburgh: Edinburgh University Press, 1969), 72.

[61] Wright, "Biology and the Philosophy of Science," 124.

[62] Sewall Wright, "Gene and organism."

[63] Whitehead, *Process and Reality*, 21.

[64] Ibid., 222.

another framework, in which the lower levels of organization are also interpreted in terms of the higher. At each level the individual entities, be they protons, cells or whole organisms, are postulated to have an inner creativity which relates to cosmic purpose.

The individual entities in cosmic and biological evolution have the potentiality to become something they are not yet. From a universe of hydrogen there eventually evolved complex molecules and eventually the most complex structure known, the brain of a human being. These were potentialities from the foundation of the universe. In contemplating this cosmic evolutionary process Whitehead argued that "the potentiality of the universe must be somewhere."[65] By "somewhere," Whitehead meant "some actual entity." Whitehead named this actual entity "the mind of God." In process theology God is not conceived as the omnipotent, supernatural, legalistic ruler of the universe. Ideas associated with omnipotence, intervention, and law-giver are not part of process thought. Concepts of persuasive divine love, ultimate concern, and infinite passion are.

Divine potentiality becomes concrete reality in the universe by means of persuasive love. A threefold creative activity characterizes God's interaction with the individual entities of the world: (i) God's primordial envisagement of the individual entities of creation, (ii) the creation of actual individual entities in the world, and (iii) the absorption into the divine life of the completed actual entities.

8.1 *The feelings of God* for *the world*

The possibilities and potentialities of the world are not conceived as some sort of storehouse of ideas in a divine pantry. They are feelings that God has for the world in the form of persuasive love. God acts by being felt by his creatures, the individual entities of creation. A note on a tuning fork elicits a response from a piano because the piano already has in it a string tuned to the same note. The individual entities of creation have within their subjective nature "strings" tuned to the persuasive influence of God for their immediate future. We know of the persuasive influence of God in human life as creatively transforming. None of us need stay as we are. For each of us there are persuasive possibilities not yet realized. In every event we are addressed by God. We are tuned to the lure of God in our lives. When God is felt by us we are transformed. This aspect of God is given many names: the light on a hill, the treasure in the field, the pearl of great price. It has also been called the divine eros to emphasize that this is a strongly felt relationship.

The first proposition of process thought is that this same influence is at work in all the individual entities of creation from protons to people. God is persuasive love, ever confronting the world as it now is with the possibilities for its future. With each successive creative advance in cosmic evolution, stabilities become established. The hydrogen atom has a stability concerning its future. The DNA molecule is more susceptible to change, as is that of the cell, and that of the human person even more so. There is a fullness of time for each set of possibilities to become concretely real.

The potentialities of the universe and its individual entities are not in the form of a blueprint of the future, so it is misleading to speak of a divine design. The term 'design' has connotations of a preconceived detailed plan, which is one reason why Darwinism dealt such a blow to the deism of Paley's natural theology. The term 'purpose' is better as it does not carry this connotation. Nothing is completely determined. The future is open-ended. One reason for this is that God is not the sole

---

[65] Ibid., 46.

cause of all happenings. God exercises causality always in relation to beings who have their own measure of self-determination. God, the source of all unactualized possibilities, is constantly creating within the universe by confronting what is with that which is possible, and this by persuasion and not by direct manipulation. Thus God, like all other beings, is in some aspects incomplete and is our companion in the creative advance toward the actualisation of possibilities not yet realized. So it would be true to have God say "I am what I am becoming," which is regarded by some scholars as the appropriate rendering of Exodus 3:14.[66]

Whitehead speaks of the appetition of God as the basis of all order. This is the appeal of God to the creation for ever greater creativity. So Whitehead says that the world lives by its incarnation of God in itself. These are images of an organic relation of God to the world, of "being present in another entity" without being identified with that other entity.

It is appropriate to conceive of providence in these terms. 'Providence' is a difficult word with a number of meanings. The meaning in the present context is that God provides the possibilities. In so doing God is forever active and never need be persuaded to act. Providence does not mean a divine planning by which everything is pre-determined, as in an efficient machine. Rather, as Tillich understands the meaning of providence, there is a creative and saving possibility in every situation which cannot be destroyed by any event.[67] The use of persuasion as opposed to coercion is not to be conceived as based on a voluntary self-limitation of God. We might think a surer way to create would be to combine a bit of persuasion with coercive manipulation from time to time. I know of no evidence to support such a view. The world does not appear to be made that way.

Some traditional theists have said to me "you make God limited if he does not have omnipotent control." But is God limited if he cannot work any nonsense in the world when he wants to, such as to create a stone so heavy that he could not carry it? The imagery leads to absurdity. It is as absurd to say that we have our own power and freedom (which we all presuppose) but that God can step in and control our actions. It is absurd to suppose that to do what has to be done God cannot work with the order of nature as we have to, but has to destroy the creation to do that.

On the other hand, the nature of the world is consistent with the concept of God as persuasive love that is never coercive. That in the end is the only sort of power that matters. The form of power that is admirable and creative is that which empathizes with others and empowers them. Some events in the history of the cosmos, including human history, have more significance than others. These are peak events. This is not because God intervenes in these events and not in others. To interpret significant events as special acts of God is to turn God into an agent of mechanical intervention or into a magician. It is to replace persuasive love by fiat. This unfortunately seems to be an increasing trend in current "folk" theology in much of the western world.

This view of divine action is not only a view of the nature of goodness but also of the nature of evil. Cobb says:

> If God is understood as that factor in the universe which makes for novelty, life, intensity of feeling, consciousness, freedom, and in man for genuine concern for others, and which provides that measure of order which supports these, we must

---

[66] Birch, *On Purpose*, 101.

[67] Paul Tillich, *The Shaking of the Foundations* (New York: Charles Scribner's Sons, 1953), 106.

recognize that he is also responsible in a significant way for the evil in the world. If there were nothing at all or total chaos, or if there were only some very simple structure of order, there would be little evil—there would instead be the absence of both good and evil. Earthquakes and tornadoes would be neither good nor evil in a world devoid of life. Only where there are significant values does the possibility of their thwarting, their conflict, and their destruction arise. The possibility of pain is the price paid for consciousness and the capacity for intense feeling. Sin exists as the corruption of the capacity for love. Thus God by creating good provides the context within which there is evil.[68]

In this view evil springs not from providence but from chance and freedom. Because of chance, freedom, and struggle there are misfits, suffering, and what is called the evil in nature .

In the evolutionary process mutation of genes, natural selection, and self-organization, as discussed in the first part of the paper, make sense within this context of divine action. Genes have their own degree of creativity through mutation and through diversity of action, which may involve choice. Natural selection is a natural consequence of the production of individuals that are not all perfectly adapted to their environment as earlier theories of evolution supposed. The weak go to the wall. The antelope needs the possibility of becoming swifter to escape from the cheetah. The cheetah needs the possibility of becoming swifter to catch up with the speeding antelope. Likewise the bacterium *Staphylococcus aureus*, if it is to survive in a world of antibiotics, needs the capacity to mutate to resistant forms at least as fast as humans make new antibiotics. One organism is pitched against another in the struggle to survive.

Where then is God in this struggle? God is on the side of the antelope and the cheetah, of the *Staphylococcus* and us. It makes no sense to suppose that God directs this mutation rather than that one, or that God determines the direction of natural selection. No such intervention is necessary. The possibilities of mutant change are enormous, so sooner or later the appropriate one occurs. The evidence now indicates that the *Staphylococcus* bacterium was producing mutations resistant to the modern antibiotics before these ever existed. The appropriate mutations do not arise in response to the presence of antibiotics in its environment. They arise whether the antibiotics are present or not. In a similar way insects were producing mutants that made them resistant to DDT before DDT, which is a man-made chemical, was invented. But these particular mutants had no special value in an environment free of DDT.

8.2 *The world's response to God's feelings for the world*

The second proposition is that creativity is the response of the entities of creation to God's feelings for the world. We recognize this most clearly in the proposition that the only adequate human response to God's persuasive love is infinite passion. Infinite passion is the phrase Paul Tillich borrowed from Kierkegaard to express the only adequate "with all" response of humans to God's persuasive love. In process philosophy every individual is both self-created and created by others. Hartshorne puts it this way:

> The divine imperative is to be creative and foster creativity in others. An artist, scientist, or statesman, a husband, wife or friend... are all creating experiences in themselves and helping others to do the same. Even an atom of uranium is not just deciding whether or not at a given moment to change into an atom of lead. That is but

---

[68] John B. Cobb, *God and the World* (Philadelphia, Penn.: Westminster, 1968), 96.

a partial aspect of what is being decided. Whitehead soared above all previous thought when he said. "The many become one and are increased by one." The new unity is not a mere rearrangement of old units, it is a new single actuality.[69]

We are free only as we are co-creators. This applies to all individual entities, though the element of creative freedom may be slight for many individual entities such as hydrogen atoms. God takes a chance on what free creatures will do. Contrary to what Einstein said, God does throw dice. Some kind of appetition of the individual, freely chosen toward further possibilities, some way in which the individual entities of the world lay hold of an order which lies beyond the present, must belong to the individual entities of the world.

8.3 *God's feelings of the world*

The third proposition of process theology is that God responds to the world with infinite passion. It is as true to say that the world experiences God as the world is created, as to say that God experiences the world as the world is created. God is both cause (in creating the world) and effect (in experiencing the world). God is conceived as in process of becoming rather than as static being. In process theology the creator, like other individual entities, is in process of being created, not simply self-created or simply created by the creatures, but the two together. By contrast, in the classical view, God is said to be loving, yet without anything like emotion, feeling, or sensitivity to the feelings of others. Aristotle said it first: "God is mover of all things, unmoved by any." So says the first of the thirty-nine articles of the Church of England, listed in the back of the Anglican Prayer Book to this day, which articles I was required to study in preparation for confirmation.

The alternative proposition of process theology is that whatever we do makes a difference to God. Furthermore, whatever any individual entity of creation does makes a difference to God. That includes the sparrow who falls to the ground. A love that leaves the lover unaffected by the joys and sufferings of the one who is loved is not worthy of being called love at all. There is biblical testimony to a God who is deeply involved with the creation and even with its suffering. The denial of God as one who feels the world's joys and sufferings was largely due to the Greek notion that perfection involves immutability—if God is perfect then God cannot be changed in any way by what happens in the world. Such a God is invulnerable to the world's suffering. On the contrary, to be enriched by the enrichment of the world is to be responsive to the world and therefore to be more loving. Responsiveness, not immutability, is the nature of perfection.

In Whitehead's language, "What is done in the world is transformed into a reality in heaven."[70] In the sentence that follows he adds "and the reality in heaven passes back into the world." This is an interesting speculation, that the world not only makes a difference to God but that the difference to God floods back into the world as new possibilities for the world.

Evolution involves creativity of the individual entities of creation, from protons to people, acting under the persuasive influence of divine creativity. The creative process involves chance, accident, freedom, purpose, self-determination, and self-organization, as discussed in earlier sections of this essay. Creativity is not simply a rearrangement of parts as in a new model of a motor car. It involves being

---

[69] Charles Hartshorne, "A Reply to My Critics," in Edwin Hahn, ed., *The Philosophy of Charles Hartshorne*, The Library of Living Philosophers, vol. xx, (La Salle Il. 1991), 585.

[70] Whitehead, *Process and Reality*, 351.

members one of another. It involves the anticipation and move to possibilities not yet realized.

Acknowledgment. John B. Cobb read earlier drafts of this manuscript and I am grateful to him for a number of helpful suggestions. I also had helpful discussions with Paul Davies and Charles Taylor about self-organization, and I value the criticisms and suggestions from other contributors to this consultation.

III

RELIGIOUS INTERPRETATIONS OF BIOLOGICAL THEMES

THE THINKING UNDERLYING THE NEW 'SCIENTIFIC' WORLD-VIEWS

George F. R. Ellis

1 *Issues*

Various versions of an atheistic view of the universe, each based on some aspect or other of modern science, have appeared in the popular literature in the past, and more are being published at present (by Jacques Monod, Carl Sagan, Heinz Pagels, Edward O. Wilson, Richard Dawkins, Peter W. Atkins, Daniel C. Dennett, to name some of the authors concerned). While they each are based in a serious and sustained investigation of some specific aspect of modern science (usually evolutionary biology or modern cosmology) they also have the characteristic of moving from these subjects to cosmic absurdity (essentially, the claim that the whole affair happened by pure chance and there is nothing to understand except the laws of physics and evolutionary molecular biology). In so doing, they usually mount a strong attack on the possible validity of a religious world-view. Furthermore, that move is usually not made in a spirit of scientific or philosophical enquiry, but rather is presented as if it were an indubitably established fact—although there is no way that this certainty can in fact be attained. In many cases these authors seem to present something akin to a scientific religion in that they purport to give an overall world-view, sometimes including ethical or pseudo-ethical statements.

I consider that move in a number of cases, first showing that in each case it is based on some combination of scientifically unjustified assumptions and rhetorical or emotional appeal, together either with undue restrictions on the scope or method of enquiry, or with attempts to apply the scientific method to domains where it is inapplicable. A hidden metaphysical agenda underlies what is presented to the public as a pure and neutral scientific rendition of nature. Indeed, something of this kind has to be so, for the move cannot be made on analytically justifiable grounds. In each case there is the attempt to claim the authority of scientific status for the results attained, when they are in fact based on hidden philosophical or metaphysical assumptions that are not and cannot be based on strictly scientific methods. Thus their foundations and justification are not what is (explicitly or implicitly) claimed. The methods or limits of science are not respected in these arguments.

Second, the authors concerned sometimes misrepresent conclusions in their public statements about their own scientific area, claiming as definitively established scientific results what are in fact ill-supported hypotheses. This indicates a lack of care in presenting evaluations of evidence in the scientific area of expertise of the author, where the scientific method is properly applicable. This seems to be done to bolster the hidden metaphysical agenda.

The third problem is that in order to counter opposing world-views, the authors concerned deny their validity without taking their arguments seriously. This is not done on the basis of adequately studied consideration, for that cannot lead to such a total dismissal of themes that have been of major concern to some of the most perceptive thinkers in the world continuously for thousands of years. Rather it is done either by 1) an invalid argument from evolutionary origins, or by a combination of 2) denying the meaning of the domain of issues dealt with (or simply ignoring it), 3) denying the validity (or even the very existence) of the data that support the opposing view, and 4) creating straw men—simply unbelievable theories supposed to represent the viewpoint of the opposition—that are trivial to demolish.

Now it is true that some strongly held religious viewpoints are indeed of such an intellectually indefensible nature. However, there are other serious positions that have to be considered before religion *per se* can be discarded. It appears that the basis of approaches that ignore such positions is pre-existing prejudice which prevents a serious considerations of the issues. The authors concerned do not seriously contemplate considered modern presentations of the scientific-religious alternatives,[1] which cannot so easily be dismissed. Rather the argument tends to proceed on the basis of authoritative dogmatic statements. Sometimes this is disguised by being presented as ridicule, but it is still without proper analytic or intellectual support. Associated with this may be an implied or explicit rejection of the *bona fides* of all those who take these domains of enquiry or data seriously.

Thus many scientists' rejection of religion because of its dogmatism[2] is now aiming at the wrong target: the vanguard of religious thinkers today are undogmatic and open in their attitude,[3] while some of the scientists offering scientifically based religions are not, as one can demonstrate directly from their own statements.

In contrast to this, some scientists attempting broadly scientifically based world-views (Steven Weinberg, for example) are less dogmatic and more tentative in their conclusions, also attempting a less-encompassing science-based position. This is an acceptable stance. Their conclusion is self-critical atheism rather than dogmatic atheism—a position much more in accord with the modern understanding of knowledge.

In looking at the scientific religions[4] I do not give a full analysis of all the relevant papers and books by all the authors concerned, or even by any one of them, because of time and resource limitations. Rather I consider a restricted set of their

---

[1] For example, Robert J. Russell, William R. Stoeger, and George V. Coyne, eds., *Physics, Philosophy. Theology: A Common Quest for Understanding* (Vatican City State: Vatican Observatory, 1988); Ted Peters, ed., *Cosmos as Creation* (Nashville: Abingdon Press, 1989); Ian Barbour, *Religion in an Age of Science* (London: SCM Press, 1990); John Polkinghorne, *Reason and Reality: The Relationship Between Science and Religion* (London: SPCK Press, 1991); Nancey Murphy, "Evidence of Fine Tuning in the Design of the Universe," in *Quantum Cosmology and the Laws of Nature*, ed. by Robert J. Russell, Nancey Murphy and Chris J. Isham (Vatican City State: Vatican Observatory, 1993), 407-436; Arthur Peacocke, *Theology for a Scientific Age* (Minneapolis: Fortress Press, 1993).

[2] For example, H. Bondi, Letter to *Europhysics News*, 22 (1991): 130.

[3] See for example, W. Mark Richardson and Wesley J. Wildman, *Religion and Science: History, Method, Dialogue* (New York & London: Routledge, 1996).

[4] The present paper is based on and develops themes in Nancey Murphy and George F. R. Ellis, *The Moral Nature of the Universe: Cosmology, Theology, and Ethics* (Minneapolis: Fortress Press, 1996), which in turn develops themes from Nancey Murphy, *Theology in the age of Scientific Reasoning* (Ithaca: Cornell University Press, 1990); idem,"Evidence of Fine Tuning in the Design of the Universe"; George F. R. Ellis, *Before the Beginning* (London: Bowerdean Press/Marion Boyers, 1993); idem,"The Theology of the Anthropic Principle," in *Quantum Cosmology and the Laws of Nature*, ed. by Robert J. Russell, Nancey Murphy and Chris J. Isham (Vatican City State: Vatican Observatory, 1993), 367–406. Langdon Gilkey, "What Ever Happened to Immanuel Kant? A Preliminary Study of Selected Cosmologies," in *The Church and Contemporary Cosmology: Proceedings of a Consultation of the Presbyterian Church (USA)*, ed. by J. B. Miller and K. E. McCall, (Pittsburgh: Carnegie Mellon University Press, 1990) has similar aims, but a different emphasis. I thank Wim Drees for drawing this paper to my attention.

popular works, which is where this viewpoint is presented to the public.[5] I believe this is sufficient to raise the major issues at stake and to provide a useful framework for a future full study of these kinds of works, which I suggest would be a worthwhile project. In conclusion I emphasize my criticism is not of their strictly scientific papers but of some of the philosophical positions they take and unwarranted assertions they make in their popular books and articles.

2 Foundations

To set the framework for this study, I start by looking at the foundations that underlie it.

2.1 Knowledge and Models

I assume a general understanding that our knowledge is encapsulated in models of reality of greater or lesser complexity, each with a specific domain of applicability, and supported to some degree by evidence.[6] Indeed, I take the view that all understanding is based on models of one kind or another, even a straightforward description being at base a reference to an image or set of images that is a model also, even if a very simple one. Both metaphors and analogies are models, in the sense envisaged. Models of lower complexity or scope are combined in diverse ways to form models of higher complexity or scope, that at the highest level form overall explanatory theories. Support for theories comes from hypothetico-deductive reasoning (Hempel) applied within specific research programs (Lakatos) imbedded in broad traditions of understanding (MacIntyre).[7]

These models and theories can give great insight and understanding within their spheres of application. However, each model or theory will also of necessity be partial and incomplete, illuminating in some aspects but misleading in others, for only reality itself can represent with complete fidelity all its own aspects;[8] and this partial nature of the representation of reality is of particular relevance in theories that attempt to explain its ultimate nature.[9] Furthermore, all theories are open to questioning and rational evaluation on the basis of the available evidence. Thus they should all be regarded as provisional. Some theories of specifically limited scope can attain virtual certainty in describing present laboratory experiments or engineering practice, but we can never attain complete certainty about the future even in relation to these theories both because we cannot guarantee that present physical conditions will continue to the future, and because cosmological horizons prevent us from having the data necessary to make predictions with certainty.[10]

---

[5] Indeed, it is probable the viewpoints presented would not be publishable in relevant research journals because of their philosophical inadequacies.

[6] Barbour, *Religion in an Age of Science*; Ellis, *Before the Beginning*; P. M. Senge, *The Fifth Discipline: The Art and Practice of the Learning Organization* (New York: Currency Doubleday, 1990); P. M. Senge et al., *The Fifth Discipline Fieldbook* (London: Nicholas Brealey, 1994).

[7] For a discussion, see Murphy and Ellis, *The Moral Nature of the Universe*.

[8] A. Bullock, A. Stanleybrass, and S. Trombley, *The Fontana Dictionary of Modern Thought* (London: Fontana, 1988), 537.

[9] J. Hick, *An Interpretation of Religion: Human Responses to the Transcendent* (New Haven: Yale University Press, 1992).

[10] George F. R. Ellis, "Relativistic cosmology: its nature, aims and problems," in *General Relativity and Gravitation*, ed. by B. Bertotti et al. (Reidel, 1984), 215–88.

The essential feature in the search for understanding is thus an openness of approach and a readiness to evaluate all the possibilities on an equal footing, as far as possible without prejudice.[11] This openness to possibility enables us to some extent to see what is really there—insofar as this can be captured by our models of reality—rather than what we would like to be there. It is the weapon against being beguiled by wishful thinking. On this basis, contrary to any of the extreme claims of relativism or deconstruction,[12] the systematic analytic approach of science can attain considerable knowledge of the nature of the universe, and this knowledge has the character of reliable understanding of the nature of reality within appropriate limited domains (but subject to the caveat above).

Thus an acceptable approach to knowledge is characterized by an understanding of the partial nature of models and the tentative nature of knowledge; in particular it should not confuse models with reality. It should also take account of the multiplicity of causation: the fact that the outcomes that occur are brought into being by elaborate networks of causality and constraint, and that when we talk of a 'cause' we are singling out one causal factor from an interlocking network of features which in fact were all of consequence in determining an outcome. "Life is too short to specify the whole range of constraint which has controlled an eventuality into its outcome... we therefore 'pick out' a single cause or set of causes to explain why a particular appearance or event has come into being, leaving tacit or assumed the many other relevant constraints."[13]

Finally when it comes to comparing and testing different world-views, two key questions that can be put to any advocate of any view are as follows: 1) does your theory cope with the ostensible counter evidence? 2) when we consider the criticisms you make of other peoples' world-views, how does your own position stand up to precisely those same criticisms? Each viable stance must be able adequately to answer both of these questions.

Within this broad approach to knowledge, the important feature for what follows is an understanding of some of the limits of science that arise from the nature of the scientific enterprise and its methods. Specifically I refer to two themes: the metaphysics of cosmology, and the issue of values.[14]

2.2 The Metaphysics Underlying cosmology

When science studies the nature of cosmology, it does so on the basis of the specific laws of physics that apply in the unique Universe we inhabit. It can interrogate the nature of those laws, but not the reason for their existence, nor why they take the particular form they do. Neither can science examine the reason for the existence of the Universe.

---

[11] Senge, The Fifth Discipline; Ellis, Before the Beginning.

[12] See for example, R. Scruton, Modern Philosophy: An Introduction and Survey (London: Sinclair Stevenson, 1994); P. R. Gross and N. Levitt, Higher Superstition: The Academic Left and its quarrels with science (Baltimore and London: Johns Hopkins University Press, 1994); G. Priest, Beyond the Limits of Thought (Cambridge: Cambridge University Press, 1995).

[13] J. Bowker, Is God a Virus? Genes, Culture, and Religion (London: SPCK, 1995), 96.

[14] George F. R. Ellis, "Modern Cosmology and the Limits of Science," Trans Roy Soc SA, 50 (1995): 1–26.

These are metaphysical issues, whose examination lies beyond the competence of science *per se*,[15] because there is only one Universe, and we are unable to perform experiments in which we vary its initial conditions or the laws of physics that apply in it. Neither can science investigate the issue of whether or not there is an underlying purpose or meaning to physical existence, for these are non-scientific categories. However, these issues are of significance to us; in particular they underlie examination of the anthropic issue,[16] and are importantly related to ethical choices.[17]

This is not to deny that science constrains our views on metaphysics in an important way. It does provide a context within which metaphysical viewpoints are developed; and that context does indeed limit the range of possibilities. Nevertheless, the scope of that constraint is limited to what science can legitimately achieve and does not extend to direct statements on any metaphysical issues.

2.3 Ethical Issues and Issues of Values

Issues of value are of fundamental importance to life in general, and to applied science in particular, but science itself is unable to provide a foundation for the choice of values, whether in ethics, aesthetics, or daily life.

We must distinguish this statement from two issues that are separate from it. First, there are certain values that are essential to the conduct of science, which we might call scientific virtues (honesty in relation to other scientists and to the experimental data, for example); these are taken for granted by science, forming part of the foundation necessary for it to flourish. Thus science can be said to support or demand these values.[18] Second, some values are necessary for science to continue studying particular objects of interest: if the Thunderwing Butterfly becomes extinct, we can no longer study its mating habits. Thus one might stretch things a bit and claim that zoology imposes an obligation to prevent extinction of that species. Neither of these begins to provide a basis for the value choices needed in real encounters with aesthetics or ethics. This is because there is no possible scientific test that can measure whether something is *good* or *bad*, or is *beautiful* or *ugly*; for these are non-scientific categories—they are not amenable to determination by any scientific experiment. This does not mean they are non-rational, but rather that the experimental methods of science cannot give sensible answers in this domain, nor can they result in a corresponding technology (for example, a meter that when pointed at a picture would evaluate its beauty).

What science can do, given some basic set of value choices, is to determine whether specific policies or actions are likely to promote these values or not. For example, if our values specify that the existence of whales or of humans is valuable or important, then science can help determine what courses of action are likely to promote their continued existence; but science *per se* (biology or ecology, for example) cannot determine if either whales or people are in any absolute sense 'valuable' or 'important', as these too are non-scientific categories.[19]

---

[15] George F. R. Ellis, "Major Themes in the Relation between Philosophy and Cosmology," *Mem Ital Ast Soc*, 62 (1991): 553–605; idem,"Modern Cosmology and the Limits of Science."

[16] Ellis, *Before the Beginning*; idem,"The Theology of the Anthropic Principle"; idem, "Evidence of Fine Tuning in the Design of the Universe."

[17] Murphy and Ellis, *The Moral Nature of the Universe*.

[18] Jacques Monod, *Chance and Necessity* (London: Collins/Fount Paperbacks, 1970).

[19] Ellis, *Before the Beginning*; idem,"The Theology of the Anthropic Principle"; idem, "Modern Cosmology and the Limits of Science."

I am not claiming that one cannot establish ethical values on a credible basis. On the contrary, Nancey Murphy and I claim that this is possible and indeed essential to human conduct.[20] However, this is only possible by means of discernment involving commitment and an element of faith. It is open to those who are responsive to inner leanings and willing to give up their own demands in favor of receptivity to seeing what is around them. A particular ethical view based on this foundation can be established as logically sensible and can be tested experientially with credibility. But it cannot be proved to be valid in a rigorous scientific sense by means of scientific experimentation; its status will always be open to questioning.

To put it another way, while it is plausible there is an ethical basis to the universe—the universe being grounded not only in the intelligible but also the good—this basis is not purely or fully contained in the laws of logic and physics that control the functioning of the physical world; it nonetheless has an independent and real existence. However, there is an important interaction between them: the scientific laws which enable us to attain virtual certainty at certain levels in the hierarchy of structure and explanation, to some degree shape the nature of ethical demands; and it is fundamental that they also enable conscious beings to exist and to function in an ethical manner.[21]

2.4 The Limits of Science

In both cases—metaphysics and ethical values—it is not the situation simply that science is at present insufficiently advanced, and at some future date will solve these issues. The impossibility lies in the very nature of science and the scientific method, and no future scientific advances will change this. Indeed, should someone claim to "solve" such questions—for example, by producing the aesthetics meter mentioned above—they would be laughed at. Such issues—including religious issues such as the existence or non-existence of God—lie outside the competence of science; indeed, there is a profound ambiguity of the universe relative to these issues. Science cannot solve them, and neither can philosophy give an indisputable answer. This is not to say one cannot make strong cases one way or another, or attain viable viewpoints as a basis for living. However, intellectual certainty is unattainable. This realization is the basis of some recent approaches to religious pluralism.[22]

This ambiguity was known to Hume and Kant, who both realized that no certainty can be reached by human reasoning regarding the existence or non-existence of a deity, or the nature of any deity there may be. According to Gaskin, Hume "adopted a species of 'mitigated skepticism' which is inconsistent with any positive assertion about God's nature, existence, or non-existence"; and thus, "any conclusion as confident and positive as atheism would have been inconsistent with the skepticism expressed by Hume in his philosophical works."[23] Similarly, according to Fischer, after many years of thought and study, "Kant determined for all time and for all who follow that it is impossible to prove the existence—or the non-existence—of God, and as a consequence of this situation Kant expressed the

---

[20] Murphy and Ellis, *The Moral Nature of the Universe*. The argument cannot be handled adequately here.

[21] See Davies in this volume and Murphy and Ellis, *The Moral Nature of the Universe*.

[22] Hick, *An Interpretation of Religion*.

[23] Introduction to the re-issued David Hume, *Dialogues and Natural History of Religion*, World Classics Series, foreword by J. C. A. Gaskin, (Oxford: Oxford University Press, 1993).

standing plea that no one should in the future bother him with further attempts of this sort."[24]

Given such logical uncertainty, one can still look for the weight of data and experience and use this to attain a reasonable basis of support for particular standpoints. This is a sensible and practical thing to do. However, we must then remember that whatever we may do, certainty will not be attainable.[25]

The essential feature that both Hume and Kant made quite clear is that neither philosophic nor scientific argumentation is able to solve the religious questions that have always faced us; nor can they solve the metaphysical issues that underlie physics. It may be suggested that the situation has not changed since then: arguments for the existence of God cannot give a definitive conclusion,[26] despite the renewed attacks, for example, by Alvin Plantinga.[27] The same is true for arguments for the non-existence of God.[28] The fundamental point is that our best current philosophical understanding indicates that neither science nor philosophy can establish the truth of either theism or of atheism (later I will discuss why neither the argument from evolutionary origins nor the argument from evil, including evil done by the Church, undermines this logical position). Any serious attack from either side that claims to attain certainty has to show why this hard-won position has to be abandoned.

3 The Scope of Theories and the Limits of Science

One can and indeed should see what science says about the world and the universe in the larger sense, which goes beyond the traditional scope of scientific investigation. Indeed, it is essential to see how far the understanding of science can go. The investigations of those concerned with a wider understanding are to be applauded insofar as they pursue this broader understanding in an open-minded way. But conversely there is the need for an awareness of the limits of science (see the previous section), and in particular, an awareness that one cannot deduce from science that the object of religion does or does not exist.

What can be said legitimately is that one of the old pillars of support for religion—the argument from special design—has fallen away because of the progress of scientific understanding, in particular through evolutionary theory. Given this fact, it is legitimate to see what science by itself may suggest about this issue, given this new situation; but the limits of science must be kept in mind when doing so. The thesis to be developed here is that the scientific religions, in examining this question, develop positions that go beyond the bounds of what is attainable on the basis of science itself, while attempting to attach the authority of science to those

---

[24] E. P. Fischer, "Book review of The Physics of Immortality," Die Weltwoche, no. 15, 14 April, 1994, p. 45.

[25] The word 'certainty' here means absolute certainty. Bill Stoeger suggests the idea of 'relative certainty' or 'fallible knowledge'. I am not sure what these mean. Relative certainty is not certain, and fallible knowledge is not knowledge. One can attain sound standpoints for thought and action, ones that are well based, and as certain as one can be, but that allow doubt. This is not certainty, but it is the best one can do.

[26] See for example, J. Hospers, An Introduction to Philosophical Analysis (London: Routledge, 1989); Scruton, Modern Philosophy.

[27] Alvin Plantinga, God, Freedom, and Evil (Grand Rapids, Mich: W. Eerdmans, 1974).

[28] Hick, An Interpretation of Religion. This discussion does not mean I accept Kant's general anti-metaphysical conclusions, but simply the inability to attain a once-hoped for certainty.

positions. They move beyond the legitimate domain of scientific evidence without taking adequate account of the epistemological hazards involved in doing so. Indeed, if the foundations put forward in the previous section have been correctly identified, then this must be so: wherever we find authors proposing 1) purely physically based metaphysics, or 2) purely scientifically motivated values, or 3) a denial of the ambiguity of the metaphysical basis of the universe, something of this kind must be happening. A series of examples follow.[29] Note that I am not here attempting to consider or critique the scientific writings of the author's concerned, but rather their popular presentations of their understandings, which go beyond the scientific domain into philosophical and metaphysical realms.

3.1 Carl Sagan

Carl Sagan writes with elegance and persuasiveness about the discoveries of modern astronomy. However, he goes beyond that: his writings proposes a mix of science and what can properly be called "naturalistic religion." He begins with biology and cosmology but then uses concepts drawn from science to fill in what are essentially religious categories that fall into a pattern surprisingly isomorphic with the Christian conceptual scheme.[30] He has a concept of ultimate reality: "The Universe is all that is or ever was or ever will be." He has an account of ultimate origins: Evolution with a capital 'E'. He has an account of the origin of sin: the primitive reptilian structure in the brain, which is responsible for territoriality, sex drive, and aggression. His account of salvation is gnostic in character in that it assumes salvation comes from knowledge. The knowledge in question is scientific knowledge, perhaps advanced by contact with extraterrestrial life forms who are more advanced than we. Sagan's account of ethics is based on the worry that the human race will destroy itself. Morality consists in overcoming our tendencies to see others as outsiders; knowledge of our intrinsic relatedness as natural beings—we are all made of the same star dust—can overcome our reptilian characteristics.

Clearly here Sagan is proclaiming an ethic that cannot be based solely on scientific reasoning, and so cannot follow simply from either astronomy or evolutionary history. His argument has a metaphysical/philosophical basis that goes way beyond what science can say. One may perhaps agree with his ethic as a partial ethical statement, but on what ultimate basis does it rest? It is simply an *ex cathedra* statement. One could equally propose the opposite, that the way for survival is to eliminate the weak and unintelligent who drag down the rest of the human race; this would have at least as much 'justification' from our evolutionary history.

Sagan in fact relies on simple emotional response to establish his 'ethic' as well as his basic metaphysical view. This is clearly shown in his book *Pale Blue Dot*, where his prime argument against a Creator is simply one of size: the Earth is a very

---

[29] We do not consider Frank J. Tipler's *The Physics of Immortality: Modern Cosmology, God, and the Resurrection of the Dead* (New York: Doubleday, 1994) and his Omega-point theory, despite the fact it shows most of the errors discussed here, because it is different in aim and character from the rest in that it ostensibly takes religious issues seriously. For an analysis of the major problems with this approach, see William R. Stoeger and George F. R. Ellis, "A response to Tipler's Omega-Point Theory," *Science and Christian Belief* 7 (1995): 163-172.

[30] T. M. Ross,"The Implicit Theology of Carl Sagan," *Pacific Theological Review*, 18.3 (1985): 24–32.

small planet compared to the size of the universe.[31] How then could the Universe be created in order to enable life to exist?[32]

This confusion of importance with size is remarkable; for example, it is clear that microbes are important in biological functioning and evolution even though they are so small. But Sagan intends to refer to value in a metaphysical rather than biological sense; and his central theme that human beings are not physically dominant in the universe at large, nor situated at any special place, and therefore are unimportant,[33] can have no scientific basis, as "important" is not a scientific category; and the emotional argument has no logical power for there is no reason that size and importance should be directly related.[34] Consequent on this, his argument against the anthropic principle is basically emotional rhetoric (his use of the label "conceit," for example), and does not begin to address the issues raised in thoughtful and sophisticated modern analyses of that question.[35]

Overall, Sagan chooses not to address the metaphysical issues that underlie cosmology (and hence the entire evolutionary process); given this limitation of scope of his viewpoint, he then cannot see the purpose of the anthropic argument. Apparently the scope of metaphysical questioning that he is prepared to entertain is very limited; in particular, he takes for granted, without question, the existence and the nature of the universe and of the laws of physics. He even proposes the old canard that if the universe is infinitely old (which in fact has no solid experimental or theoretical support), "then it was never created and the question of why it is as it is, is rendered meaningless."[36] This does not acknowledge that there are other possibilities: specific conditions exist, rather than others that might have been.

Within this limited framework that ignores the most important issues, his argument is based on emotion rather than on rational analysis that touches the real issues. However, he is not shy in making his claims: Kant's views on the religious basis of the universe are "self-indulgent folly."[37] His limited metaphysical view is proclaimed as unquestionable absolute truth.

3.2 Richard Dawkins

Richard Dawkins is a brilliant expositor of the nature and power of evolutionary theory. However, he similarly simply takes for granted the conditions that allow evolutionary processes to take place.[38] The fact that Dawkins believes science is in competition with religion, and seeking to answer the same questions,[39] suggests an exaggerated view of the domain of science, namely, that it pertains to issues of good and evil and of ultimate meaning. Any such proposal simply ignores the limits on what science can and cannot do. However, his public persona is that of the scientist

---

[31] Carl Sagan, *Pale Blue Dot* (London: Headline, 1995), 11.

[32] He conflates this with the presumption that the anthropic argument only makes sense if it is taken to be applied to one particular species on one planet, which is obviously not necessary or even plausible; see Ellis, *Before the Beginning*, idem,"The Theology of the Anthropic Principle"; and section 6.2 of this essay.

[33] Sagan, *Pale Blue Dot*, 48.

[34] W. E. Paden, *Interpreting the Sacred* (Boston: Beacon Press, 1992), 121–22.

[35] Ibid., 37–38. For a more careful analysis, see Murphy, "Evidence of Fine Tuning in the Design of the Universe."

[36] Sagan, *Pale Blue Dot*, 35.

[37] Ibid., 39.

[38] Richard Dawkins, *The Blind Watchmaker* (London: Penguin, 1988), 15.

[39] Richard Dawkins, interview in *Sunday Times*, London, 30 July, 1995.

proclaiming on the nature and metaphysical underpinnings of the universe, again without any shadow of doubt.

His conclusion is "The theory of evolution by cumulative natural selection is the only theory we know of that is in principle capable of explaining the existence of organized complexity."[40] This is challenged by present day complexity theory, but that is not the issue at present. The point is that he does not query the origin of the structure and the complexity of physical principles and laws, as understood by modern physics, which makes all of this possible: quantum field theory, symmetry principles and symmetry breaking, variational principles, hierarchical structuring and the renormalization group,[41] and so on. He simply takes all this for granted and treats it as unquestionable.

In doing so, he conveys the message that scientists know the truth about everything, not just science. His strongly proclaimed atheistic stance takes the statement above and makes metaphysical deductions that simply do not follow; this step ignores the boundaries of scientific method. In making his claims, he character-izes religion as a virus, because of the way it (as a thought system) replicates and spreads. The question as to why evolutionary biology should not also be regarded as a virus in exactly the same way then arises and does not receive a satisfactory answer.[42]

3.3 Edward O. Wilson

Sociobiology, initiated by E. O. Wilson, attempts to give an account of morality—a bottom-up account, based in evolutionary terms. Originally this kind of view mainly explained aggressive behavior, but current interest focuses on explaining altruistic (that is, "moral") behavior as well.[43] It involves an attempt to show that ethics can be entirely reduced to the biological level, and, at least in the case of Wilson, that religion is misguided.[44]

If it were merely an attempt to show that there are biological or genetic factors that affect a person's ability to behave in moral ways, it would be entirely unexcep-tionable. The central problem with typical sociobiological projects is that they attempt to give an exhaustive account of moral behavior. But were they to show that "moral" behavior is genetically programed and represented nothing more than this, they would thereby show that it is not moral at all.[45] The very meaning of morality involves its voluntary, intentional nature (whether the decision process is genetically determined or not). If it were possible to show that all of the characteristics our culture praises as admirable or condemns as blameworthy were simply the result of evolutionary selection rather than the expression of a character shaped by deliberate choices, the consequence would be the denial of the realm of the moral. If so-called moral behavior could be shown to be genetically programed on the basis of

---

[40] Dawkins, *The Blind Watchmaker*, 317.

[41] Silvan S. Schweber, "Physics, Community, and the Crisis in Physical Theory," *Physics Today* (November, 1993): 34–40.

[42] A detailed analysis of the problems with his approach is given in Bowker, *Is God a Virus?*, 61–63.

[43] T. H. Clutton-Brock and Paul H. Harvey, eds., *Readings in Sociobiology* (San Francisco: W. H. Freeman, 1978).

[44] See Bowker, *Is God a Virus?*, 37.

[45] See Nancey Murphy's article in this volume.

Darwinian evolution and had no other content or meaning, this would undercut moral prescriptions.[46]

Wilson goes further, in his attempt to show that religion is wrong. Like Dawkins, he purveys the fallacy of evolutionary origins (see section 5.1 below). In his argument he also takes for granted the issue of why conditions exist that allow evolution to take place, thus ignoring the metaphysical issues at stake, which cannot be tackled by his limited line of reasoning with its restricted domain of concerns (if they could, Hume and Kant would be wrong). However, no doubt is apparent in his claim that religious myths are known to be absurd and false. This is based on the assertion that "Scientific materialism presents the human mind with an alternative mythology that until now has always, point for point in zones of conflict, defeated traditional religion."[47] He does not, however, adequately treat the metaphysical issues raised in the modern science-religion debate.[48]

3.4 Daniel C. Dennett

Similar difficulties emerge in Daniel Dennett's enthusiastic writing on evolution.[49] An example of not appreciating the limits of science is his claim that through Darwinian evolution, the values of excellence and worth can emerge out of mindless, purposeless forces. But 'excellence' and 'worth' are words that only attain meaning at higher levels of the hierarchy of explanation,[50] and cannot be based on a scientific viewpoint alone. To be ontologically meaningful, they demand a separate existence independent of their evolutionary origin or of any scientific "justification." Thus his scheme, in claiming to give an account of the evolution of values, introduces concepts that can have no place in a purely scientific world-view, which is what he claims to be presenting. His view, which is within the sociobiology tradition, reduces ethics to something less than ethics.[51]

He also fails to address the major metaphysical issues underlying the existence of a world where evolution is able to take place. His starting point is evolution, which he sees as the key to everything. His scope of enquiry simply fails to question how any evolutionary process at all can be possible. He takes the existence and nature of the world and the universe and the laws of physics for granted. However, he claims a certainty about the non-existence of God.[52] There is no doubt or openness here, despite the fact that his domain of argument is so restricted that he does not begin to tackle the serious metaphysical issues in his writing.

3.5 Jacques Monod

A writer in a similar mold (albeit writing earlier) is Jacques Monod, who proclaims at the outset of his major manifesto, "the ultimate aim of science is, I believe, to

---

[46] Thomas Tracy points out that there is a long-standing and vigorous philosophical debate about the relation between incompatibilist (indeterministic) freedom and moral responsibility. I believe, however, that the point made here remains.

[47] E. O. Wilson, *On Human Nature* (Cambridge: Harvard University Press, 1978), 192.

[48] A detailed analysis of the problems with Wilson's line of argument is given in Bowker, *Is God a Virus?*, chapter 7.

[49] Daniel C. Dennett, "Darwin's Dangerous idea," *The Sciences*, 35 (1995): 34; idem, *Darwin's Dangerous Idea* (New York: Simon and Schuster: 1995).

[50] Murphy and Ellis, *The Moral Nature of the Universe*.

[51] See the analysis of such approaches by Murphy in this volume.

[52] Daniel C. Dennett, interview in *Times Higher Educational Supplement*, 29 September, 1995, p. 15.

clarify man's relationship to the universe."[53] This is a worthy aim but science *per se* cannot carry us very far down this road. It can legitimately do so in physical terms, pointing out for example the small size of humans relative to the universe—the point Sagan makes so much of—and their evolutionary origins. This sets the ground for full explorations of our relationship to the universe, but it cannot determine the results of such an exploration. Monod claims that biology is the ultimate science; in so doing, he ignores the basis of biology in physics and chemistry, and, in evolutionary terms, in the origin and evolution of the universe and of the world. So the scope of his enquiry is in fact rather limited; his conclusions are therefore correspondingly narrow.

Monod's book contains a good summary of biological ideas up to the time of writing. Much of his metaphysical effort is devoted to a defense against Marxism. Although this may no longer be quite so necessary, it certainly was in Monod's time, and it still responds to the dogmatism of the Marxist thought-system. The problem from the present viewpoint is his proposal, imbedded in an argument that is presented in highly emotional terms, that science can be a source of truth and of ethical premises rooted in objective knowledge.[54]

He correctly states that ethics is central to action, which brings knowledge and values—essentially separate domains which must not be confused with each other—simultaneously into play. He suggests as the basic proposition of an ethical system, the ethic of knowledge, where "it is the ethical choice of a primary value that is the foundation."[55] On this view, humankind makes the choice, imposing the ethic on themselves, rather than discovering it as an external reality. From my point of view, this is an inadequate account of ethics: it removes the essential core of what ethics is about, precisely because it no longer has the character of an external demand. He proposes that the ethics which underlies the enterprise of pure science, is the universal ethical principle:

> Where then shall we find the source of truth and the moral inspiration for a really scientific socialist humanism? Only, we suggest, in the sources of science itself, in the ethic upon which knowledge is founded, and which by free choice makes knowledge the supreme value—the measure and guarantee for all other values. An ethic which bases moral responsibility upon the very freedom of that axiomatic choice.[56]

He puts this forward as the ethical foundation for social and political institutions. However, in reality this does not begin to cope with the complex issues that arise in applied science, namely, the use of nuclear weapons, issues in biotechnology (see Ted Peters, in this volume), problems in information technology, and so on. It is a valid part of the full story, but is very inadequate indeed when proposed as the whole, which is his strongly held viewpoint. It simply is not realistic.

Having said that, it must be said that at least his tract realizes and considers seriously the central importance of ethics, and the problem of the source of ethics. Even if I disagree with the conclusion, I applaud the attention Monod gives to this central issue.

---

[53] Monod, *Chance and Necessity*, 11.
[54] Ibid., 159.
[55] Ibid., 163–64.
[56] Ibid., 166.

3.6 *Peter Atkins*

Atkins is a talented writer of physical chemistry textbooks, who also has written an excellent text on entropy. However, he believes in "the limitless power of science."[57] He writes,

> Scientists, with their implicit trust in reductionism, are privileged to be at the summit of knowledge, and to see further into truth than any of their contemporaries... there is no reason to expect that science cannot deal with any aspect of existence.... Science, in contrast to religion, opens up the great questions of being to rational discussion... *reductionist science is omnicompetent* [my italics]... science has never encountered a barrier that it has not surmounted or that we can at least reasonably suppose it has the power to surmount.... I do not consider that there is any corner of the real universe or the mental universe that is shielded from [science's] glare.[58]

This is a very clear statement of the belief that science can answer questions that are in fact outside its domain of competence. The useful question we can ask is, Is Atkins in fact claiming that science can deal with everything of importance to humanity, or rather, that anything outside the limited scope of science is unimportant? It appears that the latter is his true position, for he throws out of the window not only theology but also all philosophy, poetry, and art:

> ...although poets may aspire to understanding, their talents are more akin to entertaining self-deception. Philosophers too, I am afraid, have contributed to the understanding of the universe little more than poets.... I long for immortality, but I know that my only hope of achieving it is through science and medicine, not through sentiment and its subsets, art and theology.[59]

His frame of reference thus excludes all the highest understandings of the human predicament that have been attained throughout history; he defines reality to be only that which can be comprehended by his narrow view of reductionist science. Indeed, he frames his viewpoint so narrowly that it even excludes psychology, all the social sciences, and behavioral biology, for he states, "A gross contamination of the reductionist ethic is the concept of purpose. Science has no need of purpose."[60] This is the framework within which he claims to consider "the great questions of being." The conclusions he attains are dictated by the self-imposed, extraordinarily narrow limits of his analytic scheme.

Atkins is motivated by the power of science to show connections between the disparate, and its ability to show that at its foundations the world is simple. He rightly wants to see how far this approach can go. However, this then becomes a dogma that drives all before it irrespective of how inadequate it is in some spheres of understanding, and he raises reductionism to a first principle to be adhered to even when it cannot deal with the issues at hand (unlike for example Campbell's superb text on biology[61]). Anything that does not fit into this narrow framework is claimed to be self-deception or delusion. His argument against religion, which he characterizes as only being based on ignorance and fear,[62] is essentially that it does not fit into his restrictive scheme. His whole approach is based on an *a priori* metaphysical

---

[57] P. W. Atkins, "The Limitless Power of Science," in *Nature's Imagination: The Frontiers of Scientific Vision*, ed. by J. Cornwell (Oxford: Oxford University Press, 1995), 122–32.

[58] Ibid., 123, 125, 129, 131.

[59] Ibid., 123, 131.

[60] Ibid., 127.

[61] N. A. Campbell, *Biology* (San Francisco: Benjamin Cummings, 1991).

[62] Atkins, "The Limitless Power of Science," 124.

position of a fundamentalist nature, which is his ground and starting point. This viewpoint is claimed to be the only metaphysical position compatible with science, which is simply untrue; he can maintain this supposition only by ignoring both the rationally based arguments that carefully consider all the other metaphysical options, and all reasonably sophisticated versions of religious explanation.

One might ask[63] what is the pay-off of this impoverished world-view, which consigns to the dustbin *inter alia* Plato, Aristotle, Kierkegaard, Shakespeare, Dostoyevsky, Tolstoy, Victor Hugo, T. S. Eliot? It appears to be two-fold: firstly, one claims absolute certainty (even if this is not attainable)—yet another manifestation of the human longing to be free of the metaphysical doubt that in fact faces us. Second, given this view, scientists become the high priests of this barren religion— they are the people with privileged access to omnicompetent knowledge.[64] It is their prerogative to judge and dispense truth in this desolate landscape. Thus the temptation to scientists to promote this view is the same as has throughout history been the temptation to those claiming absolute knowledge of truth: they can see themselves as superior to their contemporaries. However, this no longer works: while a few will follow, the main result is to convince the majority of the public that scientists have little understanding of the real world. This kind of exaggerated position (whether explicit or implicit) is one of the reasons why anti-science views are presently on the increase in the populace at large.

3.7 The Issue of Doubt and Certainty

In the end, the biggest concern in all of these cases is the lack of a tentative approach and/or presentation in relation to some of the most difficult issues that face us in our attempts to understand the universe. Sometimes this lack of doubt is explicit, in some cases it is implicit and conveyed through the presentation, but it is none the less real for that (remembering here Marshall McLuhan's dictum that "the medium is the message"):

> ... these books manifest what can only be called 'unyielding dogmatism' on issues relevant to teleology and religion, *i.e.* on 'ontological' and 'metaphysical' issues. Here, clearly, the characteristic tentativeness of scientific statements, which each well represents when scientific theories are under discussion, ceases and utter certainty enters the scene... one of the problems of the unphilosophical consciousness of many scientists—as is also the case with theologians—is that the former become as rigid and dogmatic as have religionists on subjects that transcend science and so are essentially unprovable, and yet, thinking they are still speaking in the discourse of science, the scientists claim the authority of science—in fact much more authority than real science ever claims.[65]

The authors concerned are tackling important, indeed vital, issues, and that is certainly to be commended. What is lacking is some appreciation of the uncertainty that must accompany our best attempts to resolve them. The progress of modern science has not in any way changed that essential uncertainty, and any claim that it has is seriously misleading to the reader.

---

[63] See Mary Midgley, "Reductive megalomania," in *Nature's Imagination: The Frontiers of Scientific Vision*, ed. by J. Cornwell, (Oxford: Oxford University Press, 1995), 122–32.

[64] See the quotes above, and the analysis in Margaret Wertheim, *Pythagoras' Trousers* (New York & Toronto: Times Books, Random House, 1995).

[65] Gilkey, "What Ever Happened to Immanuel Kant? A Preliminary Study of Selected Cosmologies," 157.

4 *Adequacy of Theory Within its Scope*

The next point is that writers concerned with establishing a scientific world-view that goes beyond what science can legitimately achieve, often prepare the way by moving to questionable ground even in presenting the legitimate arena of science, but usually without alerting their readers to the speculative or contested status of the specific scientific theories being considered. These speculations are in fact aimed at extending science in metaphysical directions, but without revealing openly that this is what is happening. I give some examples to illustrate this.

4.1 *Carl Sagan*

By and large, the scientific part of Sagan's *Pale Blue Dot* is excellent. However, he cannot resist pushing things beyond where they can properly go, in order to make philosophic points. For example, his argument that we do not have a preferred position in the universe is supported by the claim that there do not exist preferred velocities in the universe.[66] This is simply wrong; there is indeed such a preferred velocity in the standard cosmologies, and nowadays we test for it by observing the cosmic background radiation dipole, which reveals our motion relative to this standard of rest. A preferred rest frame is fundamental to the Robertson-Walker geometries that underlie the standard Friedmann-Lemaitre universe models.

However, this is not too important, as its contribution to his supposed metaphysical deductions is minimal. More to the point is that he would like a universe without beginning, because of the mistaken belief that this alters the need for a Creator (apparently being unaware of contrary conclusions in the debate on this issue, which has ranged from St. Augustine to the present time[67]). To support this view, he espouses the chaotic inflationary cosmology of Andrei Linde, without warning the reader that this model has no solid physical basis (there is at present no physical candidate for the underlying "inflaton"), and that there is no observational evidence that it is right.[68] It is an intriguing speculation that lies beyond the bounds of presently well-established physics.

Most importantly, he tries to argue that there is no basis to anthropic hypotheses, because all laws of physics that are possible allow life.[69] There are very good reasons for denying this argument.[70] To support it, Sagan argues that Newtonian gravity is necessary rather than contingent (because space is three-dimensional, which is itself a contingent fact). This claim is simply untrue, for there are numerous relativistic theories of gravity that do not have a Newtonian limit.

4.2 *Richard Dawkins*

Dawkins' approach to evolution is hotly contested because it is so centered on the gene alone, without allowing for the effects of culture and the top-down effects whereby evolution modifies the gene, the effects, in other words, of the causal chain

---

[66] Sagan, *Pale Blue Dot*, 30.

[67] See the discussion in Robert J. Russell, Nancey Murphy and Chris J. Isham, eds., *Quantum Cosmology and the Laws of Nature* (Vatican City State: Vatican Observatory, 1993).

[68] Sagan, *Pale Blue Dot*, 38.

[69] Sagan, *Pale Blue Dot*, 36–37.

[70] John D. Barrow and Frank J. Tipler, *The Anthropic Cosmological Principle* (Oxford: Oxford University Press, 1986); Y. V. Balashov, "Resource Letter AP-1: The Anthropic Principle," *American Journal of Physics*, 59 (1991): 1069–76.

acting in the direction opposite to that which is central to Dawkins' thesis.[71] Central
to his approach to religion is his idea of a genetic unit of cultural inheritance—a
"meme"[72]—which, while a useful analogy, would seem to be an idea that goes
beyond what can be scientifically established. The point is that while it may be a
useful concept in principle, it has no scientific standing unless one can specify what
will characterize its survival or not, apart from the mere fact of survival; but the very
complex causal features that allow or prevent this cannot easily be characterized.
Without some such criteria, it is unclear what conclusions based on this concept can
be experimentally tested—it simply serves as a new name for those notions which
in fact continue to exist and exert an effect on society. Its relation to concepts of
"fitness" and "hereditary" (and hence to Darwin's theory) remains inexact and
unproven. Indeed, this may be a case where Popper's charge of evolutionary ideas
being tautological is true.

The key point is that "many theories of gene-culture co-evolution simplify the
way in which genes and culture cause or produce phenotypical outcomes."[73] These
theories choose restricted aspects of causation in situations where many complex
factors interact, and hence arbitrarily pick out one causal chain from all those in
action. This applies to Dawkins' general use of the idea of a meme.[74] From the
viewpoint of this article, Dawkins' attempt to use an argument based on memes to
undermine the basis of religious understanding is a version of the argument from
evolutionary origins, discussed below (section 5.1).

Turning to broader issues, there is an obvious temptation for those promoting
science as the answer to everything to underplay the problems encountered by
science and overstate its successes. Dawkins' presentation in his new *Evolution of
Life* CD-Rom is a specific example. There he states "Our existence was once the
greatest mystery. It is no longer. It has now been solved." As the interviewer in
Personal Computer World commented, "Well, evolution does answer many
questions, but it is a long way from solving the mystery of life."[75] Indeed, Dawkins'
statement is a substantial misrepresentation of the scientific position. Quite apart
from all the issues raised in discussions of the anthropic issue,[76] he is ignoring the
fact that major steps in the evolution of life—in particular the creation of the first
cell—are still not understood; the various proposed solutions are very speculative
and unproven.[77] Presumably Dawkins' metaphysical concerns have led him to this
misleading presentation of the present scientific position.

4.3 *Daniel C. Dennett*

Similar difficulties emerge in Dennett's writing on "Darwin's Dangerous Idea,"[78]
which puts forward a scientism that claims to explain all. An example is his central

---

[71] See for example, N. Eldredge, *Reinventing Darwinianism: The Great Evolutionary
Debate* (London: Weidenfeld and Nicolson, 1995).

[72] Richard Dawkins, *The Extended Phenotype* (Oxford: Oxford University Press, 1982).

[73] Bowker, *Is God a Virus?*, 83.

[74] Problems with use of this concept in relation to religion in particular are tackled in
depth by Bowker, *Is God a Virus?*

[75] C. Akass, "Life's work" (interview with Richard Dawkins) *Personal Computer
World*, (December, 1996): 158–60.

[76] Barrow and Tipler, *The Anthropic Cosmological Principle*; Balashov, "Resource
Letter AP-1: The Anthropic Principle."

[77] A far more reliable statement of the position will be found, for example, in A. Scott,
The Creation of Life (Oxford: Blackwell, 1986).

[78] Dennett, "Darwin's Dangerous idea"; idem, *Darwin's Dangerous Idea*.

claim that algorithms are foolproof recipes that always work ("all algorithms are guaranteed to do whatever they do"[79]); this is the infallible underpinning for his expansive claims. However, this is not true; in fact major parts of modern computer science are concerned precisely with the issue of when algorithms do and do not work (the "halting problem": does the algorithm in fact succeed in giving an answer, or does it grind on forever, possibly in some endless loop, without ever completing its task?).

He also strongly bases his analysis on memes, despite the lack of evidence to support them (see above). In fact he interestingly discusses the difficulties in making this concept scientific,[80] but then continues to give it a major role in his philosophy, indeed giving memes an extraordinary personalized power. Thus he states, "some memes definitely manipulate us into collaborating on their replication,"[81] reminding one more than anything else of the spirits of animism.

4.4 *Peter Atkins*

In order to bolster his metaphysical position, Atkins claims that science has never encountered a barrier it cannot surmount. Quite apart from the lack of appreciation of the limits of the scientific domain displayed here, this is signally untrue even within the proper area of competence of science. Considering Atkins' own domain of physical chemistry, he does not tell his reader that there are no secure logical foundations for the quantum mechanics that underlies chemistry; indeed, it is not even a truly coherent theory because the quantum measurement problem has not been solved.[82] Even more striking, given his presentation of the Second Law of Thermodynamics as one of the greatest insights of science, is that he does not mention that it is precisely here that one of the greatest failures of the reductionist project lies, because we do not yet have any really viable solution to the arrow of time problem. There is as yet no fully adequate resolution of how it is that the underlying fundamental laws of physics are time symmetric but their effects in the macroscopic world are not.[83] So he hides from his readers a major failure of his own program in the area of science that he claims is one of its greatest triumphs.

Equally striking is the extravagance of his hopeful claim that science can provide a full description of the creation of the universe—it is precisely here that science reaches its limits.[84] Atkins blissfully ignores all the problems of testing and confirmation that underlie any proposals for theories of the creation of the universe. He claims, with great emphasis, that a theory of creation can proceed from nothing at all, without even attempting to explain why on this basis we should expect any laws of physics to hold, let alone the particular laws that in fact obtain. He has no apparent view on whether the laws of physics are Platonic or descriptive, whether

---

[79] Dennett, *Darwin's Dangerous Idea*, 57.

[80] Ibid, 352–68.

[81] Ibid., 363.

[82] See for example, Roger Penrose, *Shadows of the Mind* (Oxford: Oxford University Press, 1994), 237–339; Chris J. Isham, *Lectures on Quantum Theory* (London: Imperial College Press, 1995).

[83] See for example, Roger Penrose, *The Emperor's New Mind* (Oxford: Oxford University Press, 1989); H. D. Zeh, *The Physical Basis of the Direction of Time* (Berlin: Springer, 1989).

[84] Atkins, "The Limitless Power of Science," 131. On scientific limits, see for example, Ellis, *Before the Beginning*.

they precede the universe or were created with it.[85] He claims science can account for cosmogenesis but does not tell his readers that quantum cosmology is regarded with considerable suspicion by many physicists—even the concept of "the wave function of the universe" is under dispute. He presents this untested proposal as if it were solid science, and then compares it with what he calls "the soft flabbiness of non-scientific argument."[86] Because he presents as solid science unconfirmed incomplete proposals that lie beyond the bounds of testability, he himself is indulging in fantasy.

4.5 Scientific Metaphysics and Presentation

In many cases the pursuit of scientific themes (for example, the steady-state universe, the no-boundary idea in quantum cosmology, the idea of action at a distance, the pursuit of a fundamental physical "Theory of Everything") has to a considerable degree been based on the pursuit of some metaphysical idea that it was hoped would be embodied in a particular scientific theory or hoped-for synthesis.

There is nothing wrong with this,[87] provided the resulting theory is carefully assessed for internal consistency and in terms of its relation to data. The problem arises when wishful thinking—based on the hoped-for metaphysical conclusions— leads to inadequate assessment of the resulting theory and the relevant data. Each of the writers referred to pursues interesting and important themes in an interesting way. However, I have presented evidence that sometimes they allow their metaphysical hopes to distort their evaluation of the evidence, resulting (in their popular writings) in fanciful extensions of science within its own domain, but presented with certainty and apparent authority. I suggest that particularly in popular presentations of scientific themes, it is important to ask: Is the presentation of the scientific work clear about the uncertainties and difficulties in the theory presented? It is clear this is not always the case; but this should be seen as one of the major responsibilities of popularizers of science, for otherwise we will indeed deserve the attacks by those hostile to the claims that science is able to reliably detect any invariant aspects of reality.

5 Flawed Arguments

In order to present one's position as absolutely certain, as in some of the works mentioned, it is necessary *inter alia* to undermine the gravamen of the opposition's arguments, attempting to demonstrate that it cannot be seriously held by a reasonable person even though there cannot be a solid analytic basis for this view (see the discussion in section 2). The main attempt at doing this by logical argumentation is the argument from evolutionary origins; however, this is illusory (section 5.1). The other primary methods are denying the meaning of the domain of issues dealt with (section 5.2), or in various ways denying the data supporting the position (section 5.3).

---

[85] Indeed, he presents no coherent position on any of the associated fundamental problems discussed in depth by Paul Davies in this volume, or in the papers in Russell, Murphy, and Isham, eds., *Quantum Cosmology and the Laws of Nature*.

[86] Atkins, "The Limitless Power of Science," 131.

[87] And indeed, it is to be expected if one takes into account that the rise of physics in Western culture has been fundamentally a religiously inspired enterprise; see Wertheim, *Pythagoras' Trousers*.

5.1 *The Argument by Evolutionary Origins*

This is the initially plausible argument that because one can in some way or another explain religious origins in functional and evolutionary terms as meeting some economic, social, or psychological human need,[88] perhaps related to a particular stage of evolution,[89] one has explained religion away. *Inter alia*, Dawkins and Wilson are authors who argue this viewpoint.

However, this is a spurious argument, because if it applies to religion, then it applies equally to science as a whole, as well as its specific branches such as evolutionary biology (and to literary criticism, sociology of knowledge, Marxist theory, and so on); for each subject can be analyzed in terms of its evolutionary origins and the human needs that it meets.[90] Of course the argument may not apply equally to all these subjects, for the degree of destructiveness depends on how strongly coupled are the content of the discipline and the explanation of the discipline's origin. Nevertheless, in principle it applies to all subjects, for each has an origin that can be explained in historical and sociological terms.

In fact this argument *per se* does not tell one anything about the truth value of the theory concerned.[91] While each subject has a sociological role and a historical origin, each also has its own logic of testing and confirmation (which is imbedded in a larger tradition of understanding).[92] One has to examine each theory in the light of this logic to test its truth-value.[93] Such analysis of ultimate issues ends in ambiguity, in that logical certainty cannot be attained (see section 2).

This is not to say that study of the social and psychological origins of scientific, philosophical, and theological theories and world-views is without interest. At least as interesting as the study of those tendencies and situations that lead to religious world-views is the study of those that lead to the atheistic positions. One might guess, for example, that the extreme scientism of some of these views could only arise if one's world-experience were strongly limited to an "academic" environment far removed from the pressures and responsibilities of ordinary life.

5.2 *Denying the Meaning of the Domain*

Aside from direct logical argument, there are various ways of denying the meaning of the domain that underpins it, in this case the domain of religious experience and argumentation. In effect one defines the problem in a way that excludes the opposition view by restricting the scope of argument and enquiry. This may be done implicitly by simply omitting that domain from consideration (see some of the examples in section 3), but it may alternatively be done directly. Failing a logical basis for attaining certainty, if one is to deny the opposing view as a rational position then one has to create an illusion of certainty. There are various ways this can be done.

---

[88] Paden, *Interpreting the Sacred*.

[89] J. H. Barkow, L. Cosmides, and J. Tooby, *The Adapted Mind: Evolutionary Psychology and the Generation of Culture* (Oxford: Oxford University Press, 1992).

[90] Paden, *Interpreting the Sacred*.

[91] Peter Berger, *A Rumor of Angels* (Garden City, New York: Anchor Books/Doubleday, 1990); A. Flew, *Thinking about Social Thinking* (London: Fontana, 1991). If it were valid, it would speak against the conclusions presented in section 2.

[92] See Alasdair MacIntyre, *Whose Justice? Which Rationality?* (Notre Dame: University of Notre Dame Press, 1988); Murphy and Ellis, *The Moral Nature of the Universe*.

[93] Murphy, *Theology in the age of Scientific Reasoning*; idem,"Evidence of Fine Tuning in the Design of the Universe."

The main way is by appealing to emotion or use of rhetoric, perhaps in the form of ridicule. Thus Monod writes, "'liberal' societies of the West still pay lip-service, and present as a basis for morality, a disgusting farrago of Judeo-Christian religiosity, scientific progressivism, belief in the natural rights of man, and utilitarian pragmatism."[94] This simply substitutes emotional invective for logical argument. Dawkins has said that religious people confronted with science are "know nothings" and "no contests," and dismisses metaphysical questions as questions that should never be put and don't deserve an answer.[95] Atkins writes emotionally about how theologians have no right to claim that the existence of God is a useful explanatory hypothesis.[96] Because their proposal does not fit his restricted explanatory scheme, pursuing it is "an abnegation of the precious human power of reasoning." This blustering—it has a threatening element in the way it is presented—fails to discredit reasoned theological positions; his reference to his opponents as "the underinformed or the wily"[97] is that oldest of rhetorical tricks, the resort to personal abuse because a more cogent argument is lacking.

Another method is the creation of "straw men"—simply unbelievable theories supposed to represent the viewpoint of the opposition—that are trivial to demolish; or, equivalently, one can simply ignore the strong arguments on the other side. Slightly more subtle, but essentially equivalent, is to lump together as a whole the vast variety of religious possibilities, and reject them all because some extremes that are included in the set are unpalatable or clearly unviable. Thus Atkins describes religious people as "not merely the prejudiced but also the underinformed" and presents religion as simply founded on ignorance.[98] Sagan valiantly fights off Brian Appleyard and (quite rightly) belabors the Church for the Galileo affair.[99] But neither considers for example the reasoned scientific-religious views of Peacocke, Barbour, and Polkinghorne;[100] their highly developed arguments are much more difficult to dismiss. While it may well be that the crude religious views belabored in this way are what Dawkins, Dennett, Atkins, and Sagan encountered in their youth—and rightly rejected—they should by now be aware that other much more sophisticated and palatable options are available.

Again, Monod lumps Jewish and Christian viewpoints together as "rooted in animism" and "fundamentally hostile to science"—a caricature of some of the modern religious viewpoints.[101] Sagan rightly fights and discounts all kinds of superstition and occult practice, and then almost as an aside includes seriously thought out religious positions in the same category, without any attempt to distinguish the possibly true from the palpably false. It is as if science were to be rejected because it is undeniable that pseudo-science exists, and indeed has numerous supporters. Again, Dennett apparently believes that discounting the view of religion contained in a child's song, "Tell me why the stars do shine" (also held dear by many adults), is sufficient to completely discount theism: indeed, "God is,

[94] Monod, *Chance and Necessity*, 159.

[95] Dawkins, interview in *Sunday Times*.

[96] Atkins, "The limitless power of science," 128.

[97] Atkins, "The limitless power of science," 129.

[98] Ibid., 125.

[99] Sagan, *Pale Blue Dot*.

[100] Peacocke, *Theology for a Scientific Age*; Barbour, *Religion in an Age of Science*; Polkinghorne, *Reason and Reality*.

[101] Monod, *Chance and Necessity*, 159.

like Santa Claus, not anything a sane, undeluded adult could literally believe in."[102] Thereby he casually dismisses the life's work of many of the greatest minds that have ever existed. It is possible his analytic ability is so much better than that of Augustine, Aquinas, Newton, Kant, Kierkegaard, Barbour, Polkinghorne, Peacocke, to name a few, that they can indeed be casually dismissed in this way; but then again his underlying presupposition of infallible superiority may be wrong.

These are examples of the point made in section 1, namely, the resort to dogmatic *ex cathedra* statements, and the denigration of the opposition instead of rational argument. Further, there is no shadow of doubt in these writer's minds; they have attained absolute certainty about metaphysical issues that have puzzled some of the most able thinkers ever since the dawn of consciousness. The essential element of openness of approach (section 2.1) is missing. The caution that characterizes true scientific endeavor has been abandoned.

These methods all avoid facing the issues that are taken seriously by the other side. They are expressions of a mind closed to issues which are of crucial importance to many. This could in fact be interpreted as a form of intellectual or personal arrogance, for it seems to imply a lack of ability or even of integrity in those with opposing views, as well as depreciation of the commitment made and suffering undergone of many on behalf of what they believe. It is an undeniable fact that many people have undergone great hardship on behalf of their religious or metaphysical beliefs. Now that does not show these beliefs are right, but it does give them a certain weight, a right to be considered seriously and not just dismissed with contempt.

Indeed, in dealing with moral and religious issues, one is entitled to query what moral wrestling has been undertaken by those making major intellectual claims in these areas, and how that has been expressed in their lives, for this is what morality is about. This does not determine the correctness or the incorrectness of their views, but it does give an idea of the depth of experience incorporated, and the depth of commitment involved. Where that commitment is deep, in particular when serious sacrifices have been made on behalf of metaphysical, religious, or moral views, a serious dialogue is required to attempt to see what glimpses of understanding may be incorporated in these views. An approach to understanding with integrity cannot simply dismiss such views out of hand.[103]

5.3 Denying the Validity of the Full Data

When denying the validity of a domain of concern, one has of necessity also to deny the validity (or even the very existence) of the data that support the opposing view. I have on a number of occasions been told that there are no data supporting a religious world-view. This is an astonishing claim given the existence of the Church and many other religious institutions; the highly developed philosophical systems based in them and resulting from their historical experience;[104] the countless lives of people who have devoted their whole existence to religious endeavors such as hermits, monks, priests, nuns, and so on; and the even more numerous lives of

---

[102] Dennett, "Darwin's Dangerous idea," 40.

[103] There is no implication here that only those on the religious side have engaged in such wrestling; Carl Sagan for one certainly has done so in regard to the environment, nuclear weapons, etc. The argument applies in all cases.

[104] G. S. Wakefield, *A Dictionary of Christian Spirituality* (London: SCM Press, 1983); Huston Smith, *The World's Religions: our great wisdom traditions* (San Francisco: HarperSanFrancisco, 1991).

ordinary believers of many faiths, many of whom are completely convinced of the reality of their experience of God.[105] The one thing that is quite clear is that there is a huge body of data supporting a variety of religious world-views. Denying the existence of this body of data is intellectually insupportable. If the aim rather is to reinterpret this data reductionistically, then that should be made clear, for it is a very different position.

Indeed, the real problem is rather the opposite: there is too much data. Contained in this body of potential evidence is much conflicting data, supporting opposing religious views. There is clearly a real question as to how to handle this. Accepting the existence of this data and then debating its validity or otherwise opens up an important area to analysis and debate.

In particular—and here I support what is contained in some of the writings I am criticizing—there is a real problem for supporters of a religious world-view in the manifest evil that has been wrought by many churches and other religious institutions over the years. They are right in raising this issue, and it has to be dealt with very seriously.

One approach is to admit relevant data exists but to avoid the problem of seriously assessing it by selection of only that which favors your view and neglecting that which does not, thereby attaining the desired result irrespective of what the full data set actually indicates. In the case of religion, the prime example is considering only negative evidence. For example, often in history the behavior of particular sections of the Christian Church has denied its proclaimed tenets of faith; this can be interpreted as evidence against the truth-claims of that particular group, but some take it as sufficient to show that religion as a whole cannot be true. More generally, the problem of innocent suffering and human evil may be taken as a denial of the proposition that a loving God could have created the universe.

In both cases the problem from the viewpoint of epistemology is that drastic selection has taken place: all the positive evidence has been ignored. For while the evil done in the name of religion is undeniable, so is the good: the high profile acts of saints and reformers, the major campaigns by churches to improve conditions for millions, as well as the ordinary lives of many millions of the faithful of many beliefs, humbly doing their best to improve the world around them and live according to their understanding of a loving God. Similarly, as well as evil, there is demonstrably tremendous love and good in the world: generous acts of all kinds, kindness on behalf of others, self-sacrifice and love, which requires an explanation just as much as does the evil in the world. Ignoring all this positive data is a method of avoiding a true conclusion by failing to take the full set of evidence seriously. It can function as a way of imposing a predetermined conclusion on the debate, irrespective of what the full body of evidence says.

6 Taking the Data Seriously

These strategies are devices that avoid the actual analysis needed. There is nothing wrong with selection of data when constructing a theory; indeed, that is essential; however, when theory is complete it must then handle both positive and negative data.[106] The real task, which takes the data seriously, is to take into account the vast

---

[105] A. Frossard, *God Exists: I have Met Him* (London: Collins-Doubleday, 1970); D. Edwards, *Human Experience of God* (New York: Paulist Press, 1983); J. Punshon, *Encounter with Silence* (Richmond, Indiana: Friends United Press, 1987).

[106] For a discussion, see Murphy and Ellis, *The Moral Nature of the Universe*.

range of material included in this full data set of religious and moral experience, and to consider seriously methods of discernment within it of true and false evidence (or simply noise), just as one must do in the hard sciences.

In the latter,[107] a sophisticated process of testing takes place, involving not only 1) an assessment of the analytic and experimental procedures involved in some truth claim, and 2) a testing of whether the data obtained represents a signal or pure noise, but also 3) an implicit or explicit assessment of the *bona fides* of the research worker involved, and 4) an evaluation of whether the result makes sense in terms of fitting into the current scientific world-view as embodied in theory, authority figures, textbooks, practice, and tradition.[108] This testing 5) takes place within an overall framework of understanding that assigns basic meaning to concepts used in setting up and analyzing experimental data; in this sense, data is theory-laden. Furthermore, 6) this overall framework itself is also subject to testing and revision.[109]

By and large this process works well, and there is indeed a body of scientific knowledge that can be regarded as established with effective certainty, within its domain of application, through this process (for example, Newton's laws of motion and Maxwell's equations for electromagnetism). However, there are always outliers (that is, individuals who question one or other aspect of the present orthodoxy) and cases where the method does not work at the present time (a significant group of scientists maintaining certainty has been attained, and others denying this). As time progresses, these cases may or may not be embodied in the mainstream (cosmological examples include: Lemaitre's models of the expanding universe, Arp's observations of unusual galaxy-QSO associations and consequent questioning of the nature of redshift, and the lack of solar neutrinos as predicted.) There are also experiments that contradict the current understanding but whose results come to be rejected for a variety of reasons relating to the particular experiment in question (for example, data relating to the speed of light, to existence of a fifth force, to detection of gravitational radiation, and to existence of massive neutrinos.)

The claim then, from the religious side, essential for making sense of the relevant body of data, is that in the religious area also there is a process of discernment and testing needed to determine which data is good evidence and which is not.[110] This process involves testing data against current understanding, as embodied in practice, doctrine, scripture, tradition, authority, and community perception, with the aim of detecting charlatans and weeding out false perceptions and the insidious processes of self-deception and projection, while at the same time subjecting the current orthodoxy to questioning and testing in the light of this data. The same kind of testing of data is essential in the area of morality and ethics, which may be claimed to be closely related.[111]

---

[107] W. H. Newton-Smith, *The Rationality of Science* (London: Routledge, 1981).

[108] J. Ziman, *Reliable Knowledge: An Exploration of the Grounds for Belief in Science* (Cambridge: Cambridge University Press, 1991).

[109] Murphy, *Theology in the age of Scientific Reasoning*, idem, "Evidence of Fine Tuning in the Design of the Universe."

[110] Jonathan Edwards, *The Religious Affections* (Carlisle, Pennsylvania: Banner of Truth Trust, 1746/1991); E. Trueblood, *The Trustworthiness of Religious Experience* (Richmond, Indiana: Friends United Press, 1979); K. Yandell, *The Epistemology of Religious Experience* (Cambridge: Cambridge University Press, 1993).

[111] Berger, *A Rumor of Angels*; idem, *A Far Glory* (Garden City, New York: Anchor Books/Doubleday, 1992); Murphy and Ellis, *The Moral Nature of the Universe*.

The result is a process that in effect classifies data as good (it can be used as evidence) or bad (it should be rejected). It does so on the basis of the best current overall method of understanding within which the data is tested. This must be compatible with the overall tradition of understanding employed in the area of scientific testing, but modified and extended to be applicable to and properly take cognizance of the broader area of data that arises in examining metaphysical, philosophical, and ethical issues, as well as the themes of daily life, which all lie outside the competence of the scientific method proper.

Two comments are in order here. First, faith in the end is based on personal experience and revelation;[112] nevertheless, it is possible to some degree to apply an experimental approach to the religious sphere.[113] The problem again is how to evaluate the result of the experiment. One can argue it can be reliable if approached in the proper way.[114]

Second, as has been implied in the discussion above, I assume that religious traditions and practices should be judged by their fruits—their outcomes—as well as by their overall religious stance, explanatory satisfactoriness, and support from data.[115] The overall reason for this standpoint is the intimate link between ethics and religion, with religion providing the basis of ethics,[116] which in turn is a primary element in the practice of religion and test of its validity (see Matthew 7:15–20). Thus the ethical stance adopted in practice is a touchstone for the true nature of the religion. Religious beliefs have caused both good and bad. My position here is that we can use this as a crucial test of the validity of particular religious beliefs within the broad spectrum of religion and of particular sects within each faith. This is in contrast to those views that reject all religious positions because some of them have manifestly caused bad.[117]

In applying this three issues arise.[118] First, one must distinguish carefully between cohesive cultural or social subgroups who believe and act in a certain way (this group of Christians, that faction of Moslems) and the generally approved or recommended behavior genuine (Christian or Moslem) belief encourages. Any person or group may engage in evil behavior and appeal to their religious beliefs in an illegitimate or distorted way; or they may simply ignore the generous, self-sacrificing behavior their beliefs insist upon. Their behavior should be condemned unquestionably, but the religious beliefs they espouse should only be condemned or called into question if a genuine and authentic interpretation of them promotes evil behavior. Most Christians do not follow Christ in any way perfectly, nor Moslems

---

[112] J. Scott, *What Canst Thou Say?* (London: Quaker Home Service, 1980).

[113] See for example, G. Hubbard, *Quaker by Convincement* (London: Penguin, 1976), ch. 6 entitled, "The Spiritual Experiment."

[114] Yandell, *The Epistemology of Religious Experience.*

[115] I am not here claiming this is the only criterion. For example, I am not rejecting Ted Peters' central thesis that out of divine love God forgives sinners. But I am saying that any branch of faith which is manifestly and clearly leading to evil—say lynching parties of church men hanging blacks, or a church giving major public support to the Apartheid system—can for that reason and on those grounds alone be rejected, unless manifest and obvious steps are being taken to correct the situation, this itself being a sign that enlightenment is present.

[116] This is because ethical requirements are arguably based on the nature of God; see Murphy and Ellis, *The Moral Nature of the Universe.*

[117] It is also in contrast to those views that reject all scientific world-views because science has enabled massive environmental destruction and made possible a nuclear holocaust.

[118] I am grateful to Bill Stoeger for the following comments.

Mohammed. The question regarding the overall faith is, What does living out that faith in an authentic way approximating the ideal lead to?

Second, one must distinguish among the various interpretations of a given authentic belief system. One interpretation of a given "doctrine" within the Christian belief system may wreak havoc, and thus should be condemned on that basis (being death-dealing rather than life-giving); another interpretation may bring just the opposite: life, reverence, and respect. Certain erroneous interpretations may have been considered authentic in the past—or may be considered authentic now—but on the basis of such testing they have been, or hopefully are being, revealed as seriously "out of step" with the overall character of the Good News. In this way, one avoids dismissing an entire belief system on the basis of distortions or weaknesses which have crept into it. In all faiths, continual conversion and purification must go on, at every level—personal, communal and institutional—because any realization of the faith community and its comportment is always imperfect, often subject to cooptation or distortion, and in some circumstances may be downright evil. Insofar as it is evil, or insufficiently critical of what is evil and death-dealing, it must be challenged and condemned. But that does not necessarily mean that the belief system those people are proclaiming to try to follow is wrong or misguided. It may be simply a distorted interpretation which is being forced upon them; or perhaps they are not living out what they proclaim. Perhaps they are trying but are not succeeding very well. Perhaps they are badly misinterpreting certain aspects of the faith they espouse, or have their priorities within it scrambled.

Thirdly, there may as a matter of fact be whole belief systems which are intrinsically flawed. It would take considerably more evaluation and testing to determine that it wasn't due to distortions, erroneous interpretations, ideologizing of the religion, or domination by authoritarian and repressive factions, but rather to the intrinsically evil or inadequate character of the religion or belief system itself.

This is an important and difficult area. Given all the above, there are religious interpretations that survive the acid test. Of course the same test applies to atheistically based or "scientific" ethical systems, no matter what their supposed theoretical origins. They too can be tested in this way.

7 Theories of Ultimate Reality

I now briefly consider in turn three views that seem viable, in terms of taking the data seriously. I do not claim they are the only ones; indeed, one of the major challenges facing us is to investigate how large the range of realistic possibilities is.

7.1 The Kenotic Moral-theistic Position

A position that seems satisfactory in terms of being able to meet all the criteria is one where some religious viewpoints and data are taken as preferred over others and are accepted as valid viewpoints and evidence, their compatibility with ethical understanding being a key requirement for acceptance. On this view, a subset of the religious data is accepted and the associated understanding is taken to represent correctly, insofar as models can do so, the real nature of the universe, which has a theistic character. Others are rejected because they are either demonstrably ethically inadequate—those that have been responsible for evil actions are excluded unless manifest action is being taken to correct the situation—or because they cannot be accommodated in an adequately broad framework of understanding (as outlined above); that is, they are intellectually inadequate.

The problem here is to identify this acceptable subset by identifying the ethical and religious data pointing in the correct direction, as opposed to that which does not. One way to proceed is to use the concept of kenosis as the key identifier in both ethics and theology. Kenosis here is taken to be *self-emptying and voluntary sacrifice on behalf of others, based on genuine and freely given love of others, and resulting in the generosity and respect that flow from it.*[119] As in the case of metaphysics, there is an ambiguity in ethics too in that one cannot prove beyond questioning that this is indeed the ethical under-pinning of the universe, perhaps justice being the key contender. However, one can make a very good case for it based on both theory and practice. While this view has rich Christian associations, and is probably most strongly expressed in that tradition, it is certainly not restricted to this heritage.[120] For example, one of the most effective practitioners in recent times was Mahatma Gandhi. The experiences in his life and in that of Martin Luther King Jr., together with, for example, recent studies in restorative justice[121] and a wealth of experience in non-violent political action and cooperative economic movements provide evidence this is a practical policy worth pursuing.

The proposal will be that the ethics underlying the universe—having the same ontological status as the laws of logic and physics—has a kenotic nature, derived from and expressing the kenotic nature of God as experienced by human beings who are fully open to and receptive to that nature.

Adopting this view, one will select data that is meaningful—it is not fanciful, hallucination, or self deception—and then use it to separate out religious practices and understanding into those that are kenotic (based on love, generosity, and freely given assent) and those that are opposite in spirit (based on fear, authority, and coercion). These criteria cross the lines of religious traditions. Both occur in Christianity, in Judaism, and so on; and in many cases both occur also in different branches of these traditions. Any attempt to use kenotic theology to legitimate ideological positions which involve keeping indigenous people in slavery or the submission of women to men would be rejected; similarly it must be distinguished from forms of masochism and self-destruction by a positive intention and method. The practical test of true kenosis is not always easy, but considerable experience is available and has been thoughtfully analyzed by practitioners such as Gandhi and King.[122]

---

[119] W. Temple, *Readings in St. John's Gospel* (London: MacMillan, 1961); Peacocke, *Theology for a Scientific Age*; Ellis, "The Theology of the Anthropic Principle"; idem,"God and the Universe: kenosis as the foundation of being," *CTNS Bulletin* 14 (1994): 1–14; Murphy and Ellis, *The Moral Nature of the Universe*; John Haught, in this volume.

[120] See for example Philippians 2:6–11 and William Temple's interpretation of the temptations in the desert in his *Readings in St. John's Gospel*, which provide a strong theoretical foundation for its use as a basis for ethics. The focus here is not on Christ's kenosis, which has many dimensions I cannot pursue here, but rather on its practical and ethical (*i.e.* behavioral) dimensions for the believer. However, for the Christian those are founded on Christ's attitudes and behavior. The few sentences leading into the Philippians kenotic hymn help to expresses this.

[121] J. Consadine, *Restorative Justice: Healing the Effects of Crime* (Lyttelton, New Zealand: Ploughshares Publications, 1995).

[122] See for example, D. Dalton, *Mahatma Gandhi: Non-Violent Power in Action* (New York: Columbia University Press, 1993); J. Ansbro, *Martin Luther King Junior: The Making of a Mind* (Maryknoll, New York: Orbis, 1992).

As far as the ethical test is concerned, the first kind of faith is tentatively accepted, and the second is rejected because while ethical implications certainly do not constitute the whole of a religious tradition, or provide a full validation, on this view they do constitute a necessary test for any faith. Thus this view initially accepts the faiths, traditions, or branches of religion that are open, loving, and kenotic; and it rejects those that are dogmatic, authoritarian, or based on fear. Those that are accepted in terms of this criterion are then considered further in purely religious terms: looking at data of the other kinds supporting the view (the usual kinds of justification in terms of experience, authority, tradition, scriptures), and at the overall satisfactoriness of the viewpoint both as an explanatory scheme (is it intellectually viable?) and as a guide to living. We do not pursue these further criteria here.

Thus in this particular case, we identify kenotic religious data and practice as valid evidence supporting a theistic view in which God is creator and sustainer of the Universe, immanent and transcendent, and kenosis is a fundamental expression of the telos of His/Her being and so of the moral nature of the created universe. An important feature of this view is its ability to deal with much of the apparent counter-evidence, including the existence of opposing views and of evil behavior; in essence, both are inevitable in a situation where freewill is given free reign in order to enable a truly free and voluntary response.[123] It is supported by evidence of a religious kind in a variety of traditions; the kenotic nature of God is supremely evidenced, in the Christian tradition, by God's entering into the resulting suffering in the world.[124]

On this view, the creation process is aimed at creating intelligent beings capable of a freely given kenotic response, but that does not mean it focuses solely on the human beings on this planet (or some subset of them who are 'chosen people'). Rather, conditions are created where such evolution will take place on many planets in many galaxies (Davies, in this volume) and the Creator's purpose then lies in this wider creation of self-conscious ethical beings throughout the Universe, not just those on the particular planet we inhabit.[125]

This view accepts fully the role of evolution in creation of self-conscious human beings, enabled and guided by laws of physics that promote self-organization (see Davies' discussion in this volume). God's creative action lies in defining, implementing, and maintaining in being these laws,[126] but God's action also encompasses providing intimations of reality to the faithful and others without any transgression of the established physical laws, through the freedom allowed by quantum indeterminacy.[127]

It is then fundamental that this kenotic view of the universe requires to be true, for its self-consistency, the key element on which this paper focuses, namely, the hidden nature of reality, which is needed in order to allow a free response to reality through the resulting metaphysical ambiguity. No particular faith is forced on us by incontrovertible evidence. Rather, for full coherence in this approach there needs to be a basic epistemic uncertainty.[128] Irrespective of any other reasons for why

[123] J. Hick, *Evil and the God of Love* (San Francisco: Harper, 1977).

[124] Temple, *Readings in St. John's Gospel*; Thomas Tracy, in this volume.

[125] Ellis, "The Theology of the Anthropic Principle."

[126] Ellis, Murphy, and others in Russell et al., *Quantum Cosmology and the Laws of Nature*.

[127] Murphy, Tracy, Ellis in Robert J. Russell, Nancey Murphy, and Arthur R. Peacocke, eds., *Chaos and Complexity* (Vatican City State and Berkeley: Vatican Observatory and the Center for Theology and the Natural Sciences, 1995); see also Paul Davies in this volume.

[128] Hick, *Evil and the God of Love*; Ellis, "The Theology of the Anthropic Principle";

development of self-conscious beings should take place through an evolutionary process, something like evolution based on the regularities of physical laws is needed in order that creation of humans satisfy this requirement of metaphysical ambiguity. Special creation would not do so, for if there were no natural process that could lead to "design" of living beings,[129] the argument from design would be more or less overwhelming, and provide a proof that would remove this uncertainty. On the other hand it is of course important that evolution does not prove whether or not God exists.

Overall, this proposal provides a viable viewpoint.[130] It might be true; and yet it might not. This is the fundamental point of the metaphysical ambiguity of the universe. Personal experience may persuade one that this viewpoint is correct— indeed, one's experience may be so strong as to leave one with no doubts. One can know it through what one has lived through (just as one can have no doubt one is loved by another), and one can make a powerful circumstantial case for it. And yet, epistemologically the case will remain ambiguous. The certainty may be the result of illusion and self-deception, or it may not.

7.2 Self-critical Atheism

At the other end of the scale, one may adopt a self-critical atheistic position; namely, that after careful review of all the data, and after taking into account all the conflicting evidence embodied in various religious views, none of them is established; and indeed, none of them can be correct, because there is too little commonality between them, and there is no reason to prefer any one over the others. The variety of religious viewpoints in the end are all just wishful thinking. The negative evidence wins over the positive, and the true situation is cosmic absurdity.

This is a possible view; it could be correct. However, it may also be wrong; some of the religious data may be truly pointing at the nature of reality. The difference between this view and the dogmatic atheistic position of the scientific religions (see sections 3, 4, and 5 above) is the openness of approach and the proper appreciation of the metaphysical issues involved. The scientific method is not employed beyond its proper domain, and there is a willingness to consider all the relevant data carefully: as well as strictly scientific data, one considers the personal moral and religious experience of numerous individuals, the lives of the saints, the evidence from morality and from intimations of transcendence, and so on. One can in the end decide the proper course, on the balance of evidence, is to reject all this, while acknowledging that there remains the possibility it is in fact what it claims to be: evidence of a transcendent existence that has a moral and theistic nature. The move to cosmic absurdity is made, but it is neither rapid nor dogmatic.[131]

This is a coherent position, but is problematic in terms of ethical choices and decisions that face us. It might be taken as saying they have no real basis or value; cosmic absurdity gives no meaning to life or morality. This could be true, but if so it is a very bleak position, and is usually belied by the stances people take in their daily lives, irrespective of their stated metaphysical stance.

Murphy and Ellis, *The Moral Nature of the Universe*.

[129] See Dawkins, *The Blind Watchmaker*.

[130] It is laid out in some depth in Murphy and Ellis, *The Moral Nature of the Universe*.

[131] See Steven Weinberg, *Dreams of a Final Theory* (London: Vintage, 1988), chapter XI, for a thoughtful exposition of this position.

7.3 *Agnosticism*

Here one reviews the evidence and cannot decide between the different competing views; and so takes no stand and regards the question as open. Metaphysical uncertainty wins; no choice is made. This is a philosophically viable position; but again, it leaves a considerable problem as regards ethical choices. It provides no solid basis or guide for real choices facing us in the world. Inevitably such choices will in fact be made, probably implying a more definite metaphysical position even if it is not made explicit. Thus there is a probability of a dual philosophical position being employed, with the philosophy underlying daily life being disjointed from that proposed as a mental picture of the universe at large.

7.4 *Viable Alternatives*

The argument, then, is that there is a set of viable, alternative theories that can handle the data and the problem of uncertainty. These cover the range from theism to atheism and include agnosticism.

It can be argued that a kenotic ethical-religious position is satisfactory in terms of overall coherence and ability to explain the data, while a self-critical atheist (as opposed to dogmatic atheist) position is also intellectually defensible, as is agnosticism. Other positions that take seriously the full data set in a tentative approach to knowledge may be equally defensible.[132] It is ultimately personal experience and predilection that chooses between them, strongly influenced by one's cultural heritage and community belief, rather than the force of logic or indisputable evidence (for some people logical coherence will be important in making this choice, and for others it will not). The fundamental point is that despite the problems in each approach, a choice is available to the rational person.

One can try to build up a cumulative argument for one or other that becomes strong, and can to some extent succeed; but the fact of the matter is that even then, some people will find it compelling and others will not. Given all the data and in particular their own experience, there are convinced Christians, Jews, Muslims, Hindus, atheists, agnostics; each finds their own position convincing. The choice remains. This does not mean that truth is irrelevant in the matter of religion, nor that logic has no part to play, but rather that indubitable certainty is unattainable.[133] I can reach a position which I am convinced is true and can support in depth through logic and analysis. This can provide a satisfactory basis for living life based on belief, but like all other knowledge of ultimate issues, it is provisional and open to correction.

This feature is coherent with and indeed central to the proposed theological-ethical position of my book with Nancey Murphy. The requirement that ultimately we need to give up intellectual certainty about religion as well as other ultimate issues, acknowledging here too the provisional nature of knowledge, is in the end a liberating and necessary move.[134] It acknowledges the reality of the epistemic freedom which can be claimed to be one of the essential features of the kenotic religious nature of the universe.

---

[132] The scientific religions, however, are not in this set, as they are not tentative in their approach to fundamental knowledge.

[133] Logically, alternative positions can always be argued and cannot be conclusively disproved; emotionally and experientially I may reach conviction, but millions of others equally will have reached opposite positions with similar conviction.

[134] R. Bellah, *Beyond Belief* (New York: Harper and Row, 1976).

8 *Conclusion*

The ultimate metaphysical ambiguity of the universe is an essential element in the scheme outlined here. The scientific religions have not provided any adequate counter-arguments to the epistemic uncertainty detected by Hume and Kant, nor have scientific or philosophic advances since their day conclusively undermined this conclusion; so the basic uncertainty remains.

Murphy comments that, "two vastly important developments in epistemology separate us from Hume. One is the recognition of the role of hypothetico-deductive reasoning; the other is holism: the view that our beliefs face the tribunal of experience not singly but as a body."[135] Given these changes, we still cannot attain certainty, but we can attain a sound logical basis of belief drawn both from our personal experience and from the lives of others whom we regard as wise. We cannot prove our view is right, but we can show it is not logically impossible or intellectually feeble. And that applies whether our view is basically religious or basically atheist (provided it respects the guidelines for argumentation and analysis outlined here; and in each case that is possible).

Implicit or explicit claims by the authors of scientific religions that deny this basic uncertainty are evidence either of ignorance of the philosophic and epistemological issues at hand, or of some kind of attempt at replacing solid argument by illusion. This is the mark of pseudo-science rather than of true science, which must surely reject such methods. Nevertheless there is one feature here that should be applauded, despite these negative comments: it is positive and worthwhile that those engaged in science step back and see what overall world-view this endeavor might support (rather than forever being engaged only in their scientific study without considering its broader implications). However, this too must be done with the care and attention to detail that is lavished on science. The rules of evidence and argument must be followed carefully here too.

If there is openness, acknowledgment of what the other side is trying to do, and tricks of rhetoric are abandoned on both sides, then there is room for a real debate that can further understanding. This paper will have been useful if it helps clear the way for scientific world-views that are less dogmatic and more open than the majority on the market today. Clearly an equal openness is required of those wanting to partake in the debate from the religious side; I have briefly indicated that there are viewpoints of this kind that are open in approach and are intellectually acceptable. If both sides can agree that the other can have a rational and defensible position, then we will be able to move the debate to the deeper areas, such as ethical understanding, that really matter.

Acknowledgment. I thank Lewis Wolpert for drawing my attention to Hume's writings and Nancey Murphy for discussions that have helped shape this article. This version has benefited from comments received from conference participants on previous drafts. I thank Wesley Wildman and Ian Barbour for their support in this project, and Bill Stoeger for very helpful comments and proposals as to how to improve the manuscript.

---

[135] Murphy, "Evidence of Fine Tuning in the Design of the Universe."

DARWIN'S REVOLUTION IN *THE ORIGIN OF SPECIES*: A HERMENEUTICAL STUDY OF THE MOVEMENT FROM NATURAL THEOLOGY TO NATURAL SELECTION

Anne M. Clifford

> When the views advanced in this volume, and by Mr. Wallace, or when analogous views on the origin of species are generally admitted, we can dimly foresee that there will be a considerable revolution in natural history.
>
> Charles Darwin (1809–82)[1]

1 *Introduction*

The views Charles Darwin advanced in *The Origin of Species* precipitated not only a revolution in the study of natural history, but also the development of modern biology which assumes the self-replicating metabolic organization of matter and energy in terms of a one-way, irreversible process in time. This process generates novelty, diversity, and higher levels of organization in living organisms. Darwin was not the first to write about biological evolution. Prior to Darwin evidence indicated to some scholars, such as Jean Baptiste de Lamarck (1744–1829), that species were not immutable. But Darwin, more than any of his predecessors, provided a mechanism for explaining the gradual transformation of one species into another.[2]

Because of the theory of natural selection, which Darwin and Alfred Russel Wallace developed independently, Darwin believed that the gradual transformation of species would be comprehended with the same clarity that a mechanical invention is understood by the workers who use it. About natural selection Darwin reflects:

> A grand and almost untrodden field of inquiry will be opened on the causes and laws of variation, on correlation, on the effects of use and disuse, on the direct action of external conditions, and so forth.... Our classifications will come to be, as far as they can be so made, genealogies; and will then truly give what may be called the plan of creation.[3]

This passage with its focus on "causes and laws" is written in a genre that has the marks of a scientific treatise, but it also ends with an overtly religious expression of personal conviction: the outcome of the inquiry in this "almost untrodden field" will reveal "the plan of creation."[4] Some of Darwin's contemporaries saw his work

---

[1] Charles Darwin, *The Origin of Species by Means of Natural Selection or the Preservation of Favored Races in the Struggle of Life* [Hereafter *The Origin of Species*], 1st ed., 1859. 6th ed, 1872. Reprint, with forward by Julian Huxley (New York: The New American Library of World Literature, 1958), 447. The first edition did not mention Alfred Russel Wallace by name. A joint paper on natural selection by Darwin and Wallace had been delivered at a meeting of the Linnaean Society in 1858.

[2] When Darwin was writing *The Origin of Species*, the notion of biological evolution was already being debated. Perhaps to avoid confusion with the ideas of others Darwin did not use the term "evolution."

[3] Ibid., 148.

[4] As Langdon Gilkey has argued, using "creation" in reference to any scientific theory "represents the epitome of religious speech." Langdon B. Gilkey, "The Creationist Issue: A Theologian's View" in *Cosmology and Theology, Concilium*, vol. 166, David Tracy and Nicholas Lash, eds. (New York: Seabury Press, 1983), 61.

in this light. For example, Frederick Temple, who would become Archbishop of Canterbury in 1896, supported Darwin's theory of species transformation, arguing that it was compatible with Christianity because it presupposed an original Creator.[5] But for others, including many influential naturalists who were also clergy, *The Origin of Species* could not be accepted because it conflicted with the tenets of natural theology which stressed God's purposeful design and benevolent agency in the world.[6] Among some of the proponents of natural theology were those who rejected Darwinian evolution because of the doubt it cast on both the supernatural causality and the design of creation.

At the same time, scientists who were not favorably disposed toward natural theology hesitated to embrace Darwin's explanation of evolution on the grounds that Darwin had not adequately applied the scientific model of inquiry that was widely accepted in the scientific community. They argued that Darwin's methodology did not comply with the canons of the inductive method of Bacon and the principles of Newton's natural philosophy.[7] Darwin's revolution was not immediate. It took over twenty years after the publication of *The Origin of Species* before the majority of scientists accepted his theory. Today, the neo-Darwinian synthesis (that is, the combination of evolution by natural selection and the discovery of DNA as the mechanism of variation) is widely considered the best explanation of the increase of biological complexity by the scientific community.

Darwinian evolution, however, is not exempt from criticism and outright rejection in the United States where creationism is not only a religious commitment but also a socio-political movement promoted with missionary fervor by fundamentalist Protestants. This group continues to regard Charles Darwin's theory of evolution by natural selection as a threat to biblical Christianity. Unwilling to relinquish the position that the Bible provides inerrant truth about the universe, they have developed "creation science," a discipline which, while avoiding any explicit

---

[5] See for example, Frederick Temple, "Apparent Collision between Religion and the Doctrine of Evolution," lecture VI of his Bampton Lectures published as "The Relations between Religion and Science," 1884, reprinted in Tess Cosslett, *Science and Religion in the Nineteenth Century*, (Cambridge: Cambridge University Press, 1984), 192–204. Although Temple is in fact interested in healing the conflict between theologians and scientists, his lectures nonetheless exemplify a division between these two communities and the ways of knowing that they represent. For more details on the range of responses to Darwin's theories see Arthur Peacocke, "Biological Evolution—A Positive Theological Appraisal," in this volume.

[6] One of the earliest exponents of natural theology was Raymond de Sabunde who published *Theologia Naturalis* in 1438. Natural theology rose to importance in England in the seventeenth century in conjunction with the rise of the empirical sciences. Robert Boyle (1627–1691) and Isaac Newton (1642–1727) were natural theologians as well as scientists; both interpreted the world's operations by mechanical principles within a theistic framework.

[7] Among the critics of Darwin were the empiricist philosophers John Stuart Mill and John Herschel. Mill was positively disposed toward Darwin's theory of natural selection because he believed it conformed with the strict principles of the logic of induction, but was reluctant to surrender adaptations in nature to an explanation other than creation by intelligence. Herschel, a philosopher whom Darwin greatly admired, referred to Darwin's theory as the "law of higgledy-piggledy" and refused to grant that it had any scientific merits at all; see David L. Hull, "Charles Darwin and Nineteenth-Century Philosophies of Science," in *Foundations of Scientific Method: the Nineteenth Century*, edited by Ronald N. Giere and Richard S. Westfall (Bloomington: Indiana University Press, 1973), 115–32.

mention of God, is clearly founded on the initial chapters of Genesis and belief in God's "special creation" of each distinct species.

This essay proposes to examine *The Origin of Species* with attention to how Darwin's conception of scientific theorizing is related to a major current of nineteenth-century British thinking known as "natural theology," considered at that time to be a form of science. Like the creation science of the twentieth century, late eighteenth and early nineteenth-century natural theology, forged a unique union between discoveries about nature and Christian belief in a creator. But natural theology, unlike the creation science of today, did not look to the initial chapters of Genesis for the major warrants for its truth claims. The study of nature provided the natural theologian with a reasonable foundation independent of biblical revelation for affirming the existence of God. God's handiwork was evident in nature. Although natural theology in fact confirmed the notion of the "special creation" of distinct species reflective of a literal interpretation of Genesis 1:1–2:4a, its methodology was primarily concerned with providing evidence from nature to substantiate God's sovereignty over and purposeful design of the universe.

Natural theology is summarily dismissed as "bad science" today, but a perfunctory dismissal of William Paley (1743–1805)[8] and the authors of the *Bridgewater Treatises*[9] ignores the limitations inherent in taking what is currently accepted as "normal science"[10] as the standard for examining the historical development of scientific rationality. If one attends only to the positions that are regarded as right due to the hegemony that Darwin's theory of natural selection now enjoys, then one's understanding of Darwin's revolution will be impoverished. The examination of Darwin's revolution in this essay is informed by an insight expressed by Herbert Butterfield, who over forty years ago argued that when one examines scientific positions,

> it is necessary on each occasion to have a picture of the older systems—the type of science that was overthrown.... Little progress can be made if we think of the older studies as merely the case of bad science or imagine that only the achievements of the scientist in very recent time are worthy of serious attention at the present day.[11]

To explore the extent to which *The Origin of Species* precipitated a revolution, Darwin's positions will be examined in relationship to the theoretical framework exemplified by representative natural theologians, especially William Paley, whose work Darwin studied while he was a seminary student at Cambridge. This analysis of *The Origin of Species* is undertaken mindful that the relationship between Christianity and evolutionary biology cannot be resolved exclusively in the domain

---

[8] William Paley's most important contribution to the field of natural theology is his *Natural Theology; or, Evidences of the Existence and Attributes of the Deity, Collected from the Appearances of Nature*, 1802, (Reprint. Hartford: S. Andrus and Sons, 1847).

[9] The *Bridgewater Treatises* were eight volumes commissioned by the Reverend Francis Henry Egerton, eighth Earl of Bridgewater, to demonstrate how God's design was manifested in new scientific findings. For a helpful account of the authorship and content of these volumes, see Charles Coulston Gillispie, *Genesis and Geology, A Study in the Relations of Scientific Thought, Natural Theology and Social Opinion in Great Britain, 1790–1850* (Cambridge: Harvard University Press, 1951), 209–16.

[10] "Normal science," a term coined by Thomas Kuhn, is the practice of science accepted by a scientific community at a particular time; see Thomas S. Kuhn, *The Structure of Scientific Revolutions*, 2nd ed. (Chicago: University of Chicago Press, 1970), 10.

[11] Herbert Butterfield, *The Origins of Modern Science: 1300–1800* (New York: Free Press, 1957), 9–10.

of history. History is relevant for an analysis of Darwin's revolution, but also necessarily limited because history has the potential of providing justification for particular positions. From a hermeneutical standpoint, there are many ways to read texts such as Darwin's *The Origin of Species* and Paley's *Natural Theology*. Every historical text is open to a plurality of possible readings; any reading is affected by the presuppositions and social location of the reader.

Although hermeneutics is closely associated with the interpretation of texts, it encompasses much more. One of the tasks of hermeneutics germane to this study is how language, whether it is the language of the theories of science or the doctrines of theology, is used to make claims about reality. In the formation of these claims what is already understood from a totality of references which we call our "world" come into play. In both science and theology, it is widely recognized that the use of metaphors is inevitable and, therefore, fundamental to the scientific and theological enterprises. Both science and theology draw from references that are beyond their ordinary boundaries in the formation of their metaphorical language. Theological metaphors not only express religious faith; they also articulate interpretations of certain aspects of communal experience, including its cognitive aspects as they are expressed in other disciplines. This is evident in William Paley's choice of metaphors for God as he envisioned God related to nature. By the same token, scientists choose metaphors that not only explain the data of observation, experimentation, and prediction; they also articulate an understanding of this data in ways that are influenced by forces at work in the broader society in which the scientist participates. In Darwin's attempt to make intelligible his observational data, his choice of metaphors relates to broader social currents of thought, as we shall see.

Darwin's position on the descent of species through modification by natural selection presented enormous challenges to the natural theology positions of William Paley. These will be given attention. Since among them is how to speak about God, a proposal for a metaphorical naming of God responsive to Darwin's challenges and rooted in the biblical tradition will be offered.

2 Darwin and Nineteenth-Century Natural Theology

To throw light on a particular phenomenon, at times a hermeneutics of suspicion must be applied to uncover distortions. For the reader who presumes that warfare metaphors aptly describe the relationship of Darwin's theory of biological evolution to Christianity, one must ask, Is this assumption related directly to Darwin and the positions of his contemporaries, or does it come from other possible post-Darwinian interpretations? For North Americans, a likely source for the warfare depiction is Andrew Dickson White's *A History of the Warfare of Science with Theology in Christendom*.[12]

---

[12] Andrew Dickson White, *A History of the Warfare of Science and Theology in Christendom*, 1896 (Reprint. New York: The Free Press, 1965). An earlier work of lesser importance is John William Draper, *History of the Conflict between Religion and Science* (London and New York: D. Appleton and Company, 1874). Draper's work is largely a diatribe against the Roman Catholic Church and its *Syllabus of Errors* (1864). Protestantism is spared his criticism and science is even said to be the "twin-sister of the Reformation" (353). For more on these two works see Claude Welch, "Dispelling Some Myths about the Split between Theology and Science in the Nineteenth Century," in *Religion and Science*, W. Mark Richardson and Wesley J. Wildman, eds. (New York: Routledge, 1996), 29–31.

White liberally used military metaphors to describe "the struggle between Science and Dogmatic Theology."[13] White's focus on "dogmatic theology" rather than the Christian religion *per se* was founded on his differentiation between theology which makes dogmatic statements about the world and treats the Bible as a scientific text, and religion which recognizes "a Power in the universe" and, thereby, fosters science.[14] White did not apply his many metaphors of warfare to religion, including Christianity, but to interference with science by Christian theologians whom he describes as foolishly indulging

in exhortations to "root out the wicked heart of unbelief," in denunciation of "science falsely so called," and in frantic declarations that "the Bible is true"—by which they meant that the limited understanding of it which they happened to inherit is true.[15]

For White, it is narrow-minded theologians who have caused the war that resulted in direct harm both to religion and to science.

One of the persons whom he highlights as engaging in such behavior in response to Darwin's theory is Bishop Samuel Wilberforce who encountered Darwin's self-appointed "bull dog," Thomas H. Huxley, at a meeting of the British Association for the Advancement of Science at Oxford in 1860. Is White correct in his highly negative assessment of Wilberforce? He provides no documentation for his account of the debate, and ignores Huxley's own admission that his motivation for engaging Wilberforce was an unabashed opposition to "the ecclesiastical spirit, that clericalism which in England, as everywhere else... is the deadly enemy of science."[16]

The encounter of Huxley and Wilberforce is often cited as a paradigmatic symbol for the conflict between Christianity and Darwinian evolution and of Christianity's defeat at the hand of science. However, this depiction is based on what is quite possibly an apocryphal story that portrayed Wilberforce asking Huxley whether he traced his monkey ancestry through his grandfather or grandmother. Huxley supposedly responded that he preferred to be a descendent of a humble monkey rather than of a man who employs his knowledge in misrepresenting those who have given their lives to the pursuit of truth.[17]

From a late twentieth-century North American perspective, particularly one colored by the warfare rhetoric of White and of creation scientists who occupy the opposite end of the spectrum, it is easy to stereotype Bishop Wilberforce as a "Bible thumping" fundamentalist who opposed scientific investigation into God's creation

---

[13] White, *A History of the Warfare of Science and Theology in Christendom*, ix.

[14] Ibid., xii.

[15] Ibid., 48.

[16] Thomas Henry Huxley, *Autobiography*, 1889, Reprint and edited by Gavin de Beer (London: Pitman Press, 1974), 109.

[17] Unfortunately, no detailed contemporary account of the affair exists. The descriptions of the meeting that are usually cited are either by T. H. Huxley's son, written forty years later, see Leonard Huxley, *Life and Letters of Thomas Henry Huxley*, vol. 1 (London: Oxford University Press, 1900), 179–89, or Isabella Sidgwick, "A Grandmother's Tales," *Macmillan's Magazine*, 78 (1898): 433–34. Both have been critically examined by J. R. Lucas, "Wilberforce and Huxley: A Legendary Encounter," *The Historical Journal*, 22 (1979): 313–30. His analysis of these sources as well as the proceedings of the British Association published in *The Athenaeum*, 1703–7, led him to conclude that the data attributing a masterful defeat of Wilberforce by Huxley to be extremely thin. The accounts written at the turn of the century tell the reader more about the currents of thought at that time than what actually happened in Oxford on June 30, 1860.

on biblical grounds. This stereotype misses the mark, however, for several reasons. The obvious one is the fact that Wilberforce antedates the advent of fundamentalist Protestantism.[18] The more important reason, however, is that Wilberforce, although an Anglican bishop, was also a naturalist who specialized in ornithology, or to be more exact, ornithotheology,[19] and served on the Council of the Geological Society.[20] In Wilberforce's review of *The Origin of Species* which appeared in print shortly after the Oxford meeting,[21] the primary basis on which he attacked evolution was not that it opposed biblical revelation, but that it contained scientific errors.[22] He argued against Darwinian evolution on scientific grounds, maintaining that this is the basis on which such arguments should be judged. He further asserts:

> We have no sympathy with those who object to any facts or alleged facts in nature, or to any inference logically deduced from them, because they believe them to contradict what it appears to them is taught by revelation.[23]

Focusing on the weaknesses in Darwin's theory of evolution from the standpoint of the widely accepted scientific methodology of his time, Wilberforce pointed out the absence in the geological record of any convincing data that showed that one species had developed into another. Wilberforce was critical of Darwin, but he also praised him for important contributions to science in *The Origin of Species*, including natural selection which he believed was simply God's plan for weeding out the unfit. However, to prove to Wilberforce that natural selection brought changes that could transform individuals within a species to a new form, Darwin would have to provide far more evidence.

Unwilling to accept evolution, Wilberforce's opposition was rooted both in the lack of sufficient fossil evidence and in his commitment to harmony between science and theology which natural theology strongly affirmed, a harmony which Darwin's theory of biological evolution challenged. The approach adopted by Wilberforce in his critique of *The Origin of Species* reflects a distinction between natural theology and divine revelation that can be traced to another Englishman, Francis Bacon (1561–1626), and his frequently cited division: the "book of God's word" and the "book of God's works." The latter was distinct from the former in that its positions were not dependent on Scripture; it was to include only what was accessible to human reason and observation. The difference between the two books did not constitute opposition, for both were united in one God who was the author of both.

---

[18] Fundamentalism is an interdenominational movement upholding Protestant orthodoxy against the challenge of modernity, especially Darwinian evolution which challenges the inerrancy of the Bible. The movement gets its name from the twelve-volume paperback series, *The Fundamentals*, published between 1910 and 1915.

[19] Herbert James Paton points out that because of English naturalists' concerns about the details of God's purpose in nature, natural theology in the nineteenth century was broken into a series of subordinate disciplines. For example, hydrotheology studied God's purpose in water in all of its forms and ornithotheology studied God's purpose in the great variety of birds; *The Modern Predicament, 1950–51 Gifford Lectures,* 20–21, cited in Bernard L. Jones, *Earnest Enquirers after Truth, A Gifford Anthology* (London: George Allen & Unwin Ltd., 1970), 33.

[20] Standish Meacham, *Lord Bishop, The Life of Samuel Wilberforce* (Cambridge: Harvard University Press, 1970), 212.

[21] Samuel Wilberforce, "[Review of] Darwin's *The Origin of Species,*" *The Quarterly Review* 108 (July 1860): 225–64.

[22] Ibid., 226.

[23] Ibid., 256.

God's sovereignty over the world remained in tact. What the division did provide for those engaged in the study of nature, however, was freedom from making their findings conform rigidly to current interpretations of biblical texts. Instead, their discoveries could aid exegetes in their future interpretations of the Bible.

At the time of his writing *The Origin of Species*, Darwin himself evidently accepted the concept of Bacon's two books. On the frontispiece of *The Origin of Species*, Darwin cites these words from Bacon's *Advancement of Learning*:

> To conclude, therefore, let no man out of a weak conceit of sobriety, or an ill-applied moderation think or maintain, that a man can search too far or be too well studied in the book of God's word, or in the book of God's works; divinity or philosophy; but rather let men endeavor endless progress or proficiency in both.[24]

It may be dangerous to read too much into Darwin's choice of this quote, but its inclusion does not give the impression that *The Origin of Species* was envisioned by Darwin as a polemic against the book of God's word.

In his many books and notebooks, Darwin gives evidence that he was a man well schooled in the books written on God's works. Among these were the texts of William Paley[25] whose *Natural Theology* presented scientific evidence from observations of nature for the existence of a grand design and an intelligent designer. Darwin was attracted to Paley's works because Paley accounted for the adaptation of species to their environments in a clear and methodical way.[26]

In developing the argument for design in the first part of *Natural Theology*, Paley employed an inductive method whereby he drew on observations of the adaptation of organisms to their environments. These observations provided him with a basis for arguing that there was a direct and perfect correspondence between the characteristics of animals and their environments which showed that there was an evident design at work in the creation of each species for its own geographic province. This intricate design pointed to a divine designer. In this regard, Paley exemplifies Bacon's definition of natural theology as that knowledge concerning God which may be obtained by the contemplation of his creatures, which may be truly divine in respect of the object and natural in respect of the light.[27]

Yet, Paley's argument from design detailed not only nature's order, but also nature's purpose. In this regard, Paley broke with Bacon who argued against the intrusion of teleological reasoning into science because "the handling of final causes... intercepted the severe and diligent inquiry of all real and physical causes."[28] Nearly everything Paley wrote in *Natural Theology* is connected by interlocking analogies that bind the evidence drawn from extensive scientific inquires into a single teleological argument which united science and religious faith. Paley carefully

---

[24] Darwin, *The Origin of Species*, n. p.

[25] Darwin notes that while he was at Cambridge his course of study required passing an exam on the works of William Paley. About this he reflects: "I did not at that time [1830–31] trouble myself about Paley's premises; and taking these on trust I was charmed and convinced by the long line of argumentation"; in Charles Darwin, *The Autobiography of Charles Darwin, with Two Appendices*, edited by Francis Darwin (Reprint. London: Watta & Co., 1929), 22. Paley's *Natural Theology* remained on the examination list at Cambridge until 1920; see M. L. Clarke, *Paley: Evidences for the Man* (London: SPCK; Toronto: University of Toronto Press, 1974), 129.

[26] Darwin, *The Autobiography of Charles Darwin*, 22.

[27] Francis Bacon, *The Advancement of Learning*, 1605 (Reprint. London: Oxford University Press, The World's Classics, 1951), 103.

[28] Ibid., 113.

crafted arguments to defeat skeptics who allowed that nature manifested a design, but not necessarily a designer.[29] He provided his readers with evidence for the existence of a designer by rational demonstration: as the complexity of the watch proved the existence of a watchmaker, so the various animal species and parts of the human body, such as the eye, required a designer of supernatural intelligence.[30]

His design argument is developed further in the second part of *Natural Theology* in which Paley demonstrates that the design so evident in nature absolutely requires a designer, and that this is none other than God. Paley's God, however, is not merely a principle of order and purpose operating in nature. Logic requires that that "which can design, must be a person," because design is an act of intelligence and "the seat of the intellect is a person."[31] He further argues that this personal God manifests the attributes which Christianity has traditionally associated with God, namely, omnipotence, omniscience, omnipresence, eternity, unity, and goodness.

To the last of these Paley devotes an entire chapter, in which he deduces from observations of nature a God of incomparable goodness. He writes:

> It is a happy world after all. The air, the earth, the water teem with delighted existence... Species are running about, with an alacrity in their motions, which carries with it every mark of pleasure.[32]

Later he asserts: "the deity has added pleasure to animal sensations, beyond what was necessary for any other purpose."[33] No experience of evil could weaken Paley's depiction of a benevolent God. Evil was but an aberration. Pain, though violent and frequent, was only seldom both violent and of long duration.[34]

---

[29] The refutation of the objections of skeptics is the major goal of the first six chapters of *Natural Theology*. Among them is likely David Hume, who in the *Dialogues on Natural Religion* argued that it was "merely an arbitrary act of the mind" to claim that, because everything within the universe has a cause, the universe itself must have a cause. The reasoning is false, according to Hume because the universe is of a logically different class than its parts; see David Hume, *Dialogues concerning Natural Religion*, 1794, reprint with forward and editing by Norman Kemp Smith (New York: Bobbs-Merrill Company, Inc., 1947), 190. Paley counters this argument by denying that the universe was of a distinct logical class. Appealing to infinite regress, he proposed the analogy of a linked chain. A chain of causes that is finite differs from one that is infinite only in length and not in kind; see Paley, *Natural Theology*, 8.

[30] Paley, *Natural Theology*, 1–2. The book opens with this famous analogy:

Suppose I found a watch upon the ground, and it should be enquired how the watch happened to be in that place.... When we come to inspect the watch, we perceive that its several parts are framed and put together for a purpose... This mechanism being observed, the inference, we think, is inevitable; that the watch must have had a maker, that there must have existed, at some time and at some place or other, an artificer or artificers who formed it for the purpose we find it actually to answer...

This analogy was not unique to Paley; it can be traced to Cicero, who compared the world to a sun dial or water clock and praised the craftsperson who created the incredible numerous and inexhaustibly varied species of nature; see Cicero, *De natura Deorum*, trans., H. Rackman (Cambridge: Harvard University Press, 1956), 217, 219. At the time in which Paley wrote *Natural Theology*, the watchmaker metaphor for God was commonly used by deists. Although Paley was not a deist himself, he did adapt the rationalistic metaphor of deism which depicts God making the world to be a finely tuned machine to account for the design of living creatures.

[31] Ibid., 284.
[32] Ibid., 278.
[33] Ibid., 281.
[34] Ibid., 303.

Paley's apologetic arguments were directed not only to skepticism about the existence of a divine designer, but also to the theory of biological evolution which was emerging when Paley composed *Natural Theology*. Paley, not blind to the threat that theories of evolution, such as that of Erasmus Darwin, Charles' grandfather, posed to his own,[35] assumed that God designed all of the details related to plant and animal species. There was no room for random occurrence in Paley's *Natural Theology*. Against those who argued that the living plants and animals were only so many of the possible varieties, which "the lapse of infinite ages has brought into existence" and that millions of other species perished, "being by the defect of their constitution incapable of preservation," Paley argues: there is "no foundation at all," no empirical evidence whatsoever to substantiate such a theory.[36] Nature manifests an orderly division of plants and animals into genera and species which, by its design, contradicts an arbitrary selection process.

Paley's popular *Natural Theology* provided the inspiration for the *Bridgewater Treatises*, written to further demonstrate the "Power, Wisdom and Goodness of God, as Manifested in the Creation" by drawing on the findings of scientific observation. Among the contributors to this series was William Whewell (1794–1866) whom Darwin also quotes in the frontispiece of *The Origin of Species:*

> But with regard to the material world, we can at least go so far as this—we can perceive that events are brought about not by insulated interpositions of Divine power, exerted in each particular case, but by the establishment of general laws.[37]

What Darwin cites from Whewell, however, has a somewhat ironic significance for what is to follow. The quote is taken from the beginning of Whewell's chapter entitled, "On the Physical Agency of the Deity," in which he argues that the view of the universe proper to science combines itself harmoniously with the doctrines of natural theology.[38] Included among these doctrines are the fixity of nature and the exercise of God's sovereign power in maintaining it.[39] Although Darwin shared with

---

[35] Although it may be accidental that Paley's book appeared shortly after Erasmus Darwin's writings on development, M. J. S. Hodge argues that Paley set about defeating the positions taken by Erasmus Darwin in his *Zoonomia*; see M. J. S. Hodge, "Darwin as a Lifelong Generation Theorist," in *The Darwinian Heritage*, David Kohn, ed., (Princeton, NJ: Princeton University Press, 1985), 211.

[36] Paley, *Natural Theology*, 39.

[37] *The Origin of Species*, n. p. Darwin knew Whewell well; in his *Autobiography* he describes walks taken with him, 66. Yet, this citation from Whewell *(Astronomy and General Physics, Considered with Reference to Natural Theology*, vol. 3, *Bridgewater Treatises* (Philadelphia: Carey, Lea & Blanchard, 1836), 182) is somewhat curious because in a notebook entry written twenty years prior to the publication of *The Origin of Species* Darwin indicates that he found little of merit in Whewell's positions; see "Notebook C, 1838," #72, in Charles Darwin, *Charles Darwin's Notebooks, 1836–44, Geology, Transmutation of Species, Metaphysical Enquiries*, edited and arranged by David Kohn (London: British Museum [of Natural History], 1987), 62.

[38] Whewell, *Astronomy and General Physics*, 182–83.

[39] Ibid., 185–87. Given Whewell's commitment to natural theology and the physical agency of God in nature, it is not surprising that he did not accept Darwin's theory of evolution. Author of several books on the history of science, Whewell was a Kantian idealist who argued that science is interpretation. The goal of inductive logic practiced by scientists was to show that one's science is the right interpretation. As far as Whewell was concerned, Darwin's theory of natural selection was not the right interpretation. He wrote to Darwin, saying that he had not been won over by his theory of evolution by natural selection. "But there is much of thought and of fact in what you have written... it is not to be contradicted

Whewell the conviction that nature's processes operate in accord with general laws, Whewell's ideas on God's agency and the fixity of nature are the very positions that Darwin's theory of natural selection will argue are untenable. Yet, Whewell does indicate that one cannot expect to learn from physical investigation exactly how God acts upon the universe and its many varied components; in this regard, Darwin is in agreement. Darwin also agreed with Whewell that teleological explanations should be excluded from the investigations of the naturalist.[40]

3 Gradual Descent with Modification by Natural Selection

Darwin's intellectual conversion from the natural theology he studied while a seminary student emerged gradually. What he discovered in his five years of travel on the Beagle (1831–1836) led him to raise the question of the origin of the many species no British naturalist had yet investigated. Of particular importance was the data he gathered on the Galápagos Islands, located several hundred miles west of South America. There Darwin found different species on the individual islands that seemed to result from adaptations to the islands' distinct environments. A conflict in interpretations presented itself to him. The variations in finches on islands separated by just a few miles, for example, begged for an explanation; they could not have resulted from an unbroken chain of causality designed from all eternity. Darwin took from Paley's natural theology answers that he had mastered as a student and converted them into questions. Did God purposefully design each and every distinct species, or was some natural mechanism of adaptation at work?[41]

For some twenty years after his voyage, Darwin sought an answer to this question in the data he gathered in his many notebooks. These notebooks provide us with clues into his abandonment of Paley's ideas on the perfect adaptation of species to their environments and the development of his theory of natural selection which accounted for the variation and mutability of species. The formulation of his theory involved the convergence of several ideas. The first was the idea of a gradually evolving world over a long span of time. In this area, Charles Lyell's (1797–1875) *Principles of Geology* (1830–33), which Darwin read while on the Beagle, was a major influence.[42] Of particular importance was the sequence of different species found in the fossil records of successive geological eras.

Lyell disputed both Jean-Baptiste de Lamarck's account of evolving life forms and the catastrophism of natural theologians, among whom was one of Darwin's teachers, Adam Sedgwick. Lamarck speculated that nature produced all the species of animals in a progressive succession beginning with the least perfect and ending with the most perfect in order to create a gradually increasing complexity in their

---

without careful selection of the ground and manner of the dissent." Cited in I. Bernard Cohen, *Revolution in Science* (Cambridge: The Belknap Press of Harvard University Press, 1984), 546.

[40] For more on Whewell and the relationship of his positions to those of Darwin, see Dov Ospovat, *The Development of Darwin's Theory, Natural History, Natural Theology, and Natural Selection, 1838–59* (London: Cambridge University Press, 1981), 12 and passim.

[41] In "Notebook C" #184, written between February and July 1838, Darwin indicates that he wants to study further plant and animal forms in Africa. In this context he raises this question: "Did Creator make all new yet (*sic*) forms like neighboring Continent?" He continues, "my theory [likely referring to his theory of transmutation] explains this, but no other will"; Darwin, *Charles Darwin's Notebooks, 1836–44*, 296.

[42] Darwin, *The Autobiography of Charles Darwin*, 36–37.

organization.[43] Lyell rejected Lamarck's proposal that in the gradual modification of species no organism ever became extinct, because it conflicted with the geological data.

Like the catastrophists, Lyell affirmed that living creatures had become extinct, but he refused to accept the proposal that the former fauna had perished through some catastrophe, such as Noah's flood, and were replaced by more progressive fauna, due to God's direct intervention. Lyell, a uniformitarian who believed that past occurrences must be studied in analogy with the present, proposed that individual species became extinct one by one, rather than entire fauna, as environmental conditions changed. In time, the gaps in nature were filled by the introduction of new species through some supernatural means operative in nature itself.

The question of how these new species originated was left unanswered by Lyell. The specifics of the process of species' origination became the problem that Darwin would address. He shifted the focus from Lyell's concern about survival among species (inter-species competition) to a contest among individual members of a species (intra-species competition).

In forming his theory of natural selection in 1838, Darwin brought together data he had gathered, insights on uniformitarianism and the transmutation of species, observations about animal breeding, and *An Essay on the Principle of Population* by Thomas Malthus (1766–1834), an English political economist.[44]

Darwin noted that breeders were able to make desired changes in their livestock by selectively breeding individuals who possessed a characteristic they wished to enhance.[45] This was "artificial selection." It provided him with evidence from which he was able to deduce a type of "natural selection" at work in nature as a whole.

The major catalyst for the development of the theory of natural selection, however, was Thomas Malthus' "principle of population" which gave Darwin a way to make intelligible the process of species' adaptation. Following Malthus' logic, especially his argument for the exponential increase in population reproduction exceeding the amount of resources, Darwin reasoned that there was an inevitable struggle for existence.[46] To the problem of the origin of species, Darwin had found an answer. Populations of species have a tendency to increase beyond the means of subsistence. Nature provides a mechanism that keeps populations in bounds. In the course of competition among individuals, only those best suited to an environment are able to produce healthy offspring.[47] By inheritance their characteristics

---

[43] Lamarck's *Philosophie Zoologique* 1809, cited in *Evolution, A Book of Readings*, George E. Brosseau, ed., (Dubuque: Wm. C. Brown Co, 1967), 13. Lamarck explained the process of the evolution of species as the inheritance of environmentally induced modifications. His positions had considerable influence on Herbert Spencer's theory of social evolution, published two years prior to Darwin's *The Origin of Species*.

[44] Darwin indicates that he read *The Principle of Population* for amusement on September 28, 1838; see "Notebook D," #135 in Darwin, *Charles Darwin's Notebooks, 1836–44*, 375; idem, *The Autobiography of Charles Darwin*, 57.

[45] Darwin, *The Origin of Species*, chap. 1.

[46] The sentence from Malthus that likely sparked Darwin's theory of natural selection was "It may safely be pronounced, therefore, that population, when unchecked, goes on doubling itself every twenty-five years, or increases in a geometrical ratio;" in *Evolution, A Book of Readings*, Brosseau, ed., 18.

[47] Ernst Mayr argues that "it is primarily the insight that competition is among individuals rather than species that is clearly a Malthusian contribution" to Darwin's theory of Natural Selection: see Ernst Mayr, *One Long Argument, Charles Darwin and the Genesis of Modern Evolutionary Thought* (Cambridge: Harvard University Press, 1991), 85.

predominate in the next generation with the effect of a gradual transmutation of a species. Survival belonged to the fittest. In the struggle for survival among individuals, favorable variations would be preserved and the unfavorable ones lost. In summarizing his position, Darwin states:

> This principle of preservation, or the survival of the fittest, I have called Natural Selection. It leads to the improvement of each creature in relation to its organic and inorganic conditions of life; and consequently, in most cases to what must be regarded as an advance in organization.[48]

This principle of preservation Darwin referred to as "descent with modification," a term which he preferred over evolution.[49] Darwin recognized that many objections against the theory of modification through variation and natural selection were likely to be raised. He reflects:

> Nothing at first can appear more difficult to believe than the more complex organs and instincts have been perfected not by means superior to, though analogous with human reason, but by the accumulation of innumerable slight variations, each good for the individual possessor.[50]

Darwin points out that a person can choose to believe that a divine creator purposefully constructed all the animals and plants "on a uniform plan, but that this is not a scientific explanation."[51] Paley's depiction of a designer God is dismantled by Darwin who challenged one of Paley's major arguments in *Natural Theology*: the complexity of the eye. Although he never mentions Paley by name, Darwin summarizes Paley's analogy which can be expressed as a simile: as the telescope is to the eye, so is the human intellect to the divine designer. He tersely raises this question: "Have we any right to assume that the Creator works by intellectual powers like those of man?"[52] Darwin's response to his own question is in the form of an observation that illustrates natural selection. During the process of millions of years of a countless number of species a living optical instrument developed that is far superior to any human-made telescope. He believes that the only way in which his natural selection argument could be refuted is, "if it could be demonstrated that any complex organ... could not possibly have been formed by numerous, successive slight modifications..."[53] Darwin was convinced that the theory of natural selection explains functional adaptation without recourse to an external designer. Thus, Darwin uses the same evidence that Paley had used to defeat one of Paley's central arguments.

Later in his *Autobiography* Darwin's criticism of Paley's conception of the designer is more pointed:

> The old argument from design in nature as given by Paley, which once seemed to be so conclusive, fails now that the law of natural selection has been discovered. We can no longer argue that, for instance, the beautiful bivalve shell must have been made by an intelligent being. There seems to be no more design in the variability of organic beings and in the action of natural selection, than in the course which the wind blows.[54]

---

[48] Darwin, *The Origin of Species*, 128.
[49] Ibid., see in particular chaps. V and VI.
[50] Ibid., 426.
[51] Ibid., 404.
[52] Ibid., 171.
[53] Ibid.
[54] Darwin, *The Autobiography of Charles Darwin*, 145.

For Darwin, adaptation is the *sine qua non* of survival of the fittest without design being the *sine qua non* of adaptation. Nor, for that matter, is adaptation due to the external agency of a designer.

Darwin's theory of natural selection disputed not only the concept of God as designer, but also two other core doctrines of natural theology: the direct creation and immutability of species, and the benevolence of God. Darwin points out that descent with modification makes it impossible to draw a line of demarcation between species, commonly supposed to have been produced by special acts of creation. Convinced of the gradual transmutation of species by adaptation, Darwin also dismissed natural theology's belief in a God who directly and deliberately created each species. In commenting on "the ordinary view of each species having been independently created,"[55] Darwin argues that, although the geological record is imperfect, "the facts which the record does give strongly support descent with modification."[56] He further argues:

> To my mind it accords better with what we know by the laws impressed on matter by the Creator, that the production and extinction of the past and present inhabitants of the world should have been due to secondary causes, like those determining the birth and death of the individual. When I view all beings not as special creation, but as the lineal descendants of some few beings which lived long before the first bed of the Cambrian system was deposited, they seem to me to become ennobled.[57]

The secondary causality about which he writes is carried out, not in the mode of catastrophism, as so many natural theologians claimed, but by a uniformitarian process of speciation by gradual descent with modification.

Darwin's difficulties with natural theology were not limited to its depictions of God as designer and of the special creation of each species; they extend also to the portrayal of a benevolent God who displays unbounded goodness in nature. For Darwin, evil in nature was no aberration; there was abundant evidence for waste, pain, famine, and death in the struggle for survival.[58] Although Darwin recognized that good ends sometimes resulted from suffering in nature, he took exception to the optimistic caricature of nature that Paley used to describe nature's processes. While natural theologians enthusiastically sought to highlight the perfection of God which they discerned from nature's harmony and perfect adaptations, they also failed to describe nature as it is.

In sum, the three major areas of Darwin's revolution prompted by his theory of natural selection were: 1) design and the accompanying questions of teleology in nature, and of God as the designer, 2) "special creation" of immutable species, and 3) ontic evil and its companion problem of theodicy. These areas, to a greater or lesser degree, continue to be concerns for theologians as we approach the dawn of the twenty-first century.[59]

4 *Darwin's Metaphors*

An important vehicle whereby Darwin's revolution in science was facilitated was the metaphors he chose to explicate his theory. In science, the metaphorical redescrip-

---

[55] Darwin, *The Origin of Species*, 437.

[56] Ibid., 438.

[57] Ibid., 449.

[58] Ibid., chap. III.

[59] For examples of responses to these concerns in this volume, see the contributions by Robert Russell and Thomas Tracy.

tion of reality is central to any theoretical revolution which results in a major paradigm change. Metaphors are figures of speech that enable a person to speak about one thing in terms which are suggestive of another in a new way that enables the person to express adequately what can be expressed in no other way.[60] As Paul Ricoeur points out, this new way is a redescription of reality that facilitates the emergence of meaning in ways impossible without this redescription.[61]

Metaphors are not merely poetic ornamentation. A metaphor is a type of comparison in which there is an inherent tension between what might be questionable or even absurd on a literal level and the reality to which it points. Sallie McFague emphasizes that a metaphor finds similarity in the midst of dissimilars and therefore not only says "is not" but also "is," not only "no" but also "yes."[62]

In *The Origin of Species*, Darwin used metaphors to disclose new insights into the data that indicated to him that "species had not been independently created, but had descended, like varieties, from other species."[63] These metaphors had to be carefully crafted by Darwin in order for him to show how the innumerable species inhabiting Earth had been modified and are the descendants of other now extinct species. In the metaphors that Darwin used, he adapted ideas and insights from other contexts and applied them in ways that radically recast commonly held perceptions of nature's many life forms, including the human species.

When Darwin's *The Origin of Species* was first read by his contemporaries many experienced a conflict of meaning in the interaction of the terms that constituted his metaphors. Since the seventeenth century the educated persons of Great Britain had accepted a compromise between the naturalistic and theistic viewpoints that reserved mechanical laws for physics, chemistry, and geology on the one hand, and divine purposes for biology on the other. No natural laws could explain living organisms, including and especially the human species. Darwin's explanation of the origin of species by natural selection shifted the lines of demarcation among the sciences and inserted life forms into the thought-world of impersonal natural laws. This is a major reason why Darwin rightly believed his theory to be revolutionary.

To limn further Darwin's revolution, two key metaphors, "the origin of species" and "natural selection," will be examined. At first glance, the expression, "the origin of species," may not appear to be a revolutionary metaphor to today's reader. But from the standpoint of the theological state of the biological sciences[64] many of the first readers of *The Origin of Species* expected Darwin to give attention to God, the ultimate origin and designer of living beings. Many of his former teachers did not waste any time condemning his theory. His Cambridge mentor Adam Sedgwick expressed his absolute sorrow. Darwin had replaced the theological reasoning proper to true method of induction, as explained by Whewell, with the mechanism of natural selection.[65] Even Darwin's closest friends and supporters, such as Lyell,

---

[60] For this helpful definition of "metaphor" I am indebted to Janet Martin Soskice, *Metaphors and Religious Language* (Oxford: Clarendon Press, 1985), 15, 31 and passim.

[61] Paul Ricoeur, *The Rule of Metaphor*, trans. Robert Czerny with Kathleen McLaughlin and John Costello (Toronto: University of Toronto Press, 1977), 7.

[62] Sallie McFague, *Metaphorical Theology, Models of God in Religious Language* (Philadelphia: Fortress Press, 1983), 19.

[63] Darwin, *The Origin of Species*, 28.

[64] Darwin recognized zoology's theological status, see "Notebook N," October 2, 1838–April 28, 1840, in Darwin, *Charles Darwin's Notebooks, 1836–44*, 566.

[65] David J. Depew and Bruce H. Weber, *Darwinism Evolving, Systems Dynamics and*

found his ideas difficult to accept because they believed that affirmation of the divine origin of species was integral to the work of a scientist.

To diffuse the tension surrounding Darwin's unorthodox position, Asa Gray, a Harvard University professor, compiled some of his own articles on Darwin's work that had appeared in the *Atlantic Monthly* and gave the monograph the unwieldy title, "Natural Selection Not Inconsistent with Natural Theology."[66]

At no point does Darwin undertake to clarify the issue of *the origin* of species. Rather, given his observations of a variety of plant and animal species, Darwin responds to the questions: How did these species become as they are? How might these species have changed over time through the process of adaptation? The shortened form of the title of his book, therefore functions as a metaphor with an "is," "is not" character to it. It points to the emergence of new species, but does not address the question of their ultimate beginnings. This question is treated tangentially, however, in references to nature as creation in *The Origin of Species* and in the book's concluding sentences in which he alludes to Genesis 2: 7,

> There is grandeur in this view of life (that is, evolution by natural selection) with its several powers, having been *originally breathed by the Creator into new forms or into one* [emphasis mine]; and that whilst this planet has gone cycling on according to the fixed laws of gravity, from so simple a beginning endless forms most beautiful and most wonderful have been, and are being evolved.[67]

Although God is not given a central focus in *The Origin of Species*, Darwin does not explicitly reject God, as the following statement makes clear.

> I see no good reasons why the views given in this volume should shock the religious feelings of any one.... A celebrated author and divine (that is, minister) has written to me that "he has gradually learnt to see that it is just as noble a conception of the Deity to believe that He created a few original forms of self-development into other and needful forms, as to believe that he required a fresh act of creation to supply the voids caused by the action of His laws."[68]

Elsewhere, Darwin indicated that he considered the theory of evolution to be compatible with belief in God, but that one must remember that different persons have different definitions of what they mean by God.[69] What was clear to him was that Paley's conception of God must be rejected.[70]

Even many of those whose approach to science was quite different from the natural theologies of Paley and the authors of the *Bridgewater Treatises* could not accept Darwin's treatment of origins. Although uniformitarians of Darwin's time, such as Lyell, were at odds with the catastrophism of many natural theologians, they nevertheless shared with natural theologians the belief that the work of the scientist was to give witness to a divine plan for nature. Science was not to detract from the honor of the divine author and governor of the world. Darwin's volume shocked the religious sensibilities of Lyell, who argued that although species may have been created in succession at such times and at such places as to enable them to multiply

---

the Genealogy of Natural Selection (Cambridge: MIT Press, 1995), 142.

[66] Adrian Desmond and James Moore, *Darwin, the Life of a Tormented Evolutionist* (New York: W. W. Norton and Co., 1991), 502.

[67] Darwin, *The Origin of Species*, 450.

[68] Ibid., 443. The person to which Darwin is referring is the Rev. Charles Kingsley.

[69] Francis Darwin cites his father in Appendix II of *The Autobiography of Charles Darwin*, 142.

[70] Ibid., 145.

and endure for an appointed period, and to occupy an appointed space on the globe, this was still in keeping with a pre-existing divine plan.[71]

Darwin's theory of evolution, therefore, transformed the understanding of the origin of nature's life forms and removed the question of the ultimate origin from the study of natural history. About such questions Darwin reflects:

> I cannot pretend to throw the least light on such abstruse problems. The mystery of the beginnings of all things is insoluble by us; and I for one must be content to be an agnostic.[72]

The new meanings mediated through Darwin's conception of origins, however, precipitated only a gradual shift in horizons in his own thinking. There is considerable ambiguity about God and God's discernability in nature in Darwin's writings. Only three years prior to declaring himself an agnostic, Darwin wrote that it was impossible for him to conceive "that this grand and wondrous universe with our conscious selves, arose through chance;" this "seems to me to be the chief argument for the existence of God..."[73]

In company with the gradual transformation of the understanding of origins, the understanding of species was also transformed. No longer were species discrete kinds identified as collections of organisms whose descendants necessarily belonged to the same species as their ancestors as natural theologians taught. Rather, species have a fluidity about them. They possess an inherent capacity to change, to become transmuted into new species through adaptation to changing environments, and even to multiply into several species.

Darwin's notion of species and their origin by descent through modification, cannot be dissociated from "natural selection." The full title of his monograph was *On the Origin of Species by Means of Natural Selection*. Although Darwin stressed that natural selection was not the exclusive means of modification,[74] it is the most important metaphor in *The Origin of Species*. The combination of terms in this metaphor produces new meaning by means of an inherent conflict in the literal meanings of the two terms. In "natural selection" Darwin brought together terms from two distinct epistemic fields into a metaphor that is capable of both engendering and organizing a network of emergent meaning: animal breeding, a humanly directed selective process, and nature in the wild, as observed by naturalists.

During Darwin's time, animal breeding and the study of natural history were separate fields with very little interaction.[75] In the metaphor "natural selection" Darwin articulates a phenomenon in nature analogous to the selection done by animal breeders to produce certain desired results.[76] Darwin reflects:

---

[71] Charles Lyell, *Principles of Geology*, vol. 3 (London: J. M. Dent, 1833), 99–100.

[72] Darwin, *The Autobiography of Charles Darwin*, 149. Francis Darwin indicates that Charles Darwin wrote this in 1876. James Moore argues that Charles Darwin's agnosticism may have had more to do with the death of his beloved daughter Annie than with his scientific positions; see James Moore, *The Darwin Legend* (Grand Rapids, Mich.: Baker Books, 1994), 61.

[73] Ibid., 142. Francis Darwin indicates that his father wrote these words in response to a letter from a Dutch student in 1873.

[74] Darwin, *The Origin of Species*, 442.

[75] James Secord, "Darwin and the Breeders: A Social History," in Kohn ed., *The Darwinian Heritage*, 519–42.

[76] Darwin, *The Origin of Species*, 74.

As man can produce, and certainly has produced, a great result by his methodical and unconscious means of selection, what may not natural selection effect?[77]

The tension in this metaphor lies in its depiction of selection in nature as if it were by choice in the same way that breeders choose. There may be unconscious elements in the selecting process as carried out by animal breeders in Darwin's era—a formal study of genetics played no role in animal breeding. But breeders did know how to purposefully change livestock to achieve desired results. The unpredictable elements of natural selection may be analogous to "unconscious means," but they can scarcely be compared to the exercise of methodological choice without inserting a purposeful vitalism into the content of natural selection that Darwin's observations do not substantiate, and which he rejected in the positions of Lamarck. Yet, natural selection is a metaphor with great explanatory power for what it affirms: there is a mechanism in nature that regulates species evolution. This metaphor, therefore, serves as a heuristic aid for discovering what could not be directly observed.

The "is" and "is not" tension in "natural selection" is made evident in Ernst Mayr's critique of Darwin's choice of this metaphor. To say that nature somehow selects species' evolution suggests purposeful agency in nature. Mayr writes:

> This, of course, is not what natural selection does. The term simply refers to the fact that only a few of all the offspring of a set of parents survive long enough to reproduce.[78]

It may very well be the case that there is no selective force in nature, nor a selecting agent, but this is not clear from how Darwin writes about this metaphor. Darwin's use of the term "selection" connotes purpose and implies teleology in nature. The connotations of agency and teleology are intensified when one considers the quasi-divine status that Darwin conferred on natural selection.

In response to nameless critics, Darwin notes that some claimed that he attributed to natural selection "an active power or Deity." He points out that his critics ignore the metaphorical character of his speech and claims that it is difficult to avoid personifying the word "nature." He indicates that by nature he means only the aggregate action and product of many natural laws.[79] But in the light of Darwin's recurring linguistic patterns, the question, "Does he deify natural selection?" is not sufficiently addressed by him in this perfunctory response. As Adrian Despond and James Moore point out in their biography of Darwin, even some of Darwin's old friends were reluctant to accept his theory. Selection implied a selector.[80] They wondered why Darwin did not see the divine properties of natural selection.

Given the many revised editions of *The Origin of Species*, it is puzzling that Darwin did not rewrite the troublesome passages on natural selection. Darwin's descriptions of natural selection as "acting" remain. His personification of natural selection with divine-like qualities is evident in many passages, particularly in this one:

> Metaphorically speaking, natural selection is daily and hourly scrutinizing, throughout the world, the slightest variations; rejecting those that are bad, preserving and adding up all that are good; silently and insensibly working, *whenever and wherever*

---

[77] Ibid., 90.

[78] Mayr, *One Long Argument*, 86.

[79] Ibid., 88.

[80] Desmond and Moore, *Darwin, the Life of a Tormented Evolutionist*, 492.

opportunity offers, at the improvement of each organic being in relation to its organic and inorganic conditions of life.[81]

The same pattern is found in the final chapter:

> What limit can be put to this power [natural selection], acting during long ages and rigidly scrutinizing the whole constitution, structure and habits of each creature—favoring the good and rejecting the bad?[82]

A power without limit which favors the good and rejects the bad does seem to have attributes that Christians have traditionally ascribed to God. Further, natural selection preserving, silently and insensibly working, appears to exhibit divine qualities. The difference between Darwin's deified natural selection and the God of Paley and Whewell is that natural selection governs the world from within nature's selective processes rather than from a locus of transcendent agency.

Darwin's retention of speech patterns in which nature and natural selection are given a quasi-divine status in the six editions of *The Origin of Species* underscores Darwin's depiction of natural selection as a self-regulating mechanism for species evolution. Mechanistic interpretations of nature's processes antedate Darwin, but are given a new perspective by him. This perspective is reflective of the marked increase in self-regulating machines in nineteenth-century industry and in the conception of society as self-regulatory, made prominent in Adam Smith's laissez-faire economics. The relevance of the latter for understanding Darwin's interpretation of the data he collected is not surprising, if one considers that the discipline of political economy enjoyed a much higher status in the mid 1800s than we are inclined to grant it today.

Michael Polanyi has astutely pointed out that the metaphors chosen by scientists for their theories do not mirror nature. Rather, scientific theories are statements of probability which utilize metaphors that are intellectually satisfying to the participants of a common culture.[83] One perspective that did seem to be intellectually satisfying to many of Darwin's contemporaries was Malthus' "principle of population," a major influence on Darwin in the formation of the theory of natural selection, and on Wallace as well. In Malthus' "principle" Darwin and Wallace found a mechanism that would account for nature's process of population change. Malthus' principle, intended as a rationale for resolving the problem of insuring the economic soundness of England at a time in which the number of poor people was escalating, called for limiting the number of births of the less fit lower class. Malthus argued that generosity to the poor on the part of the rich allowed the unfit to reproduce, indeed to produce children at a greater rate than the more fit upper class which showed more moral restraint. Without definitive measures by the upper class, there would be dire consequences.

Malthus' ideas contrasted with those of his contemporary, William Paley. As John Durant points out, for Malthus human and nonhuman nature were governed by providential laws that set limits to what could be achieved and awarded Christian virtue with worldly success, while in Paley's schema, human and nonhuman nature demonstrated the wisdom, justice, and benevolence of the creator.[84] In Paley's

---

[81] Ibid., 90. [emphasis his] In the first edition of *The Origin of Species*, 1859, the qualifying term: "metaphorically" did not appear.

[82] Ibid., 434.

[83] Michael Polanyi, *Personal Knowledge, Towards a Post-Critical Philosophy* (Chicago: University of Chicago Press, 1958), chap. 2 and passim.

[84] John Durant, *Darwinism and Divinity, Essays on Evolution and Religious Belief* (Oxford: Basil Blackwell, 1985), 14.

chapter in *Natural Theology* on "The Goodness of the Deity," he gives evidence of familiarity with Malthus' principle of population, but casts it in a more optimistic light than Malthus by arguing that the systems of superfecundity and destruction are part of nature's economy as perfectly designed by God.[85] In *The Essay on the Principle of Population*, Malthus argued against the thesis of perfectibility of nature (proposed by Condorcet and Godwin), pointing out that nature's harmony is far from perfect: by a struggle for life nature regulates the multiplication of living beings and both assures the survival of the fittest and the elimination of the unfit.

It is not possible, however, to know the full extent of Malthus' influence on Darwin's thinking.[86] By the same token, how the class struggle in Great Britain affected Darwin in his formulation of his theory of natural selection cannot be explained. What can be surmised is that his theory is not unaffected by prevailing nineteenth-century British political ideology.

What is clear is that Darwin's choice of metaphors has profoundly affected the course of the history of evolutionary biology and the relationship of science to theology. The meaning that has emerged in front of his chosen metaphors disclosed a vision of Earth's life forms, including *Homo sapiens*, and raises a question: Who is God, if God is not the designing watchmaker?

5 *A Proposal for a Metaphorical Naming of God Consonant with Darwin's Revolution*

Darwin's prediction that his theory of natural selection would result in a revolution has been fulfilled. Darwin's revolution was one among many in a century of revolutions. *The Origin of Species* was published a little over a decade after a wave of political revolutions swept over Europe. This was also a time in Great Britain and in the rest of the Western world, in which industry and the economy were undergoing sweeping changes that were profoundly transforming the life of humans, and of nonhuman nature as well. In addition, authors such as Mill and Whewell, whose works on science Darwin read, used the term "revolution" in reference to the new developments in the sciences.[87] The use of "revolution" by Darwin in reference to his own work, therefore, is not surprising given his historical context.

As we have seen, Darwin's own revolution was not a violent upheaval in the sense of a war directed against Christianity, as it has sometimes been stereotyped. His revolution was a personal intellectual conversion from the natural theology in which he had been educated to a science that he believed better reflected what he knew of the laws impressed on matter that resulted in both the production and extinction of an enormous variety of species.

In *The Origin of Species*, Darwin rejected neither the authority of the Bible, nor the doctrine of creation. What he did reject was the union of a highly rationalistic Christian theism and a limited body of scientific observations exhibited in the writings of natural theologians. When evidence was discovered to contradict the immutability of species and the perfection of adaptation affirmed by the natural theology of Paley and authors of the *Bridgewater Treatises*, the concept of God they

---

[85] Paley, *Natural Theology*, 289–93.

[86] For an analysis of the influence of Malthus on Darwin see Depew and Weber, *Darwinism Evolving*, 113–39.

[87] See I. Bernard Cohen, "Darwin's Awareness of Revolution," supplement 19.1, in *Revolution in Science* (Cambridge, Mass.: The Belknap Press of Harvard University Press, 1984), 546.

affirmed had to be abandoned also. Darwin's revolution was the overthrow of a specific type of "religious science" and its replacement by a science committed to consistent scientific positions, not limited by specific theistic doctrines and not accountable to Christian institutional authorities.

"Natural selection" as a metaphor used by Darwin to make the common descent of species with modification intelligible does not rule out the affirmation that nature is the creation. Contrary to the claim of Julian Huxley, who argued on the occasion of the centenary celebration of *The Origin of Species*, that "Darwin removed the whole idea of God as the creator of organisms from the sphere of rational discussion,"[88] this is not what Darwin did. What he did was to demonstrate the unintelligibility of Paley's highly artificial watchmaking analogy and the accompanying watchmaker and designer metaphors for God that envisioned God creating countless numbers of discrete species perfectly adapted to their unchanging environments at the beginning of time. Darwin rendered Paley's God, a God who exercised radical sovereignty over a passive and static world-machine, untenable.

Paley's divine designer, the great watchmaker, serves as a reminder that in more ways than is possible to account for, God has been understood or misunderstood by different human beings, in different places, and at different times. Paley's metaphors for God contributed to misunderstanding and became mere period pieces that inhibited rather than facilitated the pursuit of scientific and theological truth and meaning. Paley's unfortunate over-emphasis on God as designer also points out the dangers inherent in God-language when it claims to affirm too much.

In the writings of Thomas Aquinas (1225–74) one is reminded of the inherent limitations of language for God. In the *Summa Theologica* he wrote:

> God can be named by us from creatures, yet not so that the name which signifies Him expresses the divine essence in itself...[89]

Effects in nature point to God's existence and suggest metaphors for God, but do not define God. In any naming of God there is a tension between affirmation and negation. No name for God tells us who God is in God's self; every metaphor falls short of a full representation of the divine. The process of naming God, even in its affirmative, cataphatic moment inexorably leads to an apophatic moment of silent wonder, because God exceeds anything we might say. In the use of any metaphors for God the apophatic dimension must always be affirmed because, as Janet Soskice points out, "we are most in danger of theological travesty when we forget that this is so."[90] When we claim to know too much about God we risk idolatry.

Although Darwin's theory of natural selection does lend itself to the idea that Earth's life forms make themselves without transcendent agency, there is nothing in the theory that necessarily rules out God's existence, as Darwin himself recognized. What is at stake is the definition, or more correctly, since God as ineffable mystery can never be defined, the appropriate metaphors for God that are both consonant with accepted scientific theory and rooted in the biblical tradition.

Sallie McFague has written extensively on models for God that express God's relationship not only to humans but also to the cosmos as a whole. She has proposed

---

[88] Julian Huxley with Sol Tax et al. " 'At Random,' a Television Preview," in *Issues in Evolution*, vol. III of *Evolution After Darwin*, edited by S. Tax and C. Callender (University of Chicago Press, 1960), 45–46.

[89] Thomas Aquinas, *Summa Theologica*, I-I, vol. 1 (London: Blackfriars, 1965), q. 13, a. 1.

[90] Soskice, *Metaphors and Religious Language*, 160.

a model of the universe as God's body as a way of speaking of God that does not depict God as transcendent over the universe in the sense of being apart from its many processes and material forms.[91] My proposal is for a variation of the model of God as embodied in the form of the metaphor of a mother giving birth. This maternal metaphor is proposed on the grounds that it is an apt way of affirming that God is the source and power of life in all living species.

A mother giving birth as an image applied to God is far less familiar to Christians than depictions of God as totally other than creatures. But as Paul Tillich has pointed out, the symbolic dimension of God as "the ground of being" affirms "the mother-quality of giving birth, carrying and embracing…"[92] God as a mother giving birth is reality-depicting without claiming to prove God's existence or to define God in relationship to evolutionary biology in a definitive way. This metaphor is closer to our experience of life as embodied beings than Paley's watchmaker-designer, that distant sovereign who made a mechanistic world by divine fiat.

The maternal metaphor for God has an inherent tension; it brings together two distinct fields of meaning: 1) a reproductive element in nature central to descent with modification and therefore of the evolution of species, and 2) the biblical doctrine of God as the creator. God giving birth evokes the mystery of creative and generative love selecting and encircling new life, making possible its existence and nurturing it in the face of the threat of nonexistence.

This maternal metaphor for God is not an argument for a pantheistic God. The metaphor "God giving birth" does not identify God and the world monistically. Rather it conceives of God's involvement with life engendering evolutionary processes as their inner modification. But this immanence of God in the life of creatures does not eliminate God's transcendence. The Mother God transcends the creatures to which she gives life. What is proposed, therefore is a panentheistic depiction of God in which all living things have their origins in God and nothing exists apart from God's involvement. As Elizabeth Johnson argues, the metaphor of God giving birth is the paradigm without equal for a panentheistic notion of the coinherence of God and the world.[93] To see the world dwelling in God and being continually birthed by God is reflective of the biblical insight that in God, we (and by implication all creatures who share a common evolutionary history) live and move and have our being (cf., Acts 17:28).

These words allude to what is more explicitly expressed in Deuteronomy. Near the end of the book in a song sung after Moses had finished writing the law, words directed to an unfaithful people explain their need for the law: "You were unmindful of the Rock that bore you; you forgot the God who gave you birth" (Deut. 32:18). Mindfulness of the Mother, the rock-like dependable source from whom all of creation receives life is basic to a relationship with God as creator.

In the biblical prophetic literature the maternal image of God giving birth appears in the context of words of consolation. Second Isaiah (40:1–55:13) emphasizes that God is the creator of the entire world. It is God who created the heavens and stretched them out, and who spread out the earth and what comes from

---

[91] Sallie McFague, *The Body of God, An Ecological Theology* (Minneapolis: Fortress Press, 1993), 20.

[92] Paul Tillich. *Systematic Theology*, vol. 3 (Chicago: University of Chicago Press, 1963), 293–94.

[93] Elizabeth A. Johnson, *She Who Is, The Mystery of God in Feminist Theological Discourse* (New York: Crossroad, 1992), 234.

it (Is. 42:5). Isaiah's creation imagery focuses on God as a mother giving birth. Crying out like a woman in labor: gasping, and panting (Is. 42:14), there is anticipation. Life is emerging.

Does this maternal metaphor for God respond to Darwin's problems with 1) God the Designer, 2) the immutability of species, and 3) the suffering inherent in the struggle for existence? The biblical metaphor—God giving birth—expresses, to borrow from the words of McFague, an "immanental transcendence or transcendent immanence."[94] God is not totally other than creation, designing it from afar. God is present within the natural processes of emerging life. God acts in and through the highly complex evolutionary processes that began millions of years ago. McFague's reflection on her preferred panentheistic metaphor, "the world as God's body," is applicable to God giving birth: "this does not mean that God is reduced to the evolutionary process, for God remains as an agent, the self, whose intentions are expressed in the universe."[95]

Where the immutability of species and their special creation are concerned, a selectivity in Earth's life-giving processes is implied in this metaphor. There is risk and unpredictability in childbirth. Bringing new life into the world is a process that is open to novelty. From a religious perspective all of the diverse species that comprise the web of life are creative expressions of natural selection which is ultimately empowered by the divine mother giving birth. Through natural selection, God vivifies self-organizing nature, indwelling each species, quickening and setting free, while weaving bonds of solidarity among all in complex eco-systems.

In this maternal metaphor, suffering and its accompanying vulnerability are acknowledged as part of nature's processes. God is no stranger to suffering and to struggle, the Spirit of God groans with creatures in their travail (Rom 8:22–23). Suffering, though an undeniable reality, is not given the final word. Although God imaged as a woman in labor puts God in the vulnerable position of sharing the pain and suffering of creation, this metaphor also puts God in the position of power in bringing forth new life.

The maternal metaphor God-giving-birth is an appropriate theological response to Darwin's revolution. It has consonance with Darwin's theory of evolution by natural selection while affirming the sacredness of Earth's complex processes, for the God at work in evolution is over all and through all and in all (Eph. 4:6), indwelling and unfolding creation.

Realistically speaking, for Darwin as a mid-nineteenth-century Englishman to have imaged God as a mother giving birth to all the forms of life that fascinated him is beyond the range of possibilities. His intellectual conversion was not accompanied by a major religious conversion to a new understanding of God that emphasized God's immanent involvement in creation rather than transcendent sovereignty over it. But there is nothing to prevent us from embracing a religious conversion that lifts up the metaphor of God giving birth to limn a world that is made intelligible by Darwinian evolution. This proposal for a maternal metaphor for God giving birth is made with the cognizance that no metaphor applied to God is ever fully adequate, even if it does seem to respond to a particular set of critical concerns at a particular juncture in history.

---

[94] McFague, *The Body of God*, 21.

[95] Sallie McFague, *Models of God, Theology for an Ecological, Nuclear Age* (Philadelphia: Fortress Press, 1987), 73.

EVOLUTIONARY NATURALISM AND RELIGION[1]

Willem B. Drees

A meme that occurs in many guises in the world's folklore is the tale of the initially terrifying friend mistaken for an enemy. "Beauty and the Beast" is one of the best-known species of this story. Balancing it is "The Wolf in Sheep's Clothing." Now, which meme do you want to use to express your judgment of Darwinism?

Daniel C. Dennett[2]

1 *Introduction*

Why would anybody use either metaphor to judge Darwinism? I cannot imagine persons evaluating fluid dynamics as a Wolf-in-Sheep's-clothing or as a Beast-Friend-of-Beauty. Evolutionary theory is subject to more intense responses than other parts of science. This intensity stems from consequences assumed, anticipated, and feared—consequences for major aspects of our own self-understanding and for our religious convictions.

There are at least three debates about (real or apparent) implications of an evolutionary view. First, an evolutionary view is a challenge to a literalist understanding of the first chapters of Genesis. In my opinion, this kind of controversy has to be resolved by a better appreciation of the historical nature of religious scriptures; creationism treats the biblical narrative as if it purported to be a scientifically accurate account.[3] Second, in a theistic metaphysical view of the world, God plays some role in the processes of the world.[4] Thus, some Christian thinkers oppose an evolutionary view since it seems to leave insufficient room for divine involvement, at least not of the kind they would like to see. A third area of controversy is about the implications of an evolutionary view for our ideas about human nature and culture. This discussion is of interest to orthodox and liberal religious believers and humanists alike. What would be lost from our moral, social, and religious life if one accepts an evolutionary account?

The overwhelming majority of biologists have accepted an evolutionary view of natural history and of its underlying mechanisms. However, in discussing consequences for our self-image, we go beyond the domain of evolutionary biology

---

[1] This paper draws in many places on Willem B. Drees, *Religion, Science and Naturalism* (Cambridge: Cambridge University Press, 1996). I want to thank the participants in the Vatican–Center for Theology and the Natural Sciences conference of June 1996 for their criticisms, comments, and suggestions.

[2] Daniel C. Dennett, *Darwin's Dangerous Idea: Evolution and the Meanings of Life* (New York: Simon and Schuster, 1995), 521.

[3] For creationism and science, see Philip Kitcher, *Abusing Science: The Case Against Creationism* (Cambridge: MIT Press, 1982); for creationism and religion, see Langdon B. Gilkey, *Creationism on Trial: Evolution and God at Little Rock* (San Francisco: Harper & Row, 1985); for the rise and development of creationism, see Ronald L. Numbers, *The Creationists: The Evolution of Scientific Creation* (New York: A. A. Knopf, 1992).

[4] I here use theism as distinct from deism; the latter would do without divine activity at later stages. For a far more elaborate overview of positions, see Robert J. Russell, "Introduction," in *Chaos and Complexity*, Robert J. Russell, Nancey Murphy, and Arthur R. Peacocke, eds. (Vatican City State: Vatican Observatory & Berkeley: Center for Theology and the Natural Sciences, 1995; distribution by University of Notre Dame Press), 11.

in a restricted sense. One can imagine that various wider perspectives might be developed. Are there frameworks which are consistent with evolutionary biology and other parts of science which grant a major place to teleology? Is process philosophy in the Whiteheadian tradition a good example? In my view an extension which stays as close as possible to insights offered and concepts developed in the sciences is to be preferred; I label this position "naturalism," and explicate and defend it below.

The main question for this paper is what the consequences would be if a naturalist evolutionary view is right. To run ahead of the argument, the view defended is that morality and the richness of experience can be understood without many losses but that the challenges for religion are more serious. The difference arises since a functional and immanent understanding of morality need not be as problematic to moral persons as a similar understanding of religion may be to religious believers. Thus, some fears, especially from religious persons, are to the point. However, I think that the grounds for accepting a naturalist evolutionary view of the world, ourselves included, are strong. Hence, rather than backing away, I prefer to reflect on perspectives for religion within an evolutionary framework. Thus, in this paper I will defend the following three theses.

1. In the area of anthropology and ethics fearful consequences come from a simplistic understanding of the implications of an evolutionary view; upon a more nuanced understanding of what the scientific theory achieves and of how human nature is seen, a naturalist evolutionary perspective can do justice to the richness of experience and to our sense of morality.[5]
2. In the area of religion, similar arguments do not work as well.
3. However, there nonetheless is in a naturalist view room for religion, at least as a way of life and as a response to "limit questions" regarding the framework within which science operates and within which we live and move and have our being.

The order of the paper is as follows. First, I will explicate my version of a naturalist view of the world. At the end of this section I briefly discuss the claim that such a naturalism need not be atheistic and compare the naturalism defended here with other views which seek to accommodate contemporary science, such as process philosophy. Then, we come to the thesis about understanding morality and the richness of experience (sections 3, 4). Section 5 explicates why the same defenses do not work as well for religion (the second thesis) and the final section explores the room for religion upon such an evolutionary-naturalist view (the third thesis).

2 Naturalism

2.1 A Few Characteristics of Science

Science does not discover *the* true story of the world by using *the* scientific method. Studies of actual science have shown that successes have been achieved by violating officially acknowledged methods, that subsequent scientific accounts exhibit substantial discontinuities, and that social relations among scientists and between scientists and the wider community have been important to the development of science. However, this is not a reason to dismiss science. The point is that:

---

[5] Aside of the richness of experience and of ethics, other issues could be considered as well, especially issues in the philosophy of mind (intentionality, consciousness, etc.); it is assumed that they too could be dealt with naturalistically.

Flawed people, working in complex social environments, moved by all kinds of interests, have collectively achieved a vision of parts of nature that is broadly progressive and that rests on arguments meeting standards that have been refined and improved over centuries. Legend does not require burial but metamorphosis.[6]

Let me give a few characteristics of science.

1. Science is *realist* in the sense that it studies a reality which is to a large extent independent of humans, and even more independent of human attempts to find out about reality. However, science is not restricted to phenomena which are independent of humans; superconducting materials made by humans and human mental processes can be studied as well.

2. Such a realism does not carry us very far in debates on *scientific realism*, which are not debates about the existence of "reality out there" but debates about the quality of our knowledge. Which theories, or which elements in our theories, can we take seriously as "depicting" the way reality is, and to what extent? What criteria should we apply when we attempt to answer such questions? Unqualified realism is too much; scientific explanations and concepts are provisional human constructs organizing the natural world; they are not wholly independent of human intellectual capacities, social interactions, and contingencies of history.

Debates about realism sometimes become heated due to conflation of this debate about the quality of our knowledge with the debate about the existence of reality out there. This happens especially when it comes to religious issues, since the consequence of a certain view of our knowledge may be that one has a low regard for the belief that religious terms refer to a particular existent with the characteristics ascribed to God without, however, challenging the existence of reality as such; perhaps the religious terms are understood psychologically or sociologically. To meet the challenge of a non-realist understanding of religious terms, a general defense of realism in the first sense is insufficient.

3. A major characteristic of the sciences is their wide *scope*; their domain seems to be without boundaries. In the course of history, terrestrial physics turned out to be applicable to heavenly phenomena as well, and chemistry can be applied to all processes in living beings. The domain of the sciences extends from the smallest objects to the universe at large, from extremely brief phenomena to the stability of rocks, and from heavy objects to massless light.

4. Correlated with the extension of science is the inner *coherence* of our scientific knowledge, which has proved to be a heuristically fruitful guide in the development of the sciences, and, if temporarily seriously violated, has at least reestablished itself as a result of later scientific developments. Coherence has become a criterion which makes us consider with suspicion purported knowledge which stands in splendid isolation, even if it does not conflict with the rest of our knowledge.

5. Science *enlarges and changes* our view of the known world.[7] In science there is more risk involved than in formal demonstrations (as in mathematics) since theories may postulate entities and concepts of a kind not found in the data; theories are more than generalizations. Whatever we think of the realist status of scientific

---

[6] Philip Kitcher, *The Advancement of Science: Science without Legend, Objectivity without Illusions* (New York: Oxford University Press, 1993), 390.

[7] The expression is inspired by Ernan McMullin, "Enlarging the known world," in *Physics and Our View of the World*, Jan Hilgevoord, ed., (Cambridge: Cambridge University Press, 1994).

theories, they offer us, as Wilfrid Sellars described it, *scientific images* of the world which *differ* from our *manifest images*.[8] This is especially relevant when we consider religion and personal life, since religion and our concept of person (with an inner life, emotions, responsibilities, etc.) are intimately related to manifest images and ordinary practices.

6. Contemporary natural science is *stable and provisional*. It is stable in the sense that many branches of science build upon knowledge acquired in the last few centuries. It seems extremely unlikely that physicists and chemists ever will abandon belief in atoms and in the periodic table arranging the various elements or that biologists will abandon evolution as a view of the natural history of organisms and as a theory explaining this natural history in terms of transmission of properties and differences in reproduction. However, science is also provisional. Not only may we extend our knowledge into new domains, but we can also reach a further understanding of domains already known, and in consequence have to modify our views. Without affecting the periodic table, our understanding of the particles that make up the core of an atom has changed; now they are taken to consist of quarks and gluons. If one probes further, the physics is very speculative, and certainly not as stable as our belief in atoms.

2.2 *Varieties of Naturalism*

The label "naturalism" is used in various ways. One distinction, introduced by Strawson, is between "soft" or "nonreductive" and "hard" or "reductive" naturalism. Upon the "soft" understanding, naturalism refers to what we ordinarily do and believe as humans, say about colors, feelings, and moral judgments. When a painting is considered "naturalist," it is so in this "soft" sense. The "hard" version, according to Strawson, attempts to view human behavior in an "objective," "detached" light as events in nature.[9] This distinction corresponds to some extent with the distinction made above between "manifest" and "scientific" images. I am of the opinion that in light of the successes of science we have to give "hard naturalism" priority over "soft naturalism" in intellectual, cognitive projects; science in many instances on good grounds *corrects* our (soft) "natural" understanding of reality.

Among "hard naturalists" some emphasize the scientific method, whereas others emphasize science as basis for ontological assertions. Clarity is served, I think, if one distinguishes these two versions of naturalism. Let me first describe the epistemological one.

Epistemological naturalism. Arthur Danto defines naturalism in the *Encyclopedia of Philosophy* as "a species of philosophical monism according to which whatever exists or happens is *natural* in the sense of being susceptible to explanation through methods which, although paradigmatically exemplified in the natural sciences, are continuous from domain to domain of objects and events." Such a naturalism is "ontologically neutral in that it does not prescribe what specific kinds of entities there must be in the universe or how many distinct kinds of events we must suppose to take place.... it is a methodological rather than an ontological monism..., a monism leaving them [philosophers] free to be dualists, idealists, materialists, atheists, or non-atheists, as the case may be."[10]

---

[8] Wilfrid Sellars, *Science, Perception and Reality* (London: Routledge and Kegan Paul, 1963).

[9] P. F. Strawson, *Skepticism and Naturalism: Some Varieties* (New York: Colombia University Press, 1985), 95.

[10] Arthur C. Danto, "Naturalism," in *The Encyclopedia of Philosophy*, vol.5, Paul

Such a methodological understanding of naturalism, which is quite common among philosophers, seems to be potentially friendly towards religion by abstaining from ontological claims. However, I see two major disadvantages.[11]

1. If "continuity of methods" is given some discriminating sense, it may exclude too many relevant intellectual enterprises. Methodological or epistemological naturalism does not accommodate "higher," more metaphorically laden forms of discourse, such as that which is characteristic of the humanities and of religious narratives. It shares this problem with an epistemologically reductionist form of ontological naturalism, although an ontological understanding of naturalism need not assume such an epistemological reductionism.

2. If continuity with scientific methods is the main criterion, questions which cannot be answered by these methods will be dismissed as meaningless, whereas if naturalism is understood in ontological terms, there is no ground to dismiss such questions *a priori*. This is especially relevant when we consider questions regarding the framework of existence which is not explained but assumed by the sciences. The methodological naturalist has to dismiss such questions, whereas an ontological naturalist can be more open-minded with respect to such limit-questions.

Ontological varieties of naturalism. Ontological naturalism or materialism comes in varieties.[12] *Reductive materialists* take it that regularities about "higher" phenomena such as mental states correlate with regularities at the level of physical processes (type-type identity). *Nonreductive materialists* reject a strict correlation between physical and psychological properties and regularities, even though each actual mental event is embodied physically (token-token identity). Money exists as precious metals, paper, electronic codes, and shells; it would be neither feasible nor helpful to deal with economic processes in terms of the physical characteristics of money. The taxonomy of a science describing higher level phenomena need not carve up the world in the same way as physics. Besides, the "higher" level may depend on a particular environment (see also Nancey Murphy's contribution on "supervenience" in this volume). Hence, successful reduction of regularities and theories is not to be expected, without thereby denying that the higher phenomena are embodied as material entities and processes.

Eliminative materialists believe that we should eliminate "higher level" notions. For instance, Paul Churchland wrote in 1981 that our "folkpsychology" will fade out of existence once it is replaced by a more adequate neurophysiological vocabulary.[13] This is a gross overstatement. Once one understands how a concept from a higher level of description is understood in terms of the lower level the original term may be superfluous (in the rare case of an exhaustive type-type reduction) but is not thereby dismissed. If the temperature of a gas can be identified with the mean kinetic energy of the molecules, there is nothing wrong with saying that the temperature of the air in my room is currently 291 Kelvin. Reduction is not elimination of

---

Edwards, ed., (New York: Macmillan, 1967), 448.

[11] Methodological naturalism is also criticized in Steven J. Wagner and Richard Warner, eds., *Naturalism: A Critical Appraisal* (Notre Dame, Ind.: Notre Dame Press, 1993). On the first page they define naturalism as "the view that only natural science deserves full and unqualified credence." They trace this use of "naturalism" to W. V. O. Quine, *Word and Object* (Cambridge: MIT Press, 1960).

[12] Paul K. Moser and J. D. Trout, eds. *Contemporary Materialism: A Reader* (London: Routledge, 1995), 5.

[13] Paul Churchland, "Eliminative materialism and the propositional attitudes," *Journal of Philosophy* 78 (1981): 67–90.

phenomena, but rather the opposite.[14] Once genes are understood in terms of DNA there is less reason to doubt their existence, not more. Pain does not become less painful when its physiological basis is understood. However, in many cases reduction comes with modification; pure examples of successful reduction are rare, and even then there is information hidden in bridge laws. Hence, I think that some variety of *nonreductive ontological naturalism* is the most adequate one.

2.3 Major Aspects of Ontological Naturalism

Non-material aspects of reality, such as music, science, and social meanings, are not studied as such by any of the natural sciences, but they are embodied in forms which are in the domain of the natural sciences, whether as ink on paper, sound waves in the air, or neural patterns in a brain, and only as embodied they do seem to be causally efficacious. If we call the domain of the natural sciences the "natural world" (living and non-living, stable and ephemeral, physical objects and embodied mental and cultural entities), the core of *ontological naturalism* can be articulated as follows:

1. The natural world is the whole of reality that we know of and interact with; no supernatural or spiritual realm distinct from the natural world shows up *within* our natural world, not even in the mental life of humans.

Some seek to interpret the impressive *coherence* in our knowledge as an indication of an holist or organic unity of the world. However, the actual coherence of the sciences, which resembles a hierarchy with more fundamental sciences describing the behavior of the constituents of the more complex systems described by "higher" sciences, suggests that the coherence of the sciences arises due to the fact that different entities are constituted from the same basic stuff, say atoms, forces and spatial relations—a *constitutive reductionism*.

2. Our natural world is a unity in the sense that all entities are made up of the same constituents. Physics offers us the best available description of these constituents, and thus of our natural world at its finest level of analysis. This I label *constitutive reductionism*.

Upon this view, every biological, mental, or social change is at the same time a change in the physical state of the system (token-token identity). However, regularities described in processes at higher levels need not correspond to regularities at the physical level. Constitutive reductionism does not necessarily imply conceptual reductionism, or type-type identity, though occasionally there may be successful reductions of this kind. Naturalism need not exclude as meaningless and superfluous concepts which are involved in explanations in sciences other than physics.

This view of irreducibility can also be argued in relation to ideas that have arisen in physics: "The ideas of symmetry breaking, the renormalization group and decoupling suggest a picture of the physical world that is hierarchically structured in quasi-autonomous domains, with the ontology and dynamics of each layer essentially quasi-stable and virtually immune to whatever happens in other layers."[15]

---

[14] This conclusion is argued convincingly in J. Schwartz, "Reduction, elimination, and the mental," *Philosophy of Science* 58 (1991): 203–20.

[15] Silvan S. Schweber, "Physics, community and the crisis in physical theory," *Physics Today* 46 (November 1993): 36. Cf. P. W. Anderson, "More is different: Broken symmetry and the nature of the hierarchical structure of science," *Science* 177 (4047, 4 August 1972): 393–96.

This view could be labeled ontological nonreductionist, in the sense that successful causal explanations at a certain "level" use terms which relate to real causal dependencies and natural kinds at that level. Protons attract electrons. The fact that at a different level of description protons are understood as being constituted of quarks and gluons is no reason to deny the ontological status of protons. Nor does the role of "protons" in certain causal explanations imply that they are fundamental rather than constituted of other particles. Similarly, green plants are green plants, even though at the atomic level all life forms are carbon, phosphorus, hydrogen, and other well-known atoms. Desires and emotions may perhaps be fundamental concepts in a psychological analysis, even when affective and mental phenomena are rooted in physiological processes, which are, "further down," identical with physical processes.[16]

The acceptance of the reality of "higher level" entities may mislead some to forget the physical realities underlying such entities. The conclusion of the relative independence of various sciences is conceptual and explanatory rather than ontological in the constitutive sense. Therefore, I prefer to label the next claim *conceptual and explanatory nonreductionism*:

3. The description and explanation of phenomena may require concepts which do not belong to the vocabulary of fundamental physics, especially if such phenomena involve complex arrangements of constituent particles or extensive interactions with a specific environment.

With respect to living organisms evolutionary biology provides a powerful pattern of explanation which is not primarily in terms of constituents and laws, as in physics, but in terms of interactions between organisms and their environments. Its explanatory schemes are primarily functional: within the constraints due to natural history, traits which contribute to the functioning of an organism (or, more precisely, to the propagation of that trait in its particular environment) are likely to become more abundant than competing traits which are functionally less advantageous.

Biology is more clearly a "higher level" compared to physics than, for instance, chemistry. The reason is that whereas both physics and chemistry classify phenomena primarily in terms of their causal powers and microstructure, biology also classifies by function.[17] Hence, in biology, and beyond biology in psychology and the social sciences, there is a greater variety of types of explanations. One may explain in functional terms what happens, in causal terms how it happens, and in evolutionary terms why the organism is structured so that this behavior can happen.

That traits arose via biological evolution does not imply that every trait must have been optimal in its original context; contingencies of natural history may have determined to a large extent which traits developed. Capacities may have been

---

[16] A way to articulate causality for entities "at higher levels" is offered by Paul Humphreys, "How properties emerge," *Philosophy of Science*, 64 (March, 1997): 1–17, and in a paper by him on "Natural Emergence," presented at a conference on naturalism at Elizabethtown College in July 1997.

[17] Thus, Anne R. Mackor, "The alleged autonomy of psychology and the social sciences," in *Logic and Philosophy of Science in Uppsala*, Dag Prawitz and Dag Westerståhl, eds., (Dordrecht: Kluwer Academic Publishers, 1994), 542, developing arguments from Ruth G. Millikan, *Language, Thought and Other Biological Categories* (Cambridge: MIT Press, 1984); idem, *White Queen Psychology and Other Essays for Alice* (Cambridge: MIT Press, 1993).

deployed for novel tasks; perhaps the ability to read animal tracks endowed us with brain structures which we now deploy for reading texts. An evolutionary view does not imply a particular position in the so-called "nurture versus nature" debates regarding the role of the environment versus the role of the constitution of an organism. We need to avoid too simplistic views of evolutionary explanations if we want credible arguments about the implications of evolution for our view of human nature and culture.

4. Evolutionary biology offers the best available explanations for the emergence of various traits in organisms and ecosystems; such explanations focus on the contribution these traits have made to the inclusive fitness of organisms in which they were present. Thus, the major pattern of evolutionary explanation is functional.

2.4 *Ontological Naturalism Is Not Atheistic*

Reductionistic explanations within a naturalist framework do not explain the framework itself, as a thumbnail sketch of the sciences may illustrate. Concerning the properties of genes a biologist may refer to the biochemist in the next office. When asked "When and where did the ninety-two elements arise?" the chemist can refer to the astrophysicist. The astrophysicist might answer in terms of processes in stars, supernovas, and the early universe, referring for further explanations to the nuclear physicist and the cosmologist. This chain of referring to "the person in the next office" ends, if successful at all, with the cosmologist and the elementary particle physicist, the one on the ultimate historical questions and the other on the most basic structural aspects of reality.[18] Since physicists and cosmologists cannot refer to a "person in the next office" they more easily engage in philosophical and theological speculation than scientists from other disciplines, though not necessarily with greater competence.

5. Fundamental physics and cosmology form a boundary of the natural sciences, where speculative questions with respect to a naturalist view of our world come most explicitly to the forefront. The questions which arise at the speculative boundary I will call *limit questions*.

The questions left at the metaphorical "last desk" are questions about the existence and structure of the world as a whole, and not only questions about its beginning. Such limit questions are persistent, even though the development of science may change the shape of the actual ultimate questions considered at any time. Naturalism does not imply the dismissal of such limit questions as meaningless, nor does it imply one particular answer to such limit questions. Religious views of reality which do not assume that a transcendent realm shows up *within* the natural world but which understand the *natural world as a whole* as a creation which is dependent upon a transcendent Creator—a view which might perhaps be articulated with the help of a distinction between primary and secondary causality, or between temporal processes in the world and timeless dependence of the world on God—are consistent with such a materialism.

---

[18] Charles W. Misner, "Cosmology and theology," in *Cosmology, History, and Theology*, Wolfgang Yourgrau and Allen D. Breck, eds., (New York: Plenum Press, 1977), 97; on this image, cf. Steven Weinberg, *Dreams of a Final Theory* (New York: Pantheon Books, 1992), 242.

To make the argument differently, we may distinguish between four views of God's relation, if any, to natural reality and its regularities.

1. *God against natural regularities*. When God acts, God can do so against any laws of nature. It is like shifting from the automatic pilot to manual control; whereas on the basis of natural processes one would expect A to happen, God makes event B (non-A) to happen. A disadvantage of such a view is, in my opinion, that it adversely affects our esteem for God's creation.

2. *God in the contingencies of natural processes*. One could also argue that God need not interfere in a way which suspends laws of nature, since there is enough looseness (contingency) in the web God created in the first place. This looseness might perhaps be found in complex and chaotic systems[19] or at the level of quantum physics.[20] Upon such a view natural processes could result in different outcomes, say A, B, C, and D, and God makes C rather than any of the other outcomes happen. This view depends on a proper role of an ontological kind of contingency *in* nature, whether at the quantum level or elsewhere.

3. *God as the creator of a world with integrity*. Naturalism need not deny the existence of such contingencies *in* nature as emphasized in (2); perhaps natural reality is to some extent hazy and under-determined. However, it abstains from supplementing natural reality with additional (supra-natural) determining factors. Chance can be understood as chance, not as hidden determination. Naturalism accepts that nature is, when considered at the level of causal interactions, complete, without religiously relevant holes. The natural world has an integrity which need not be supplemented within its web of interactions. However, this integrity is not to be confused with self-sufficiency; it does not imply that the natural world owes its existence to itself or is self-explanatory. If a religious believer accepts naturalism as integrity, it is still possible to see God as the creator of this framework, the ground of its existence. This is best understood, in my opinion, as a non-temporal notion, rather than as a creator who started it all a long time ago.

4. *A self-sufficient world without God*. Naturalism can also be understood differently, not as emphasizing *integrity* but as claiming the *self-sufficiency* of the natural world, without any role for God as creator or ground of natural reality. However, an argument about self-sufficiency has to be quite different from any argument about explanations within the natural world, since here we have to do with the contingency *of* existence rather than contingency *in* existence; the first is best modeled in terms of logic, whereas the second has to do with causality (and hence with time). This is a difference that a polemical atheist like Peter Atkins slid over when he claimed in 1981 that science is about to explain everything.[21] He can trace back everything to a beginning of utmost simplicity, but he cannot do without assuming existence, without assuming a framework where certain rules apply, where mathematics applies, etc.

A naturalist need not assume (4) the self-sufficiency of the framework when seeing (3) the framework itself as a whole which has integrity.

---

[19] One advocate is John C. Polkinghorne, *Reason and Reality: The Relationship Between Science and Theology* (London: SPCK Press, 1991); see also the discussions in Russell, Murphy, and Peacocke, eds., *Chaos and Complexity*.

[20] A recent advocate is Robert J. Russell, "Theistic evolution: Does God really act in nature?" *Center for Theology and the Natural Sciences Bulletin* 15.1 (1995): 19–32; see also his contribution in this volume.

[21] Peter. W. Atkins, *The Creation* (Oxford: Freeman, 1981).

2.5 *Why Not a "Richer Naturalism"?*

Naturalism as presented here is a metaphysical position. It goes beyond the details of insights offered by the sciences in an attempt to present a general view of the reality in which we live and of which we are a part. However, it is a "low level" metaphysics in that it stays close to the insights offered and concepts developed in the sciences rather than imposing certain metaphysical categories on the sciences or requiring a modification of science so that it may fit a metaphysical position taken *a priori*.

Hermeneutical approaches that are at odds with an epistemological variant of naturalism may concur with the five aspects mentioned. However, there are views which are at odds with this naturalism. By emphasizing the integrity of the natural world, this view excludes ontological dualisms (except for a dualism between creator and creature). Angels, ghosts or other non-embodied minds acting in natural reality are excluded. However, there is no need to deny that humans discover certain invariant truths about geometry and thus construct or discover timeless truths.

The label "religious naturalism" or "empirical theology" is used by a variety of thinkers who hold positions similar to the view taken here. For instance, in an essay on science and "empirical theology," Karl E. Peters wrote, "Human fulfilment and the ultimate source of fulfilment are to be found not beyond the spatial-temporal world but within it. If there are realms of being other than space-time nature and history (as in supernaturalism), they are beyond our ken and have no relevance to life today."[22]

However, within this stream of thought, many seem to dislike the prominent place given to physics in "constitutive reductionism"; they prefer "organic" metaphors. For instance, Nancy Frankenberry claims that "the fundamental image of nature in terms of interpenetrating fields of forces and organically integrated wholes has replaced that of self-contained, externally related bits of particles of inert matter."[23] Frederick Ferré saw a new image of the world:

> If the image of the Garden, in which humanity and nature interact with balance and mutual benefit, becomes a fundamental image of our world, it will of course be easier to see how the Machine can fit—as an inorganic simplification and servant of the organic—than it is now to understand how a Garden could come to grow in the cosmic Machine.[24]

Ferré believes that postmodern sciences "have broken sharply with the ideals and assumptions that have been identified with modern science for long centuries." Ferré refers briefly to quantum physics, but his main example is ecology which "includes and transcends analysis in a holistic way that is essential to its conceptual task."[25]

I am not convinced by the view of science suggested here. Quantum physics, on some interpretations, introduces non-local correlations, but it does not thereby introduce into our view of the world holism in a sense related to subjectivity or values. I consider as revealing Ferré's statement that it is easier to understand how the Machine fits in the Garden than the reverse. He thereby largely abandons

---

[22] Karl. E. Peters, "Empirical theology and science," in *Empirical Theology: A Handbook*, Randolph. C. Miller, ed., (Birmingham, Alabama.: Religious Education Press, 1992), 63.

[23] Nancy Frankenberry, "Major themes of empirical theology," in *Empirical Theology*, Miller, ed., 39.

[24] Frederick Ferré, *Hellfire and Lightning Rods* (Maryknoll: Orbis, 1993), 95.

[25] Ibid., 93f.

mainstream theories of evolution, which see more complex entities as products rather than as initial states. Claims about a transition from modern to post-modern science underestimate the success and the potential for further development of modern science in the way it has progressed over the last few centuries. There are interesting changes in science, which have triggered various debates in the philosophy of physics and elsewhere. Ideas relating to space, time, substance, and determinism have acquired a new shape. However, neither these changes in science nor these philosophical discussions warrant the claim that there has been a "reintegration of understanding with valuational intuition."[26] Science is not modified by our "valuational intuitions," but rather seems to offer the possibility of understanding the origins of our "valuational intuitions."

The place of physics in the order of disciplines is an important area of disagreement in science-and-religion discussions. Can one offer an account of our world which is in its fundamental ontology radically different from the way it is viewed on the basis of contemporary physics?

If the focus is on current physics, the answer must be positive. Underlying the level of particle theory there might be a quite different theory, formulated perhaps in terms of superstrings, twistors, or quantized building blocks of spacetime in a yet-unknown theory of quantum gravity. Such changes may well have consequences for our concepts of object, space, time, substance, and force, and for ideas on issues such as determinism and causality. However, such a change in physics would respect the hierarchical structuring of phenomena, and of the corresponding sciences, which is more or less the backbone of contemporary science, from quarks, to nuclei, to atoms and molecules, to macromolecules, and on to living organisms, followed by consciousness and culture. We might change our understanding of the foundation of reality, ontologically speaking. But fundamental physics is a kind of pinnacle of the building of knowledge. If physics were to change, this building would not collapse, though it might need some reorganizing.

A more radical alternative would be one which would in some way reject this overall pattern of the natural sciences and the difference between the order of knowing and the order of being (with physics being fundamental in the latter though not in the former). In discussions about the relationship between science and religion the most prominent example of such an alternative is process philosophy, which draws on the categorial scheme developed by Alfred N. Whitehead in his *Process and Reality* (1929). On this view, "values" and "choices" are relevant at the most fundamental level of reality. Physics is adequate for uninteresting entities, such as electrons or stones, while features which show up most clearly in human relations are characteristic of the most fundamental structure of reality; the "Garden" has priority over the "Machine."

The attempt to develop such an alternative view of the fundamental structure is legitimate. It would be a remarkable change in the history of ideas if such an alternative organization of scientific knowledge would replace the consensus view, but it is not to be rejected *a priori*. Such accounts should be at least be comparable in detail and precision to those of the currently dominant view. With respect to process philosophy, I am not convinced that the categorial scheme which gives a metaphysically basic role to values and choices can be developed in sufficient quantitative detail, nor do I expect it to be more adequate than the standard view. I

---

[26] Ibid., 95.

thus see no reason to abandon a physicalist version of naturalism for an organismic one.

3 Could Evolved Morality Be Moral?

Evolutionary views have gotten bad press when applied in the social arena. Political dimensions surrounded the theory from its very beginning[27] up to the present time.[28] Given the impression that scientific approaches to human nature result in dubious politics, it is not surprising that many well-intending citizens reject science. There is also concern about the view of human morality that evolutionists have presented. For instance, E. O. Wilson wrote in his monumental book *Sociobiology*, that "ethical philosophers intuit the deontological canons of morality by consulting the emotive centers of their own hypothalamic-limbic system."[29] Despite the fact that Wilson has called for us to choose consciously "among the alternative emotional guides we have inherited,"[30] he has been heard by many as someone who has reduced morality to emotional drives, and thus as playing down the relevance of moral deliberation.

I will argue here that one should not judge the implications of an evolutionary view by the claims of the most "greedy reductionists"—a label Dennett used for those who think that everything can be explained without "cranes," that is, without many intermediate steps of various kinds.[31] One issue is whether we can envisage an evolutionary explanation of human moral behavior. This will be treated briefly before we turn to the main issue: can we consider morality to be moral if it has arisen by evolution?

Humans with their bodies and social behavior have arisen in the course of a long evolutionary process which we shared with the other great apes until a few million years ago.[32] One author who has presented an evolutionary account of human morality is Richard Alexander.[33] In brief, he argues that the evolution of cultures with moral codes may have been driven by two major factors: group cohesion against other groups and indirect reciprocity related to status within a group.

---

[27] E.g. Adrian Desmond, *The Politics of Evolution: Morphology, Medicine, and Reform in Radical London* (Chicago: University of Chicago Press, 1989).

[28] E.g., Richard J. Herrnstein and Charles Murray, *The Bell Curve: Intelligence and Class Structure in American Life* (New York: Free Press, 1994). Some of the responses have been brought together in Russell Jacoby and Naomi Glauberman, eds., *The Bell Curve Debate: History, Documents, Opinions* (New York: Random House, 1995). Many critics have pointed out that the empirical and statistical basis of *The Bell Curve* is shaky while the basic argument from "inherited" to "nothing to be done about" is totally mistaken—bad eyesight may be inherited but nonetheless corrected with the help of glasses.

[29] Edward O. Wilson, *Sociobiology: The New Synthesis* (Cambridge: Harvard University Press, 1975), 563; see also p. 3; controversies over sociobiology are documented in Arthur. L. Caplan, *The Sociobiology Debate: Readings on the Ethical and Scientific Issues Concerning Sociobiology* (New York: Harper & Row, 1978).

[30] E. O. Wilson, *On Human Nature* (Cambridge: Harvard University Press, 1978), 6.

[31] Dennett, *Darwin's Dangerous Idea*, 82.

[32] On the evolution of hominids and of human characteristics, see Camilo. J. Cela-Conde in this volume; on evolution and morality, see also Cela-Conde, *On Genes, Gods and Tyrants* (Dordrecht: Reidel, 1987). On research on primates and insights into human behavior, see Frans De Waal, *Good Natured: The Origins of Right and Wrong in Humans and Other Animals* (Cambridge: Harvard University Press, 1996) and J. Diamond, *The Rise and Fall of the Third Chimpanzee* (London: Radius 1991).

[33] Richard D. Alexander, *The Biology of Moral Systems* (New York: De Gruyter, 1987).

Recently, David Sloan Wilson and Elliot Sober have argued more extensively for selection operating at the level of groups rather than of individuals; this approach strengthens the case for an evolutionary understanding of social behavior within large groups of unrelated individuals.[34] Earlier accounts of evolution in terms of group selection had the problem that cooperative behavior seemed to be unstable against an individual who profits without contributing—say, who does not give an alarm call (for the benefit of the group) when being the first individual to spot a predator. Genes for asocial behavior would be advantageous and thus become more widely spread.[35] However, Wilson and Sober argue that the emphasis on individual genes is inadequate. They are the units which replicate, but they only do so in coordination with other genes—just as humans can come out first in a rowing contest only by cooperating as members of a team. Thus, the (more encompassing) "vehicles of selection" are the units favored (or not) by the selective processes; the "essence of the vehicle concept is *shared fate*."[36] When applied to humans, "human cognitive abilities provide other mechanisms for concentrating natural selection at the group level, even when the groups are composed of large numbers of unrelated individuals."[37]

This is not the place to pretend to decide on the best theory in contemporary biology. However, the examples of Alexander and of Wilson and Sober indicate that there are resources within evolutionary theory which seem to be able to understand human social behavior as a product of evolutionary processes. And, unlike expositions in terms of, for instance, "selfish genes" (Dawkins), these more recent expositions treat social behavior not so much as a "cover" hiding selfishness, but rather as an element in the rise of culture. Our cognitive capacities and cultural traditions complicate a biological analysis. It would be a case of greedy reductionism if one pretended to explain moral behavior without reference to cultural and mental aspects. And it is precisely these cultural and mental aspects which make it possible to consider evolved morality to be moral.

Michael Ruse seems to argue that an evolutionary understanding of morality implies that the ethical norms and moral attitudes we hold are selected for their survival value. However, Francisco Ayala suggests that it is intelligence that has been selected for, delivering as a byproduct the conditions necessary and sufficient for ethical behavior. Elliot Sober is even more outspoken on the role of mind: "cultural selection can be more powerful than biological selection. The reason for this is not some mysterious metaphysical principle of mind over matter. When cultural selection is more powerful than biological selection, the reason is humble and down to earth: *thoughts spread faster than human beings reproduce*."[38] In that

---

[34] D. S. Wilson and Elliott R. Sober, "Reintroducing group selection to the human behavioral sciences," *Behavioral and Brain Sciences* 17 (1994): 585–654.

[35] An earlier group-selectionist account is Vero C. Wynne-Edwards, *Animal Dispersion in Relation to Social Behavior* (Edinburgh: Oliver & Boyd, 1962); a major critic has been George C. Williams, *Adaptation and natural selection: A critique of some current evolutionary thought* (Princeton: Princeton University Press, 1966); Williams emphasized the genes as the level where almost all selection takes place. The gene-centered approach became widely known through Richard Dawkins, *The Selfish Gene* (Oxford University Press, 1976).

[36] Wilson and Sober, "Reintroducing group selection," 591.

[37] Ibid., 598 and 605f.

[38] Michael Ruse, "Evolutionary Ethics: A Defense," in *Biology, Ethics and the Origins of Life*, Holmes Rolston, III, ed., (Boston: Jones and Bartlett, 1995), 93–112; Francisco Ayala, "The difference of being human: Ethical behavior as an evolutionary byproduct," in *Biology,*

sense, humans have some measure of independence from the evolutionary process that produced us. Even more, a better understanding of our biological drives gives us the possibility of opposing them. According to Sober, a "brain is able to liberate us from the control of biological evolution because it has given rise to the opposing process of cultural evolution."[39] As a consequence of the difference in speed, the prejudice that biological influences must overwhelm the more superficial cultural ones is wrong; cultural influences may have a greater impact on the outcome than biological influences.

If one accepts the possibility of an evolutionary explanation of the origins of human social behavior, the habit of making moral pronouncements on the behavior of others included, a further question is whether morality thus understood could still be considered moral. In debates on human sociobiology, one fear—apart from an unwarranted elevation of evolution to the status of a moral principle—seems to be that an evolutionary understanding of morality would undermine the specific moral character of such behavior. There are at least four sources for such a fear.[40] It can be seen as 1) the fear that our moral language is a screen for hiding amoral motives. It can be understood as 2) the fear that, given their humble ultimate origins, moral considerations are not as worthy as we take them to be. The fear can also be related to "ontological" issues, such as 3) the apparent denial of human freedom (and thereby responsibility) or 4) of objective values.

The first dispute concerns proximate mechanisms, such as the nature of our motives. The suggestion is that we use moral language to serve and hide our interests. We may think that we are driven by moral considerations, but we are mistaken about the mechanisms that drive our behavior. Instead of springing from concern for the well-being of others, our actions aim at increasing inclusive fitness. In order to evaluate this suggestion, let us briefly consider another phenomenon: the feeling of pain. When I have my hand too close to a flame, I will quickly move my hand away. The proximate mechanism is neurological, partly automatic (reflexes) and partly conscious (feeling pain). I would not explicate my behavior by saying that I withdrew my hand to maximize inclusive fitness. An evolutionary account of the origin of a neurological mechanism does not in any way deny its reality, or the reality of the sensation of pain. The situation is similar with motives for moral behavior. Motives and feelings are not covering up a supercomputer which calculates which behavior is most profitable, but they are the means by which we have come to cooperate. As proximate mechanisms, our motives and moral pronouncements may well be sincere.

A second source of concern about the implications of the evolutionary view for morality is the expectation that the recognition of the evolutionary origins of our motives would undermine our assessment of their worth. Consider a childless couple spending time and energy caring for children with birth defects. To point out that this behavior derives from the propensity to care for offspring, a propensity which arose to maximize inclusive fitness, in no way diminishes the personal sacrifices made by

---

Ethics and the Origins of Life, Rolston, ed., 117–35; Elliott R. Sober, "When natural selection and culture conflict," in *Biology, Ethics and the Origins of Life*, Rolston, ed., 156.

[39] Sober, "When natural selection and culture conflict," 151.

[40] Philip Kitcher, *Vaulting Ambition: Sociobiology and the Quest for Human Nature* (Cambridge: MIT Press, 1985), 395–434.

the couple. The remote action of evolutionary forces "is irrelevant to the assessment of moral worth."[41]

But—a third source of concern—perhaps the childless couple is driven by an innate propensity to care for children. If their behavior is the consequence of evolution, they were externally or internally coerced to behave the way they did. Would their actions—though perhaps morally good if judged by the results—still be moral? The suggestion is that their behavior is determined by the evolutionary past and/or by the environment in which they live; reasons which include moral considerations would be irrelevant. But why would they be irrelevant? Our deliberations, and thus our values, etc., are part of the conditions which determine our behavior.[42] Freedom is not the opposite of determination. Rather, freedom is self-determination, that is determination by my character and desires, controlled by my rational reflection on my past actions and on potential consequences of various options, by my second order desires with respect to my life plan, and by my values. A morally relevant sense of freedom is at odds with a crude version of an evolutionary understanding of human nature, which does not pay sufficient attention to the complexity of culture and mind which generate the possibility for self-reflection, deliberation, and acting on the results of deliberations. However, a more sophisticated evolutionary view may well allow for freedom as morally relevant self-determination.[43]

A fourth fear seems to be that sociobiology undermines the possibility of objective values with respect to which we evaluate moral behavior. It is indeed at odds with an evolutionary perspective to consider human values as revealed or imposed by religious authorities or as entities residing in some timeless realm. Thus, we seem to be left with a subjectivist view of values as rooted in the emotions of the individual, as E. O. Wilson suggested, or with an evolutionary view which grants the presence of values which are shared by various organisms, but only in as far as these organisms share a common evolutionary past or have similar interests in similar situations; Michael Ruse has argued that rape would perhaps not be wrong for a species with a quite different biology, say one which evolved on another planet.[44]

The focus on emotions, such as E. O. Wilson's emphasis on the limbic system, is odd. Why would a sociobiologist consider the higher structures of the brain—including the capacity to reason about consequences and principles on the basis of which certain behavior could be justified publicly—to be superfluous, neither affecting the functioning of the limbic system nor affecting canons of morality?

Human morality has many aspects related to our biological make-up which seem to be accidental. Consider, for instance, widely spread double standards with respect to extramarital sexual activity by males or females. As the philosopher Peter Singer points out, even if we accept a sociobiological explanation for this phenomenon, we still may consider such a double standard an example of sexism which is morally unacceptable; there is room for a considered moral judgment which differs from conventional moral sentiments. And "by explaining the widespread acceptance of the double standard, we also remove any lingering idea that this standard is some

---

[41] Ibid, 404.

[42] See, for instance, the analysis in Daniel C. Dennett, *Elbow Room: Varieties of Free Will Worth Wanting* (Cambridge: MIT Press, 1984).

[43] Kitcher, *Vaulting Ambition*, 405–17.

[44] Ruse, "Is rape wrong on Andromeda?" in Ruse, *The Darwinian Paradigm: Essays on its History, Philosophy and Religious implications* (London: Routledge, 1989), 237.

sort of self-evident moral truth. Instead it can be seen as the result of the blind evolutionary process and, as such, something about which we should make a more deliberate decision, now that we have understood it."[45] A sociobiological explanation does not offer a justification, but rather reveals the contingent character of many moral practices, and thus offers an opportunity to reconsider behavior. But then the question arises, By what standards do we evaluate our "natural" moral sentiments?

The committed sociobiologist might say that we do not escape our biology; we only bring into play further values which are also part of our biology. Upon a naturalist view, there seem to be no resources for substantial moral judgments except for the heritage of our biological and cultural past. There is no room for the justification of ethical decisions in relation to entities in some Platonic realm, as if we come to hold moral principles by intuiting an absolute moral order. However, a procedural view of moral justification such as offered by Rawls may be compatible with an evolutionary view.[46] It does not justify claims about categorical objective moral truth, but such an absolute, "rational intuitionist notion of objectivity is unnecessary for objectivity," and can be replaced by a social one.[47] Procedures of consensus-formation and public justification are a valuable complement to and corrective of our moral intuitions as rooted in our biology. Ethical objectivity "involves the existence of a standard beyond personal wishes, a standard in which the wishes of others are given their place."[48] A procedural form of ethical justification may offer us ways to cope with the conflicting interests of individuals.

The claim is not that ethical values could be deduced by thinking and public fiat alone. Rather, we reflect upon our moral intuitions, and thus consider whether they have certain general features which we consider desirable. In our reflection, we test moral judgments by criteria such as generality and disinterestedness, coherence, contribution to happiness and to the reduction of suffering, etc. We owe our intuitions to the evolutionary past, but they can be considered and corrected, since we have the ability to evaluate our primary responses and to act upon such evaluations, though we do not act easily upon them, as the apostle Paul observed (Romans 7:19). The difficulty of acting morally shows that genuine moral behavior does not come to us "by nature," but rather requires moral effort; ethics is not prediction of what is most likely to happen. Whatever the origin of the human capacity for reflection, we now have this capacity and can use it for a reconsideration of our moralities. The way we reason is itself also subject to change. However, this does not assume a higher norm by which we would adjudicate changes in procedures. Rather, it is a piecemeal process of improvement, just as in the development of precision technology from cruder forms.

The criteria which we use in moral evaluations, such as the requirement of disinterestedness, may also be seen as the product of our evolutionary past. At some moment in the past one of our hominid ancestors asked a fellow hominid the equivalent of the question "Why did you do that?" in the presence of a third party,

---

[45] Peter Singer, "Ethics and sociobiology," *Zygon* 19 (1984): 141–58. Cf. Singer, *The Expanding Circle: Ethics and Sociobiology* (Oxford: Clarendon Press, 1981), 154.

[46] John Rawls, *A Theory of Justice* (Cambridge, Mass.: MIT Press, 1971); Richard D. Alexander, "Biological considerations in the analysis of morality," in *Evolutionary Ethics*, Matthew H. Nitecki and Doris V. Nitecki, eds., (Albany: SUNY Press, 1993), 180ff.

[47] John Rawls, "Kantian constructivism in moral theory," *Journal of Philosophy* 77 (1980), 570.

[48] Kitcher, *Vaulting Ambition*, 432.

and the answer was couched not in terms of emotions (I like to do that) or in terms of self-interest, but in terms which were sufficiently general to be recognizable and acceptable to all bystanders, and thus, perhaps, brought the others to similar behavior.[49] The point is that we developed the habit of evaluating and justifying behavior in terms which are acceptable to the whole group. Even the criteria delineating the relevant group can change.

Formal analysis, the application of criteria such as disinterestedness and coherence, and the moral deliberation of many people together are important for the credibility of morality, precisely because they surpass and may correct the conclusions of our ordinary biological and psychological mechanisms. One might include all these elements in a sociobiological description, but then ethical considerations would not so much have been eliminated, but rather would have been included in a modified sociobiology which includes consideration of the mechanisms by which we override the psychological processes explained by traditional sociobiology. Or one could say that our moral intuitions are explained by sociobiology, but that these intuitions need not be our best ethical conclusions, since we can reconsider them.[50]

4 Scientific Images and the Richness of Experiences

Another reason for hesitations about evolutionary views of reality, ourselves included: Can they do justice to the richness of reality or of our experiences of it? As the novelist John Fowles expressed it:

> Ordinary experience, from waking second to second, is... hopelessly beyond science's powers to analyze. It is quintessentially "wild," in the sense my father disliked so much: unphilosophical, uncontrollable, incalculable. In fact it corresponds very closely—despite our endless efforts to "garden," to invent disciplining social and intellectual systems—with wild nature. Almost all the richness of our personal existence derives from this synthetic and eternally present "confused" consciousness of both internal and external reality, and not least because we know it is beyond the analytical, or destructive, capacity of science.[51]

Understanding human experiences and religions in the context of our scientific image of the world should not be achieved by pruning away complex, "wild" experiences for the sake of simplicity. On the other hand, appealing to our immediate experience or our inner lives may also be less safe ground than it seems. We have learned all too well that we may be deluded and delude ourselves, offer rationalizations for behavior we exhibit, etc. An adequate scientific image should not only make intelligible the phenomena, but also our experience of the phenomena. If "wild" nature is to be understood in the framework generated by the sciences, we have to understand why it appears so "wild" as to be beyond science's power to analyze.

---

[49] Singer, *The Expanding Circle*, 92ff.

[50] A similar conclusion could be defended with respect to the status of epistemology: either we expand psychology by including scientific procedures (double blind experiments, etc.) by which we correct our ordinary belief-forming processes, or we acknowledge the difference between psychology and epistemology: there is no need to say that epistemology is eliminated in a naturalist view.

[51] John Fowles, *The Tree* (St. Alban's: Sumach Press, 1979), 40f.

4.1 *Why Does Nature Seem "Wild"?*

As I see it, this wildness is related to various limitations which manifest themselves almost everywhere in nature. For example, with respect to human nature: we do not monitor our inner states, nor could we if we intended, and the causal webs of responding to the environment are the product of such a long, convoluted history as to be beyond detailed analysis. We need to be aware of such limitations when using science for understanding human nature. We often use metaphors which are too simple, such as metaphors borrowed from technology when talking about experiences; "letting off steam" and "being under pressure" are metaphors which depend on nineteenth-century technology; "recharging batteries" and "tuning in" reflect the earlier electromechanical technology, and the personal computer era has generated a whole new set of metaphors. Such metaphors derived from current technology are fine as long as the metaphorical character is kept in mind. However, if the analogy between humans and technical artifacts, from clocks to computers, is made too tight, it tends to become ridiculous; "wild nature," including human nature, is richer than such technological metaphors can express.

We are also limited with respect to detailed explanations of particular events. Chaos theory has made clear what could have been obvious to students of historical evidence: we never have sufficient knowledge of all details to provide a full account of the course of events. Besides, a full account would be so cumbersome as to be totally unmanageable and inaccessible to us. As long as the concept of "explanation" is not used in an overdemanding way (which would make it hard to find any cases where anything is explained), such limitations do not imply that the phenomena which actually happen in a complex, chaotic, or evolving system are unexplained or inexplicable in some deep, religiously significant sense;[52] phenomena are explained when underlying causes or mechanisms are discerned (an ontic notion of explanation) and when they are located in a wider theoretical framework (an epistemic notion of explanation).[53] Quantum physics may be interpreted as showing that there are limitations not only to the determinateness of our knowledge but also of reality. However, even then we may consider the outcome of a quantum event to have been explained when it is understood as one of the possible outcomes given the setup of the situation. Those with small probabilities are understood just as good as those with higher probabilities.

While scientists are in many cases able to understand how a particular phenomenon fits into the larger picture, this insight comes at a price; often, the actual process is understood to be the consequence of processes which cannot be traced in full detail, either because it concerns a history about which we have insufficient data (for example, evolutionary histories of species) or because it concerns a system about which we know that we cannot know the actual state of affairs at a single moment (both for apparently ontological reasons, as in quantum physics, and for epistemological reasons, as in chaotic systems).

---

[52] Here I disagree with John Polkinghorne who has used the unpredictability of chaotic systems as a model for divine action; see Willem B. Drees, "Gaps for God?" in *Chaos and Complexity*, Russell, Murphy, and Peacocke, eds.; idem, *Religion, Science and Naturalism*, 99–100.

[53] Wesley C. Salmon, *Four Decades of Scientific Explanation* (Minneapolis: University of Minnesota Press, 1990).

4.2 A Brief History of Science: via the Simple to the Complex

Aristotelian physics was in many ways more immediately adequate to our experiences than modern physics. The first phase of modern physical and chemical science has been to study simple phenomena, either those with obvious regularity such as motions of the planets or those which were artificially created or approximated in experiments, such as balls rolling along inclined planes or reactions between homogeneous volumes of chemicals, passing over the complexities of friction and of surface phenomena in chemistry. An enormous amount of abstraction and simplification (compared to the real world) has taken place in order to develop science in some depth. It should be obvious that science in this phase was woefully inadequate in so far as it was used to deal with complex phenomena from the real world.

Science has not ended with the analysis of simple phenomena. Especially in this century, the scope of science has been extended enormously, studying all kinds of more complex entities and complicated processes such as those which are not in equilibrium, processes in thin layers, etc.; the area referred to as the study of complexity and insights about "chaos" exemplify (but do not exhaust) this development. Increased computing power and powerful techniques in molecular biology and physics have joined forces with interest in details of particular processes. Thus, science is more and more able to study systems which match or approximate the complexity of the real world.

To visualize the process: from the richness of our manifest images we have first gone down to the study of gross simplifications, and then reconstructed from insights about these simplifications an understanding of more complex phenomena. Perhaps one should rather envisage a reiterated process, with more than one consecutive cycle of simplification and building up again towards an understanding of complex phenomena. As a consequence of the detour through the study of simpler systems, science now more fully understands "wild" reality in its variety and at the same time its own limitations in explanatory and predictive power.

The conviction that science is unable to account for "wild" nature is to some extent true: it is not able to predict events in full detail. However, science has over time become better and better at understanding complex phenomena (and not merely simple systems which are obviously inadequate models for understanding complex experiences). If we intend to be fair to science, we should not dismiss science on the basis of straw men, that is, simple models which can be dismissed too easily. Besides, we need to be realistic about phenomena in the world; if one has too lofty a view of human nature, for instance with respect to free will, rationality, perception and the like, then we cannot see how humans would fit into the scientific image.[54]

At some places the distance between our manifest and scientific images may be minimal; at other places it may be more significant. If we find ourselves with two images, which one is more important? That depends on the purpose. It may well be that the "wild" richness of experiences is more important when we deal with one another as humans, when we long for consolation, for a sense of beauty, etc. The scientific image, however, has gone through a critical process of articulation with precision and testing, and is therefore more adequate when we are after "intellectual

---

[54] Both Mary Midgley, *The Ethical Primate* (London: Routledge, 1994) and Dennett, *Elbow Room* make clear that an analysis of human free will requires a realistic assessment of human nature; whether angels possess rationality and free will is quite a different question from the question of whether humans do.

adequacy" since that is what it has been selected for. In that case, the scientific image need not be recognizable to the original subjects. In and through the sciences we have come up with all kinds of scientific images of the world which differ significantly from the way we experience the world, that is, our manifest images.

5 *Religion as a Phenomenon*

Science results in a view of the world which differs from the images generated by religious traditions. In this context, one may speak of a conflict between science and religiously motivated beliefs (for example, creationism). One could also attempt to accommodate theological notions such as divine action to scientific insights and insights from historians, anthropologists, etc. (for example, Peacocke[55]). Or one could attempt to reinterpret scientific insights to fit better the theological view (for example, process thought). It is in this context too that debates about theological realism and the similarities and dissimilarities between theological and scientific claims flourish. However, religion is not only a partner of science in understanding and explaining the world but also a phenomenon, and thus an object of study and explanation. How does that affect our understanding of religion?

In *The Selfish Gene* Richard Dawkins suggested seeing religion in terms borrowed from pathology: The God-meme is a virus. Belief in God is a meme (an idea which functions as a unit) which has been able to induce organisms infected with it to produce multiple copies of it (by psychological appeal, suggested rewards, and punishments). Compared with such a greedy dismissal of religion, Dennett is far more appreciative; religion has been valuable:

> Long before there was science, or even philosophy, there were religions. They have served many purposes (it would be a mistake of greedy reductionism to look for a single purpose, a single *summum bonum* which they have all directly or indirectly served). They have inspired many people to live lives that have added immeasurably to the wonders of our world, and they have inspired many more people to lead lives that were, given their circumstances, more meaningful, less painful, than they otherwise could have been.... At their best, religions have drawn attention to love, and made it real for people who could not otherwise see it, and ennobled the attitudes and refreshed the spirits of the world-beset. Another thing religions have accomplished, without this being thereby their *raison d'être*, is that they have kept *Homo sapiens* civilized enough, for long enough, for us to have learned how to reflect more systematically and accurately on our position in the universe. There is much more to learn. There is certainly a treasury of ill-appreciated truths embedded in the endangered cultures of the modern world, designs that have accumulated details over eons of idiosyncratic history, and we should take steps to record it, and study it, before it disappears, for, like dinosaur genomes, once it is gone, it will be virtually impossible to recover.[56]

Dennett's approach is in the end a more serious challenge than Dawkins' dismissal as pathology. Religions are seen as phenomena within reality, which can be studied just like other human phenomena. The natural sciences in a restricted sense do not have much to contribute to the study of religions; this level of complexity and intractability requires approaches which may be less fine-grained and precise but are thereby able to take some of the richness of social interactions into account. However, even though the specific study of religions may be the

---

[55] Arthur R. Peacocke, *Theology for a Scientific Age: Being and Becoming—Natural, Divine and Human* (London: SCM & Minneapolis: Fortress, 1993).

[56] Dennett, *Darwin's Dangerous Idea*, 517–19.

business of anthropologists, sociologists and psychologists, the perspective arising out of the natural sciences offers some outlines for views of religions.

Within an evolutionary perspective, one would primarily explain the emergence of religions along lines similar to the explanations one would advance for social phenomena such as political institutions and languages. One looks for their contributions to the inclusive fitness of the individuals or communities in which they arose. Alternatively, one could claim that religions arose as a side-effect with the emergence of some other trait. Perhaps with the rise of consciousness questions about the origin and meaning of the world and of one's individual existence could arise, and as long as other explanations were not available, explanations in terms of spirits and personal powers in and beyond the world were attractive.

As for the functional role of religion, various proposals may be considered, such as Ralph Burhoe's view that religions made the cooperation of larger groups of hominids, beyond close-kin, possible.[57] Another proposal places less emphasis on the role of religion in contributing to human cooperation, and more on its role in living with intractable, apparently contingent features of our environment.[58] Such proposals need further specification. One might test their credibility by analyzing in greater detail how religions may have arisen and may have been sustained in the environments of various epochs, and what their adaptive value (via culture) may have been. The functions of religions may have changed over time as well. Here I will not defend one particular view of the function of religions, but rather reflect on some general implications of such naturalist views of religions as functional, cultural practices. When they arose, since they were functional, the major question is whether we should take seriously religious rituals, myths, narratives, conceptualities, etc., as cognitive claims.

To say that religions are, or were, functional is not necessarily to deny that their central terms refer to realities. However, if a religious claim purports to be about a supernatural reality, such as one or more gods, one might raise the question of whether the way in which this claim is rooted in evolved reality makes it more or less likely that it is right or wrong. We may consider as analogies our access to the world around us by perception.

On an evolutionary view, the adequacy of our language about trees, with notions such as bark, leaves, firewood, etc., is intelligible since the language has been modified in a long history of interactions of humans with trees and with each other in conversations about trees. This web of causal interactions lies behind the adequacy of our language about trees. If one came across a culture with no past experiences with trees, it would be a very surprising coincidence if they had an adequate vocabulary for trees. We refer to trees, and we seem to do so in fairly adequate ways, because our language has arisen and been tested in a world with particular, ostensible trees.

Now back to religions and reference. On a naturalist view there is no locus for particular divine activities in a similar ostensible way. Thus, it is extremely unlikely that our ideas about gods would conform to their reality.[59] Hence, an evolutionary view of religions challenges them not only by offering an account of their origin but

---

[57] Ralph W. Burhoe, *Towards a Scientific Theology* (Belfast: Christian Journals Ltd., 1981).

[58] Niklas Luhmann, *Funktion der Religion* (Frankfurt am Main: Suhrkamp, 1992), 26.

[59] Robert A. Segal, *Religion and the Social Sciences: Essays on the Confrontation* (Atlanta: Scholars Press 1989), 79.

thereby also by undermining the credibility of their references to a reality which would transcend the environments in which the religions arose.

One might propose a different analogy, not between religion and sense perception (seeing trees) but between religious claims and our more abstract activities such as mathematics and ethics. For instance, Philip Kitcher wrote, in the midst of a discussion on sociobiology and ethics: "Even if [E. O.] Wilson's scenario were correct, the devout could reasonably reply that, like our arithmetical ideas and practices, our religious claims have become more accurate as we have learned more about the world."[60] An account of origins—how we have come to a certain conviction—does not in itself decide on the truth of that conviction. To argue otherwise, conflating issues of origins of beliefs and of their justification, is to commit what is called the "genetic fallacy."

However, there are relevant differences between the status of mathematics and ethics, and the status of religious ideas. Mathematics may be seen as a second-order activity, growing out of the analysis of human practices such as counting and trading. Similarly, ethical considerations involve a second-order reflection upon procedures or standards which may be fruitful in resolving conflicts of interests with reference to an (unavailable) impartial perspective. As second order activities, they aim at norms of universal validity, but these universal, "transcendent" claims may be construed without reference to a realm of abstract objects apart from the natural realm with all its particulars. Moral intuitions and judgments may be considered first-order phenomena, but they do not need a "supernatural" realm, either for their explanation or for their justification.

In contrast, religions are first-order phenomena in which there is, in most cases, some form of reference to transcendent realities, denizens of another realm. Whereas such references in morality and mathematics may be reconstructed in terms of procedures for justification (and of some insights about human nature and the world in which we act), religions are much more tied to an ontological view of those realities: gods are either supernatural realities or they would be considered unreal, non-existent. In this sense, an account of the evolutionary origins and adaptive functions of religion is a much stronger challenge to the truth of religious doctrines than is a similar understanding of the origin and function of arithmetic or morality, since mathematical and moral claims need not be seen here as truth claims about reality, say about causally efficacious entities, whereas religious claims are often taken to be of such a kind. Thus, whereas an account of ethics which avoids reference to a non-natural realm would not affect morality, a similar account in theology would have more radical consequences, because it would undermine the referential character of statements which purport to be about a non-natural God.

6 Religions For Wandering And Wondering Humans

Religion is not only related to explanatory enterprises and their limit questions but also to ways of life. Any naturalist, whether religious or atheistic, should acknowledge the existence of religious traditions as phenomena within reality. They are there, just as languages and bodily features, as products of a long bio-cultural evolution. We have a physical "memory" of the past in our genes and an implicit and explicit heritage in our cultures, languages, and moral and religious traditions, passed on in brains and books. The fact that these traditions arose and were passed

---

[60] Kitcher, *Vaulting Ambition*, 419.

on from generation to generation (unlike others which disappeared) implies that surviving traditions embody well-winnowed practical wisdom which deserve attention, though in new circumstances not necessarily uncritical allegiance.

Among those who have sought to articulate an understanding of theology in the context of such a naturalist view of religion, Ralph Burhoe has been the one who emphasized most strongly their character as "well-winnowed wisdom." For him, the overwhelming power of the evolutionary process relates to our images of God's sovereignty. Gerd Theissen emphasizes the variety of adaptations which arose through evolution; he underlines tolerance or grace as the main characteristic of "ultimate reality."[61] Philip Hefner also relates "the way things are" to God; altruism and love are interpreted by Christian theology as expressions of "basic cosmological and ontological principles."[62]

Aside from those who seek to identify characteristics of the evolutionary process and of reality as a whole, there are also some who have sought to see religious language as language about some aspect of reality. Thus, Lindon Eaves sees it as dualistic language which we use to speak about features "lost in the mists of evolution and hidden from language and logic in the genetic code"[63] and which we also use to articulate an "ought" in a world of "is." Another proposal is Charley Hardwick's "valuational theism."[64] My own view, which seeks to take evolution not only seriously as a feature of the world (an issue theologically developed by Arthur Peacocke and many others), but also as the explication of human religious traditions, resembles these latter approaches in seeking to understand the function of religions *within* the evolutionary process rather than looking for some general characteristic *of* the process.

6.1 *Wandering Humans: a Variety of Particular Traditions*

Humans do not speak language, but particular languages; they are immersed in some culture. They also relate to particular religious traditions. Some people are totally immersed in a tradition; others are confronted with a variety of traditions, and seek to respond to that variety. There are different ways of responding intellectually to the variety of religious traditions. One might attempt to find common features in underlying processes or common first principles. However, this does not seem to do sufficient justice to the variety of religions. A more promising approach is to consider religious practices in relation to their own contexts and their own history, and to see how, for instance, cosmogonic beliefs, conceptions of ethical order, and social circumstances interact.[65] Such an approach in the study of religion is an obvious extension of an evolutionary approach, where organisms are also studied in relation to (the history of) their environments.

Religious traditions are complex entities. A *way of life* may be suggested by parables, as for instance that of the Good Samaritan helping a stranger from another

---

[61] Gerd Theissen, *Biblical Faith: An Evolutionary Approach* (Philadelphia: Fortress, 1985; translation of *Biblischer Glaube in evolutionärer Sicht*, München: Kaiser, 1984).

[62] Philip Hefner, *The Human Factor: Evolution, Culture, and Religion* (Minneapolis: Fortress Press, 1993), 197.

[63] Lindon B. Eaves, "Adequacy or orthodoxy? Choosing sides at the frontier," *Zygon* 26 (1991): 499.

[64] Charley D. Hardwick, *Events of Grace: Naturalism, Existentialism and Theology* (Cambridge: Cambridge University Press, 1996).

[65] Robin W. Lovin and Frank E. Reynolds, eds. *Cosmogony and Ethical Order: Studies in Comparative Ethics* (Chicago: University of Chicago Press, 1985).

culture (Luke 10:29–37), by historical narratives (such as various accounts of prophets protesting against injustice, or of Jesus forgiving those who persecuted him) and it may be articulated more explicitly, as in the Ten Commandments (Deuteronomy 5:6–21). Such a way of life need not always strengthen the conformity of the believer to the expectations of the larger community; it may also emphasize individual responsibility even where the individual goes counter to the interests of others. Such a way of life is not only a practical matter. It is oriented by an *ultimate ideal* which surpasses any achievable goal or situation. Thus, religious traditions include elements such as "the Kingdom," "Paradise," "Heaven," "Nirvana," immortality, emptiness, openness, perfection, or unconditional love. Such notions function as regulative ideals with which actual behavior is contrasted in order to evaluate it. A tradition's way of life is affirmed and strengthened by the particular *forms of worship* and devotion of that religious tradition. Worship and other forms of ritual behavior express and nourish the individual and communal spirituality in relation to the joys, sorrows, and challenges of life and to the conceptions and ultimate ideals of good life. Believers see their religious way of life as *rooted* in certain claims about historical events, ultimate destiny, or authoritative command-ments. These claims are supposed to justify the way of life espoused by a tradition as corresponding to the way one should live one's life—justified because they derive from an authoritative source, because they deliver future happiness, or because they correspond to the way reality is intended to be or, deep down, really is.

A particular tradition (or stage of, or element in a tradition) may, of course, have to be rejected as outdated. The circumstances may have changed, and hence models of good life or forms of worship may have to change. For instance, we stand in a different relation to nature, we are more powerful than before, and we are confronted with neighbors across the globe. In relation to such changes, traditional models and metaphors may be employed differently, or they may be understood as they always were but this may now be inappropriate to the circumstances (for example since they fuel the exploitation of natural resources). Not only have circumstances changed, but so have our moral and spiritual sensitivities, for example with respect to conflicts between ethnic or religious groups, slavery, or cruelty to animals. To this process of change have contributed religious traditions, changing circumstances, a wider encounter with other cultures, and philosophical insights; changes in moral sensitivity have also changed our religious traditions. We evaluate traditions also by the moral and spiritual life they support. One more reason for change, though not the most important one, is the cognitive credibility of a tradition. If the images which support the way of life are not recognizable, or if the claims by which the way of life is justified have become incredible (and thus no longer support it), then that too challenges the religious tradition, though more indirectly than challenges to the appropriateness of the circumstances of the way of life and to its moral and spiritual adequacy.

Granted that we may have to discard some traditions or may have to modify them, why would one keep alive any such tradition? The reason, in my opinion, is that they are useful and powerful. They are so even for reflective and well-informed persons. No human is only a rational being who could entertain all his motives and desires consciously and intentionally; the structure of our brains is such that much goes on which is not dealt with consciously. This is the risk involved in religious forms of behavior (since so much cannot be scrutinized consciously) and the reason for their importance: through religious metaphors and forms of behavior we address reality especially in a way which confronts us with ideals, with what ought to be, with a vision of a better world, or with images of a paradisiacal past or an ultimate comforting presence.

6.2 *Wondering humans: Limit Questions*

There is another aspect of religious traditions. Humans have, with the development of consciousness and communication, contemplated questions about the world in which they found themselves. Many of their speculative answers may have been functional; creation myths and other cosmogonies are not merely speculative attempts at explanation, but ways of presenting and justifying moral imperatives and social structures.[66] However, some speculations may well be useless, or at least reach beyond what is sufficient for the circumstances of the moment. In earlier ages, answers to speculative questions may have been closely allied with the way the world was experienced, which is still to a large extent reflected in our manifest image of the world. In this manifest image, persons are the major agents from which action proceeds. Hence, it is not very amazing that animist ways of speaking about the world have become widespread; experiences with many phenomena are modeled after experiences with human agents. Sometimes, such agents are understood as residing in the phenomena, say as spirits, and sometimes, the agent is thought of as a god who transcends the phenomena but acts through them.

Such models are still with us; animist ways of speaking about cars or computers are common, and many persons discern intentions behind bad luck such as being struck by a disease. Such ways of speaking are not credible given our knowledge of cars and diseases. However, even though earlier answers have lost their credibility and questions may have changed their appearance, humans can still be wondering persons, contemplating questions that transcend our current answers. Religious traditions offer answers to such questions, but—and more importantly, in my view—they are also ways of nourishing sensitivity to such questions.

7 Conclusions

In this paper we have considered the existence of particular religious traditions, each functional in its own way in certain circumstances, and each offering speculations on philosophical limit-questions about the world as a whole. In my view, these two approaches can complement each other. I do not mean that together they result in a complete view, but we may see them as independent contributions which can be brought together in a larger world-view. The openness expressed in the limit-questions may induce a sense of wonder and gratitude about the reality to which we belong. Such a cosmological approach might primarily be at home with a mystical form of religion, a sense of unity and belonging, as well as dependence upon something which surpasses our world. The functional view of religion offers some opportunities for a prophetic form of religion, with a contrast between what is the case and what is believed ought to be the case. The contrast might be seen as a consequence of our evolutionary past, which has endowed us with wisdom that is encoded in our constitution and in our culture (including religious traditions). Another way to articulate a prophetic element is to argue that evolution has endowed us with the capacity for imagination, for reconsidering our situation from a different perspective. This capacity has as its limit the regulative ideal of an impartial view transcending all our perspectival views. That such a point of view is inaccessible, is beneficial since it protects us from fanaticism; if one were inclined to believe that

---

[66] Ibid., 1.

one's view could be the final one, one would not be incited to self-questioning.[67] It is precisely in this role that a speculative approach, in response to limit-questions, with a radical notion of divine transcendence may be of major significance in our dealings with particular traditions. When considered in relation to the radical concept of divine transcendence all regulative ideals as they arise in particular religious traditions are relativized; they can never lay unrestricted claim to our allegiance.

We know, collectively, a great deal about our world. Our knowledge is also limited. Certain phenomena may be intractable, even though they fit into the naturalist framework. Limit-questions regarding the whole naturalist framework can be posed but will not be answered. Our knowledge and our capacity for knowledge have arisen in the midst of life, and if we are to use them anywhere at all, it will have to be there. They allow us to wonder about that which transcends and sustains our reality, but all the time we wander in the reality in which we live, move, and have our being. To its future we contribute our lives.

---

[67] Stewart R. Sutherland, *God, Jesus and Belief: The Legacy of Theism* (Oxford: Basil Blackwell, 1984), 110.

BIOCULTURAL EVOLUTION: A CLUE TO THE MEANING OF NATURE

Philip Hefner

1 *The Argument*

The argument follows a straightforward line of thought: 1) In the emergence of its biocultural phase, evolution has given rise to a creature, *Homo sapiens,* who is a meaning-seeking and meaning-shaping creature, indeed, meaning-seeking and shaping describes the distinctiveness of this creature; 2) culture becomes increasingly significant in this phase of evolution, and in the present epoch, culture is accompanied by a crisis that is defined by its inadequacy in sustaining and directing human life on this planet; 3) a careful reading and interpretation of the scientific understandings that underlie both the evolutionary emergence of humans and their culture and the dynamics of their ongoing development, provide a basis for speaking of the meaning of nature and also of the action of God; 4) classical Christian theology offers a framework for this interpretation.

To summarize: On the basis of scientific understanding of the emergence and character of *Homo sapiens,* and contextualizing this understanding within classical Christian theology, we state the meaning of nature (including human nature) as follows: With the appearance of the human being, having reached the present level of culture, nature has evolved on Earth to the point where adaptation requires the self-aware interpretation of the nature of the natural realm and free intentional cultural behavior that is in accord with that nature. The human being is presently the bearer of that behavior. In our present times, this behavioral requirement poses a challenge to the human community that has reached the proportions of a global civilizational crisis. Nature is revealed to possess the character of a project, and in the human species this character is currently most clearly revealed; this species bears the burden of the short-term future of the project on Earth. We propose to interpret the human creature and its situation within evolution with the concept of *created co-creator*. The Christian faith proposes that nature's project is God's project. Since the intentionality of God and the meaning of the project of nature is embodied in Jesus Christ, who is the normative expression of the image of God in the creation, the crisis of human culture to which we have referred will best be met by behavior that is in accord with the intentions of the creator God, instantiated by Jesus Christ. The task and meaning of human nature is to determine the concrete import of Jesus Christ for the conduct of human culture today and to shape human culture accordingly. Hence we can conceptualize the human being as *God's* created co-creator.

2 *Evolution in Its Biocultural Phase and the Emergence of Homo sapiens*

The scientific understandings that are most pertinent to my theme begin with a biocultural evolutionary model within the physical ecosystem. Although here I devote relatively little attention to the physico-chemical dimensions of evolution, they are, of course, the presupposition of the biocultural. This biocultural model has been described by Solomon Katz as follows:

> Biocultural evolution consists of a series of interactions among: the biological information resident within individuals and populations in the form of the genetic constitution (namely, the DNA); the cultural information which is the sum of the knowledge and experience which a particular society has accumulated and is available

for exchange among its members; and thirdly, a human central nervous system (CNS), which is of course a biologically based system, whose principal evolved function with respect to this model is to facilitate the communication or storage of individually and socially developed knowledge and awareness.[1]

This model is fruitful for our thinking about the human animal.

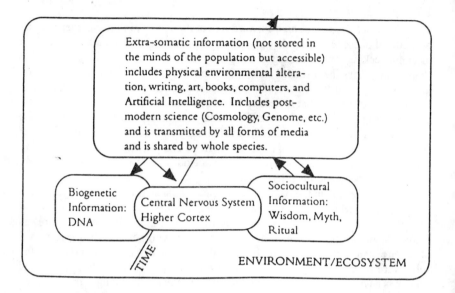

Extra-somatic information (not stored in the minds of the population but accessible) includes physical environmental alteration, writing, art, books, computers, and Artificial Intelligence. Includes postmodern science (Cosmology, Genome, etc.) and is transmitted by all forms of media and is shared by whole species.

Biogenetic Information: DNA

Central Nervous System Higher Cortex

Sociocultural Information: Wisdom, Myth, Ritual

TIME

ENVIRONMENT/ECOSYSTEM

A Basic Model of Biocultural Evolution (After Solomon H. Katz, 1993)

2.1 *The Two-Natured Animal*

One of the most provocative perspectives which this biocultural model gives us is what I shall call the "two-natured" character of the human. This may indeed be the most interesting and deeply significant insight that these sciences have given us into human being. *Homo sapiens* is itself a nodal point wherein two streams of information come together and co-exist. The one stream is inherited genetic information, the other is cultural information. Both of these streams come together in the CNS. Since they have coevolved and coadapted together, they are one reality, not two.[2] To speak of them as "two" is metaphorical. Even though such language of duality may be unavoidable and even necessary for heuristic purposes, it is certainly as misleading as it is useful. The confluence of these two streams of evolutionary information in the human being has been noted for a long time; it has given rise to the perennial discussion about "nature" and "nurture" as two building blocks of human life. We are far from understanding adequately how these two dimensions of human life and its evolution are related. We do know, however, that—as inseparable as they are—they often appear to be quite different from one another, and we require different sets of dynamics and principles for understanding each of these two

---

[1] John Bowker, "Editorial," *Zygon* 15 (December, 1983): 356; Solomon H. Katz, "Biocultural Evolution and the Is/Ought Relationship," *Zygon* 15 (June, 1980): 155–68.

[2] Timothy H. Goldsmith, *The Biological Roots of Human Nature: Forging Links between Evolution and Behavior* (New York: Oxford Univ. Press, 1991).

dimensions. We also understand that they have been selected to mix sufficiently well to differentiate humanity from other forms of life and that, even though they flow in different channels (remembering that this is a metaphorical manner of speaking), these channels merge in the human brain. The relation between them is at times a tense one. The cultural reality can easily put the biological to death, just as the latter can apparently withhold its cooperation from the former. It is the cultural agency that makes life interesting; culture lifts human existence to its heights, and it also plunges us into the depths. Nevertheless, for humans the genetic agent has both mandated the necessity and provided the possibility for the cultural reality, just as it holds the final cards in the game of life; if those cards are played in a fatal manner, culture is obliterated. The cultural and the genetic have coadapted to one another and to their common environments, so as to coevolve, in a relationship that may be termed symbiotic, again speaking metaphorically.

The information that we call culture does not remain stored in the human CNS. Rather it soon is transferred to other forms of storage—the designated "rememberer" or shaman, drawings, writing, eventually libraries and computer data bases. In other words, the information that comprises culture is stored outside the human body, in what is called *extrasomatic information*. Biocultural evolutionary models include this extrasomatic information, both as stimulant to cultural evolution and also as that which does evolve. The diagram above (page 330) illustrates these aspects of the biocultural model.

2.2 Burhoe's Theory of Symbiosis versus the Selfish Gene

How these two dimensions of human being—genes and culture—are related is the subject of intense scientific scrutiny just now. No theory can claim wide consensus in explaining their interrelationship. Ralph Wendell Burhoe's suggestion, that they constitute two "organisms" that are co-adapted in *Homo sapiens* and that they co-evolve within the species in a symbiosis, is one of the most interesting of the proposed theories. His theory underscores that in an important sense the management of this symbiosis, within the larger system of reality in which we live and its requirements, is what human existence is all about.[3]

Burhoe's suggestion rests on the striking hypothesis that "humanity is not a single species but a new kind of symbiotic community."[4] The significant symbiosis that he is talking about is between the biological creature *Homo* and "a new creature such as the earth had never seen before, a creature that is only partly biological, only partly programed by genetic information"[5] Biological humans (Burhoe calls them *Homo,* without the *sapiens sapiens*) and sociocultural systems are living in symbiosis and undergoing the process of natural selection as a coevolving and coadapting supraorganism. The evolutionary processes by which creatures who are "only partly programed by genetic information" have emerged are a long time in elaborating themselves. Consequently, the symbiotic creature we call human has forebears, and not only among the other higher primates.[6]

---

[3] Ralph Wendell Burhoe, "The Source of Civilization in the Natural Selection of Coadapted Information in Genes and Culture," *Zygon* 11 (1976): 263–303; idem, "Religion's Role in Human Evolution: The Missing Link between Ape-Man's Selfish Genes and Civilized Altruism," *Zygon* 14 (1979): 135–62.

[4] "The Source of Civilization in the Natural Selection of Coadapted Information in Genes and Culture," 282.

[5] Ibid.

[6] Frans de Waal, *Good Natured: The Origins of Right and Wrong in Humans and*

In one sense, Burhoe's theory underscores the tension between genes and culture, as a number of scientists have done, including the celebrated "selfish gene" imagery of Richard Dawkins and Richard Alexander.[7] Burhoe's basic understanding is of the cooperation and reciprocity between genetic and cultural streams of information, however, in the image of symbiosis. A successful symbiosis is necessary for the survival of both human genes and culture. Later he suggests that religion plays a special role in culture's contribution to this symbiosis.[8]

2.3 Freedom as Emergent

The earthly career of this two-natured creature, the human being, is characterized through and through by the marks of being both conditioned and free. The conditionedness of the human being is rooted in its evolutionary development. *Homo sapiens* has emerged from a deterministic process that extends back to the origins of the universe. Furthermore, humans are ecologically situated so as to be perched on a delicate balance within a specific planetary ecosystem. Both its emergence from the evolutionary past, with the rich heritage which that entails, and also its present ecological placement define the human being within very finite limits.[9]

Within this deterministic evolutionary process, freedom has emerged, and with it our reflection is provoked in a deeply dialectical mode. That freedom should be produced from determinism seems startling, until we understand that freedom apparently is in the interest of the deterministic evolutionary system.[10] What we call freedom is rooted in the genetically controlled adaptive plasticity of the human phenotype.[11] This plasticity takes a particular and unique form in the human being, and it is constituted by 1) exploration of the environment to consider appropriate behaviors, 2) self-conscious consideration of alternative decisions and behaviors, 3) a supportive social matrix which both allows for exploration by the individual and at the same time demands that group relationships and the welfare of other individuals as well as of the whole society be respected. The last-named element is a biological ground of values and what we term morality.

Freedom in the sense that I have just pictured it is linked with responsibility. The essence of freedom in human affairs is that human beings can take the kind of deliberative and exploratory action that I have just referred to, while at the same time they and they alone must finally take responsibility for the action. They must take responsibility for the future developments that are entailed in any actions, in the sense that they are the creatures who must live with the consequences and respond adequately to those consequences. If the consequences are undesirable, and the human actors are not extinguished by their mistakes, they and they alone have the

---

Other Animals (Cambridge: Harvard University Press, 1996).

[7] Richard Dawkins, *The Selfish Gene* (London: Granada, 1978); Richard D. Alexander, *The Biology of Moral Systems* (New York: Aldine de Gruyter, 1987).

[8] Burhoe, "Religion's Role in Human Evolution: The Missing Link between Ape-Man's Selfish Genes and Civilized Altruism."

[9] Philip Hefner, "Nature's History as Our History: A Proposal for Spirituality." in *After Nature's Revolt: Eco-Justice and Theology*, Dieter T. Hessel, ed. (Philadelphia: Fortress Press, 1992).

[10] Ted Peters, "Review of *Free Will and Determinism*, Viggo Mortensen and Robert C. Sorensen, eds." *Zygon* 26 (March, 1991): 178–80.

[11] Theodosius Dobzhansky, *The Biological Basis of Human Freedom* (New York: Columbia University Press, 1956), 68.

responsibility and the possibility for compensating for errors. This evolutionarily-fashioned context reinforces conditions that are suitable for the emergence of values.

The fact that conditionedness and freedom have emerged together, reaching the point of development that we find in *Homo sapiens,* is of no little significance. This phenomenon is an evolutionary preparation for values and morality, which indicates that the dimensions of "oughtness" and value are built into the evolutionary process and need not be imported from the outside.[12] Recent developments in a number of sciences, including Ilya Prigogine's work on the thermodynamics of non-equilibrium processes and Manfred Eigen's work on biochemical processes, have taught us to recognize how determinism and indeterminacy, conditionedness and freedom, are built into the basic structure of all natural processes.[13]

It appears that as the evolution of the biosphere has continued through the animals that have more complexly developed CNSs, oughtness and values are intrinsic to the process—it is a value-driven process.[14] The genetically controlled adaptive plasticity of the human phenotype is confronted with the intrinsic demand that it make choices between alternatives. Among the alternatives are those that are not presently actual, but which can be imagined and made actual in the future. Survival hinges upon these choices. The choices have consequences, and choices must be made as to how those consequences will be dealt with. The chain of choice, feedback, consequence, and response to feedback, is endless.

2.4 *Evolution as the Ambience of Freedom and Value*

What I have just described is the ambience of values—the emergence of values as a requirement for life and its evolution, the clarification of values, the achieving of consensus about specific values, and the taking of responsibility for actualizing the values. This theme will play a major role in later stages of this discussion. Most recently, scientific research into this constellation of issues has been the focus of the emerging disciplines of *evolutionary psychology* and *human behavioral ecology.* Each of these disciplines, the successors of what was formerly termed *sociobiology,* is in fact interdisciplinary, including researchers from genetics, biochemistry, evolutionary biology, ethology, primatology, psychology, anthropology, game theory, and the various neurosciences.

These disciplines carry the investigations of Burhoe, Pugh, and the sociobiologists into wider ranging and more complex directions. Two items are of particular importance: 1) The hypothesis that adaptive behaviors are shaped by critical moments in evolutionary history, and that these are transmitted in bundles of adaptations. One of the leading suggestions is that the Pleistocene was a particularly significant epoch, in which many basic human behaviors evolved, including those pertaining to child rearing, cooperative strategies, and relation to other animals. 2) The focus on the evolution of cooperative behaviors, including morality, although still at a relatively early stage of development, is producing a large literature on the genetic, neurobiological, and cultural interactions that inform these behaviors.[15]

---

[12] Jeffrey Wicken, "Toward an Evolutionary Ecology of Meaning," *Zygon:* 24 (June, 1989): 153–84.

[13] Arthur Peacocke, *Creation and the World of Science* (Oxford: Clarendon Press, 1979); Ilya Prigogine and Isabelle Stengers, *Order out of Chaos* (New York: Bantam, 1980); Manfred Eigen and Ruthild Winkler, *Laws of the Game* (New York: Harper and Row, 1981).

[14] George Edwin Pugh, *The Biological Origin of Human Values* (New York: Basic Books, 1977); Wicken, "Toward an Evolutionary Ecology of Meaning."

[15] Donald T. Campbell, "The Conflict between Social and Biological Evolution and the

2.5 *Summary*

The current state of research suggests the following concerning the issues we have discussed in this section.

1. The distinction between the genetic and the cultural dimensions of human nature are suggested by the most recent scientific research, and these distinctions are of great heuristic value. At the same time, it would be highly misleading if this distinction were to be read into human nature in a way that suggests a dualism.

2. Genetics and the neurosciences are foundational disciplines for understanding human culture and the behaviors it produces. The significance of the human CNS can hardly be overstated, inasmuch as without it there would be no possibility for culture as humans know it. The brain has emerged, however, on the basis of the genetic information that allowed it to emerge and shaped its emergence.

3. The evolution of behavior, specifically of those behaviors that pertain to morality, is constituted by several factors, and one of the most important is the human ability to fashion frameworks of meaning that contextualize the concrete data of experience that suggest the meaning of those data and the their implications for human behavior. Morality may be considered to be highly dependent on the construction of such frameworks and bestowing normative meaning through them. The human CNS is highly competent in forming such frameworks. It is this ability, along with the adequacy for interpreting the world and its challenges, that underlies the spectacular adaptive success of the human species. Religions are also bodies of such frameworks of meaning. These three components—the CNS, morality, and religion—are bound together in this particular dimension of human nature.[16] Scientists recognize this interrelatedness, even if they do not approach it in ways that theologians do. Up to this point in our discussion, we have not given attention to this element that I term "fashioning frameworks of meaning" and "contextualizing behavior within meaning"; the next section will introduce the discussion of these matters.

3 *Meaning, Transcendence, and Project*

The question of meaning, as we began to note at the end of the previous section, forms an intrinsic element of the human being. *Homo sapiens* has evolved as a creature who seeks meaning. Its CNS is a highly competent instrument for seeking and describing meaning, and its survival depends on the adequacy of its search for meaning. These three elements—human evolutionary history, the human brain, and human survival—are enveloped by the reality of what we call *meaning* and the search for it. My concept of meaning will become clearer as I elaborate these three elements.

Concept of Original Sin," *Zygon* 10 (1975): 234–49; Robert Boyd and Peter J. Richerson, *Culture and the Evolutionary Process* (Chicago: University of Chicago Press, 1985); William Irons, "Where Did Morality Come From?" *Zygon* 26 (March, 1991): 49–90; idem, "Morality, Religion, and Human Evolution," in *Bridges between Science and Religion*, Wesley Wildman and Mark Richardson, eds. (New York: Routledge, 1996); Jerome H. Barkow, Leda Cosmides, and John Tooby, *The Adapted Mind: Evolutionary Psychology and the Generation of Culture* (New York: Oxford Univ. Press, 1992).

[16] Irons, "Where Did Morality Come From?" idem, "Morality, Religion, and Human Evolution."

3.1 *Genes, Culture, and the Locus of the Search for Meaning*

The fact that meaning is fundamental to human nature seems correlative to the dual nature of *Homo sapiens,* constituted by both genetic and cultural information. Culture is defined here as learned and taught patterns of behavior and the symbol systems that contextualize those behaviors, both interpreting and justifying them. Unlike genetic information, cultural information is not programed, but rather discovered, learned, taught, interpreted, and justified. Even in the earliest phases of human history, a sense of appropriateness was required in the selection and application of these behaviors. The intentionality and consequences of the behaviors had in some sense to be in mind, particularly with behaviors that pertain to child-rearing, status and authority in the community, cooperative behaviors, especially among non-kin, and hunting and gathering behaviors. These considerations of appropriateness, intentionality, and consequence are all components of what we call the *meaning* of behavior and of the world in which that behavior takes place.

3.2 *The Brain and the Search for Meaning*

The human brain is an appropriate organ for the creature who is cultural and who is faced with the challenge of meaning that I have just described. The brain makes culture possible and is also the organ that conducts culture. Neuroscientists tell us that the human brain is distinguished by its ability to process information, which includes constructing frameworks that interpret the data those brains receive. Our brains do this so successfully that they enable the human creature to survive as the dominant species of the planet.[17]

This process of constructing frameworks and interpretations is not in all respects a moral process; some would call it "pre-moral." It does inevitably become moral, whether directly or indirectly. The important thing to note is that constructing frameworks and interpretations for our experience is not tangential to what our brain does, nor peripherally relevant to our survival, not an add-on, but rather intrinsic, a basic element of our brains and our survival.

In doing this, our brains specialize, as anthropologist Solomon Katz has noted, in ascribing what isn't there to surface observation, but what turns out upon further consideration to be real.[18] The frameworks and interpretations we construct supply meaning to the data of our experience, and thereby augment our experience by conjoining what is actually present in our experience with interpretations that require deeper discernment of that experience. We do not experience indigestion or stomach cancer; rather, we experience sensations of pain and discomfort that we and our doctors interpret, by adding what is not there in the original experience, the diagnosis—and our survival depends on how well we discern this interpretive element that "isn't there." Having made this diagnosis and acted upon it, we proceed not only to say that the diagnosis "is" there, but also to say that it is part of what is "most real" about the experience.

---

[17] Rodney Holmes, "*Homo Religiosus* and Its Brain: Reality, Imagination, and the Future of Nature," *Zygon* 31 (1996): 441–55; Terrence W. Deacon, "Brain-Language Coevolution," in *The Evolution of Human Languages*, J. A. Hawkins and M. Gell-Mann, eds. (Santa Fe Institute Studies in the Sciences of Complexity, Proc. Vol. X, 1990); idem, "Rethinking Mammalian Brain Evolution," *American Zoologist*, 30 (1990): 629–705.

[18] Solomon H. Katz, "Evolution and the Human Brain," paper delivered at the American Academy of Religion, Philadelphia, Penn., on 20 November 1995; Robert Ornstein and Paul Ehrlich. *New World, New Mind* (New York: Doubleday, 1989).

3.3 *Transcendence*

What we have, based on what we have just discussed, is an encounter between humans and the rest of nature, consisting of the mutual and reciprocal impact of *Homo sapiens* upon its environment and of the environment upon *Homo sapiens*. In this reciprocal relationship, possibilities are elicited in both parties, engendering new actualities. The human species has been unleashed within the ecosystem of terrestrial evolution as a species that is bounded by that ecosystem and defined by its working, but the species, bestowed with self-awareness and the ability to act upon it, must actively seek to *discover* its bounds and its definition—thus it is driven to self-definition within the context of its already having been defined. The human being has emerged as a creature that can understand itself, and that possesses the ability both to define nature and itself and to act responsibly upon that definition. In this discovery, self-definition, and action, humans inevitably alter what is, even in the act of discerning and defining what is, and thereby make an impact on the rest of the ecosystem. Human survival requires this act of self-defining and the responsible action that flows from it. As we shall discuss in greater length later, this is what is described in the phrase, *humans as the created co-creators*.

This self-definition is both reflective, in that it leads to deeper self-understanding, and also political, in that it is accompanied at every point by action that is directed by the self-definition. As such, self-definition configures the encounter with transcendence in our lives. This encounter can be spoken of in five aspects: 1) The evolutionary process and the contemporary ecosystem are transcending themselves when they question their purpose through *Homo sapiens*. Humans do, in fact, engage in reflection and discernment on behalf of the natural processes that have produced them. 2) The act of humans naming the world and its elements and correlating to those names a set of uses brings a dimension of transcendence to the non-human world. 3) Defining individuals in relation to groups and groups in relation to their individuals carries with it a transcendence with respect to our self-understanding and behavior. 4) The task of self-definition in the context of the global village requires that we bring together in our self-understanding nature, individual, group, and global humankind. This confluence brings with it still another dimension of transcendence, as we see ourselves relativized and yet integral to even larger and more complex communities. 5) In our self-defining, we transcend our own future by opening up new futures.

> To sum up, we encounter transcendence in the self-defining process, inasmuch as in the attempt to know and to actualize ourselves we find at every point that from within what seems to be a "natural" thrust that is every bit "ours," which we own as ours, at every moment we come to the brink of what is MORE, the transcendent, and although the process is very much this-worldly, we find ourselves in touch with the MORE than this world, in its depths. We are ourselves continually altered and enlarged as we define ourselves. We surpass ourselves, and thus we are pulled and pushed towards newness. This is an unsettling state. Yet, being unsettled by the push and pull of transcendence is also very profoundly sustaining and consoling, because the disturbance that belongs to our attempt *to be who we are*, and the effort to be what we are called to be is the most comforting and sustaining moment of life…. because there is no greater comfort than to be struggling to actualize our self-definition in accord with what we feel moved to become.[19]

---

[19] Philip Hefner, "The Foundations of Belonging in a Christian Worldview," in *Belonging and Alienation: Religious Foundations for the Human Future*, Hefner and Schroeder, eds., 161–80 (Chicago: Center for the Scientific Study of Religion, 1976), 175.

The import of the term *meaning* should be obvious now, even if it has not been defined precisely. Meaning has to do with a sense of the nature of the world in which we live and how it relates to us. Meaning includes an understanding of our purposes and their relation to the world in which we live, as well as an intelligent grasp of our possibilities for conducting our culture in this world and the impact of our cultural strategies upon ourselves and our world.

Project enters here as a companion concept to meaning. This term 'project' occurs in the work of certain social philosophers, such as Alfred Schutz and Gibson Winter.[20] It is rooted in the philosophy of Edmund Husserl and Martin Heidegger. Project refers to "what it is all about"—the "it" in question may be an individual, in which case we speak of the individual's *project*; it can also be a community, such as a social or ethnic community, or even an entire society. I am suggesting that we can also speak of the project of the human species and, since that species is fully natural, nature's project. Heidegger's point is that the world encounters us in the form of possibilities to which we must relate and in which we must invest ourselves. "The world presents to me not so much actualities as possibilities. Rather than being free because I consider possibilities, I consider possibilities because I am free."[21] Both freedom and possibility are deepened in the awareness that we may invest ourselves in the possibilities in various ways.[22] In this vein, Winter writes, "the *project* is the total intentionality with which subjectivity as a totality is stretched toward the world as possibility—shaping the future."[23]

This being the case, we experience the world and seek to understand it in terms of *what it is all about* and *what it is for*. To understand what the world is all about involves also understanding what we, as individuals, as groups and societies, and as a species are all about. To be engaged in the search for understanding *what it is all about* and in the behavior that constitutes our self-investment in the possibilities that are thus presented to us, is to *have* a project or to be *engaged* in a project. The concept of project brings a number of dimensions of our lives together into an interrelated whole: knowing, understanding, intentionality, and behavior.

We have come to what is perhaps the central and most important insight that is delivered by the argument we are pursuing here. The human creature and human culture intrinsically bear the character of *project*. The human brain, the biological basis of culture, has been described as an organ which itself bears a project-character. The search for "what it is all about" is what Katz refers to as the element "that isn't there." The human brain, in both its evolutionary development and in its structure and function, appears to be geared for survival in a natural world that is project. The distinction of the human brain lies in its fitness to discern and construct projects with respect to nature; its survival rests on the fact that what it takes to be the projects of nature do indeed work. Furthermore, the cultural situation in which we live today and the crisis it poses suggest that we are losing our grasp on nature's project; we require a deeper understanding of what our project is as humans, and how that relates to the project of the natural and social world in which we live. We

---

[20] Alfred Schutz, "Choosing among Projects of Action," in *Collected Papers*, vol. 1, Arvid Brodersen, ed. (The Hague: Martinus Nijhoff, 1962); Gibson Winter, *Elements for a Social Ethic* (New York: Macmillan, 1966), 119–56.

[21] Michael Gelven, *A Commentary on Heidegger's "Being and Time"* (New York: Harper and Row, 1970), 86.

[22] Ibid.

[23] Winter, *Elements for a Social Ethic*, 135.

suggest that since this human being and its culture have emerged within the continuum of nature, there is in fact a project-character to nature. We believe that evolution to this point on Earth has revealed that nature itself is a project. Just as the diagnosis we spoke of earlier is first discerned and then asserted to be "what is truly the case," so we discern that nature is a project, and this project-character is what nature "is really like." The challenge for us is to discern just what the content of this project is.

3.4 Homo sapiens *as Created Co-Creator*

The understanding of nature and the human species within nature that we have just sketched is best captured in an interpretation of *Homo sapiens* under the concept of the *created co-creator*. The term gathers together the data that we have surveyed and refers to the emergence of the human creature, who, on the one hand is thoroughly a creature of nature and its processes of evolution—hence the term *created*—and who at the same time is created by those very processes to be a creature of freedom and culture—hence *co-creator*. It is the experience of this creature that forms the locus of what we have called the search for meaning, the encounter with transcendence, and nature (including human nature) as project.

4 *The Emergence of Culture and Its Crisis*

In order to comprehend the depth and scope of the human situation, we must attend to the challenge of culture. Earlier, we defined culture as learned and taught patterns of behavior and the symbol systems that contextualize those behaviors, both interpreting and justifying them. Culture is a vast system of information that is constructed to guide our behavior and interpret it. In our present globally dominant position, the human species has in fact conditioned all of the other natural systems with systems of human culture, chiefly those that we call *technology*.

Herein lies the great challenge, the challenge *to* human culture and at the same time the challenge *posed by* human culture. In our present condition, human culture must interface successfully with all of the non-cultural systems of information. These non-cultural systems include the systems that are at work in all of the rest of the natural world—oceans and rivers, mountains and flood plains, atmosphere, plants and animals, and all the rest. The survival of the human species and that of the other planetary subsystems depends on a viable symbiosis between human culture and the rest of nature.

Furthermore, even though it is in many respects a unified system of information, human culture itself is not a single monolithic system. Human culture is a constellation of many diverse cultural strategies for living. These strategies have accessed somewhat different pools of available information and organized those pools in a multitude of ways to meet the challenges of living in diverse settings. This diversity is also reflected in the myths and rituals that inform the strategies. In addition to interfacing with all of the non-human systems of information that drive our planet's life, global human culture is challenged to intermesh its diversity of cultures in ways that are life-giving and life-sustaining for the whole of the human community and for the planetary ecosystems of which it is a part.

In describing this situation, we are also in fact delineating a crisis that faces us today. It is a crisis of human culture that pertains to the manner in which we direct our learned patterns of behavior. The entirety of human culture is caught up in this crisis, but some of its elements come into special focus: science, technology, human freedom, and our capability of interpreting these other elements. Science and its technology are perhaps the most dramatic forms of human culture today, and have in a sense created the world in which we live. It is because of them that freedom has assumed both its contemporary urgency and present shape. Perhaps it would be more accurate to say that it is because evolution on Earth took the course it has taken—namely, the emergence of the human creature who is an animal of culture, which takes the form of science and technology—that we face the challenge of freedom as I outline it here. Or, to state the matter still differently, the condition of freedom is a phase of nature's evolution on planet earth, the phase of *Homo sapiens* and its science and technology. Let us consider briefly just why this is so.

Freedom describes the condition of life in which we must make decisions, evaluate those decisions, justify them, and take responsibility for them. Freedom is an urgent, life-or-death matter today, first of all, because we inhabitants of the Earth are essentially one interdependent global human community, existing in symbiosis with one finite, planetary ecosystem; secondly, because the human community is stretching the carrying capacity of the global ecosystem to its very limits; and thirdly, because in the present density of population, the technology of war intensifies violence, oppression, and injustice. All three of these factors—the interrelated planetary community (with all of its psycho-social-political-economic dimensions), the pressure on the planet's ecosystem, and the thirst for violence—have come about under conditions that are enabled by science and its technology. Without the sciences of medicine and agriculture, and their related technologies, the population of the planet would not have reached its present state, including the intensity with which it covers the surface of the earth. Similarly, the science and technology that undergird transportation, communications, and armaments have contributed to the scope and intensity of the relationships within Earth's human sector. Without the size of the population, its global spread, and the science-based technologies upon which that population depends, the carrying capacity of the ecosystem would not be in jeopardy.

In one sense, these developments are not new. The challenge of relating to other tribes or human groups has been a prime factor in human evolution since at least 30,000 years ago. Also, the allocation of natural resources, including the food derived from plants and animals, has been part of the challenge posed by the growth of human population and the necessary interrelations that go with it.[24] What is new about the situation of our epoch are the factors that 1) the carrying capacity of the planetary ecosystem is approaching its limits and 2) there is virtually no access for humans to Earth's resources that are essential for survival except through the very forms of culture that have come to this state of crisis. There are no new herds to search out, no new wildernesses to develop, and no new fields of limitless mineral resources to exploit. Indeed, were the present level of technological overlay to be sizeably cut back, the world's human population would have to be trimmed by many millions of persons. Clearly, science and technology are inseparable from issues of war and peace, social justice, ethnic diversity (and other expressions of human diversity) and conflict, as well as environmental issues. Perhaps the most novel aspect of the current situation is that we are the first epoch to be aware of the planetary condition and our role in it.

---

[24] John Pfeiffer, *The Creative Explosion: An Inquiry into the Origins of Art and Religion* (Ithaca, N.Y.: Cornell University, 1982), chaps. 3, 8, 12; Richard Klein, *The Human Career: Human Biological and Cultural Origins* (Chicago: University of Chicago Press, 1989), chap. 8.

Against this background, we can say that science and technology form the context in which human life will find either fulfillment or catastrophe, or some tense homeostasis of the two. This context is that of freedom—decisions must be made, and we have to be prepared to live with the consequences of these decisions. Science and technology not only provide the context, but their development is also the spur for motivating humans to face up to the necessity of decisions and their complexity. There is still more, however: since science and technology are now essential for human life as we know it, they thoroughly condition the future of the planet. Since, barring catastrophe, the future is one in which science and technology will, even more, condition the basic circumstances of life on our planet, we can say that they also constitute the medium through which the future of life on earth will be expressed.

A first awesome consideration becomes clear when we recognize the magnitude of the task that culture faces. Its *function* is comparable to that of physico-bio-genetic systems of information that surround us, namely, to guide behavior. The magnitude of this task dawns upon us when we consider how thoroughly the behavior of the other members of the biosphere—plants and animals—is guided by their physico-bio-genetic inputs. Furthermore, those other inputs, lacking cultural supplements, are inadequate for humans. Our definition of culture underscores this function of guiding behavior: a system of learned patterns of behavior and the symbol-systems that contextualize and interpret those actions. Culture's functioning is not yet complete when it provides symbol systems that put one's world together; those symbol systems must also motivate and guide behavior.

Culture's *scope* becomes clear when we recognize that the cultural systems of information interface directly or indirectly with virtually all of the planetary physico-bio-genetic systems. Beginning with our own bodies and extending to the entire ecosphere of the planet and to the other human cultures that we live with, the range of that with which culture must interface is almost beyond comprehension. Here the urgent struggle that is inherent in the quest for human identity comes into full view. The formation of culture is not just of private concern, either to the individual or to the local community that comprises the culture. That formation must synchronize with the ecosystem and with other human cultures and subcultures. In our present epoch, the crises that threaten our lives grow out of the failure to synchronize as we must with these other systems of information. Human survival can be defined for us today as the process of understanding and actualizing congruence with these other systems.

The *intellectual and spiritual* dimension of culture's challenge is the interpretive task that culture must perform. How are we to understand the purpose of being human and the significance of our human venture within our segment of the evolutionary history of nature? I interpret our own era as the time when a human cultural overlay has been imposed upon virtually all of the natural systems that comprise the planet Earth. Or perhaps a more apt image would be that human culture has permeated all of Earth's natural systems, with the consequence that all of those natural systems, including human beings and their societies, are substantially—even life-threateningly—dependent upon human decisions and the ability of humans to conceive and execute those decisions adequately. Since our era is the time in which it is impossible to retreat to a time or place where natural processes still operate on their own, apart from or untouched by human decisions, there is little or no possibility for humans to take up residence in locales where they can live off non-human natural processes without human manipulation of those processes. The time is long past when we can let those processes simply take their course and sustain us. There is no wilderness that is really untouched by human decisions, no foods that are "natural" in a pre-human sense, very little water or atmosphere that is in such a non-humanified state. Even though it is possible to fantasize to the contrary, such fantasies are misleading and even dangerous.

This state of affairs is what Teilhard de Chardin and a previous generation of thinkers called the condition of *hominization*—humanification of nature and its evolution. The hybridized navel orange, the automobile, the asphalt parking lot, the computer—these are nature. We call them *techno-nature*, recognizing that techno-nature is, in a real sense, the only nature there is now, on our planet. All of nature is in some sense *Cyborg*, to cite Donna Haraway's analysis.[25] Such terminology is offensive to many persons, and that offense is rooted in our alienation from nature, from our own nature, and from the distinguishing feature of human nature in our time: culture and the technological form it has taken. Unless this alienation is overcome, humans will not be capable of assuming their authority over their own nature and directing it into genuinely humane channels. There is essentially no difference between the phenomenon of the bee producing honey and the human being fashioning a fast-food burger. The technological overlay that characterizes the production of the burger is the same nature as in the honey bee, performing in a manner appropriate to the evolutionary context of human culture. That we may attach differing value judgments to the work of the bee and the work of technologically advanced food processing should not cloud our sense of the fundamental sameness of the two activities.

The term for designating this historical situation is 'Technological Civilization'. Technology has been interpreted from a number of perspectives. My hermeneutic for interpreting it is rooted in the fact that human culture and the challenges it faces have emerged within the evolutionary processes of nature. Technology participates in the same challenges as culture, because technology is part of culture, perhaps the most luminous facet of culture in our time, since it impinges so significantly upon the future of *Homo sapiens* and the planet. Technology must be interpreted; it is a behavior to be guided; and it must interface in a wholesome or well-synchronized manner with the full range of Earth's physico-bio-genetic information systems.

My approach for interpreting technology, drawn from evolutionary history, is non-dualistic: technology is a form or segment of nature, and its emergence is grounded in the same neurobiological matrix as human beings and the rest of their culture. Furthermore, my interpretation asserts that the purpose of technology, along with the purpose of *Homo sapiens* and culture, is to be referred to on the basis of the natural order whence it emerged. This angle of vision on technology gives a quite different interpretation of its significance than some others, especially those that set it off dualistically from nature and from the supposedly "more authentic" character of human being.[26]

My interpretation does not, however, weaken our sense of urgency at the crisis that technology poses for humans and the planet. Furthermore, I use the term Technological Civilization to suggest that, unlike the Sorcerer's Apprentice, we are

---

[25] Donna Haraway, *Simians, Cyborgs, and Women: The Reinvention of Nature* (New York: Routledge, 1991); William Grassie, "Donna Haraway's Metatheory of Science and Religion: Cyborgs, Trickster, and Hermes," *Zygon* 31 (1996): 285–304.

[26] Martin Heidegger, *The Question Concerning Technology* (New York: Harper & Row, 1977).

confronted not simply with tools that have gotten out of control, but rather with a fundamental, all-permeating condition of existence that threatens to turn against itself and the rest of nature. This phenomenon of the good turning against itself is one description of the demonic element of life.[27] Furthermore, technology does tend to focus our attention on causation in nature, to the exclusion of a concern for the meanings of nature. In the long-run, the suppression of meaning in favor of causation awakens dissatisfaction in human experience.[28] Our vantage point also allows us to throw different light on the nature of the crisis, its potential disastrous consequences, and on the appropriate response to the crisis. The nature of the crisis is identical to the crisis that attends human culture, as such: Technology must be interpreted (our theory is providing one such interpretation) and it must interface with the rest of nature's systems so as to meet, on the one hand, the pragmatic criterion of wholesomeness, and, on the other hand, to function within the larger purposes of nature. A re-organization of consciousness is called for, if we are to respond adequately to this crisis.[29] Humans do use technology, and they are technology, in the same sense that we not only use our hands and eyes, but we are our eyes and hands. Eyes and hands are not good or evil in and of themselves, but they are good or evil depending on us and the agency we exercise through them. Our technology can gain control over us, almost demonically, just as our physical bodies can gain control over us if we devote most of our time and resources to body-building strategies or to sexual prowess or to comfortable lifestyles. In comparable ways, any part of us can go berserk, so to speak, and take us over—our intellectual dimensions, our sexuality, or our gregariousness.

The appropriate response to Technological Civilization is to recognize that it is human culture, that it is an emergent from what we have described as human freedom, and that it is constituted by our self-consciousness, our constructions, and our decisions, for which we take responsibility. The response, in order to be appropriate, must be the response of creatures who are themselves natural creatures, and who understand that they are responding to the natural world in the form that it has taken commensurate with our particular epoch in evolutionary history.

5 Classical Christian Theological Resources for Interpreting Nature: Creation

We now move into the final and most constructive phase of the discussion, in which we sketch a presentation of classical Christian theology in a form that can serve to interpret the preceding analysis. This theological interpretation proceeds under the rubric of the doctrine of creation.

5.1.1 Creation: The Natural World Interfaces with the Character of God

Christian faith invests the natural world with meaning by proposing that that world is conditioned by the character of God. The proposal is grounded in the traditional

---

[27] Paul Tillich, "The Demonic." in *The Interpretation of History*, Paul Tillich, ed., 77–122 (New York: Charles Scribner's Sons, 1936).

[28] Paul Tillich, *The Spiritual Situation in Our Technical Society* (Macon. Ga.: Mercer Univ. Press, 1988); Charles D. Laughlin, "On the Relationship between Science and the Life-World: A Biogenetic Structural Theory of Meaning and Causation," in *New Metaphysical Foundations of Modern Science*, Willis Harman and Jane Clark, eds. (Sausalito, Calif.: Institute of Noetic Sciences, 1994).

[29] Mihaly Csikszentmihalyi, "On the Relationship between Cultural Evolution and Human Welfare," unpublished paper, delivered at the American Association for the Advancement of Science, in Chicago, February 1987.

theology of creation. This theology orbits around two technical phrases: *creation-out-of-nothing* (*creatio ex nihilo*) and *continuing creation* (*creatio continua*). These two doctrines finally mean the same thing, that the relation of God to nature at the point of nature's origins continues as God's relation to nature forever.

In the vocabulary of the disciplines that study culture, these affirmations and doctrines are "memes" that the Christian tradition has unleashed for our consideration, memes that compete for our attention and that claim to be beneficial for us.[30] I will finally describe these memes under the rubrics of Creation, Sin, and Love. I call these memes Christian, even though I recognize that they are in large part present also in Jewish and Muslim traditions.

5.1.2 *What the Doctrine of Creation Does Not Mean*

The creation-out-of-nothing affirmation is particularly critical for our understanding. This particular meme is, to my way of thinking, greatly misunderstood. I summarize the major misunderstandings as follows.

1. Contrary to what is often said, the term "out of nothing" is not primarily a statement of chronology. Its chief affirmation is not that, looking backwards over a long chain of cosmic events, we come finally to God and creation, the unmoved mover of the continuum. Such ideas may be implied, at least in the minds of many people, and they are not necessarily mistaken, but they are not what the doctrine of creation-out-of-nothing mainly is about. Chronological speculations may not be wrong, but they are difficult and in part inappropriate. Since time or chronology is a characteristic of nature, it cannot have existed as we think of it before there was a world.

2. Contrary to what some thinkers have suggested—even the great Nicholas Berdyaev—the term "out of nothing" does not intend to depict that the world emerged out of a state of "nothingness" or chaos. "Nothing" does not refer to a state prior to the emergence of the world, out of which God made the world.

3. Although the concept of Creation-out-of-nothing emerged in the late second century in response to misunderstandings and attacks on Christian faith by Hellenistic thinkers, its meaning cannot be reduced to the polemical situation in which it took shape. Rather, it grew out of Christian awareness that the faith expressed in the Bible required this conceptual development if it were to be articulated in the Hellenistic situation.[31]

5.1.3 *What the Doctrine of Creation Does Mean*

What can we say concerning the positive, constructive meaning of the basic Christian affirmations about creation? The affirmations mean to assert, first of all, that there is but one ultimate grounding for the natural world that is described by the sciences, and this grounding is the God affirmed by the Christian tradition. Secondly, these primal affirmations say something about the conditions under which this grounding exists. Let us examine these assertions in more detail, because they are the foundations for everything that the Christian religion asserts about nature.

---

[30] Mihaly Csikszentmihalyi, *The Evolving Self* (New York: HarperCollins, 1993), 119–46; Dawkins, *The Selfish Gene*.

[31] Robert Wilken, *The Christians as the Romans Saw Them* (New Haven: Yale Univ. Press, 1994), chap. 4; Wesley Fuerst, "Old Testament Views of the World's Beginnings," in *The Epic of Creation*, Thomas Gilbert, ed. (in press); Edgar Krentz, "The Greco-Roman Context for New Testament Ideas of Cosmogony and Cosmography," in *The Epic of Creation*, Gilbert, ed.

The out-of-nothing affirmation establishes that there is one and only one source or grounding for the natural world: the God of the Hebrew-Christian tradition. This is the core of the presentations in the two Genesis creation stories, as well as the traditions that are embedded in the Book of Job, the nature Psalms, and the New Testament traditions that we find in the Gospels and the writings of Paul, the Epistle to the Hebrews, and the Book of Revelation.

We grasp the significance of this motif of a single grounding for the natural world when we compare it with the creation story that most Hellenistic intellectuals accepted, the story reported in Plato's *Timaeus*. Plato reports that God, the demiurge, was constrained by a second grounding element of nature, pre-existent chaos. Consequently, God could not create the world that was most desirable, a perfect world, but had to work within the possibilities that chaos would allow. The end result was a natural world that was defective, capable only of partial conformance to the creator's intention. If perfection were to be attained, it would require separation from the world—the soul would have to be liberated from the prison-house of the body.

In addition to the motif of a single ground of the natural world, the creation-out-of-nothing doctrine also eliminates all mediation between the creator God and the natural world. As Gerardus van der Leeuw points out, the Hebrew story of creation is a very simple equation, comprised of three factors: God, what God creates, and the future that God brings about for the creation.[32] The interface of God and the natural world is one of immediacy. The significance of this immediacy comes to the fore when we compare the Hebrew story with myths that require some sort of testing or ordeal before God's creation work can be completed. Such myths include those that depict creation as the conquest of chaos. The Babylonian *Enuma Elish* is an example of such a story. Marduk must struggle in a violent contest with chaos, killing Tiamat, before the natural world can emerge.

5.1.4 *Nature Conditioned by the Character of God*

The logic that comes into play from this theological base is of great significance for understanding nature and the fundamental stance of Western religion toward nature. This significance is not often perceived, even by Western theologians. By this double affirmation, that God is the sole ground of nature and that God stands in an unmediated relation to the creation, these traditional memes assert that the character of God conditions the nature of nature, in a fundamental way.

An analogy may help us understand this point, even though it is finally an inadequate analogy. Think of a person who wishes to give a special gift to a beloved friend or lover or family member, perhaps for a special occasion or as special sign of love and respect. If the person buys a gift, we recognize immediately that the giver is dependent upon what is available, and must accept the fact that what is available was not made specially with the beloved, the occasion, or the special relationship between the giver and recipient in mind. The giver must make do, in other words, in the context of a formidable array of constraints that are in fact obstacles to the intention of the giver. Consider, however, the difference if the giver makes the gift personally—designing it with particular reference to the relationships involved, the occasion, and the beloved. We also assume that the giver can acquire whatever

---

[32] Gerardus van der Leeuw, "Primordial Time and Final Time," in *Man and Time*, vol. 3 of *Papers from the Eranos Yearbooks*, J. Campbell, ed., 346 (New York: Pantheon Books, 1957).

materials and tools are desired; here the analogy breaks down, to a degree, because the giver must accept the inherent properties of the available materials. Given these factors, when the gift is bestowed, we know that it is exactly what the giver desired, and that it is precisely what the giver believes is appropriate to convey the giver's intentions on this specific occasion, for this particular beloved person. We know that the nature of the gift is conditioned only by the character and intentions of the giver.

The upshot of the double affirmations of God as sole ground of nature, creating in an unmediated relationship to the creation, is to place God in the position of the giver who designs and makes the gift, with the exception that God could create whatever materials were desired. God, therefore, worked under no constraints except those of God's own character and choosing. The end result is that nature is not only the creation of God, when perceived from within the Christian frame of vision, but it is also the creation that God intended.

Viewed in the perspective of our entire argument, we may say that the doctrines of creation assert that nature is God's project.

5.1.5 Pantheism: A Disclaimer

We should be clear that pantheism is not a reasonable conclusion to be drawn from what I have described as the logic of the Christian affirmations about nature. To say that there is an unmediated relationship between God and what God has created does not establish an identity between God and nature. This point is conveyed by our analogy. A work of art may well convey the intention of the artist, as the gift may convey the intention of the giver, but there is no question that the artist is the work of art, nor that the giver is the gift. Nor is the Protestant principle, that there is an infinite qualitative difference between God and the world of nature, violated in the least.

5.1.6 The Conditioning Character of God: Freedom, Intentionality, and Love

What is the character of God that conditions the natural world, just as the character of the hypothetical giver conditions the gift? The Christian tradition forcefully enunciates three aspects of God's character that are important for our reflection: freedom, intentionality, and love.

Freedom asserts that God was under no constraints in the creating work. God could create whatever world was desired. The polemic of the second century C.E. was particularly concerned to assert this freedom. The much-discussed impassibility of God should be interpreted in this light. Impassibility means that God does not suffer. This is not intended to depict a God who is cold and lacking in compassion. In our context, it means that God did not suffer the coercion of any other being in creation. Plato's God, the demiurge, was passible precisely in this sense, that he could not gainsay the impact of preexistent chaos in his creating work. One might suggest that the consequence of this freedom for nature is the *evolutionary character* of the natural world and the ways evolution proceeds.[33]

Intentionality asserts not only that God willed to create the world, but also that the world is created in accord with the particular intentions of God, that God had purposes in creating, and that the creation embodies those purposes. In the same way, our hypothetical gift was the consequence of the giver's purposes and embodies

---

[33] John Hick, "An Irenaean Theodicy," in *Encountering Evil*, Davis, Stephen, ed., 69–100 (Atlanta, GA: John Knox, 1981); see also Rudolf Brun, "Integrating Evolution: A Contribution to the Christian Doctrine of Creation," *Zygon* 29 (September, 1994): 275–96.

those purposes. So, too, Christian faith asserts that, however inscrutable they may be, *nature has purposes.*

Love asserts that the creation was motivated by God's love and is governed in its workings most fundamentally by that love. Love, then, becomes for the Christian the *operating principle* of nature. The classic tradition of theology asserted this at several points, such as Augustine's image of the created universe as a *carmen Dei* (song of God); Fr. Zachary Hayes, who is influenced by Bonaventure, utilizes such traditional resources to argue that we must recognize the "sacred nature" of the universe.[34] Antti Raunio has demonstrated that Martin Luther also took up this theme.[35]

All three of these elements of the divine character—freedom, intentionality, and love—must be discerned and interpreted. Discernment and interpretation are undertaken on the basis of what the Christian tradition understands about God and God's intentions and love. The final criterion of both God's intentions and love is set forth, for Christians, in Jesus Christ, just as for Jews it is set forth in the Torah and for Muslims in the Qu'ran.

In light of the basic logic of the Christian affirmations, we must conclude that the natural world is the world that God intended, and that the nature of nature is conditioned by God's freedom, intentionality, and love. The doctrine of creation out-of-nothing asserts this for the natural world in its origins, while the doctrine of continuing creation asserts this for the natural world in every moment of its existence.

5.2.1 *Excursus: The Centrality of Meaning*

The Christian affirmations about nature weave God's meaning into nature inseparably from God's agency. The Christian faith, as well as the Jewish and Muslim faiths, have no real interest in any conceptions of nature that do not include nature's meaning. In technical theological terms, we say that creation is subsumed under soteriology, the thatness of nature is inseparable from its meaning. In these considerations, we observe how thoroughly Christian faith and theology share the notion that the world and nature are project, that they are to be interpreted by a hermeneutic of intentionality. For Christian theology, quite obviously, nature is God's project, the subject and object of God's intentionality.

Discussions between science and theology often overlook this "thick" religious perspective. Arguments about the existence of God are finally unsatisfying to the theological mind, as are discussions about whether and how God might be present in nature. Such arguments and discussions are not unimportant—to the contrary, they are vital—but they are not enough. Christian theology is more concerned to say that nature has meaning, and this implies finally that nature reflects the character of the creator God. Two thousand years of discussion between theologians, natural scientists, and philosophers on the question of whether there is a God and whether God could be present in nature has pretty much come to an impasse, in the sense that the outcome of this discussion leads one to say that it is equally intelligent either to believe in God or to disbelieve. The issue of meaning is another question, however. Without meaning, human life is impossible, but all proposals for meaning are just

---

[34] Zachary Hayes, O.F.M. "Christology and Cosmology," in *The Epic of Creation*, Thomas Gilbert, ed. (in press).

[35] Antti Raunio, *Summe des christlichen Lebens: Die "Goldene Regel" als Gesetz der Liebein der Theologie Martin Luthers von 1510 bis 1527* (Helsinki: Department of Systematic Theology, 1993).

that, proposals, hypotheses. In theological jargon, all meaning involves a leap of faith. In his "Ars Poetica,"Archibald MacLeish spoke memorably in the statement, "A poem should not mean, But be."[36] We appreciate this wisdom, and we know why he said this, to remind overconfident critics that a work of art cannot be reduced to any single reader's interpretation, nor to all of the possible interpretations put together. The theologian would be inclined to restate MacLeish's aphorism thus: "Nature never simply is, it also means." For the Christian faith, this translates into the assertion that nature reflects, or conforms to, the character of the creator God: freedom, intentionality, and love.

5.2.2 *Meaning Is Woven from Thatness*

To say that the *thatness* of the natural world is inseparable from its meaning cannot, however, encourage a dualistic perspective composed of facts and interpretation, or causes and meaning. It is just as true to say that there is no meaning apart from the way nature is described by the sciences, as it is to say that nature cannot be experienced in isolation from frameworks of meaning. It cannot be overemphasized that the work of the sciences in describing nature is constitutive for the work of interpreting the meaning of the natural world. The meaning of nature grows out of our experience of it, as much as it is imposed upon that experience, and this experience of nature includes our scientific experience, as described by the scientists.

If intentionality is ascribed to nature, that intentionality is experienced under the forms of randomness and natural selection, as well as under the forms of genetic predisposition. If freedom is posited of nature, it is freedom that works according to the various realms of evolutionary process—cosmic, biological, cultural. This is true even if the intentionality and freedom are understood to be rooted in God's character, because apparently God chose to work out freedom and intentionality within the evolutionary context. It is even more essential that we understand love within this perspective. I contrast such a view with conventional philosophical or theological moves that define God's love in ways that cannot be accommodated to the natural order and then conclude that nature is not governed by love.

Such moves flow from an anthropocentrism that prefers to conceptualize intentionality, freedom, and love solely with reference to human affairs. The force of an interpretation like mine, which insists that the logic of the Christian faith conceives of nature as conditioned by the character of God, suggests that if we define freedom, intentionality, and love solely with respect to humans, then we are in fact forfeiting any possibility of relating God's essential character (as revelation claims to know that character) to the natural world, just as we are also in the deed separating humans from the natural order. Since it is absurd to commit either of these separations—God from nature or humans from nature—we have no choice but to conceptualize the characteristics of God (freedom, intentionality, and love) with reference to the entire natural order and then submit to the incredibly difficult task of discerning how those characteristics actually work themselves out in the creation, that is, what their content really is.

We may summarize this argument by paraphrasing Paul Tillich: *scientific description provides the form of nature's meaning, meaning provides the substance of scientific description of nature.* Or, in other terms: *evolution provides*

---

[36] Archibald MacLeish, "Ars Poetica," in *Modern American Poetry*, Louis Untermeyer, ed., 472–73 (New York: Harcourt, Brace, and Co., 1950).

the form of God's presence in nature, God's presence in nature is the substance of evolution. My focus in this paper upon the foundations of meaning in the conditioning of nature by God's character, gives attention to the *substance* of the meaning that takes form in the concepts provided by scientific descriptions.

5.3 *The Image of God*

That human beings are created in the *image of God* (*imago Dei*) is the chief Hebrew-Christian-Muslim assertion about *Homo sapiens.* Although this assertion has had a long and rich tradition of interpretation, there is no consensus on exactly what it means; there is no single official or even standard interpretation of the concept of the image of God. There is a great deal at stake in this classical theological symbol, since it moves toward making explicit what it means that God's character conditions nature, by locating within nature a place where God's intentionality for nature is represented. By designating the human being as the image of God, human being is thereby said to be a possible metaphor for the meaning of nature. In what follows, I will offer an interpretation of this symbol that contributes to the overall theological argument that is here proposed.

I start from the straightforward assumption that the image of God refers to that which portrays or sets forth God. To say that humans are the image of God means, therefore, that humans are somehow to be a portrayal or representation of God in the creation. Human beings were created, according to Christian faith, for this purpose: to represent or portray God in the creation. I am interpreting the concept of the image of God, therefore, to speak of the purpose of human being. Such an interpretation stands in sharp contrast to many past efforts that have in some way used the concept of the image of God to undergird the notion that humans are of greater value than the rest of nature, or that, since they are superior to the rest of nature, they are destined to dominate it and exploit it for their own benefits.

What does it mean to represent or portray God within the natural world? I refer to the way in which I have spoken of the character of God that conditions nature, including three elements: freedom, intentionality, and love. These characteristics of God condition all of nature, but in humans they come to explicit representation or portrayal. Since *Homo sapiens* is thoroughly a creature of nature, we may say that in humans nature itself becomes aware of itself as free, intentional, and loving. In this sense, *human being is a metaphor of the meaning of nature.* The form of this metaphor lies in the project character of human life, its intentionality. The content of the metaphor is constituted by love.

This way of interpreting human being relies heavily on my own concept of the human being as the created co-creator.[37] In this concept, I incorporate an understanding of the human creature as a symbiosis of genes and cultures. Our genetic heritage marks us as ineradicably creatures of nature. We are also creatures of culture, which I define as learned and taught patterns of behavior and the symbol systems that interpret and justify those behaviors. This character as cultural creatures is the seat of our freedom, our intentionality, and our self-awareness. What it means to be human is spelled out in the text that we produce as we conduct ourselves culturally, a text that is written in our social forms, our technology, our conduct of relations within the human community and also within the natural environment—and in all of

---

[37] Philip Hefner, *The Human Factor: Evolution, Culture, and Religion* (Minneapolis: Fortress Press, 1993).

the other aspects of our culture. We produce this text under the conditions of the civilizational crisis that I described in detail above.

5.4 Sin and Evil: Secondary to God's Conditioning of Nature

The traditional theology of Fall, Sin, Original Sin, and Evil are among the most misunderstood and maligned items of the theological repertoire. They are also critical for the Christian understanding of nature and human origins. I want to suggest that they take on very important meaning when interpreted within the framework I have sketched thus far. We do not properly understand these doctrines if we wrench them from their context of reflection upon nature, thereby referring them only to human beings in isolation. Their import is seriously misunderstood, for example, if they are simply taken to mean that humans are weak or bad.

In summary form, I suggest that these doctrines of Fall and Original Sin serve to underscore that when it comes to interpreting the natural world, human error and sin, as well as the evil they cause, are *secondary* to nature's status as conditioned by the character of God. What I mean by the term "secondary" is clarified if we refer to Plato's *Timaeus* and the Babylonian *Enuma Elish*. In both of those myths, nature is constituted by error, defect, sin, and evil; there can be no talk about creation unless defect and evil are placed in the center of this discussion. The Hebrew-Christian-Muslim myths differ strikingly. The discussion is completely turned on its head: *error, defect, sin, and evil cannot be properly understood unless we first consider that in creation, nature is conditioned by the character of God.* As we have observed earlier, this conditioning includes the action of the loving God who said of the original creation, "It is good." In the Hebrew myth, this is articulated in creation coming first, before error, defect, and sin, are brought on stage—the act of creation is a simple, uncluttered divine act, requiring no qualification or prior history, as is the case with the Platonic and Babylonian accounts.

What we need to understand is that these doctrines of Fall and Original Sin serve to create an epistemic distance between the ambiguity of the actual human condition and the primordial intentionality and love that mark the Creation. This distance or gap, which is extraordinarily significant, is characterized by the image of Fall, inasmuch as that image suggests that something intervenes between what we are essentially meant to be as human beings and what we are in actuality . The key here is the discrepancy between what Tillich called our essential and our existential states. The term "original sin" adds to this the insight that the discrepancy goes far back in our history; it is real today, but it did not happen only yesterday.

This discrepancy or epistemic distance is critical to the mythic assertions of Christian faith (and also Hebrew-Jewish and Muslim faiths), because it preserves the grounding of creation in goodness and meaningfulness from being negated by the obvious condition of ambiguity in which we live as humans, a condition of fallibility and vulnerability, marked by attitudes and behaviors that are almost always morally ambiguous and often simply shameful and evil. At the same time it allows that primordial grounding to serve as the norm for what humans are called to, for their destiny—both in terms of judgment and inspiration. This is accomplished by insisting that humans are not in actuality what they are called and created to be. Humans are creatures who live in deviation from their own essential character. If our actuality were taken as the clue to the essential status of the natural world, we could hardly sustain the affirmation of goodness and intentionality.

It may seem strange to identify these doctrines as liberating, but in fact they are, because they assert that as basic as sin and ambiguity are to human existence, they are not the norm for reality, and they do not finally disqualify humans from sharing in the essential goodness and purpose of the Creation. If it were not for the distance that these myths create between the empirical actuality of human life and essential goodness, we would have no recourse but to take sin and ambiguity as ontologically normative.

What we have in the Christian mythic assertions about nature as Creation and human beings as created in the image of God, yet fallen and sinful, is a massive ideological structure that aims to preserve the fundamental reading of the natural world as rooted in meaningfulness, goodness, and love. The logic of the Christian myth is such that it will tolerate just about anything—including great embarrassment concerning evil—except the weakening of this fundamental conviction about the Creation. This conviction, in turn, is what we should first focus upon in our efforts to relate Christian faith to the scientific world-views. This conviction governs the Christian attitude toward the world and it forms the foundation for understanding how humans should conduct their lives.

When we seek to relate science and Christian faith, the focus should be upon the fundamental Christian affirmations of goodness and meaningfulness. However, remember that I do not mean that issues of causality, randomness, quantum physics, and the like are not important. As I have already explained, they are highly important, because they constitute the loom on which nature's meaning is woven. I simply want to make it clear that the challenge to Christian faith is not solely a question of whether or not there is a God, or whether or not ancient Scriptures can speak to us today, or whether random natural processes can give rise to purpose. The challenge is whether we believe that deep down in the fabric of nature, what governs the world and what counts most is meaningfulness expressed in love. Perhaps this is what Gerard Manley Hopkins meant in saying that

> there is a freshness deep down things;
> And though the last lights off the black West went
> Oh, morning, at the brown brink eastward, springs—[38]

If goodness and meaningfulness are not deep down the essential principles of the world we live in, then the existence of the God of the Christians is a moot and uninteresting question. If we are able to discern goodness and meaningfulness as the basic elements that make this world tick, then the reality of God as described in Christian faith will be in view, even if the name "God" is never spoken.

5.5 The Challenge of Meaninglessness and Evil

The alternatives to the Christian affirmations ought to be brought into the discussion, such as the basic conviction that nature is grounded in evil and hatred, or that it is neutral to all values, lacking intentionality altogether. These are real challenges to the fundamental Christian-Hebrew-Muslim conviction. Placed in the context of these challenges, the sciences play an interesting and important role. For example, the theories of the "selfish gene" raise the question of whether altruism or love really are central to the natural world, at least in the biosphere. Hence the importance of E. O. Wilson's famous dictum that the existence of altruism beyond the kin group is the great intellectual problem facing sociobiology.[39] In *The Selfish Gene* Richard

---

[38] Gerard Manley Hopkins, "God's Grandeur," in *The Poems of Gerard Manley Hopkins,* 4th ed., W. H. Gardner and N. H. MacKenzie, eds. (London: Oxford Univ. Press, 1877/1967).

[39] E. O. Wilson, *Sociobiology: The New Synthesis* (Cambridge: Harvard Univ. Press,

Dawkins argues for the selfishness of our biology, but then implores his readers to construct cultures that can counter the genes.[40] Scientists like William Irons are in a sense working off Wilson's dictum; I interpret their work as the attempt to understand how altruism really is rooted in the basic workings of the genes-culture symbiosis, not counter-genetic, as Dawkins would have it. Other scientists simply reject Dawkins' notion that genetic selfishness determines animal behavior, especially humans and other higher primates.[41]

Some theologians consider chance and randomness to be great alternatives to Christian faith. This is no doubt true, if chance and randomness are both ontological and final. Over the long haul of scientific cosmology, I see no conclusive answers here, even though if one takes some short term slice of cosmic history, like the Burgess Shale or certain physical processes, it may appear that one or the other—randomness or determinism—is clearly dominant.[42] Certainly chance and necessity seem to be interwoven in nature's processes. Directionality seems to be constructed epigenetically.

The presence of evil, of which meaninglessness is the mirror image, or the flip side, seems in some ways to be more challenging to Western religion. Any religious or philosophical perspective that is both monistic and which posits goodness to be the originating essence of things will be greatly embarrassed by the actuality of evil, because there seem to be no grounds for evil to emerge. It is for this reason that I consider natural selection to be the major intellectual and moral challenge that the scientific world-views pose to the Christian faith. I say this, because the process of selecting out or being selected out appears on the surface to include evil, at least where humans are involved. Natural selection poses the issues of theodicy. Theodicy is defined as the problem of how one can reconcile a good and powerful God with the actuality of evil. The primary way that Christians have dealt with the actuality of evil is in terms of the epistemic distance or gap that I have described—evil is a phenomenon that occurs after the initial articulation of the essential goodness of Creation, and it is a phenomenon that will be transmuted into goodness eternally. The question is, then, how to deal with the evil that marks the natural world in which we live.

5.6.1 Incarnation, Sacraments

I have already suggested that the theologoumena of creation and the image of God are significant for interpreting God's intention and action in the world, since they establish nature as a whole, evolution in particular, and the human being as a realm of God's will and creation. Our attention is now turned to those theologoumena that establish the viability and capability of nature, specifically human nature, to be vessels of God's intention and action. These include the Incarnation, sacramental theology, and the undergirding doctrine of the finite as capable of the infinite (*finitum capax infiniti*).

The upshot of these theological motifs is an extraordinarily high valuation of nature and its possibilities. Notwithstanding the reputation of the Christian tradition in the minds of critics and the actual practice of many Christians to the contrary, the

---

1975).

[40] Dawkins, *The Selfish Gene*.

[41] de Waal, *Good Natured*.

[42] Steven J. Gould, *Wonderful Life: The Burgess Shale and the Nature of Reality* (New York: W. W. Norton and Co., 1989).

classical theological tradition regards the created order as an amazing realm that is a suitable vessel to bear the presence and intentional action of God.

The Incarnation of the Word of God in the flesh of a human being, Jesus of Nazareth, is a powerful statement of the Christian valuation of nature as God's creation. It would be a misinterpretation of the classical theological tradition to think that the doctrine of the Incarnation demeans human nature in favor of a superhuman intrusion into the natural realm. On the contrary, the tradition asserts that the human nature and the divine presence can each be fully itself while retaining their respective integrities as divine and human.[43] This is the strongest possible affirmation of the viability of the created order (in its human form, at least) as an instrument of the will of God.

Traditional interpretation of the sacraments takes place within the discussion of both nature and Christ. On the one hand, the sacraments convey the grace of God in Christ and are closely tied to the work of Christ, and on the other hand they are constituted by fully natural elements. The result of the sacramental theology is that nature, as in the form of water, bread, and wine, when it is intimately associated with Christ and interpreted in the light of Christ, performs its God-given purpose. Thus, the purposes of nature as revealed in the sacramental context become metaphors for the ways in which the purposes of nature are to be considered and in which nature is to be employed.

In the course of reflection upon the sacraments, particularly the Eucharist, the viability of the natural elements of bread and wine to carry the presence and grace of Christ's body and blood has been vigorously discussed. Although there is diversity of opinion on this point among various Christian communities, the overwhelming consensus is that the natural elements are, in analogy to Christ's human body, fit vessels of that presence and grace. My own Lutheran tradition has favored the terminology of the "finite capable of the infinite"—that is, that finite natural elements can be the means of God's grace (*finitum capax infiniti*). This is yet another strong testimony to the traditional valuation of nature.

The chief point of this discussion of incarnation and sacraments deserves emphasis: Within the classical Christian theological perspective, nature is considered to be fully capable of participating in the meaning that God has conceived for nature in the creation. Nature need not be destroyed nor overwhelmed in order for it to serve its Creator's purposes.

5.6.2 *Another Note on Evil and Sin*

Does not the presence of evil and sin seriously qualify the strong affirmation of nature's capabilities that I have located within the classic Christian tradition? There is no unanimity among Christian theologians and churches on this point. The response that I offer here does, however, stand firmly within the tradition.

The response is complex, in two parts. First, it is clear that nature now, including human nature, is not in conformity with God's intention at creation. The symbols of Fall, Original Sin, and Actual Sin make that clear. Furthermore, if there were not a dramatic discrepancy between what nature was created to be and what it actually is, there would be no crucifixion in the story of redemption.

Nevertheless—and this is the second part of my response—fallenness, sin, and finitude do not bar nature from its God-given intentionality. The sinful creature is

---

[43] Edward Rochie Hardy, *Christology of the Later Fathers* (Philadelphia: Westminster Press, 1964), 359–74.

nevertheless the one created in the image of God. The theologoumena of forgiveness, justification, sanctification, and consummation all point to the acceptability of nature, including human nature, in the sight of God. Moreover, the work of Christ and the sacraments underscores that actual nature can be the vessel of God's continuing creation and grace, so as to fulfill God's intentions for the creation. The way of the Cross becomes, not the negation of nature, but a portrayal of the deeper reaches of nature's career under the conditions of grace.

5.7 *Christ*

In the Christian system of thought, Jesus Christ provides the core of meaning. This is what it means to call him *Logos*, the Word of God. It is in the meanings associated with Christ that we discover the content of several basic concepts that I have emphasized, particularly the image of God and the character of God in terms of freedom, intentionality, and love (see sections 5.1–2 above). My argument requires a full-scale Christological elaboration at this point. Since that is obviously impossible, I will underscore three elements of the Christ symbol that are essential: the focus on self-giving love, the Cross and Resurrection, and cosmic redemption.

In speaking of the human being called to live for the sake of the created natural order in which humans have emerged and which has shaped them, the Love command defines formally the governing principle of human behavior in and toward nature. The norm and example of that love is found in Jesus' life and teaching.

The crucifixion and resurrection testify that the nature of nature includes, in a mysterious fashion, entry into evil and non-being, critique of surface expectations, and surprising emergences. In this respect, New Testament scholar Gerd Theissen has utilized Burhoe's theory (and other related ones) brilliantly to interpret the biblical faith. Theissen's insight is that Jesus actually constitutes a new proposal for cultural evolution that will, via the human being, lead evolution into a qualitatively new phase of existence. This new phase gives content to the concept of God's will for humanity. As cultural evolution negotiates the passage into this new phase, it transcends certain dimensions of biological inheritance, those which are obstacles to the new phase; at the same time it ushers the biological into a new set of circumstances that qualify as the fulfillment of the genetic heritage.[44]

The traditions of the Cosmic Christ, present in the New Testament and continuous throughout Christian history, take seriously the interrelatedness of all of nature. Those traditions understand that Christ's significance reaches out far beyond the human form in which the Incarnation took place. The symbol of the Cosmic Christ reinterprets the Incarnation as a statement about all of nature as a place for God's indwelling, just as it interprets redemption to pertain to all of creation. These assertions are already present, for example, in the Epistles to the Ephesians (1:3–10) and the Colossians (1:13–20), and subsequent theology and philosophy has elaborated them in detail. We are still far, however, from possessing an interpretation of the Cosmic Christ that is intelligible in terms of contemporary evolutionary science. Teilhard de Chardin has provided the most adequate interpretation to date, but it certainly needs updating and revision.[45]

---

[44] Gerd Theissen, *Biblical Faith: An Evolutionary Approach* (Philadelphia: Fortress, 1972).

[45] Pierre Teilhard de Chardin, *Science and Christ* (New York: Harper & Row, 1965); George A. Maloney, S.J. *The Cosmic Christ from Paul to Teilhard* (New York: Sheed and Ward, 1968); Joseph Sittler, *Essays on Nature and Grace* (Philadelphia: Fortress Press, 1972); J. A. Lyons, *The Cosmic Christ in Origen and Teilhard de Chardin* (Oxford: Oxford

In summary, in the meanings that are associated with Christ, we find the content of evolution's meaning in love, and the assertion that the Christ-meanings are cosmic in import.

6 The Created Co-Creator as Metaphor for Meaning: The Meaning of Nature and God's Intentions for the World

6.1 Summary

One-half of the preceding discussion has focused upon certain scientific understandings of the evolutionary emergence of *Homo sapiens,* with commentary on their significance, while the remaining half has elaborated traditional theological concepts pertaining to creation, image of God, incarnation, and sacraments. What are we to make of this discussion? What does it amount to? We propose the following, in sum: On the assumption that *Homo sapiens* is thoroughly and without remainder a natural creature, an emergent within the processes of evolution as they have taken shape on planet earth, nature elaborates itself as a project, and the human creature emerges as the first evolved species that can begin to take the measure of that project, define it, and act freely to behave in accordance with that project; that is to say, in *Homo sapiens* nature reveals itself to be a realm of intentionality, and the human species is evolved as the creature that can participate in that intentionality under the conditions of self-awareness and freedom. The Christian doctrines of creation, image of God, incarnation, and sacrament provide an interpretation of how this natural world can be an expression of nature's project as God's own project, and how the natural human creature can express that divine intentionality as God's created co-creator, under the criterion of meaning that is presented in the person and life of Jesus of Nazareth, whom Christians interpret as the Christ. The emergence and evolution of the human creature can thus be understood to be a clue to the meaning of nature and of God's intentions for nature. Hence the title of this piece, *biocultural* evolution as clue to meaning. In concluding, I shall itemize in summary form specific considerations, derived from this discussion, that are pertinent to the argument as I have just summarized it.

6.2 Created Co-Creator as Metaphor for the Meaning of Nature: Scientific

The scientific descriptions that our pertinent to our argument have raised the following considerations that are essential for our understanding.

1. *The created co-creator is a natural entity, and its placement is fully within nature*. This means that when we look at the created co-creator and reflect upon its character and function, we are reflecting upon nature. Knowledge concerning the human being is knowledge of nature.

2. *The reference or context for the created co-creator and its activity is nature*. Whatever purpose we ascribe to humans, whatever goals we set for their activity is to be referred to the rest of nature. The purpose of humans is a purpose within and for nature. The goals of human activity are in the service of nature. The contribution that humans make is a contribution to nature. Any purposes, goals, and contributions that are ascribed to humans are simultaneously statements about nature, about what nature is, and about its possible purposes. This is another way of asserting that the human project is inseparable from nature and nature's project.

3. *That the created co-creator has appeared as a fully natural creature is a statement about what nature has come to, what nature is capable of, and what*

Univ. Press, 1982).

nature itself has produced or allowed to appear. Here we touch on the question of whether awareness, intentionality, and even agency and personhood can be imputed to nature and its processes. I will not deal with these issues here, but it does seem clear that whatever we do impute to human beings is also thereby imputed in some sense to nature—and unavoidably so.

4. *The experience of transcendence.* The human created co-creator is caught up in a process of eliciting possibilities from its environment, and in turn the environment elicits possibilities from the human being. This process is termed a reality of transcendence, in that new possibilities form the context for new adaptations that are more capable than preceding ones.

5. *If transcendence and freedom are characteristics of human being, they are* ipso facto *characteristics of nature.* It is nature that encounters transcendence in human being, and it is nature that experiences freedom. It would be important here to ask whether the non-human sciences, such as physics, astrophysics, cosmology, and biochemistry can identify any correlates to transcendence and freedom as I have described them in this essay. Do relativity theory, chaos and complexity theory, and theories of biochemical autocatalysis, for example, involve pre- and non-human correlates to what I have called transcendence and freedom?

6. *The appearance of the created co-creator inclines us to speak of nature as "project," and to ask what the appearance of this creature suggests concerning "nature's project."* In the human species, we observe nature's own behavior of being aware of itself, interpreting itself, and conceiving its own project. In the human species we observe most vividly nature's attempts to take the measure of its possibilities that invest itself in them. The pragmatic criterion of wholesomeness that we applied to human behavior (see above, p. 342) is defined in the context of the human project interacting with nature's project.

Summing up. When the created co-creator is understood as a metaphor for the meaning of nature, it is stated in this way: The appearance of *Homo sapiens* as created co-creator signifies that nature's course is to participate in transcendence and freedom, and thereby nature makes the transition into the condition in which it interprets its own essential nature and takes responsibility for acting in accord with that nature; nature thus shares in freedom. In other words, *the appearance of the created co-creator within the evolutionary processes and its history up to the present moment reveals that nature is a project.* This is the meaning that is to be derived from the biocultural human creature, the created co-creator. This is the meaning for which the human created co-creator is the metaphor.

6.3 *Created Co-Creator as Metaphor for God's Intentions for the World: Theological*

The classical Christian theological themes that are pertinent to our argument have raised up the following considerations that are essential for our understanding.

1. *Nature's project is God's project.* Consequently, nature's struggle to discern its project and carry it out is an encounter with transcendence and with God.

2. *God intends nature to be project.* This implies that God intends the created natural order to discern its purposes and values freely and likewise to behave freely in accord with those purposes

3. *The human project is inseparable from nature's project as a whole.* Thus, the human project must be in the service of nature's project. To the extent that the two projects are separated or alienated from one another, the projects have not been properly understood.

4. *The human being is created in the image of God, which is to say that in the human struggle to take the measure of the human project, which in turn is part of nature's project, God's intentionality is present as the norm of the project.*

5. *Christ as the norm of the image of God gives content to God's intentionality.* This content includes love.

6. *To assert the doctrines of sin and evil is to say that humans experience their nature as the image of God both as gift that has been received and as mandate yet to be fulfilled, since it presently falls short of what God intends.*

7. *The purpose of the created co-creator and, hence, of nature is to carry out the project instantiated in Jesus Christ.* Humans must recognize that this Christ-project is nature's project and cannot therefore be carried out against nature or apart from it.

Summing up. Taking the scientific understandings of *Homo sapiens* as interpreted in the preceding section, and asserting that the species, as nature's created co-creator, is a metaphor for the meaning of nature, classical Christian theology deepens the interpretation to include the following: that nature's created co-creator is the creation of God, that the project of nature as exemplified in the human being is God's own project, and that the criterion and content of this project is set forth in Jesus of Nazareth. To interpret Jesus as the Christ is to make the assertion that he offers the criterion and content of God's project.

BIOLOGICAL EVOLUTION—A POSITIVE THEOLOGICAL APPRAISAL

Arthur Peacocke

> Darwinism appeared, and, under the disguise of a foe, did the work of a friend. It has conferred upon philosophy and religion an inestimable benefit, by shewing us that we must choose betwen two alternatives. Either God is everywhere present in nature, or He is nowhere. (Aubrey Moore, in the 12th edition of *Lux Mundi*, 1891, p. 73).

1 *Introduction*

It[1] would, no doubt, come as a surprise to many of the biologically-cultured "despisers of the Christian religion," to learn that, as increasingly thorough historical investigations are showing,[2] the nineteenth-century reaction to Darwin in theological and ecclesiastical circles was much more positive and welcoming than the legends propagated by both popular and academic biological publications are prepared to admit. Furthermore, the scientific reaction was also much more negative than usually depicted, those skeptical of Darwin's ideas including initially *inter alia* the leading comparative anatomist of his day, Richard Owen (a Cuverian), and the leading geologist, Charles Lyell. Many theologians deferred judgment, but the proponents of at least one strand in theology in nineteenth-century England chose to intertwine their insights closely with the Darwinian—that "catholic" revival in the Church of England of a stress on the doctrine of the Incarnation and its extension into the sacraments and so of a renewed sense of the sacramentality of nature and God's immanence in the world. I have summarized elsewhere[3] some of this history, not only in Britain but also in France, Germany, and America—suffice it to say that more of the nineteenth-century theological reaction to Darwin was constructive and reconciling in temper than practically any biological authors today will allow.

That is perhaps not surprising in view of the background to at least T. H. Huxley's aggressive propagation of Darwin's ideas and his attacks on Christianity, namely that of clerical restriction on, and opportunities for, biological scientists in England in the nineteenth century. His principal agenda was the establishment of science as a profession independent of ecclesiastical control—and in this we can sympathize. So it is entirely understandable that the present *Zeitgeist* of biological science is that of viewing "religion" as the opposition, if no longer in any way a threat. This tone saturates the writings of the biologists Richard Dawkins and Stephen Gould and many others—and even philosophers such as Daniel Dennett. Indeed, the strictures of Jacques Monod in his 1970 publication *Le hasard et la*

---

[1] General and theological reflections on biological evolution will appear below in this font, scientific accounts in this font. Some of the phrasing in this paper follows sections of my *Theology for a Scientific Age* [henceforth *TSA*] (London: SCM Press. Minneapolis: Fortress Press, 2nd enlarged ed., 1993), and my *Creation and the World of Science* [henceforth *CWS*] (Oxford: Clarendon Press, 1979).

[2] E.g., J. R. Moore, *The Post-Darwinian Controversies: a study of the Protestant struggle to come to terms with Darwin in Great Britain and America* (Cambridge: Cambridge University Press, 1979); J. R. Lucas, "Wilberforce and Huxley: a legendary encounter," *The Historical Journal* 22.2 (1979): 313–30; J. V. Jensen,"Return to the Huxley-Wilberforce Debate," *Brit. J. Hist. Sci.* 21 (1988): 161–79.

[3] Arthur Peacocke, "Biological Evolution and Christian Theology—Yesterday and Today," in *Darwinism and Divinity*, ed. John Durant (Oxford: Blackwell, 1985), 101–30.

nécessité, especially in its English translation *Chance and Necessity,* could be said to be one of the strongest attacks in this century, in the name of science, on belief in God and in the universe having any attributable meaning and purpose. His remarks on this are well known and widely quoted.

Perhaps this has polarized the scene, but what I find even more surprising, and less understandable, is the way in which the "disguised friend" of Darwinism, and more generally of evolutionary ideas, has been admitted (if at all) only grudgingly, with many askance and sidelong looks, into the parlors of Christian theology. I believe it is vital for this churlishness to be rectified in this last decade of the twentieth century if the Christian religion (indeed any religion) is to be believable and have intellectual integrity enough to command even the attention, let alone the assent, of thoughtful people in the beginning of the next millennium.

2 Biological Evolution and God's Relation to the Living world

We shall consider in this section various features and characteristics of biological evolution and any theological reflections to which they may give rise.

2.1 Continuity and Emergence

A notable aspect of the scientific account of the natural world in general is the seamless character of the web that has been spun on the loom of time: the process appears as continuous from its cosmic "beginning," in the "hot big bang," to the present and at no point do modern natural scientists have to invoke any non-natural causes to explain their observations and inferences about the past. Their explanations are usually in terms of concepts, theories and mechanisms which they can confirm by, or infer from, present-day experiments, or reasonably infer by extrapolating from principles themselves confirmable by experiment. In particular, the processes of biological evolution also display a *continuity*, which although at first a conjecture of Darwin (and, to be fair, of many of his predecessors), is now thoroughly validated by the established universality of the genetic code and by the study of past and present species of DNA nucleotide sequences and of amino acid sequences in certain widely-distributed proteins.

The process that has occurred can be characterized also as one of *emergence*, for new forms of matter, and a hierarchy of organization of these forms, appear in the course of time. These new forms have new properties, behaviors and networks of relations which necessitate not only specific methods of investigation but also the development of new epistemologically irreducible concepts in order to describe and refer to them. To these new organizations of matter it is, very often, possible to ascribe new levels of what can only be called "reality": that is, the epistemology implies at least a putative ontology. In other words new kinds of reality may be said to "emerge" in time. Notably, on the surface of the Earth, new forms of *living* matter (that is, living organisms) have come into existence by this continuous process—that is what we mean by evolution.[4]

---

[4] Even the "origin of life," that is the appearance of *living* matter on the surface of the Earth some four and a half billion years ago, can be subsumed within this seamless web of the operation of processes at least now intelligible to and entirely conformable with the sciences, if inevitably never entirely provable by repeatable experiments—the situation with all the *historical* natural sciences (cosmology, geology, evolutionary biology). Studies on dissipative systems (the Brussels school) have shown how interlocking systems involving feedback can, entirely in accord with the second law of thermodynamics, undergo transitions to more organized and more complex forms, provided such systems are open, non-linear and far from equilibrium. All these conditions would have been satisfied by many systems of chemical reactions present on the Earth during its first billion years of existence. Furthermore, with our increasing knowledge of how molecular patterns can be copied in present living systems, it is

What the scientific perspective of the world, especially the living world, inexorably impresses upon us is a *dynamic* picture of entities and structures involved in continuous and incessant change and process without ceasing. Any static conception of the way in which God sustains and holds the cosmos in being is therefore precluded, for new entities, structures, and processes appear in the course of time, so that God's action as Creator is both past and present: it is continuous. The scientific perspective of a cosmos, and in particular that of the biological world, as in development all the time must re-introduce into our understanding of God's creative relation to the world a dynamic element which was, even if obscured by the assigning of "creation" to an event in the past, always implicit in the Hebrew conception of a "living God," dynamic in action. Any notion of God as Creator must now take into account that God is continuously creating, continuously giving existence to, what is new; God is *semper Creator*, and the world is a *creatio continua*. The traditional notion of God *sustaining* the world in its general order and structure now has to be enriched by a dynamic and creative dimension—the model of God sustaining and giving continuous existence to a process which has a creativity built into it by God. God is creating at every moment of the world's existence in and through the perpetually-endowed creativity of the very stuff of the world. God indeed makes "things make themselves," as Charles Kingsley put it in *The Water Babies*, and aptly quoted by Charles Birch at the head of his paper in this volume.

Thus it is that the scientific perspective, and especially that of biological evolution, impels us to take more seriously and more concretely than hitherto the notion of the immanence of God-as-Creator—God is the Immanent Creator *creating in and through the processes of the natural order*. I would urge that all this has to be taken in a very strong sense. If one asks where do we see God-as-Creator during, say, the processes of biological evolution, one has to reply: "The processes themselves, as unveiled by the biological sciences, *are* God-acting-as-Creator, God *qua* Creator."[5] God gives existence in divinely-created time to a process that itself brings forth the new: thereby God is creat*ing*. This means we do not have to look for any extra supposed gaps in which, or mechanisms whereby, God might be supposed to be acting as Creator in the living world.

The model of musical composition for God's activity in creation is, I would suggest, particularly helpful here. There is no doubt of the "transcendence" of the composer in relation to the music he or she creates—the composer gives it existence and without the composer it would not be at all. So the model properly reflects, as do all those of artistic creativity, that transcendence of God as Creator of all-that-is which, as the "listeners" to the music of creation, we wish to aver. Yet, when we are actually listening to a musical work, say, a Beethoven piano sonata, then there are times when we are so deeply absorbed in it that, for a moment we are thinking

---

now possible to make plausible hypotheses concerning how early forms of nucleic acids and/or proteins might have formed self-replicating molecular systems (e.g., the "hypercycle" of the Göttingen school). Such systems can be shown to multiply at the expense of less efficient rival ones (q.v., Peacocke, *The Physical Chemistry of Biological Organization* (Oxford: Clarendon Press, 1983), chap.5). These studies indicate the inevitability of the appearance of more organized self-replicating systems, the properties of atoms and molecules being what they are; but what form of organization would be adopted is not strictly predictable (by us, at least) since it depends on fluctuations (M. Eigen, "The Self-Organization of Matter and the Evolution of Biological Macromolecules," *Naturwissenschaften* 58 (1971): 465–523).

[5] This is not pantheism, for it is the *action* of God that is identified with the creative processes of nature, not God's own self.

Beethoven's musical thoughts with him. In such moments the

> Music is heard so deeply
> That it is not heard at all, but you are the music
> While the music lasts[6]

Yet if anyone were to ask at that moment, "*Where* is Beethoven now?"—we could only reply that Beethoven-*qua*-composer was to be found only in the music itself. The music would in some sense be Beethoven's inner musical thought rekindled in us and we would genuinely be encountering Beethoven-*qua*-composer. This very closely models, I am suggesting, God's immanence in creation and God's self-communication in and through the processes by means of which God is creating. The processes revealed by the sciences, especially evolutionary biology, are in themselves God-acting-as-Creator. There is no need to look for God as some kind of *additional* factor supplementing the processes of the world. God, to use language usually applied in sacramental theology, is "in, with, and under" all-that-is and all-that-goes-on.

2.2 *The Mechanism of Biological Evolution*

There appear to be no serious biologists who doubt that natural selection is a factor operative in biological evolution—and most would say it is by far the most significant one. At one end of the spectrum authors like Dawkins argue cogently for the all-sufficiency of natural selection in explaining the course of biological evolution. Certainly he has illustrated by his "biomorph" computer program how the counter-intuitive creativity of evolution could be generated by the interplay of chance events operating in a law-like framework—this is, in the case of biological evolution, the interplay between mutational events in the DNA of the genome with the environment of the phenotype to which it gives rise. This was well illustrated by Dawkins[7] with his program in which two-dimensional patterns of branching lines, "biomorphs," are generated by random changes in a defined number of features combined with a reproduction and selection procedure. It was striking how subtle, varied and complex were the "biomorph" patterns after surprisingly few "generations," that is, reiterations of the procedure. Such computer simulations go a long way toward making it clear how it is that the complexity and diversity of biological organisms could arise through the operation of the apparently simple principles of natural selection.

However, other biologists are convinced that, even when the subtleties of natural selection are taken into account, it is not the whole story; and some even go so far as to say that natural selection alone cannot account for speciation, the formation of distinctly new species.

What is significant about all these proposals[8] is that they are all operating entirely

---

[6] T. S. Eliot, "The Dry Salvages," *The Four Quartets* (London: Faber & Faber, 1944), ll. 210–12, p. 33.

[7] Richard Dawkins, *The Blind Watchmaker* (Harlow: Longmans, 1986).

[8] Some considerations other than natural selection which, it is claimed, are needed to be taken into account are thought to be:

1. The "evolution of evolvability" (Daniel C. Dennett, *Darwin's Dangerous Idea* (London and New York: Allen Lane, Penguin, 1995) , 222 and *n.* 20, for references.), in particular the constraints and selectivity effected by self-organizational principles which shape the possibilities of elaboration of structures and even direct its course (Stuart A. Kauffman, *The Origins of Order: Self-organization and Selection in Evolution* (Oxford Univ. Press, New York and London, 1993); idem, *At Home in the Universe* (Penguin Books, London, 1995); B. C. Goodwin, *How the Leopard Changed Its Spots: the evolution of complexity* (New York: Scribner's Sons, 1994));

2. The "genetic assimilation" of C. H. Waddington (*The Strategy of the Genes: a*

within a naturalistic framework—to use Dennett's graphic designation, they are all "cranes" and not "skyhooks"[9]—and, moreover, they assume a basically Darwinian process to be operating, even when they disagree about its speed and smoothness (e.g., the presence or absence of the sudden "saltations" proposed by Gould). That being so, it has to be recognized that the history of life on Earth involves chance in a way unthinkable before Darwin. There is a creative interplay of "chance" and law apparent in the evolution of living matter by natural selection. However what we mean by "chance" in this context first needs closer examination.

Events are unpredictable by us in basically two ways:

1. They can be unpredictable because we can never possess the necessary detailed knowledge with the requisite accuracy at this microlevel of description. In such cases talk of the role of "chance" can mean either: (A) we cannot determine accurately the microparameters of the initial conditions determining the macroevents (e.g., the forces on a tossed coin) while often knowing the overall constraints that must operate on the system as a whole (e.g., the symmetry constraints making for equal probabilities

---

discussion of some aspects of theoretical biology (London: Allen and Unwin, 1957);

3. How an organism might evolve is a consequences of itself, its state at any given moment, historical accidents, as well as its genotype and environment (R. C. Lewontin, "Gene, Organism and Environment," in *Evolution from Molecules to Man*, D. S. Bendall, ed. (Cambridge: Cambridge University Press, 1983), 273–85);

4. The innovative behavior of an individual living creature in a particular environment can be a major factor on its survival and selection and so in evolution (A. Hardy, *The Living Stream* (London: Collins, 1985), 161ff., 189ff.);

5. "Top-down causation" operates in evolution (D. T. Campbell, "Downward Causation in Hierarchically Organized Systems," in *Studies in the Philosophy of Biology: reduction and related problems*, F. J. Ayala and T. Dobzhansky, eds. (London: MacMillan, 1974), 179–86) and does so more by a flow of information between organism and environment and between different levels (q.v., *TSA*, 59) than by any obvious material or energetic causality;

6. The "silent" substitutions in DNA are more frequent than non-silent ones (with an effect on the phenotype); in other words, the majority of molecular evolutionary change is immune to natural selection (M. Kimura, "The Neutral Theory of Evolution," *Sci. Amer.* 241(1979): 98–126);

7. The recent re-introduction of *group* (alongside that of *individual*) selection in a unified theory of natural selection as operating at different levels in a nested hierarchy of units, groups of organisms being regarded as "vehicles" of selection (D. S. Wilson and E. Sober, "Reintroducing group selection to the human behavioral sciences," *Behavioral and Brain Sciences* 17 (1994): 585–654);

8. Long-term changes in the genetic composition of a population resulting from "molecular drive," the process in which mutations spread through a family and through a population as a consequence of a variety of mechanisms of non-reciprocal DNA transfer, thereby inducing the gain or loss of a variant gene in an individual's lifetime, leading to non-Mendelian segregation ratios (G. A. Dover, "Molecular Drive in Multigene Families: how biological novelties arise, spread and are assimilated," *Trends in Genetics* 2 (1986): 159–65).

9. An emphasis on the context of adaptive change (or, in many cases, non-change) in species regarded as existing in interlocking hierarchies of discrete biological entities (genes, populations, species, eco-systems, etc.) in a physical environment (N. Eldredge, *Reinventing Darwin: the great evolutionary debate* (London: Wiedenfeld and Nicolson, 1996).

[9] A crane "is a sub-process or feature of a design process that can be demonstrated to permit the local speeding up of the basic, slow process of natural selection, *and* that can be demonstrated to be itself the predictable (or retrospectively explicable) product of the basic process" (Daniel C. Dennett, *Darwin's Dangerous Idea* (London & New York: Allen Lane, Penguin, 1995), 76). Cranes include sex and the Baldwin Effect. Dennett means by a skyhook "a 'mind-first' force or power or process, an exception to the principle that all design, and apparent design, is ultimately the result of mindless, motiveless mechanicity" (ibid.).

of heads and tails); or (B) the observed events are the outcome of the crossing of two independent causal chains, accurate knowledge of which is unattainable both with respect to the chains themselves and to their point of intersection.

2. At the subatomic level events can also be inherently unpredictable because of the operation of the Heisenberg Uncertainty Principle.

Events of type (1) and (2), unpredictable as they are, can produce effects at the macroscopic level which operate in a law-like framework which delimits the scope of the consequent events or provides them with new and unexpected outcomes. Both ways of viewing the matter are pertinent to biological evolution, which depends on a process in which changes occur in the genetic-information carrying material (DNA) that are random with respect to the biological needs of the organisms possessing the DNA—in particular, random with respect to its need to produce progeny for the species to survive. What we call "chance" is involved both at the level of the mutational event in the DNA itself (1A and/or 2), and in the intersecting of two causally unrelated chains of events (1B). The biological niche in which the organism exists then filters out, by the processes of natural selection, those changes in the DNA that enable the organisms possessing them to produce more progeny in an entirely law-like fashion.

The interplay between "chance," at the molecular level of the DNA, and "law" or "necessity" at the statistical level of the population of organisms tempted Jacques Monod, in *Chance and Necessity*[10] to elevate "chance" to the level almost of a metaphysical principle whereby the universe might be interpreted. As is well known, he concluded that the "stupendous edifice of evolution" is, in this sense, rooted in "pure chance" and that *therefore* all inferences of direction or purpose in the development of the biological world, in particular, and of the universe, in general, must be false. As Monod saw it, it was the purest accident that any particular creature came into being, in particular *Homo sapiens*, and no direction or purpose or meaning could ever be expected to be discerned in biological evolution. A creator God, for all practical purposes, might just as well not exist, since everything in evolution went on in an entirely uncontrolled and fortuitous manner.

The responses to this thesis—mainly from theologically informed scientists and some philosophers, rather than from theologians—have been well surveyed and analyzed by Bartholomew.[11] I shall pursue here what I consider the most fruitful line of reflection on the processes that Monod so effectively brought to our attention—a direction that I began to pursue[12] in response to Monod and which has been further developed by the statistically-informed treatment of Bartholomew.

There is no reason to give the randomness of a molecular event the metaphysical status that Monod attributed to it. The involvement of "chance" at the level of mutation in the DNA does not, of itself, preclude these events from displaying trends and manifesting inbuilt propensities at the higher levels of organisms, populations, and eco-systems. To call the mutation of the DNA a "chance" event serves simply to stress its randomness with respect to biological consequence. As I have earlier put it (in a response later supported and amplified by others):

> Instead of being daunted by the role of chance in genetic mutations as being the manifestation of irrationality in the universe, it would be more consistent with

---

[10] Jacques Monod, *Chance and Necessity* (London: Collins, 1972).

[11] David J. Bartholomew, *God of Chance* (London: SCM Press, 1984).

[12] Peacocke, "Chance, Potentiality and God," *The Modern Churchman*, 17 (New Series 1973): 13–23; idem, in *Beyond Chance and Necessity*, ed. J. Lewis (London: Garnstone Press, 1974), 13–25; idem, "Chaos or Cosmos," *New Scientist*, 63 (1974): 386–89; and *CWS*, chap. 3.

observation to assert that the full gamut of the potentialities of living matter could be explored only through the agency of the rapid and frequent randomization which is possible at the molecular level of the DNA.[13]

This role of "chance," or rather randomness (or "free experiment") at the microlevel is what one would expect if the universe were so constituted that all the potential forms of organization of matter (both living and non-living) which it contains might be thoroughly explored. Indeed, since Monod first published his book in 1970, there have been the developments in theoretical and molecular biology and physical biochemistry of the Brussels and Göttingen schools. They demonstrated that it is the interplay of chance and law that is in fact creative within time, for it is the combination of the two which allows new forms to emerge and evolve—so that natural selection appears to be opportunistic. As in many games, the consequences of the fall of the dice depend very much on the rules of the game.[14] It has become increasingly apparent that it is chance operating within a law-like framework that is the basis of the inherent creativity of the natural order, its ability to generate new forms, patterns and organizations of matter and energy. If all were governed by rigid law, a repetitive and uncreative order would prevail; if chance alone ruled, no forms, patterns or organizations would persist long enough for them to have any identity or real existence and the universe could never be a cosmos and susceptible to rational inquiry. It is the combination of the two which makes possible an ordered universe capable of developing within itself new modes of existence (cf. Dawkins' bio-morphs). The "rules" are what they are because of the "givenness" of the properties of the physical environment and of the other already evolved living organisms with which the organism in question interacts.

This givenness, for a theist, can only be regarded as an aspect of the God-endowed features of the world. The way in which what we call "chance" operates within this "given" framework to produce new structures, entities and processes can then properly be seen as an eliciting of the potentialities that the physical cosmos possessed *ab initio*. Such potentialities a theist must regard as written into creation by the Creator's intention and purpose and must conceive as gradually being actualized by the operation of "chance" stimulating their coming into existence. One might say that the potential of the "being" of the world is made manifest in the "becoming" that the operation of chance makes actual. God is the ultimate ground and source of both law ("necessity") and "chance."[15]

---

[13] *CWS*, 94.

[14] R. Winkler and M. Eigen, *Laws of the Game* (New York: Knopf, 1981. London: Allen Lane, 1982).

[15] D. J. Bartholomew in his *God of Chance* has urged that God and chance are not only logically compatible, as the foregoing has argued, but that there are "positive reasons for supposing that an element of pure chance would play a constructive role in creating a richer environment than would otherwise be possible" (97). He argues that "chance offers the potential Creator many advantages which it is difficult to envisage being obtained in any other way" (97). Since in many natural processes, often utilized by human beings, chance processes can in fact lead to determinate ends—for many of the laws of nature are statistical—"there is every reason to suppose that a Creator wishing to achieve certain ends might choose to reach them by introducing random processes whose macro-behavior would have the desired character" (98). Thus the determinate ends to which chance processes could lead might well be "to produce intelligent beings capable of interaction with their Creator" (98). For this it would be necessary, he suggests, to have an environment in which chance provides the stimulus and testing to promote intellectual and spiritual evolution.

For a theist, God must now be seen as acting to create in the world *through* what we call "chance" operating within the created order, each stage of which constitutes the launching pad for the next. The Creator, it now seems, is unfolding the divinely-endowed potentialities of the universe, in and through a process in which these creative possibilities and propensities (see next section), inherent by God's own intention within the fundamental entities of that universe and their inter-relations, become actualized within a created temporal development shaped and determined by those selfsame God-given potentialities.[16]

2.3 *Trends in Evolution?*

Are there any trends or discernable directions in evolution? This is a notoriously loaded question, which human beings are only too ready to answer on the basis of their own believed significance in the universe grounded on their own importance to themselves! Is there any objective, non-anthropocentrically biased evidence for directions or at least trends in biological evolution? Biologists have been especially cautious not to answer this question affirmatively, not least because they do not wish to give premature hostages to those seeking to gain a foothold for claiming some kind of "skyhook," such as divine action, intervention even, directing the course of evolution. Evolution is best depicted biologically not as a kind of Christmas tree, with *Homo sapiens* accorded an angelic position crowning the topmost frond, but rather as a bush—"Life is a copiously branching bush, continually pruned by the grim reaper of extinction, not a ladder of predictable progress."[17] Nevertheless, G. G. Simpson can affirm that "Within the framework of the evolutionary history of life there have been not one but many different kinds of progress."[18] While admitting that such lines of "progress" can be traced in the evolutionary "bush," other biologists would be more neutral.

The question of more general significance that is being addressed in relation to the biological evolutionary story is, it would seem, Are there any particular properties and functions attributable to living organisms which could be said to be in themselves helpful for evolution to occur because they are advantageous in natural selection (for survival of progeny) of organisms possessing them? Simpson's suggested list seems to fulfil this criterion and so it could be said that the evolutionary process manifests what Karl Popper[19] has called a "propensity" in nature for such properties to appear. He argued that a greater frequency of occurrence of a particular kind of event may be used as a test of whether or not there is inherent in a sequence of events (equivalent to throws of a die) a tendency or propensity to realize the event in question. He has pointed out

---

[16] Cf., *CWS*, 105–6.

[17] S. J. Gould, *Wonderful Life: the Burgess shale and the nature of history* (London and New York: Penguin Books, 1989), 35. Gould's principle thesis, on the basis of his interpretation of the Burgess shale, of the role of contingency in evolution as rendering impossible any generalizations about trends in evolution, has recently been strongly contraverted and refuted by Simon Conway-Morris, F.R.S., Professor of Evolutionary Palaeobiology at the University of Cambridge, England, who has spend most of his research life on the contents of the Burgess shale, in *The Crucible of Creation* (Oxford: Oxford University Press, 1998).

[18] G. G. Simpson, *The Meaning of Evolution* (New Haven: Bantam Books, Yale Univ. Press, 1971), 236. He instances the kinds of "progress" (prescinding from any normative connotation) as: the tendency for living organisms to expand to fill all available spaces in the livable environments; the successive invasion and development by organisms of new environmental and adaptive spheres; increasing specialization with its corollary of improvement and adaptability; increase in the general energy or maintained level of vital processes; protected reproduction-care of the young; individualization, increasing complexity, and so forth.

[19] Karl Popper, *A World of Propensities* (Bristol: Thoemmes, 1990).

that the *realization of possibilities*, which may be random, *depend on the total situation within which the possibilities are being actualized* so that "there exist weighted possibilities which are *more than mere possibilities*, but tendencies or propensities to become real"[20] and that these "propensities in physics are properties of *the whole situation* and sometimes even of the particular way in which a situation changes. And the same holds of the propensities in chemistry, in biochemistry, and in biology."[21] I suggest that there *are* propensities, in this Popperian sense, in evolution towards the possession of certain characteristics, propensities that are inherently built into an evolutionary process based on natural selection of the best procreators. Such properties naturally enhance survival for procreation in certain widely-occurring environments.

Among the plethora of such properties of living organisms which might in some circumstance or another be advantageous in natural selection, there are a number which characterize *Homo sapiens* and are pertinent to our wider concerns in this paper (and volume). They are as follows.

1. *Complexity.* The human brain is the most complex natural system known to us in the universe. Is there a propensity to complexity in *biological*[22] evolution? There certainly seems to be, and "increasing complexity" was included in Simpson's list (see n. 18) as characteristic of it. What significance is to be attributed to this? Is it simply that biological "evolution is a process of divergence and wandering rather than an inexorable progression towards increasing complexity" so that evolution "*permits* the emergence of new complexity, but does not in any particular case necessitate it"?[23]

The fact is that there *has* been, taking biological evolution as a whole, an emergence of increasingly complex organisms, even if in some evolutionary lines there has been a loss of complexity and so of organization. So, on Popper's criterion enunciated above, we would be correct in saying that there is a propensity towards increased complexity in the evolution of living organisms. Saunders and Ho[24] identify the

---

[20] Ibid., 12.

[21] Ibid., 17.

[22] There certainly seems to be such a propensity in non-living matter, for in those parts of the universe where the temperature is low enough for molecules to exist in sufficient proximity to interact, there is a tendency for more and more complex molecular systems to come into existence and this process is actually driven, in the case of reactions that involve association of molecules to more complex forms, by the tendency to greater overall randomization, that is, as a manifestation of the Second Law (q.v., Peacocke, *The Physical Chemistry of Biological Organization* (Oxford: Clarendon Press, 1983), section 2.7). Such systems, if open and if they also exhibit feedback properties, can become "dissipative" and undergo sharp changes of regime with the appearance of new patterns in space and time. In other words, even in these non-living systems, there is an increase in complexity in the entities involved in certain kinds of natural process. This appears to be an example of "propensity" in Popper's sense.

[23] W. McCoy, "Complexity in Organic Evolution," *J. Theor. Biol.*, 68 (1977): 457. J. Maynard Smith, *Towards a Theoretical Biology, vol. 2, Sketches*, C. H. Waddington, ed., (Edinburgh: Edinburgh University Press, 1969), 88–89, has pointed out, "All one can say is that since the first living organisms were presumably very simple, then if any large change in complexity has occurred in any evolutionary lineage, it must have been in the direction of increasing complexity... 'Nowhere to go but up'... Intuitively one feels that the answer to this is that life soon became differentiated into various forms, living in different ways, and that within such a complex ecosystem there would always be *some* way of life open which called for a more complex phenotype. This would be a self-perpetuating process. With the evolution of new species, further ecological niches would open up, and the complexity of the most complex species would increase."

[24] P. T. Saunders and M.-W. Ho, "On the Increase in Complexity in Evolution," *J. Theor. Biol.*, 63 (1976): 375–84. But see also W. McCoy, "Complexity in Organic Evolution"

basis of this tendency to be the process by which a self-organizing system optimizes its organization with respect to locally defined requirements for fitness. Even if it cannot be predicated as inevitable in any particular evolutionary line, there has been an overall trend towards and an increase in complexity along particular lines in biological evolution, so that it is right to be speak of a propensity for this to occur.

The need for *organization* for survival was beautifully demonstrated by H. A. Simon[25] who showed that the simplest modular organization of, say, the structure of a watch, so that each module had a limited stability, led to an enormous increase in survivability during manufacture in the face of random destructive events. Hence the increases we observe during evolution in complexity and organization (subsumed under "complexity" from now on) in the biological world are entirely intelligible as contributing to success in natural selection and are not at all mysterious in the sense of requiring some non-naturalistic explanation.

2. *Information-processing and -storage ability.* The more capable an organism is of receiving, recording, and analyzing signals and using the information to make predictions useful for survival about changes in its environment, the better chance it will have of surviving under the pressures of natural selection in a wide variety of habitats. In other words, there is a propensity towards the formation of systems having the functions we now recognize in nervous systems and brains. Such ability for information-processing and -storage is indeed the necessary, if not sufficient, condition for the emergence of consciousness.

3. *Pain and suffering.* This sensitivity to, this sentience of, its surroundings inevitably involves an increase in its ability to experience pain, which constitutes the necessary biological warning signals of danger and disease, so that it is impossible readily to envisage an increase of information-processing ability without an increase in the sensitivity of the signal system of the organism to its environment. Hence an increase in "information-processing" capacity, with the advantages it confers in natural selection, cannot but have as its corollary an increase, not only in the level of consciousness, but also in the experience of pain. Insulation from the surrounding world in the biological equivalent of three-inch nickel steel would be a sure recipe for preventing the development of consciousness!

Each increase in sensitivity, and eventually of consciousness, as evolution proceeds inevitably heightens and accentuates awareness both of the beneficent, life-enhancing, and of the inimical, life-diminishing, elements in the world in which the organism finds itself. The stakes for joy and pain are, as it were, continuously being raised, and the living organism learns to discriminate between them, so that pain and suffering, on the one hand, and consciousness of pleasure and well-being, on the other, are emergents in the world. Thus there can be said to be a propensity for them to occur. From a purely naturalistic viewpoint, the emergence of pain and its compounding as suffering as consciousness increases seem to be inevitable aspects of any conceivable developmental process that would be characterized by a continuous increase in ability to process and store information coming from the environment, for this entails an increase in sensitivity, hence in vulnerability, and consequently in suffering as consciousness (minimally the sum of the brain states reflecting it all) ramifies. In the context of natural selection, pain has an energizing effect and suffering is a goad to action: they both have survival value for creatures continually faced with new problematic situations challenging their survival.[26] In relation to any theological reflections, it must be emphasized that pain

---

and C. Castrodeza, "Evolution, Complexity and Fitness," *J. Theor. Biol.*, 71 (1978): 469–71, for different views.

[25] H. A. Simon, "The architecture of complexity," *Proc. Amer. Phil. Soc.*, 106 (1962): 467–82.

[26] Holmes Rolston (*Science and Religion: a critical survey*, New York: Random

and suffering are present in biological evolution as a necessary condition for survival of the individual long before the appearance of human beings on the scene. So the presence of pain and suffering cannot be the result of any particular human failings, though undoubtedly human beings experience them with a heightened sensitivity and, more than any other creatures, inflict them on each other.

4. *Self-consciousness and language.* If an information-processing and -storage system can also monitor its own state at any moment, then it has at least the basis for communicating what that state is to other similar systems. Hence, provided the physical apparatus for communicating has also evolved, the capacity for language becomes possible, especially in the most highly developed of such systems. In other words, there is an inbuilt propensity for the acquisition of language and so for developing the necessary basis for *self*-consciousness. This would be an advantage in natural selection, for it is the basis of complex social cooperation in the creatures that possess it (apparently supremely *Homo sapiens*) with all the advantages this gives against predators and in gaining food.[27]

Given an immanentist understanding of God's presence "in, with, and under" the processes of biological evolution adopted up to this point, can God be said to be implementing any purpose in biological evolution? Or is the whole process so haphazard, such a matter of happenstance, such a matter of what Monod and Jacob called *bricolage* (tinkering), that no meaning, least of all a divinely intended one, can be discerned in the process?

I have given reasons above for postulating that there are propensities in evolution towards the possession of certain characteristics, propensities that are inherently built into an evolutionary process based on natural selection, for they naturally enhance survival for procreation in a wide range of environments. Thus it is that the evolutionary process is characterized by propensities towards increase in complexity, information-processing and -storage, consciousness, sensitivity to pain, and even self-consciousness (a necessary prerequisite for social and development and the cultural transmission of knowledge down the generations). *Some* successive forms, along *some* branch or "twig" (à la Gould), have a distinct probability of manifesting more and more of these characteristics. However, the actual physical form of the organisms in which these propensities are actualized and instantiated is

---

House, 1987) has developed this characteristic of biological evolution in what he calls "cruciform naturalism" (289ff). Sentience, he argues, evolves with a capacity to separate the "helps" from the "hurts" of the world: with sentience there appears caring (287). With the appearance of life, organisms can now view events as "pro-" or "anti-life" and values and "dis-values" appear—the world becomes a "theater of meanings" and nature may be variously judged as "hostile," "indifferent," and "hospitable" (244). "The step up that brings more drama brings more suffering" (288). But "pain is an energizing force" so that "where pain fits into evolutionary theory, it must have, on statistical average, high survival value, with this selected for, and with a selecting against counterproductive pain" (288). "Suffering... moves us to action" and "all advances come in contexts of problem solving, with a central problem in sentient life the prospect of hurt. In the evolution of caring, the organism is quickened to its needs" (288). "Suffering is a key to the whole, not intrinsically, not as an end in itself, but as a transformative principle, transvalued into its opposite" (288).

[27] It is interesting to note that Richard Dawkins, too, in his *River Out of Eden* (London: Wiedenfeld and Nicolson, 1995) includes (151ff) amongst the thresholds that will be crossed *naturally* in "a general chronology of a life explosion on any planet, anywhere in the universe.... thresholds that any planetary replication bomb can be expected to pass," those for high-speed information-processing (no. 5, achieved by possession of a nervous system), consciousness (no. 6, concurrent with brains), and language (no. 7). This list partly corresponds to the "propensities" referred to in the text.

contingent on the history of the confluence of disparate chains of events, including the survival of the mass extinctions that have occurred (96% of all species in the Permo-Triassic one[28]). So it is not surprising that recent re-interpretation of the fossils of very early (*circa* 530 million years ago) soft-bodied fauna found in the Burgess shale of Canada show that, had any larger proportion of these survived and prevailed, the actual forms of contemporary, evolved creatures would have been very much more disparate in anatomical *plans* than those now observed to exist—albeit with a very great diversity in the few surviving designs.[29] But even had these particular organisms, unique to the Burgess shale, been the progenitors of subsequent living organisms, the same propensities towards complexity, etc., would also have been manifest in *their* subsequent evolution, for these "propensities" simply reflect the advantages conferred in natural selection by these features. The same considerations apply to the arbitrariness and contingency of the mass extinctions, which Gould also strongly emphasizes. So that, providing there had been enough time, a complex organism with consciousness, self-consciousness, social and cultural organization (that is, the basis for the existence of "persons") would have been likely eventually to have evolved and appeared on the Earth (or on some other planet amenable to the emergence of living organisms), though no doubt with a physical form very different from *Homo sapiens*. There can, it seems to me *(pace* Stephen Gould[30]) be overall direction and implementation of divine purpose through the interplay of chance and law without a deterministic plan fixing all the details of the structure(s) of what emerges possessing personal qualities. Hence the emergence of self-conscious persons capable of relating personally to God can still be regarded as an intention of God continuously creating through the processes of that to which God has given an existence of this contingent kind and not some other. It certainly must have been possible since it actually happened—with us!

I see no need to postulate any *special* action of God—along the lines, say, of some divine manipulation of mutations at the quantum level (as proposed by others in this volume)—to ensure that persons emerge in the universe, and in particular on Earth. Not to coin a phrase, "I have no need of that hypothesis!"[31] If there are any

---

[28] Gould, *Wonderful Life*, 306, citing David M. Raup.

[29] Ibid., 49.

[30] Ibid., 51 and passim.

[31] My basically theological and philosophical objections to the location of divine action in quantum events—in evolution, and elsewhere in the natural world (including that of the human brain)—may be summarized as follows.

1. This hypothesis assumes that if God does act to alter quantum events (e.g., in the present context, quantum events in DNA that constitute mutations), this would still be a "hands on" *intervention* by God in the very processes to which God has given existence, even if we never, in principal, could detect this divine action. It would imply that these processes without such intervention were inadequate to effect God's creative intentions if operating in the way God originally made and sustains them in existence.

2. Yet one of the principal reasons, certainly for a scientist and those influenced by the scientific perspective, for adducing from the nature of these processes the existence of a Creator God is their inherent rationality, consistency *and* creativity in themselves.

3. If one does not assume, with most physicists, that there are "hidden variables," that quantum events are indeed ontologically indeterminate within the restrictions of deterministic equations governing their probability—then God cannot know definitely the precise outcome of any quantum event because God can only know that which it is logically possible to know (and God knows everything in this category—that is what constitutes God's omniscience).

such influences by God shaping the direction of evolutionary processes at specific points—for which I see no evidence (how could we know?) and no theological need—I myself could only envisage them as being through God's whole-part constraint on all-that-is affecting the confluence of what, to us, would be independent causal chains. Such specifically-directed constraints I would envisage as possible by being exerted upon the whole interconnected and interdependent system of the whole Earth in the whole cosmos which is in and present to God, who is therefore its ultimate boundary condition and therefore capable of shaping the occurrence of particular patterns of events, if God chooses to do so.[32]

2.4 The Ubiquity of Pain, Suffering and Death.

The biological inevitability of the experience of pain in any creature that is going to be aware of—and can gain information from—its environment and thereby avoid dangers has already been emphasized. The pain associated with breakdown of health due to general organic causes also appears to be simply a concomitant of being a complex organized system containing internal as well as external sensors. When pain is experienced by a conscious organism, the attribution of "suffering" becomes appropriate and a fortiori, with self-consciousness the suffering of others also becomes a burden. The ubiquity of pain and suffering in the living world appears to be an inevitable consequence of creatures acquiring those information-processing and - storage systems, so advantageous in natural selection, which we observe as nerves and brains in the later stages of evolution.

New patterns can only come into existence in a finite universe ("finite" in the sense of the conservation of matter-energy) if old patterns dissolve to make place for them. This is a condition of the creativity of the process—that is, of its ability to produce the new—which at the biological level we observe as new forms of life only through death of the old. For the death of individuals is essential for release of food resources for new arrivals, and species simply die out by being ousted from biological "niches" by new ones better adapted to survive and reproduce in them. Thus, biological death of the individual is the prerequisite of the creativity of the biological order, that creativity which eventually led to the emergence of human beings. At this biological level we discover

---

Ontological indeterminacy at the quantum level precludes such precise knowledge *for* God to have. Thus God could not know (logically could not know) the outcome of the interference by God in the quantum events which this hypothesis postulates and could not effect the divine purposes thereby.

4. For the overall probabilistic relationships which govern statistically the ontologically indeterministic quantum events to be obeyed, if God were to alter one such event in a particular way, then many others would also have to be changed so that we, the observers, detected no abrogation of the overall statistics, as the hypothesis assumes. So it is certainly no tidy, neat way to solve the problem and one wonders where the chain of necessary alterations would end.

5. Finally, in any case, the hypothesis is otiose if God is regarded as creating in evolution, as elsewhere, *through* the very processes, themselves creative, to which God gives existence and which God continuously sustains in existence.

This is why I think there is no need of this hypothesis in this evolutionary context—or indeed in that of any other (e.g., as a way God might affect human brain states and so thoughts).

[32] For a fuller exposition of this approach, see TSA, 157–65; and, more particularly and recently, "God's Interaction with the World: The Implications of Deterministic 'Chaos' and of Interconnected and Interdependent Complexity," in *Chaos and Complexity*, R. J. Russell, N. Murphy, and A. Peacocke, eds., (Vatican City State: Vatican Observatory, Berkeley, Calif.: Center for Theology and the Natural Sciences, 1995), 263–87.

the process to be one of "natural selection," but it is possible to discern cognate processes occurring also at other levels.

For complex living structures can only have a finite chance of coming into existence if they are not assembled *de novo,* as it were, from their basic subunits, but emerge through the accumulation of changes in simpler forms, as demonstrated by H. A. Simon in his classic paper.[33] Having come on to the scene, they can then survive, because of the finitude of their life spans, only by building pre-formed complex chemical structures into their fabric through imbibing the materials of other living organisms. For the chemist and biochemist there is the same kind of difficulty in conceiving how complex material structures, especially those of the intricacy of living organisms, could be assembled otherwise than from less complex units, as there is for the mathematician of conceiving of a universe in which the analytic laws of arithmetic were inapplicable. So there is a kind of *structural* logic about the inevitability of living organisms dying and preying on each other—for we cannot conceive, in a lawful, non-magical universe, of any way whereby the immense variety of developing, biological, structural complexity might appear, except by utilizing structures already existing, either by way of modification (as in biological evolution) or of incorporation (as in feeding).[34] The statistical logic is inescapable: new forms of matter arise only through the dissolution of the old; new life only through death of the old. It would seem that the law of "new life through death of the old" (J. H. Fabre's "sublime law of sacrifice"[35]) is inevitable in a world composed of common "building blocks" (atoms, etc.).

But death not only of individuals but of whole species has also occurred on the Earth during the periods of mass extinction which are now widely attributed to chance extraterrestrial collisions of the planet with comet showers, asteroids or other bodies. These could be cataclysmic and global in their effects and have been far more frequent than previously imagined. This adds a further element of sheer contingency to the history of life on the Earth.

The theist cannot ignore these features of the created order. Any theodicy has to come to terms with the obliteration of far more species than now exist on the Earth. The spontaneity and fecundity of the biological world is gained at the enormous price of universal death and of pain and suffering during life.[36] Yet individual living creatures scarcely ever commit suicide in any way that might be called intentional. Let us pay attention to this positive aspect first.

The natural world is immensely variegated in its hierarchies of levels of entities, structures, and processes, in its "being," and abundantly diversifies with a cornucopian fecundity in its "becoming" in time. From the unity in this diversity and the richness of the diversity itself, one may adduce,[37] respectively, both the essential

---

[33] H. A. Simon, "The architecture of complexity."

[34] The depiction of this process as "nature, red in tooth and claw" (a phrase from Tennyson that actually pre-dates Darwin's proposal of evolution through natural selection) is a caricature, for, as many biologists have pointed out (e.g., G. G. Simpson in *The Meaning of Evolution* (New Haven: Bantam Books, Yale University Press, 1971 edition), 201), natural selection is not even in a figurative sense the outcome of struggle, as such. Natural selection involves many factors that include better integration with the ecological environment, more efficient utilization of available food, better care of the young, more cooperative social organization—and better capacity for surviving such "struggles" as do occur (remembering that it is in the interest of any predator that their prey survive as a species!).

[35] Quoted by C. E. Raven, *Natural Religion and Christian Theology,* 1951 Gifford Lectures, Series 1, *Science and Religion* (Cambridge: Cambridge University Press, 1953), 15.

[36] Cf., Dawkins' epithet, "DNA neither cares nor knows. DNA just is. And we dance to its music" (*River out of Eden,* 133).

[37] *TSA,* chap. 8, section 1.

oneness of its source of being, namely the one God the Creator, and the unfathomable richness of the unitive being of that Creator God. But now we must reckon more directly with the diversity itself. The forms even of non-living matter throughout the cosmos as it appears to us is even more diverse than what we can now observe immediately on the Earth. Furthermore the multiply branching bush of terrestrial biological evolution appears to be primarily opportunist in the direction it follows and, in so doing, produces the enormous variety of biological life on this planet.

We can only conclude that, if there is a Creator, least misleadingly described in terms of "personal" attributes, then that Creator intended this rich multiformity of entities, structures, and processes in the natural world and, if so, that such a Creator God takes what, in the personal world of human experience, could only be called "delight" in this multiformity of what he has created—and not only in what Darwin, called "the most exalted object which we are capable of conceiving, namely the production of the higher animals."[38] The existence of the *whole* tapestry of the created order, in its warp and woof, and in the very heterogeneity and multiplicity of its forms must be taken to be the Creator's intention. We can only make sense of that, utilizing our resources of personal language, if we say that God has something akin to "joy" and "delight" in creation. We have a hint of this in the satisfaction attributed to God as Creator in the first chapter of *Genesis*: "And God saw everything he had made, and behold, it was very good."[39] This naturally leads to the idea of the "play" of God in creation on which I have expanded elsewhere,[40] in relation to Hindu thought as well as to that of Judaism and Christianity.

But now for the darker side. The ubiquity of pain, suffering and death as the means of creation through biological evolution entails, for any concept of God which is morally acceptable and coherent, that if God is immanently present in and to natural processes, in particular those that generate conscious and self-conscious life, then we cannot but infer that God suffers in, with, and under the creative processes of the world with their costly unfolding in time.

Rejection of the notion of the impassibility of God has, in fact, been a feature of the Christian theology of recent decades. There has been an increasing assent to the idea that it is possible to speak consistently of *a God who suffers eminently and yet is still God, and a God who suffers universally and yet is still present uniquely and decisively in the sufferings of Christ.*[41]

---

[38] C. Darwin, *The Origin of Species* (London: Thinkers Library, Watts, 6th ed.), 408. As Charles Darwin himself put it in a famous passage at the end of one edition of this work: "It is interesting to contemplate a tangled bank, clothed with many plants of many kinds, with birds singing on the bushes, with various insects flitting about, and with worms crawling through the damp earth, and to reflect that these elaborately constructed forms, so different from each other, and dependent upon each other in so complex a manner, have all been produced by laws acting around us.... There is grandeur in this view of life, with its several powers, having been originally breathed by the creator into a few forms or into one; and that, whilst this planet has gone cycling on according to the fixed law of gravity, from so simple a beginning endless forms most beautiful and most wonderful have been, and are being evolved."

[39] *Genesis* 1:31.

[40] *CWS*, 108–11.

[41] Paul S. Fiddes, *The Creative Suffering of God* (Oxford: Clarendon Press, 1988), 3 (emphasis in the original).

As Paul Fiddes points out in his survey and analysis of this change in theological perspective, the factors that have promoted the view that God suffers are new assessments of "the meaning of love [especially, the love of God], the implications of the Cross of Jesus, the problem of [human] suffering, and the structure of the world."[42] It is this last-mentioned—the "structure of the world"—on which the new perspectives of the biological sciences bear by revealing the world processes to be such, as described above, that involvement in them by the immanent Creator has to be regarded as involving suffering on the Creator's part. God, we find ourselves having to conjecture, suffers the "natural" evils of the world along with ourselves because—we can but tentatively suggest at this stage—God purposes *inter alia* to bring about a greater good thereby, namely, the kingdom of free-willing, loving persons in communion with God and with each other.[43]

Because sacrificial, self-limiting, self-giving action on behalf of the good of others is, in human life, the hallmark of love, those who believe in Jesus the Christ as the self-expression of God's own self have come to see his life as their ultimate warrant for asserting that God is essentially "Love," insofar as any one word can accurately refer to God's nature. Jesus' own teaching concerning God as "Abba," Father, and the conditions for entering the "Kingdom of God" pointed to this too, but it was the person of Jesus and what happened to him that finally, and early, established in the Christian community this perception of God as self-offering Love.

On their, and subsequent Christians', understanding Jesus the Christ is the definitive communication from God to humanity of the deep meaning of what God has been effecting in creation—and that is precisely what the Prologue to John's Gospel says in terms of God the Word/*Logos* active in creation and as now manifest in the person of Jesus the Christ.

As we saw above, it may be inferred, however tentatively, from the character of the natural processes of creation that God has to be seen as suffering in, with, and under these selfsame processes with their costly unfolding in time. But if God was present in and one with Jesus the Christ, then we have to conclude that *God* also suffered in and with him in his passion and death. The God whom Jesus therefore obeyed and expressed in his life and death is indeed a "crucified God,"[44] and the cry of dereliction can be seen as an expression of the anguish also of God in and through creation. If Jesus is indeed the self-expression of God in a human person, then the tragedy of his actual human life can be seen as a drawing back of the curtain to unveil a God suffering in and with the sufferings of created humanity and so, by a natural extension, with those of all creation, since humanity is an evolved part of it. The suffering of God, which we could glimpse only tentatively in the processes of creation, is in Jesus the Christ concentrated into a point of intensity and transparency which reveals it to all who focus on him.

---

[42] Ibid., 45 (see also all of chap. 2).

[43] I hint here at my broad acceptance of John Hick's "Irenaean" theodicy in relation to "natural" evil (q.v., "An Irenaean Theodicy," in *Encountering Evil*, ed. Stephen T. Davis (Edinburgh: T. & T. Clark, 1981), 39–52; and his earlier *Evil and the God of Love* (London: MacMillan, 1966), especially chapters 15 and 16); and the position outlined by Brian Hebblethwaite in chapter 5 ("Physical suffering and the nature of the physical world") of his *Evil, Suffering and Religion* (London: Sheldon Press, 1976).

[44] The title of Jürgen Moltmann's profound book, *The Crucified God* (London: SCM Press, 1974).

2.5 *Human Evolution*

1. *Homo sapiens a late arrival.* Remarkable and significant as is the emergence of self-conscious persons by natural processes from the original "hot big bang" from which the universe has expanded over the last ten to twenty billion years, this must not be allowed to obscure humanity's relatively recent arrival in the universe, even on a time-scale of the history of the Earth. How recent this is can be realized if one takes the age of the Earth as two days of forty-eight "hours" (one such "hour" equals 100 million years)—then *Homo sapiens* appears only at the last stroke of midnight of the second day. Other living organisms had existed for some two billion or more years (which equals over twenty "hours" on the above scale) before our relatively late arrival.

2. *The emergence of humanity.* The biological-historical evidence is that human nature has emerged only gradually by a continuous process from other primates. No sudden breaks of any substantial kind in the sequences are noted by paleontologists and anthropologists. This is not to say that the history of human culture is simply a smooth rising curve. There must have been, for example, key turning points or periods in the development of speech and so of social cooperation and of rituals for burying the dead, with provision of food and implements, testifying to a belief in some form of life after death. These apparently occurred among the Neanderthals of the middle Paleolithic even before the emergence of *Homo sapiens* some 100,000 or so years ago, when further striking developments occurred.[45] However there is *no* past period for which there is reason to affirm that human beings possessed moral perfection existing in a paradisal situation from which there has been only a subsequent decline. All evidence points to a creature slowly emerging into awareness, with an increasing capacity for consciousness and sensitivity and the possibility of moral responsibility and, the religions would affirm, of response to God (especially after the "axial period" around 500 B.C.[46]). So there is no sense in which we can talk of a "Fall" from a past perfection. There was no golden age, no perfect past, no individuals, "Adam" or "Eve," from whom all human beings have now descended and declined and who were perfect in their relationships and behavior. We appear to be rising beasts[47] rather than fallen angels!

With regard to (1): the fact that human beings represent only an extremely small, and very recently-arrived, fraction of living organisms that have populated the Earth raises the question of God's purposes in creating such a labyrinth of life. To assume it was all there simply to lead to us clearly will not do. Hence the attribution to God of a sheer exuberance in creativity for its own sake, to which we have already referred. Since biological death was present on the Earth long before human beings arrived on the scene and was the pre-requisite of our coming into existence through the processes of biological evolution, when Saint Paul says that "sin pays a wage, and the wage is death"[48] that cannot possibly mean for us now *biological* death. It can only mean "death" in some other sense, such as the death of our relation to God

---

[45] See Karl J. Narr, "Cultural achievements of early man" in *The Human Creature,* ed. G. Altner (Garden City, New York: Anchor Books, Doubleday, 1974), 115–17. "...a marked evolutionary expansion manifests itself after around 30,000 B.C., at the beginning of the upper Paleolithic. The new picture that emerges can be characterized by such terms as accumulation, differentiation and specialization. There is an increase and concentration of cultural goods, a more refined technology with greater variety in the forms of weapons and tools produced and corresponding specialization of their respective functions, more pronounced economic and general cultural differentiation of individual groups."

[46] K. Jaspers, *The Origin and Goal of history* (London: Kegan Paul, 1953), chap.13, section 2.

[47] Rising from an a-moral (and in that sense) innocent state to the capability of moral, and immoral, action.

[48] *Rom.* 6:23.

consequent upon sin. I can see no sense in regarding biological death as the consequence of that very real alienation from God that is sin, because God had already used biological death as the means for creating new forms of life, including ourselves, long before we appeared on the Earth. This means those classical Christian formulations of the theology of the redemptive work of Christ that assume a causal connection between biological death and sin urgently need replacing.

With regard to (2): although there was no perfect past, no original, perfect, individual "Adam" from whom all human beings have now declined, what *is* true is that humanity manifests aspirations to a perfection not yet attained, a potentiality not yet actualized, but no "original righteousness." Sin as alienation from God, humanity, and nature is real and appears as the consequence of our very possession of that *self-consciousness* which always places us at the egotistical center of the "universe" of our consciousness. Sin is about a falling short from what God intends us to be and is part and parcel of our having evolved into self-consciousness, freedom, and intellectual curiosity. The domination of Christian theologies of redemption by classical conceptions of the "Fall"[49] urgently needs to be rescinded, and what we mean by redemption to be rethought if it is to make any sense to our contemporaries.

Now, we all have an awareness of the tragedy of our failure to fulfil our highest aspirations, to come to terms with finitude, death and suffering, to realize our potentialities, and to steer our path through life. Freedom allows us to make the wrong choices, so that sin and alienation from God, from our fellow human beings, and from nature, are real features of our existence. So the questions of not only "Who are we?" but even, "What should we be becoming—where should we be going?" remain acute for us. To be brief, I find the clue to the answers to these questions in the person of Jesus of Nazareth and what he manifested of God's perennial expression in creation (as the *Logos* of God incarnate). Hence I am impelled to consider specifically *Christian* affirmations in the light of the foregoing reconsiderations of God's creating through an evolutionary process.

3 The significance of Jesus the Christ in a Christian evolutionary perspective

> ... [I]n scientific language, the Incarnation may be said to have introduced a new species into the world—the Divine man transcending past humanity, as humanity transcended the rest of the animal creation, and communicating His vital energy by a spiritual process to subsequent generations... (J. R. Illingworth, in the 12th edition of *Lux Mundi*, 1891, p. 132.)

Jesus' resurrection demonstrated to the disciples, notably to Paul, and now to us, that it is the union of *his* kind of life with God which is not broken by death and capable of being taken up into God. For he manifested the kind of human life which can become fully life with God, not only here and now, but eternally beyond the threshold of death. Hence his imperative "Follow me" now constitutes for us a call for the transformation[50] of humanity into a new kind of human being and becoming. What happened to him, Jesus saw *could* happen to all.

---

[49] The myths of Adam and Eve and of the "Fall" have long since been interpreted non-historically and existentially by modern theologians and biblical scholars; see, for example, Alan Richardson, "Adam, Man," in *A Theological Word Book of the Bible*, Alan Richardson, ed., (London: SCM Press, 1957), 14, quoted in TSA, 249.

[50] But we cannot, today, use for this transformation Illingworth's phrase, "a new species," in any literal sense, for the concept "species" is a purely biological one.

In this perspective, Jesus the Christ, the whole Christ event, has shown us what is possible for humanity. The actualization of this potentiality can properly be regarded as the consummation of the purposes of God already incompletely manifested in evolving humanity. In Jesus there was a *divine* act of new creation because the initiative was *from God* within human history, within the responsive human will of Jesus inspired by that outreach of God into humanity designated as "God the Holy Spirit." Jesus the Christ is thereby seen, in the context of the whole complex of events in which he participated (the "Christ event"), as the paradigm of what God intends for all human beings, now revealed as having the potentiality of responding to, of being open to, of becoming united with, God.

But how can what happened in and to him then, happen in us now? Can what happened in and to him be effectual, some 2000 years later, in a way that might actually enable us to live in harmony with God, ourselves and our fellow human beings—that is, can we experience the fulfilment for which human nature yearns?

I can outline here only sketchily how such questions might be answered in the affirmative.[51] *Any* credible answer today will have to be grounded on our sharing a common humanity with Jesus. Certain features in the evolutionary perspectives we have been delineating will now properly constrain this response, namely:

1. biological death of the individual, as the means of the evolutionary creation of new species by natural selection, cannot now be attributed to human "sin";
2. the evidence is against human beings ever in the past having been in some golden age of innocence and perfection from which they have "fallen."

The Nicene Creed simply affirms that Christ "was crucified *for us* under Pontius Pilate. He suffered and was buried." This reticent "for us" encompasses a very wide range of interpretations. Although the church in its many branches has never officially endorsed any one particular theory of this claimed at-one-ment, yet a number have become widely disseminated doctrinally, liturgically, and devotionally. Most (with the exception of the Abelardian) propose a change in God's relation and attitudes to humanity because of Jesus' death on the Cross. These purportedly "objective" theories of the atonement also rely heavily on presuppositions that are contrary to (1) and (2) above. Moreover they fail to incorporate our sense derived from the vista of evolution unfolded by the sciences of humanity as *emerging* into an individual and corporate consciousness and self-consciousness, an awareness of values, social cooperation, culture, and of God. The classical theories of the atonement fail to express any dynamic sense of the process of human *becoming* as still going on. They also fail to make clear how the human response which is an essential part of the reconciliation between God and humanity is evoked.

Let us now put the question again as: How can what happened in and to Jesus the Christ actually evoke in us the response that is needed for our reconciliation to God and enable us to live in harmony with God and humanity *now*? This question may be answered most effectively by seeing the life, suffering and death of Jesus the Christ as an act of love, an act of love *of God*, an act of love *by God*.

In the suffering and death of Jesus the Christ, we now also perceive and experience the suffering, self-offering love of *God* in action, no more as abstract knowledge, but actually "in the flesh." For the openness and obedience of the human

---

[51] For a fuller treatment, see *TSA*, 319ff. A more fully argued exposition of this whole theological stance is given in *TSA*, part III, chaps. 14–16.

Jesus to God enabled him, as *the* God-informed human person, to be a manifest self-expression in history, within the confines of human personhood, of God as creative self-expressing Word/*Logos*/"Son." Thereby is uniquely and definitively revealed the depths of the divine Love for humanity and the reality of *God*'s gracious outreach to us as we are, alienated from God, humanity, and ourselves, that is, as "sinners." As such this love of God *engages* us, where "to engage" means (OED): "to attract and hold fast; to involve; to lay under obligation; to urge, induce; to gain, win over." The Cross is a proposal of God's love and *as such* engages our response. Once we have really come to know that it was God's love in action "for us" which was manifest in the self-offering love and obedience of Jesus the Christ, then we can never be the same again. God, in that outreach to humanity we denote as God the Holy Spirit, united the human Jesus with God's own self and can now kindle and generate in us, as we contemplate God in Christ on the Cross, a love for God, for the humanity for whom Jesus died, and for the creation in which God was incarnated.

I am proposing here that this action of God as Holy Spirit in us engages our response and this itself effects our at-one-ment, is itself salvific, actually making us whole, making us "holier." *Such an understanding of the "work of Christ" coheres with our present evolutionary perceptions that the specifically human emerged and still emerges only gradually and fitfully in human history*, without a historic "Fall."

For since God took Jesus through death into his own life, it is implicit in this initiation and continuation of this process in us, that we too can thereby be taken up into the life of God, can be "resurrected" in some way akin to that of Jesus the Christ. Since Jesus was apprehended as having been taken through death with his personhood and identity intact and as having been "taken up" into the presence of God, it *could* happen to us and that is the ground of our hope for our individual future and that of humanity corporately.[52]

Furthermore the interpretation of the death and resurrection of Jesus as manifesting uniquely the quality of life which can be taken up by God into the fullness of God's own life implicitly involves an affirmation about what the basic potentiality of all humanity is. It shows us that, regardless of our particular human skills and creativities—indeed regardless of almost all that the social mores of our times applaud—it is through a radical openness to God, a thoroughgoing self-offering love for others and for God's creation, and obedience to God that we grow into such communion with the eternal God that *God* does not allow biological death to rupture that potentially timeless relation. Irenaeus, in accord with the Eastern Christian tradition, says it all:

The Word of God, our Lord Jesus Christ
Who of his boundless love
became what we are
to make us what even he himself is.[53]

---

[52] The virtue of being agnostic about the relation between the "empty tomb" and the risen Christ here becomes apparent. For, within a relatively short time after our own biological death, our bodies will lose their identity as their atomic and molecular constituents begin to disperse through the earth and its atmosphere, often becoming part of other human beings. (See the discussion in *TSA*, 279–88.)

[53] Irenaeus, *Adv. Haer.*, v, praef. (author's translation).

ORIGINAL SIN AND SAVING GRACE IN EVOLUTIONARY CONTEXT

Denis Edwards

1 *Introduction*

A Christian theology which embraces the theory of biological evolution will involve a rethinking of the doctrines of original sin and saving grace. How can theology talk meaningfully of original sin and grace in an evolutionary world? I see this as a classic instance of the need for the kind of interaction between science and theology which Ian Barbour has called "doctrinal reformulation."[1]

This paper will be an attempt to contribute to such a reformulation. First, I will survey the work of three theologians who have recently made contributions on this issue, Gerd Theissen, Sallie McFague, and Philip Hefner. Then, in a second step, I will attempt to build on their work in a constructive reformulation of the theology of original sin and grace.

Since Christian theologies of sin and grace are diverse, it will be helpful to begin by naming my own starting point, which is the reformulation of these doctrines already undertaken by Karl Rahner. For Rahner, grace is the self-communication of God. It is the Spirit of God, always freely and graciously present in self-offering love to every human person. God creates in order to give Godself in love. This self-offering love which is always present can be either rejected (as sin) or accepted (as justification). God surrounds and embraces every human person in radical nearness, so that this divine self-offer is a constitutive dimension of human existence. We live in a world of grace. Rahner's theology of grace is summed up in his saying that the human person is "the event of a free, unmerited and forgiving, and absolute self-communication of God."[2]

But grace is not the only reality that confronts the freedom of the human person. There is also "original sin." Rahner understands this as having to do with the fact that we actualize ourselves as free subjects in a situation which is always determined by other persons and by history. The sin of others is intrinsic to and partly constitutive of the situation of our human freedom. The sin of others is a universal and permanent part of the human condition and is in this sense original. This determination of our inner situation by the sin of others is called sin only in a loose and analogous sense. Strictly speaking, sin is personal and actual; it is the free and deliberate rejection of God. It is only for this kind of personal sin that we are morally responsible. Original sin consists of the fact that human beings have a history of refusal and radical rejection of God's self-communicating love, and this history of personal and communal sin enters into and becomes an inner dimension of each person's situation.

I will ask what difference biological evolution makes to such a theology of sin and grace. Arthur Peacocke has pointed out that the link between biological evolution and human cultural history provides a "new brief" and an exciting role for contemporary theology.[3] He points particularly to the fact that in human beings

---

[1] Ian Barbour, *Religion in an Age of Science* (London: SCM Press, 1990), 26–28. Doctrinal reformulation fits within the "theology of nature" approach in Barbour's typology.

[2] Karl Rahner, *Foundations of Christian Faith* (New York, Seabury Press, 1978), 116.

[3] Arthur Peacocke, *God and the New Biology* (London: J. M. Dent, 1986), 115.

genetic evolution gives rise to a new form of information transfer—that of culture. In this paper, I will attempt to contribute to a rethinking of the theology of original sin and grace by bringing this theology into dialogue with contemporary insights into both genetic and cultural evolution.

2 A Survey of Three Theologies

I have chosen to begin this reformulation with a survey of the contributions of Gerd Theissen, Sallie McFague, and Philip Hefner, because their work is at least partially interconnected. McFague and Hefner make reference to Theissen's ideas,[4] and both Theissen and Hefner make frequent references to the works of Donald T. Campbell and Ralph Wendell Burhoe.[5] I will conclude the summary of each theologian's views by raising critical questions of their work. These questions will then be taken up and developed in the constructive sections on original sin and grace.

2.1 Gerd Theissen

Gerd Theissen is well known for using insights from social science in his study of the Scriptures. In 1984, with his *Biblical Faith: an Evolutionary Approach*, he took up the scientific framework of evolutionary theory and applied it in a thoroughgoing way to the interpretation of biblical faith.[6] Theissen, using the "evolutionary epistemology" of thinkers like Karl Popper, argues that what is common to science and theology can be discovered through categories drawn from the theory of

---

[4] See Sallie McFague, *The Body of God: An Ecological Theology* (Minneapolis: Fortress Press, 1993) where she refers to Philip Hefner and acknowledges indebtedness to Theissen in n. 10, p. 257. See Philip Hefner, *The Human Factor: Evolution, Culture and Religion* (Minneapolis: Fortress, 1993) where he refers to Theissen on pp. 30, 190, 207, 247, 249, 252, 253, 274.

[5] Campbell argues that urban society has been made possible by cultural evolution and that religion plays a crucial role in this cultural history. He points out that unlike social insects, human beings who cooperate together are genetically competitors. In order to succeed in their environment, humans must develop behaviors that enhance the welfare of those who are not their kin. Human cities depend upon cooperation which is made possible by social evolution and, in this kind of evolution, groups which indoctrinate members into religious commitments could have a social-evolutionary advantage. He speaks of the value of religious "counter-hedonic" traditions, which allow human beings to serve other values than those of immediate satisfaction; see Donald T. Campell, "On the Conflicts Between Biological and Social Evolution and Between Psychology and Moral Tradition," *Zygon* 11 (1976): 167–208. Burhoe understands the human being as a symbiosis of two sources of information, genetic and cultural. This symbiotic structure means that human beings can occupy an evolutionary niche which is radically different from that of any other form of life. This human symbiosis has been selected for during the last million years as a unity. Genes and culture co-evolve and co-adapt. Like Campbell, Burhoe sees religion playing a vital function in human evolution, fostering altruism and cooperation and enabling human society to survive and flourish; see Ralph Wendell Burhoe, "The Source of Civilization in the Natural Selection of Coadapted Information in Genes and Culture," *Zygon* 11 (1976): 263–303; idem, *Toward a Scientific Theology* (Belfast: Christian Journals Limited, 1981), 15.

[6] Gerd Theissen, *Biblical Faith: An Evolutionary Approach* (London: SCM Press, 1984).

evolution.[7] He sees both science and religious faith as attempts to adapt to reality. For faith, the "central reality" to which it must adapt is understood as God.

Theissen's argument is that culture is a process which "reduces selection" and that religion is the heart of human culture. He sees religion as a protest against the principle of selection. He makes his argument by considering what he calls three great "mutations" of Christian faith: 1) biblical monotheism, in which the prophets called for a far reaching change of behavior towards the sojourner and the widow;[8] 2) New Testament Christology, in which the proclamation and activity of Jesus reveal the nature of divine love and act as a "protest against the pressure of selection";[9] 3) the experience of the Holy Spirit, in which those seized by the Spirit "are incorporated into the history of protest against selection."[10]

Theissen suggests that in Jesus "the principle of solidarity replaced the principle of selection... [which] brought about a new era in the cosmos and completed the history of the God who revealed himself in the experiences of Israel." He claims that the new mutation that has appeared in Jesus is love, a love defined as "a solidarity with the weak, a contradiction of the processes in nature and history which are oriented on selection."[11] While selection means aggression against one's enemies, the Gospel calls us to love our enemies (Matt 5:43). While selection involves solidarity with our genetic kin, the Gospel calls for solidarity with outsiders. While selection benefits those with power, the Gospel calls for a form of leadership modeled on the slave (Mark 10:44). While selection means preference for the strong and healthy, the Gospel teaches identification with the weak.[12] The good news of the Gospel is offered in the name of God, and reveals the true nature of the "central reality."

While Theissen does not like the language of original sin, he sees humans as having a "natural" inclination to sin. The pull in us towards anti-social behavior springs from a biological foundation. We have pre-programmed biological tendencies which are held in check by strong cultural control, but which can collapse with disastrous consequences. Theissen writes that "we experience the tension within us between biological and cultural evolution as 'guilt'."[13] It is the capacity for symbol-making, Theissen argues, which enables true altruism. By virtue of a symbolic act

---

[7] See Theissen, *Biblical Faith*, 18. He mentions alongside K. Popper the work of K. Lorenz, G. Vollmer and R. Riedl. He refers to those who have interpreted religion with the help of evolutionary ideas: D. Campbell, A. Hardy, R. W. Burhoe and H. V. Ditfurth. Though I admire Theissen's contribution, in the end it seems to me that evolutionary categories do not deal with Christian faith in an adequate way. Ian Barbour points to the inadequacy of evolutionary epistemology applied to science, and notes that "here again the differences between cultural and biological evolution are more significant than their similarities." See Barbour, *Religion in an Age of Science*, 194. Hans Schwarz, in his "Review of *Biblical Faith*," *Zygon* 21 (1986): 540–42, suggests that "to a large degree Theissen forces the Judeo-Christian faith into the straitjacket of evolutionary conceptuality." Reviewers differed on this aspect of Theissen's work. For a positive response see Walter Wink, "Review of *Biblical Faith*," *Zygon* 21 (1986): 543–45, and for a more negative critique see Luke T. Johnson "Review of *Biblical Faith*," *The Catholic Biblical Quarterly* 49 (1987): 161–63.

[8] Thus Theissen sees the breakthrough of biblical monotheism as an "evolution of evolution." See Theissen, *Biblical Faith*, 80–81.

[9] Ibid., 112.

[10] Ibid., 141.

[11] Ibid., 87.

[12] For other examples see Theissen, *Biblical Faith*, 115.

[13] Ibid., 147.

human beings can see "brothers" and "sisters" in others even when this is not true in a genetic sense. When our symbol-making capacity is actualized in a pro-social way it is "our spirit being grasped by the 'Holy Spirit' of anti-selectionist motivation."[14]

Theissen's work is vulnerable to criticisms from biology. The sharp opposition he sets up between natural selection and culture does not take into account more recent interactionist models, which are based on a cooperative and coadaptive relation between genes and culture. Theissen's view of natural selection is further challenged by the ecological model of evolution proposed by Charles Birch and John Cobb, where what comes into view is the inter-relationship and cooperation between organism and environment.[15]

A critical theological question to be addressed to Theissen concerns the opposition he sets up between natural selection and the way of Jesus Christ. Does Theissen mean to suggest that natural selection is evil? What does this do to the biblical teaching of the goodness of creation? Is divine action only through opposition to natural selection? Is the way of Jesus really opposed to natural selection?

2.2 Sallie McFague

In her recent book, *The Body of God: An Ecological Theology*, Sallie McFague develops a theology of God's engagement with creation by using the model of the universe as God's body. Here, I will not attempt a response to this model, but simply focus on the Christological section of this book where McFague discusses the relationship between natural selection and the inclusive praxis of Jesus.[16]

McFague's overall argument is that it is in the story of Jesus that we can find the pattern for divine immanence in creation. We find in Jesus something that is not evident simply from evolutionary history, namely that the universe is directed "toward inclusive love for all, especially the oppressed, the outcast, the vulnerable."[17] She points to three aspects of Jesus' ministry: his "destabilizing" parables that side with the outcast; the healing stories which point to the importance of bodies; and his practice of eating with sinners which points to radical inclusivity. All of this together forms what McFague calls the "Christic paradigm." The distinctive feature of the Christic paradigm is the inclusion of the outcast and the oppressed, and McFague argues that because of human ecological destruction, this must include as well the "new poor" of nature. This argument, along with her fundamental concept that the universe can be understood as God's body, provides the foundations for her ecological ethics.

She asks: What consonance can there be between the Christian message of inclusion of the oppressed and biological evolution, in which millions of lives are wasted? What is the relationship between solidarity with the oppressed and natural selection? Her answer is that there is in one sense consonance and in another sense dissonance between Christianity and natural selection.

---

[14] Ibid., 144.

[15] Charles Birch and John Cobb, *The Liberation of Life: From the Cell to the Community* (Cambridge: Cambridge University Press, 1981), 79–96.

[16] Sallie McFague's use of the metaphor of creation as God's body is careful and subtle, but I have some reservations about the central metaphor because it seems to me that the metaphor itself can easily move in the direction of compromising both divine transcendence and creaturely integrity and of suggesting a dualistic view of God.

[17] McFague, *The Body of God*, 160.

McFague sees solidarity with the oppressed as consonant with evolutionary history, in the sense that she sees cultural evolution as the current stage of evolution, and she understands this stage as one in which it is necessary for human culture to help all life forms to share the goods of the planet. But she finds dissonance in the sense that the radical inclusiveness at the heart of Christian faith runs counter to the wastage and suffering of natural selection. Even within human cultural evolution, she points out, we find a tendency to construct our worlds to benefit ourselves. Christian solidarity with the oppressed has some not only counter-evolutionary and but also counter-cultural characteristics. It involves resistance to evil and action aimed at liberation of the oppressed. But it can also mean suffering with the oppressed as God suffers, suffering in hope based on God's salvific action.[18]

God cannot set aside the laws of nature, but suffers with suffering creation. Because McFague sees the world as God's body, she argues that nothing happens in creation that does not happen to God. But she points out that sin rather than natural evil is the principle form of evil endangering our planet. Sin is the desire to have everything for oneself, the centering of everything on ourselves. Human selfishness and greed constitute the real ecological problem, a problem that is gender, race, and class specific.

It seems to me that the following questions need to be put to this part of Sallie McFague's work: Is there a tendency to be too negative (moralistic) about natural selection? Is solidarity truly opposed to natural selection? Is the Christic paradigm opposed to natural selection and evolution, or does the Christic paradigm define not only God's action in Jesus Christ but also God's creative action in and through natural selection?

2.3 Philip Hefner

In his *The Human Factor: Evolution, Culture, and Religion*, Philip Hefner presents a challenging synthesis of his understanding of Christian anthropology. He sees human beings as "God's co-creators whose purpose is to be the agency, acting in freedom, to birth the future that is most wholesome for the nature that has birthed us."[19]

Following Burhoe, Hefner sees *Homo sapiens* as "a two-natured creature, a symbiosis of genes and culture."[20] He argues that the "pre-eminent and essential means for human adaptation are now cultural—that is, not genetically programed behaviors, but learned behaviors."[21] Religion, which embraces myth, ritual, and praxis, is a central dimension of culture. The origin of religious myths "lies in the primordial human reading of the world and our place in it."[22] At some point in evolutionary history human beings became aware of transcendence, and this sense of religious transcendence expressed in myth and ritual enables humans to live in kinship with other creatures and to act responsibly towards them. Today we face a crisis because human beings have polluted and damaged nature. We need to use all the resources of our religious tradition to respond to this crisis. A key resource, Hefner argues, will be the theology of the human being as created co-creator.

In developing this argument, Hefner covers many issues and argues his case in a somewhat circular fashion. I will focus on two aspects of his argument: his

---

[18] Ibid., 174.

[19] Hefner, *The Human Factor*, 27.

[20] Ibid., 102.

[21] Ibid., 120.

[22] Ibid., 125.

treatment of original sin, and his treatment of altruism. In his treatment of original sin, Hefner asks what light scientific understandings can throw on this doctrine. He offers two answers to this question.

Hefner's first suggestion is that original sin can be understood in terms of discrepancy between the information coming to us from our genes and our culture. He sees *Homo sapiens* as a creature dependent upon and formed by these two kinds of information, genetic and cultural. Genes and culture have formed a symbiosis or supraorganism, which flourishes precisely because the two sources of information have coadapted both to each other and to the environment. Although the human person is a real unity, there is dissonance and tension between the two sources of information. Hefner argues that the concepts of the fall and original sin may be considered as "mythic renditions of this biologically grounded sense of discrepancy."[23] This sense of discrepancy is deeply rooted in our being and primordial in our self-awareness and in this sense "original."

There are a number of aspects of this discrepancy which Hefner names. First, there is the loss of the immediacy which the creatures who preceded *Homo sapiens* had with their own nature and with their environment. Second, we find that some of the information from our genetic past is unacceptable (such as territoriality).[24] Third, there is a sense of finitude and guilt about our incapacity to satisfy our biologically given motivators (such as the drive to excel).[25] Fourth, there is the clash in us between cooperation and altruism on the one hand and our genetic selfishness on the other. Hefner refers to Campbell's claim that urban living is made possible by culture which overcomes the genetic urge to competition.[26] Campbell and Hefner both see this state of affairs as related to the doctrine of original sin.

In a second contribution to the discussion on original sin, Hefner points to human fallibility, which he says, shows that sinfulness is *intrinsic* to the human condition. Cultural evolution, like biological evolution, occurs only through the process of trial and error. But this process is accompanied by a consciousness of fallibility. Human freedom emerges in probing forward, but this is always connected with a sense of inadequacy and guilt. The fallibility that gives rise to error and evil is intrinsic to the human being and is basic to the process that gives rise to life and enables its enrichment. So, this process is good, but as a good process it is always accompanied by failure and sin.[27]

In his contribution on altruism, Hefner begins with the question that E. O. Wilson sees as the central problem of sociobiology: "how can altruism, which by definition reduces personal fitness, possibly evolve by natural selection?"[28] Wilson understands altruism as increasing the fitness of another at the expense of one's own fitness.[29] Hefner seeks to show the links between the scientific discussion of altruism and the theological tradition of love of neighbor. He aligns himself with the view of Campbell, that the religious traditions are the chief carriers of "counter-hedonic" and

---

[23] Ibid., 132.

[24] Ibid.

[25] Hefner is here following the work of George Edwin Pugh, *The Biological Origin of Human Values* (New York: Basic Books, 1977), 284–88.

[26] Donald T. Campbell, "The Conflict Between Social and Biological Evolution and the Concepts of Original Sin," *Zygon* 10 (1975): 234–49.

[27] Hefner, *The Human Factor*, 240.

[28] Edward O. Wilson, *Sociobiology: The New Synthesis* (Cambridge: Harvard University Press, 1975), 3.

[29] Ibid., 117.

altruistic values.[30] He also agrees with Burhoe that "trans-kin altruism" is a distinguishing mark of humanity, and that religious tradition is the chief carrier of this dimension of the human.[31]

Hefner takes it to be the general opinion of scientists working in this area that trans-kin altruism cannot be accounted for on the basis of genes alone. He agrees with Burhoe that genes and culture must coevolve, coadapt, and be coselected.[32] Hunting, gathering, food preparation and preservation, and child-rearing required a fine balance between cultural and genetic systems in *Homo sapiens*. He argues that since Neanderthal times, 100,000 years ago, religious myth and ritual have been significant dimensions of the cultural system which enabled the first humans to live in human conditions. Religious myths have enabled and encouraged social behavior.

It is in this context that Hefner considers the biblical commands to love one's neighbor (Matt 22:37–40), to love with mutual love (John 13:34), and to love one's enemy (Matt 5:43–48). In this last and most radical formulation, the love command is grounded not in family, nor in kin, but only in God. Altruism is thus claimed to be intrinsic to the way things are. The Christian claim is that "altruistic love holds the status of a cosmological and ontological principle."[33] Altruistic love is grounded in the fundamental character of reality.

There are two critical questions that I would put in the light of Hefner's insights. 1) With regard to original sin, can discrepancy between information coming from genes and culture and intrinsic human fallibility *themselves* be equated with sin? 2) With regard to altruism, is the *ultimate* principle of divine and human life altruistic love?

3 *Towards a Further Reformulation*

Theissen, McFague, and Hefner are far from being the only theologians working on these issues,[34] but they form a significant interrelated group which has something

---

[30] See Donald T. Campbell, "The Conflict Between Social and Biological Evolution and the Concepts of Original Sin," 234–49; idem, "On the Conflicts Between Biological and Social Evolution and Between Psychology and Moral Tradition," 167–208.

[31] See Ralph Wendell Burhoe and Solomon H. Katz, "War, Peace and Religion's Biocultural Evolution," *Zygon* 21 (1986): 439–72.

[32] Hefner, *The Human Factor*, 200.

[33] Ibid., 208–9.

[34] Birch and Cobb, for example, develop the idea of a fall "upwards," suggesting that new levels of evolutionary order and freedom (animal life, human life, cultural evolution) are bought at the price of suffering. Each liberation brings new forms of enslavement. See Birch and Cobb, *The Liberation of Life*, 117–22. Gabriel Daly also sees the fall as a movement from one level of evolution to another, as an alienation from peace at one level in order to attain it at a higher level. Instincts and drives from the pre-human level are not sin, but they can become the instruments of sin if they are not shaped by grace. See Gabriel Daly, *Creation and Redemption* (Dublin: Gill and Macmillan, 1988), 131–47. Sebastian Moore also sees the fall in terms of the birth of consciousness, both of the species and the individual psychic life, in which there can be an arrest at an early ego stage, so that the ego of the adult remains compulsively self-securing. See Sebastian Moore, *Jesus the Liberator of Desire* (New York: Crossroad, 1989), 25–30. Moore's ideas have been developed by Neil Ormerod. He sees original sin as original self-disesteem, a felt rejection of our human nature prior to the exercise of our freedom. See Neil Ormerod, *Grace and Disgrace* (Newtown, N.S.W.: E. J. Dwyer, 1992), 151–62. Marjorie Hewitt Suchocki argues that original sin should be seen as rebellion against creatures. God is the co-sufferer in this rebellion. She finds structures that predispose us to sin in 1. the human bent towards aggression, 2. the interrelationships of humans which

important to offer in the re-thinking of sin and grace in the light of evolutionary science. In what follows, I will attempt to offer critical qualifications and developments of their positions.

3.1 *Original Sin in Evolutionary Context*

Gerd Theissen sees human beings as predisposed to sin. There is a pull in us towards anti-social behavior which springs from pre-programed biological tendencies. Philip Hefner offers a much more developed theology of original sin, which incorporates and goes beyond Theissen's view. Hefner sees original sin first in terms of the discrepancy we experience between information coming from our genes and from our culture, and second in terms of the fallibility and limitation that are essential to human evolution.

In my view, Hefner's insights genuinely illuminate the human condition. Great thinkers like Paul, Augustine, and Luther have described the existential struggle which characterizes human existence in the world, the gap between what we would do and what we do, the sheer weight which opposes our best intentions, our seeming inability to give ourselves completely, our incapacity to love with our whole hearts. Our appetites and passions erupt in unruly disorder, our wills seem fettered, our minds cluttered and confused. This inner state of affairs has been called concupiscence in the theological tradition. It has generally been understood as the result of original sin, and Luther identified it as sin. What Hefner does is to help us understand this inner disorder in the light of our evolutionary history as creatures who are a symbiosis of genes and culture, and as fallible creatures who move forward only through trial and error. To my mind, this evolutionary insight does help us understand ourselves. It offers genuine insight into what the Christian tradition has called concupiscence.

3.2 *Discrepancy, Fallibility and Sin*

There are, however, two suggestions I would make. First, I would argue that the discrepancy and fallibility described by Hefner are real, but they are not of themselves sin. Karl Rahner has offered a major clarification of the theological concept of concupiscence.[35] He distinguishes between the disorder that springs from sin and the disorder that is intrinsic to the human. The disorder of original sin comes from the history of the sinful rejection of God which partially shapes us and which is the context for our free decisions. But, he suggests, there is disorder which is not the result of sin but which is intrinsic to being a spiritual creature who is at the same time radically bodily and limited. Because of bodiliness and finitude human beings are never fully autonomous, integrated, or in control. Rahner, however, sees this is not as something to be overcome, but as part of God's good creation. This form of concupiscence is not sinful but morally neutral. In fact our bodily and psychic selves can and do resist us not only when we make good decisions but also when we make bad ones—as when a blushing face betrays a lie.

What Rahner offers us, Stephen Duffy has said aptly, is a clear distinction between concupiscence as "the child of sin" and concupiscence as the "the

---

involve us in a world of violence, 3. the social structures that shape and influence us. See Marjorie Hewitt Suchocki, *The Fall to Violence: Original Sin in Relational Theology* (New York: Continuum, 1994).

[35] Karl Rahner, "The Theological Concept of Concupiscentia," *Theological Investigations* I, 347–82 (Baltimore: Helicon Press, 1961), 347–82.

companion of finitude."[36] I see this distinction as absolutely essential in rethinking original sin in the light of evolution. What Hefner offers us is fundamentally a good description of the human being as the companion of finitude. We are creatures who experience discrepancy between information coming from our genes and our culture and we are evolutionary creatures who are intrinsically fallible. This is the way that God has created us. It is not sin.

But we are also children of sin because the fallible symbiosis that is the human has rejected the Creator. Original sin has to do with God, and with our relationship to God. It has to do with the human stance before the God who is both the source of genetic inheritance and the one who in grace calls us into a life centered on love. Original sin involves the inner effect on a person of the history of human rejection of God and of our creaturely status before God. It is not the structure of the human (as a fallible symbiosis of genes and culture) which constitutes original sin, but the inner impact on each human person's free situation of previous human rejection of God. By way of an aside, it is important to note that if original sin is understood as springing from the history of human free decisions against God and God's creation, then non-human nature is not "fallen," except insofar as nature is damaged by human sin.[37] The ecological crisis makes it all too apparent that nature is damaged by human decisions. But the fundamental Christian insight about creation is not that creation is "fallen," but that all of creation will be transformed in Christ (Rom 8:18–25; Col 1:15–20).

Our existential state then is constituted 1) by our evolutionary structure as fallible symbiosis of genes and culture, and 2) by the additional fact that the history of human sin is an inner constitutive element in our own free acts. This second element twists, compounds, and distorts the complexity and fallibility that is part of our evolutionary makeup.

3.3 *Genes, Culture and Sin*

My second suggestion is that it is important to avoid any tendency to identify selfishness and sin with the biological side of the human being and unselfish behavior with the cultural side. On the one hand, our genetic inheritance carries messages that are necessary for human life, and these are to be seen, according to biblical tradition, as part of God's good creation. On the other hand, culture, and religion as part of culture, can carry not only messages of altruistic love, but also messages of systemic evil.

Reinhold Niebuhr long ago warned against the idea that we can pinpoint the source of moral evil either in nature or in human culture. He pointed out that it is must be located in the responsible human being who sins against God.[38]

Original sin, as the impact of accumulated history of rejections of God which enters into the inner place of our freedom, may involve a lack of acceptance and integration of either side of our humanity before God. When the biological side dominates, it is possible that powerful self-serving urges from our genetic side can obscure relational tendencies coming from the cultural inheritance. But at other times particular forms of cultural conditioning can so predominate that healthy genetic

---

[36] Stephen J. Duffy, *The Dynamics of Grace: Perspectives in Theological Anthropology* (Collegeville: The Liturgical Press, 1993), 226.

[37] On this, see Holmes Rolston, III, "Does Nature Need to be Redeemed?" *Zygon* 29 (1994): 205–29.

[38] Reinhold Niebuhr, *The Nature and Destiny of Man* (New York: Charles Scribner's Sons, 1941).

drives and life-giving impulses may be obscured or driven underground. Rejection of God may involve the distortion, denial, repression from consciousness, or the lack of integration of either side of our heritage. Both forms of denial can twist and distort the human before God.

It is essential to recall that culture can carry not only altruistic information, but also messages that legitimate the dominance of the powerful and the oppression of the poor, of women, and of those who are "other." As Langdon Gilkey has said:

> Culture is the locus of the social institutions that pass on systemic injustice; it was culture's information system that perpetuated and justified slavery, and also class, gender, and racial domination. Culture is also the locus of the mores and morals that encourage, defend and justify those unjust (and cruel) institutions. Culture is the site of *ideology*, whether religious or secular, which incites, increases, and excuses, in fact justifies through its myths and rituals, these injustices.[39]

Gilkey argues that sin itself is a "compound" or a "symbiosis" of both genetic and cultural influences. I would suggest that there are instances where it seems apparent that genes and culture together are involved in structural sin, which is sin carried in culture. Patriarchal culture, for example, with its support for the position of dominant males and for structures of oppressive power over others, is obviously sin carried in culture. But this form of cultural sin may spring, in some part, from genetic tendencies inherent in the sexual relations of our prehuman ancestors. Extreme militarism may be understood, in part, as a radically destructive culture, but it also seems that it gives expression to genetic urges we share with our prehuman ancestors about dominance and territoriality.

The intellectual and technological capacity of human culture greatly enlarges the genetic/cultural capacity in human beings to do damage to other creatures. Competition for survival and predation in nature involve the death of many creatures. But when competitive instincts are enfleshed in unintegrated and technologically powerful and sinful human beings, the capacity for death and destruction involves the whole planetary community. Self-serving genetic tendencies can be "amplified" by the enormous capacities of human culture and human technology. The self-serving actions of a chimp may have negative effects on a small number of other creatures, but the unintegrated self-serving human being, with all the power and resources of human culture, can destroy the rain forests of the Earth and drive the uncounted species that inhabit them to extinction.

Genetic and cultural inheritances are profoundly inter-related in human evolutionary development. They are so closely connected that we cannot properly attribute behavior patterns simply to one side of our inheritance. We cannot locate sin in our genetic inheritance. We are not genetically determined even though we may well be subject to genetic tendencies. Original sin is not to be understood as springing from one side of the human, but rather from the history of human free decisions which enters into the inner place of our own decision-making.

4 Grace in Evolutionary Context

Theissen, McFague, and Hefner all suggest that the message of inclusive love is at the heart of the Christian vision. They understand Jesus' words and deeds as revealing God as liberating, healing, and inclusive love (McFague's Christic paradigm). They suggest that it is the Spirit of this God which enables us to relate

---

[39] Langdon Gilkey, "Evolution, Culture, and Sin: Responding to Philip Hefner's Proposal," *Zygon* 30 (1995): 305.

to others in self-transcending love. This same Spirit calls us to solidarity with the global community. The experience of the Spirit is the experience of barriers being broken down and communion being established. While I very much agree with these ideas, I have questions about the ultimate place given to altruism, particularly in Gerd Theissen's and Philip Hefner's work, and will suggest that ultimacy belongs instead to mutual relations. I will argue that the concept of God as mutual relations leads towards a relational ontology. Finally, I will point to the need to affirm that the God of mutual relations is the God who creates through the process of natural selection.

4.1 *The God of Mutual Relations*

It is undeniable that the Gospel calls for love of the "other" and that the Cross of Jesus is the central Christian symbol of self-sacrificing love. Altruism is clearly a radical dimension of the Christian understanding of divine and human love. But is "altruism" a sufficient description of this love? If altruism is understood in Wilson's terms as increasing the fitness of another at the expense of one's own fitness, then obviously this does not express the highest ideals of Christianity. But even if altruism is understood in Christian terms as self-sacrificing love, there are two reasons why I think it is better to look beyond altruism in order to express the ultimate Christian vision of love.

The first reason comes from theological anthropology. Feminist scholars have argued that Christian theologies of sin and salvation show the effect of having been constructed by men. In this traditional theology there has been a tendency to identify sin with pride, self-assertion, and self-centeredness. But does this reflect universal human experience? The admonition to a more self-sacrificial love may well offer a corrective to a dominant form of sinfulness operative in powerful people, but it may exacerbate an oppressed person's sin of "hiding" from freedom and self.[40] Altruism may be essential learning for dominant groups, but some oppressed persons may be thought of as "altruistic" to a fault. Admittedly this latter form of altruism does not represent the Christian concept, which involves self-love as well as love of the other. But the point is that indiscriminate calls to altruism and self-sacrifice can function in some circumstances to maintain oppression. This argument, in my view, does not undermine the significance of the sacrificial love of the Cross, nor the importance of altruistic love, but offers an important critique of indiscriminate, undifferentiated, and uncritical calls to self-sacrifice and altruism.

The more fundamental reason for reserve about the ultimacy of altruism comes not from anthropology but from the doctrine of God. From the perspective of trinitarian theology, it seems clear that, even while the Cross of Jesus points to altruism and self-sacrifice as essential components of divine and human love, love is revealed most radically in the trinitarian relations of mutual, equal, and ecstatic friendship. The Christian ideal of love is undeniably altruistic, modeled on the Cross of Jesus, but it is more than altruistic. It concerns self-possession as well as self-giving, love of self as well as love of the other. In Christian trinitarian theology altruism is understood within a vision of mutual and equal relations. So, while Philip

---

[40] See, for example, Valerie Saiving Goldstein, "The Human Situation: A Feminine View," *Journal of Religion* 40 (1960): 100–12; Judith Plaskow, *Sex, Sin and Grace: Women's Experience and the Theologies of Reinhold Niebuhr and Paul Tillich* (New York: University of America, 1980); Susan Nelson Dunfee, "The Sin of Hiding," *Soundings* LXV (1982): 316–27.

Hefner sees altruistic love as holding the status of "a cosmological and ontological principle," I believe that it is Persons-in-Mutual-Relations that has this status.

The good news of Christianity is that God is not simply a God of self-sacrifice, but a God of reciprocal giving and receiving, a God of perichoretic relations of mutual love. John Damascene (675–749) first used the word *perichoresis* to refer to this trinitarian communion. *Perichoresis* describes the being-in-one-another, the mutual dynamic indwelling of the trinitarian Persons (John 10:30; 14:9; 17:21). The word comes from *perichoreo*, meaning to encompass. It distinct from *perichoreuo*, meaning to dance around, but the image of the divine dance captures the vitality of the idea of *perichoresis*.[41] The word describes reciprocal relations of intimate communion. It describes a communion in which diversity and unity are not opposed. In this type of unity individuality finds full expression. *Perichoresis* expresses the ecstatic presence of divine person to the others, the being-in-one-another in supreme individuality and freedom. Each person is present to the other in a joyous and dynamic union of shared life.

4.2 An Ontology of Being-In-Relation

I think it matters a great deal that the "cosmological and ontological" principle of the universe is personal, relational, and communal. The late twentieth-century retrieval of the doctrine of the Trinity is suggesting precisely a relational ontology. One of the seminal thinkers pointing in this direction is Orthodox theologian John Zizioulas. He refers often to the intellectual breakthrough made by the Cappadocians, Basil the Great (c. 330–79), Gregory of Nazianzus (330–89), and Gregory of Nyssa (330–95). Their insight, he says, is that "the being of God is a relational being." God's being *is* communion. This communion is a *"primordial* ontological concept," not a notion added to the divine substance, or something which follows substance.[42] Thus, communion rather than substance is understood as the fundamental ontological concept. It is communion that makes things be. Nothing exists without it. Zizioulas, faithful to the ancient tradition of the East, insists that everything that is exists because of a person, the Father. But the First Person exists only in a communion of Persons. Reality springs from Persons-In-Relation. For Zizioulas "God" has no ontological content without communion. Nothing is conceivable as existing only by itself. There is no true being without communion.[43]

Many other theologians join Zizioulas in arguing that this approach to trinitarian theology leads to a radically different ontology. I will mention only two of them. Walter Kasper points out that the doctrine of the Trinity means that "the final word belongs not to the static substance, the divine self-containment, but to being-from-another and being-for-another."[44] He writes that "the development of the doctrine of the Trinity means a breaking out of an understanding of reality that is characterized by the primacy of subject and nature, and into an understanding of reality in which

---

[41] Bonaventure translated *perichoresis* into Latin as *circumincessio*, from *circumincedere*, meaning to move around. Others used the word *circuminsessio*, from *circuminsedere*, meaning to sit around—suggesting presence in repose.

[42] John Zizioulas, *Being as Communion: Studies in Personhood and the Church* (Crestwood, N.Y.: St. Vladimir's Seminary Press, 1985), 17.

[43] Ibid. See also John Zizioulas, "The Doctrine of the Holy Trinity: The Significance of the Cappadocian Contribution," in *Trinitarian Theology Today*, ed. Christoph Schwöbel, 44–60 (Edinburgh: T&T Clark, 1995).

[44] Walter Kasper, *The God of Jesus Christ* (London: SCM Press, 1983), 280.

person and relation have priority."[45] Catherine LaCugna sees the essence of God as relational and personal: "God's To-Be is To-Be-in-relationship, and God's being-in-relationship-to-us *is* what God is."[46] In the light of this view of God, LaCugna's whole book, *God For Us*, becomes an argument for an ontology of relations.

If the essence of God is relational, if the very foundation of all being is relational, if everything that is, springs from Persons-in-Relation, then I would argue that this points towards an ontology which might be called an ontology of "being-in-relation." In such an ontology, each creature can be understood as a being-in-relation.[47] Of course, there is an infinite difference between created being-in-relation and the divine communion; but what continuous creation means is that created being-in-relation always springs from, depends upon, and in a creaturely way participates in, the being of divine Persons-in-Relation.[48]

It seems clear that an ontology of being-in-relation can be seen as having some congruence with the insights of evolutionary biology. Biology suggests a world of cooperative, coadaptive, symbiotic, and ecological relations. An ecological biology, along with other areas of twentieth-century science, points towards a philosophy of nature that is fundamentally relational. It seems that biology and theology both point towards a view of reality in which relationships have a primary place. Trinitarian theology and ecological biology can meet in an ontology which understands the being of things as being-in-relation.

4.3 *The Relational God Creates Through Natural Selection*

The view of God I have developed here, of God as Persons-In-Mutual-Relations, is highly dependent on Christian sources and is intended to be faithful to those sources. But it also seeks to make sense of Christian faith within an evolutionary world-view. One complex issue that arises at the point of intersection between Christian theology and evolutionary biology is the apparent gap between the positive Christian view of God and the negative dimensions of evolutionary history—the dead ends, the extinctions, the suffering, and the death that are part of the process.

In the light of what we know about the "tooth and claw" of nature, it is understandable that in the work of Gerd Theissen, and to less extent in Sallie McFague, there is a tendency to set up an opposition between natural selection and the Gospel of Jesus Christ. I want to argue that this approach is not finally helpful. It sets up a dichotomy between divine action in natural selection and divine action in the Christ event. It can lead to a dialectical opposition between the theology of creation and redemption.

I would want to maintain that, for Christian theology, insight into God's action in creation is necessarily related to the liberating and inclusive view of God that is found in Jesus of Nazareth. This Christic paradigm points to inclusive love as the meaning of the universe. Theological coherence demands, it seems to me, that this

---

[45] Ibid, 310.

[46] Catherine Mowry LaCugna, *God For Us: The Trinity and Christian Life* (San Francisco: HarperSanFrancisco, 1991), 250.

[47] Anthony Kelly's theology of the Trinity as divine Being-in-Love would lend support to this line of thought. See his *The Trinity of Love: A Theology of the Christian God* (Wilmington, Delaware: Michael Glazier, 1989). See also Colin Gunton's trinitarian principles of freedom-contingency, relation, and energy which, he suggests, apply in science, in his *The Promise of Trinitarian Theology* (Edinburgh: T&T Clark, 1991), 142–61.

[48] On this concept of creaturely communion as communion by participation, see Zizioulas, *Being as Communion*, 94.

same Christic paradigm is understood as defining the God who creates through the process of natural selection.

If we maintain the goodness of the God of mutual relations, and the conviction that this God creates through natural selection, and if we are aware at the same time of the struggle and pain of evolutionary history, then it must be acknowledged that we are confronted with the issue of theodicy, a demanding and urgent issue which Thomas Tracy deals with directly and helpfully.[49] I would like to offer only two lines of thought on this issue: first, to suggest that natural selection be understood in a non-anthropomorphic way, and second to argue that the God of natural selection is to be understood as a relational God who freely accepts the limitations of natural processes.

For Theissen natural selection has a negative, almost an evil, character. Instead, I think it needs to be seen as a non-anthropomorphic and non-moral process. It is not an issue of human morality and cannot be evaluated in human terms as morally good or bad. Like the law of gravity, it simply is. It is one of the ways in which the universe works. In terms of theology, however, it can be understood (like gravity) as an aspect of the process whereby God creates what is good. In response to those who question divine morality, I would argue, with Thomas Tracy that we are not in a position to make a judgment on God's morality because of our limited perspective and information. Those who accept biblical revelation will wager that God can be trusted in the process of creating through natural selection, and will find nothing intellectually incoherent in this position.

Darwin himself has described natural selection soberly enough as the "preservation of favorable variations and the rejection of injurious variations."[50] Charles Birch offers a definition of natural selection as "the differential survival and reproduction of individuals in a population."[51] Francisco Ayala notes that while Darwin formulated natural selection primarily as differential survival, the modern understanding is formulated in genetical and statistical terms as differential reproduction:

> Natural selection implies that some genes and genetic combinations are transmitted to the following generations on the average more frequently than their alternatives. Such genetic units will become more common in every subsequent generation and their alternates less common. Natural selection is a statistical bias in the relative rate of reproduction of alternative genetic units.[52]

Natural selection, understood in such a realistic and non-mythological fashion, does not stand opposed to goodness or to God. The process itself is not to be judged in anthropomorphic terms any more than the process of stellar nucleosynthesis or the big bang itself. Like them, it is a part of the pattern of divine action in creation, a pattern which respects the intrinsic properties of things.

If natural selection is approached in a non-anthropomorphic way, then I would argue that the issue of theodicy is no more intense with regard to natural selection than it is with regard to other dimensions of existence, above all with regard to death itself. The problem of evil is not specific to natural selection. In fact, once death is accepted as essential to biological life, it seems to me that natural selection can be understood as a positive process whereby the negativity of death is subsumed, in

---

[49] Thomas Tracy, in this volume.

[50] Charles Darwin, *On The Origin of Species* (London: Unit Library, 1902), 76.

[51] Charles Birch, in this volume.

[52] Francisco J. Ayala, in this volume.

some circumstances, into a process which leads to wonderfully creative new possibilities for life.

If one's view of God is of a being who is completely omnipotent, unencumbered by any limits of any kind whatsoever, then it is difficult to reconcile such a God with the suffering that accompanies natural selection and still affirm divine goodness. I want to suggest that God is not completely unlimited, but rather freely accepts the limits of loving finite and created beings.

The creative trinitarian God must be understood not only in relational terms but also in terms of the limitations that are freely accepted in loving relationships. This trinitarian view of the God-world relation differs from Whiteheadian process theology because of its emphasis on the Trinity, divine transcendence, and divine freedom in creation. But it shares two very significant emphases of process theology.[53] The first is commitment to a relational theology. The second is the insistence that this relational theology involves a real, two-sided, but differentiated, relation between God and creatures.[54] God really relates to creatures and in the relating becomes vulnerable. The divine act of creation is an act of love, by which the trinitarian persons freely make space for creation, and freely accept the limits of the process. God respects the integrity of nature, its processes, and its laws. And in creating and relating with human creatures, God freely accepts the vulnerability of interpersonal love, and enters into love with a divine capacity for self-giving love. God accepts the limits of physical processes and of human freedom. The theology of incarnation and the theology of the Cross point to a God of unthinkable vulnerability and self-limitation. It is this concept of God, I believe, which needs to be brought into relation with natural selection.

If God is to be understood as consistent and faithful, then theological logic demands that the Christic paradigm applies to the God who creates. The God of natural selection *is* the liberating, healing, and inclusive God of Jesus. I would argue that in spite of all the suffering, death, and loss of species that is associated with evolutionary history, it is necessary to make the theological claim that natural selection is a way of divine action. God creates in and through this process.

This suggests a God who freely accepts the limits of the process of emergence through chance and lawfulness, a God who creates through the losses and gains of evolutionary history. It suggests a God engaged with creation, a God who respects

---

[53] For the development of these themes in Process Theology see Alfred North Whitehead, *Process and Reality*, corrected edition, D. R. Griffin and D. W. Sherbourne, eds. (New York: Free Press, 1929/1978); Charles Hartshorne, *The Divine Relativity: A Social Conception of God* (New Haven: Yale University Press, 1948/1964); idem, *A Natural Theology For Our Time* (La Salle, Ill.: Open Court, 1967); John B. Cobb and David R. Griffin, *Process Theology: An Introductory Exposition* (Philadelphia: Westminster Press, 1976). See also Barbour, *Religion in an Age of Science*, 218–42; Charles Birch, *On Purpose* (Kensington: New South Wales University Press, 1990), 87–109, and, for a theology of divine kenosis influenced by process thought, John F. Haught, *Mystery and Promise* (Collegeville: Liturgical Press, 1993).

[54] I have developed this line of thought in *Jesus the Wisdom of God: An Ecological Theology* (Maryknoll: Orbis Press, 1995), 122–30. See also the work of Arthur Peacocke, *Theology in a Scientific Age* (Minneapolis: Fortress Press, 1993), 87–183; John Polkinghorne, *Science and Christian Belief* (London: SPCK, 1994), 71–87; Jürgen Moltmann, *God in Creation* (London: SCM Press, 1985), 185–214; Ted Peters, *God as Trinity: Relationality and Temporality in Divine Life* (Louisville: Westminster/John Knox Press, 1993), 179–82.

the process, who suffers with creation, a God whose ongoing action is adventurously creative in and through the unfolding of evolutionary history.

The God of mutual relations is the God of continuous creation. Continuous creation is itself a relation. It is the intimate relation between each creature and God by which the creature exists. "Reality is intrinsically relational," Stephen Happel says, "because God is present as inner relationality."[55] The relation of creation means that the transcendent God, Persons-In-Mutual-Communion, is immanent to every creature, with a wonderfully differentiated interior relation to each of them, constantly luring each to be and become. As William Stoeger puts it, "God acts immanently in nature—in every 'nook and cranny' of nature, at the core of every being and the heart of every relationship—to constitute and maintain it just as it is and just as it evolves."[56] And from the side of the creature, this relationship of creation is the creature's own finite and specific participation in God's own being and trinitarian relationships. This two-sided relation is not distinct from divine action but is itself divine action.

The relation of grace is not a new presence of God to human creatures over and above the divine presence in ongoing creation, but simply the distinctive relation that the ever present God freely chooses to have with human creatures. According to Christian theology the divine action of grace incorporates us even now into the divine perichoretic relations (John 14:20, 23). The being-in-one-another of the divine life is extended to human beings. All of this demands a new attitude of mutual love on the part of human beings towards one another (John 13:34; 15:12,17). This experience of being embraced in the divine relations of mutual love, imperfectly realized in us this side of the eschaton, is the healing and freeing of our humanity and its genetic and cultural heritage.

The insight that God is Persons-in-mutual-relations suggests a world-view and a praxis in which relations are primary. If the central religious message is that relations are primary then this grounds our commitment to other humans, which is not only altruistic but a commitment to equal and mutual relations. It also grounds our ecological commitment to other creatures. We are caught up with them in a relational world as fellow creatures before a relational God.

---

[55] Stephen Happel, "Divine Providence and Instrumentality," in *Chaos and Complexity*, Robert John Russell, Nancey Murphy, and Arthur R. Peacocke, eds. (Vatican City: Vatican Observatory Publications, 1995), 200.

[56] William R. Stoeger, "Describing God's Action in the World in Light of Scientific Knowledge of Reality," in *Chaos and Complexity*, Russell, Murphy, and Peacocke, eds., 256–57.

DARWIN'S GIFT TO THEOLOGY

John F. Haught

1 *Introduction*

In a recent book, philosopher Daniel Dennett presents Darwin's notion of natural selection as a "dangerous idea." By "dangerous" he means, among other things, that Darwinian science is contrary to all human hope that nature follows a divine plan. The universal and impersonal working of natural selection, he argues, decisively refutes religious and theological assertions that any divine agency is effectively active in the world. He claims that "evolution is, in the end, just an algorithmic process," one that unfolds in the same way as a computer program.[1] Evolution requires nothing more than purely random mutations, the deterministic laws of "natural selection" and enormous spans of time. Resorting to non-natural causation, or what Dennett calls "skyhooks," to account for evolution's emergent complexity is altogether unnecessary. Evolution's prodigious creations are not the result of forces that miraculously pull matter toward life and consciousness from above. Rather they are simply products of a sequence of "cranes," one superimposed upon the other, all firmly planted on the ground of purely mechanical causation, lifting only from below.

We may think of evolution as taking place, Dennett proposes, in an open-ended "Design Space" consisting of all the logically possible forms of life. Or we may imagine it as occurring in a limitless "library" made up of all the mathematically conceivable chains of DNA. In this "Library of Mendel" evolution wanders aimlessly about, experimenting with countless potential assemblages of the molecules of life, until it finds those that actually "work." The workable or "fit" combinations are the ones whose phenotypic expressions happen to be adaptive to their environments and are thus able to survive and reproduce. Through selection of adaptive changes in organisms over a period of several billion years this totally blind process can bring about all the diversity of life, including sight and consciousness.

If we look closely at what actually goes on in biological evolution we must conclude, Dennett insists, that no purposeful divine causation has ever played any role in the production of species, including human beings. Those evolutionists "who see no conflict between evolution and their religious beliefs," therefore, are simply not looking carefully enough at the facts.[2] Although Dennett's deposition is not new, the thoroughness, passion, and conviction with which he sets it forth render it worthy of comment by theologians whose lives and work he considers evolutionary biology to have rendered altogether futile.

Dennett espouses a view similar to that of the well-known evolutionist Richard Dawkins. In *The Blind Watchmaker*[3] and again in his more recent *River Out of Eden*[4] Dawkins has argued that blind chance and natural selection, deployed over long periods of time, can account for the wonderful creativity in evolution all by themselves. He claims that the fundamental units of evolution are the coded

---

[1] Daniel C. Dennett, *Darwin's Dangerous Idea: Evolution and the Meaning of Life* (New York: Simon & Schuster, 1995), 266.

[2] Ibid., 310.

[3] Richard Dawkins, *The Blind Watchmaker* (New York: Norton & Co., 1986).

[4] Richard Dawkins, *River Out of Eden* (New York: Basic Books, 1995).

segments of DNA that we call genes. So all we need in order to understand natural selection is a law that describes the recurrent and predictable activity of these genetic units. According to Dawkins this law can be formulated as follows: *it is the nature of genes to maximize opportunities for survival and reproduction.* In other words, genes are somehow driven by blind physical necessity to do everything possible to secure their immortality. Their blind and indifferent striving to perpetuate themselves, Dawkins admits,

> ...is not a recipe for happiness. So long as DNA is passed on, it does not matter who or what gets hurt in the process. It is better for the genes of Darwin's ichneumon wasp that the caterpillar should be alive, and therefore fresh, when it is eaten, no matter what the cost in suffering. Genes don't care about suffering, because they don't care about anything.[5]

The same impersonal physical necessity that governs stars and atoms also rules the relentlessly "selfish" units of evolution. An utterly unintelligent momentum to survive at all costs, and by whatever clever ploys they can find, will even lead genes to create, among their many "vehicles," intelligent beings who unknowingly but inexorably carry these genes on to subsequent generations. Natural selection impersonally sifts out the reproductively "fit" from the "unfit" genes in each generation of living organisms as they pass through time.[6] Maximization of the "utility function" of *DNA survival* is sufficient to explain all the products of evolution.[7] Thus nature has no need for an active, ordering deity. In fact, according to Dawkins, Darwin has given atheism a solid intellectual foundation, perhaps for the first time.[8]

Dennett fully agrees. He is not himself a biologist, but a scientifically literate philosopher who has taken on the role of being Dawkins' "bulldog." He presents Dawkins' controversial adaptationist interpretation of Darwinian selection not only as scientifically adequate and unrevisable, but also as the definitive refutation of theism.

The use of evolutionary ideas by Dawkins and Dennett in the service of an atheistic manifesto is an obvious challenge to contemporary theology. Their rigorously adaptationist leanings are by no means universally accepted by biologists, but their sleek "hyper-Darwinist" version of evolution is sufficiently influential that theology cannot profitably avoid the task of responding to its claim that an accurately understood evolutionary science renders superfluous the action of God in nature.

This means that any theology in dialogue with contemporary biological science has to be, in part at least, apologetic. Its apologetic character, however, need not take the same shape as the natural theology of the past. After all, the picture of nature given to us by contemporary evolutionary biology is much more subtle and complex than those presupposed by the classical arguments from design to which apologetics formerly appealed. Theology can no longer appropriately pretend that the human mind will move swiftly and automatically from a grasp of scientific pictures of nature to the conclusion that God exists or is active in nature. What theology may do at the very least, however, is demonstrate that the purely natural causes science is obliged to look for in explaining evolutionary process do not logically rule out the simultaneously effective action of God.

---

[5] Ibid., 131.

[6] "Fitness" in evolution is understood as a measure of the probability of reproducing.

[7] Dawkins, *River Out of Eden*, 95–133.

[8] Dawkins, *The Blind Watchmaker*, 6.

And yet theology must also do much more than this. It is essential for its integrity that theology today go beyond merely demonstrating that the data of evolutionary science are "consonant" with an understanding of a God who acts in the world. It must also show that a theologically appropriate[9] understanding of God's action, even though beholden to science for concrete information about the story of life, can in a certain sense logically anticipate, systematically ground, and render more coherent than otherwise, what evolutionary science is now discovering about the nature of life. Rather than merely presenting theism as a logically credible option with respect to interpreting Darwin's "dangerous idea," theology should seek to develop the notion of God in such a way that evolutionary science not only poses no obstacle to faith, but instead contributes to the fruitful conceptual unfolding of the very character of that faith.

In this essay, therefore, I shall propose that the discoveries of evolutionary science, including those that Dennett and Dawkins consider to be fatal to theology, are not only logically consistent with, but are in a certain sense "anticipated" by, the disturbing image of God and God's relation to the world given to Christian faith. This means that it is not necessary for a theology of evolution to begin by safeguarding the emaciated idea of God or "divine design" that Dennett, Dawkins, and other scientific skeptics consider to have been debunked by Darwin. Rather, a theology of evolution should *first* set forth, however inadequately, what *faith* means by "God," and only *after* doing this, should it present any relevant arguments that evolution does not logically rule out God's creative action in nature. Thus, this essay places the apologetic component of evolutionary theology in a secondary, subordinate position.

The challenge by Darwin to theology, I shall argue, will in the end prove not to be a danger, but a gift, for the understanding of nature implied in Darwinian evolution demands that we depart, perhaps more decisively than ever before, from all notions of God that have suppressed the self-giving, and absolutely relational character of the divine mystery. More specifically, by inviting us to abandon once and for all the idea that divine action *forcefully* determines the course of nature, evolutionary biology now allows theology to lay out in unprecedented relief the image of a compassionate, suffering God—a God who, precisely by virtue of the divine humility, is powerfully active in and fully related to the world.

Our thoughts about God in a post-Darwinian age can no longer comfortably bear all of the features that theism assumed before we became aware of evolution. After Darwin we may still think of God as powerfully effective in the world, but we may have to do so in a manner quite distinct from that implied in much theology of the past. Of course, in order to approximate theological integrity, our new thoughts about God must be continuous with the authoritative scriptural and traditional sources of faith. But our attending to the details of evolutionary science invites us to think of God in fresh ways today—precisely in order to remain in touch with the classic sources of faith. We shall still be able to confess the creativity, effectiveness, and redeeming presence of God in the world, but we shall be able do so in a way that takes the data of evolutionary science more fully into account.

How, though, can we accomplish such an objective in view of the position represented by Dawkins' and Dennett's skepticism on the one hand, and the suspicion in which so many Christians have held evolution on the other? Can a

---

[9] By "appropriate" I mean consistent with the biblical and traditional witnesses of faith as well as with the demands of contemporary experience, including especially that of scientists.

theology after Darwin become not just an accommodational retreat, but instead a genuine, and perhaps even aggressive, advance in religious understanding? Here I shall propose that the core of Christian "revelation" provides us with an image of God that is not only religiously satisfying, but which may fruitfully illuminate the evolutionary character of the world as well. When viewed in the light of this understanding of revelation, evolutionary biology will appear not only consistent with theism, but also enriching of it.

By "revelation" I mean here what H. Richard Niebuhr refers to as the *gift of an image* that brings intelligibility to aspects of our experience which would otherwise remain obscure to us.[10] More precisely, revelation is "that special occasion which provides us with an image by means of which all occasions of personal and common life become intelligible."[11] I extend this definition, as in principle Niebuhr would also, to include not just the rendering of human life and history intelligible, but the natural world as well. To Christian faith, "revelation" is not entirely unlike those new models or theories in science which suddenly light up the world in such a way as not only to solve previously unexplained problems, but also to promise further discovery and richer understanding in the future. Just as an imaginative breakthrough in science can bring hitherto impenetrable dimensions of nature into the explanatory purview of a single integrating picture, so also the image of God given to us through what we take to be "revelation" can allow us to view our experience of the entire world, including nature, in an entirely new light. An authentically revelatory "image" will bring new coherence to what we have already experienced, as well as an opening to further discovery.[12]

Does Christian faith provide us with such a revelatory image, one that can bring wider intelligibility to our relatively new and "dangerous" understanding of evolution through natural selection? Does it have the capacity not only to permit us to make some sense of the strange ways of nature, but also to bring us to a deeper understanding of the story of life on Earth? Is there a revelatory image in the authoritative Christian sources that might function in such an illuminating, integrating, and heuristically fruitful way?

Such an image, I would argue, is present in the biblical picture of the humility of God made manifest in Jesus. This portrait of God as self-abandoning, self-giving mystery, while always implicit in Christian faith, has recently become more prominent than ever in shaping the character of theology.[13] It has by no means penetrated the sensibilities of all believers—and it is completely absent from scientific skeptics' caricatures of "God"—but a growing number of theologians now consider it to be central to any genuinely Christian understanding of ultimate reality. The modern anguish over God's absence, the enormous scale of suffering and sorrow in our century, and the widespread destruction of Earth's life-systems—all of these and numerous other tragedies have compelled theology to bring before us with unprecedented intensity the image of God as self-emptying, or *kenotic*, love.

---

[10] A fuller treatment of the notion of revelation, of course, would address not only the question of revelation's meaning but also that of its truth. I have dealt with the latter at length in *Mystery and Promise: A Theology of Revelation* (Collegeville: Liturgical Press, 1993), esp. 199–214.

[11] H. Richard Niebuhr, *The Meaning of Revelation* (New York: the Macmillan Co., 1960), 80.

[12] Ibid., 69.

[13] See William C. Placher, *Narratives of a Vulnerable God* (Louisville: Westminster John Knox Press, 1994), 3–26.

The image of God's humility has always been present in Christian tradition,[14] but it has taken the travails of the present century to move it toward the center of theological reflection. Recent scientific clarifications of the nature of evolution now encourage us to bring this image of God even more sharply into focus. It is in this sense that evolution may prove to be not so much a danger as a gift to theology.

At the very center of Christian faith lies a trust that in the passion and crucifixion of Jesus we are presented with the mystery of a vulnerable God who pours the divine selfhood into the world in an act of complete self-abandonment. This image has led some theologians even to speak provocatively of the "powerlessness" of God. Dietrich Bonhoeffer, for example, wrote from prison that only a "weak" God can be of help to us. Edward Schillebeeckx, on the other hand, has proposed that we speak not of divine weakness and powerlessness, but rather of the "defenselessness" or "vulnerability" of God. In other words, when we emphasize the divine humility there is no need to deny the teaching that God is at the same time powerfully effective. After all, as Schillebeeckx says, those who make themselves vulnerable can also powerfully and effectively disarm evil. Since divine "power" implies the capacity to bring about significant effects, it is given a new definition for faith as the consequence of God's defenselessness in the face of our own human use of "force." Schillebeeckx writes:

> ... the divine omnipotence does not know the destructive facets of the human exercising of power, but in this world becomes "defenseless" and vulnerable. It shows itself as power of love which challenges, gives life and frees human beings, at least those who hold themselves open to this offer.[15]

Theological reflection on this image of divine defenselessness (which, it must be emphasized, is not the same as powerlessness), can also help faith appropriate and make sense of the ways of evolution, especially as these are depicted by neo-Darwinian biology.

The image of a self-emptying God lies at the heart of Christian revelation and the doctrine of the Trinity.[16] And it is the same picture of God that illuminates the world's creation as well. Drawing on both Christian and Jewish themes, Jürgen Moltmann explains:

> God "withdraws himself from himself to himself" in order to make creation possible. His creative activity outwards is preceded by this humble divine self-restriction. In this sense God's self-humiliation does not begin merely with creation, inasmuch as God commits himself to this world: it begins beforehand, and is the presupposition that makes creation possible. God's creative love is grounded in his humble, self-humiliating love. This self-restricting love is the beginning of that self-emptying of God which Philippians 2 sees as the divine mystery of the Messiah. Even in order to create heaven and earth, God emptied himself of all his all-plenishing omnipotence, and as Creator took upon himself the form of a servant.[17]

---

[14] See, for example, the studies by Donald G. Dawe, *The Form of a Servant* (Philadelphia: The Westminster Press, 1963); Lucien J. Richard, O.M.I., *A Kenotic Christology* (Lanham, Md.: University Press of America, 1982); Jürgen Moltmann, *The Crucified God*, trans. by R.A. Wilson and John Bowden (New York: Harper & Row, 1974); and Hans Urs Von Balthasar, *Mysterium Paschale*, trans. by Aidan Nichols, O.P. (Edinburgh: T & T Clark, 1990).

[15] Edward Schillebeeckx, *Church: The Human Story of God*, trans. by John Bowden (New York: Crossroad, 1990), 90.

[16] See Eberhard Jüngel, *The Doctrine of the Trinity: God's Being is in Becoming*, trans. by Scottish Academic Press Ltd. (Grand Rapids: Eerdmans Press, 1976).

[17] Jürgen Moltmann, *God in Creation*, trans. by Margaret Kohl (San Francisco: Harper

What does the image of this humble God contribute to our new knowledge of evolution? In faith's response to it there lurks, I think, a way of bringing new meaning not only to our perplexity at the broken state of social existence or individual human suffering, but also to our more recent intellectual puzzlement over the unfathomed epochs of struggle, apparent waste, and suffering that occur as the result of "evolution by natural selection." The ways of nature take on a distinctively new significance when we view them in light of the defenselessness of God. Theology's suppression of—or at least its failure to take seriously—the startling image of divine vulnerability has only nourished the skeptical conviction that the existence of God is incompatible with an evolving world. A theology of evolution, on the other hand, must make central the idea of a self-emptying God who opposes the domineering kind of power that religion often projects onto the divine. The same loving self-withdrawal of God that in Moltmann's portrayal makes creation initially possible (*creatio originalis*) also allows for the ongoing creation (*creatio continua*) of the world through evolution.

A theology of evolution need not lose sight of other aspects of revelation, but it makes central the theme of divine suffering love. The biblical notions of God's word and promise, of exodus, redemption, covenant, justice, Wisdom, the Logos made flesh, the Spirit poured out on the face of creation, and the trinitarian character of God—these are also indispensable elements in any Christian theology. But they disclose to us the nature of God only when they are closely joined to the theme of divine self-abnegation which for faith assumes explicit expression in the life, death, and resurrection of Jesus.[18]

Of course, as the world's cumulative religious wisdom cautions us, we have to qualify very carefully anything we say about God. So I do not intend to argue here that the notion of kenosis or humility adequately summarizes the ineffable depths of the divine mystery. In fact, taking the notion of divine vulnerability as exclusively sufficient for our thoughts about God would hardly do justice to Christian tradition and religious experience, for to faith God is also the one who creates the world, who raises Jesus from the dead, and who promises to bring everything to fulfillment in the New Creation. God, as Paul Tillich emphasizes, is the "Power of Being." Religious trust would be impossible unless we thought of God as a truly effective presence capable of bringing about the fulfillment of all creation. Thus I would agree with John Macquarrie when he suggests that our theism must always be dialectical: everything we say about God needs to be qualified by subsequent statements which may seem at first to contradict earlier ones, but which can have the long-run effect of pulling our theology away from any deadening literalism.[19]

I am not arguing, therefore, that we should dispense with the notion of divine power, much less that of divine ecstasy, but only that we are clearly encouraged by the Christian sources to transfigure these dialectically with images of God's self-emptying. The result of such a dialectic would be a radically different understanding of divine power from the one that religion has typically projected onto the sacred, and which has also made its way into Christian theology.

& Row, 1985), 88.

[18] See John B. Cobb, Jr. and Christopher Ives, editors, *The Emptying God* (Maryknoll: Orbis Books, 1990); also Haught, *Mystery and Promise*.

[19] John Macquarrie, *In Search of Deity: An Essay in Dialectical Theism* (New York: Crossroad, 1985).

Moreover, we must keep in mind the fundamentally relational vision of God expressed especially in trinitarian theology. In fact a fruitful way of thinking about "kenotic" power in an evolving world is in terms of divine love's longing to bring about and intensify relationships. God's kenosis can be seen as the supreme exemplification of the restraint resident in any truly loving relationship. And the urge to enter into relationship will prove in the end to be a much more effective kind of power than the domineering forcefulness that overwhelms, divides, and excludes. According to Christian faith, when Divine Love moves into a situation defined by "earthly power," and feels the relationless caused by brute force, it proves to be transformative by first taking its place among the victims of such circumstances. By thus making itself vulnerable to victimization, it begins to restore and redeem.[20] Such vulnerability is not the antithesis but the expression of genuine power.

The picture of an incarnate God who suffers along with creation is offensive, of course, to our customary sense of what should pass muster as ultimate reality. But it deserves the name "revelation" precisely because it breaks through our pedestrian projections of the absolute, and in doing so brings new meaning to the world's experience of suffering, struggle, and loss. This new meaning consists, in part at least, of the unexpected intuition that the suffering of living beings is not undergone in isolation from God, but is taken up into the very "life-story" of God. Because in the minds of skeptics, and of many theists as well, the concept of God scarcely connotes such participatory, empathetic love, it appears incompatible with evolution by natural selection. But, as Donald Dawe writes, we easily forget that "belief in the divine self-emptying or condescension in Christ" is in fact "basic to Christian faith."

> According to Christian faith, God in his creation and redemption of the world accepted the limitations of finitude upon his own person. In the words of the New Testament, God had "emptied himself, taking the form of a servant." God accepted the limitations of human life, its suffering and death, but in doing this, he had not ceased being God. God the Creator had chosen to live as a creature. God, who in his eternity stood forever beyond the limitations of human life, had fully accepted these limitations. The Creator had come under the power of his creation. This the Christian faith has declared in various ways from its beginning.[21]

Unfortunately, however, as Dawe laments, "the audacity of this belief in the divine kenosis has often been lost by long familiarity with it."

> The familiar phrases "he emptied himself [*heauton ekenosen*], taking the form of a servant," "though he was rich, yet for your sake he became poor," have come to seem commonplace. Yet this belief in the divine self-emptying epitomizes the radically new message of Christian faith about God and his relation to man.[22]

---

[20] Some theologians are sensitive to the possibility that a one-sidedly kenotic theology could provide religious sanction for acquiescence in subordinate social roles and psychically unwholesome passivity. Perhaps, however, such understandable apprehension can be avoided if we emphasize that the impetus of divine love is *primarily* to constitute and affirm relationship, as the doctrine of the Trinity implies. It is *relationality* and not suffering or self-abasement that the Christian God idealizes. However, when relational power enters into situations where coercive power has already vitiated appropriate relationships, it clearly cannot be effective if it merely imitates the kind of domination that already divides, but apparently only if it first participates in the relationlessness of a broken world. Thus it makes itself vulnerable to such a situation in order to overcome it, not to sanction it.

[21] Dawe, *The Form of a Servant*, 13.

[22] Ibid., 15.

An evolutionary theology, it goes without saying, extends the picture of God's suffering beyond the human sphere to embrace also the struggles of the entire evolutionary process. God's empathy enfolds the whole of creation, and this can only mean that the billions of years of evolutionary travail and creativity are also God's own suffering and enjoyment. Nothing that occurs in creation, including all of the tribulation that it involves, can appropriately be understood by faith and theology as taking place outside of God's compassionate experience. Because God suffers along with creation in order to deliver it from relationlessness, faith can look toward a final redemption that would consist, in part at least, of the restoration of all the broken relationships in the cosmos and human history.[23]

In the words of St. Paul, of course, such a picture of God's vulnerability amounts to "foolishness" in comparison with our conventional wisdom (I Cor 1:25). John Macquarrie writes:

> That God should come into history, that he should come in humility, helplessness and poverty—this contradicted everything—this contradicted everything that people had believed about the gods. It was the end of the power of deities, the Marduks, the Jupiters... yes, and even of Yahweh, to the extent that he had been misconstrued on the same model. The life that began in a cave ended on the Cross, and there was the final conflict between power and love, the idols and the true God, false religion and true religion.[24]

Religious thinkers in this century have become very appreciative of Sigmund Freud's and Friedrich Nietzsche's perception that an exaggerated feeling of human "weakness and fear" can be "normalized" by coarse ways of thinking about divine power.[25] Many would now agree with Alfred North Whitehead that the idea of God in traditional Christian theology has been modeled too often along the lines of Caesar rather than the humble shepherd of Nazareth.[26] Moreover, from the point of view of theology's attempts to come to grips with Darwin, despotic notions of divine power—often enfolded in sophisticated theological wrappings—have only made the data of evolutionary science all the more difficult for faith to digest.[27] We continue to resist the image of God's humility, even though, as Karl Rahner asserts, "[t]he primary phenomenon given by faith is precisely the self emptying of God..."[28] Today evolutionary science invites us to appropriate more solemnly than ever this way of

---

[23] I do not have the space here to develop the ways in which such an understanding of God must also be part of any attempt by theology to respond to the question of theodicy. I have made a brief attempt to do so in *The Cosmic Adventure* (New York: Paulist Press, 1984).

[24] John Macquarrie, *The Humility of God* (Philadelphia: The Westminster Press, 1978), 34.

[25] A good example of such analysis is Paul Ricoeur's essay "Religion, Atheism, and Faith," in *Conflict of Interpretations*, edited by Don Ihde (Evanston: Northwestern University Press, 1974), 440–67.

[26] See Alfred North Whitehead, *Process and Reality*, corrected edition, edited by David Ray Griffin and Donald W. Sherburne (New York: the Free Press, 1978), 342.

[27] As Jürgen Moltmann notes, despotic and Oedipal features have been painted even over the symbol of the Cross; but the *pathos* of God requires that we cleanse these from our theologies. *The Crucified God*, 306.

[28] Karl Rahner, *Foundations of Christian Faith*, trans. by William V. Dych (New York: Crossroad, 1978), 222. In the light of Rahner's consistent reference to the divine mystery as self-emptying love, it is surprising that he expresses reservations about Jürgen Moltmann's theology of divine suffering.

thinking about the divine mystery. At the same time, faith's sense of the divine humility can open both our minds and our hearts more fully to the ways of evolution.

Sallie McFague rightly notes that the traditional notion of an "Almighty" God does not *necessarily* entail a controlling kind of divine power, but that in fact "power as domination has been and still is a central feature of the Western view of God."[29] In light of this bias, the real stumbling block to reconciling evolution with God is not the "dangerous" features of evolution that Dennett dwells upon, but the scandalous image of God's humility latent in the Christian sources themselves. In debates about "God and evolution" theologians have usually focused on the question of how to reconcile God's "power" and "action" with the autonomous, random, and impersonal features of nature's evolution. In doing so they have quite often tacitly understood God's "power" as "power of domination"—to use McFague's words—and have thus failed to think of nature in terms of how evolution might appear when viewed in terms of God's self-sacrificing love. In many of its encounters with evolutionary science theology has assumed a notion of God already clouded over with the same motifs of patriarchal domination that have persistently vitiated Western theism. In spite of the New Testament's own images of a vulnerable God, theology's predilection for a God of "power and might" has more than occasionally led it either to deny evolution altogether or at least to ignore features of evolution that skeptics such as Dawkins and Dennett consider to be most injurious to faith.[30] But the randomness of variations, the autonomous impersonality and creativity of natural selection, and even the enormous "waste" and suffering that accompany evolution through time, can all be illuminated by the notion of a vulnerable God.

The main reason theology has resisted thinking of evolution in terms of the revelatory image of God's humility is that such an image seemingly gives God too little "power," and perhaps even no power at all, to act in nature. And since any coherent faith demands that God be actively involved in nature, a vulnerable deity may not seem capable of providing adequate foundations for our hope in redemption, resurrection, and new creation. Perhaps it is for this reason, as Macquarrie observes, that "[t]he God of Jesus Christ, like Yahweh before him, has been turned back again and again into a God of war or the God of the nation or the patron of a culture."[31]

The main task of a theology of evolution, however, is to tell the story of nature in the light of faith in the God who renounces despotic exercise of force, whose creative and loving concern for the world's being is the ultimate ground of nature's evolution, and whose participation in evolution restores relationship and redeems all the suffering and struggle that the process involves. But is such an understanding of God both religiously adequate and consistent with post-Darwinian science?

---

[29] Sallie McFague, *Models of God: Theology for a Nuclear Age* (Philadelphia: Fortress Press, 1987), 16.

[30] I have found this to be the case especially in some instances of "theistic evolution" which remain content with the bald assertion that "evolution is God's way of creating," and that fail to look very closely at the pain, the elimination of the weak, and the enormous struggle and waste involved in such a "clumsy" process as evolution appears to be. See, for example, John A. O'Brien, *God and Evolution: The Bearing of Evolution Upon the Christian Faith* (Notre Dame: University of Notre Dame Press, 1961). In its time O'Brien's treatment of Darwin may have appeared sophisticated and fair, but I think it is hard to sustain its benign version of theism in the light of contemporary scientific discussions of evolution.

[31] Macquarrie, *The Humility of God*, 34.

2 *What is Really at Issue?*

Our scientific understanding of the autonomously creative resourcefulness of natural selection working over the course of immense epochs of time raises questions about whether and how God could be meaningfully operative in nature. Why do we need to call upon the notion of "divine action" at all if nature can so autonomously create itself? The main support for Dawkins' and Dennett's atheism is the impression evolution gives them that all the creativity in the biosphere (and indeed in all of nature[32]) can be fully accounted for by the automatic algorithmic filtering process known as "natural selection." Evolutionary science does not actively rule out God's action; rather, it simply appears to make God *superfluous*. For if nature is *self-creative*, where is the need for a providential, personal, intelligent God to act or intervene in evolution?[33] Moreover, do not other branches of recent science, especially non-equilibrium thermodynamics and chaos theory, support the idea that natural processes may all be spontaneously "self-organizing" and thus in need of no organizing divine intelligence?

Science, of course, is methodologically constrained from the very outset to explain all natural occurrences without reference to supernatural intervention, design, or final causes. Thus when biologists today seek to explain the evolution of complexity, life, and even consciousness by means of the purely natural "laws" of evolution alone, they are not doing anything unexpected. In principle, then, there is no more theological "danger" involved in scientists' accounting for evolution by appealing to purely natural causes than there is in any other branch of science. We simply do not expect scientists *qua* scientists *ever* to resort to supernatural "skyhooks." In fact, the more science itself becomes "secularized" and purges itself of all such appeal to miracles or any other adventitious assumptions, the less confusing are its conversations with theology.

However, when scientists go beyond their disciplinary concern for methodological purity and take the additional step, as Dawkins and Dennett clearly do, of asserting that science *leaves no room* for alternative or complementary ways of grasping and articulating the intelligibility of nature, they have already left the arena of science and are giving expression to what would better be called "beliefs." In such extra-scientific excursions they are claiming, at least implicitly, that the scientific method is alone capable of giving us a completely adequate understanding and knowledge of the real world. And they generally add to this a metaphysical assumption that *physical reality as graspable by science* constitutes the only realm of objective being available to human understanding and knowledge.

Neither Dawkins nor Dennett provides a truly *scientific* justification for these commitments. Thus, logically speaking, the opposition to theology by some evolutionists is not directly scientific, but ideological.[34] It is clear that scientism and scientific naturalism (especially what Whitehead called "scientific materialism") contradict theism, but it is not immediately self-evident that evolution by natural selection does so. It would appear to do so only if science can firmly establish on

---

[32] See John Gribbin, *In the Beginning: After COBE and Before the Big Bang* (Boston: Little, Brown, 1993) for an application of the notion of natural selection to the entire universe's coming into being.

[33] To the obvious question of how to account for why there exists anything at all in the absence of God, Dennett's reply is simply "why not?" *Darwin's Dangerous Idea*, 180–81.

[34] See the paper by George Ellis in this volume for a similar critique of the metaphysical assumptions employed by Dennett and Dawkins.

purely scientific grounds that evolutionary biology constitutes the *only* appropriate method for making sense of the fascinating story of life.

It is difficult, of course, to imagine how science itself could ever provide such confirmation. Yet, it is common practice for evolutionary scientists quietly to overlay their accounts of evolving nature with materialist ideology and then present this mixture to the public as though it were pure science. This "conflation" is openly promoted not only by the "ultra-Darwinist" expositions of Dennett and Dawkins, but also by their expert evolutionary adversary, paleontologist Stephen Jay Gould. In his earlier work *Ever Since Darwin* Gould expressed a view that he has never abandoned: he persistently argues that Darwin's theory is "dangerous" (to use Dennett's term) because it inevitably carries with it a "philosophical message" that conflicts with "Western" (by which he means both progressivist and theistic) assumptions. According to Gould, Darwin's "message"—allegedly inseparable from Darwin's science—is that the universe is inherently purposeless and that "matter" is all there is to reality. Gould writes:

> ...I believe that the stumbling block to [the acceptance of Darwin's theory] does not lie in any scientific difficulty, but rather in the philosophical content of Darwin's message—in its challenge to a set of entrenched Western attitudes that we are not yet ready to abandon. First, Darwin argues that evolution has no purpose. Individuals struggle to increase the representation of their genes in future generations, and that is all.... Second, Darwin maintained that evolution has no direction; it does not lead inevitably to higher things. Organisms become better adapted to their local environments, and that is all. The "degeneracy" of a parasite is as perfect as the gait of a gazelle. Third, Darwin applied a consistent philosophy of materialism to his interpretation of nature. Matter is the ground of all existence; mind, spirit and God as well, are just words that express the wondrous results of neuronal complexity.[35]

Even though Dennett complains that Gould's evolutionary ideas are still vitiated by a nostalgia for skyhooks,[36] it is clear that Gould shares with him and Dawkins a more fundamental belief that in order to accept evolution *as science* we must first make an *ideological* commitment to materialism and all of its consequences. That is, we must situate all biological data within the framework not only of Darwinian theory, but also of a materialist metaphysics, for otherwise we run the risk of contaminating the "objective facts" of evolution with theistic assumptions.

It would be logically impossible, of course, to reconcile theology with evolutionary theory if the latter is in fact inextricably implicated with the "philosophical message" of materialism that claims the allegiance of scientists like Gould, Dennett, and Dawkins, for the only choice such a state of affairs would leave us is either to renounce theism on the one hand or evolutionary science on the other.[37] However, there remains for an increasing number of theologians today the option of situating the data of evolutionary science within the framework of a metaphysical theology shaped by the story of the suffering love of God. Without in any way

---

[35] Stephen Jay Gould, *Ever Since Darwin* (New York: W. W. Norton, 1977), 12–13. In his most recent book, *Full House: The Spread of Excellence from Plato to Darwin* (New York: harmony Books, 1996), 17–21, Gould repeats his conviction that it is not Darwin's science as such, but the philosophical message that comes with it, that Westerners find so offensive.

[36] Dennett, *Darwin's Dangerous Idea*, 262–313.

[37] It is not surprising that when evolutionary science is presented as though it were inevitably and inherently anti-theistic, those who hold belief in God to be the most important aspect of their lives will be repelled by it.

contradicting what *science* quite naturally takes to be the undirected character and spontaneous creativity of natural selection, such a theology may simultaneously discern in evolutionary accounts of nature the lineaments of a momentous story of divine self-giving to the universe. In order to do this, of course, it needs to take its lead from faith in the power and effectiveness of God as understood in terms of the revelatory image of divine vulnerability.

Theology has not always articulated an understanding of divine power that is completely consistent with faith in the infinite depths of divine compassion. But unless it does so, I doubt that it will ever have the right ingredients for a coherent theology of evolution. For if we focus only on a God of "power and might" we cannot help but look in perpetual disillusionment for actual divine actions in nature that will "correct" nature's profligate and lazy impersonality. Since such interventions obviously do not take place in the manner we would expect, the evolutionary scientist is often compelled to become a skeptic about the existence of God.

Obviously, then, it is difficult if not impossible to reconcile God's power with evolution as long as we understand "power" in a crudely mechanical and interventionist sense.[38] Moreover, it is hard for many compassionate human beings to embrace religiously a God who is said to be able at any point to intervene in the cosmic process, where so much suffering occurs, and who yet refuses to do so. For this reason it may be too much to expect those who are fully aware of the troubling ways of evolution to accept ideas of God that have not yet been tempered by the image of divine suffering love. Ironically, the central image of Christian revelation has in principle already vanquished all notions of divine omnipotent apathy and arbitrariness, but theology and Christian spirituality have not yet been fully transformed into faithful conveyors of this image. Indeed, we might say that the evolutionary objections to theism are less of a challenge to theology than is Christian faith's own implicit awareness of the depths of God's compassion. But a theology that has begun to be transformed by profound meditation on the image of God's self-effacing love should prove quite capable of rendering intelligible, and even systematically (obviously not chronologically) anticipating, the evolutionary picture of the world as it is presented by neo-Darwinian science.

The seeds of such a theology have already been sown—though they have yet to blossom fully—in several forms of contemporary theology. In my opinion, however, no contemporary systematic theology has confronted the reality of evolution with more sophistication and metaphysical acumen than the one known as "process theology." Process theology has not yet been embraced by all Christian theologians, partly because it requires further development in terms of its consistency with traditional teachings about divine creation, but also because its insistence on the scandalous notion of a compassionately suffering God seems unduly to diminish the effectiveness of divine power to create and redeem.[39] Yet another reason why it has not become more widely accepted is that theologians have so far paid only scant attention to nature and the problematic character of its evolution. Focusing primarily on questions about the meaning of human life, politics, economics, history, and

---

[38] For this reason, some theologians today also seek to understand divine power and action in more eschatological terms, consistent with the "unfinished" character of the universe. We shall return to this important emphasis below.

[39] See David Basinger, *Divine Power in Process Theism: A Philosophical Critique* (Albany: State University of New York Press, 1988).

personal freedom, modern theology has generally ignored the natural world,[40] especially as it has been pictured by evolutionary science. Ironically, though, it seems to me that theology's scanty reflection on evolution has in the long run crippled its capacity to respond convincingly to questions raised by the ambiguities of human life and historical existence as well.

Process theology has always been apprised of the very features of evolution that Dennett, Dawkins, and Gould consider to be so "dangerous" to religion, and it has been molded in accordance with that awareness. Thus, an appropriate place to begin fashioning an evolutionary theology may be to scan the contributions process theology can make to such an effort. I shall first summarize some of these, and then suggest very briefly how a theology of hope and its implied "metaphysics of the future" can complement a process theological understanding of evolution.[41]

3 Process Theology and Evolution

Process theology reflects on God and nature in the light of concepts developed especially in the philosophy of Alfred North Whitehead. Other philosophers, such as Charles Hartshorne, have contributed to process thought, but most of its main themes were already anticipated by Whitehead. I shall not attempt here to respond to all of the difficulties some traditional theists have with Whiteheadian thought and process theology. Nor is it my intention to defend every aspect of what is in fact a rather diversified body of religious thought. Instead, I shall simply focus on features of process thought that I consider especially helpful in the task of constructing an evolutionary theology consistent with and illuminative of Darwinian science.

Process thought acknowledges that all of reality, and not just life, is in process, and that this pervasive cosmic becoming is itself comprised of temporal moments, occasions, events, or happenings. Nothing ever stands frozen in time, or it would cease to be at all. Temporality is woven into the very actuality of things.[42] In the course of the world's evolution an enormous variety of perishable patterns has emerged, responding to an adventurous restlessness that has always been present in the universe. Process thought does not take this restless cosmic dynamism for granted, but considers it fundamentally important to ask *why* nature has a temporal character and orientation toward novelty in the first place.

The universe, after all, is not, as many classical thinkers supposed, a "cosmos" in the sense of an eternally fixed order. As science now shows, nature is always becoming or evolving in a process that combines order with novelty in partially predictable but also partially surprising ways. This dynamic novel patterning, however, needs to be metaphysically accounted for. In its attempts to discover such an explanation, process thought turns to the religious idea of "God" as the most appropriate name for the necessary source of novel forms of order required by evolution.

---

[40] See Stanley L. Jaki, *Universe and Creed* (Milwaukee: Marquette University Press, 1992), 54.

[41] I do not subscribe to all aspects of Whitehead's metaphysics, nor do I agree with every aspect of "process theology" (a tag which itself covers a considerable variety of positions). I see the need for this kind of theology's continuing development, but I have found in its processive, organismic, and especially relational metaphysics indispensable concepts for thinking about the universe, life, mind, and God in an evolutionary context.

[42] Whitehead, *Science and the Modern World* (New York: The Free Press, 1967), 50. Time, therefore, is not an abstraction. Temporal occasions or events are the irreducibly concrete constituents of nature. Events rather than abstract, spatialized chunks of "matter" can be said to be the ultimate "stuff" out of which the cosmos is composed.

The God of process theology is active and involved in the cosmos, not in an intrusive or coercive way—since this would be incompatible with God's character as love—but persuasively, offering to the universe relevant new forms of order at every instant of its becoming. God, however, does not force the cosmos to respond to the divine proposals for new being. By remaining everlastingly present to it in the hidden mode of providing new possibilities for its self-actualization, God is deeply efficacious (or powerfully active) in the universe, yet without making the evolving cosmos cease to be radically distinct from its divine source. Moreover, God is intimately involved with the cosmic process in another sense: God experiences, suffers, and "remembers" forever all that occurs in the story of the evolving universe. In this sense God "redeems" the world and saves it from absolute perishing.[43]

Although, like every other philosophical system that has been taken into the service of theology, Whiteheadian process thought is not without difficulties, I think that it still provides indispensable conceptual ingredients for a theology capable of responding to the claims, such as those of some contemporary neo-Darwinians, that evolution decisively rules out any divine influence. Process theology's notion of a persuasive God appears to me to be completely faithful to the revelatory image of God as suffering love, while providing a much more secure metaphysical accounting for the novelty in evolution and the temporality of the world than either classical forms of theism or mechanistic materialism have offered.

Theologically speaking, God's character as persuasive love may itself be religiously grounded in the revelatory image of divine *kenosis*. The persuasive nature of divine "power" would then be a function of God's absolute self-sacrificing love. It is God's own loving "self-withdrawal" that not only allows the otherness of new creation to spring into being, but also actively invites the cosmos to swell forth continually, through immense epochs of time, into an always free and open future, and to do so in the relatively autonomous mode of "self-creation" that we discover in cosmic, biotic, and cultural evolution. As such, God intends the independence and freedom of the creature,[44] for the compassionate divine concern for the world's (relative) autonomy allows a considerable degree of self-creativity on the part of the created universe. Indeed, process theology argues that the independence of creation contributes much to God's own enjoyment of the world.

In this vision of things, the extensive reservoir of possibilities that Dennett calls "Design Space" is derived from and made available to natural selection by the generosity of a God who, in a manner consistent with the nature of love, does not insist on stamping all of these possibilities immediately and directly into actuality.

---

[43] We may understand this "redemption," at least fragmentarily, in terms of the process notions of relationship, patterning, and beauty. The world's constituent occasions are always open to a divinely inspired patterning that can give continually novel meaning (e.g., beauty and value) to the past, including its sufferings. Cosmic purpose resides precisely in the world's aim toward beauty, that is, toward novel patterning that can resolve local clashes and contradictions, such as life's suffering and evolution's wanderings, into an ever richer "harmony of contrast." Thus the world-process may be thought of as having an eternal meaning or destiny, without this destiny necessarily having to be a fixed and final stopping point.

[44] A theme emphasized by Wolfhart Pannenberg, *Systematic Theology*, Vol. II, trans. by Geoffrey W. Bromiley (Grand Rapids: Eerdmans, 1994), 127–36. "Theologically, we may view the expansion of the universe as the Creator's means to the bringing forth of independent forms of creaturely reality" (127). "Creaturely independence cannot exist without God or against him. It does not have to be won from God, for it is the goal of his creative work" (135).

The extensiveness of Dennett's "Mendelian library" may be seen here as the overflowing gift to the world of God's compassionate concern that the world be given a virtually unrestricted scope and time to "become itself." The world, after all, must be allowed to *become itself* if God is to communicate the divine selfhood to it. Nicholas of Cusa's prayer to God was: "How could you give yourself to me if you had not first given me to myself."[45] Analogously, there can be no self-giving of God to the universe unless this same universe is first allowed in some sense to actualize "on its own," some of the possibilities proposed to it by God. Theologically we may locate the enormous epochs of cosmic time, the apparently autonomous evolution of life by random variation and natural selection, and the self-organizing character of other physical systems as well, within the conception of a universe that is always being challenged to "become itself" so as eventually (and perhaps only eschatologically) to be transformed into a fully receptive "other," completely open to God's self-communication. The history of cosmic evolution, including all of those features that seem to render it absurd when interpreted in terms of our human standards of right order, receives its meaning as an adventure toward a freedom and "otherness" *vis-a-vis* the Creator that can in turn allow for the deepest intimacy of God with the world. As Karl Rahner would have it, a graced world is radically distinct from God, but it differs from God in direct, not inverse, proportion to its union with God.[46]

God's gift of Self to the world, therefore, does not dissolve the creation but invites it to become ever more distinctly itself. The world's emergent self-creativity and self-coherence is entirely consonant with an intensification of its intimacy with God. And so, in light of the axiom that its union with God actually differentiates the world from God, theology is not surprised at the undirected experimentation with ways of adapting, the spontaneous creativity, or the enormous spans of time involved in evolution. From the point of view of God's incarnation, a world forcefully directed in every detail by an imagined divine potentate would in any case be theologically incoherent, the reason being that such a world would in fact imply the absence of divine love, and so would not be a truly graced universe. A hypothetical "creation" imagined to be in every respect the direct and "perfect" implementation of the divine will, could never be a world unto itself. It could not be so distinct from its Creator as to be meaningfully open to the divine self-donation.[47]

In summary, then, our kenotic understanding of God in alliance with some important contributions of process theology allows: 1) that God is the sole ground

---

[45] Cited in John J. O'Donnell, *Hans Urs Von Balthasar* (Collegeville: Liturgical Press, 1992), 73.

[46] See, for example, *Foundations of Faith*, trans. by William V. Dych (New York: Crossroad, 1984), 116–37. Or, as Teilhard de Chardin says, "true union always differentiates." See *The Future of Man*, trans. by N. Denny (New York: Harper Colophon Books, 1969), 55ff.

[47] Theologically speaking, a deterministic theism is "incoherent" since infinite love posits the otherness of the finite beloved, which would mean that the finite other must have at least some degree of indeterminacy *vis-a-vis* God's vision of the world's future. If theism entails God's radical transcendence of the world, then a hypothetical "world" which in every respect were nothing more than the immediate and direct implementation of a divine "plan" would be reducible to an emanation of God's being rather than a distinct reality endowed with some degree of autonomy. Moreover, God can only *transcend* a world that is truly distinct from the divine being. Distinctness, however, does not entail separateness. Thus a panentheistic understanding of God and the world allows for their utter differentiation, and hence for God's transcendence, while still stressing the redemptive intimacy of God with the evolving world.

of the world's being; 2) that God's eternal self-restraint, by grounding the world's (relative) autonomy and allowing for its self-creation, shows God to be more intimately involved with and powerfully effective in the world than a more immediately directive divine agency would be; 3) that God acts effectively in the world by offering to it a wide range of autonomously realizable possibilities within which it can "become itself"; 4) that God simultaneously gives the divine self away completely to the world which has by God's will been encouraged to develop as something radically "other" than God; 5) that the phenomena of life's evolution, including the randomness, the wandering prodigality, and the enormous amount of time required for the emergence of complexity and consciousness, become theologically intelligible when seen in the light of God's self-limiting and persuasive love; and finally, 6) that the sufferings, struggles and achievements of the evolving world nonetheless take place within God's own experience, not outside of it: God's compassionate feeling and remembering of the sufferings, struggles, and achievements of the entire story of cosmic and biological evolution redeem and give meaning to everything, though in an always partially hidden way.

Such a theological stance is not only consistent with, but ultimately explanatory of the world as it is seen in the light of current evolutionary science.[48] A theological understanding of God as the hidden but inexhaustible source of all possible forms of ordered novelty gives a grounding to evolution that scientific materialism is incapable of providing. For materialism allows no real novelty to enter into the cosmos, since it "explains" all evolutionary developments as nothing more than the consequence of what is always already there.[49] According to Dennett, for example, in order to account for evolutionary creativity, we must think of living and conscious beings as the products of a series of purely mechanical "cranes" working entirely out of the dictates of deterministic laws and physical stuff already fully available.

---

[48] In positing God as the sole ground of the world's being, I am affirming the intention of the doctrine of *creatio ex nihilo* to secure the distinction of God from the world. The deep religious meaning in this classic teaching is that the world is a free gift of God. I see no reason why process theology cannot be "modified" in order to accommodate this indispensable religious conviction. For that reason, then, I am uncomfortable with Whitehead's own wording, that the world also "creates God." At the same time, however, the notion of divine vulnerability demands that we think of the world and its evolution as affecting and even "changing" God's being. This changing or becoming in God is not a threat to the notion of divine transcendence and eternity as long as we understand the latter in terms of the biblical sense of God's everlasting fidelity. In fact, it would appear that an absolute, eternal fidelity could actualize itself concretely only by becoming deeply sensitive or vulnerable to the events in the finite world. It is in this sense, I think, that most process theologians want us to understand the Whiteheadian notion that God not only creates the world, but the world also "creates" God by being taken into God's "Consequent Nature."

[49] If, as Dennett argues, evolution is just an algorithmic process, then nature already contains within itself *in potentia* all the possible entities that can ever emerge. All evolution can do, in this view, is "process" what is already given. This would mean that the world's entire future is already fully coiled up in the most remote past. The alternative I shall propose on the basis of process thought is that the past world cannot alone be the reservoir of all the possibilities to be actualized in the future, as a materialist metaphysics would have to imply. Rather the world's future states are always a synthesis, on the one hand, of data received by the present from the objectively given past, and, on the other, of genuinely new possibilities that emerge not from the past as such but from each occasion's anticipation of a truly novel future. It seems to me that biblical faith and eschatology *require* a conception of the universe in which the dead past is not the source of all apparent novelty. For that reason a theology of evolution needs something like a "metaphysics of the future."

Nothing truly new can come into such a world if the explanation for evolution resides in the inanimate cosmic material and motion fully resident in the past.[50]

In this commonly accepted "scientific" view, the "surprising" developments that take place in biological evolution, therefore, can be accounted for by three constituents: 1) "contingent" or chance occurrences, which by definition have no intelligible explanation beyond themselves; 2) the "necessary" laws of physics, chemistry, and natural selection; and 3) enormously protracted periods of time.

Process theology, however, credibly allows that "contingent" events may have a deeper meaning, since it is through contingencies that novelty, which ultimately has its origin in God, can enter into the world in such a way as to give our universe the character of temporal evolution. Nor is the "necessity" of nature so inexorably rigid and impersonal that physical and chemical processes cannot be taken up into an indefinitely extended range of novel and momentous patterns. Nature at every instant is enfolded by a field of possibilities ("Design Space," if you prefer) graciously offered to it in a manner relevant to its particular developmental phase. The ground or source of these possibilities for novel patterning is one of the things an evolutionary theology would mean by "God."

The question remains, of course, whether this understanding of God as the self-concealing source of evolutionary novelty is adequate not only to the scientific understanding of evolution but also to the religious doctrine of divine creation. In order to respond suitably to this question an evolutionary theology may have to go beyond Whitehead's ambiguous interpretation of divine creativity.[51] And at the same time, in order to approximate a more biblical understanding of God, evolutionary theology would do well also to open itself to a profoundly *eschatological* understanding of divine creativity. While process theology's strength has been its attunement to scientific discovery, it has not always demonstrated to everyone's satisfaction that its interpretation of nature is completely consistent with traditional Christian teachings on creation and eschatology.

In my own judgment process theology can unfold its potential contributions to a Christian evolutionary theology if it articulates more explicitly its own contributions to a "metaphysics of the future,"[52] by which I mean that the fullness of being is not found in the past or present, but in what is yet to come. In such a vision new and unprecedented natural occurrences arise primarily because novel possibilities are being offered to the present from the temporal realm that we refer to as the future. Accordingly, God can be thought of as continually creating the universe not by compelling it from the past, but by persuading it to actualize itself in conformity with a range of relevant possibilities proposed to it by the future, the arena from which the biblical God is pictured as coming into the present. God acts in the world not by forcing it to conform unbendingly to the pathways of a dead past—although this kind of "causality" is what natural science is methodologically habituated to discern—but by entering into it (via contingent occurrences) out of an unpredictable future.

---

[50] The question of how to account in any ultimate way for this "given" past is an issue that Dennett and Dawkins dismiss as either unanswerable or as downright silly.

[51] Whitehead's claim, for example, that creativity is distinct from God may be difficult for Christian theology to appropriate, even though his intention was the legitimate one of emphasizing that all entities are invited to participate in the process of their own creation.

[52] Process theology already seems to allow for the causal role of the future inasmuch as each occasion is subjectively attuned to a realm of novel possibilities proposed to it ultimately by God. I am suggesting that a theology of evolution needs to make this point even more explicit.

Revelation, as Wolfhart Pannenberg says, is the "arrival of the future."[53] It is the entrance of a divine future into the present that *ultimately* makes evolution possible.

I am not arguing, of course, that science needs to see things in this way. Rather, I am proposing only that cosmic process and biological evolution can be *fundamentally* grounded by a theological metaphysics according to which the world's future concrete actualizations are not exhaustively derived from the past. I realize that in an intellectual climate shaped dominantly by scientific notions of efficient and mechanical causation, the idea of a metaphysics of the future may not seem initially intelligible. Science's (or, perhaps more precisely speaking, scientism's) understanding of causation is tacitly rooted in what might be called a "metaphysics of the past," an ideology that seems to rule out the possibility of any genuinely new creation. On the other hand, biblical experience assumes something like a "metaphysics of the future" according to which the fullness of creation can at present be grasped only fragmentarily, and only through acts of hope in what is yet to come. Faith posits a God "who makes all things new" (Isaiah 42:9; 43:19; Revelation 21:5) and "whose very essence is future."[54] Hence in its attempts to bear witness conceptually to faith's experience of the *Novum* it seems to me that theology requires an understanding of causation inclusive of but also surpassing that used by natural science. A metaphysics of the future would not only respond to faith's need to ground its hope, but also to our attempts to explain, in any profound way, how the universe is capable of evolving at all.

An obvious objection to the idea of a "metaphysics of the future" is that the future, which is not yet actual, cannot be "causal" in any sense. Doesn't such a proposal amount to an incomprehensible "ontologizing" of the future? In response I would suggest first that we recall Paul Tillich's counsel that the attempt to be too specific about any conceivable causal relationship between God and the world runs the risk of implicitly "naturalizing" the divine by embedding it within a finite causal series. Hence we must never too literally apply the notion of causation to God's action, or look obsessively for a "causal joint" between the world and God's action. Causation, after all, is a notion derived from our experience of occurrences in the realm of finite entities in a finite world. As such, it is literally inapplicable to divine action. If it is used at all, as Tillich argues, it must only be in a metaphorical sense.

Acknowledging our religious need to speak of God's action in a metaphorical rather than literal manner, Tillich suggests that we use the term "Ground" rather than "Cause" when making reference to God's creative and sustaining relationship to the

---

[53] "By contemplating Jesus' resurrection, we perceive our own ultimate future," Pannenberg writes. And he adds: "The incomprehensibility of God precisely in his revelation, means that for the Christian the future is still open and full of possibilities." *Faith and Reality*, trans. by John Maxwell (Philadelphia: Westminster Press, 1977), 58–59.

[54] See Ernst Bloch, *The Principle of Hope*, Vol. I, trans. Neville Plaice, Stephen Plaice and Paul Knight (Oxford: Basil Blackwell, 1986). See also Jürgen Moltmann, *The Experiment Hope*, edited and translated by M. Douglas Meeks (Philadelphia: Fortress Press, 1975). The writings of Teilhard de Chardin, Pannenberg, and Rahner also imply a metaphysics of the future. Hans Küng, *Eternal Life*, trans. by Edward Quinn (Garden City, New York: Doubleday & Co., 1984), 213–14, acknowledges the debt that Christian theology owes to the Marxist philosopher Ernst Bloch and to Bloch's main Christian theological follower, Jürgen Moltmann, for retrieving the Bible's futurist understanding of God. For centuries it had been buried beneath the non-eschatological categories of Greek philosophy that still determine in great measure the shape and character of scientific as well as theological understanding.

world.[55] But, as Tillich was especially aware, no metaphor can be taken as exhaustively adequate without becoming idolatrous, and so theology is obliged to experiment with a diversity of them. For example, process theology finds the notion of "Persuasion" to be especially illuminative of God's "causal" relation to the world, for it bears nuances that the term "Ground" does not. And likewise, I would suggest, theology can appeal to faith's sense of being grasped by what is yet to come, and make this the experiential base for yet another powerful metaphor, that of "Future," as a way to characterize God's creative relationship to the universe.

Tillich describes biblical faith's sense of the power of the future in these terms:

> The *coming* order is always coming, shaking *this* order, fighting with it, conquering it and conquered by it. The coming order is always at hand. But one can never say: "It is here! It is there!" One can never grasp it. *But one can be grasped by it.*[56]

Faith's experience of *being grasped* by "that which is to come" encourages theology to attribute genuine efficacy to the future at least as far as human experience is concerned. However, "being grasped by the future" must pertain to more than just our human experience since, as process thought correctly emphasizes, human experience cannot be artificially separated from that of the whole universe. Thus theology can legitimately claim, along with St. Paul (Rom 8:22), that the entire universe is under siege by the redemptive future.

The sense of being grasped by the power of the future is palpable to Christian religious experience, but it cannot be translated into scientifically specifiable concepts for the obvious reason that science typically attributes efficacy only to what lies in the causal past. Nevertheless, if we follow Whitehead's great insight that human experience may be the source of metaphysical categories that by analogy we can assume to characterize the experiential events that make up the rest of nature, then theology can infer that the same "power of the future" that grasps us in faith must also be effectively and persuasively present to the entire cosmos.

If this sounds too metaphorical to be scientifically palatable, we might note here the significant role that analogy and metaphor also play in scientific explanation. Evolutionary biology in particular has to resort to metaphorical language, and even a scientist as materialistic as Richard Dawkins complains of the "physics envy" among some of his fellow biologists who seek to escape metaphor by collapsing the hierarchy in nature down to a level of analysis where only mathematics is appropriate.[57] The explanatory success of Darwinian biology, Dawkins insinuates, is not in spite of, but *because* of its reliance on foggier but still illuminating metaphors such as adaptation, cooperation, competition, survival, and selection. Biologists are completely justified in resisting any request that they abandon these metaphors in the interest of more mathematical precision. All the more then, it seems to me, must theology likewise resist the temptation to forsake the relatively indistinct but powerful realm of metaphor and analogy for the sake of a more precise or atomistic

---

[55] See especially Paul Tillich, *The Courage to Be* (New Haven: Yale University Press, 1952).

[56] Paul Tillich, *The Shaking of the Foundations* (New York: Charles Scribner's Sons, 1996), 27, emphasis added.

[57] See Dawkins' comments in John Brockman, *The Third Culture* (New York: Touchstone Books, 1996), 23–24. However, as Niles Eldredge rightly points out, Peter Medawar's label "physics envy" is no less applicable to Dawkins who, in spite of his and Dennett's claims that he is not a greedy reductionist, borrows from physics the atomistic ideal of explanation and applies it univocally to the genetic monads that, for him, are sufficient to explain life and evolution. See ibid., 122.

account of divine causality in nature and evolution. By employing metaphors for God's influence such as "Ground of Being" or "Absolute Future,"[58] theology can in principle account for the fact of evolutionary novelty at a deeper if less precise level of explanation than the scientific. And such metaphorical explanation would not contradict or compete with evolutionary biology any more than evolutionary biology's own metaphorical "explanations" at its own level conflict with the chemical or physical explanations at another.

To attribute explanatory efficacy to the future, then, means that the world can *truly* evolve only if novelty comes into it in such a way as to allow it to become clearly distinct from its past. But if, as materialist evolutionism asserts, the full explanation of everything in evolution can be found only by tracing events backward in time through a series of mechanical causes (or Dennett's cranes), we would be compelled to say that past events are the only constituents of the real. Certainly biblical faith rules out such a metaphysics, and so theology is obliged to search for an alternative kind of explanation, one commensurate with the biblical intuition that the fullness of being lies up ahead, in the future. For this reason, in spite of the apparent philosophical difficulties it raises for those of us still attuned to the Greek metaphysics of an eternal present or to the scientific ideal of explaining things in terms of past causes, a theology faithful to the biblical witness must in some way affirm the power of the future as the deepest explanation of evolution. Such a metaphysical explanation, of course, can exist side by side with purely scientific accounts of evolution which *at their own abstract level of understanding* may have a "completeness" into which it makes no sense to insert theological categories.

Materialist interpretations of evolution, as I have already noted, implicitly appeal to a "metaphysics of the past" in order to explain present states of natural activity. In doing so they fail to acknowledge that it is only the constant emergence of new present moments that thrusts events into the "past" in the first place. A metaphysics which looks for explanation only in the dead past will inevitably suppress a sense of the future out of which new moments make their appearance. It will not realize that the very possibility of things being fixed in the past clearly depends upon the constant and faithful coming of a boundlessly resourceful future. If there were no "arrival of the future" there could be no past either, and therefore no science, which appeals exclusively to an explanatory domain of past series of efficient and mechanical causes. Science, of course, is appropriately constrained to follow a method of inquiry that leaves out consideration of any "power of the future," but at the level of metaphysics such constraint may be dropped.

Process theology is inherently open to the notion of the "power of the future," where the term "future" refers to that which brings renewal not only to human experience, but also to the rest of the evolving world to which we belong. In its temporal understanding of reality, process theology posits a realm of yet unrealized possibilities that can be synthesized into each present moment so as to bring about evolutionary novelty. The ultimate source of these possibilities it calls God. Past events may, of course, still be said to act efficaciously on the present, but not so much because they have a massive momentum that thrusts them forward dumbly and deterministically into the present (in the fashion of Dennett's cranes). Rather, the

---

[58] The notion of God as the "Absolute Future" comes from Karl Rahner, *Theological Investigations*, Vol. VI, trans. by Karl H. Kruger and Boniface Kruger (Baltimore: Helicon Press, 1969), 59–68. See also Ted Peters, *God—The World's Future: Systematic Theology for a Postmodern Era* (Minneapolis: Fortress Press, 1992).

past is causal only because every present event actively synthesizes or assimilates the given world into its own unique actuality in a manner that remains simultaneously open to the realm of new possibilities that arrive, so to speak, from the future.

The interaction of God and world, therefore, need not be imagined only in terms of one substance bumping up against another, as science sometimes pictures efficient causation. Rather, God interacts with the world in a manner structurally analogous to the way in which any present moment of actuality receives the cumulative past into itself and actively orders or patterns that given past in a way that also lets in the future and its wealth of possibilities. Thus any "special" actions faith attributes to God are not to be thought of as interruptions, but as intensifications, of structures characteristic of all natural occurrences.

By combining a "metaphysics of the future" with Whitehead's understanding of nature-in-process, an evolutionary theology can metaphorically point to the "causal" role of God in an evolving world without stepping competitively onto the explanatory "level" at which neo-Darwinian biology accounts for the same world through its own metaphors. Such a theological approach, however, is *not* compatible with the implied "metaphysics of the past" upon which evolutionist materialism is based. Not everything that occurs in evolution can be *fully* explained by what has already been. A metaphysics of the past is almost compelled to interpret the contingent emergence of novelty as absurd, since contingencies simply do not make sense when viewed in terms of what has already been. A metaphysics of the future allows us to interpret the contingent occurrences in evolution as the in-breaking of a freshness whose depth and meaning, by definition, cannot be unfolded in terms of any past or present scheme of understanding. In this sense it is more *fundamentally* explanatory of evolution than is the materialist metaphysics of the past.

4 Is Evolutionary Science Open to Theological Interpretation?

I have just sketched one way in which we might begin to move toward a consistently Christian theology of evolution. However, I have postponed until now a fuller discussion of what many would consider to be a prior question: does evolutionary theory leave any room for theology at all. Such a way of proceeding seemed necessary for the elementary reason that before we can talk about the relationship of the idea of God to Darwinian evolution we needed first to arrive at some specificity as to what we mean by "God." I have stated above that Christian faith is shaped by a kenotic understanding of God and by the experience of the advent of a new and powerful future. Hence it would not be in the best interest of genuine dialogue between theologians and evolutionary scientists for us to argue the merits of a more philosophically abstract and religiously innocuous concept of God, one that has very little connection with the experience of Christian faith. Nevertheless, in its conversations with scientific skeptics, theology today must also give more logically explicit reasons for its conviction that evolutionary science cannot provide us with a *complete* account of the story of life. It must be able to show that in principle there is still a powerfully explanatory role for its idea of God as far as life's evolution is concerned, and that such an explanation in no way interferes with those of biologists.

Many scientific writings today overtly claim that the phenomena of life can be explained sufficiently in purely naturalistic terms. Even in an encyclopedia, where one might expect more neutrality, an article on evolution claims that "Darwin did two things: he showed that evolution was a fact contradicting scriptural legends of creation and that its cause, natural selection, was automatic with *no room for divine guidance or design.*"[59] It is incontestable, of course, that science illuminates the phenomena of life in ways that religion cannot. Moreover, it goes without saying that science has no business searching for non-natural explanations of living phenomena, no matter how recalcitrant the latter may be to present modes of scientific understanding. However, the significant question is whether Darwinian science leaves any room for religious explanation *at all.* Whether deliberately intended or not, scientific writings often claim that the concept of God has *no explanatory value* concerning the evolution of life's diverse forms of order.[60] If this means only that we should not introduce theological ideas into specifically scientific inquiry, then obviously the theologian cannot object. Or if evolutionary scientists are only seeking to avoid the notorious god-of-the-gaps, then one can only applaud. However, it is often suggested that theology has no legitimately explanatory role *whatsoever* with respect to the events that have occurred in the evolution of life, implying that explanatory adequacy belongs to science alone. To claim that Darwinian selection adequately accounts for the design of organisms, and that such an explanation *excludes* God, amounts to asserting that the idea of God is irrelevant to the fact of nature's intelligibility.

Subscribing to the belief that natural science can elucidate life in an exhaustive way, many biologists now share the view that Darwinian evolution leaves no room whatsoever for theological attempts at explanation. They consider it no longer acceptable after Darwin to seek any deeper meaning in the story of life than what biological science provides in its elegantly simple recipe for evolution. Allegedly only three ingredients are necessary to explain life and evolution: 1) first, you need accidental or *contingent* occurrences. These include, for example, the highly improbable chemical accidents required for the origin of life, the random genetic mutations that make possible the diversification of life, and many other fortuitous events in natural history that shape the course of evolution (for instance, the mass extinctions caused by famines or meteorite impacts); 2) a second ingredient is the determinism or *necessity* implied in the remorseless "law" of natural selection and

---

[59] Gavin de Beer, "Evolution" in *The New Encyclopedia Britannica*, 15th ed. (London: Encyclopedia Britannica, 1973–74), emphasis added. I owe this reference to Holmes Rolston III, *Science and Religion: A Critical Survey* (New York: Random House, 1987), 106. Francisco Ayala is the author of the article on "Evolution" in the 1987 edition of the same encyclopedia. Ayala does *not* take the position defended by de Beer, as is clear from his articles in this volume.

[60] Even as lucid and careful an evolutionary witness as Francisco Ayala can easily give this impression to readers when he writes: "It was Darwin's greatest accomplishment to show that the directive organization of living beings can be explained as the result of a natural process, natural selection, without any need to resort to a Creator or other external agent." Ayala, "Darwin's Revolution," in John H. Campbell and J. William Schopf, ed., *Creative Evolution?!* (Boston: Jones & Bartlett, 1994), 4. He also states that "Darwin's theory encountered opposition in religious circles, not so much because he proposed the evolutionary origin of living things (which had been proposed many times before, even by Christian theologians), but because his mechanism, natural selection, *excluded* God as accounting for the obvious design of organisms" (5, emphasis added). While theologians will agree that at the level of purely scientific explanation it is inappropriate to bring in any idea of God, Ayala's wording could possibly be read as a *philosophical* claim that divine influence plays no explanatory role *at any level* of our understanding of life (though this is not his intention). As in many discussions of science and religion the fundamental issue is whether we may allow for a multi-leveled hierarchy of explanations that includes theology, or must instead submit to the non-scientific claim that science alone gives a complete account of living beings.

in the inflexible laws of chemistry and physics;[61] 3) and finally, evolution requires enormous spans of *time*, for without the immensity of the universe's temporal duration the many improbable events in evolution could never have occurred, especially in the absence of a divine "designer."

It might seem then that the mixing and brewing of these three components—contingency, necessity, and time—is sufficient to account for the entire evolutionary stew. However, not only does this simple formula leave out countless particularities that comprise the unique story of life on Earth,[62] it also leaves ample room for theological comment. For the topics of contingency, necessity, and time are theology's perennial concerns. Indeed, making sense of them is its main and persistent reason for being. In case such an assertion does not seem immediately obvious, we need only recall that the human experience of nature's and history's contingencies (especially those that cause pain) is what prods our religions to seek out a framework wide enough to give such events an intelligibility that presently eludes our best intellectual efforts. Likewise, theology also asks about the source of the intelligibility, order, predictability, and lawfulness—or what some scientists tendentiously refer to as the "necessity"—resident in the universe. Since science itself only assumes, and can in no sense explain the fact of nature's intelligibility, it is part of theology's distinctive mission to do so. And so it understands the word "God" as a symbol whose meaning consists, at least in part, of pointing us toward the *Ground* of the world's intelligibility. Finally, the fact of "time" is a persistent stimulus to theology, especially since the "perpetual perishing" it brings with it leads us to wonder what possible meaning might reside in this vanishing world.

Therefore, even though theology does not seek to supply a *scientific* account of the contingency and order that occur in the temporal process of evolution, it can still fruitfully raise significant and interesting questions about contingency, order, and time as such. These questions are of an entirely different logical order from those of science itself, and so there is no danger that any answers to them will intrude into gaps left open in the evolutionary accounts. Theology will leave free reign to science, but it has every right to raise questions about the claims of *scientism*. Three pertinent examples of such theological questions are discussed below.

1. *Must we interpret the contingency (or non-necessity) of natural occurrences, such as genetic mutations, as purely aimless accidents arising out of a dead and blind past, or may we understand contingency as essential for the inbreaking of new creation, for the arrival of a future that by definition lies beyond the horizon of science with its habitual orientation toward the causal past?* Nothing in the scientific accounts of life rules out a more fundamental "explanation" of the contingency required by evolution. Within the framework of a process theology attuned to a metaphysics of the future, for example, the events that evolutionary scientists call "contingent"—in the sense of absurd or meaningless—can at another level be viewed as signals of the fundamental openness of the cosmos to new forms of order. Such an interpretation, of course, cannot be validated by looking only at the relationship any particular contingent event has to past patterns of physical activity

---

[61] I shall use the term "necessity" to stand for what scientists refer to as the lawful, determined, and predictable aspects of nature. In fact, however, I think it is a most inappropriate term, for what it obliquely refers to is not only the predictability or apparent determinism in nature, but above all to its intelligibility.

[62] For a careful study of just how much biological science does leave out of its accounts of life, see John Bowker, *Is God a Virus?* (London, 1995).

as these are understood by science, for science methodologically restricts itself to looking at nature in abstraction from unanticipated possibilities and genuinely new future developments.[63] Hence, whenever evolutionary scientists insist *a priori* that contingency in nature is absurd and inherently meaningless, this is not a scientific but a metaphysical claim— one to which there may be theological alternatives which are completely consistent with the reality of evolution. Evolutionary science has by no means clearly ruled out the possibility that nature's contingency is the consequence of God's longing for the world to grow toward increasing independence.

2. *Does the ingredient in evolution sometimes referred to as necessity leave us with such a closed and determined universe that there is no room for divine action?* Or is the apparent "necessity" in natural selection an indication—especially at the level of life and evolution—of nature's pervasive *intelligibility* and *reliability* rather than of any godless impersonality? May we even be permitted to understand it as the expression of God's faithfulness?[64] Whenever nature's "necessity" is taken in abstraction from its contingency, we may be tempted to universalize and eternalize it as evidence of a fated and indifferent cosmos. But as John Polkinghorne has pointed out, the notion of natural necessity is a misleading abstraction that does not exist in any pure state. Nature *in the concrete* is dominated neither by rigid necessity nor by blind contingency. Rather it is a dynamic blend of reliability and openness.[65] It is the kind of universe, in other words, that one might expect from a creator who selflessly seeks to give it a degree of internal self-coherence and consistency, and who also endows it with the capacity for self-transcendence and consciousness.

3. *Finally, how are we to make sense of the fact of time?* Evolution requires unimaginably prolonged periods of temporal duration as one of its three main ingredients. Given the allegedly unguided character of evolution implied in its "contingency" and "necessity," there would scarcely have been sufficient opportunity to produce life's improbable concoctions without seemingly endless epochs of time. And so, any attempt to explain the outcomes of evolution parsimoniously by appealing to utterly random mutations and impersonal natural selection also requires, as a third indispensable constituent, enormous spans of terrestrial and cosmic time. If "divine guidance" is not a factor in evolution, then apparently only the immensity of time and the latitude it allows for experimentation can adequately compensate.

Note that the enormous amount of time that evolution requires is one of the main reasons for scientific skepticism about the explanatory relevance of the idea of God. Gould, Dawkins, Dennett, and numerous other skeptics repeatedly observe that an intelligent "designer" would never have fooled around and tinkered so clumsily with creation in the way the evolutionary record depicts. Hence the sheer immensity of time that has transpired during life's evolutionary meanderings places the idea of God in as much doubt as do the elements of randomness and impersonal necessity.

However, if we think of God's "design" for the universe in terms of a concern for maximizing its opportunities for independence, self-creativity, and emergent freedom, then the world's struggle to realize such possibilities during the course of

---

[63] New reflections on the nature of science in the light of the phenomena of "chaos" and "complexity" may eventually bring about a less restrictive understanding of scientific explanation.

[64] Wolfhart Pannenberg, *Toward A Theology of Nature*, edited by Ted Peters, (Louisville: Westminster/John Knox Press, 1993), 109.

[65] See, for instance, John Polkinghorne, *The Faith of a Physicist* (Princeton: Princeton University Press, 1994), 25–26, 75–87.

an enormous span of time poses no theological difficulties. The vastness of cosmic and evolutionary time contributes no less than spatial enormity to the richness and narrative beauty of the cosmic whole which, to the religious imagination, can be an invigorating sacrament of God's infinity. However, in addition to our religious awe at the aesthetic grandeur implicit in the billions of years of cosmic emergence that science has laid out before us, theology can also make the following comments on the evolutionary biologist's appeal to time as a "causal" ingredient in evolution.

Evolutionary science can appeal to and build its ideas upon the fact of the universe's irreversible temporality, but it cannot give an adequate account of why the universe has such a temporal character at all. Just as evolutionary science cannot say exactly why the universe is open to contingency at all, or why it is in some respects intelligibly ordered and sufficiently reliable to allow for the emergence of organic complexity, neither can it ground the fact of cosmic temporality. At its own level of explanation biological science can only take the great mystery of time as a brute datum on which it has no light of its own to shed.

Of course, like all other sciences evolutionary science, *qua* science, is not obliged to account for the fundamental features of nature that its recipe requires. It can quite legitimately limit itself to "explaining" what occurs at the logically distinct level of organisms, populations and their achievements, and can use explanatory notions different from those of physics or theology. Biology, in other words, must have at least some degree of autonomy if it is to be fruitful. But even though there is a methodological need for each of the sciences to fence itself in and develop its own kind of illumination, there still remains considerable room for a more penetrating look at the totality of things from which each regional discipline inevitably abstracts.

Time is inseparable from the existence and nature of the universe as such. And because of the inherent connection of time with the cosmos as a whole, it turns out that evolutionary biology still leaves a large opening for theological explanation, not as a filler of gaps in the theory itself, but as part of the quest to understand the encompassing cosmic system within which evolution occurs. Thus any *adequate* account of the creativity in evolution may eventually lead us toward reflection on the fundamental conditions that give the natural world the character of temporal becoming in the first place.

Evolutionists typically take time for granted, and in the selective manner characteristic of scientific specialization, they generally leave out any consideration of the complex linkage time has to the cosmos as such.[66] Dennett, for example, postulates time's irreversible and generous open-endedness as the capacious arena for evolution to take its random walk through the "Library of Mendel." But the library acquires its countless volumes only by virtue of an assumption that time is a virtually limitless arena. The implication is that if time were much shorter, the boundaries of evolutionary creativity would be too restrictive. Take away the enormity of time, and the recipe for evolution falls one ingredient short.

Like many other neo-Darwinian thinkers, Dennett does not take the time to dwell on the wonder of time itself, even though without time evolution by natural selection would be inconceivable.[67] According to most versions of evolutionary

---

[66] A point convincingly made by Jeffrey S. Wicken, *Evolution, Thermodynamics, and Information* (New York: Oxford University Press, 1987), 59ff.

[67] Dennett briefly looks at the connection between the fact of life and the physics of the early universe, but he abruptly sidesteps any serious consideration of the possible wider implications of this connection. *Darwin's Dangerous Idea,*100ff.

theory, vast amounts of time must already be available for life's continuing and usually unsuccessful stabs at adaptation. Given enough time, almost anything can happen, but time must first be given. So fundamental is time for evolutionary science that biologists often implicitly and misleadingly give to time's *immensity* a quasi-causal role in the creation of new forms of life. A deeper inquiry, however, must at some point do more than simply postulate temporal duration as part of evolution's astonishingly simple recipe. It must inquire into *all* the conditions that make a universe of irreversible temporal becoming possible in the first place. I would suggest that if we pursue these conditions far enough down, there will be plenty of room for theological inquiry alongside (not mixed in with) other sciences.

Unless we wish arbitrarily to cut short our longing for understanding, it is not enough to view time merely as a preexisting co-ingredient, ready to be mixed together with random mutations and natural selection so as to produce evolutionary novelty. A theological inquiry into the deepest conditions of evolution—and here metaphors are no less appropriate than in evolutionary biology—would follow faith's claim that the world's temporality is itself a gift stemming from the influx of a future which in its absolute depth may be called God. The world's "being grasped" by this future would allow for the break from routine we call "contingent" events in nature, among which are the many serendipitous occurrences necessary for creative evolution. The coming of the future into the present is ultimately what allows us to become aware of time. Biblically speaking, what endows the universe with its temporality is not just change or motion, as Aristotle thought, but the arrival of the divine *Novum*. Without the constant in-pouring of new creation there could be no temporal becoming and no evolution.[68] Thus it is finally the arrival of the future that "explains" the fact of contingency, gives time its irreversibility, allows the cosmos to expand, and eventually arouses the emergent adventure of life and consciousness.

Such an interpretation not only fits the promissory vision of biblical faith, but is also completely consistent with what science tells us about a universe open to the evolution of life by natural selection. This theological explanation does not compete with evolutionary science but instead looks for *ultimate* reasons why our universe has the contingency, necessity, and temporal openness that allow evolution to take place at all. It also leaves open to physics, biology, and other sciences the opportunity to explain things at their own levels. It can still claim for itself the responsibility of looking beneath all scientific inquiries for even deeper understanding. If it remains aware of the fact that its own accounts will themselves always be only fragmentary, it will ensure not only a future for itself, but it will also lay out before the sciences the limitless horizon of a cosmic intelligibility that gives them a future as well.

---

[68] Classical mechanics was unable to detect the irreversible character of time, for by abstracting considerably from the *actual* realms of physical activity, it was enabled to conceive of physical processes as "reversible." Such abstractness still perdures in much "scientific" thinking, and it is especially this habit of mind that continues to seal off so much scientific consciousness from any sense of *new* being.

FIVE MODELS OF GOD AND EVOLUTION

Ian G. Barbour

Is evolutionary theory compatible with the idea that God acts in nature? Through most of Western history it had been assumed that all creatures were designed and created by God in their present forms, but Darwin claimed that they are the product of a long process of natural selection. His theory of evolution not only undermined the traditional version of the argument from design; it also explained the history of nature by scientific laws that seemed to offer no opportunity for God's providential guidance. However several themes in the biological sciences offer promising new ways of conceiving of divine action in evolutionary history without intervention or violation of the laws of nature.

The first section of this essay traces the development of evolutionary theory from Darwin himself to molecular biology and recent hypotheses about complexity. The second explores four themes in recent writing about biological processes: self-organization, indeterminacy, top-down causality, and communication of information. Subsequent sections examine theological models of God's action in nature based on analogies with each of these four characteristics of organic life. I will suggest that a fifth model from process theology avoids some of the problems arising in other models of God's relation to nature.

1 Darwinism Evolving

Evolutionary theory has undergone significant reinterpretation and modification since Darwin. First, the growth of population genetics and molecular biology is briefly described. Then the expansion of Darwinism is discussed, particularly the recognition that other factors in addition to natural selection influence the direction of evolutionary change. Finally, recent theories of complexity and self-organization are considered.

1.1 From Darwin to DNA

In Darwin's day, Newtonian mechanics was looked on as the form of science which other sciences should emulate. The Newtonian viewpoint was atomistic, deterministic and reductionistic. It was believed that the behavior of all systems is determined by a few simple laws governing the behavior of their smallest components. Change was thought to be the result of external forces, such as gravity, acting on bodies which are themselves essentially passive. Darwin agreed with the philosophers of science who held that Newtonian physics represented an ideal for all the sciences, and his theory of evolution shared many of its assumptions.[1]

Darwin held that evolutionary change is caused by natural selection acting on variations among individual members of a species. Under competitive conditions, those individuals with a slight adaptive advantage will survive better to reproduce and pass on that advantage to their offspring. His viewpoint was "atomistic" in assuming that selection acts on separate traits in individual organisms. For him, as for Newton, change was the result of external forces; he held that the direction of change is determined by natural selection, not by the efforts of organisms themselves

---

[1] Michael Ruse, *Philosophy of Biology Today* (Albany, N.Y.: State University of New York Press, 1988), 6; idem, *The Darwinian Paradigm* (New York: Routledge, 1989).

as Lamark had believed. The assumptions which Darwin shared with Newton are explored in detail in a recent volume by Depew and Weber.[2]

By the end of the nineteenth century, *probability* was an important concept in several areas of physics. Maxwell and Boltzmann showed that the probability of different configurations of gas molecules can be calculated even when the motions of individual molecules are too complicated to describe mathematically. Statistical averages can be used to predict the relationship between large-scale variables such as pressure, volume, temperature, heat flow, and entropy. In statistical mechanics and classical thermodynamics, equilibrium macrostates can be calculated without knowing the initial distribution of molecules.

Probabilistic reasoning was also important in the merging of *population genetics* and evolutionary theory early in the twentieth century in the theories of Fisher, Wright, and Dobzhansky. Fisher acknowledged the influence of nineteenth-century physics on his ideas about calculating gene probabilities in individual organisms and gene frequencies in populations. The *"modern synthesis"* in which Julian Huxley, G. G. Simpson and Ernst Mayr were prominent, continued the Darwinian belief that the evolution of species was the result of a gradual accumulation of small changes. If some members of a population are geographically or reproductively isolated from other members, accumulated changes may result in a new species that can no longer interbreed with the original population. In a very small isolated population, gene frequencies may differ, purely by chance, from those in the larger population; the direction of evolutionary change ("genetic drift") would then be the result of chance rather than natural selection. But natural selection was still viewed as the principal agent of evolutionary change.[3]

The discovery of *the structure of DNA* in 1953 led to the identification of the molecular components of the genes which population genetics had postulated. The "central dogma" of molecular biology asserted that information is transferred in one direction only, from the sequences of bases in DNA to the sequences of amino acids assembled by the DNA to form proteins. It was claimed that the environment has no direct effect on genes except to eliminate or perpetuate them through selective pressures on the organisms that carried them. Molecular biology has been immensely fruitful in illuminating almost every aspect of evolutionary history, but some of the assumptions initially associated with it have more recently been questioned.

1.2 *The Expansion of Darwinism*

Most of the challenges to the modern synthesis in recent decades should be seen as part of an expanded Darwinism (or neo-Darwinism), rather than as a rejection of earlier insights. For example, it has been claimed that *selection occurs at many levels,* and not just on the level of organisms in populations. Dawkins speaks of selection at the level of genes; he views organisms as mechanisms by which genes perpetuate themselves. E. O. Wilson speaks of kin selection and others defend group selection. Both philosophers and biologists have argued that selection occurs also at the species level. Whereas an organism produces other organisms by reproduction, and it perishes by death, a species produces other species by speciation, and it perishes by extinction. The speciation rate of a species may be as important in the long run as the reproduction rate of individual organisms. Variation and selection

---

[2] David P. Depew and Bruce H. Weber, *Darwinism Evolving* (Cambridge: MIT Press, 1995), Part I.

[3] Ibid., Part II.

occur at several levels at once, and of course changes at one level will influence those at other levels.[4] Darwin himself stressed the struggle and competition for survival, but more recent interpretations point to a larger role for cooperation and symbiosis.

The idea of *punctuated equilibrium* defended by Gould and Eldredge challenged the earlier assumption that macroevolution is the result of the gradual accumulation of many small changes. They point to fossil records that show millions of years with very little change, interspersed with bursts of rapid speciation in relatively short periods, especially in the early Cambrian period when all the known phyla and basic body plans appeared in a very short period. They postulate that alterations in developmental sequences produced major structural changes. Their view is *holistic* in directing attention to polygenic traits, the genome as a system, and the role of regulatory programs in development, rather than to small changes due to mutations in single genes governing separate traits that might be subject to selection. The directions of change are determined by the possibilities of developmental reorganization as well as by selective forces acting on organisms.[5]

Gould and Lewontin hold that evolutionary change arises from *many differing causes,* and they criticize explanation by natural selection alone ("panadaptationism"). They point out that one can always postulate a possible "selective advantage" for any trait by making up a "just-so story" of how it might be adaptive, even in the absence of independent evidence for such an advantage.[6] But most biologists probably follow Stebbins and Ayala in claiming that all the known data are consistent with an expanded and enriched version of neo-Darwinism in which variation and natural selection are still the main factors in evolutionary change.[7] The communication of information from DNA to proteins is indeed crucial, as the "central dogma" asserted, but other sources of information are significant in determining how genes are expressed in living organisms. Some of this information is in the cytoplasm outside the cell nucleus, and some comes from elsewhere in the organism or wider environment. A complex feedback and regulatory system turns particular genetic programs on and off. Outside influences can also affect the transposition of genes.[8]

Some biologists have noted that the *internal drives* and novel actions of organisms can initiate evolutionary changes. The environment selects individuals, but individuals also select environments, and in a new niche a different set of genes may contribute to survival. Some pioneering fish ventured onto land and were the ancestors of amphibians and mammals; some mammals later returned to the water

---

[4] R. N. Brandon and R. M. Burian, eds., *Genes, Organisms, Populations: Controversies over the Units of Selection* (Cambridge: MIT Press, 1984); Niles Eldredge and Stanley Salthe, "Hierarchy and Evolution," in *Oxford Surveys of Evolutionary Biology* (Oxford: Oxford University Press, 1985).

[5] Stephen Jay Gould, "Darwinism and the Expansion of Evolutionary Theory," *Science* 216 (1982), 380–87; S. J. Gould and Niles Eldredge, "Punctuated Equilibrium Comes of Age," *Nature* 366 (1993), 223–27.

[6] S. J. Gould and Richard C. Lewontin, "The Spandrels of San Marco and the Panglossian Paradigm: A Critique of the Adaptionist Programme," *Proc. of Royal Society of London B* 205 (1979), 581–98.

[7] G. Ledyard Stebbins and Francisco Ayala, "Is a New Evolutionary Synthesis Necessary?" *Science* 213 (1981), 967–71. See also Ayala, in this volume.

[8] John Campbell, "An Organizational Interpretation of Evolution," in *Evolution at the Crossroads,* David P. Depew and Bruce H. Weber, eds. (Cambridge: MIT Press, 1985).

and were the ancestors of dolphins and whales; some forest woodpeckers began to hunt in the mountains. In each case organisms themselves took new initiatives; genetic and then anatomic changes followed from their actions through "genetic assimilation" (the Baldwin effect). The changes were not initiated by genetic variations. Lamark was evidently right that the purposeful actions of organisms can eventually lead to physiological changes, though he was wrong in assuming that physiological changes occurring during an organism's lifetime can be inherited directly by its offspring.[9]

Finally, some biologists, including Mayr, Gould, and Lewontin, consider themselves exponents of an expanded Darwinism but insist on *the autonomy of biology* from physics. They say that even the probabilistic physics of classical thermodynamics cannot serve as a model for evolutionary biology because chance and contingent historical contexts play such crucial roles. We can describe evolution through a unique historical narrative but we cannot deduce its path from predictive laws. These authors also defend the distinctiveness of biological concepts and their irreducibility to the concepts of physics and chemistry, as I will note later.[10]

1.3 Beyond Darwinism?

Darwin's theory shared many of the assumptions of Newtonian physics; the modern synthesis was influenced by the probabilistic reasoning of statistical mechanics. Future understanding of evolution may be enhanced by recent work on *chaos* and *complexity* in the physical sciences. Whereas the linear systems of classical thermodynamics are insensitive to small initial differences and attain predictable equilibrium states, nonlinear thermodynamic systems far from equilibrium are extremely sensitive to very small initial differences and are therefore unpredictable. Prigogine and others have described the emergence of new types of order in dissipative systems far from equilibrium. An infinitesimal difference in initial conditions will lead to alternative end-states and new levels of order described by system-wide relationships rather than by interactions at the molecular level.[11]

Stuart Kauffman draws from theories of complexity in arguing that evolution is the product of *self-organization* as well as chance and selection. He looks at the common properties of diverse systems, for example those in embryonic development, neural networks and computer networks. As we will see in the next section, he argues that dynamical systems can achieve new ordered states without any external selective pressures.[12] Jeffrey Wicken has insisted that we cannot understand evolutionary history without looking at the entropy, order, and flow of energy in the wider ecosystems within which organisms co-evolve. Moreover, he says, structural and thermodynamic constraints drastically limit the stable combinations when amino

---

[9] C. H. Waddington, *The Strategy of the Genes* (New York: Macmillan, 1957); Robert J. Richards, *Darwin and the Emergence of Evolutionary Theories of Mind and Behavior* (Chicago: University of Chicago Press, 1987), chap. 10.

[10] Ernst Mayr, *The Growth of Biological Thought* (Cambridge: Harvard University Press, 1982); idem, "How Biology Differs from the Physical Sciences," in *Evolution at the Crossroads*, Depew and Weber, eds.

[11] Ilya Prigogine and Irene Stengers, *Order Out of Chaos* (New York: Bantam Books, 1984).

[12] Stuart Kauffman, *The Origins of Order: Self-Organization and Selection in Evolution* (New York: Oxford University Press, 1993); idem, *At Home in the Universe: The Search for Laws of Self-Organization and Complexity* (New York: Oxford University Press, 1995).

acids are randomly assembled to form proteins. These authors adopt a holistic approach that attempts analysis at a variety of levels, avoiding the reductionism evident in much of evolutionary theory. They claim that natural selection works on a field of already self-organized systems.[13]

In the past, the phenomena of *embryology* and *developmental biology* have been poorly understood and have been difficult to incorporate into neo-Darwinism. How do cells differentiate so that the right organs are formed at the right place in the growing organism? Some biologists postulated a "morphogenic field" which imposes a pre-existing plan that guides cells in their differentiation. Others postulated "developmental pathways" which direct growth toward specific anatomical forms. These hypotheses appear increasingly dubious in the light of recent research on genetic and molecular mechanisms in embryological development. Regulatory genes produce proteins that act as "switches" to turn on secondary genes, which in turn control the tertiary genes responsible for protein assembly in cells, tissues, and organs. In recent experiments, the master control gene that initiates the program for the development of an eye in the fruit fly was introduced into cells on its wings, legs, and antennae, and complete eyes developed at these sites. If the control gene for eye development in a *mouse* is inserted in cells of a fly's wing, *a fly's eye* will develop, suggesting that the control genes for eyes in the two species are virtually unchanged since a common evolutionary ancestor, even though the eye structures of insects and mammals evolved in radically different directions.[14] Our understanding of such processes is still very limited, but research on the molecular basis of development holds great promise for broadening our understanding of evolutionary history. For example, the Cambrian explosion of new phyla may well have been caused by changes in the genetic networks that regulate very early development.

Even after recognizing the power of molecular explanations, however, one can argue that developmental patterns are constrained by principles of *hierarchical organization* and the possible forms of physiological structures. The variability of phenotypes is limited by the architecture and dynamics of developmental systems. Goodwin, Ho, and Saunders have defended a structuralism in which a relatively autonomous developmental dynamic is the main source of macroevolution.[15] Their ideas are controversial and outside the mainstream of current biological thought, but should not be dismissed if they might be able to account for observed phenomena more adequately than neo-Darwinist theory.

These authors see themselves as having moved beyond even an expanded Darwinism. If these ideas prove fruitful they may lead to what Kuhn would call a *paradigm shift*, in which the basic assumptions of Newtonian and nineteenth-century physics will be replaced by an alternative set of assumptions. Or perhaps we could say, in Lakatos' terms, that the core of Darwinism (the importance of variation and natural selection) will have been preserved by abandoning some of its *auxiliary*

---

[13] Jeffrey Wicken, *Evolution, Thermodynamics, and Information: Extending the Darwinian Program* (New York: Oxford University Press, 1987).

[14] George Halder, Patrick Callaerts, and Walter Gehring, "Induction of Ectopic Eyes by Targeted Expression of the *Eyeless* Gene in *Drosophila,*" *Science* 267 (1995), 1788–92.

[15] Mae-Won Ho and Peter Saunders, eds., *Beyond Neo-Darwinism* (New York: Harcourt Brace Jovanovich, 1984); Brian Goodwin and Peter Saunders, eds., *Theoretical Biology: Epigenetic and Evolutionary Order from Complex Systems* (Edinburgh: Edinburgh University Press, 1989); see also Robert Wesson, *Beyond Natural Selection* (Cambridge: MIT Press, 1991).

hypotheses (such as gradualism and the exclusive role of selection as a directive force). We could also follow the philosophers of science who hold that in studying complex phenomena we should seek *limited models* applicable to particular domains, rather than universally applicable predictive laws. Natural selection may be more important in some contexts than in others. As a minimum we can say that we should consider other factors in addition to variation and natural selection, and that we should look at what is going on at a variety of levels. In the discussion that follows, I will be drawing primarily from the advocates of "the expansion of Darwinism," but I will refer to the work of Kauffman, who considers himself "beyond Darwinism."

2 *Philosophical Issues in Biology*

Four concepts in recent biological thought require more careful analysis: self-organization, indeterminacy, top-down causality, and communication of information. Each of these concepts is crucial in one of the theological interpretations explored in the subsequent section.

2.1 *Self-organization*

Evolutionary history does indeed show a *directionality*, a trend toward greater complexity and consciousness. There has been an increase in the genetic information in DNA, and a steady advance in the ability of organisms to gather and process information about the environment and respond to it. The emergence of life, consciousness, and human culture are especially significant transitions within a gradual and continuous process. But evolution does not display any straight-line progressive development. For the majority of species, opportunistic adaptations led to dead ends and extinction when conditions changed. The pattern of evolution does not resemble a uniformly growing tree so much as a sprawling bush whose tangled branches grow in many directions and often die off. Nevertheless, there is an overall trend. Who can doubt that a human being represents an astonishing advance over an amoeba or a worm?

Some authors have argued that if the amino acids in primeval oceans had *assembled themselves by chance* to form protein chains, the probability of being assembled in the right order to form a particular protein would be fantastically small. It would be highly unlikely to occur even in spans of time many times longer than the history of the universe.[16] The argument is dubious because amino acids do not combine by chance with equal probability, for there are built-in affinities and bonding preferences and structural possibilities. Some combinations form stable units which persist, and these units combine to form larger units. Organic molecules have a capacity for self-organization and complexity because of structural constraints and potentialities.

Other authors have used *hierarchy theory* to indicate how advances to a higher level of organizational complexity are preserved. Imagine a watchmaker whose work is disrupted occasionally. If he has to start over again each time, he would never finish his task. But if he assembles groups of parts into stable sub-assemblies, which are then combined, he will finish the task more rapidly. Living organisms have many stable sub-assemblies at differing levels which are often preserved intact and only loosely coupled to each other. The higher level of stability often arises from functions

---

[16] Fred Hoyle and Chandra Wickramasinghe, *Evolution from Space* (London: Dent, 1981).

that are relatively independent of variations in the microscopic details. Evolution exhibits both *chance* and *directionality* because higher levels embody new types of order and stability that are maintained and passed on.[17]

Let us examine Kauffman's thesis that evolution is a product of *self-organization* as well as of random variation and natural selection. He finds similar patterns in the behavior of complex systems that appear very different—for example, in molecules, cells, neural networks, ecosystems, and technological and economic systems. In each case feedback mechanisms and nonlinear interactions make cooperative activity possible in larger wholes. The systems show similar emergent *systemic* properties not present in their components. Kauffman gives particular attention to the behavior of networks. For example, an array of 100,000 light bulbs, each of which goes on or off as an adjustable function of input from its four neighbors, will cycle through only 327 states from among the astronomical number of possible states. Genes are also connected in networks; in the simplest case, gene A represses gene B and vice versa, so only one of them is turned on. Kauffman notes that there are only 256 cell types in mammals, and suggests that this may be the result of system principles and not merely an historical accident.[18]

Many of Kauffman's ideas are speculative and exploratory, but they reflect a new way of looking at evolution. He finds that *order emerges spontaneously* in complex systems, especially on the border between order and chaos. Too much order makes change impossible; too much chaos makes continuity impossible. We should see ourselves not as a highly improbable historical accident, but as an expected fulfilment of the natural order. In his book, *At Home in the Universe,* Kauffman calls for awe and respect for a process in which such self-organization occurs.

2.2 Indeterminacy

Many features of evolutionary history are the product of *unpredictable events.* The particular pair of organisms that mate and the particular combination of genes that are inherited by their offspring cannot be predicted; genetic laws can only be expressed probabilistically for individuals in large populations. Many mutations and replication errors seem to occur at random. A few individuals may form a small isolated population which happens to differ genetically from the average of the larger population, leading to "genetic drift." Such unpredictability is compounded when co-evolving species interact competitively or cooperatively in historically contingent ecosystems and environments. An asteroid collision at the end of the Permian period may have drastically altered Earth's climate and its evolutionary history. We can only describe evolution by a historical narrative; we could not have predicted its course.

Many of these "chance events" seem to represent the unpredictable *intersection of separate causal chains.* Two causal chains may each be determinate, but if they are completely independent of each other, no lawful regularity describes their intersection in time and space. The idea of a causal chain is of course an abstraction. When we speak of "the cause" of an event we are selecting from among the many necessary and jointly sufficient conditions the one to which we want to direct attention in a particular context of inquiry. But our ignorance of the immensely complicated and ramifying web of causal influences in evolutionary history does not in itself imply that it is not determined.

---

[17] Stanley Salthe, *Evolving Hierarchical Systems* (New York: Columbia University Press, 1985).

[18] Kauffman, *At Home in the Universe,* chap. 4.

But an indeterminacy in nature itself seems to be present *at the quantum level.* In quantum theory, predictions of individual events among atoms and subatomic particles give only probabilities and not exact values. A particular radioactive atom might decay in the next second or a thousand years from now, and the theory does not tell us which will occur. Some physicists think that this unpredictability is attributable to the limitations of current theory; they hope that a future theory will disclose hidden variables that will allow exact calculations. But most physicists hold that indeterminacy is a property of the atomic world itself. Electrons and subatomic particles apparently do not have a precise location in space and time; they are spread-out waves representing a range of possibilities until they are observed.[19]

Among large groups of atoms in everyday objects, indeterminacy at the atomic level averages out statistically to give predictable large-scale behavior. However, in some biological systems, especially in the genetic and nervous systems, changes in a small number of atoms can have *large-scale effects.* A mutation could arise from a quantum event in which a single molecular bond in a gene is formed or broken, and the effects would be amplified in the phenotype of the growing organism, and might be perpetuated by natural selection. Such evolutionary unpredictability would reflect indeterminacy in nature and not merely the limitation of human knowledge.[20]

In *chaos theory* and *nonlinear thermodynamic systems* far from equilibrium, an infinitesimally small uncertainty concerning initial conditions can have enormous consequences. In chaotic systems, a very small change may be amplified exponentially. This has been called "the butterfly effect" because a butterfly in Brazil might alter the weather a month later in New York. The effect of moving an electron on a distant galaxy might be amplified over a long period of time to alter events on Earth.[21] Deterministic laws can be applied only to closed systems; they are an approximation to reality because actual systems that are extremely sensitive to initial conditions can never be totally isolated from outside influences.

According to Stephen Kellert, the *unpredictability of chaotic systems* is not merely a reflection of temporary human ignorance. Prediction over a long time period would require more information than could be stored on all the electrons of our galaxy, and the calculations would take longer than the phenomena we were trying to predict. Moreover chaotic systems would amplify the quantum indeterminacies that set limits to the accurate specification of initial conditions in both theory and practice. Kellert also notes that in classical physics the behavior of a larger whole is deduced from predictive causal laws governing interactions of its constituent parts. Chaos theory, by contrast, studies the qualitative form of *large-scale patterns* that may be similar even when the constituents are very different. Chaos theory examines holistic geometrical relationships and systemic properties rather than seeking microreduction to detailed causal mechanisms. Order is a

---

[19] Ian G. Barbour, *Religion in an Age of Science* (San Francisco: Harper Collins, 1990), 96–104.

[20] On the topic of quantum indeterminacy and its possible role in mutations, see Ellis, Murphy, Tracy, and Russell in this volume and in *Chaos and Complexity: Scientific Perspectives on Divine Action,* Robert John Russell, Nancey Murphy, and Arthur R. Peacocke, eds. (Rome: Vatican Observatory, and Berkeley, CA: Center for Theology and the Natural Sciences, 1995).

[21] James Gleick, *Chaos: Making a New Science* (New York: Penguin Books, 1987); John Holte, ed., *Chaos: The New Science* (Lanham, MD: University Press of America, 1993).

broader concept than law because it includes formal, holistic, historical, and probabilistic patterns.[22]

2.3 Top-down Causality

Living organisms exhibit a many-leveled hierarchy of systems and sub-systems. A *level* identifies a unit which is relatively integrated, stable, and self-regulating, even though it interacts with other units at the same level and at higher and lower levels. One such hierarchy is identified structurally: particle, atom, molecule, macromolecule, organelle, cell, organ, organism, and ecosystem. Other hierarchies are identified functionally: the reproductive hierarchy (gene, genome, organism, and population), or the neural hierarchy (molecule, synapse, neuron, neural network, and the brain with its changing patterns of interconnections). Human beings also participate in all the social and cultural interactions studied by the social sciences and humanities. A particular discipline or field of inquiry focuses attention on a particular level and its relation to adjacent levels.

We can distinguish three kinds of *reduction between levels*.

1. *Methodological reduction* is a research strategy that studies lower levels in order to better understand relationships at higher levels. Analysis of molecular interactions has been a spectacularly successful strategy in biology, but it is not incompatible with multi-level analysis and the study of larger systems.

2. *Epistemological reduction* claims that laws and theories at one level of analysis can be derived from laws and theories at lower levels. I have argued that biological concepts are distinctive and cannot be defined in physical and chemical terms. Distinctive kinds of explanation are valid at differing levels. But inter-level theories may connect adjacent levels, even if they are not derivable from the theories applicable to either level alone. A series of overlapping theories and models unifies the sciences without implying that one level is more fundamental or real than another.[23]

3. *Ontological reduction* is a claim about the kinds of reality or the kinds of causality that exist in the world. It is sometimes asserted that an organism is "nothing but organized molecules," or that "only physical forces are causally effective." I have defended ontological pluralism, a multi-leveled view of reality in which differing (epistemological) levels of analysis are taken to refer to differing (ontological) levels of events and processes in the world, as claimed by critical realism. In evolutionary history, novel forms of order emerged which not only could not have been predicted from laws and theories governing previously existing forms, but which also gave rise to genuinely new kinds of behavior and activity in nature. We can acknowledge the distinctive characteristics of living organisms without assuming that life is a separate substance or a "vital force" added to matter, as the vitalists postulated.

Bottom-up causation occurs when many sub-systems influence a system. *Top-down causation* is the influence of a system on many sub-systems. Higher-level events influence chemical and physical processes at lower levels without violating

---

[22] Stephen Kellert, *In the Wake of Chaos: Unpredictable Order in Dynamical Systems* (Chicago: University of Chicago Press, 1993).

[23] For analyses of reduction, see Ian G. Barbour, *Issues in Science and Religion* (Englewood Cliffs, NJ: Prentice-Hall, 1966), 324–37 and *Religion in an Age of Science*, 165–69; Francisco Ayala, "Reduction in Biology" in *Evolution at the Crossroads*, Depew and Weber, eds.; Arthur Peacocke, *God and the New Biology* (London: J. M. Dent & Sons, 1986), chaps. 1 and 2.

lower-level laws.[24] Microproperties are not referred to in the specification of the macrostate of the system. Network properties may be realized through a great variety of particular connections. Correlation of behaviors at one level does not require detailed knowledge of all its components. The rules of chess limit the possible moves but leave open an immense number of possibilities that are consistent with but not determined by those rules. So, too, the laws of chemistry limit the combinations of molecules which are found in DNA, but do not determine them. The meaning of the message conveyed by DNA is not given by the laws of chemistry.

The *holistic* and *anti-reductionistic* character of chaos theory has been described by one of its best-known exponents, James Gleick:

> Chaos is anti-reductionist. This new science makes a strong claim about the world, namely, that when it comes to the most interesting questions, questions about order and disorder, decay and creativity, pattern formation, and life itself, the whole cannot be explained in terms of the parts. There are fundamental laws about complex systems, but they are new kinds of law. They are laws of structure and organization and scale, and they simply vanish when you focus on the individual constituents of a complex system—just as the psychology of a lynch mob vanishes when you interview individual participants.[25]

We know little about how memories are preserved in the brain, but computer simulations of neural nets suggest that memory may be stored in *distributed patterns* rather than at discrete locations. In some computer networks with parallel distributed processing, the nodes in a series of layers can be connected by links whose strength can be varied. In one experiment, the inputs are groups of letters, and the outputs are random sounds in a voice synthesizer. Every time the correlation between an input and the correct output is improved, the strongest links are strengthened, so the network gradually improves its performance. The network can be taught to pronounce written words. The connective patterns involve the whole network and they are learned by experience rather than by being directly programed. Patterns develop in the whole without prior specification of the parts; the readjustment of the parts can be considered a form of top-down causation.[26] We should also note that the brain of a baby is not finished or "hard-wired" at birth. The neural pathways are developed in interaction with the environment and are altered by the baby's experiences.

Of all the sciences, ecology is the most *holistic* in its outlook. No part of an ecosystem can be considered in isolation because changes in one component often have far-reaching ramifications elsewhere in the system. The participants in an ecosystem are linked by multiple connections and cycles. The oxygen inhaled by animals is exhaled as carbon dioxide which is in turn taken in by plants and converted back to oxygen. The food chain connects various life forms. Predator and

---

[24] On top-down causation, see Donald Campbell, "'Downward Causation' in Hierarchically Ordered Biological Systems" in *The Problems of Reduction*, Francisco Ayala and Theodosius Dobzhansky, eds. (Berkeley: University of California Press, 1974); Michael Polanyi, "Life's Irreducible Structures," *Science* 160 (1968), 1308–12; Elizabeth Vrba, "Patterns in the Fossil Record and Evolutionary Processes" in *Beyond Neo-Darwinism*, Ho and Saunders, eds.

[25] James Gleick, address at 1990 Nobel Conference, Gustavus Adolphus College, quoted in Steven Weinberg, *Dreams of a Final Theory* (New York, N.Y.: Pantheon Book, 1992), 60.

[26] C. Rosenberg and T. Sejnowski, "Parallel Networks That Learn to Pronounce English Text," *Complex Systems* 1 (1987), 145–68.

prey are dependent on each other in maintaining stable populations. A holistic approach is also used in the field of systems analysis which studies the dynamics of urban, industrial, and electronic systems. In all these cases, there are of course lawful relations among the parts, but their behavior is analyzed in relation to a larger whole.

Holism is both a rejection of ontological reductionism and a claim that the whole influences the parts. Attention is directed to the parts of a particular whole, even though it is in turn a part of a larger whole. The whole/part distinction is usually structural and spatial (for example, a *larger* whole). *Top-down causality* is a very similar concept, but it draws attention to a hierarchy of many levels characterized by qualitative differences in organization and activity (for example, a *higher* level). Levels are defined by functional and dynamic relationships. Patterns in time are emphasized, though of course they are inseparable from patterns in space.

2.4 The Communication of Information

Information has been an important term in many fields of science. In the thermody-namics of gases, systems of *low entropy* are highly improbable molecular configura-tions, which tend to degrade into the more probable configurations of uniform equilibrium states. This entails a loss of order and pattern that is also a loss of information. *Information theory* was first developed in World War II in studies of the communication of messages by radio. Communication is more reliable if the signal-to-noise ratio is high and if a coded message contains regularities and redundancies which allow the detection of errors. With the advent of computers, instructions could be encoded in a binary representation (0/1 or off/on) and quantified as *"bits" of information.* The computer responds to the instructions in the program which specify the connections in its electrical circuits. It manipulates the electrical representations of the symbols fed into it ("information processing") and then activates some form of output. The letters on a printed page are of course the classical case of the communication of information to a reader.[27]

Information is *an ordered pattern* (of alphabetical letters, auditory sounds, binary digits, DNA bases, or any other combinable elements) which is one among many possible sequences or states of a system. Information is *communicated* when another system (reader, listener, computer, living cell, etc.) responds selectively— that is, when information is coded, transmitted, and decoded. The meaning of the message is dependent on a wider *context of interpretation.* It must be viewed dynamically and relationally rather than in purely static terms as if the message were contained in the pattern itself.

The information in DNA sequences in genes is significant precisely because of its context in a larger organic system. In the growth of an embryo, a system of time delays, spatial differentiation, and chemical feed-back signals communicates the information needed so that the right proteins, cells, and organs are assembled at the right location and time. Complicated developmental pathways, with information flowing in both directions, connect genes with molecular activities and physiological structures. A genome contains an immense number of possible developmental scenarios, of which only a few are realized. In *The Ontogeny of Information,* Susan Oyama argues that the meaning and informational significance of genetic instructions depend on what cells and tissues are already present, and on the actual functioning

---

[27] Jeremy Campbell, *Grammatical Man: Information, Entropy, Language, and Life* (New York: Simon and Schuster, 1982).

of the developmental system. In place of a one-way flow of information we must imagine interactive construction in a particular context.[28]

An enzyme speeds the interaction of two molecules by recognizing them (by shape and chemical affinity) and holding them at adjacent sites where they can react with each other. Molecules of the immune system recognize an invading virus, which is like a key that fits a lock, and they are activated to release a specific antibody. The communication between molecules is dependent on properties of both the sender and the receiver. A receptor is part of an embodied action system that implements a response to signals.

Stored in the DNA is a wealth of *historically acquired information* including programs for coping with the world. For example, a bird or animal uses specific visual or auditory clues to recognize and respond to a dangerous predator which it has not previously encountered. Individuals in some species are programed to communicate warning signals to alert other members of the species. Higher primates are capable of symbolic communication of information, and human beings can use words to express abstract concepts. Human information can be transmitted between generations not only by genes and by parental example, but also in speech, literature, art, music and other cultural forms. The storage and communication of information is thus an important feature of biological processes at many levels and it must always be understood dynamically and relationally rather than in purely static and formal terms. Even at low levels, reality consists not simply of matter and energy, but of matter, energy, and information.

3 Models of God's Action in Nature

What models of God's relation to nature are compatible with the central affirmations of the Christian tradition and also with a world which is characterized by self-organization, indeterminacy, top-down causality, and communication of information? I will examine theological proposals that draw from each of these four characteristics.

All four models reject the idea of *divine intervention* that violates the laws of nature. In none of them is God invoked to fill particular gaps in the scientific account (the "God of the gaps" who is vulnerable to the advance of science). God's role is different from that of natural causes. In each case, a feature of current scientific theory is taken as a model (that is, a systematically developed analogy) of God's action in nature.[29] Some authors in the first group below do propose a new version of *natural theology* in which evidence from science is used as an argument in support of theism, even if it does not offer a proof of God's existence. The other authors are proposing ways in which a God who is accepted on other grounds (such as religious experience in a historical interpretive community) might be reconceived as acting in nature. I have called such an approach *a theology of nature* rather than a natural theology.[30]

---

[28] Susan Oyama, *The Ontogeny of Information: Developmental Systems and Evolution* (Cambridge: Cambridge University Press, 1985).

[29] Barbour, *Myths, Models, and Paradigms* (New York: Harper and Row, 1974).

[30] Barbour, *Religion and Science: Historical and Contemporary Issues* (San Francisco: Harper Collins, 1997), chap. 4.

3.1 *God as Designer of a Self-organizing Process*

Until the nineteenth century, the intricate organization and effective functioning of living creatures were taken as evidence of an intelligent designer. After Darwin, the argument was reformulated: God did not create things in their present forms, but designed an evolutionary process through which all living forms came into being. Today we know that life is possible only under a very narrow range of physical and chemical conditions. We have seen also that in the self-organization of molecules leading to life there seems to have been considerable built-in design in biochemical affinities, molecular structures, and potential for complexity and hierarchical order. The world of molecules seems to have an inherent tendency to move toward emergent complexity, life, and consciousness.

If design is understood as a detailed pre-existing plan in the mind of God, *chance* is the antithesis of *design*. But if design is identified with the general direction of growth toward complexity, life, and consciousness, then both law and chance can be part of the design. Disorder is sometimes a condition for the emergence of new forms of order, as in thermodynamic systems far from equilibrium, or in the mutations of evolutionary history. We can no longer accept the clockmaker God who designed every detail of a determinate mechanism. But one option today is a revised deism in which God designed the world as *a many-leveled creative process of law and chance*. Paul Davies is an exponent of this position in this volume and elsewhere.[31]

A patient God could endow matter with diverse potentialities and let the world create itself. We can say that God respects the integrity of the world and allows it to be itself, without interfering with it, just as God respects human freedom and allows us to be ourselves. Moral responsibility requires that the world have some openness, which takes the form of chance at lower levels and choice at the human level. But responsible choice also requires enough lawfulness that we have some idea of the probable consequences of our decisions.

An attractive feature of this option is that it provides at least partial answers to the problems of *suffering and death* which were such a challenge to the classical argument from design. For competition and death are intrinsic to an evolutionary process. Pain is an inescapable concomitant of greater sensitivity and awareness, and it provides a valuable warning of external dangers. My main objection to a reformulated deism is that we are left with a distant and inactive God, a far cry from the active God of the Bible who continues to be intimately involved with the world and human life.

One could still argue that God has an ongoing role in *sustaining* the world and its laws. Some theologians maintain that the world does not stand on its own but needs God's continual concurrence to maintain and uphold it in what is known today to be a dynamic rather than a static process. According to neo-Thomists, God as *primary cause* works through the matrix of *secondary causes* in the natural world. William Stoeger argues that there are no gaps in the scientific account on its own level; God's action is on a totally different plane from all secondary causes.[32] Many

---

[31] Paul Davies, *The Cosmic Blueprint: New Discoveries in Nature's Creative Ability to Order the Universe* (New York: Simon & Schuster, 1988); idem, *The Mind of God: The Scientific Basis for a Rational World* (New York: Simon & Schuster, 1992); idem, in this volume.

[32] Austin Farrer, *Faith and Speculation* (London: Adam & Charles Black, 1967), chaps. 4 and 10; William R. Stoeger, "Describing God's Action in the World in the Light of Scientific

neo-Thomists maintain that divine sovereignty is maintained if all events are foreseen and predetermined in God's plan. God does not have to intervene or interfere with the laws of nature; divine action occurs indirectly and instrumentally through natural processes. This view respects the integrity of science and the transcendence of God, whose action is not like causality within the world. Some theologians hold that God sees all events in timeless eternity without determining them, but I would argue that predestination is not compatible with human freedom or the presence of chance, evil, and suffering in the world.

3.2 God As Determiner Of Indeterminacies

I suggested earlier that uncertainties in the predictions made by quantum theory reflect indeterminacy in nature itself, rather than the inadequacy of current theory. In that interpretation, *a range of possibilities* is present in the world. Quantum events have necessary but not sufficient physical causes. If they are not completely determined by the relationships described by the laws of physics, their final determination might be made directly by God. What appears to be chance, which atheists take as an argument against theism, may be the very point at which God acts.

Divine sovereignty would be maintained if God *providentially controls* the events that appear to us as chance. No energy input would be needed, since the alternative potentialities in a quantum state have identical energy. God does not have to intervene as a physical force pushing electrons around, but instead actualizes one of the many potentialities already present—determining, for example, the instant at which a particular radioactive atom decays.[33]

We have seen that under some conditions the effects of very small differences at the microlevel are greatly amplified in *large-scale phenomena*. In nonlinear thermodynamics and chaos theory, an infinitesimal initial change can produce dramatic changes in the larger system. Similar trigger effects occur in evolutionary mutations and in genetic and neural systems today. Scientific research finds only law and chance, but perhaps in God's knowledge all events are foreseen and predetermined through a combination of law and particular divine action. Since God's action would be scientifically undetectable, it could be neither proved nor refuted by science. This would exclude any proof of God's action of the kind sought in natural theology, but it would not exclude the possibility of God's action affirmed on other grounds in a wider theology of nature.

If we assume that *God controls all indeterminacies,* we could preserve the traditional idea of predestination. This would be theological determinism rather than physical determinism, since nothing happens by chance. But then the problems of waste, suffering, and human freedom would remain acute. Nancey Murphy has proposed that God determines all quantum indeterminacies but arranges that law-like regularities usually result, in order to make stable structures and scientific investigation possible, and to ensure that human actions have dependable consequences so that moral choices are possible. Orderly relationships do not constrain God, since they are included in God's purposes. God grants causal powers to created entities. Murphy holds that in human life God acts both at the quantum level and at

---

Knowledge of Reality," in *Chaos and Complexity*, Russell, Murphy, and Peacocke, eds.

[33] William Pollard, *Chance and Providence* (New York: Charles Scribner's Sons, 1958); Donald MacKay, *Science, Chance, and Providence* (Oxford: Oxford University Press, 1978).

higher levels of mental activity, but does it in such a way that human freedom is not violated.[34]

An alternative would be to say that most quantum events occur by chance, but *God influences some of them* without violating the statistical laws of quantum physics. This view has been explained by Robert Russell, George Ellis and Thomas Tracy, and it is consistent with the scientific evidence.[35] A possible objection to this model is that it assumes *bottom-up causality* within nature once God's action has occurred, and thus seems to concede the reductionist's claim that the behavior of all entities is determined by their smallest parts (or lowest levels). The action would be bottom-up even if one assumed that God's intentions were directed to the larger wholes (or higher levels) affected by these quantum events. However most of these authors also allow for God's action at higher levels which then results in a *top-down influence* on lower levels, in addition to quantum effects from the bottom up. The model can thus be combined with one of the models discussed below.

3.3 *God as Top-down Cause*

The idea of levels of reality can be extended if God is viewed as acting from an even higher level than nature. Arthur Peacocke holds that God exerts a *top-down causality* on the world. God's action would be a constraint on relationships at lower levels that does not violate lower-level laws. Constraints may be introduced not just at spatial or temporal boundaries, but also internally through any additional specification allowed by lower-level laws. In human beings, God would influence their highest evolutionary level, that of mental activity, which would affect the neural networks and neurons in the brain.[36] Within human beings, divine action would be effected down the hierarchy of natural levels, concerning which we have at least some understanding of relationships between adjacent levels. (Peacocke gives a table showing the hierarchy of academic disciplines, from the physical sciences to the humanities, which study successively higher levels, with some disciplines addressing inter-level questions.[37]) His use of top-down causality seems to me more problematic in the case of divine action on inanimate matter; we would have to assume direct influence between the highest level (God) and the lowest level (matter) in the absence of intermediate levels—which has no analogy within the natural order.

Peacocke also extends to God the idea of *whole-part* relationships found in nature. He proposes that God as "the most inclusive whole" acts on "the-world-as-a-whole." But this spatial analogy seems dubious because the world does not have spatial boundaries, and it has no temporal ones if we accept Stephen Hawking's version of quantum cosmology. Moreover the rejection of universal simultaneity in relativity theory makes it impossible to speak of "the-world-as-a-whole" at any one

---

[34] Nancey Murphy, "Divine Action in the Natural Order: Buridan's Ass and Schrödinger's Cat," in *Chaos and Complexity*, Russell, Murphy, and Peacocke, eds; Nancey Murphy and George F. R. Ellis, *On the Moral Nature of the Universe: Theology, Cosmology and Ethics* (Minneapolis: Fortress Press, 1996).

[35] Thomas F. Tracy, "Particular Providence and the God of the Gaps," and George F. R. Ellis, "Ordinary and Extraordinary Divine Action: The Nexus of Interaction," in *Chaos and Complexity*, Russell, Murphy, and Peacocke, eds.; R. J. Russell, in this volume.

[36] Arthur Peacocke, *Theology for a Scientific Age: Being and Becoming—Natural, Human, and Divine*, enlarged edition (Minneapolis: Fortress Press, 1993), chap. 3, and his "God's Interaction with the World" in *Chaos and Complexity*, Russell, Murphy, and Peacocke, eds.; idem, in this volume.

[37] Peacocke, *Theology for a Scientific Age*, 217.

moment. The whole is a spatio-temporal continuum with temporal as well as spatial dimensions. In such a framework God's action would presumably have to be more localized in space and time, interacting more directly with a particular part rather than indirectly through action on the spatio-temporal whole.

One version of top-down causality uses the relation of mind to body in human beings as an analogy for God's relation to the world. Some authors urge us to look on *the world as God's body*, and God as the world's mind or soul. In using the analogy, we can make allowance for the human limitations that would not apply to God. We have direct awareness of our thoughts and feelings, but only limited awareness of many other events in our bodies, whereas God would be directly aware of all events. We did not choose our bodies and we can affect only a limited range of events in them, whereas God's actions are said to affect all events universally. From the pattern of behavior of other people we infer their intentions which cannot be directly observed; similarly, the cosmic drama can be interpreted as the expression of God's intentions.[38]

But the analogy breaks down if it is pressed too far. The cosmos as a whole lacks the intermediate levels of organization found in the body. It does not have the biochemical or neurological channels of feedback and communication through which the activities of organisms are coordinated and integrated. To be sure, an omnipresent God would not need the cosmic equivalent of a nervous system. God is presumably not as dependent on particular bodily structures as we are. However, we would be abandoning the analogy if we said that God is a disembodied mind acting directly on the separate physical components of the world. It appears that we need a more pluralistic analogy allowing for interaction among a community of beings, rather than a monistic analogy that pictures us all as parts of one being. The world and God seem more like a community with a dominant member than like a single organism.

3.4 *God as Communicator of Information*

In radio transmissions, computers, and biological systems, the *communication of information* between two points requires a physical input and an expenditure of energy (the Brillouin-Szilard relationship). But if God is omnipresent (including presence everywhere at the microlevel), no energy would be required for the communication of information. Moreover, the realization of alternative potentialities already present in the quantum world would convey differing information without any physical input or expenditure of energy.

Arthur Peacocke has used a rich variety of analogies in addition to top-down causality. Some of these involve the communication of information. God is like the choreographer of a dance in which much of the action is left up to the dancers, or the composer of a still unfinished symphony, experimenting, improvising, and expanding on a theme and variations.[39] Peacocke suggests that the purposes of God are *communicated through the pattern of events* in the world. We can look on evolutionary history as the action of an agent who expresses intentions but does not follow an exact predetermined plan. Moreover, an input of information from God could influence the relationships among our memories, images and concepts, just as our

---

[38] Grace Jentzen, *God's World, God's Body* (Philadelphia: Westminster Press, 1984); Sallie McFague, *The Body of God: An Ecological Theology* (Minneapolis: Fortress Press, 1993).

[39] Peacocke, *Creation and the World of Science* (Oxford: Clarendon Press, 1979), chap. 3, and *Theology for a Scientific Age*, chap. 9.

thoughts influence the activity of neurons. Peacocke maintains that Christ was a powerfully God-informed person who was a uniquely effective vehicle for God's self-expression, so that in Christ God's purposes are more clearly revealed than in nature or elsewhere in history.[40]

John Polkinghorne proposes that God's action is *an input of "pure information."* We have seen that in chaos theory an infinitesimally small energy input produces a very large change in the system. Polkinghorne suggests that in imagining God's action we might extrapolate chaos theory to the limiting case of zero energy. (This differs from quantum theory in which there actually is *zero* energy difference between alternative potentialities, so no extrapolation is needed). Polkinghorne holds that God's action is a nonenergetic input of information which expresses holistic patterns. God's selection among the envelope of possibilities present in chaotic processes could bring about novel structures and types of order exemplifying systemic higher-level organizing principles.[41]

The biblical idea of divine Word or *Logos* resembles the concept of information. In Greek thought, the *Logos* was a universal rational principle, but biblical usage also expressed the Hebrew understanding of Word as creative power. The Word in both creation and redemption can indeed be thought of as the communication of information from God to the world. As in the case of genetic information and human language, the meaning of the message must be discerned within a wider context of interpretation. God's Word to human beings preserves their freedom because it evokes but does not compel their response.[42] But the divine *Logos* is not simply the communication of an impersonal message since it is inseparable from an ongoing personal relationship. The *Logos* is not a structure of abstract ideas like Plato's eternal forms, or like a computer program that exists independently of its embodiment in a particular medium or hardware system. If we believe that one of God's purposes was to create loving and responsible persons, not simply intelligent information processors, we will have to draw our analogies concerning the communication of information primarily from human life, rather than from the genetic code or computer programs.

4 God's Action in Process Theology

Process theology shows similarities with each of the four models above, but also differs because it adds a fifth idea, that of interiority. Christian process theology combines biblical thought with process philosophy, the attempt of Alfred North Whitehead and his followers to develop a coherent set of philosophical categories general enough to be applicable to all entities in the world. Process theology is advocated by Charles Birch and John Haught in the present volume.[43]

---

[40] Peacocke, *Theology for a Scientific Age*, chap. 9.

[41] John Polkinghorne, *Reason and Reality* (Philadelphia: Trinity International Press, 1991), Chap. 3; idem, "The Metaphysics of Divine Action" in *Chaos and Complexity*, Russell, Murphy, and Peacocke, eds; idem, *The Faith of a Physicist* (Princeton, NJ: Princeton University Press, 1994), pp. 77–78.

[42] John Puddefoot, "Information Theory, Biology, and Christology," in *Religion and Science: History, Method, Dialogue,* W. Mark Richardson and Wesley J. Wildman, eds. (New York: Routledge, 1996).

[43] Charles Birch and John Haught, in this volume.

4.1 Biology and Process Philosophy

Many features of contemporary science are strongly represented in process philosophy. Whitehead was indebted to quantum physics for his portrayal of the discrete, episodic, and indeterminate character of all events. He was indebted to relativity for his view that all entities are constituted by their relationships. Process thought is evolutionary in stressing temporality and change. Becoming and activity are considered more fundamental than being and substance. The continuity of evolutionary history implies the impossibility of drawing absolute lines between successive life forms historically, or between levels of reality today.[44] Each of the four themes outlined earlier can be found in process philosophy:

1. *Self-organization* is a characteristic of the basic units of reality, which are momentarily unified events (Whitehead called them "actual occasions," but I will refer to them simply as "events," which reminds us of their temporal character). No event is merely a passive product of its past. All events are also products of present creative activity in which organization is realized—that is, pattern and structure which are temporal as well as spatial. But self-organization is analyzed by process thought in a distinctive way. Interiority is postulated in every event, providing a unifying center for the organizing activity.

2. *Indeterminacy* is assumed by process thought not only in the quantum world but at all levels of integrated activity. Both order and openness are present at all levels. At lower levels, order predominates, while at higher levels there is more opportunity for spontaneity, creativity, and novelty.

3. *Top-down causality* is defended in process writings. Process thought is *holistic* in portraying a network of interconnected events. Every event is a new synthesis of the influences on it; it occurs in a context which affects it and which it in turn affects. This can be called a relational or ecological view of reality. Not even God is self-contained, for God's experience is affected by the world. More specifically, reality is taken to be *multi-leveled*. Events at high levels of complexity are dependent on events at lower levels. But genuinely new phenomena emerge at higher levels which cannot be explained by the laws describing lower-level phenomena. Charles Hartshorne's version of process philosophy makes extensive use of the concept of hierarchical levels with differing characteristics, and he gives a careful critique of reductionism.[45]

4. *The communication of information* is not prominent in early process writings, which is not surprising since its scientific importance was not recognized prior to World War II. However the idea that a concrescing event takes other events into account resembles the contextual and relational character of information in action. James Huchingson notes that information always involves selection from among possible states; he proposes that Whitehead's "actual occasions" are information-processing entities that select from among the possibilities provided by God and previous events. Moreover information from the world feeds back to God; this feedback leads to relevant readjustment, as in cybernetic systems. Huchingson finds holism and top-down causality in the role of information in both process thought and systems theory. A system works as a whole to restrict the ability of its

---

[44] Alfred North Whitehead, *Science and the Modern World* (New York: Macmillan, 1925); *Process and Reality* (New York: Macmillan,1929). See Barbour, *Religion in an Age of Science*, chap. 8, or *Religion and Science: Historical and Contemporary Issues*, chap. 11.

[45] Charles Hartshorne, *Reality as Social Process* (Glencoe, IL: Free Press, 1953).

components to realize all possible states. New forms of order are generated at higher levels of organization, according to both process and systems thinking.[46]

4.2 Interiority

Interiority is the most controversial theme in process thought. Reality is construed as a network of interconnected events which are also *moments of experience,* each integrating in its own way the influences from its past and from other entities. The evolution of interiority, like the evolution of physical structures, is said to be characterized by both continuity and change. The forms taken by interiority vary widely, from rudimentary memory, sentience, responsiveness and anticipation in simpler organisms, to consciousness and self-consciousness in more complex ones. Human life is the only point at which we know reality from within. If we start from the presence of both physical structures and experience in human life, we can imagine simpler and simpler structures in which experience is more and more rudimentary. But if we start with simple physical structures totally devoid of interiority, it is difficult to see how the complexification of external structures can result in interiority.[47]

The approach and avoidance reactions of bacteria can be considered elementary forms of perception and response. An amoeba learns to find sugar, indicating a rudimentary memory and intentionality. Invertebrates seem to have some sentience and capacity for pain and pleasure. Purposiveness and anticipation are clearly present among lower vertebrates, and the presence of a nervous system greatly enhances these capacities. The behavior of animals gives evidence that they suffer intensely, and even invertebrates under stress release endorphins and other pain-suppressant chemicals similar to those in human brains. Some species exhibit considerable problem-solving and anticipatory abilities and a range of awareness and feelings. Conceptualizing interiority requires that we try to look on an organism's activities from its own point of view, even though its experience must be very different from our own.[48]

We noted earlier that evolutionary change can be *initiated by the activity of organisms* in selecting their own environments (the Baldwin effect). Their diverse responses and novel actions may create new evolutionary possibilities. Among the creatures who were the common ancestors of bison and horses, some charged their enemies head on, and their survival would have been enhanced by strength, weight, strong skulls and other bison-like qualities. Others in the same population fled from their enemies, and their survival depended on speed, agility, and other abilities we see in horses. The divergence of bison and horse may have arisen initially from different responses to danger, rather than from genetic mutations related to anatomy. Emotions and mental responses are not uniquely determined by the genes, though they occur in nervous systems which are the product of an inherited set of genes.

---

[46] James Huchingson, "Organization and Process: Systems Philosophy and Whiteheadian Metaphysics," *Zygon* 11.4 (1981): 226–41.

[47] Charles Birch, *A Purpose for Everything* (Mystic, CT: Twenty-Third Publications, 1990); Birch and Cobb, *The Liberation of Life: From the Cell to the Community* (Cambridge: Cambridge University Press, 1981).

[48] Donald Griffin, *The Question of Animal Awareness: Evolutionary Continuity of Mental Experience* (Los Altos, CA: William Kaufmann, 1981); Charles Birch and John B. Cobb, Jr., *The Liberation of Life.*

Organisms participate actively in evolutionary history and are not simply passive products of genetic forces from within and environmental forces from without.[49]

In the study of human beings, psychology was once dominated by behaviorists who correlated observable stimuli and responses and claimed that mental life is inaccessible to science. But the more recent *cognitive psychologists* talk about perception, attention, memory, intention, mental representation and consciousness. These issues are highly disputed today, but some authors have been trying to relate data from three sources: phenomenological self-description, neurological research on the brain, and computer simulations of neural nets.[50] Others insist that *subjectivity*, which always involves a particular perspective or point of view, cannot be represented in the objective framework of science.[51]

We are each aware of our experience despite the difficulty of studying it scientifically. It is this direct awareness that leads us to attribute subjectivity to other humans, animals, and even to lower forms of life. While the terms consciousness and mind should be restricted to organisms with a nervous system, it is reasonable to attribute rudimentary forms of perception and experience to organisms as simple as the amoeba. I would argue that in the light of evolutionary continuity and in the interest of metaphysical generality we should take *experience* as a category applicable to all integrated entities, even if consciousness appears only in higher life forms.

4.3 Christianity and Process Theology

In process thought God is the source of order and also the source of novelty. God presents new possibilities to the world but leaves alternatives open, eliciting the response of entities in the world. God is present in the unfolding of every event but never exclusively determines the outcome. This is a God of persuasion rather than coercion. For process theologians, God is not as an omnipotent ruler but the leader and inspirer of an interdependent community of beings. John Cobb and David Griffin speak of God as "creative-responsive love" which affects the world but is also affected by it. God's relation to human beings is used as a model for God's relation to all beings.[52]

Process theologians stress God's *immanence* and participation in the world, but they do not give up *transcendence*. God is said to be temporal in being affected by interaction with the world, but eternal and unchanging in character and purpose. Classical ideas of omnipresence and omniscience are retained, but not even God can know a future which is still open. Compared to the traditional Western model, God's power over events in the world is severely limited, especially at lower levels where events are almost exclusively determined by their past. The long span of cosmic history suggests a patient and subtle God working through the slow emergence of novel forms. Christian process theologians hold that the life and death of Christ are the supreme examples of the power of God's love and participation in the life of the world. The Cross is a revelation of suffering love, and the resurrection reveals that even death does not end that love.

---

[49] C. H. Waddington in *Mind in Nature*, John B. Cobb, Jr. and David Griffin, eds., (Washington DC: University Press of America, 1977).

[50] Owen Flanagan, *Consciousness Reconsidered* (Cambridge: MIT Press, 1992).

[51] Thomas Nagel, *The View from Nowhere* (New York: Oxford University Press, 1986); Colin McGinn, *The Problem of Consciousness* (Oxford: Blackwell, 1991).

[52] John B. Cobb, Jr. and David Ray Griffin, *Process Theology: An Introduction* (Philadelphia: Westminster Press, 1976).

Process thought shares insights with each of the theological models described earlier, but it differs at crucial points.

1. Like God *the designer of a self-organizing process,* the God of process thought is the source of order in the world. But the process God is also directly involved in the emergence of novelty through the interiority of each unified event. Deism is avoided because God has a direct and continuing role in the history of the world.

2. Like those who say that God *determines quantum indeterminacies,* process thinkers hold that God influences systems that are not fully determined by past events. It is never an absolute determination, for God always works along with other causes. In process thought God's activity occurs at higher levels of organization in addition to the quantum level. This avoids a reliance on quantum events alone which would perpetuate the reductionist's assumption that only bottom-up causality operates within natural systems..

3. Like those who postulate God as *top-down cause,* process thinkers stress God's immanence and participation in an interdependent many-leveled world. But process thought has no difficulty conceptualizing the interaction between the highest level (God) and the lowest (inanimate matter) in the absence of intermediate levels, because God is present in the unfolding of integrated events at all levels. Hartshorne has indeed used the analogy of *the world as God's body,* though we must remember that in the process scheme the body is itself a community of integrated entities at various levels. Most process theologians, however, insist on a greater divine transcendence and greater human freedom than the analogy of a cosmic body suggests. Using a social rather than organic analogy they imagine us, not as cells in God's body, but as members of a cosmic community of which God is the preeminent member.

4. The idea that God *communicates information* to the world is consistent with process thought. God's ordering and valuation of potentialities is a form of information within a larger context of meaning. God also receives information from the world, and God is changed by such feedback. The communication of information occurs within the momentary experience of integrated events at any level, rather than by bottom-up causality through quantum phenomena alone, or through the trigger points of chaos theory, or by top-down causality acting on the whole cosmos. God, past events, and the event's present response join in the formation of every event. Process thought uses a single conceptual representation for divine action at all levels, whereas some of the authors mentioned earlier assume very different modes of divine action at various levels in the world. At the same time, process thought tries to allow for differences in the character of events that occur at diverse levels.

The idea of God's *self-limitation* or *kenosis* in recent theology is in many ways similar to the assertions of process theology. Some theologians have suggested that God voluntarily set omnipotence aside in creating a world. They hold that the life and death of Christ reveal a God of love who participates in the world's suffering. They suggest that, like a wise teacher or the parent of a growing child, God respects the integrity of the created world and the freedom of human beings, but does not abandon them. They balance the classical emphasis on transcendence, eternity, and impassibility with a greater emphasis on God's immanence, temporality, and vulnerability.[53] Feminist authors have urged that patriarchal images of power as

---

[53] W. H. Vanstone, *Love's Endeavor, Love's Expense* (London: Dartmon, Longman, and Todd, 1977); Jürgen Moltmann, *God in Creation* (San Francisco: Harper & Row, 1985),

coercive control be replaced by the images of empowerment, nurture, and cooperation that are associated with women in our culture. They propose the image of God as Mother to balance the traditional image of God as Father.[54] Many feminists are sympathetic to the idea of *kenosis*, but with the caveat that divine vulnerability and suffering love must not be cited to support the submission and self-abnegation of women. Power as control is a zero-sum game: the more one party has, the less the other can have. Power as empowerment is a positive-sum situation and does not imply weakness in either party. Empowerment and the nurturing of growth and interdependence also seem to be appropriate features of a model of God in an evolutionary world.

Proponents of *self-limitation* hold that God is in principle omnipotent but voluntarily accepts a limitation of power in order to create a community of love and free response. The goal is relationship and transformation, not *kenosis* in itself. Moreover, the use of personal images of the relation between God and the world suggests that God might influence events in the world without controlling them, so we do not end up with a powerless or deistic God. God's dialogic relation to human beings serves as a model of divine activity throughout nature. Process thought agrees with many of these assertions. However, it holds that the limitations of God's power are not voluntary and temporary but metaphysical and necessary—though they are integral to God's essential nature and not antecedent or external to it.

The role of God in process thought has much in common with the biblical understanding of the Holy Spirit. Like the process God, the Spirit works from within. In various biblical passages, the Spirit is said to indwell, renew, empower, inspire, guide, and reconcile. According to Psalm 104, the Spirit creates in the present: "Thou dost cause the grass to grow for the cattle and plants for man to cultivate.... When thou sendest forth thy Spirit, they are created; and thou renewest the face of the ground." The Spirit represents God's presence and activity in the world. This is an emphasis on immanence which, like that in process theology, does not rule out transcendence. Moreover, the Spirit is God at work in nature, in human experience, and in Christ, so creation and redemption are aspects of a single activity.[55] Process thought similarly applies a single set of concepts to God's role in human and nonhuman life, and it is not incompatible with the idea of particular divine action and human response in the life of Christ. The Holy Spirit comes to us from without to evoke our response from within. It is symbolized by the dove, the gentlest of birds. Other symbols of the Spirit are wind and fire, which can be more overpowering, but they usually represent inspiration rather than sheer power. I have elsewhere tried to show that the process view of God is consistent with other aspects of the biblical message.[56]

4.4 *Some Objections*

Let me finally note some possible objections to process thought.

86–93; Paul Fiddes, *The Creative Suffering of God* (Oxford: Clarendon Press, 1988); Murphy and Ellis, *On the Moral Nature of the Universe.*

[54] Sallie McFague, *Models of God: Theology for an Ecological, Nuclear Age* (Philadelphia: Fortress Press, 1987); Elizabeth A. Johnson, *She Who Is: The Mystery of God in Feminist Theological Discourse* (New York: Crossroads, 1992).

[55] G.W. H. Lampe, *God as Spirit* (Oxford: Clarendon Press, 1977); Alisdair Heron, *The Holy Spirit* (Philadelphia: Westminster Press, 1983).

[56] Barbour, *Religion in an Age of Science*, 235–38.

1. *Is panexperientialism credible?* Process thinkers attribute rudimentary experience, feeling, and responsiveness to simple entities. They hold that mind and consciousness are present only at higher levels in more complex organisms, so they are not panpsychists as the term is usually understood. Rocks and inanimate objects are mere aggregates with no unified experience. There are no sharp lines between forms of life in evolutionary history or among creatures today. It appears that for matter to produce mind, in evolution or in embryological development, there must be intermediate stages or levels, and mind and matter must have some characteristics in common. No extrapolation of physical concepts can yield the concepts needed to describe our subjective experience. Process thought interprets lower-level events as simpler cases of higher-level ones, rather than trying to interpret higher-level events in terms of lower-level concepts or resorting to dualism.

However, Whitehead himself was so intent on elaborating a set of metaphysical categories applicable to all events that I believe he gave insufficient attention to the radically different ways in which those categories are exemplified at different levels. In that regard, Hartshorne, Griffin, and other more recent process thinkers are more helpful. I have also questioned whether Whitehead's understanding of the episodic character of moments of experience provides an adequate view of human selfhood. I would argue that we can accept more continuity and a stronger route of inheritance of personal identity, without reverting to traditional categories of substance.

2. *Is this a God of the gaps?* In earlier centuries, God was invoked as an explanation for what was scientifically unexplained. It was held that God intervened at discrete points in an otherwise law-abiding sequence. This was a losing strategy when the gaps in the scientific account were successively closed. According to process philosophy, by contrast, God does not intervene unilaterally to fill particular gaps. God is already present in the unfolding of every event, but no event is attributable to God alone. God and the creatures are co-creators. The role filled by God is not a gap of the kind that might be filled by science, which studies the causal influence of the past. The contribution of God cannot be separated out as if it were another external force, for it operates through the interiority of every entity, which is not accessible to science. God's influence on lower-level events would be minimal, so it is not surprising that the evolution of new forms has been such a long, slow process.

3. *Can we worship a God of limited power?* The God envisaged by process thought is less powerful than the omnipotent ruler of classical theology. But different kinds of power are effective in different ways. The power revealed in Christ is the power of love to evoke our response, rather than the power to control us externally. Moreover, the God of process thought is everlasting, omnipresent, unchanging in purpose, knows all that can be known, and has a universal role and priority in status reminiscent of many of the traditional divine attributes. But I would grant that the numinous experience of the holy and the Christian experience of worship seem to require a greater emphasis on transcendence than we find in Whitehead himself. We can adapt Whiteheadian categories to the theological task of interpreting the experience of the Christian community without accepting all of his ideas.

4. *Is process thought too philosophical?* Metaphysical categories seem abstract and theoretical, far removed from the existential issues of personal life which are central in religion. Some process writers use a technical vocabulary which is understandable only after considerable study, though process ideas can be expressed in a more familiar vocabulary. No theologian can avoid the use of philosophical categories in the systematic elaboration of ideas. Augustine drew from Plato,

Aquinas from Aristotle, Barth from Kant, and so forth. However, we do always need to return to the starting point of theological reflection in the formative events and characteristic experiences of the Christian community. Imaginative models are more important than abstract concepts in the daily life of the church. No model is a literal or exhaustive representation, and we can use different models to imagine different aspects of God's relation to the world. In our search for universality we must be in dialogue with people in other social locations, since economic interests, cultural values, and gender affect all our interpretive categories.

Perhaps, after all, we should return to the biblical concept of the Holy Spirit. This will help us to avoid the separation of creation and redemption that occurred in much of classical Christianity. It is free of the male imagery so prominent elsewhere in Christian history. It will help us recover a sense of the sacred in nature that can motivate a strong concern for the environment today. The Spirit is God working from within, both in human life and the natural world, which is consistent with process thought. The theme of the 1991 assembly of the World Council of Churches in Canberra, Australia, was a prayer in which we can join: "Come, Holy Spirit, renew thy whole creation."

IV

BIOLOGY, ETHICS, AND THE PROBLEM OF EVIL

BEYOND BIOLOGICAL EVOLUTION: MIND, MORALS, AND CULTURE[1]

Camilo J. Cela-Conde and Gisele Marty

1 *Introduction*

Since 1985, we have been working on a model of human evolution that could explain how our species evolved and developed the traits that characterize it at present. Some of these traits (such as large brains, and small teeth with thick enamel) are morphological. Others (such as speech and sophisticated tool-making) are functional. However, the human phenomenon does not stop at the confluence of morphological and functional traits, which we may observe in any individual of our own species. Some important characteristics of human beings go beyond the field of the individual and must be considered in a collective perspective. Language is one of them. Culture, of course, is as well. On the border that separates (or maybe links) language and culture, moral codes are another example. Even the mind has been defined in a collective way.[2]

How these three phenomena (evolutionary processes, individual characteristics, and collective achievements) end up converging in the body/mind of each member of our species is the main topic that anthropology should explain. Furthermore, it constitutes a necessary reference to any theological discussion on divine action and human evolution. Let us discuss some speculative aspects of human evolution, such as moral behavior, that are very often discussed when arguing about the theological consequences of Darwinian theory.

Any theological interpretation of the origin of morphological modern human beings (MMHB) should take into account the gradual evolution of our family, with several lines of adaptation (such as the gracile vs. robust lines; see Cela-Conde's introductory essay in this volume). Consequently, two of the three debates mentioned by Willem B. Drees (in this volume) about the theological implications of an evolutionary view should definitely be put aside; both a straightforward literalist understanding of Genesis and a strong metaphysical view of God's action in the world may be discarded, as they collide with evidence coming from evolutionary theory. For instance, the creationist assumption of a clear-cut gap between great apes and human beings with no intermediate species cannot anymore be held.[3]

However, the third debate mentioned by Drees concerning the implications of an evolutionary point of view for our ideas about human nature and culture is still worth pursuing. Apart from the evolution of human morphology, there are conspicuous non-physical traits that characterize human beings. What has evolutionary theory to say about their origin and appearance? If we remember that theological assumptions normally refer to non-physical traits, the need to examine

---

[1] This study has been supported by the Comisión Interministerial de Ciencia y Tecnología, Ministerio de Educación (Spain), project no. PS95-0059.
    [2] R. Fischer, "Why the Mind is not in the Head but in the Society's Connectionist Network," *Diogenes*, 151 (1990): 1–24.
    [3] For a more detailed discussion, see J. Foley, "Fossil Hominids," *The Talk Origins Archive*, available from http://earth.ics.uci.edu:8080/faqs/fossil-hominids.html, and M. I. Vuletic, "Frequently Encountered Criticisms in Evolution vs. Creationism: Revised and Expanded," available from http://icarus.uic.edu/~vuletic/cefec.html.

how these traits appeared during human evolution is clear. When discussing them, we shall refer to some of the theological approaches given by contributors to this volume, such as Clifford, Drees, Haught, Hefner and Murphy, who explicitly mention the evolutionary human process.

2 *Nature vs. Culture*

In almost all of the theological approaches to human evolution,[4] there is a tendency to counterbalance "biology" and "culture," treating the evolution of culture as a contrary alternative to the evolution of morphological traits. Drees gives an example of this kind of thought when he discusses the sociobiological theory of human evolution. To refute the existence of a strong influence of biology on human behavior, Drees quotes Elliott Sober's idea of how culture influences human evolution. According to Sober a "brain is able to liberate us from the control of biological evolution because it has given rise to the opposing process of cultural evolution."[5] What Sober is trying to refute is the strong biological determination of some behaviors, such as the reaction to an enemy violating the nest. The response, highly automatic in some species of invertebrates, is considered altruistic behavior. Soldier ants, for instance, sacrifice their lives to protect the colony against an intruder. As Sober says, this is not the way human beings act—but we will come back to this later. However, it is very easy to misunderstand Sober's arguments. In the biology/culture clash, the metaphor of the brain liberating us from biological control could assign to the brain itself a Promethean role, seeing it as a God who decides that the human species should be liberated from slavery. Some aspects of the metaphor do not fit well with what we know about human evolution. Firstly, this kind of liberation occurred among the great apes as well. Secondly (and mainly), the brain itself is a result of biological evolution. Therefore, the evolutionary process should be seen as the real protagonist of the liberation. This kind of argument leads us to the discussion about the teleology of "evolutionary progress," a topic that other scholars have dealt with. For instance, Philip Hefner (in this volume) gives a very interesting—and unusual—non-dualistic account for what technology is, considering it to be as "authentic" ("natural" would be an equivalent) as other human character-istics.

Cultural evolution cannot be seen as something that collides head-on with biological evolution because the former is a result of the latter. Interactionist models that can help us understand the relationships between culture and morphology during the evolution of hominids are needed. The so-called "second sociobiology"[6] tried to give us this kind of integrated model of how culture and morphological traits could have evolved altogether. This is not a metaphor. Early cultural traditions, such as Oldowan culture, existed during a span of time long enough to allow hominids to evolve, that is, to undergo morphological changes.[7]

---

[4] And, to be fair, this would true in almost all the evolutionary approaches as well.

[5] E. R. Sober, "When natural selection and culture conflict," in *Biology, Ethics and the Origins of Life*, Holmes Rolston, III, ed., (Boston: Jones and Bartlett, 1995), 151.

[6] For instance, C. J. Lumsden and E. O. Wilson, *Genes, Mind and Culture: The Coevolutionary Process* (Cambridge: Harvard University Press, 1981).

[7] See F. Kort, "An Evolutionary-Neurobiological Explanation of Political Behavior and the Lumsden-Wilson 'Thousand-Year Rule'," *J. Social. Biol. Struct.*, 6 (1983): 219–30.

3 *Human Characteristics*

To hold that cultural traits and morphological traits are sufficiently related, obliging us to study them in a coordinate way, does not mean that we have found the correct way to study them. We have said nothing yet of how non-physical human traits evolved.

The main characteristics that we normally identify with MMHB are:

1. articulate language
2. highly developed culture, implying a complex brain which is capable of performing high cognitive functions, such as self-consciousness

Both characteristics have a long and gradual history, probably going back as far as 2.5 million years. According to the interactionist model that we are describing, culture, language, and a complex brain are all related, and evolved in a coordinated way. All those characteristics imply large quantitative differences when comparing human beings to great apes, but it is very difficult to say when these differences became manifest. Let us consider, for instance, the case of language. Almost all anthropologists think that human language implies a phenomenon of emergence. But, when did it take place?

The idea of an almost instantaneous and isolated emergence of language is supported by Lieberman's idea of its very late appearance, related to morphological changes of the larynx in our own species, *Homo sapiens*. Lieberman rejected the notion that Neandertals could speak.[8] However, this is hard to accept, if we take into account that Neandertals were responsible for such a developed culture as the Mousterian one. If Lieberman is right, it is possible to hold that language itself suffices to identify a MMHB. This does not help us very much when trying to find empirical proof of when and where human beings appeared—proof that might be used to either support or to reject theological assumptions—since language, obviously, does not fossilize. However, Lieberman compared the larynx of great apes to those of human beings in order to identify that part of the anatomy of the muscular mass that allows humans to speak. From the marks left by them on fossilized bones, he concluded a late and very rapid appearance of language.

This idea collides head-on with Phillip V. Tobias' model of a long, gradual, and early development of language, that may have started with *Homo habilis*. Tobias relates the emergence of language with the development of the brain—that is, with some parts of the brain, such as Wernicke's and Broca's areas, which are related to the use of speech. In Tobias' words, we do not speak with the larynx, but with the brain. Tobias expressed this thesis a long time ago,[9] but it had little support until scholars such as Falk claimed that there was a need to consider the existing relationship between the brain's evolution and the appearance of language.[10] Tobias has since published numerous papers developing his ideas on the evolution of language.[11] As he relates a functional characteristic (language) to a morphological

---

[8] P. Lieberman, "On the Evolution of Language: A Unified View," *Cognition*, 2 (1973): 59–94; idem, *The Biology and Evolution of Language* (Cambridge: Harvard University Press, 1984).

[9] P. V. Tobias, *The Brain in Hominid Evolution* (New York, N.Y.: Columbia University Press, 1971).

[10] D. Falk, "Cerebral Cortices of East Africa Early Hominids," *Science*, 221 (1983): 1072–74.

[11] These include: P. V. Tobias, "The Brain of *Homo habilis*: A New Level of Organization in Cerebral Evolution," *Journal of Human Evolution*, 6 (1987): 741–61; idem,

one (development of some areas of the neocortex), it is possible to examine fossil specimens to look for support of his hypotheses. Endocasts, for instance, sometimes show the imprints of the brain's surface, allowing us to determine its evolution. On the basis of such endocasts, Tobias rejected the possibility of speech in the *Australopithecus africanus* (Taung specimen), but considered as plausible the beginning of language development in *Homo habilis*. Since *Homo habilis* is widely considered as the maker of early implements belonging to the Oldowan culture as well, Tobias' thesis receives further support.

As the Chomskian model holds, language implies an enormous functional difference existing between great apes and human beings that completely separates the cognitive capacities of each.[12] However, if we take into account the phylogenetic process, it seems impossible to say where the qualitative barrier is. Primates like Vervet monkeys have syntactic language and complex social relationships based on this means of communication.[13] This is also a strong argument in favor of Tobias' ideas.

What can we say about other mental abilities, apart from language? They may give some important clues when talking about the theological implications of human evolution. When dealing with consciousness, Francis Crick talks about "the scientific search for the soul."[14] However, consciousness, like language, is a characteristic shared, to a greater or lesser degree, with other primates as well. Chimpanzees have self-consciousness[15] and perform political strategies.[16] When we talk about our closest ancestors, the continuity seems unavoidable. Even if the mitochondrial Eve theory (mtEve) may be considered a plus when dealing with the appearance of human beings as a relatively unique episode (remember that the original population of "Eve" may have had as many as 10,000 individuals), Neandertal and MMHB shared, for a long span of time in Europe, the same Mousterian culture,[17] and the

---

"The Emergence of Spoken Language in Hominid Evolution," in *Cultural Beginnings: Approach to Understanding Early Hominid Life-Ways in the African Savanna*, J. D. Clark, ed., (Bonn: Dr. Rudolf Habelt GMBH, 1987), 67–78; idem, "Some Critical Steps in the Evolution of the Hominid Brain," in *The Principles of Design and Operation of the Brain*, J. C. Eccles and O. Creutzfeldt, eds., (Civitate Vaticana: Dr. Rudolf Habelt GMBH, 1990), 67–78; idem, "Relationship Between Apes and Humans," in *Perspectives in Human Evolution*, A. Sahni and R. Gaur, eds., (Delhi: Renaissance Publishing House, 1991), 1–19; idem, "The Craniocerebral Interface in Early Hominids," in *Paleoanthropological Advances in Honor of F. Clark Howell*, R.S. Curriccini and R. L. Ciochon, eds., (Englewood Cliffs, N.J.: Prentice-Hall, 1994), 185–203; idem, *The Communication of the Dead: Earliest Vestiges of the Origin of Articulate Language* (Amsterdam: Stichting Nederlands Museum voor Anthropologie en Praehistorie, 1995).

[12] E. H. Lenneberg, *Biological Foundations of Language* (London: John Wiley & Son, 1967).

[13] R. M. Seyfarth and D. L. Cheney, "Grooming, Alliances and Reciprocal Altruism in Vervet Monkeys," *Nature*, 308 (1984): 541–43; idem, "Meaning and Mind in Monkeys," *Scientific American*, 267 (1992): 122–28; D. L. Cheney and R. M. Seyfarth, "Précis of *How Monkeys See the World*," *Behavioral and Brain Sciences*, 15 (1992): 135–82; D. Cheney, R. Seyfarth, and B. Smuts, " Social relationships and social cognition in nonhuman primates," *Science*, 234 (1986): 1361–66.

[14] F. Crick, *The Astonishing Hypothesis* (New York, N.Y.: Scribner, 1994).

[15] G. G. Gallup Jr., "Towards a comparative psychology of mind," in *Animal Cognition and Behavior*, R. L. Mellgren ed., (Amsterdam: North-Holland, 1983), 473–510.

[16] R. Byrne, *The Thinking Ape: Evolutionary Origins of Intelligence* (New York, N.Y.: Oxford University Press, 1995).

[17] G. A. Clark and J. M. Lindly, "The Case for Continuity: Observations on the

Châtelperronian culture as well,[18] thus providing proof of the continuity of the cognitive abilities.[19] Why MMHB populations displaced Neandertal populations may have an adaptive explanation that does not force us to regard Neandertals as animals.

Whether divine action had a hand or not in the appearance of MMHB is a problem that has been widely posed in relation to these "high" characteristics of language and cognition. We do not know how divine action is to be considered in this continuous evolutionary field. However, we think that any theological study compatible with state-of-the-art paleoanthropology must take cognitive continuity into account.

Are language and culture the only differences that we must take into account when separating human beings from great apes? Let us refer to a different trait, mentioned by Charles Darwin, that could perhaps be considered as an exclusive property of human beings, namely, our moral sense.

It is well known that Darwin pointed out the difficulty of distinguishing between the similar reactions of baboons and humans when each is giving aid to a wounded mate.[20] Darwin concluded by referring to a special sense, the "moral sense," as the sole characteristic capable of distinguishing between animals and human. Thanks to this special sense, it is possible to talk about "heroes" when referring to the human species, but not in the case of an ape, even if both of them are performing what, in biological terms, is considered "altruistic behavior." What the biological concept of altruism means will be discussed below. Let us now underline that Darwin was the first to speak of a biological mechanism capable of conferring morality on the human species. And even though there are authors[21] who maintain that Darwin makes no attempt to derive substantive moral principles from biological phenomena, the truth is that Darwin, on examining moral progress, links both entities.[22] Let us see how.

Biocultural Transition in Europe and Western Asia," in *The Human Revolution: Behavioral and Biological Perspectives on the Origins of Modern Humans*, P. Mellars and C. Stringer eds., (Princeton, N.J.: Princeton University Press, 1989), 626–76; R. L. Holloway, "The Poor Brain of *Homo sapiens neanderthalensis*: See What You Please...," in *Ancestors: The Hard Evidence*, E. Delson ed., (New York, N.Y.: Alan R. Liss, 1985), 319–24; E. Trinkaus and P. Shipman, *The Neandertals: Changing the Image of Mankind* (New York, N.Y.: Alfred A. Knopf, 1993).

[18] J. J. Hublin, F. Spoor, M. Braun, F. Zonneveld, and S. Condemi, "A Late Neandertal associated with Upper Paleolithic Artefacts," *Nature*, 381 (1996): 224–26.

[19] J. A. J. Gowlett, "Early Human Mental Abilities," in *The Cambridge Encyclopedia of Human Evolution*, S. Jones, R. Martin and D. Pilbeam eds., (Cambridge: Cambridge University Press, 1992), 341–45. On the contrary, the latest finds of Arcy-sur-Cure (France) are not conclusive enough to disprove the mtEve theory, as the newly discovered derived Neandertal morphological features (bony labyrinth) do not exist in Upper Paleolithic MMHB specimens from Western Europe.

[20] Charles Darwin, *The Descent of Man and Selection in Relation to Sex* (1871), 2nd. ed., modified, 1874. Quoted from the N.Y. Modern Library ed., 1936 (numerous editions) which follows the 2nd. edition of 1874 and also contains the *Origin of Species*, chap. IV.

[21] J. G. Murphy, *Evolution, Morality and the Meaning of Life* (Totowa, N.J.: Rowman & Littlefield, 1982).

[22] Narrowly identifying "motive" with "biological phenomena," and "criterion" with "substantive moral principles," presupposes a reduction because, in fact, there are biological aspects (emotions, for example) that are influential in moral judgment, and there are also ethical aspects (certain values) which count in the motivation to act. But even granting the existence of these influences, it is useful to distinguish between these two areas. This analysis will be amplified below.

Two years after the end of the voyage of the *Beagle*, Darwin obtained the most modern bibliography available concerning human moral behavior, including the latest works of Martineau, Mackintosh and Abercrombie, together with the more classical works of William Paley.[23] Some of these authors (Martineau, Paley) defended the merely conventional character of ethics, utilizing an argument that has often been exploited, namely, the diversity of moral codes. Darwin observed the vast variety of customs and moral norms of South American Indians; but this diversity did not bewilder him. On the contrary, he saw in it a response to what we would call today, "adaptation to environmental conditions."

Thus an adaptive response could very well proceed from certain deeper capacities, a common substratum, unique to all human species; and it could become oriented in a variety of directions. Darwin, following a previous tradition,[24] assumed that mankind relied on a special sense, the "moral sense," which allowed for the making of ethical decisions, and which, in fact, constituted the main difference between humans and animals. But in Darwin's work it is not clear what this special sense consists of—there are general references to the possibility of human beings able to anticipate the consequences of their actions, and that is all. The moral sense does not go beyond an *ad hoc* concept, inserted to fill a gap in the evolutionary framework. Its explanatory weakness stands out when biological mechanisms (such as the sympathetic instinct) are related to moral evolution.

No matter how much the moral sense guarantees a certain universality of human behavior, this could not be, naturally, eternal. It would be subject, after all, to evolution by natural selection, and Darwin was well aware that different cultures manifested successive stages of a "positive" moral evolution. Human beings, by means of a nature that includes the moral sense,[25] go on constructing societies in which ethical conduct and codes of approval of such conduct are present. Initially, the group that benefits from this complex of actions and codes is small, but slowly, by virtue of intellectual, material, and moral progress, the radius of moral action widens. Primitive humans respect and help their closest relatives; they then extend their fellow-feeling to the tribe, and later to the entire population.[26] In time, Darwin concluded, it will be the whole human race that forms a unique body of morality expressed in a universal code and a generalized feeling of fellowship. Will it be by means of an instinct or by a perfected rationalization? Darwin's view of human morality relies on the union of these two factors.[27]

---

[23] Regarding the reading material of the young Darwin, see E. Manier, *The Young Darwin and His Cultural Circle* (Dortrecht: Reidel, 1978), and R. J. Richards, *Darwin and the Emergence of Evolutionary Theories of Mind and Behavior* (Chicago, Ill.: University of Chicago Press, 1987).

[24] Authors such as Hutcheson, Shaftesbury, Hume and, later, Adam Smith, based the foundation of morality in general, and justice and private property in particular, on a psychological mechanism (sympathy).

[25] Such a nature would also include the sympathy instinct. In spite of Darwin's reading in his youth, however, as well as his friendship with authors like James Mackintosh (a distant relative of his), thirty years had to pass for Darwin to establish, in *The Descent of Man*, solid connections between sympathy and ethical codes.

[26] This idea has been developed in a very different context, of course, in P. Singer, *The Expanding Circle* (New York, N.Y.: Farrar, Strauss & Giroux, 1981).

[27] For a discussion of the meaning of "progress" in Darwinism, and its axiological sense, see Francisco J. Ayala, "The Concept of Biological Progress," in *Philosophy of Biology*, F. J. Ayala and T. Dobzhansky, eds., (London: MacMillan, 1974), 339–55., and C. Castrodeza, *Ortodoxia darwinista y progreso biológico* (Madrid: Alianza Editorial, 1988).

Darwin's implicit theory of heredity includes, as is well known, the incorporation of phenotypic transformations or, to put it in today's fashionable terms, the inheritance of acquired characteristics from the genetic pool. The formula, proceeding from Lamarck, is a guarantee for the combined progress of sympathy instincts and ethical codes, since later generations benefit through a parallel and compatible line from every new discovery. In one sense, as *The Descent of Man* tells us, it is ethics that shape the human person, transforming the savage pre-human into the modern citizen. Nevertheless, the explanatory nature of the example is still seriously affected by the demands of Lamarckian inheritance. After the definitive implantation of Weissman's theory of heredity, the neo-Darwinists were obliged to reject this structure of progress and harmony.[28]

To go beyond the "moral sense" approach of Darwin, we cannot use for our analysis a simple model of moral behavior like that of an automatic response to biological tendencies. That is to say, we must clarify what "altruistic behavior" means, and what difference, if any, exists between the altruistic behavior of one of our ancestors and the altruistic behavior of some other species. To say that human beings (and their ancestors, as well) have "moral" altruism but all other species have "biological" altruism, has, at the moment, no meaning. To endow these phrases with content, we must explain what the exclusive human characteristics of moral behavior are.

4 What is Human Morality?[29]

During the course of a famous workshop given by the Dahlen Konferenzen and chaired by Gunther Stent almost twenty years ago, Norbert Bischof presented a paper entitled *On the Phylogeny of Human Morality*[30] in which he proposed the discussion of four key points:

1. What phenomena are implied when we speak of "human morality"?
2. What kind of analogs to human morality can be found on the animal level?
3. How can the emergence of these analogs be explained phylogenetically?
4. How may human morality have developed in the framework of these analogs?

This proposed scheme, and the entire workshop, was in line with the expectations created by the unfolding of sociobiology as an autonomous science. Sociobiologists were able to offer fascinating theories to explain group conduct, which could scarcely be overlooked by those interested in moral phenomena. In the introduction to his work, Bischof pointed out that none of the theories in these areas

---

[28] Among neo-Darwinists there were very different positions regarding the relationship of evolutionary process and ethics, positions which are not going to be mentioned in this brief résumé. For example, there is considerable distance between C. H. Waddington, "The Relations between Science and Ethics," *Nature*, 148 (1941): 270–74, and A. Keith, *Evolution and Ethics* (New York, N.Y.: Putnam's Sons, 1947)—almost as much as between E. O. Wilson, *Sociobiology: The New Synthesis* (Cambridge: Harvard University Press, 1975) and Science for the People, *Biology as a Social Weapon* (Minneapolis, Minn.: Burgess, 1977). A clear analysis of different neo-Darwinist positions figures can be found in Richard D. Alexander, *The Biology of Moral Systems* (New York, N.Y.: Aldine De Gruyter, 1987).

[29] In this section, we have used material freely from C. J. Cela-Conde, "On the Phylogeny of Human Altruism," *Human Evolution*, 5 (1990): 139–51.

[30] N. Bischof, "On the Philogeny of Human Morality," in *Morality as a Biological Phenomenon*, G. S. Stent, ed., (Berkeley, Calif.: University of California Press, 1978), 48–66.

achieved the usual standards of scientific verification, but all of them were concerned with mere heuristic propositions. Even allowing for that important specification, sociobiology seemed to offer a solid base for ethical naturalism.

In present times, a great deal of that enthusiasm has died down. But it would be unfair to deny that more precise and fertile examples of moral human behavior, thanks to sociobiology, can be established. We are trying to examine here how, ten years later, the debate may continue. Therefore, we shall follow up some points proposed by Bischof, especially regarding what we mean by "human morality" and its phylogenetic development. Except for isolated cases, the propositions are still heuristic, but we trust that the integrated model of the biological base of human morality presupposes some progress on what was previously found.

In spite of all kinds of reductionism that have been suggested by biological interpretations of ethical behavior, human morality seems to be a very complex phenomenon. One of the most prolific authors on the sociobiology of human species, Richard D. Alexander, starts his book *The Biology of Moral Systems* by agreeing with a philosopher, Alasdair MacIntyre, about the mystery of ethics and the frustration of the search for its roots. Why is this so? In principle, it should be relatively easy to find the biological roots of such conspicuous behavior. If morality is a phenomenon that appeared early in our phylogenetic history, it must necessarily have been caused by some genetic mechanisms capable of initiating it. But taking for granted that this is so, which of the different aspects of moral phenomenon are the behaviors controlled by these kind of mechanisms? In order to answer this question it is essential to examine the structure of moral action, that is to say, its diverse characteristics.

In his 1978 paper, Bischof, on analyzing the characteristics of human morality, distinguished between "form characteristics" and "content characteristics." Nancey Murphy (in this volume), quoting Francisco Ayala's point of view,[31] refers to this distinction (the "capacity" for moral thought versus the "content" of moral codes). It is an old and pertinent distinction, and we will take the liberty of discussing its meaning in greater depth.

The difference between the two was explained several centuries ago, when the Enlightenment Scottish philosophers proposed to distinguish, within moral conduct, between the *motive* for action and the *criterion* of moral action. This difference makes it possible for us to realize, for example, that regret or remorse might exist, or that a person might believe that it is his duty to help a wounded person on the highway or run from the scene in fright when faced with possible complications.[32] The difference between motive and criterion is also important in dealing with a common confusion regarding what the evolution of morality is. When Drees asks (in this volume) whether "evolved morality" could be moral, he mentions the fear "that an evolutionary understanding of morality would undermine the specific moral character of such behavior." This fear relies on the sociobiological explanations of some moral behavior, such as taking care of offspring. It is clear that the more biologically determined this behavior is, the less moral it may be considered. Why worker ants take care of the queen's eggs is something that is explained very well

---

[31] Francisco J. Ayala, "The Biological Roots of Morality," *Biology & Philosophy*, 2 (1987): 235–52.

[32] For a deeper discussion see C. J. Cela-Conde, *On Genes, Gods and Tyrants* (Dortrecht: Reidel, 1986).

by the sociobiological theory of kin selection. However, does the same explanation allow us to consider this behavior "moral"?

It is important to realize that the behavior in itself does not solve the dilemma. Ants do take care of their offspring. Human beings do as well. However, it is normal to talk about moral behavior only in the latter case. Why? Let us consider the example given by Michael Ruse (and quoted by Drees, in this volume). A human-like species, with quite a different biology, that evolved on a distant planet in the Andromeda galaxy may, perhaps, consider rape as not being wrong. We are not interested in the arguments about differences in the biology that, in Ruse's words, could justify the morality of rape. What we mean is that Andromeda's and Earth's evolutionary processes are so similar in Ruse's *gedankenexperiment* that, in both cases, they lead the "human" species to discuss what is wrong and what is right. This is the key-point of moral behavior—not taking care of offspring but qualifying whether a mother who takes care of children (or who rapes them) acts in a correct way.

Nevertheless, one may ask which of these two aspects would affect the genetic causation in early morality? Is it the mere inclination to act in favor of other members of the group, or is it the rising of genetically driven elemental norms of morality that should be considered as some kind of initial ethical criteria? In Lorenz's view,[33] going beyond the Kantian categorical imperative by means of a "biological imperative" already contains certain norms, such as not killing members of the group.[34] But this is mere speculation, while the biological mechanisms that intercede in the control of such norms are explicit.

5 Moral-analogous Behavior in Humans and Animals, and How to Explain It

During the last half of this century a new naturalist topic has surfaced, that of the sociobiological theory of altruism, with a far greater explanatory scope and capability for including humans and animals in an integrated model. Instead of taking refuge in an *ad hoc* concept like that of moral sense, the models of group selection[35] and kin selection[36] propose strict laws and models based on population genetics to explain the adaptive paradox of altruism.

Let us examine the general features of the problem. Altruistic action would not exist if we relied on the classical scheme of evolutionary theory. Natural selection, as it is widely known, strives to maximize the fitness of individuals. In such a way, the most apt individual is finally selected. So, in accordance with the model, it is to be hoped that everywhere we meet up with individuals capable of furthering their aptitudes.

But altruistic behavior seems not to be included in the evolutionary model. Far from augmenting individual fitness, it does the contrary: it *diminishes* it. An altruist

---

[33] K. Lorenz, "Kant's Lehre von Apriorischen im Lichte gegenwärtiger Biologie," *Blatt für Deutsche Philosophie*, 15 (1941): 94–125.

[34] For an interesting discussion on the functionalist explanation of the role of morality in phylogeny and the existence of "ethical innate knowledge," see G. S. Stent, "The Poverty of Scientism and the Promise of Structuralist Ethics," in *Knowledge, Value and Belief*, H.T. Engelhardt Jr. and D. Callahan, eds., (Hastings-on-Hudson: Institute of Society, Ethics and the Life Science, 1977), 225–46.

[35] V. C. Wynne-Edwards, *Animal Behavior in Relation to Social Behavior* (Edinburgh: Oliver and Boyd, 1962).

[36] W. D. Hamilton, "The Genetical Evolution of Social Behavior," *J. Theoret. Biol.*, 7 (1964): 1–52.

squanders nourishing resources, shares territory, and can even come to the point of risking death by warning others in the group, for example, of the approach of a predator. It seems difficult to understand how such an individual would be capable of transmitting characteristics to the next generation in order to guarantee the future existence of altruists among the population. No matter how much the group benefits in global terms from the presence of a altruistic member, it does not explain the altruist's adaptive success. The neo-Darwinist theory of evolution by natural selection insists on an *individual* behavior capable of assuring the transmission of genetic qualities. The presence of a selfish mutant within a group of altruists would very quickly (in a few generations) lead to the whole group being composed of egoist members because they would enjoy far superior possibilities of producing offspring. Therefore any group selection theory, such as that developed from Wynne-Edwards' ideas,[37] must get rid of the challenge of the egoists' presence which threatens the evolutionary stability of this kind of altruistic behavior.[38]

David Sloan Wilson and Elliott Sober have recently given a new impulse to the group selection theory, denying the pertinence of selection at the individual level (and, more strongly still, at the gene level, as the kin selection theory holds—see below).[39] Drees (in this volume) discusses Wilson and Sober's theory, so we will not repeat their arguments. Suffice it to say that Wilson and Sober's group selection needs highly developed cognitive mechanisms to avoid the presence of individuals performing selfish behavior, and therefore connects with the models of reciprocal altruism such as those of Trivers. Let us first examine how a very different theory of the evolution of altruism—the kin selection theory—may apply to human beings.

The theory of kin selection[40] explains reasonably well the evolution of social insect populations and their reproductive systems (haplo-diploidy). Sister workers have a great number of genes that are common to them all. Thus, the queen's capacity to transmit their genetic content to their offspring, the *inclusive fitness*, may be considered as the individual fitness of the queen, plus the advantage that workers give to her and to her developing eggs. The closer the relatives and the higher the amount of shared genes, the larger the probability of an "altruistic gene" being selected. This kind of altruism has been considered as selfishness at the genes' level.[41] As regards Drees' discussion (in this volume) of Dawkins' ideas, we will comment very briefly. Putting aside the question of the kind of "genes" which lead to ants' and bees' behavior, taking for granted this theory, and projecting it onto the human species (something that the sociobiologists certainly often do), we have established a much firmer naturalist link than we could have done earlier. A

---

[37] Wynne-Edwards, *Animal Behavior in Relation to Social Behavior*.

[38] The concept of an "evolutionary stable strategy" (ESS) was developed, within the theory of games by Maynard Smith; see J. Maynard Smith, "Evolution and the Theory of Games," *American Scientist*, 64 (1976): 41–55; idem, "Models of Evolution of Altruism," *Theoretical Population Biology*, 18 (1980): 151–59; J. Maynard Smith and G. R. Price, "The Logic of Animal Conflict," *Nature*, 246 (1973): 15–18. ESS refers to a strategy that, in some detailed conditions, cannot be overcome by a different one. Computer-developed models of group selection may be an ESS, but only under very special conditions (see Wilson, *Sociobiology*).

[39] D. S. Wilson and E. Sober, "Reintroducing Group Selection to the Human Behavioral Sciences," *Behavioral & Brain Sciences*, 17 (1994): 585–654.

[40] Hamilton, "The Genetical Evolution of Social Behavior."

[41] For instance, see R. Dawkins, *The Selfish Gene* (Oxford: Oxford University Press; Hay ed. Castellana, 1976).

naturalist method of justification is thereby provided for Hume's intuition about how benevolence gradually takes on a weaker form when comparing our relationships with friends to those with strangers.

But such extrapolations, those that carry over from social insects to the human species, lead us to many doubts. How can the fact that insects of the *Hymenoptera* order sacrifice themselves be affirmed as moral conduct if we are employing the term in the same way that philosophers do? Some authors, such as Bertram and Voorzanger,[42] have put forth the idea of distinguishing "biological altruism" from "moral altruism." Biological altruism (which Voorzanger suggests calling "bioaltruism") would be that behavior, the presence of which involves a paradoxical diminution of the individual's genetic fitness in favor of another member (a relative) of the group. But, what about moral altruism? How can it be described? Does it involve a behavior which, besides being altruistic in a biological sense, has other added connotations? Or, to the contrary, is it something altogether different—a phenomenon that cannot be clearly seen, not even partially, under the sociobiological microscope? Voorzanger is unyielding in this regard: "Whatever we know about the one [bioaltruism] *is irrelevant* to the other [moral altruism]."[43]

If we opt for this solution, our position will be far removed from any ethical naturalism, at least in this concrete aspect. We shall be at one with Kant in his first *Kritik*—the world of moral values and the world of facts (whether they be biological, social or even political) are irreversibly separated, and not even a common frontier exists between them. But if we find in moral altruism certain shared aspects with biological altruism, then naturalism will come into play, more or less strongly, depending on the degree of overlap.

The latter is, of course, the position most often ascribed to Darwin, neo-Darwinists, ethologists, and sociobiologists. But it would be worth insisting on something, obvious as it is, that is often forgotten. To maintain that moral altruism is biological altruism (and something more on which we have not yet touched) is not necessarily to reduce ethics to biology.[44] Nancey Murphy offers (in this volume) strong arguments for non-reducibility. Even taking into account characteristics such as the supervenience of moral language, however, it seems clear that between one (biological altruism) and the other (moral altruism) there are connections impossible to bypass. And if the concept of biological altruism which is being used relates to a behavior which produces a diminution in biological fitness, it would prove difficult to deny that moral altruism leads, in fact, to the same kind of handicap. This is fairly evident, no matter how allergic we happen to be to biological determinism.

Within sociobiology many attempts have been made to bring forth a refined model of altruism—one more suitable than the theory of kin selection— to approach both philosophical and biological altruism. Darlington Jr., for example, proposed combining the group selection model and the kin selection model in an effort to explain the individual quality, as well as the collective one, in the evolution of

---

[42] B. C. R. Bertram, "Problems with altruism," in *Current Problems in Sociobiology*, King's College Sociobiology Group, ed., (Cambridge: Cambridge University Press, 1982), 251–67; B. Voorzanger, "Altruism in Sociobiology: a Conceptual Analysis," *Journal of Human Evolution*, 13 (1974): 33–39.

[43] Voorzanger, "Altruism in Sociobiology," emphasis mine.

[44] The initial framework of Wilson's *Sociobiology* could be confusing in this regard, but those who have followed his later arguments will clearly find that such a simplification, as regards the human species, will never come true.

altruism.[45] Trivers' "reciprocal altruism"[46] also tries to better adjust human moral behavior. In Trivers' model the individual must evaluate situations, analyze past (and future) actions, and make complex decisions. It is a considerable advance because, for Darwin, ethical conduct is rooted in certain mental processes of enormous complexity. In fact, Trivers has utilized his model to postulate the appearance of human morality during the Pleistocene as part of the social conduct of small, stable, and mutually dependent groups.[47]

The biological relationship, still frequent in these groups, is not necessary for the model of reciprocal altruism; the key of its existence, in human groups, is a "sense of fairness" linked both to emotional reactions and to a standard of common equity. Among the requisites for this kind of altruistic conduct there are certain psychological mechanisms necessarily involved in such complex decisions as those which are needed in moral choice. Trivers maintains that human altruism, during our phylogeny, was a factor of selective pressure towards the development of complex cognitive systems.[48] But he offers no explanation whatsoever of how such systems might be understood. His references in this area are those of Krebs and, in general, the studies in the psychology of altruism.[49] Unfortunately, such studies were carried out when cognitive science was very underdeveloped, and there was no way to integrate emotional phenomena into a theory of knowledge production.

Parallel demands regarding the need to consider complex cognitive processes (such as those implicated in the internalization of social norms) have been required in the explanatory models of human altruism by means of "ultrasociality"[50] and "sociocultural fitness."[51] Thus, there is a plethora of interpretations of altruistic behavior, in which innate tendencies are complemented either by means of evaluation processes of another's conduct or by the presence of norms (or criteria) whose significance is well-defined within the group. We are not aware, however, of any model that includes all these elements, from innate tendencies to empirical moral norms, and which explains the two classical levels of "motive" and "criterion" as affected by biological human nature. It seems that any attempt to explain human moral phylogeny should establish a model of this sort as its goal.

6 An Extended Model of Moral Action

An interesting consequence of the theory of reciprocal altruism (or any other model of altruism which involves the presence of complex cognitive processes) is that the classical distinction between the two areas (motive and criterion) is insufficient.

---

[45] P. J. Darlington Jr., "Altruism: Its characteristics and evolution," *Proc. Natl. Acad. Sci. USA*, 75 (1978): 385–89.

[46] R. L. Trivers, "The Evolution of Reciprocal Altruism," *Quarterly Review of Biology*, 46 (1971): 35–57.

[47] R. L. Trivers, "The Evolution of Cooperation," in *The Nature of Prosocial Development*, D. L. Bridgeman, ed., (New York, N.Y.: Academic Press, 1978), 43–69.

[48] Trivers, "The Evolution of Reciprocal Altruism."

[49] D. Krebs, "Altruism: An Examination of the Concept and a Review of the Literature," *Psychol. Bull.*, 73 (1970): 258–302.

[50] D. T. Campbell, "The Two Distinct Routes beyond Kin Selection to Ultrasociality: Implications for the Humanities and Social Sciences," in *The Nature of Prosocial Development*, Bridgeman, ed.

[51] J. Hill, "Human Altruism and Sociocultural Fitness," *J. Social Biol. Struct.*, 7 (1984): 17–35.

With kin-selection theory one could view biological causation of moral behavior in terms of this dualism. Within this theory, "motivation" must be largely (or totally) directed by the genetic code—only in that way would a moral behavior whose adaptive finality is the survival of certain alleles make sense. With "criterion," the weight of biological causation would be variable: from total causation, in which case the very notion of "criterion" is absurd (what are the "moral values" of termites?), to a more or less partial causation, which buttresses propositions such as those of J. H. Crook, who justifies values of certain societies in terms of population genetics.[52] This distinction between motive and criterion underlies present arguments about the "thousand-year rule"[53] and about the adaptive value of the incest taboo.

But, if we keep in mind the complex cognitive processes implied in the evaluating and making of decisions, as reciprocal altruism demands, the dual model is altogether insufficient. The motive to act, for example, in a situation like this, includes anticipating the consequences of the act in accordance with the existence of a certain criterion (that of the group). The prospect of committing a crime can be seen as being influenced, to a greater or lesser degree, by the existence of laws condemning it. Consequently, the "criterion" seems not to be a simple and integrated area, but a multiple one where both individual and collective prospects are involved. The ethical criterion can be understood collectively, as the confluence of all the values and norms present in a group. But it also has an individual form, a kind of molding of all that entire universe of values in the actual ethical criterion utilized by an individual at any given moment. Such considerations have no importance in the theory of kin selection, because the distinction between behavior in individual terms and behavior in collective terms does not play any role there.

The complexity of the criterion area compels us to extend the old distinction to which we have referred by adding the following important difference between collective and individual aspects when relating them to an ethical criterion. The extended model includes three areas (or domains) as follows:

1. the motive to act
2. the personal ethical criterion
3. the set of collective values and norms[54]

How do collective and individual aspects of the criterion relate? Bearing in mind the general features of the process by which an individual is brought into a group, it can be considered a unique actualization of information proceeding from the universe of social values, accumulated by means of an apprenticeship process.

The collective complex of information is not permanent. It *evolves* precisely by means of its actualization in different individuals. It is obvious that any new ethical value comes, finally, from an individual's special initiative but does not acquire meaning, as a part of the moral pool of society, until shared by a reasonable number of individuals. It has an individual aspect as well as a collective one, and it is the interaction of both that is able to explain the evolutionary phenomenon.

---

[52] J. Crook, *The Evolution of Human Consciousness* (Oxford: Clarendon Press, 1980).

[53] This rule refers to the presence of genetic mutations which are selected in relation to political changes in this interval of time; see Kort, "An Evolutionary-Neurobiological Explanation of Political Behavior."

[54] Cela-Conde has identified these domains as alpha-moral level (the motivation level), beta-moral level (personal criterion) and gamma-moral level (social values), adding a fourth level reserved to certain collective values which are considered by the group, from a subjective point of view, as "ultimate values" (see Cela-Conde, *On Genes, Gods and Tyrants*).

Murphy (in this volume) says that the moral level of analysis definitely depends (logically, non-motivationally) on top-down justification, given her claim that all moral systems are dependent, explicitly or implicitly, on beliefs regarding ultimate reality. We agree with her, since any level of categorical analysis relies on top-down mechanisms as a result of the way human knowledge functions.

However, beliefs regarding ultimate reality are in some relevant aspects a particular case of social knowledge. The main aspect of the discussion is, in our opinion, how all these aspects of human moral behavior—psychological motivation, the personal ethical criterion, and the set of collective values and norms—interact.

7 Possible Phylogenetic Roots of Moral Categories: Cognitive Naturalism

If it is difficult to accept the discussions about early moral altruism, due to the highly speculative character of the hypothesis, the situation is even worse when considering what kind of altruistic behavior our more ancient ancestors between human beings and chimpanzees had. The only way to comment on this is, perhaps, to examine the behavior of our present close relatives.

Almost all the characteristics of early human social groups have been deduced from the assumption that they were cooperative groups very similar in size and behavior to present-day pongid groups. Non-sharing of food among primates is an important difference from the present-day human species, but there are many other cooperative behavior patterns documented regarding the greater apes. Goodall determined that tool-using, strong mother-child relationships and complex social structures are usual traits of chimpanzee groups.[55] De Waal goes beyond this by reporting a high grade of altruistic reciprocity.[56] Consequently, it is reasonable to assume the presence of a well-developed biological altruism in our ancestors, and, consequently, to think that moral altruism is a characteristic of the human species which emerges from previous biological altruistic behavior.

This phylogenetic point of view places biological altruism and moral altruism neither as opposed polar concepts nor as distinctive peculiarities of different species, but as two successive levels of evolution in the human being's evolutionary journey. If that is so, it should be possible to trace, within our present moral behavior, the characteristics of our less developed altruistic past.

The majority of authors inquiring into moral behavior, above all those interested in its philosophical or juridical aspects, tend to overvalue the aspects related to the ethical criterion. That is understandable, because of the importance of this domain of thought in present human groups. But ethologists such as Lorenz opposed a "biological imperative" to the "categorical" one. The biological imperative is capable of justifying the highly automatic conduct so often displayed when somebody copes with a dangerous situation. Rescuing a child who has fallen onto the tracks while a train is approaching, as some scholars say ironically, does not allow much time to decide between opposite alternatives, means and ends. Once more, unfortunately, the "biological imperative" approach is proposed as an alternative to the rational approach. But some authors, like Alexander,[57] analyze human morality as a phenomenon complex enough to permit both approaches.

---

[55] J. Goodall, *In the Shadow of Man* (London: Collins, 1971).

[56] F. De Waal, *Chimpanzee Politics: Power and Sex among Apes* (New York, N.Y.: Harper & Row, 1982).

[57] Alexander, *The Biology of Moral Systems*.

In terms of the extended model used here, the biological altruistic traits, caused in some way by the genetic code, belong to the "motive domain." On the other hand, moral altruistic traits, less related to the genetic leash, form part either of the "personal ethical domain" or of the "values of the group," depending on their individual or collective dimension. None of these domains alone is able to explain the whole of human moral conduct; rather, one hopes that these considerations will together be useful as a model for analyzing it.

8 Cognitive Aspects of Moral Behavior

The importance of the mutually beneficial relationship between biological (genetically driven) altruism and the emergence of a more complex moral conduct at a time when the human species was developing its cognitive capacities, seems to be beyond doubt. The evolutionary hominid transformations likely shaped primitive moral conduct and made use of it for the appearance of groups whose survival depended on generalized altruistic conduct. The byproducts of such adaptive strategies, based on the cognitive complexity of the human being, are those that brought about our enormous moral wealth.

The burden of phylogenetic adaptations in the combined development of cognitive capacity and moral behavior is a theme that naturalism has dealt with, if not exhaustively, at least with a certain frequency. Authors such as Boehm, Maxwell, Campbell, and Alexander, all from their own theoretical perspectives, have held that biological and cultural evolution should be understood in terms of "co-evolution."[58] Although the origin of morality is viewed in different ways,[59] it is widely acknowledged that human cognitive abilities have an important role in co-evolution.

The most developed model of this co-evolutionary process was proposed by Lumsden and Wilson,[60] who propound the influence of genetic endowment in moral conduct (forming part of the larger area of social conduct) as something that cannot be expressed in terms of "all or nothing." There is a gradual influence, changing in accordance with each type of social behavior, which can only be analyzed through probability curves. In the presence of a determined adaptive problem—for example, food gathering by hunting—certain predators utilize genetically developed, highly instinctive strategies, while others are capable of accumulating "traditions" in which opportune techniques are transmitted to each generation by the learning process. The necessary biological resources for genetically developing certain very complex strategies would be prohibitive, so an elemental principle of parsimony brings us to

---

[58] C. Boehm, "The Evolutionary Development of Morality as an Effect of Dominance Behavior and Conflict Interference," *J. Social Biol. Struct.*, 5 (1982): 413–21; M. Maxwell, *Human Evolution: A Philosophical Anthropology* (London: Croom Helm, 1984); D. T. Campbell, "On the Conflicts Between Biological and Social Evolution and Between Psychology and Moral Tradition," *American Psychologist*, 30 (1975): 1103–26; idem, "The Two Distinct Routes beyond Kin Selection to Ultrasociality," in *The Nature of Prosocial Development*, Bridgeman, ed.; R. D. Alexander, *Darwinism and Human Affairs* (Seattle, Wash.: University of Washington Press, 1979); idem, *The Biology of Moral Systems*.

[59] Moral development has been characterized as stemming from: a deliberate interference in conflicts between cooperating groups (Boehm); a reflection of various tendencies toward kin altruism (Maxwell); a way to ultrasociality through mutual monitoring, forcing altruism on fellow group members (Campbell); a beneficent behavior which follows rules of "give" and "do not give" depending on costs, reciprocity and nepotism (Alexander).

[60] Lumsden and Wilson, *Genes, Mind and Culture*; idem, *Promethean Fire* (Cambridge: Harvard University Press, 1983).

the need of understanding that causation cannot be extended too far. But a totally "free" behavior would be overcome by another in which certain lines of genetic influence would dominate, given that the path of decision making would be too long and complex. Thus, genes and performance link up by means of tendencies that lead to more probable forms of behavior.[61]

Going back to the example of hunting, there exists an entire spectrum of cultural solutions, from simple, bare-handed, individual harassment to complicated group-cooperation strategies employing sophisticated instruments. Lumsden and Wilson conclude that, among all the possible and multiple solutions, there are some that human beings tend toward, thereby accumulating traditions. The tendency is genetically guided, but only in a statistical sense—the "selected behavior" only appears if we rely upon the law of the greatest number. If problems and solutions are relatively simple it is very possible, and even likely, to have a strong genetic link. But as cultural solutions became more complicated, the necessary genetic resources for closely determining them became enormous, and utopian.

Unfortunately, the Lumsden and Wilson model, despite its formalization in differential equations, contains little reference as to how the cognitive process is to be understood. *Promethean Fire* abounds in technical quotations, but the memory model used there is that of Quillian's network and seems to be inadequate for the purposes of supporting complicated problems and probabilistically directed strategies.[62] Quillian's model was far surpassed by a propositional model of a wider spectrum, namely, schemata theory, at the time when the Lumsden and Wilson book was published.[63] And yet it seems logical that if moral conduct is a byproduct of certain factors appearing through evolution, which transform human knowledge into something very complex, some interesting clues should be found.

There are two ways of dealing with the cognitive approach to the biological roots of human morality by means of the classification of moral phenomena already suggested. It is of some interest to discuss how collective and individual aspects of the phenomenon can be related and how the aspects of motive and criterion connect.

If the relationship between the individual and the collective domains (those mentioned when we were speaking of the different aspects of the ethical criterion) is the result of the development of mental schemes, it is possible to make use of a way already opened by the field of cognitive science (from Piaget and Kohlberg to the latest studies of Mandler on the emotional aspects of schemata theory).[64]

---

[61] These lines of probable behavior are expressed through "epigenetic rules," which develop the ancient epigenetic valley concept. For a criticism of the epigenetic rules, see Cela-Conde, *On Genes, Gods and Tyrants*.

[62] M. R. Quillian, "Semantic Memory," in *Semantic Information Processing*, M. Minsky ed., (Cambridge: MIT Press, 1968).

[63] In *Promethean Fire*, Lumsden and Wilson nevertheless allude to the building of knowledge by means of schemes when quoting the Tversky and Kahneman classification through stereotypes; A. Tversky and D. Kahneman, "Judgement under Uncertainly," *Science*, 185 (1974): 1124–31. On the psychology of memory, however, *Promethean Fire*, quotes P. H. Lindsay and D. A. Norman, *Human Information Processing* (New York, N.Y.: Academic Press, 1977). It is a pity that Lumsden and Wilson did not fully appreciate the role played by Lindsay, Norman and the LNR Group in the development of the schemata theory; see G. Marty, *Psicología de la memoria* (Palma de Mallorca: I.C.E., 1987).

[64] G. Mandler, *Mind and Body: Psychology of Emotions and Stress* (New York, N.Y.: Norton, 1984).

Bridgeman suggested that it would be useful to examine infants' and young children's behavior when we are searching for origins of prosocial behavior.[65]

The latter way is more difficult to assess, because we still know very little about the functioning of the mind but, even if they are hypothetical, some lines of interpretation have been suggested. Campbell and Wilson (1978) concur in the amazing fact of the human's ultrasociality: we are creatures that are very easy to instruct and convince.[66] The altruistic tendency could well be justified simply by our apparent necessity to achieve social acceptance.[67] Danielli has even suggested what the mechanism for this tendency to ultrasociality could possibly consist of—the releasing of endogenic opiates (endorphins) into the subcortical centers connected to the cerebral cortex when altruistic behavior is performed. The model, such has been established by Danielli, includes three hypotheses:

1. Children, from a very early age, learn to participate in their social surroundings by means of conduct that is both acceptable and stimulated.
2. All individuals have a system of rewards available to them which, in children, is activated and conditioned by their social environment but which can also be activated later on when individual acts fit in with those early conditionings.
3. The activation of the internal system of rewards releases substances that produce a degree of euphoria favoring the survival of certain types of conduct.[68]

There is nothing novel in Danielli's first hypothesis—it forms the basis of all current psychology—which explains learning by conditioning. The third refers to the existence of endorphins and their role as euphoria inducers. It is the second that favors a naturalistic interpretation of human morality. When discussing the internal system of rewards, Danielli indicates that the type of behavior capable of producing euphoria comes about through religious rituals, acting before public crowds, integration into small communities and, in general, conduct labeled as altruistic.

If biological altruism, like that of a rat defending its nest, is explained in this way, there is no reason to reject similar mechanisms in the case of moral altruism. But, what could be the actual behavior rewarded by the releasing of endorphins? Can we suppose the existence of some universal values (such as the taboo of incest) extensively followed because of their link with brain opiates? The idea of universal norms directing moral behavior and, consequently, having an important role in the phylogeny of the human species is typical of the first stages of sociobiological thought, when simple models of biological altruism (such as kin selection) were directly used to explain human morality, and biological causation of morals was thought of in terms of a direct relation between a set of genes and a controlled behavior. The problem with this line of thought (leaving aside the difficulties in founding universal values in their strongest sense) is, as we have already pointed out,

---

[65] D. L. Bridgeman, "Benevolent Babies: Emergence of the Social Self," in *The Nature of Prosocial Development*, Bridgeman ed.

[66] Campbell, "The Two Distinct Routes beyond Kin Selection to Ultrasociality," in *The Nature of Prosocial Development*, Bridgeman, ed.; Edward O. Wilson, *On Human Nature* (Cambridge: Harvard University Press, 1978).

[67] See Alexander, *The Biology of Moral Systems*, for the analysis of altruistic behavior as a reflection of the need for social acceptance.

[68] J. F. Danielli, "Altruism and the Internal Reward System or the Opium of the People," *J. Social. Biol. Struct.*, 3 (1980): 87–94.

the need for expensive and complicated biological equipment to support the link between gene information and behavior. As moral conduct becomes more complex, the biological mechanisms which are supposed to control it become excessive.

It is important to state that the endorphin hypothesis does not lead necessarily to this kind of determinist model. According to the evidence that we have at our disposal, endorphin release rewards not a particular value but a certain behavior which has had adaptive success. So the fact of accepting and following the moral codes of the group (Campbell's "ultrasociality") may be enough to provoke the endorphin discharge. In that case there is no "universal code of ethics," but rather a "universal tendency to accept moral codes," which is, incidentally, more compatible with the sophisticated models of mature sociobiology, such as the one suggested by Lumsden and Wilson. The relationship between genetic control and moral conduct must be now understood in a weaker sense. If we extend the idea to its limits, we are very close to a neo-Darwinist's position (like Waddington), which supposes that genes limit their control only to a pure disposition for performing moral conduct.[69]

The endorphin theory is a step forward in the search for the roots of morals. But, even limiting the role of endogenous opiates to inside the motivation domain, we must admit that there are certain problems to be solved. How can religious conversions be explained, or changes of political allegiance, or asocial behavior? Hoebel suggested we consider social behavior as a process, the first step of which would consist of innate baggage related to the internal reward mechanisms, with the addition of a second step of apprenticeship depending on the social surroundings.[70] The confluence of innate and acquired traits in human behavior seems to be out of question, but Hoebel's model does not explain non-functional conduct.

To do so, and to establish models of moral behavior that have a certain degree of accuracy, it is essential to rely on more and better knowledge of the mind's information processing methods. Both the role of previous knowledge when processing new information, and the relationship between knowledge production and emotive phenomena, are topics of great interest which have been studied by Mandler through schemata theory and the lack of accommodation between hoped-for events and real events.[71] The biological substratum in the form of tendencies towards moral conduct, the actualization of such tendencies by insertion into a social group, and the role of emotions regarding the maintenance of moral behavior are factors that must be dealt with in order to account for such a complex phenomenon as that of the human morality. State-of-the-art neurobiological theory, such as that of Damasio, strongly links emotional and rational brain circuits as collectively responsible for human action.[72] This means that no rational decision can be made in the absence of emotive phenomena. It seems clear that this is the field where any theory of what human morality is, may find a way forward.

---

[69] Lumsden and Wilson argued against these kinds of "Promethean genes," but the model finally proposed in *Genes, Mind and Culture*, while accepting the limits imposed by the parsimony rule, seems to approach that of Waddington.

[70] B. O. Hoebel, "The Neural and Chemical Basis of Reward," in *Law, Biology and Culture*, M. Gruter, M. and P. Bohannan, eds., (Santa Barbara, Cal.: Ross-Erikson, 1983), 111–28.

[71] Mandler, *Mind and Body*, idem, *Cognitive Psychology: An Essay in Cognitive Science* (Hillsdale, N.J.: Erlbaum, 1985).

[72] A. Damasio, *Descartes' Error: Emotion, Reason, and the Human Brain* (New York, N.Y.: G. P. Putnam, 1994).

SUPERVENIENCE AND THE NONREDUCIBILITY OF ETHICS TO BIOLOGY

Nancey Murphy

1 *Introduction*

The topic of this volume is the relation between evolutionary biology and divine action. This paper considers claims for the relevance of evolutionary biology to ethics, and this may not at first seem related to the topic of divine action. However, Christians and other theists have traditionally recognized a variety of means by which God influences processes within creation. One major way is through action within the sphere of nature; another is in society, influencing the development of culture by means of moral prescriptions. Some of the papers in this volume rightly defend claims regarding God's action in nature.[1] But God's action in the moral sphere has been called into question by many evolutionists as well. It will be my purpose here to show not only that morality cannot be reduced to biology, but also to participate in resuscitating the (no longer popular) view that objective moral knowledge is dependent on a theological (or at least metaphysical) framework.

George Ellis and I have argued that ethics belongs in the hierarchy of the sciences, placed above the social sciences and below theology.[2] If this is the case, then arguments pertaining to the reducibility of one science to another will be relevant to the question of the reducibility of ethics to biology. Historically, the presence of reductionism in science is but an instance of a ubiquitous pattern of thinking that has dominated the modern intellectual world—in ethics, political theory, epistemology, and philosophy of language. There have been important conceptual developments in recent philosophy on the topic of reduction, which I intend to push further in this paper by providing a clearer account of how it is possible for various levels of analysis (here the theological, the ethical, the social-scientific, and the biological) to be analyses of the same (or nested) systems, but without the higher being reducible to the lower.

To escape the pitfalls of inappropriate reductionist thinking, we need a new *concept* to describe the relations between properties, predicates, or event descriptions that belong to different levels of analysis and yet apply to roughly the same entity or event. Until recently only two concepts were available: *identity* and *causation*. For example, if we take mentalistic language to be a higher-level description of neurological events, then the two options were to say that mental states or properties are identical with brain states or that they are caused by brain states. Both options are reductionistic.

We now have a new concept, *supervenience*. Although still a matter of debate, I argue that the right definition allows us to state unambiguously the difference between reducible and nonreducible levels of analysis. It is a short step, then, to show that ethical properties are (ordinarily) not reducible to biological properties.

The discussion of supervenience might also shed light on the topic of divine action, clarifying and strengthening Arthur Peacocke's claim that God works in a

---

[1] See especially Robert Russell's account of divine action at the quantum level in evolution in this volume.

[2] Nancey Murphy and George F. R. Ellis, *On the Moral Nature of the Universe: Theology, Cosmology, and Ethics* (Minneapolis: Fortress Press, 1996).

"top-down" manner. That is, one can use these conceptual developments to explain how an event designated as an act of God could fail to be reducible to an event describable by the laws of nature.[3] The same analysis will prove useful for understanding the nature of the person, making it possible to hold a theologically respectable version of nonreductive physicalism. That is, the person *has* no such thing as a substantial soul or spirit, but this is not to deny the truly spiritual character of the person. Thus, to the question of whether or how the soul evolves, one can answer that the capacities attributed to the soul evolve as the organism evolves, but are not reducible to physical states of the organism. However, I shall not pursue these leads here.

In section 2, I will attempt to show that the logical positivists' program for the reduction of all sciences to physics was but an instance of a pattern of thought predominant throughout the modern intellectual world. Beginning in the seventeenth century, the atomist-reductionist conception of matter was used as a model to understand human reality—sensation, thought, morality, and political organization—as well as nature. The purpose of this historical survey will be to add plausibility to my claim that *a general conceptual shift* is needed to escape from reductionist thinking, and only in light of this shift can we understand the relation between biology and ethics.

Section 3 traces some of the history of "nonreductive physicalism," the view that while there are no additional metaphysical entities in the natural world besides the material, the levels of reality (the living, the conscious, the intelligent, the moral, the religious) are not reducible one to another. This philosophical position is related to recent claims, made familiar through the work of Ian Barbour and reinforced by that of Peacocke, William Stoeger, Robert Russell, Francisco Ayala, and others, that the levels in the hierarchy of the sciences are not entirely reducible to physics.[4] In this section I also mention several earlier attempts to locate ethics in the hierarchy of the sciences.

In section 4 I introduce the concept of *supervenience*, and suggest that a proper understanding of the meaning of this term allows us to state clearly when a higher-level property (predicate, description) is or is not reducible to its lower-level correlate.

In section 5 I consider the relation of ethics to science, summarizing some of the claims Ellis and I have made concerning the placement of ethics above the social sciences in the hierarchy of the sciences. The preponderance of nonreducible descriptions throughout the social sciences themselves makes a *prima facie* case against the reducibility of the ethical to the biological.

However, in section 6 I look at some specific claims for the reduction of ethics to evolutionary biology. I believe that many of these proposals exhibit a common modern failure to appreciate the true nature of the ethical itself. In section 7, drawing

---

[3] See Dennis Bielfeldt, "God, Physicalism, and Supervenience," *Center for Theology and the Natural Sciences Bulletin*, 15.3 (Summer 1995): 1–12.

[4] See Ian Barbour, *Issues in Science and Religion* (New York: Harper and Row, 1966); idem, *Religion in an Age of Science* (San Francisco: Harper and Row, 1990); Arthur R. Peacocke, *Theology for a Scientific Age*, second enlarged ed. (Minneapolis: Fortress Press, 1993); idem, "God's Interaction with the World: The Implications of Deterministic 'Chaos' and of Interconnected and Interdependent Complexity," in *Chaos and Complexity: Scientific Perspectives on Divine Action*, Robert J. Russell, Nancey Murphy, and Arthur R. Peacocke, eds. (Vatican City State and Berkeley, Calif.: Vatican Observatory and Center for Theology and the Natural Sciences, 1995), 263–87.

on the work of Alasdair MacIntyre, I present a nonreductive account of ethics, according to which ethics is essentially dependent on a yet higher level of analysis, namely theology or metaphysics. Some current proposals relating ethics to biology, then, can be seen not as the reduction of ethics to biology, but as the attempt to derive ethics from an ungrounded naturalistic metaphysic.

2 Modern Reductionism

2.1 Atomism and Reductionism in Science

When we think of the transition from medieval to modern science, the Copernican revolution is most likely to come to mind. However, the transition from Aristotelian hylomorphism to atomism has had equally significant cultural repercussions. Yet, probably because the transition happened more gradually—finally completed in biology only in the nineteenth century—this change has received less attention.

Galileo can be given as much credit for this change as for the revolution in astronomy. He was one of the first modern scientists to reject the Aristotelian theory that all things are composed of "matter" and "form" in favor of an atomic or corpuscular theory. His early version of atomism hypothesized that all physical processes could be accounted for in terms of the properties of the atoms, which he took to be size, shape, and rate of motion.

The success of the system of physics developed by Isaac Newton depended on taking *inertial mass* as the essential property of atoms, and on the development of the concept of a *force* as explaining acceleration, not velocity as Aristotle thought. Atomism was extended to the domain of chemistry by Antoine-Laurent Lavoisier and John Dalton, who made great headway in demonstrating that the phenomena of chemistry could be explained on the assumption that all material substances possessed mass and were composed of corpuscles or atoms. This was a striking triumph not only for the atomic theory of matter but also for *reductionism*, that is, the strategy not only of analyzing a thing into its parts, but also of explaining the properties or behavior of the thing in terms of the properties and behavior of the parts.[5]

The atomist-reductionist program continues to bear fruit in contemporary physics, as particle accelerators have made it possible to continue the quest for true "atoms" in the philosophical sense: the most basic, indivisible, constituents of matter. Silvan Schweber states: "Unification and reduction are the two tenets that have dominated fundamental theoretical physics during the present century." He continues:

> With Einstein the vision became all-encompassing. Einstein advocated unification coupled with a radical form of theory reductionism. In 1918 he said, "The supreme test of the physicist is to arrive at those universal elementary laws from which the cosmos can be built up by pure deduction".... The impressive success of the enterprise since the beginning of the century has deeply affected the evolution of all the physical sciences as well as that of molecular biology.[6]

The increasingly successful reduction of chemistry to physics raised the expectation that biological processes could be explained by reducing them to chemistry, and thence to physics. There have been a series of successes here,

---

[5] See Barbour, *Issues in Science and Religion*, for a much more detailed account of this history.

[6] Silvan S. Schweber, "Physics, Community and the Crisis in Physical Theory," *Physics Today*, November 1993: 35.

beginning in 1828, when the synthesis of urea refuted the claim that biochemistry was essentially distinct from inorganic chemistry, and continuing in current study of the physics of self-organizing systems and their bearing on the origin of life. The philosophical question of whether biology could be reduced to chemistry and physics, or whether the emergence of life required additional metaphysical explanation in terms of a "vital force," was hotly debated, but vitalism had almost disappeared by the end of the nineteenth century.

In the twentieth century, a variety of research projects have attempted to carry reductionism into the human sphere. There have been some impressive advances in understanding human psychology by means of biological reduction. We now have clear evidence for biochemical factors in many mental illnesses; theories regarding genetic determination or influence on a wide assortment of human traits are bound to proliferate with the progress of the Human Genome Initiative.[7] These many and valuable instances of successful reduction are taken as evidence for the philosophical thesis of reduction*ism*, the claim that such strategies can give complete accounts of human phenomena.

Thus, much of modern physics and chemistry can be understood as the development of a variety of scientific research programs, in one way or another embodying and spelling out the consequences of what was originally a metaphysical theory. It has been the era in which Democritus has triumphed over Aristotle.

In addition, the reductionist assumption has produced a model for the relations among the sciences. The general tendency of modern science has increasingly been to see the natural world as a total system, and to conceive of it in terms of a hierarchy of levels of complexity, from the smallest sub-particles (at present, quarks and leptons), through atoms, molecules, cells, tissues, organs, organisms, societies, to eco-systems and the universe as a whole. Corresponding to this hierarchy of levels of complexity is the hierarchy of the sciences, physics studying the simplest levels; then chemistry; the various levels of biology; psychology understood as study of the behavior of complete organisms; then sociology.

Causation in this view is "bottom-up": the parts of an entity are located one rung *downward* in the hierarchy of complexity, and it is the parts that determine the characteristics of the whole, not the other way around. So *ultimate* causal explanations are thought to be based on laws pertaining to the lowest levels of the hierarchy.

Francisco Ayala has helpfully distinguished among three sorts of reductionist theses: methodological, epistemological, and ontological.[8] I believe it would be useful to make additional distinctions; there are (at least) six related but distinguishable elements in the reductionist program.

1. Methodological reductionism: a research strategy of analyzing the thing to be studied into its parts.
2. Epistemological reductionism: the view that laws or theories pertaining to the higher levels of the hierarchy of the sciences can (and should) be shown to follow from lower-level laws, and ultimately from the laws of physics.

---

[7] See the paper in this volume by Ted Peters, and further references therein.

[8] See Francisco J. Ayala, "Introduction," in *Studies in the Philosophy of Biology: Reduction and Related Problems,* Francisco J. Ayala and Theodosius Dobzhansky. eds. (Berkeley: University of California Press, 1974), vii-xvi; cf. Barbour, *Religion in an Age of Science,* 165–68; and Peacocke, *God and the New Biology* (London: J. M. Dent and Sons, 1986), chaps. 1 and 2.

3. Logical or definitional reductionism: the view that words and sentences referring to one type of entity can be translated without residue into language about another type of entity.[9]
4. Causal reductionism: the view that the behavior of the parts of a system (ultimately, the parts studied by subatomic physics) is determinative of the behavior of all higher-level entities. Thus, this is the thesis that all causation in the hierarchy is "bottom-up."
5. Ontological reductionism: the view that higher-level entities are nothing but the sum of their parts. However, this thesis is ambiguous; we need names here for two distinct positions. One is the view that as one goes up the hierarchy of levels, no new kinds of metaphysical "ingredients" need to be added to produce higher-level entities from lower. No "vital force" or "entelechy" must be added to get living beings from non-living materials; no immaterial mind or soul needed to get consciousness; no *Zeitgeist* to form individuals into a society. Let us reserve the term "ontological reductionism" for this position.
6. Reductive materialism: I shall use this term to distinguish a stronger claim than the previous one, that only the entities at the lowest level are *really* real; higher-level entities—molecules, cells, organisms—are only composite structures made of atoms. It is possible to hold ontological reductionism without subscribing to this thesis. Thus, one might want to say that higher-level entities are real—as real as the entities that compose them—and at the same time reject all sorts of vitalism and dualism.

The success of methodological reductionism in modern science is what lends credence to these other sorts of reductionism. The crucial metaphysical assumption embodied in this view of the sciences is that the parts of an entity or system determine the character and behavior of the whole, and not vice versa.

2.2 Catachretical Extensions of Atomism

So the atomism that was once pure metaphysics (Democritus) has become embodied in a variety of scientific research programs—it has become scientific theory. Yet, I suggest, it continues to function *metaphysically*, though in a looser sense than that in which Democritus' thesis is metaphysical. Modern thought, not only in the sciences, but in ethics, political theory, epistemology, and philosophy of language, has tended to be *atomistic*—that is, to assume the value of analysis, of finding the "atoms," whether they be the human atoms making up social groups, atomic facts, or atomic propositions. Catachresis is the deliberate use of language drawn from one sphere in order to describe something in another. So we might say that atomism spread by catachresis into a variety of non-scientific spheres.

In addition, modern thought has been *reductionistic* in assuming that the parts take priority over the whole—that they determine, in whatever way is appropriate to the discipline in question, the characteristics of the whole. Thus, the common good is but a summation of the goods for individuals; psychological variables explain social phenomena; atomic facts provide the justifying foundation for more general knowledge claims; the meaning of a text is a function of the meaning of its parts.

It is widely recognized that the individualism of much political philosophy in the modern period is based on the attempt to extend Newtonian reasoning to the sphere of the social:

---

[9] See John R. Searle, *The Rediscovery of Mind* (Cambridge: MIT Press, 1992), 112–14.

the ideas of Newton, Hobbes, and Locke suggested to the social philosophers of the Enlightenment, like Helvétius and Holbach, that individuals in societies were not only analogous to the atomic constituents of physical wholes but were themselves intelligible in terms of a system of quasi-mechanical, hedonic attractions and repulsions. Given knowledge of the laws of human psychological mechanics, individual dispositions could be molded to a socially consistent pattern by an appropriate set of ideal institutions....[10]

While early modern physicists were developing atomist accounts of physics, English philosopher Thomas Hobbes was devising atomist accounts of ethics and politics. Hobbes is best known for his invention of social contract theory. The atomistic individualism of social contract theory is clear: Individuals are logically (if not temporally) prior to the commonwealth, which is an artificial "body." Social facts such as moral obligation and property rights only come into existence as a result of the social contract, which is motivated by the individual drive for self-preservation.

The atomist metaphors and the inherent reductionism of Hobbes' account of individual "dispositions, affections, and manners" are equally striking. Wallace Matson describes the whole of Hobbes' view of reality as follows:

> To the question "What is there to philosophize about?" Hobbes' answer is starkly simple: Bodies....Bodies move. In doing so they move other bodies; that is all that happens.[11]

Thus, in giving an account of sensation, Hobbes emphasizes the effects of the motions of matter in producing internal motions in the perceiver—motions in the head. Ratiocination is, likewise, "motion about the head"; while pleasure is "a motion about the heart." Motions within the body produce passions. All passions can be resolved into two basic forces: attraction or love, and aversion or hatred. Biology, too, is a matter of motion and mechanical causes. Here Hobbes was much influenced by Galen's work on the circulation of the blood.

Matson concludes that for Hobbes the science of motion is the only science:

> Geometry Hobbes thought of as the first and most abstract part of science, for he conceived of geometrical figures not as static entities but as paths of motions: the circle as the motion of a point at a fixed distance from another point, the sphere as the motion of a semicircle around its diameter. The second part of philosophy is the contemplation of the effects of moving bodies on one another in altering their mutual motions. Third, we must investigate the causes of seeing, hearing, tasting, smelling, and touching. Fourth, the sensible qualities—such as light, sound, and heat. These Hobbes calls in Aristotelian fashion Physics. Then we may consider more particularly the motions of the mind (that is, brain): the passions of love, anger, envy, and the like, and their causes, which comprise moral philosophy. Lastly, civil philosophy is the study of the motions of men (including their larynxes and tongues) in common-wealths.[12]

Here we see already a version of the modern conception of the hierarchy of the sciences. Here, too, is an early (reductionist) proposal for including ethics in the hierarchy of the sciences.

---

[10] Stanley I. Benn, "The Nature of Political Philosophy," *Encyclopedia of Philosophy*, vol. 6 (Macmillan: 1967), 391. The relations between science and society were not, of course, one way. Adam Smith's economics, for instance, influenced the development of Charles Darwin's thought.

[11] Wallace Matson, *A New History of Philosophy*, vol. 2 (San Diego, Calif.: Harcourt Brace Jovanovich, 1967), 288.

[12] Ibid.

There are even atomist-reductionist elements in Hobbes' epistemology and theory of language. To understand a concept one resolves (analyzes) it into its constituents. By synthesis (or composition) one can then understand causal relations. Words are marks that stand for thoughts; these marks or signs are connected together to form speech.

Hobbes' social atomism continues throughout the modern period in later versions of social contract theory. Another clear instance of atomism is the utilitarian account of ethics. Here the *only* concept of the common good is a simple summation of the goods for individuals, and individual good reduces further to a psychological variable—pleasure.

Immanuel Kant's approach to ethics brings out an important feature of modern individualism. One feature of atomic theory is the assumption that when one discovers the true atoms they will all be alike, essentially interchangeable. That is, the differences in different composite substances are not because there are ultimately different kinds of matter in the universe, but only because the identical, interchangeable atoms are arranged differently or are involved in different motions.

The contemporary scientific version of this "generic" assumption regarding basic particles is expressed by Schweber as follows:

> By "approximately stable" I mean that these particles (electrons and nuclei) could be treated as *ahistoric* objects, whose physical characteristics were seemingly independent of their mode of production and whose lifetimes could be considered as essentially infinite. One could assume these entities to be essentially "elementary" point-like objects, each species specified by its mass, spin, statistics (bosonic or fermionic) and electromagnetic properties such as its charge and magnetic moment.[13]

Alex Blair has argued that the modern view of the relation of individuals in society ought, correspondingly, to be called "*generic* individualism," to recognize not only the priority of the individual to society, but also the fact that individuals are all alike for such purposes. Thus, the whole is a mere collectivity of identical and interchangeable parts, rather than an interaction of parts with complementary functions. Individuals in society are more like marbles in a bag than like parts in a machine.[14] The generic feature of Kant's individualism is the categorical imperative, that is, the assumption that moral duty for one is, *by definition,* moral duty for each.

One factor that has complicated the modern reductionist program has been the recognition of the determinist implications for human behavior when reductionism is coupled with a deterministic account of the laws of physics: if the body is nothing but an arrangement of atoms, whose behavior is governed by the laws of physics, then how can free decisions affect it? From the very beginning of modernity, there have been two approaches. One is a materialist account of the human person that simply accepts the determinism. The other approach is to propose some form of dualism. Dualists have held that essential humanness is associated with the mind, and thus is quite independent of the workings of mechanistic nature. Then, of course, the problem of mind-body interaction arises.

Hobbes was thoroughly materialist, and accepted the determinist consequences of the reduction of thought and will to physical motions "in the head." The modern dualist account of human reality can be traced Descartes, who argued that material

---

[13] Schweber, "Physics, Community and the Crisis in Physical Theory," 35, emphasis in the original text.

[14] Alexander Blair, "Christian Ambivalence Toward the Old Testament: Corporate and Generic Perspectives in Western Culture" (Ph.D. diss., Graduate Theological Union, 1984).

substance was only part of reality. The other basic metaphysical principle was thinking substance. Human bodies were complex machines, but the person—the "I"—is the mind, which is entirely free.

From dualist roots in Descartes' philosophy has grown the distinction between the natural sciences and the *Geisteswissenschaften* (the social sciences and humanities). One strong tradition throughout the modern period has claimed, over against the logical positivists' program for the unification of science, that these are two radically different kinds of discipline with different methodologies—one dealing with a mechanistically conceived physical nature, the other dealing with the inner life of the mind and mind's cultural products. The debate over the reducibility of the human sciences continues in contemporary philosophy of the social sciences. Methodological individualists say that social events are the product of aggregates of the actions of individuals, while methodological collectivists claim that social wholes obey their own, nonreducible social laws; naturalists hold that social science should aim to reproduce the methodological features of the natural sciences, while anti-naturalists claim that their subject matter calls for a different methodology.

I believe it is fair to say that most philosophical conceptions of *language* in the modern period have also been atomistic. Before Gottlob Frege (1848–1925), the atoms were generally thought to be words. It is the words that have meaning, due to their association with individual ideas, or with objects or classes of objects, while sentence meaning is a function its constituents. After Frege, attention shifted to the sentence or proposition; thus began the search for atomic propositions.

The most extreme form of atomism in *epistemology* appeared in the logical positivists' program. The quest for suitable "protocol sentences" was an attempt to reduce the experiential foundation to its most basic constituents—atomic sentences to describe atomic facts. The manifest failure of Rudolf Carnap's attempt describe the world in terms of sense qualities at point-instants was one of the factors lying behind W. V. O. Quine's development of holist theses regarding both meaning and knowledge. It may be the very extremity and thoroughness of the positivists' program that allowed the atomist-reductionist assumptions of modernity to be recognized *as* assumptions, and thereby to be called into question.[15]

3 Nonreductive Physicalism

A major thesis of this paper is that an understanding of the proper relation between ethics and biology requires a thorough re-evaluation of the modern reductionist pattern of thought. Many of the resources are already in place.

3.1 Roy Wood Sellars

At the same time that the logical positivists were refining and promoting their reductionist program for the sciences, American philosopher Roy Wood Sellars

---

[15] What I have presented here is obviously a one-sided account of modern thought. I argue elsewhere that each of the predominant features of modern thought has produced a reaction, which is still typically modern in that it shares certain crucial assumptions with the mainline position against which it reacts. The best-known counter-modern positions in this case are absolute idealism and romanticism.

It is a judgment call, which I cannot defend here, to designate one side of these ongoing debates as the predominant view and the other as reaction. Certainly the issues will look different from an Anglo-American perspective than from the Continent. See my *Anglo-American Postmodernity: Philosophical Perspectives on Science, Religion, and Ethics* (Boulder, Colo.: Westview Press, 1997), chap. 1.

(1880–1973) was developing a nonreductionist view of the hierarchy of the sciences. He has called his view by a variety of names, including "emergent realism," "emergent naturalism," "evolutionary naturalism." A more common term today, and the one I shall employ in this paper, is "nonreductive physicalism."[16] Sellars began in 1916 to explicate a conception of the mental as an emergent property in the hierarchy of complex systems,[17] and ultimately developed a conception of nature as forming a nonreducible hierarchy of levels.

Sellars' position is expressly opposed to three competitors: Cartesian mind-matter dualism, absolute idealism (the view that the mental and its products are the only reality), and reductive materialism, as he designates the logical positivists' program for the unification of the sciences—that is, for the reduction of all sciences to physics.

According to Sellars, the natural world is one great complex system, displaying levels of complexity, which have emerged over time. In this regard he agrees with the reductive materialists as against the idealists and dualists. However, he criticizes the reductionists for having a view of nature that is overly mechanistic and atomistic. "The ontological imagination was stultified at the start by [the picture] of microscopic billiard balls."[18]

In rejecting this reductive materialism, Sellars argues that "[o]rganization and wholes are genuinely significant"; they are not mere aggregates of elementary particles. Reductive materialism overemphasizes the "stuff" in contrast to the organization. But matter, he claims, is only a part of nature. "There is energy; there is the fact of pattern; there are all sorts of intimate relations." "Matter, or stuff, needs to be supplemented by terms like integration, pattern, function."[19]

> It will be my argument that science and philosophy are only now becoming sufficiently aware of the principles involved in the facts of levels, of natural kinds, of organization, to all of which the old materialism was blind. I shall even carry the notion of levels into causality and speak of levels of causality.[20]

The levels that Sellars countenances are the inorganic, the organic, the mental or conscious, the social, the ethical, and the religious or spiritual. So here we have an early version of the claim that ethics takes its place in the hierarchy of the sciences, between sociology and theology. However, I would argue that Sellars' claims notwithstanding, his accounts of both ethics and religion are reductive. Moral values express judgments regarding the role of an object or act as it connects with human needs or desires.[21] Thus, ethical judgments are merely prudential judgments. Sellars' view of religion is purely humanist: religion is loyalty to the values of life.[22]

Despite Sellars' belief that science and philosophy were already in his day becoming adequately aware of the facts of levels and natural kinds, there are still a

---

[16] This latter term has the advantage both of being widely recognized in contemporary philosophy of mind and also, for Christians, of avoiding the atheistic connotations of "naturalism" and "materialism."

[17] See Roy Wood Sellars, *Critical Realism: A Study of the Nature and Conditions of Knowledge* (New York: Russell and Russell, 1919/1996).

[18] Sellars, *The Philosophy of Physical Realism* (New York: Russell and Russell, 1932/1996), 5.

[19] Sellars, *Principles of Emergent Realism: The Philosophical Essays of Roy Wood Sellars*, W. Preston Warren, ed. (New York: Warren H. Green, Inc., 1970), 136–38.

[20] Sellars, *The Philosophy of Physical Realism*, 4.

[21] Sellars, *Principles of Emergent Realism*, part IV.

[22] Ibid., part V.

large number of ardent reductionists, and theirs has been by far the predominant position in science and Anglo-American philosophy up to the present.[23] However, I believe that the balance is beginning to shift from reductive to nonreductive physicalism, as evidenced by developments in the philosophy of mind, as well as by the writings in the field of theology and science such as those of Barbour and Peacocke. These and a variety of other authors have begun to make it clear that despite astounding scientific advances brought about by means of reductive thinking, there is growing *scientific* evidence for the nonreducibility of higher levels. I first briefly review this evidence and then in the following section consider developments in philosophical concepts that suggest that in certain instances reduction is impossible in principle.

3.2 *Scientific Developments*

It is now recognized by (some) scientists working at a variety of levels in the hierarchy of the sciences that while analysis and reduction are important aspects of scientific enquiry, they do not yield a complete or adequate account of the natural world. In simple terms, one has to consider not only the parts of an entity, but also its interactions with its environment in order to understand it. Since the entity in its environment is a more complex system than the entity itself (and therefore higher in the hierarchical ordering of systems) this means that a "top-down" analysis must be considered in addition to a "bottom-up" analysis.

Silvan Schweber claims that recognition of the failure of the reductionist program has contributed to a crisis in physical theory:

> A deep sense of unease permeates the physical sciences. We are in a time of great change.... [T]he underlying assumptions of physics research have shifted. Traditionally, physics has been highly reductionist, analyzing nature in terms of smaller and smaller building blocks and revealing underlying, unifying fundamental laws. In the past this grand vision has bound the subdisciplines together. Now, however, the reductionist approach that has been the hallmark of theoretical physics in the twentieth century is being superseded by the investigation of emergent phenomena....
> The conceptual dimension of the crisis has its roots in the seeming failure of the reductionist approach, in particular its difficulties accounting for the existence of objective emergent properties.[24]

The demise of reductionism within physics itself can be attributed to the recognition of several related features of the relations among levels of analysis in science: emergence, decoupling, and top-down causation.

"Emergence" or "emergent order" refers to the appearance of properties and processes that are only describable by means of concepts pertaining to a higher level of analysis. The new concepts needed to describe the emergent properties are neither applicable at the lower level nor reducible to (translatable into) concepts at the lower level. The irreducibility of concepts entails the irreducibility of laws. Thus, many say that there are "emergent laws" at higher levels of the hierarchy. Barbour writes:

> The energy levels of an array of atoms in the solid state (such as a crystal lattice) are a property of the whole system rather than of its components. Again, some of the disorder-order transitions and the so-called *cooperative phenomena* have proved impossible to analyze atomistically—for example, the cooperation of elementary magnetic units when a metal is cooled or the cooperative behavior of electrons in a superconductor. Such situations, writes one physicist, "involve a new organizing

---

[23] For a current sophisticated example, see Bernd-Olaf Küppers, "Understanding Complexity," in *Chaos and Complexity*, Russell, Murphy, and Peacocke, eds., 93–105.

[24] Schweber, "Physics, Community and the Crisis in Physical Theory," 34, 39.

principle as we proceed from the individual to the system," which results in "qualitatively new phenomena." There seem to be *system laws* that cannot be derived from the laws of the components; distinctive explanatory concepts characterize higher organizational levels.[25]

"Decoupling" is a technical term in physics, but it can be used more loosely to describe the relative autonomy of levels in the hierarchy of the sciences. Schweber says:

> The ideas of symmetry breaking, the renormalization group and decoupling suggest a picture of the physical world that is hierarchically layered into quasi-autonomous domains, with the ontology and dynamics of each layer essentially quasi-stable [that is, largely stable] and virtually immune to whatever happens in other layers.[26]

In other words, the causal connections among levels in the hierarchy of complexity are being called into question in two ways. There are some changes at the microlevel that make no difference at the macrolevel. At the same time that causal relations from below are being loosened, emergent laws (laws relating variables at the higher level) are coming to be seen as significant in their own right, and not merely as special cases of lower-level laws. "A hierarchical arraying of parts of the physical universe has been *stabilized,* each part with its quasi-stable ontology and quasi-stable effective theory, and the partitioning is fairly well understood."[27]

If strict causal reductionism is denied, and autonomous, higher-level regularities among emergent properties and processes are recognized, the door is open to an even more thorough rejection of reductionism: the recognition of top-down or whole-part causation. It is now coming to be recognized in a variety of sciences that interactions at the lower levels cannot be predicted by looking at the structure of those levels alone. Higher-level variables, which cannot be reduced to lower-level properties or processes, have genuine causal impact. Biochemists were among the first to notice this: chemical reactions do not work the same in a flask as they do within a living organism. The science of ecology is based on recognition that organisms function differently in different environments. Thus, in general, the higher-level system, which is constituted by the entity and its environment, needs to be considered in giving a complete causal account.

Donald Campbell describes relations within the hierarchical orders of biology as follows:

> 1. All processes at the higher levels are restrained by and act in conformity to the laws of lower levels, including the levels of subatomic physics.
> 2. The teleonomic achievements at higher levels require for their implementation specific lower-level mechanisms and processes. Explanation is not complete until these micromechanisms have been specified.

But in addition to these reductionistic requirements, he adds:

> 3. (The emergentist principle) Biological evolution in its meandering exploration of segments of the universe encounters laws, operating as selective systems, which are not described by the laws of physics and inorganic chemistry, and which will not be described by the future substitutes for the present approximations of physics and inorganic chemistry.
> 4. (Downward causation) Where natural selection operates through life and death at a higher level of organization, the laws of the higher-level selective system determine

---

[25] Barbour, *Religion in an Age of Science*, 105, quoting Jonathan Powers, *Philosophy and the New Physics* (New York: Methuen, 1982), chap. 4.

[26] Schweber, "Physics, Community and the Crisis in Physical Theory," 36.

[27] Ibid., 38.

in part the distribution of lower-level events and substances. Description of an intermediate-level phenomenon is not completed by describing its possibility and implementation in lower-level terms. Its presence, prevalence or distribution (all needed for a complete explanation of biological phenomena) will often require reference to laws at a higher level of organization as well. Paraphrasing Point 1, all processes at the lower levels of a hierarchy are restrained by and act in conformity to the laws of the higher levels.[28]

4 *Conceptual Developments: Supervenience*

So it is fairly clear at this point that reduction in science has its limits and that a thorough-going reductionist program is misguided. What is not clear is how to *explain* this fact. That is, our multi-layered analysis of reality is all an analysis of *one* reality. How can it *not* be the case that ultimately the laws of physics govern it all? I have emphasized the ubiquity of reductionist thinking in the modern period; what we need to defeat causal reductionism, as well as to understand the independence of ethics from biology, is a new "thinking strategy."

A variety of philosophers have used the concept of *supervenience* to attempt to give naturalistic but nonreductionistic accounts of morality and of mental events. In 1952, R. M. Hare introduced the term "supervenience" as a technical term to relate evaluative judgments (including ethical judgments) to descriptive judgments. Hare says:

> First, let us take that characteristic of "good" which has been called its supervenience. Suppose that we say, "St. Francis was a good man." It is logically impossible to say this and to maintain at the same time that there might have been another man *placed exactly in the same circumstances* as St. Francis, and who behaved in exactly the same way, but who differed from St. Francis in this respect only, that he was not a good man.[29]

So the higher-level property or description "good" *supervenes* on a collection of descriptions of Francis' character traits and actions. Or, to say the same thing, these character traits and actions constitute Francis' goodness.

In 1970 Donald Davidson introduced the concept of supervenience to describe the relation between mental and physical characteristics. Davidson describes the relation as follows:

> mental characteristics are in some sense dependent, or supervenient, on physical characteristics. Such supervenience might be taken to mean that there cannot be two events alike in all physical respects but differing in some mental respect, or that an object cannot alter in some mental respect without altering in some physical respect. Dependence or supervenience of this kind does not entail reducibility through law or definition...[30]

---

[28] See Donald T. Campbell, "'Downward Causation' in Hierarchically Organized Systems," in *Studies in the Philosophy of Biology: Reduction and Related Problems*, Ayala and Dobzhansky, eds., 180; see also Roger Sperry, *Science and Moral Priority* (Oxford: Basil Blackwell, 1938), chap. 6. For a summary of the literature, see Peacocke, *Theology for a Scientific Age*, chap. 3.

[29] R. M. Hare, *The Language of Morals* (Oxford: Clarendon Press, 1952), 145.

[30] Donald Davidson, *Essays on Actions and Events* (Oxford: Clarendon Press, 1980), 214. Reprinted from *Experience and Theory*, Lawrence Foster and J. W. Swanson, eds., (University of Massachusetts Press and Duckworth, 1970).

The concept of supervenience is now widely used in philosophy of mind,[31] but there is as yet no agreement on its proper definition. Terrence E. Horgan writes:

> The concept of supervenience, as a relation between properties, is essentially this: Properties of type A are supervenient on properties of type B if and only if two objects cannot differ with respect to their A-properties without also differing with respect to their B-properties. Properties that allegedly are supervenient on others are often called consequential properties, especially in ethics; the idea is that if something instantiates a moral property, then it does so *in virtue of*—that is, as a (non-causal) *consequence of*—instantiating some lower-level property on which the moral property supervenes.[32]

Notice that there are two distinguishable notions of supervenience in this passage. In the first sentence (substituting "S" for "A" for clarity, so that S-properties are *supervenient* and B-properties are subvenient or *base* properties):

1. Properties of type S are supervenient on properties of type B if and only if two objects cannot differ with respect to their S-properties without also differing with respect to their B-properties.

But from the last sentence we can construct the following definition:

2. Properties of type S are supervenient on properties of type B if and only if something instantiates S-properties *in virtue of* (as a non-causal consequence of) its instantiating some B-properties.

These two possible definitions are not equivalent: (1) does not entail (2). The reason can be seen in Hare's original use of the term. Francis' character traits and actions (B-properties) only constitute him (or someone like him) a good man (an S-property) *under certain circumstances*. That is, it is conceivable that identical behavior in different circumstances would *not* constitute goodness. For example, we would evaluate Francis' life much differently if he had been married and the father of children.[33]

The difference between these two accounts of supervenience is absolutely crucial. If all higher-level properties are supervenient in the first sense, this ensures the reducibility of the higher level. If supervenient properties are (at least sometimes) supervenient only in the second sense, then, I claim, reduction is not a necessary consequence. Thus, I offer the following definitions (which I take to be equivalent):

3. Property S is supervenient on property B if and only if something instantiates S in virtue of (as a non-causal consequence of) its instantiating B under circumstance c.

---

[31] See, for instance, Jaegwon Kim, *Supervenience and Mind: Selected Philosophical Essays* (Cambridge: Cambridge University Press, 1993); John Heil, *The Nature of True Minds* (Cambridge: Cambridge University Press, 1992); and David J. Chalmers, *The Conscious Mind: In Search of a Fundamental Theory* (New York: Oxford University Press, 1996).

[32] Terence E. Horgan, "Supervenience," in *The Cambridge Dictionary of Philosophy*, Robert Audi, ed., (Cambridge: Cambridge University Press, 1995), 778–79.

[33] A qualification needs to be added here. Someone who wanted to argue for the reducibility of supervenient properties in all cases would point out that anyone whose life was like Francis' in *all* (nonmoral) respects, including his relations to everyone else and everything else in the universe would necessarily have the same moral properties. That is, even if moral properties do not supervene "locally" (in the first, stronger sense) it must be the case that moral properties supervene "globally" on nonmoral properties. We cannot imagine a possible world like this one in all nonmoral respects but differing only in moral respects. I believe that this claim about global supervenience is true but uninteresting for the issues at hand.

4. Property S is supervenient on property B if and only if something's being B constitutes its being S under circumstance c.

An important feature of the supervenience relation, which has long been recognized, is that supervenient properties are often *multiply realizable*. This is a term from computer science—different configurations of hardware (vacuum tubes versus circuits) can realize, constitute, the same machine considered at the functional level. So if S supervenes on B, (given circumstance c), then something's being B entails its being S, but its being S does not entail its being B. For example, goodness is multiply realizable; there are many life patterns different from Francis' that also constitute one a good person. Thus, from the statement "R. M. Hare was a good man" we cannot infer that he lived as St. Francis did. This is one respect in which supervenience relations fail to be identity relations, $S \leftrightarrow B$, since it is not the case that $S \rightarrow B$. (Here arrows represent entailment.)

My definition of supervenience recognizes another way in which supervenience relations fall short of identity. The fact that S supervenes on B does not mean that B entails S ($B \rightarrow S$) because of the dependence upon circumstances. Under c, $B \rightarrow S$, but under c' it may be the case that $B \rightarrow S$. For example, under the circumstance of having a family to support, giving away all one's money may constitute not a good act but an irresponsible one.

For the purposes of getting clear about the use of these terms, we need a very simple example, not confused by the added perplexities associated with either the mind-brain issue or moral issues. Suppose that I have a light in my window and that I have arranged with a friend to use it as a signal to let her know if I am at home or not: on means yes; off means no. I flip the switch; one state of affairs ensues, with two levels of description.

supervenient: the message is "I'm home"
subvenient: the light is on.

It is important to emphasize that there is one state of affairs, two descriptions. Turning the light on *constitutes* my sending the "at home" message under the circumstances of our having made the appropriate prior arrangement.[34]

The "at home" message is multiply realizable. We could have agreed instead that I'd leave the light off if I were home, or we could have agreed to use some other device altogether, such as leaving the window shade up or down.

We need a term to call our attention to an opposite sort of failure of the two descriptions to be identical. Not only is it the case that a variety of subvenient states can realize the same supervenient state (light on, shade up), but also, *again depending on circumstances,* the same subvenient state can constitute a variety of supervenient states. Suppose, for example, that we have agreed that on Mondays the light's being on means I'm in, but on Tuesdays the light's being on means I'm out. So depending on the circumstances of the day of the week, the same subvenient state

---

[34] The fact that it is a single event or state of affairs involved but with two descriptions makes it impossible to ask *which* event is the causative one. The correct question is, which level of description gives us the best insight into the causal connections? A consequence of this understanding for philosophy of mind will be that it becomes impossible to ask whether it is the mental event or the brain event that is the cause; we can ask only whether we can better (or only) understand the causal connections in which this single event is enmeshed if we consider it from the level of neurobiology or the level of thought. I intend to pursue this issue in the next volume in this series.

constitutes either one message or the other. I suppose we could refer to this as "multiple constitutability."

It is this latter feature of the supervenience relation that I mean to highlight by emphasizing the role of circumstances. The variability in circumstances and their role in such cases is what makes for the difference between a supervenience relation and ordinary identity relations, and thus explains why some supervenient descriptions are not reducible to the lower-level. This is the aspect that Horgan's first definition in the quotation above leaves out of account.

I have made up an example where the relations between the supervenient and subvenient levels is arbitrary in order to be able to illustrate possible variations. However, in most cases of supervenience there is no such arbitrariness. David Chalmers recognizes two sorts of supervenience, which he calls logical and natural. The former are meaning relations; for example, part of what we *mean* by "good" is manifestation of certain character traits and behavior patterns. The latter are supervenience relations that obtain in virtue of natural laws; for example, the pressure of a gas (the supervenient level) is related to the mean kinetic energy of its constituent particles (subvenient level) by Boyle's law.

4.1 *Supervenience and Reduction*

Let us now see whether these conceptual resources allow us to distinguish between cases where reduction is and is not possible. The issues that matter seem to be the following: (a) whether there are multiple circumstances such that B constitutes S in circumstance c but it is not the case that B constitutes S in circumstance c'; and if so (b) whether or not c, c' etc. are describable at the subvenient level; (c) whether S is multiply realizable; and if so (d) whether there is a finite disjunctive set of realizands.

Reduction will be possible in the limiting case where B constitutes S under all circumstances and S is not multiply realizable.

The obstacles to reduction will be somewhat different depending on whether we have an instance of logical or natural supervenience. The following table spells this out:

	logical	natural
circumstances	S cannot be defined in B-terms unless c, c', etc. can also be defined in B-terms	There can be no complete set of bridge laws (*i.e.*, laws regarding contingent equivalences) between the S-level and the B-level if there are no such equivalences between c, c', etc. and the B-level.
realization	S cannot be defined in B-terms unless there is a finite set of B's that constitute S.	There can be no complete set of bridge laws if there is no finite set of B's that constitute S.

Table 1

Having made this distinction between logical and natural supervenience, it is important to note that bridge laws are in fact definitions (based on empirical discoveries rather than convention).[35] So in both cases the failure of reduction comes down to an absence of adequate definitions of concepts from the S-level in terms of

---

[35] W. V. O. Quine's arguments for truth-meaning holism, however, suggest that this is not a hard and fast distinction.

concepts from the B-level. Thus, we can summarize as follows. Reduction will not
be possible when:

1. there are multiple circumstances that make a difference to the supervenience relation and these circumstances cannot be defined in terms of the
   subvenient level; or
2. when S is multiply realizable and there is no finite disjunctive set of
   realizands.

The example above wherein the light's being on has opposite meanings on
Monday and Tuesday is an example of the first type of nonreducibility: days of the
week cannot be defined in the language of electrical phenomena.

We cannot use agreed signals as an example of the second type of nonreducibility because this will necessarily be a finite list. Instead consider the variety of
natural signs or evidence of someone's being home: lights on, T.V. on, car in the
garage, etc. Here there is no finite list of states of affairs that constitute the
supervenient state "evidence of someone's being home"; thus there can be no bridge
laws relating the two levels. Of course there may well be cases where both types of
obstacle to reduction are present. (For a more systematic presentation of this
analysis, see the Appendix.)

4.2 *The Supervenience of Moral Language*

To illustrate the supervenience of moral descriptions, consider the event of a person
killing an animal. We can describe the event at the biological level: its throat was cut
and it bled to death. The event can only be described as an *action* at the psychological level, when we include the actor's intention. However, it is not always possible
to describe the intention without referring to still higher levels of description. Why
did he kill it? Maybe a psychological description will suffice: he is sadistic and killed
for pleasure. But perhaps we need a social description: he is the provider for his
family. Or economic: he's taking meat to market because the price is now right. We
might need a religious-level description: it is a sacrifice to the LORD. These various
levels can supervene on one another. He may be slaughtering the animal for sale, but
doing so in the manner prescribed by Jewish dietary laws, in which case the religious
supervenes on the economic, and both supervene on the psychological and biological
levels. The supervenience relation is both asymmetrical and transitive.[36]

Moral descriptions fit into the hierarchy as well: to kill for sadistic pleasure is
immoral; to kill for food, many concede, is morally justified, but even so the
particular act of slaughtering will be moral or immoral depending on circumstances
pertaining to a variety of levels: is it done humanely (biological level); is the animal
one's own property (social and economic)? This example raises the question of
where in the hierarchy of supervenient descriptions the moral level belongs.

5 *Ethics in the Hierarchy of the Sciences*

I have suggested elsewhere that the hierarchy of the sciences cannot be unambiguously ordered above biology, and that it is helpful to think of a branching hierarchy,
with one branch given over to the human sciences (psychology and the social
sciences), and the other to the physical sciences, studying increasingly more
encompassing wholes above biology.[37] In our recent book, Ellis and I concur with

---

[36] See Heil, *The Nature of True Minds*, 65.

[37] See my "Evidence of Design in the Fine-Tuning of the Universe" in *Quantum
Cosmology and the Laws of Nature: Scientific Perspectives on Divine Action*, Robert J.

Peacocke's suggestion that theology be viewed as the topmost science in the hierarchy.[38] We claim, further, that the social sciences provide an incomplete account of the human world insofar as they omit ethics, and thus we add ethics to the hierarchy above the social sciences but below theology.

I emphasize that this hierarchical ordering is a limited model for limited purposes. The natural and human sciences are distinguished not because there are no relations among them, but because it does not seem possible to solve the problem of ordering if they are mixed. One reason is that it is not clear whether the criterion for hierarchization ought to be increasing complexity of the systems studied or more encompassing systems. These criteria overlap considerably at the lower levels but diverge above biology. The human brain is probably the most complex system in the universe. Ellis and I suggest that the applied sciences, including technology, provide important connections between the natural and the human sciences, including ethics. The model is not meant to suggest, either, that sciences are related only to their nearest neighbors.[39]

The major value of the hierarchical model is that it reflects the asymmetrical relations among the sciences. One way of putting it is that states of affairs at a lower level provide necessary but not sufficient conditions for higher-level states of affairs, but the converse does not hold.[40]

Examples of supervenient but nonreducible relations among levels in the social-science hierarchy are easy to come by, as the illustration from the previous section shows. For instance, the economic judgment that it is the right time to sell meat cannot be reduced to biology (although biological factors are necessary conditions—for example, that the animal be fat). Ownership may be represented in the physical world by an artifact such as a bill of sale, but it is the social and legal systems that make the piece of paper a bill of sale, not the physical composition of the paper or the chemicals in the ink.

This is an appropriate point to mention an oversimplification that often appears in discussions of the hierarchy of complex systems. It is often said, for instance, that a society is a complex organization of individuals. In a sense it is true that a society is nothing but the individuals making it up—there is in addition no metaphysical entity such as a *Zeitgeist*. However, it is not the case that everything *pertaining to the social level* is constituted by things pertaining to the individual level. In addition to the individuals, there are social entities such as marriages, contracts, debts, rules, and so on. It will generally be these entities with no correlates at the level of individual persons whose involvement constitutes nonreducible circumstances.

It is not necessary here to repeat the argument that Ellis and I have made for the *scientific* status of ethics,[41] but only to point out that the moral does in fact supervene

---

Russell, Nancey Murphy, Chris J. Isham, eds., (Vatican City State and Berkeley, Calif.: Vatican Observatory and Center for Theology and the Natural Sciences, 1993), especially 424–25.

[38] Murphy and Ellis, *On the Moral Nature of the Universe.*

[39] See Murphy and Ellis, *On the Moral Nature of the Universe*, chap. 4 for an explanation of the ordering of the social sciences. A different way of approaching the question of order is by considering possible supervenience relations between any two disciplines. With this approach, it is clear that the social supervenes on the individual. However, the political and the economic each supervene on the social but neither upon the other in any regular way. The legal then supervenes on both the economic and the political.

[40] This seems an important topic for further reflection.

[41] See Murphy and Ellis, *On the Moral Nature of the Universe*, chap. 5. We claim that

on the biological, as well as on the individual and social levels that we place between biology and ethics. One only needs to show that properties pertaining to each of these levels can *constitute* moral properties *under proper circumstances*. This can be seen from the fact that we can ask of an intentional action, or of a social arrangement or practice, or of an economic transaction, or of a political order, or of a legal system if some aspect of it constitutes it a moral or immoral action, practice, etc. For example, killing is a moral or immoral act in virtue of its being intentional or not; an economic policy is moral or immoral in virtue of its fairness.[42] So it does appear that the moral level supervenes on these various other levels of description, and this suggests, *prima facie*, that insofar as the social level is not reducible to the biological, neither will the moral generally be reducible to the biological. However, the claims of biologists to account for morality in evolutionary terms need to be heard.

6 *Sociobiology and Ethics*

My thesis in this section is that claims to be able to reduce ethics to biology do not turn in any interesting way on new developments in biology, but can be attributed instead to metaethical ignorance. That is, the arguments depend on the absence of any clear sense of what is distinctive about the moral level of analysis—what makes the good *morally* good, as opposed to any other sort of good.

Biologists are not alone, however, in this predicament; it seems to be a pervasive feature of modern Western culture. Alasdair MacIntyre claims that our moral language is in a state of grave disorder. I quote him at length.

> Imagine that the natural sciences were to suffer the effects of a catastrophe. A series of environmental disasters are blamed by the general public on the scientists. Widespread riots occur, laboratories are burnt down, physicists are lynched, books and instruments are destroyed. Finally a Know-Nothing political movement takes power and successfully abolishes science teaching in schools and universities, imprisoning and executing the remaining scientists. Later still there is a reaction against this destructive movement and enlightened people seek to revive science, although they have largely forgotten what it was. But all that they possess are fragments: a knowledge of experiments detached from any knowledge of the theoretical context which gave them significance; parts of theories unrelated either to the other bits and pieces of theory which they possess or to experiment; instruments whose use has been forgotten; half-chapters from books, single pages from articles, not always fully legible because torn and charred. Nonetheless all these fragments are reembodied in a set of practices which go under the revived names of physics, chemistry and biology. Adults argue with each other about the respective merits of relativity theory, evolutionary theory and phlogiston theory, although they possess only a very partial knowledge of each. Children learn by heart the surviving portions of the periodic table and recite as incantations some of the theorems of Euclid. Nobody, or almost nobody, realizes that what they are doing is not natural science in any proper sense at all. For everything that they do and say conforms to certain

---

the logical structure of ethics can be conceived along the lines of Imre Lakatos' account of a scientific research program. Here the "metaphysical hard core" will be a claim about the ultimate purpose (*telos*) of human life. The auxiliary hypotheses will be claims of the form: if you are to achieve your *telos*, then *x* is an appropriate characteristic (rule, social structure) to pursue. These means-ends statements are ordinarily testable. See sec. 7 below.

[42] Of course we can ask, for instance, if it is an economically sound practice to act morally, but the *constitutive* relation does not hold here. That is, the economic soundness of the policy is not in virtue of its morality; if there is such a relationship, it is merely an empirical correlation.

canons of consistency and coherence and those contexts which would be needed to make sense of what they are doing have been lost, perhaps irretrievably.

In such a culture men would use expressions such as "neutrino," "mass," "specific gravity," "atomic weight" in systematic and often interrelated ways which would resemble in lesser or greater degrees the ways in which such expressions had been used in earlier times before scientific knowledge had been so largely lost. But many of the beliefs presupposed by the use of these expressions would have been lost and there would appear to be an element of arbitrariness and even of choice in their application which would appear very surprising to us. What would appear to be rival and competing premises for which no further argument could be given would abound. Subjectivist theories of science would appear and would be criticized by those who held that the notion of truth embodied in what they took to be science was incompatible with subjectivism.

This imaginary possible world is very like one that some science fiction writers have constructed. We may describe it as a world in which the language of natural science, or parts of it at least, continues to be used but is in a grave state of disorder....

What is the point of constructing this imaginary world inhabited by fictitious pseudo-scientists...? The hypothesis which I wish to advance is that in the actual world which we inhabit the language of morality is in the same state of grave disorder as the language of natural science in the imaginary world which I described. What we possess, if this view is true, are the fragments of a conceptual scheme, parts which now lack those contexts from which their significance is derived. We possess indeed simulacra of morality, we continue to use many of the key expressions. But we have—very largely, if not entirely—lost our comprehension, both theoretical and practical, of morality.[43]

I shall say a bit about MacIntyre's explanation for the disorder in modern moral language below, but presently wish to point out that the goal of most modern philosophical ethics has been specifically to *reduce* morality to something else. That is, in the attempt to justify moral claims, philosophers have frequently redefined morality in terms of lower levels of analysis—utility, pleasure, enlightened self-interest, feeling, social convention—and thus have lost sight of the truly moral element. The most extreme case of reduction is emotivism—the claim that moral judgments merely express one's attitudes or feelings toward an action or state of affairs. It is clear that something of traditional moral discourse has been lost here, since moral *argument* no longer makes sense if moral claims are merely expressions of preference. It is like arguing that vanilla is objectively better than chocolate.

Utilitarianism is probably the most influential form of "moral" reasoning in our pragmatic American culture. Here the moral is *defined* as that which results in the greatest good for the greatest number, and good is defined (as for Jeremy Bentham) in terms of pleasure. If there is no justification for the claim that pleasure is the ultimate good for humankind, then morality has simply been reduced to a strategy for increasing pleasure.

Another reductive understanding of morality is cultural relativism—moral norms are nothing but the conventions of the particular societies in which they are found. Thus, the moral level is reduced without remainder to the social level. More complicated social contract theories, from Hobbes to John Rawls[44] have the same result.

I suggest that in light of this history we should not be surprised to encounter attempts to reduce ethics to biology, since this is but further pursuit of the typical

---

[43] Alasdair MacIntyre, *After Virtue: A Study in Moral Theory*, 2nd edition (Notre Dame: University of Notre Dame Press, 1984), 1–2.

[44] See John Rawls, *A Theory of Justice* (Cambridge: Harvard University Press, 1971).

modern reductionist strategy. Furthermore, the *appearance* of successful reduction on the part of evolutionary biologists depends on having already redefined ethics in a reductionist manner.

E. O. Wilson is explicit about his desire to promote reduction:

> It may not be too much to say that sociology and the other social sciences, as well as the humanities, are the last branches of biology waiting to be included in the Modern Synthesis [neo-Darwinist evolutionary theory]. One of the functions of sociobiology, then, is to reformulate the foundations of the social sciences in a way that draws these subjects into the Modern Synthesis.[45]

In an often quoted passage, Wilson makes equally clear his intention to reduce ethics to biology:

> Self-knowledge is constrained and shaped by the emotional control centers in the hypothalamus and limbic system of the brain. These centers flood our consciousness with all the emotions—hate, love, guilt, fear, and others—that are consulted by ethical philosophers who wish to intuit the standards of good and evil. What, we are then compelled to ask, made the hypothalamus and limbic system? They evolved by natural selection. That simple biological statement must be pursued to explain ethics and ethical philosophers....[46]

Philip Kitcher has already offered a penetrating criticism of Wilson's metaethical position:

> Stripped of references to the neural machinery, the account Wilson adopts is a very simple one. The content of ethical statements is exhausted by reformulating them in terms of our emotional reactions.
>   ...Wilson's rush to emotivism depends on slashing the number of alternatives. Only two possible accounts of ethical objectivity figure in Wilson's many pages on the topic. One of these, the attempt to give a religious foundation for ethics, does not occur in my list of options. Wilson mentions religious systems of morality only to dismiss them; his reason is spurious: "If religion... can be systematically analyzed and explained as a product of the brain's evolution, its power as an external source of morality will be gone forever."[47]

Kitcher himself then goes on to endorse Rawls' account of justice as based on social agreement.

Richard Alexander in "A Biological Interpretation of Moral Systems," defines morality as nothing but social convention:

> A moral system is essentially a society with rules. Rules are agreements about what is permitted and what is not, about what rewards and punishments are likely for specific acts, and about what is right and wrong.[48]

The content of morality, according to Alexander, is altruism:

> Generally speaking, then, *immoral* is a label we apply to certain kinds of acts by which we help ourselves or hurt others, while acts that hurt ourselves or help others are more likely to be judged moral than immoral. As virtually endless arguments in

---

[45] Edward O. Wilson, *Sociobiology: The New Synthesis* (Cambridge: Harvard University Press, 1975). Excerpt in *Issues in Evolutionary Ethics*, Paul Thompson, ed. (New York: SUNY Press, 1995), 156.

[46] Wilson, *Sociobiology*, 153.

[47] Philip Kitcher, "The Hypothalamic Imperative," chap. in *Vaulting Ambition* (Cambridge, Mass.: MIT Press, 1985). Reprinted in *Issues in Evolutionary Ethics*, Thompson, ed., 207, 211. Quotation is from Wilson, *On Human Nature* (Cambridge: Harvard University Press, 1978), 201.

[48] Richard D. Alexander, "A Biological Interpretation of Moral Systems," *Zygon* 20. Reprinted in *Issues in Evolutionary Ethics*, Thompson, ed., 179.

the philosophical literature attest, it is not easy to be more precise in defining morality *per se.*[49]

So here Alexander is making specific reference to the lack of consensus on metaethical issues, and argues that biological knowledge can help "take the vagueness our of our concept of morality."[50]

Michael Ruse presents a more sophisticated argument for evolutionary ethics than many of his predecessors. He is specific about the metaethical problem:

A full moral system needs two parts. On the one hand you must have the "substantival" or "normative" ethical component. Here, you offer actual guidance as in, "Thou shalt not kill." On the other hand, you must have (what is known formally as) the "metaethical" dimension. Here, you are offering foundations or justification as in, "That which you should do is that which God wills." Without these two parts, your system is incomplete.[51]

Ruse is to be commended for recognizing the difference between "altruism" as a moral term and "altruism" as it is used in biology to describe animal behavior that contributes to the survival of the group. I shall use Francisco Ayala's terms: "altruism^m" for the moral concept; "altruism^b" for the biological concept.[52] Ruse suggests that whereas insects and lower animals are genetically programed for altruism^b, humans have instead been selected for a disposition toward altruism^m. Thus, he is able to argue for an evolutionary source for altruism^m without confusing it with altruism^b.

However, having properly distinguished moral behavior from superficially similar animal behavior, Ruse then goes on to argue that morality, thus properly understood, has no possible rational justification:

The evolutionist is no longer attempting to derive morality from factual foundations. His/her claim now is that there are no foundations of any sort from which to derive morality—be these foundations evolution, God's will or whatever.[53]

Since there can be no rational justification for objective moral claims, what is needed instead is a causal account of why we believe in an objective moral order. Ruse's answer is that the survival value of altruism^m does in fact provide such an explanation.

In particular, the evolutionist argues that, thanks to our science we see that claims like "you ought to maximize personal liberty" are no more than subjective expressions, impressed upon our thinking because of their adaptive value. In other words, we see that morality has no philosophically objective foundation. It is just an illusion, fobbed off on us to promote [altruism^b].[54]

So Ruse's account, while more sophisticated than Alexander's or Wilson's in that he fully appreciates the conceptual difference between morality on the one hand and sentiment, convention, and so on, on the other is most starkly reductive: moral objectivity is merely an illusion.

These are but a sampling of the positions of evolutionary ethicists, but I hope the sample is adequate to illustrate the following points:

---

[49] Ibid., 180.

[50] Ibid., 188.

[51] Michael Ruse, "Evolutionary Ethics: A Phoenix Arisen," *Zygon* 21(1986). Reprinted in *Issues in Evolutionary Ethics*, Thompson, ed., 226.

[52] See Francisco J. Ayala, "The Biological Roots of Morality," in *Biology and Philosophy* (1987). Reprinted in *Issues in Evolutionary Ethics*, Thompson, ed.

[53] Ruse, "Evolutionary Ethics: A Phoenix Arisen," 234.

[54] Ibid.

1. There is pervasive lack of clarity on metaethical issues; it is not clear at all what is the nature of the moral itself; as MacIntyre says, our moral discourse is in a state of grave disorder.
2. One important motivation for the attempt to reduce ethics to biology is the hope of clarifying or justifying moral or metaethical claims.
3. The reduction of morality to biology generally depends on a prior reduction of morality to emotion or social convention.

7 A Nonreductive Account of Ethics

If ethics is not to be reduced to the psychological (pleasure, emotion) or to the social (social conventions, rules), or to the political (social contracts), then what is its true nature? I suggest that we can answer this question best by considering again the proposal that ethics is a science (or discipline) that falls between the social sciences and theology (or metaphysics) in the hierarchy of the sciences.

It has been argued vociferously during the past decades that the social sciences are value-free. However, the burden of proof seems now to have shifted; it is widely recognized that economics, sociology, and political theory all involve what can only be called *moral* presuppositions. For example, there is the assumption in economics that self-interest, rather than benevolence, is normative and if left unchecked will lead to greater good in the end. There is the assumption in sociology that all social order is based on violence or the threat of violence and so violence is morally justifiable; in political theory there is the assumption that justice, as opposed to love, is the highest good at which government can aim; that freedom is an ultimate human good; that life, liberty, property, and the pursuit of happiness are *natural* rights.

When called into question, these assumptions all raise ethical questions, which cannot be answered by any scientific means. The social sciences are suited for studying the relations between means and ends (for example, if your goal is to avoid surpluses and shortages, the best economic system is the free market), but they are not suited for determining the ultimate ends or goals of human life. This is, instead, the proper subject matter of ethics.

Alasdair MacIntyre, whom I quoted above, argues that modern ethics is in disarray because what ethics used to be was a discipline that taught how to act in order to reach humankind's ultimate goal. The concept of humankind's *telos* was provided by either a metaphysical or a theological tradition, which informed the ethicist of the nature of ultimate reality. Two contrasting examples are: 1) the purpose of human existence is to know, love, and serve God in this life and to be happy with God in the next (cf. *The Baltimore Catechism*); or 2) a good life consists in courage in the face of the absurdity of it all, since the human race is a small-scale accident in a meaningless cosmic process.

When Enlightenment ethicists severed all ties between morality and tradition (here, largely the Christian tradition) they kept fairly traditional lists of moral prescriptions, but lost all concepts of the end for which those prescriptions were the means of achievement. So a consequence of the "autonomy of morals" was confusion about the very nature of ethics as a discipline, about the very nature of moral truth. Modern representational theories of language and knowledge led ethicists to ask what kind of "objective realities" moral prescriptions or ascriptions might represent. Failing to find any such "realities" it seemed necessary to reduce moral claims to some other kind of claims that could be verified (for example, utility); yet there was the cultural memory that moral claims must be something more than that, and so we memorize and repeat Hume's law as an incantation—"one cannot deduce an 'ought' from an 'is'"—and worry about committing "the naturalistic fallacy."

All of this confusion, taken up directly into discussions of biology and ethics, as we have just seen, is due to having forgotten that the "ought" statement—"you ought to do (or be) x"—is only half of a moral truth. The original form of an ethical claim (implicitly, at least) is, "if you are to achieve your *telos*, then you ought to do (or be) x." This latter sort of ethical claim can be straightforwardly true or false; the "ought" is no more mysterious than the "ought" in "a watch ought to keep good time." Furthermore, it can and in fact *must* be derived from certain sorts of "is" statements: about the nature of ultimate reality, about regularities in human life regarding the achievement of ends as a result of adopting certain means—the latter being amenable to empirical study.

Let us now consider the sorts of supervenience relations that obtain as we work up the scale from the personal (psychological) level to the theological. MacIntyre's own constructive ethical work involves the restoration of the ancient and medieval approach couched in the language of the virtues. The supervenience relations here are clear.

A virtue is an acquired human characteristic (psychological level). However, what constitutes a characteristic a virtue rather than a vice or a morally neutral characteristic is a set of circumstances or contexts that can only be described at a higher level. The first level of relevant context is a "social practice," which MacIntyre defines as follows:

> By a 'practice' I am going to mean any coherent and complex form of socially established cooperative human activity through which goods internal to that form of activity are realized in the course of trying to achieve those standards of excellence which are appropriate to, and partially definitive of, that form of activity, with the result that human powers to achieve excellence, and human conceptions of the ends and goods involved, are systematically extended.[55]

Practices are found throughout the hierarchy of levels of complexity studied by the social sciences, and include at the social level, games, marriage, medicine, science; at the economic level, businesses and regulatory practices; at the political level practices such as legislation and campaign management; at the legal level, the practice of law itself, the interpretation of the constitution.

An acquired characteristic necessary for attaining goods internal to a practice (for example, *fidelity* in the practice of marriage, *honesty* in business) is a *candidate* for a virtue. However, practices themselves require moral evaluation, and here we reach the level of ethics itself. Each such practice (as well as an individual's decision to participate in it) must be evaluated according to its contribution to the *telos* of human life.

So we may ask questions such as: does the practice of marriage contribute to the ultimate goal of human existence, and if so what is its proper purpose? For those who believe that the natural world is itself the ultimate reality, the only answers can be propagation of the species and personal support and pleasure. However, for those who believe that the Christian tradition provides a more adequate account of ultimate reality, marriage is a practice with additional goals, perhaps even superseding those of the naturalists' account—to witness to the faithfulness of God to God's people.

So the predicate "is a virtue" supervenes on some psychological characteristic such as fidelity or humility only in the (nonreducible) circumstances of its being

---

[55] MacIntyre, *After Virtue*, 187.

essential for attaining goods internal to social practices that in turn further the ultimate goals of human existence. Furthermore, a chosen end (survival of the race, the glorification of God) is in fact an *ultimate* goal only if it is congruent with the true (ultimate) nature of reality itself.[56]

Now, back to the question of the relation between ethics and biology. First, we can extend the subvenience relations of virtues downward in the hierarchy to the biological level. MacIntyre emphasizes that virtues are *acquired* human characteristics and so on his definition a genetically determined trait is not a candidate for a virtue. However, the habit of acting on a genetically produced capacity or propensity (say, for altruism) is a candidate for a virtue, but whether or not it is in fact a virtue depends finally on one's theological or metaphysical account of reality.

G. Simon Harak has proposed that virtues are realized physiologically or neurologically. That is, the practice of virtuous behavior actually brings about changes in one's physical makeup or functioning such that further virtuous behavior comes closer to being automatic. If this is the case it is an interesting example of top-down causation from the ethical to the biological level.[57]

This is but one instance where biological knowledge can be of use in ethics. The hierarchical ordering I have proposed, with ethics included in the hierarchy, implies that ethics (and theology) will have as much to learn from lower-level sciences (including biology) as any higher-level science does from levels below.[58]

There are some important parallels between the position developed here and the views of Francisco Ayala. First, I want to endorse his distinction between the evolution of the capacity for moral thought and behavior and the evolution of the content of moral codes. I also endorse his claim that the content of our moral systems comes not from the evolutionary process but from religious and other social traditions.

Ayala distinguishes between *religious faith* as a *motivation* to behave morally and *theology* as a resource for the *justification* of moral principles. In the latter case, a set of religious beliefs may contain propositions about human nature and the world from which the ethical norms can be derived.[59]

My claim in this essay is that, in fact, *all* moral systems are ultimately dependent *for their justification,* either explicitly or implicitly, on beliefs about the nature of ultimate reality.

Thus, I come to a second sort of claim that has been made for the relevance of evolutionary biology to ethics—claims such as those of Jacques Monod that because the evolutionary process shows no evidence of intelligent direction we must adjust our views of human nature and behavior to take account of the ultimate meaningless-

---

[56] MacIntyre recognized that the dependence of ethical systems on accounts of ultimate reality would lead to relativism if no account could be given of the rational evaluation of these larger traditions. For his answer to this epistemological problem, see MacIntyre, *Whose Justice? Which Rationality?* (Notre Dame: University of Notre Dame Press, 1988); idem, *Three Rival Versions of Moral Enquiry: Encyclopaedia, Genealogy, and Tradition* (Notre Dame: University of Notre Dame Press, 1989).

[57] See G. Simon Harak, *Virtuous Passions: The Formation of Christian Character* (New York: Paulist Press, 1993).

[58] Another instance: Paul Churchland's neurological account of learning based on prototypes rather than rules suggests that morality is best understood in terms of narratives portraying virtues, not in terms of rules. See Paul M. Churchland, *The Engine of Reason, the Seat of the Soul* (Cambridge: MIT Press, 1995), 143–50.

[59] See Ayala, "The Biological Roots of Morality," esp. 302–3.

ness of human existence. It seems clear that this is no direct argument from biology to ethics, but is an argument (a fallacious one) from biology to metaphysics or theology ("there is no intelligent creator directing the process") and then *downward* from the metaphysical view to ethics.

However, such an argument is naive when we consider the many levels of analysis involved. The shift from Aristotelian to modern biology can be described as the recognition that *intentional purpose* is not a category that applies at the purely biological level at all, but only begins to apply in the natural order at the psychological level.[60] Furthermore, God's purposes in biological processes will not be apparent to the biologist *qua* biologist, but only recognizable by seeing the sweep of evolutionary changes within the broader context of God's creation and, thus, from a theological viewpoint.[61] We might say to Monod and company: "You have not made an empirical discovery that evolution has no purpose or goal; to think that you could make such a discovery involves a category mistake. It is necessarily true (a conceptual claim) that evolution is purposeless. So this recognition cannot be a significant fact about the universe; it is not a fact, strictly speaking, at all (that is, except in the strained sense that it is a fact about language). Nothing of metaphysical interest clearly follows from it. And it is no more to be lamented than the claim that plants have no economic interests."[62]

8 *Conclusion*

It is not surprising that attempts should be made to reduce ethics to biology, since reductionism has been a central "thinking strategy" of the modern period, not only among the natural sciences but in all intellectual disciplines. Such attempts are even less surprising if we agree with MacIntyre that moderns have lost earlier conceptions of the very nature of the ethical. The rejection of reductionism in ethics requires two moves. One is the restoration of the "top-down" connections from theology (or metaphysics) to ethics. The other is a recognition of the limitations of reductionist thinking in general, which in turn depends on an adequate account of the nature of the supervenience relation.

9 *Appendix*

Here I attempt in a slightly more formal way to make good on my claim that supervenience relations, as I have defined them, do *not* automatically entail reducibility of the supervenient level to the subvenient level (although reduction will be possible in some cases).

The central issue is the following: my definition of "supervenience" involves *circumstances* as a co-determinant of a supervenience relation; S supervenes on B only in circumstance c. Therefore, in sorting out causal relations, c has to be taken into account. Let us symbolize supervenience relations as follows, where the dollar sign represents a supervenience relation (under condition c).

---

[60] See Ayala's paper in this volume (section 6, beginning on p. 109) for a set of distinctions among concepts of purpose. My claim here involves his concept of *purposeful activity*.

[61] I strongly endorse Russell's account of divine action in the evolutionary process. On this account these divine acts will appear, at the biological level, to be merely random.

[62] Before the reader objects that surely I cannot dismiss this whole area of debate so quickly, let me say that I do not deny that there are important issues regarding the consistency of a metaphysical or theological world-view with the findings of science, including evolutionary biology. My point is only that one has to recognize the differences in levels of discourse and relate them much more carefully.

$$S$$
$$\$ \text{ (c)}$$
$$B$$

Now, let us assume causal relations (symbolized by a solid arrow) at the base level. We can imagine a variety of cases such as the following:

$$S_1$$
$$\$ \text{ (c)}$$
$$B_1 \rightarrow B_2$$

but:

$$S_2$$
$$\$ \text{ (c')}$$
$$B_1 \rightarrow B_2$$

and also:

$$S_1$$
$$\$ \text{ (c'')}$$
$$B_3 \rightarrow B_4$$

Because regularity is our primary criterion for causation, we see that there are no causal laws relating prior B events to S events; in each case the relation is different. (Here outlined arrows represent mere sequence of events, not causal laws.)

$$B_1 \Rightarrow S_1$$
$$B_1 \Rightarrow S_2$$
$$B_3 \Rightarrow S_1$$

The reductionist would now want to claim that the circumstances can simply be entered into the equation to produce the requisite causal laws. If c already belongs to the subvenient level, causal reductionism is possible, so long as there is a finite and manageable set of relevant circumstances and we have the opportunity to test the effects of each. We then might get a set of causal laws relating the supervenient properties to subvenient antecedents:

$$(B_1 \text{ \& } c) \rightarrow S_1$$
$$(B_1 \text{ \& } c') \rightarrow S_2$$
$$(B_3 \text{ \& } c'') \rightarrow S_1$$

However, it is often the case that c is a factor belonging to the supervenient level or to a higher level still. For example, when relating the psychological and the neurological levels, c may pertain to the supervenient mental or the (still higher) social level. Reductionists have always claimed that no environmental variable could affect individual behavior without being *realized* neurologically. But multiple realizability may block reduction in some cases. Here the relevant issue is the *logical* reduction of the circumstances to the subvenient level. This involves translatability, which requires definition of the (supervenient) c in terms of the subvenient vocabulary, or "bridge laws" relating c to the subvenient level. Sometimes this will be possible, even if c is multiply realizable: we can define c as the disjunctive set $\{B_x \text{ or } B_y \text{ or } B_z\}$; c' as the disjunctive set $\{B_s \text{ or } B_t\}$ and c'' as the set $\{B_q \text{ or } B_r\}$. In such a case, a set of laws relating the subvenient to the supervenient levels can be established, although it will be cumbersome:

$$(B_1 \text{ \& } B_x) \rightarrow S_1$$
$$(B_1 \text{ \& } B_y) \rightarrow S_1$$
$$(B_1 \text{ \& } B_z) \rightarrow S_1$$

and:

$$(B_1 \text{ \& } B_s) \rightarrow S_2$$
$$(B_1 \text{ \& } B_t) \rightarrow S_2$$

and:

$$(B_3 \text{ \& } B_q) \rightarrow S_1$$
$$(B_3 \text{ \& } B_r) \rightarrow S_1$$

However, if there is no finite disjunctive set of realizands for c, c', and c'', then this set of laws will never be completed, and no causal account of the supervenient variable in terms of the subvenient level can be supplied. For all we know, $(B_3 \,\&\, B_n)$ will produce an entirely unexpected result. This shows clearly, I believe, that causal reduction is not always possible. There may not be laws relating subvenient antecedents to supervenient consequents.

Donald Davidson has argued for a position that he calls anomalous monism. "Anomalous monism resembles materialism in its claim that all events are physical, but rejects the thesis, usually considered essential to materialism, that mental phenomena can be given purely physical explanations."[63] I believe his position can be represented as follows:

$$S_0 \Rrightarrow S_1$$
$$\| \qquad \$ \; (c)$$
$$B_1 \rightarrow B_2$$

$$S_0 \Rrightarrow S_2$$
$$\| \qquad \$ \; (c')$$
$$B_1 \rightarrow B_2$$

$$S_3 \Rrightarrow S_1$$
$$\| \qquad \$ \; (c'')$$
$$B_3 \rightarrow B_4$$

Here, I am assuming that some supervenient (mental) events are context dependent and that others are reducible (the antecedent mental event in each case is, for simplicity, assumed to be reducible to a physical event). Still, bracketing the circumstances for the present, we fail to find supervenient laws (that is, laws relating mental events to one another):

$$S_0 \Rrightarrow S_1$$
$$S_0 \Rrightarrow S_2$$
$$S_3 \Rrightarrow S_1$$

However, there is a more important consequence. If the goal of the nonreductive physicalist is to show that it is the supervenient level that matters for explanatory purposes (in this case, the mental), then we ought to include the circumstances (which are sometimes not reducible): then we do find laws (c, c', and c'' being already defined at the supervenient level):

$$(S_0 \,\&\, c) \rightarrow S_1$$
$$(S_0 \,\&\, c') \rightarrow S_2$$
$$(S_3 \,\&\, c'') \rightarrow S_1$$

So, without assuming at all that regular causal laws fail to hold at the subvenient level, it is possible to show that these lower-level laws may not always determine sequences of events at the supervenient level. There may well be Davidson's "anomalousness" at the supervenient level. A more important point is that the actual causal regularities may only show up at the supervenient level, because it may be the case that the critical circumstances can only be described in the language of the supervenient level. That is, the circumstances will be extraneous factors invisible at the subvenient level, yet interfering with the otherwise stable regularities at that subvenient level. Recall that all scientific laws carry a tacit *ceteris paribus* clause; they hold in closed systems, barring outside interference.

---

[63] In "Mental Events," chap. in Davidson, *Essays on Actions and Events*, 214.

PLAYING GOD WITH OUR EVOLUTIONARY FUTURE[1]

Ted Peters

"The theory of evolution," writes Wolfhart Pannenberg, "has given theology an opportunity to see God's ongoing creative activity not merely in the preservation of a fixed order but in the constant bringing forth of things that are new."[2] Two things stand out as important here. First, the order of God's creation is not fixed. It changes. Second, God's creative work brings forth newness. Does newness also apply to human nature? To human creativity? To the creative self-alteration of human nature?

Karl Rahner would answer affirmatively to such questions. He describes the evolutionary history of the human race in terms of "becoming." Human becoming consists in the self-transcendence of living matter. Nature is historical in character, and its history has a direction. It is directed toward the development of freedom in the spirit. And it will not stop there. Nature will progress through and beyond the human stage toward the consummation of the cosmos as a whole. This consummation is a fulfillment inclusive of the free human spirit. Such freedom means that the human contribution to nature's history includes "an active alteration of this material world itself."[3] We human beings will apply our "technical, planning power of transformation" even to ourselves. As subject we are becoming our own object, becoming our own creator.[4]

The concept of evolution here does not apply only to the past. It applies to the future as well. Rather than a fixed reality, nature is in the process of becoming; and this becoming is subject to human influence. The ethical question we raise is this: should we influence the future course of evolutionary history, especially our own human development? Some would shout "No." When shouting, they might add a commandment: "Thou shalt not play God!" "Playing God" is the phrase invoked by many to stop the attempt by the human race to influence the future of its own evolution.

Although the phrase, "playing God," sounds theological, it is not. Or, at least it is not terminology theologians typically employ. The phrase belongs more to common parlance and is aimed at inhibiting if not shutting down certain forms of scientific research and medical therapy. In the current public debate over the impact of science on society, the proscription against "playing God" is heard particularly in the field of human genetics. More particularly, it is aimed at the prospect of germ-line intervention for purposes of human enhancement. Germ-line intervention is understood as the insertion of new gene segments of DNA into sperm or eggs before fertilization or into undifferentiated cells of an early embryo that will be passed on to future generations and may become part of the permanent gene pool.[5] Some

---

[1] This paper expands upon material in previous publications: *Playing God? Genetic Determinism and Human Freedom* (New York: Routledge, 1996) and "Playing God and Germ-line Intervention," *Journal of Medicine and Philosophy*, 19 (October, 1995): 365–86.

[2] Wolfhart Pannenberg, *Systematic Theology*, 3 vols. (Grand Rapids, Mich.: Eerdmans, 1991–96), II:119.

[3] Karl Rahner, "Christology within an Evolutionary View," *Theological Investigations*, 22 vols. (London: Darton, Longman, & Todd, and New York: Crossroad, 1961–88), 5:168; see also 21:54.

[4] Rahner, *Theological Investigations*, 5:137–38.

[5] Burke K. Zimmerman identifies "three strategies" for germ-line intervention:

scientists and religious spokespersons are trying to lock the gate to germ-line enhancement and post a no-trespassing sign reading, "Thou shalt not play God."

Our task here will be to examine the concept of playing God in conjunction with arguments favoring and opposing germ-line enhancement, especially the arguments raised by the Council for Responsible Genetics in its "Position Paper on Human Germ-Line Manipulation." We will look at both the concept of "playing God" plus these arguments in light of the Christian theology of creation. The concept of creation includes anthropology and the notion that the human race is created in the divine image.[6] I will argue that if we understand God's creative activity as giving the world a future, and if we understand the human being as a created co-creator, then ethics begins with envisioning a better future. This form of ethical thinking I dub 'proleptic' or 'anticipatory'. A proleptic ethic suggests we should at minimum keep the door open to improving the human genetic lot and, in an extremely modest way, to influencing our evolutionary future. The derisive use of the phrase, "playing God," should not deter us from shouldering our responsibility for influencing the future. To seek a better future is to "play human" as God intends us to.[7]

This work belongs within the larger field of theology and natural science. It shares the assumption made by Langdon Gilkey, when he writes: "Evolutionary science has taught us how we humans have appeared in all facets of our being *in and through* the processes of nature; hence, a theological understanding of human being must also be informed by a biological understanding."[8]

---

1) screening and selection of early stage embryos; 2) direct modification of the DNA of early stage embryos coupled with *in vitro* fertilization; and 3) genetic modification of gametes prior to conception. Although screening is not usually included in germ-line discussion, it along with the other two strategies would affect the future human gene pool. "Human Germ-Line Therapy: The Case for its Development and Use." *Journal of Medicine and Philosophy*, 16.6 (December, 1991): 593–612, esp. 594–95. See also Gregory Fowler, Eric Juengst, and Burke K. Zimmerman, "Germ-Line Therapy and the Clinical Ethos of Medical Genetics," *Theoretical Medicine*, 10 (1989): 151–65.

[6] Pope John Paul II places anthropology within creation when speaking of the humans among us as "products, knowers and stewards of creation." *Physics, Philosophy and Theology: A Common Quest for Understanding*, Robert J. Russell, William R. Stoeger, and George V. Coyne, eds. (Vatican City State: Vatican Observatory, 1988), p. M5.

[7] Paul Ramsey writes, "Men ought not to play God before they learn to be men, and after they have learned to be men they will not play God." Paul Ramsey, *Fabricated Man: The Ethics of Genetic Control* (New Haven, Conn.: Yale University Press, 1970), 138. The question of playing God in genetic intervention is only one of many reasons for inviting theological attention into this field. M. Therese Lysaught formerly at the Park Ridge Center for the Study of Health, Faith, and Ethics writes, "a Christian theological analysis of the Human Genome Project and genetics needs to examine a host of questions in addition to the question of human intervention into nature—for example, questions of theodicy, of divine agency, of theological anthropology, of social justice, of the meaning of suffering within a Christian theological framework, of the meaning of Christian community, as well as methodological questions surrounding the science/religion dialogue." M. Therese Lysaught, "Map, Myth, or Medium of Redemption: How Do We Interpret the Human Genome Project," *Second Opinion*, 19.4 (April, 1994): 83.

[8] Langdon Gilkey, "Biology and Theology on Human Nature: Ethics and Genetics," in *Biology, Ethics, and the Origins of Life*, ed. by Holmes Rolston III (Boston and London: Jones and Bartlett Publishers, 1995), 172; italics in original.

1 *Playing God?*

Should we ask our scientists to play God? Answering this questions requires answering some related ones. Does our genetic make-up represent a divine creation in such a way that it is complete and final as it is? Is our DNA sacred? Are we desecrating a sacred realm when we try to discern the mysteries of DNA, and are we exhibiting excessive human pride when we try to engineer our genetic future?

This curious phrase, "playing God," has at least three overlapping meanings. The first and somewhat benign meaning has to do with *learning God's awesome secrets*. Science and its accompanying technology are shining light where previously there was darkness. Mysteries are becoming revealed; and we sense that the scientists as the revealers are acquiring "God-like" powers. The Godlike role of the scientist has taken a familiar cultural form since the era of Isaac Newton. When the Temple of Worthies was erected in 1735 by the British government, the inscription over Newton's bust included these words:

> Sir Isaac Newton, Whom the God of Nature made to comprehend his works:
> and from simple Principles, to discover the laws never known before.

Alexander Pope says it better.

> Nature, and Nature's Laws lay hid in Night.
> God said, *Let Newton be!* and All was Light.[9]

The tie here between God and the scientist honors respectfully and humorously human genius for revealing the secrets of God's second book, the Book of Nature. At this level we do not yet have any ethical reason to object to research. Rather, what we have here is an expression of respect for science along with awe about nature.[10]

The second meaning of "playing God" applies to medical doctors working in the clinical setting with an emergency surgery. It has to do with wielding *power over life and death*.[11] A patient in critical condition feels helpless. Only the attention and skill of the surgeon stands between the patient and death. The doctor is the only door to life. The patient is utterly dependent upon the physician for his or her very existence. Regardless of whether or not doctors feel they have omnipotence in this situation, the patients impute it to them. We tell doctor jokes because some doctors confuse themselves with the true God. Cartoons depict God carrying a stethoscope—that is, God playing doctor.

The third meaning of "playing God" is the one that concerns us here, namely, the use of science to *alter life and influence human evolution*. "Playing God" here means that we—at least the scientists among us—are substituting ourselves for God in determining what human nature will be. Even though scientifically and philosophically it is difficult to conceive of what it would mean to actually alter human nature, it is a popular idea in our culture. And it is associated with placing ourselves where God and only God belongs.

---

[9] See the discussion by Margaret Wertheim, *Pythagoras' Trousers: God, Physics, and the Gender Wars* (New York: Random House, Times Books, 1995), 145.

[10] President's Commission for the Study of Ethical Problems in Medicine and Biomedical and Behavioral Research, Morris B. Abram, Chairman, 2000 K Street NW, Suite 555, Washington, D.C. 20006, *Splicing Life* (November, 1982): 54.

[11] When a surgeon picks up a scalpel and tries to save a life or when a surgeon decides to withdraw a life support system, it is a form of playing God. Leroy Augenstein, *Come, Let Us Play God* (New York: Harper and Row, 1969), 12. Whether positively or negatively construed, the simple power of life and death seems to put us on God's doorstep.

The power to alter human nature as a life form similarly evokes the question: should only God be doing this? Fifty-eight percent answered a *Time*/CNN poll saying they think altering human genes is against the will of God.[12] In a 1980 letter of warning to then President Jimmy Carter, Roman Catholic, Protestant, and Jewish spokespersons used the phrase "playing God" to refer to individuals or groups who would seek to control life forms. Any attempt to "correct" our mental and social structures by genetic means to fit one group's vision of humanity is dangerous.[13]

Why do such critics of genetic research prescribe a new commandment, "Thou shalt not play God"? The answer here is this: because human pride or *hubris* is dangerous.[14] We have learned from experience that what the Bible says is true: "pride goes before destruction" (Proverbs 16:18). And in our modern era pride among the natural scientists has taken the form of overestimating our knowledge, of arrogating for science a kind of omniscience that we do not in fact have. Or, to refine it a bit: "playing God" means we confuse the knowledge we do have with the wisdom to decide how to use it. Frequently lacking this wisdom we falsely assume we possess, scientific knowledge leads to unforeseen consequences such as the destruction of the ecosphere.[15] Applied to genetic therapy, the commandment against "playing God" implies that the unpredictability of destructive effects on the human gene pool should lead to a proscription against germ-line intervention. In light of this, "there is general agreement that human germ-line intervention for any purpose should always be governed by stringent criteria for safety and predictability."[16] This "general agreement" seeks to draw upon wisdom to mitigate pride.

A correlate to this third meaning of the phrase, "playing God," is that DNA has come to function in effect as an inviolable sacred, a special province of the divine, that should be off limits to mere mortals. New York Medical College cell biologist Stuart Newman refers to this as the "apotheosis of the gene." He calls our attention to Max Delbrück's identification of Aristotle's unmoved mover god with DNA. The "unmoved mover perfectly describes DNA. DNA acts, creates form in development, and it does not change in the process."[17]

Robert Sinsheimer, among others, suggests that when we see ourselves as the creators of life then we lose reverence for life[18] It is just this lack of reverence for life as nature has bequeathed it to us that drives Jeremy Rifkin to attack the kind of genetic research that will lead to algeny—that is, to "the upgrading of existing organisms and the design of wholly new ones with the intent of 'perfecting' their

---

[12] Philip Elmer-DeWitt, "The Genetic Revolution," *Time*, 143.3 (January 17, 1994): 46–53.

[13] President's Commission, 95–96.

[14] Jane Goodfield, *Playing God: Genetic Engineering and the Manipulation of Life* (New York: Random House, 1977), 6.

[15] Jeremy Rifkin, "Playing God with the Genetic Code," *Threshold*, 6.3 (January, 1994): 17–18. Obtain from Student Environmental Action Coalition, P.O. Box 1168, Chapel Hill, NC 27514-1168.

[16] Zimmerman, "Human Germ-Line Therapy," 606.

[17] Max Delbrück, "How Aristotle Discovered DNA," in *Physics and Our World: A Symposium in Honor of Victor F. Weisskopf*, ed. by K. Huang (New York: American Institute of Physics, 1976), 123–30; cited in Stuart A. Newman, "Carnal Boundaries: The Commingling of Flesh in Theory and Practice," in *Reinventing Biology: Respect for Life and the Creation of Knowledge*, ed. by Lynda Birke and Ruth Hubbard (Bloomington and Indianapolis, Ind.: Indiana University Press, 1995), 217.

[18] Robert L. Sinsheimer, "Genetic Engineering: Life as a Plaything," *Technology Review*, 86.14 (1983): 14–70.

performance." The problem with algeny is that it represents excessive human pride. "It is humanity's attempt to give metaphysical meaning to its emerging technological relationship with nature."[19] Rifkin's message is: let nature be! Don't try to make it better! In advocating this hands off policy, Rifkin does not appeal to Christian or Jewish or other theological principles. Rather, he appeals to a vague naturalism or vitalism, according to which nature herself claims sacred status. He issues his own missionary's call: "The resacralization of nature stands before us as the great mission of the coming age."[20]

Rifkin has garnered his share of critics. Walter Truett Anderson dubs Rifkin's hysterical attack against genetic engineering "biological McCarthyism." Anderson's own position is that the human race should become deliberate about the future of its own evolution. "This is the project of the coming era: to create a social and political order—a global one—commensurate to human power in nature. The project requires a shift from evolutionary meddling to evolutionary governance, informed by an ethic of responsibility—an evolutionary ethic, not merely an environmental ethic—and it requires appropriate ways of thinking about new issues and making decisions."[21]

What is the warrant for treating nature in general, or DNA specifically, as sacred and therefore morally immune from technological intervention? Neither molecular biologists nor Christian theologians are likely to sympathize with the implied vitalism here. Molecular biologists tend to reject vitalism when protecting reductionism—that is, they deny that DNA or even life itself is directed by a spiritual force that is more than the physics and chemistry that constitute it. This is the position of double helix discoverer Francis Crick.[22]

Theologian Ronald Cole-Turner criticizes the likes of Sinsheimer and Rifkin for making an unwarranted philosophical and theological leap from the association of DNA with life to the metaphysical proscription against technical manipulation.

> Is DNA the essence of life? Is it any more arrogant or sacrilegious to cut DNA than to cut living tissue, as in surgery? It is hard to imagine a scientific or philosophical argument that would successfully support the metaphysical or moral uniqueness of DNA. Even DNA's capacity to replicate does not elevate this molecule to a higher metaphysical or moral level. Replication and sexual reproduction are important capacities, crucial in biology. But they are hardly the stuff of sanctity.[23]

To nominate DNA for election into the halls of functional sacrality, says Cole-Turner, is arbitrary. Theologians in particular should avoid this pitfall. "To think of genetic material as the exclusive realm of divine grace and creativity is to reduce God to the level of restriction enzymes, viruses, and sexual reproduction. Treating

---

[19] Jeremy Rifkin, *Algeny* (New York: Viking, 1983), 17.

[20] Ibid., 252. Although the phrase, "play God," has been with us for some decades as a reference to the prospect of scientific creation or manipulation of life, Jeremy Rifkin thrust it before the public with his 1977 book title, *Who Should Play God?* (New York: Dell, 1977).

[21] Walter Truett Anderson, *To Govern Evolution* (New York: Harcourt, Brace and Jovanovich, 1987), 9, 135.

[22] Francis Crick, *Of Molecules and Men* (Seattle, Wash.: University of Washington Press, 1966), 16, 26. Crick defends reductionism: "The ultimate aim of the modern movement in biology is in fact to explain *all* biology in terms of physics and chemistry," 10; italics in original. Crick would reject the purposeful picture of evolution drawn by Rahner on the grounds that Crick accounts for evolutionary change by appeal to natural selection and especially to chance; ibid., 27. See also by Francis Crick, *The Astonishing Hypothesis: The Scientific Search for the Soul* (New York: Charles Scribner's Sons, 1994), 7–8.

[23] Ronald Cole-Turner, *The New Genesis: Theology and the Genetic Revolution* (Louisville, Ky.: Westminster/John Knox, 1993), 45.

DNA as matter—complicated, awe-inspiring, and elaborately coded, but matter nonetheless—is not in itself sacrilegious."[24]

What the three meanings of "playing God" raise up for us is the question of the relationship between the divine creator and the natural creation. Theists in the Jewish and Christian traditions are clear: natural life, important as it is, is not ultimate. God is ultimate.

One can argue to this position on the basis of *creatio ex nihilo*, creation out of nothing. All that exists has been called from nothing by the voice of God and brought into existence, and at any moment could in principle return to the nonexistence from which it came. Life, as everything else in existence, is finite, temporal, and mortal. The natural world depends upon a divine creator who transcends it. Nature is not its own author. Nor can it claim ultimacy, sanctity, or any other status rivaling God. This leads biologist Hessel Bouma III and his colleagues at the Calvin Center for Christian Scholarship to a pithy proposition: "God is the creator. Therefore, nothing that God made is god, and all that God made is good." This implies, among other things, that we should be careful when accusing physicians and scientists of "playing God." We must avoid idolatrous expectations of technology, to be sure; "but to presume that human technological intervention violates God's rule is to worship Mother Nature, not the creator. Natural processes are not sacrosanct."[25]

One can also argue to this position on the basis of *creatio continua*, continuous creation—that is, on the basis of the idea that creation is ongoing, one can argue for human intervention and contribution to the process. God did not just extricate the world from the divine assembly line like a car, fill its tank with gas, and then let it drive itself down the highways of history. Divine steering, braking, and accelerating still go on. The creative act whereby God brought the world into existence *ab initio*, at the beginning, is complemented with God's continued exercise of creative power through the course of natural and human history. The God of the Bible is by no means absent. This God enters the course of events, makes promises, and then fulfills them. God is the source of the new. Just as the world appeared new at the beginning, God continues to impart the new to the world and promises a yet outstanding new creation still to come.

My own way of conceiving of *creatio ex nihilo* together with *creatio continua* is this: the first thing God did was to give the world a future.[26] The act of drawing the world into existence from nothing is the act of giving the world a future. As long we have a future, we exist. When we lose our future, we cease to exist. God continues moment to moment to bestow futurity, and this establishes continuity while opening reality up to newness. Future-giving is the way in which God is creative. It is also the way God redeems. God's grace comes to the creation through creative and redemptive future-giving.

---

[24] Ibid.

[25] Hessell Bouma et al., *Christian Faith, Health, and Medical Practice* (Grand Rapids, Mich.: Eerdmans, 1989), 4–5. James M. Gustafson makes "the theological point that whatever we value and ought to value about life is at least relative to the respect owed to the creator, sustainer, and orderer of life." James M. Gustafson, "Genetic Therapy: Ethical and Religious Reflections," *Journal of Contemporary Health Law and Policy*, 8 (1992): 196. For Gustafson, the central question around which the issue of germ-line intervention is oriented is this: How do we define what is naturally normal for human life? For the theologian to answer this question, more than knowledge of biology is required. Required also is awareness of the divine ordering of human life.

[26] Ted Peters, *GOD—The World's Future* (Minneapolis, Minn.: Fortress, 1992).

God creates new things. The biblical description of divine activity in the world includes promises and fulfillments of promises. This implies two divine qualities. First, that God is not restricted to the old, not confined by the *status quo*. God promises new realities and then brings them to pass. The most important of the still outstanding divine promises is that of the "new creation" yet to come. Second, this God is faithful, trustworthy. On the basis of the past record, the God of Israel can be trusted to keep a promise. For us this means that we can trust God's creative and redemptive activity to continue in the future.

The next step in the argument is to conceive of the human being as the created co-creator. The term, "created co-creator," comes from the work of Philip Hefner.[27] The term does a couple of important things. First, the term "created" reminds us that one way God creates differs from the way we human beings create. God creates *ex nihilo*. We have been created by God. We are creatures. So, whatever creativity we manifest cannot rank on the same level as creation out of nothing, on the same level with our creator. Yet, secondly, the term "co-creator" signifies what we all know, namely, the creation does not stand still. It moves. It changes. So do we. And, furthermore, we have partial influence on the direction it moves and the kind of changes that take place. We are creative in the transformatory sense. Might we then think of the *imago dei*—the image of God embedded in the human race—in terms of creativity? Might we think of ourselves as co-creators, sharing in the transforming work of God's ongoing creation?

Human creativity is ambiguous. We are condemned to be creative. We cannot avoid it. The human being is a tool maker and a tool user. We are *homo faber*. We cannot be human without being technological, and technology changes things for good or ill. Technology is normally designed for good reasons such as service to human health and welfare, but we know all too well how shortsightedness in technological advance does damage. This is indirect evil. Direct evil is also possible. Technology can be pressed into the service of violence and war, as in the making of weapons. It is by no means an unmitigated good. Yet, despite its occasional deleterious consequences, we humans have no choice but to continue to express ourselves technologically and, hence, creatively.

We cannot not be creative. The ethical mandate, then, has to do with the purposes toward which our creativity is directed and the degree of zeal with which we approach our creative tasks.

2 *The Human Genome Project and Germ-line Intervention*

These issues come to the forefront of discussion in our time due in large part to the enormous impact of the Human Genome Project on the biological and even the social sciences. Descriptively, we know the stated purposes directing the Human Genome Project as presently conceived. First, its aim is knowledge. The simple goal that drives all pure science is present here, namely, the desire to know. In this case it is the desire to know the sequence of the base pairs and the location of the genes in the human genome. Second, its aim is better human health. The avowed ethical

---

[27] Philip Hefner, "The Evolution of the Created Co-Creator" in *Cosmos as Creation: Science and Theology in Consonance*, ed. by Ted Peters (Nashville, Tenn.: Abingdon, 1989), 212; and idem, *The Human Factor: Evolution, Culture, and Religion* (Minneapolis, Minn.: Fortress, 1993), 35–42. See also James M. Gustafson, "Where Theologians and Geneticists Meet," *Dialog*, 33.1 (Winter, 1994): 10.

goal is to employ the newly acquired knowledge in further research to provide therapy for the many genetically caused diseases that plague the human family. John C. Fletcher and W. French Anderson put it eloquently: "Human gene therapy is a symbol of hope in a vast sea of human suffering due to heredity."[28] As this second health oriented purpose is pursued, the technology for manipulating genes will be developed and questions regarding human creativity will arise. How should this creativity be directed?

Virtually no one contests the principle that new genetic knowledge should be used to improve human health and relieve suffering. Yet a serious debate has arisen that distinguishes sharply between therapy for suffering persons who already exist and the health of future persons who do not yet exist. It is the debate between somatic therapy and germ-line therapy. By 'somatic therapy' we refer to the treatment of a disease in the cells of a living individual by trying to repair an existing defect. It consists of inserting new segments of DNA into already differentiated cells such as are found in the liver, muscle, or blood. Clinical trials are underway to use somatic modification as therapy for people suffering from diabetes, hypertension, and Adenosine Deaminase Deficiency. By 'germ-line therapy', however, we refer to intervention into the germ cells that influence heredity and hopefully improve the quality of life for future generations. Negatively, germ-line intervention might help to eliminate deleterious genes that dispose us to disease. Positively, though presently well beyond our technical capacity, such intervention might actually enhance human health, intelligence, or strength.

Two issues overlap here and should be sorted out for clarity. One is the issue of somatic intervention versus germ-line intervention. The other is the issue of therapy versus enhancement. Although somatic treatment is usually identified with therapy and germ-line treatment with enhancement, there are occasions where somatic treatment enhances, such as injecting growth hormones to enhance height for playing basketball. And germ-line intervention, at least in its initial stages of development, will aim at preventive medicine. The science of enhancement, if it comes at all, will only come later.

Every ethical interpreter I have reviewed agrees that somatic therapy is morally desirable and looks forward to the advances gene research will bring for expanding this important medical work. Yet many who reflect on the ethical implications of the new genetic research stop short of endorsing genetic selection and manipulation for the purposes of improving the human species.[29] The World Council of Churches

---

[28] John C. Fletcher and W. French Anderson, "Germ-Line Gene Therapy: A New Stage of Debate," *Law, Medicine, and Health Care*, 20.1/2 (Spring/Summer, 1992): 31.

[29] W. French Anderson, for example, writes, "Somatic cell gene therapy for the treatment of severe disease is considered ethical because it can be supported by the fundamental moral principle of beneficence: It would relieve human suffering....[But] enhancement engineering would threaten important human values in two ways: It could be medically hazardous in that the risks could exceed the potential benefits and the procedure therefore cause harm. And it would be morally precarious in that...it could lead to an increase in inequality and discriminatory practices." "Genetics and Human Malleability," *Hastings Center Report*, 20.1 (January/February, 1990): 23. David Suzuki and Peter Knudtson draw a sharp ethical distinction between somatic gene therapy—which can be seen as the equivalent of an organ-transplant operation that modifies a patient's phenotype without changing the genotype—and germ-line gene therapy—which modifies "cells belonging to lineages that are potentially immortal." David Suzuki and Peter Knudtson, *Genethics: The Clash Between the New Genetics and Human Values* (Cambridge: Harvard University Press, rev. ed. 1990),

(WCC) is representative. In a 1982 document, we find

> ... somatic cell therapy may provide a good; however, other issues are raised if it also brings about a change in germ-line cells. The introduction of genes into the germ-line is a permanent alteration.... Nonetheless, changes in genes that avoid the occurrence of disease are not necessarily made illicit merely because those changes also alter the genetic inheritance of future generations.... There is no absolute distinction between eliminating "defects" and "improving" heredity.[30]

The text elsewhere indicates that the WCC is primarily concerned with our lack of knowledge regarding the possible consequences of altering the human germ-line. The problem is this: the present generation lacks sufficient information regarding the long term consequences of a decision today that might turn out to be irreversible tomorrow. Thus, the WCC does not forbid forever germ-line therapy or even enhancement. Rather, it cautions us to wait and see. In similar fashion, the Methodists "support human gene therapies that produce changes that cannot be passed on to offspring (somatic), but believe that they should be limited to the alleviation of suffering caused by disease."[31] The United Church of Christ also approves "altering cells in the human body, if the alteration is not passed to offspring."[32] On June 8, 1983 fifty-eight religious leaders issued a "Theological Letter Concerning the Moral Arguments" against germ-line engineering addressed to the U.S. Congress. The group action was orchestrated by Jeremy Rifkin of the

---

183–84. However, because the technical distinction between these two is becoming more difficult to discern, some can say "the bright ethical line separating somatic and germ-line therapy has begun to erode." Kathleen Nolan, "How Do We Think About the Ethics of Human Germ-Line Genetic Therapy," *Journal of Medicine and Philosophy*, 16.6 (December, 1991): 613. French Anderson, in a more recent work with John Fletcher, argues that the situation is changing. Whereas in the 1970s and 1980s there was a strong taboo against germ-line modification, in the 1990s that taboo is lifting. "Searches for cure and prevention of genetic disorders by germ-line therapy arise from principles of beneficence and non-maleficence, which create imperatives to relieve and prevent basic causes of human suffering. It follows from this ethical imperative that society ought not to draw a moral line between intentional germ-line therapy and somatic cell therapy." Fletcher and Anderson, "Germ-Line Gene Therapy," 31.

[30] WCC, *Manipulating Life: Ethical Issues in Genetic Engineering* (Geneva: World Council of Churches, 1982). A 1989 document reiterates this position more strongly by proposing "a ban on experiments involving genetic engineering of the human germ-line at the present time." WCC, "Biotechnology: Its Challenges to the Churches and the World" (Unpublished, Geneva: World Council of Churches, 1989), 2. Eric T. Juengst argues that the arguments for a present ban on germ-line intervention are convincing, but he argues that the risks of genetic accidents—even multi-generational ones—can be overcome with new knowledge. Germ-line alteration ought not be proscribed simply on the grounds that enhancement engineering might magnify current social inequalities. He writes, "the social risks of enhancement engineering, like its clinical risks, still only provides contingent barriers to the technique. In a society structured to allow the realization of our moral commitment to social equality in the face of biological diversity—that is, for a society in which there was both open access to this technology and not particular social advantage to its use—these problems would show themselves to be the side issues they really are." Eric T. Juengst, "The NIH 'Points to Consider' and the Limits of Human Gene Therapy," *Human Gene Therapy*, 1 (1990): 431.

[31] United Methodist Church Genetic Task Force Report to the 1992 General Conference, p.121.

[32] United Church of Christ, "The Church and Genetic Engineering," Pronouncement of the Seventeenth General Synod (Fort Worth, Tx.: 1989), 3. See Cole-Turner, "Genetics and the Church," *Prism*, 6 (Spring, 1991): 53–61; and idem, *The New Genesis*, 70–79.

Foundation on Economic Trends; and one member, James R. Crumley, presiding bishop of the then Lutheran Church in America spoke to the press saying, "There are some aspects of genetic therapy [for human diseases] that I would not want to rule out.... My concern is that someone would decide what is the most correct human being and begin to engineer the germ-line with that goal in mind."[33]

A more positive approach is taken by The Catholic Health Association. If we can improve human health through germ-line intervention, then it is morally desirable.

> ... germ-line intervention is potentially the only means of treating genetic diseases that do their damage early in embryonic development, for which somatic cell therapy would be ineffective. Although still a long way off, developments in molecular genetics suggest that this is a goal toward which biomedicine could reasonably devote its efforts.[34]

Part of the reluctance to embrace germ-line intervention has to do with its implicit association with the history of eugenics. The term 'eugenics' brings to mind the repugnant racial policies of Nazism, and this accounts for much of today's mistrust of genetic science in Germany and elsewhere.[35] No one expects a repeat of Nazi terror to emerge from genetic engineering; yet some critics fear a subtle form of eugenics may be slipping in the cultural back door.[36] John Harris may be a bit of a maverick, but he welcomes eugenics if it contributes to better human health. He makes the point forcefully: "... where gene therapy will effect improvements to human beings or to human nature that provide protections from harm or the protection of life itself in the form of increases in life expectancy... then call it what you will, eugenics or not, we ought to be in favor of it."[37]

Philosophical and ethical objections to eugenics seem to presuppose not therapy but rather enhancement. The growing power to control the human genetic make-up could foster the emergence of the image of the "perfect child" or a "super

---

[33] See: Paul Nelson, "Bioethics in the Lutheran Tradition," *Bioethics Yearbook*, Vol. 1: *Theological Developments in Bioethics: 1988–1990* (Boston, Mass.: Kluwer, 1991), 119–44. Reporting on recent developments in Scandinavian theology, Paul Nelson writes, "Biblically oriented theologians reject positive eugenics because human nature as created and willed by God rests upon a genetic foundation. Modifications aimed at making better humans would usurp the authority of the divine creation and efface the distinction between the creature and creator.... At the same time, these theologians are not opposed to negative eugenics in the form of somatic cell gene therapy.... Germ cell therapy, on the other hand, is subject to the same indictment the churches make of positive eugenics." Vol. 3: *Theological Developments in Bioethics: 1990–1992* (Boston, Mass.: Kluwer, 1993), 161.

[34] Catholic Health Association of the United States, *Human Genetics: Ethical Issues in Genetic Testing, Counseling, and Therapy* (St. Louis, Mo.: The Catholic Health Association of the United States, 1990), 19. Paulus Gregorius, Metropolitan Orthodox Bishop of Delhi, might agree. "... one cannot see anything intrinsically forbidden or evil in gene therapy, whether somatic or germ-line." Paulus Gregorius, "Ethical Reflections on Human Gene Therapy," in Zbigniew Bankowski, Alexander M. Capron, eds., *Genetics, Ethics, and Human Values.* Proceedings of the 24th CIOMS Round Table Conference (Geneva: CIOMS, 1991), 143–53.

[35] See: Peter Meyer, "Biotechnology: History Shapes German Question," *Forum for Applied Research and Public Policy*, 6 (Winter 1991): 92–97.

[36] See: Troy Duster, *Backdoor to Eugenics* (New York: Routledge, 1999); Ruth Hubbard and Elijah Wald, *Exploding the Gene Myth* (Boston, Mass.: Beacon, 1993), esp. 24–25; and Rifkin, *Algeny*, 230–34.

[37] John Harris, "Is Gene Therapy a Form of Eugenics?" *Bioethics*, 7.2/3 (April, 1993): 184.

strain" of humanity. Some religious leaders worry that the impact of the social value of perfection will begin to oppress all those who fall short. Ethicists at the March 1992 conference on "Genetics, Religion and Ethics" held at the Texas Medical Center in Houston said this:

> Because the Jewish and Christian religious world-view is grounded in the equality and dignity of individual persons, genetic diversity is respected. Any move to eliminate or reduce human diversity in the interest of eugenics or creating a "super strain" of human being will meet with resistance.[38]

In sum, with the possible exception of the Catholic Health Association, religious ethical thinking tends to be conservative in the sense that it seeks to conserve the present pool of genes on the human genome for the indefinite future.

Now the question of "playing God" begins to take on concrete form. The risk of exerting human creativity through germ-line intervention is that, though we begin with the best of intentions, the result may include negative repercussions that escape our control. Physically, our genetic engineering may disturb the strength-giving qualities of biodiversity that supposedly contribute to human health. Due to our inability to see the whole range of interconnected factors, we may inadvertently disturb some sort of existing balance in nature and this disturbance could redound deleteriously. Socially, we could contribute to stigma and discrimination. The very proffering of criteria to determine just what counts as a "defective" gene may lead to stigmatizing all those persons who carry that gene. The very proffering of the image of the ideal child or a super strain of humanity may cultivate a sense of inferiority to those who do not measure up. To embark on a large scale program of germ-line enhancement may create physical and social problems, and then we would blame the human race for its pride, its *hubris*, its stepping beyond its alleged God-defined limits that brings disaster upon itself.

Yet, there may be another way to look at the challenge that confronts us here. The correlate concepts of God as the creator and the human as the created co-creator orient us toward the future, a future that should be better than the past or present. One of the problems with the naturalist argument and the more conservative religious arguments mentioned above is that they implicitly assume the present state of affairs is adequate. These arguments tacitly bless the *status quo*. The problem with the *status quo* is that it is filled with human misery, some of which is genetically caused. It is possible for us to envision a better future, a future in which individuals would not have to suffer the consequences of genes such as those for Cystic Fibrosis, Alzheimer's or Huntington's Disease. That we should be cautious and prudent and recognize the threat of human *hubris*, I fully grant. Yet, our ethical vision cannot acquiesce with present reality; it must press on to a still better future and employ human creativity with its accompanying genetic technology to move us in that direction.

"In the not-to-distant future," says Pope John Paul II, "we can reasonably foresee that the whole genome sequencing will open new paths for therapeutic purposes. Thus the sick, to whom it was impossible to give proper treatment due to frequently fatal hereditary pathologies, will be able to benefit from the treatment

---

[38] J. Robert Nelson, "Summary Reflection Statement" of the "Genetics, Religion and Ethics Project" (1992), The Institute of Religion and Baylor College of Medicine, The Texas Medical Center, P.O. Box 20569, Houston, Texas 77225. For an analysis of the conference, see J. Robert Nelson, *On the New Frontiers of Genetics and Religion* (Grand Rapids, Mich.: Eerdmans, 1994).

needed to improve their condition and possibly to cure them. By acting on the subject's unhealthy genes, it will also be possible to prevent the recurrence of genetic diseases and their transmission."[39] To prevent the genetic transmission of disease will take human creativity operating out of a vision of better health for future generations.

3 Germ-line Therapy and Enhancement: a Closer Look

Having enunciated my own conviction that the Christian doctrine of creation with its accompanying understanding of the human being as the created co-creator leads to an ethic oriented toward striving for a better future, let me turn to a closer look at the arguments for and against germ-line intervention and manipulation. Eric T. Juengst helpfully summarizes five arguments in favor of germ-line modification for the purposes of therapy.

1. *Medical utility*: germ-line gene therapy offers a true cure for many genetic diseases.
2. *Medical necessity:* such therapy is the only effective way to address some diseases.
3. *Prophylactic efficiency*: prevention is less costly and less risky than cure.
4. *Respect for parental autonomy* when parents request germ-line intervention.
5. *Scientific freedom* to engage in germ-line inquiry.

Juengst also summarizes five arguments opposing germ-line intervention.

1. *Scientific uncertainty and risks* to future generations.
2. *Slippery slope to enhancement* that could exacerbate social discrimination.
3. *Consent of future generations* is impossible to get.
4. *Allocation of Resources*: germ-line therapy may never be cost effective.
5. *Integrity of genetic patrimony*: future generations have the right to inherit a genetic endowment that has not been intentionally modified.[40]

---

[39] Pope John Paul II, "The Human Person Must Be the Beginning, Subject and Goal of All Scientific Research," Address to the Pontifical Academy of Sciences, *L'Osservatore Romano*, N–45 (9 November, 1994): 3.

[40] Eric T. Juengst, "Germ-Line Gene Therapy: Back to Basics," *Journal of Medicine and Philosophy*, 16.6 (December, 1991): 589–90. See also Maurice A.M. De Wachter, "Ethical Aspects of Human Germ-Line Therapy," *Bioethics*, 7.2/3 (April, 1993): 166–77. Nelson A. Wivel and LeRoy Walters list four arguments against germ-line modification: 1) it is an expensive intervention that would affect relatively few patients; 2) alternative strategies for avoiding genetic disease exist, namely, somatic cell therapy; 3) the risks of multi-generational genetic mistakes will never be eliminated, and these mistakes would be irreversible; and 4) germ-line modification for therapy puts us on a slippery slope leading inevitably to enhancement. They also list four arguments favoring germ-line modification: 1) health professionals have a moral obligation to use the best available methods in preventing or treating genetic disorders, and this may include germ-line alterations; 2) the principle of respect for parental autonomy should permit parents to use this technology to increase the likelihood of having a healthy child; 3) it is more efficient than the repeated use of somatic cell therapy over successive generations; and 4) the prevailing ethic of science and medicine operates on the assumption that knowledge has intrinsic value, and this means that promising areas of research should be pursued. "Germ-Line Gene Modification and Disease Prevention: Some Medical and Ethical Perspectives," *Science*, 262.5133 (22 October, 1993): 533–38. Arthur L. Caplan believes HGI scientists may have sold their research souls too soon by

Given Juengst's classification of arguments as a general framework, I would now like to engage the issue in some detail and to test a theological commitment to the notion of the created co-creator by turning our attention to a representative case in point, namely, the position paper drafted by the Council for Responsible Genetics (CRG) in the fall of 1992.[41] The CRG proffers three types of argument in opposition to germ-line modification in humans: a technical argument, a slanderous argument, and an ethical argument.

The first argument against germ-line manipulation is technical. Although the motive for modifying germ genes may be the improvement of human well being for future generations, unexpected deleterious consequences may result. Removal of an unwanted disease gene may not eliminate the possibility that other gene combinations will be created that will be harmful. Inadvertent damage could result from biologists' inability to predict just how genes or their products interact with one another and with the environment. "Inserting new segments of DNA into the germ-line could have major, unpredictable consequences for both the individual and the future of the species that include the introduction of susceptibilities to cancer and other disease into the human gene pool."

It would seem to the prudent observer that we take a wait and see attitude, that we move cautiously as the technology develops. This is one way to interpret the WCC position.[42] The problem of unexpected consequences is one that confronts all long term planning, and in itself should not deter research and experimentation guided by a vision of a healthier humanity.[43]

The second argument appeals to guilt by association and is thereby slanderous. The CRG Human Genetics Committee says, "... the doctrine of social advancement through biological perfectibility underlying the new eugenics is almost indistinguishable from the older version so avidly embraced by the Nazis." The structure of this argument is that because germ-line modification can be associated with eugenics, and because eugenics can be associated with Nazism, therefore, we can associate proponents of germ-line enhancement with the Nazis and, on this ground, should reject it. The argument borders on the *ad hominem* (circumstantial) fallacy.

---

promising to refrain from germ-line intervention just to appease the hysteria over potential eugenic uses. There is no moral reason to refrain from eliminating a lethal gene from the human population; and there is no slippery slope from germ-line therapy to eugenics. "It is simply a confusion to equate eugenics with any discussion of germ-line therapy." "If Gene Therapy Can Cure, What is the Disease?" in *Gene Mapping*, ed. by George J. Annas and Sherman Elias (Oxford and New York: Oxford University Press, 1992), 139.

[41] CRG, "Position Paper on Human Germ Line Manipulation," Council for Responsible Genetics, 19 Garden Street, Cambridge MA 02138. Quotations here come from this paper.

[42] See page 498 above.

[43] C. Thomas Casky, like the CRG, believes that germ-line correction has little practical appeal while generating considerable ethical apprehension. Yet, he leaves the door open. "I would reserve one area for consideration of germ-line manipulation... It is conceivable that at some point in the future genetic manipulation of an individual's germ-line may be undertaken to introduce or reintroduce disease resistance." C. Thomas Casky, "DNA-Based Medicine: Prevention and Therapy," in *The Code of Codes: Scientific and Social Issues in the Human Genome Project*, ed. by Daniel J. Kevles and Leroy Hood (Cambridge: Harvard University Press, 1992), 129. John A. Robertson takes a position that would oppose the CRG, saying that "these fears appear too speculative to justify denying use of a therapeutic technique that will protect more immediate generations of offspring." *Children of Choice: Freedom and the New Reproductive Technologies* (Princeton, N.J.: Princeton University Press, 1994), 162.

One problem is that the CRG argument is too glib, failing to discern the complexities here. The eugenics movement of the late nineteenth and early twentieth centuries was originally a socially progressive movement that embraced the ideals of a better society. In England and America it became tied to ethnocentrism and the blindness of class interests, leading to forced sterilization of feeble minded prisoners. It was eventually discarded because advances in genetics proved it unscientific.[44] In Germany the eugenics movement became tied to anti-Semitism, resulting in the racial hygiene (*Rassenhygiene*) program of the Nazi SS and the atrocities of the so-called "final solution."[45] With this history in mind, the present generation must assuredly be on guard against future programs of "ethnic hygiene" which seem to plague the human species in one form or another every century. Yet we must observe that ethnocentric bias in England and America and the rise of Nazism in Germany were social phenomena that employed eugenics for their respective ends. Eugenics was not the source of injustice, even if it was a weapon in the service of injustice. The CRG's use of the volatile word "Nazi" in this discussion of germ-line enhancement is an attempt to paint their opponents in such a repulsive color that no one will open-mindedly view the matter.

The third CRG argument, the ethical argument, is much more worthy of serious consideration. The central thesis here is that germ-line modification will reinforce existing social discrimination. The position paper declares,

> The cultural impact of treating humans as biologically perfectible artifacts would be entirely negative. People who fall short of some technically achievable ideal would increasingly be seen as 'damaged goods.' And it is clear that the standards for what is genetically desirable will be those of the society's economically and politically dominant groups. This will only reinforce prejudices and discrimination in a society where they already exist.[46]

Let us look at this argument in terms of its component parts. The assumption in the first sentence is that germ-line intervention implies biological perfectibility and, on account of this, that human persons will be treated as artifacts. It is of course plausible that a social construction of the perfect child or the perfected human strain might appear in Saturday morning cartoons and other cultural forms. Yet, this does not seem to apply to the actual situation in which genetic scientists currently find themselves. They are occupied with much more modest aspirations such as protection from monogenetic diseases such as cystic fibrosis. The medical technology here is not much beyond infancy. At this point in technological history we do not find ourselves on the brink of designer children or the advent of a super strain. What is "genetically desirable" is by no means scientifically attainable. Thus, Hessel Bouma and his colleagues are less worried than the CRG because they recognize that the technological possibility of creating a genetically perfect human race is still very remote. "Things like intelligence and strength are not inherited through single genes but through multi-factoral conditions, combinations of inherited genes and numerous environmental factors. Our ability to control and to design is

---

[44] See: Daniel J. Kevles, *In the Name of Eugenics* (Berkeley and Los Angeles: University of California Press, 1985).

[45] See: Robert N. Proctor, *Racial Hygiene: Medicine Under the Nazis* (Cambridge: Harvard University Press, 1988).

[46] This statement comes directly from the position paper. It fits appropriately with what one of the drafters, R. C. Lewontin, elsewhere says critically about science and class interests: "'Science' is the ultimate legitimator of bourgeois ideology." Robert Lewontin et al., *Not In Our Genes* (New York: Pantheon, 1984), 31.

limited by the complexity of many traits, so there are seemingly insurmountable technological and economic barriers that weaken the empirical slippery-slope argument that we are sliding into the genetic engineering of our children."[47]

Continuing our analysis of the ethical argument, the CRG rightly alerts us to the social-psychology of feeling, and being treated like, "damaged goods." If a "technically achievable ideal" should become a cultural norm, then those who fail to meet the norm would understandably feel inferior.[48] Furthermore, the economically and politically advantaged groups will help to steer the definition of the ideal norm to serve their own class interests.[49] Here the CRG should be applauded for alerting us to a possible loss of human dignity.

At this point a reaffirmation of human dignity is called for, I believe, wherein each individual person is treated as having the full complement of rights regardless of his or her genes. Ethical support here comes from the Christian doctrine of creation, wherein God made men and women in the divine image and pronounced them "good" (Genesis 1:26–31). It also comes from the ministry of Jesus, wherein the Son of God sought out the outcasts, the lame, the infirm, the possessed—surely those who were considered the "damaged goods" of first century Palestine—for divine favor and healing.[50] Each human being, regardless of health or social location or genetic endowment is loved by God, and this recognition should translate into social equality and mutual appreciation. There is no theological justification for thinking of some persons as inferior to others, and new technical possibilities in genetics ought not change this.

We also note the CRG's prognostication for the future: germ-line modification "will only reinforce prejudices and discrimination in a society where they already exist." Prejudices and discrimination exist in the present, says the CRG. This is an obvious fact we readily concede. Does it follow, however, that germ-line intervention "will only reinforce" them? Is germ-line modification the cause of present prejudice and discrimination? No. Prejudice and discrimination seem to flourish quite well without germ-line manipulation, yet somehow their existence is alleged to count as an argument against the latter.

If the argument rests on the premise that germ-line enhancement will create a technical ideal achievable by some but not others, then it fails on the grounds of triviality. This could apply to countless ideals in our society. We daily confront

---

[47] Bouma, *Christian Faith, Health, and Medical Practice*, 264.

[48] "Why does the notion that medical technology might give some children an advantage elicit such a strong negative reaction?" asks Zimmerman. "Perhaps it is because the notion of fairness is well embedded in Western culture." He goes on to note that we already accept randomized differences between people and the inevitability that some individuals will excel over others. Then in support of germ-line enhancement he adds: "What about the positive side, of increasing the number of talented people. Wouldn't society be better off in the long run?" "Human Germ-Line Therapy," 606–7.

[49] We must be clear that genetic prejudice would be a cultural or social phenomenon, not a scientific one. "It is society, not biology, that turns some genetic characteristics into liabilities," writes Roger L. Shinn, *Forced Options: Social Decisions for the 21st Century* (New York: Pilgrim Press, 2nd ed., 1985), 140. If our society is serious about the fairness or justice dimension here, we could institute a sort of "Affirmative Action" public policy in which the underprivileged classes would be given privileged access to germ-line enhancement technology.

[50] Cole-Turner makes much of Jesus' healing ministry as a directive toward inspiring contemporary science and technology to continue healing and to think of this as continuing the divine work of redemption. *New Genesis*, 80–86.

innumerable ideals that are met by some but not all, whether they be athletic achievements, beauty trophies, professional promotions, or lottery winnings. These may elicit temporary feelings of inferiority on the part of those who come in second or further behind, but they are widely ignored by those who did not compete. Given the realistic prospects for what germ-line enhancement is aimed at accomplishing, the new situation would not alter the present situation in this respect. If it is technically possible to relieve some individuals from suffering the consequences of diabetes through the regular use of insulin, then the achievement of this ideal by those afflicted by diabetes leads to only gratitude on their part and on the part of those who love them. Somatic cell therapy or even germ-line modification for diabetes will only extend this gratitude. To those who are not afflicted or likely to be afflicted by diabetes, this achievement may be applauded from a distance or perhaps ignored.

One could envision a next step, of course, where germ-line intervention could, if made universally available, eliminate diabetes from the human gene pool. We would then have a future wiped clean of genetically based diabetes. If this constituted an achieved ideal for the whole human race, and if the unexpected consequences were less harmful than the diabetes, then many persons will have been spared the suffering diabetes could have caused and no reinforcement of prejudice and discrimination will have occurred.

What if we were to falter somewhere along the way? Suppose we began a worldwide program to eliminate the disposition to diabetes from the human gene pool, achieved success in some family or ethnic or class groups, and then due to lack of funding support or other factors had to abandon the project. What would happen to those individuals who still carried the deleterious gene? Would they suffer stigma or discrimination? Perhaps yes. And the CRG rightly alerts us to such a possibility. Yet, we might ask, does this prospect provide sufficient warrant to shut down the research and prohibit embarking on such a plan?

4 Inter-generational Genethics

The CRG buttresses its central ethical argument with two subarguments. One is that the present generation, presumably the one engaging in germ-line modification, cannot be held accountable by future generations for the wrongful damage we inflict on them. We, our progeny's ancestors, will not be around any more to be accountable. There may be an equivocation at work here. On the one hand, the present generation will be absent in the future and, therefore, we cannot be held accountable in the sense that we can be punished by imprisonment. On the other hand, though absent, we can be held accountable in the sense that future fingers could be waved and fists thrown into the air as our progeny express anger at our failure to assume responsibility. Just because we cannot be punished does not mean we are not accountable in a moral sense.

Yet, for the CRG somehow the concept of accountability is supposed to count against germ-line enhancement. Again the argument fails on account of triviality, because our responsibility to our progeny applies across the board to all departments of life. There is nothing special about genes. One might even make a case that environmental responsibility is of graver ethical concern. The excessive depletion of nonrenewable natural resources and pollution of the biosphere is due to the hedonism of the present generation, due to present selfishness that is willing to sacrifice the welfare of future generations for the prosperity of our own. Germ-line intervention, in contrast, could be motivated only by seeking benefit for future generations whom we may not live to see. With or without accountability, the latter at least has the virtue of altruism going for it.

The other subargument raises an interesting issue worth pondering. The CRG says: "Germ-line modification is not needed in order to save the lives or alleviate suffering of existing people. Its target population are 'future people' who have not yet even been conceived." On the face of it, this argument looks like another brand of defense for the ecological hedonists just mentioned whose interest is limited to only the present generation without any regard for future progeny. But this may be a misreading. The CRG is not eliminating our responsibility for future generations. Yet, for some unexplained reason, the CRG makes central the distinction between people who exist and people who do not yet exist. The assumption is that moral priority is given to those who exist over against those who have "not yet even been conceived." The interesting puzzle is the relative moral status of present and future, of existents and not yets.[51]

Suppose we draw up the previous concern for accountability and combine it with the concepts of rights and wrongful birth. Might future generations blame us today for their wrongful birth by damaging them through germ-line intervention? Or, in contrast, might they blame us for *not* intervening in the germ-line, thereby leaving them to suffer from diseases we could have prevented? We are on the verge of an ethical crisis—that is, on the verge of an ethical challenge where creative action is demanded—because whether we engage in germ-line intervention or not, if we are technically capable, we will be morally accountable.

Here the contrast with the environmental crisis is illuminating. We can imagine our great grandchildren living on a deforested earth, mines depleted of their minerals, lakes dead from acid rain, food supply contaminated by chemicals, skin cancerous due to excessive ozone exposure, raising their fists in anger at us. They will claim we violated their right to a life-giving environment and, despite what the CRG says, they will claim we are accountable as they burn us in effigy.

Does this apply by analogy to germ-line enhancement? We can certainly imagine a future person asserting, "My parents and great grandparents and the genetic scientists of their generation violated my rights by giving me a bad genetic endowment." It would be a variant on the wrongful birth accusation. Yet, not everyone sees the sense this makes. Hardy Jones, for example, would argue: had this

---

[51] Bioethicists dealing with genetics ask whether nonexistent future persons belong in the domain of present moral deliberation. David Heyd says no. Heyd advocates a "person-affecting approach to morality" which presupposes that only the needs, wants, interests and ideals of actual persons are the sources of value or the objects of value considerations. Value does not derive from the impersonal world, nor from as yet non-existing persons. He defines *genethics* as the field concerned with the morality of creating or procreating people. The genesis or creation or procreation of future persons marks an ethical domain for us, the procreators, but not for those yet to be brought into existence. David Heyd, *Genethics: Moral Issues in the Creation of People* (Berkeley: University of California Press, 1992). Jan Christian Heller criticizes Heyd for being anthropocentric—for deriving value only from persons and not from the world or from God—and for expelling from our ethical domain future persons whose existence is still contingent. Heller proposes a "qualified person-affecting" approach to value as part of an "impersonal theocentric" ethic in order to incorporate contingent future persons into a divine domain of value. Jan Christian Heller, *Human Genome Research and the Challenge of Contingent Future Persons* (Omaha, Neb.: Creighton University Press, 1996), 16.

individual's progenitors taken successful steps toward enhancing the genetic endowment of their offspring, then this would not be the child they actually had. Having a child with defective genes cannot be a violation of that child's right, because it is not possible to respect that right by not having the child or by bequeathing a different genetic constitution. The only child who can claim a right is one that exists, and the particular configuration of genes is definitional to the person who exists. "Genetically defective persons are not analogous to existing individuals who subsequently acquire biologically bad qualities."[52]

John A. Robertson makes a similar argument when asking about the consent or lack of consent on the part of future generations to what we do today to affect their germ-line. If no harm occurs, he argues, then this is a mere theoretical objection. If harm does occur, then the question of identity arises. "Later generations allegedly harmed without their consent may not have existed at all. Different individuals would then exist than if the germ-line gene therapy had not occurred."[53]

Perhaps the CRG position paper writers presumed this kind of distinction between existing and not yet existing persons, and this permitted them to give qualified approval of somatic modification for living persons while proscribing germ-line manipulation.[54] What this means for us here, then, is that if we are to affirm ethical responsibility for the genetic inheritance we bequeath our progeny, then the framework of rights and accountability might be inadequate. As long as the CRG works within this framework, perhaps its conclusions are understandable, even if inadequate.

A framework that includes the will of God for the flourishing of the human race could handle our present responsibility toward persons who do not yet exist. Jan Christian Heller recognizes the problem in the face of germ-line intervention: "a decision to protect a *particular* future person from predictable genetic harm will mean that a *different* person will in fact be born."[55] In order to include those *different* persons within our moral domain, Heller advocates an "impersonal theocentrism" that begins with the assumption that creation as a whole is intrinsically good independent of its ability to advance or promote human ends. Its goodness derives from God. The "impersonal" dimension here refers to the suprapersonal origin of value, God. This, Heller believes, will permit us to make moral decisions concerning both noncontingent and contingent future persons without discriminating between them.[56] It would also permit us to make moral decisions without discriminating between persons who exist and persons who do not yet exist. The point here is this: an ethic that is grounded in God and God's future orients our responsibility today toward persons we will affect (or even effect) tomorrow.

---

[52] Hardy Jones, "Genetic Endowment and Obligations to Future Generations," in *Responsibilities to Future Generations*, ed. by Ernest Partridge, (Buffalo, N.Y.: Prometheus Books, 1981), 249.

[53] Robertson, *Children of Choice*, 162.

[54] David Suzuki and Peter Knudtson promulgate a "genethic principle" that parallels the CRG: "While genetic manipulation of human somatic cells may lie in the realm of personal choice, tinkering with human germ cells does not. Germ-cell therapy, without the consent of all members of society, ought to be explicitly forbidden." *Genethics*, 163. The Suzuki and Knudtson position is obviously based upon a libertarian ethic so, to be more precise, they should be seeking the consent of those individuals involved rather than the vague "all members of society."

[55] Heller, *Human Genome Research*, 63.

[56] Ibid., 138–39.

5 The Not-yet-future and the Ethics of Creativity

Would a future-oriented theology of creation and its concomitant understanding of the human being as God's created co-creator be more adequate than what the CRG proposes? It would be more adequate for a number of reasons. First, a future-oriented theology of creation is not stymied by giving priority to existing persons over against future persons who do not yet exist. A theology of continuing creation looks forward to the new, to those who are yet to come into existence as part of the moral community to which we belong. Second, such a theology is realistic about the dynamic nature of our situation. Everything changes. There is no standing still. What we do affects and is affected by the future. We are condemned to be creative for good or ill. Third, the future is built into this ethical vision. Once we apprehend that God intends a future, our task is to discern as best we can the direction of divine purpose and employ that as an ethical guide. When we invoke the apocalyptic symbol of the New Jerusalem where "crying and pain will be no more" (Revelation 21:4), then this will inspire and guide the decisions we make today that will affect our progeny tomorrow.

The creative component to a future-oriented ethic denies that the *status quo* defines what is good, denies that the present situation has an automatic moral claim to perpetuity. Take social equality as a relevant case in point. As one can plainly see, social equality does not at present exist, nor has it ever existed in universal form. We daily confront the frustrations of economic inequality and political oppression right along with the more subtle forms of prejudice and discrimination that the CRG rightly opposes. Human equality, then, is something we are striving for, something that does not yet exist but ought to exist. Equality needs to be created, and it will take human creativity under divine guidance to establish it, plus vigilance to maintain it when and where it has been achieved. Wolfhart Pannenberg puts it this way: "The Christian concept of equality does not mean that everyone is to be reduced to an average where every voice is equal to every other, but equality in the Christian sense means that everyone should be raised up through participation in the highest human possibilities. Such equality must always be created; it is not already there."[57] An ethic that seeks to raise us to the "highest human possibilities" cannot accept the *status quo* as normative, but presses on creatively toward a new and better future. Applied to the issue at hand, Ronald Cole-Turner makes the bold affirmation: "I argue that genetic engineering opens new possibilities for the future of God's creative work."[58]

6 Conclusion

We opened with an observation of Karl Rahner regarding evolution and human openness toward the future. Self-transcendence and the possibility for something new belong indelibly to human nature. Human existence is "open and indetermined."[59] That to which we are open is the infinite horizon; we are open to a fulfillment yet to be determined by "the infinite and the ineffable mystery" of God.[60] If we try to draw any middle axioms that connect this sublime theological vision to

---

[57] Wolfhart Pannenberg, *Ethics* (Louisville, Ky.: Westminster/John Knox Press, 1981), 140.

[58] Cole-Turner, *New Genesis*, 98.

[59] Karl Rahner, *Foundations of Christian Faith* (New York: Seabury, 1978), 35.

[60] Ibid., 190.

an ethic appropriate to genetic engineering, openness to the future translates into responsibility for the future, even our evolutionary future. Such a theological vision undercuts a conservative or reactionary proscription against intervening in the evolutionary process. Rahner describes the temptation to condemn genetic research and its application as "symptomatic of a cowardly and comfortable conservatism hiding behind misunderstood Christian ideals."[61] The concept of the created co-creator we invoke here is a cautious but creative Christian concept that begins with a vision of openness to God's future and responsibility for the human future.

The health and well-being of future generations not yet born is a matter of ethical concern when viewed within the scope of a theology of creation that emphasizes God's ongoing creative work and that pictures the human being as the created co-creator. A vision of future possibilities, not the present *status quo*, orients and directs ethical activity. When applied to the issue of germ-line intervention for the purpose of enhancing the quality of human life, the door must be kept open so that we can look through, squint and focus our eyes to see just what possibilities loom before us. This will include a realistic review of the limits and risks of genetic technology.[62] But realism about technological limits and risks is insufficient warrant for prematurely shutting the door against possibilities for an improved human future. Rather than playing God or taking God's place, seeking to actualize new possibilities means we are being truly human.

---

[61] Rahner, *Theological Investigations*, IX:211.

[62] Roger Shinn's advice is salutary here. "I know of no way of drawing a line and saying: thus far, scientific direction and control is beneficial; beyond this line they become destructive manipulation. I think it more important to keep raising the question, to keep confronting the technological society with the issue." *Forced Options*, 142. Deborah Blake says it eloquently: "The risk of the nineties is the seduction of a technological fix. The challenge for the nineties is to find the moral courage necessary to guide and realize the promises made by this new genetics so that our moral wisdom is not outpaced by our technological cleverness." "Ethics of Possibility: Medical Biotechnology for the Nineties," *The Catholic World*, 234.1403 (September–October, 1991): 237.

EVOLUTION, DIVINE ACTION, AND THE PROBLEM OF EVIL

Thomas F. Tracy

> The creator's choice is an abyss, where human thought drowns.
>
> Austin Farrer [1]

1 *Introduction: Out of the Whirlwind*

When the moment finally arrives in the Book of Job for a reply to Job's passionate questioning of God's justice, the Voice from the whirlwind conducts Job on a brief but overwhelming tour of the wonders and terrors of the universe. The poet who wrote this speech presents a series of striking images of the natural world. Our attention, which has been riveted on Job's bold accusations and on the increasingly hostile dialogue with his friends, is now suddenly refocused, our frame of reference vastly expanded. We are conveyed in imagination to a time before human history, the primordial time of creation when "the morning stars burst out singing and the angels shouted for joy!" [2] We are reminded of the great forces and structures of the world, of sky and sun, sea and rain, whose powers we cannot master or comprehend. And we are shown the strange spectacle of living things on Earth. The creatures that God has made display vitalities, appetites, and habits that are surprising to us, and they both prosper and suffer in a life-drama of their own. The poet describes a natural world in which life and death, flourishing and destruction, are closely linked. The lioness hunts all day while her cubs lie in the den aching with hunger; they will eat when their mother kills. The antelope struggles in giving birth, then "her little ones grow up; they leave and never return." [3] The ostrich lays her eggs in the dirt, where they may be crushed or devoured.

> For God deprived her of wisdom
> and left her with little sense.
> When she spreads her wings to run,
> she laughs at the horse and rider. [4]

God made the ostrich stupid, but God also made her fast! This series of images culminates with a vivid description of vultures.

> Do you teach the vulture to soar
> and build his nest in the clouds?
> He makes his home on the mountaintop,
> on the unapproachable crag.
> He sits and scans for prey;
> from far off his eyes can spot it;
> his little ones drink its blood.
> Where the unburied are, he is. [5]

---

[1] Austin Farrer, *Love Almighty and Ills Unlimited* (Garden City, N. Y.: Doubleday & Co., 1961), 60.

[2] Stephen Mitchell, trans., *The Book of Job* (San Francisco: North Point Press, 1987), 79 (RSV, 38:7).

[3] Ibid., 89 (RSV, 39:4).

[4] Ibid., 89 (RSV 39: 17–18).

[5] Ibid., 84 (RSV 39:26–30).

The natural world depicted here is not a place scaled to human moral expectations. It is not readily domesticated or easily understood, and for all its familiarity, it remains strange and surprising. It reflects the workings of a creative energy that is not of our ken, a numinous vitality that generates an astonishing variety and richness of living things. We live alongside creatures that seem admirable, terrible, and pitiful all at once. This is God's world, and we find ourselves in the midst of it, in no position to say just how it came to be or to judge whether, by our lights, the whole enterprise was a good idea.

As a response to Job's anguished longing for moral intelligibility in the world, this response is of uncertain help, as has often been noted. The speech from the whirlwind leaves Job speechless with wonder. It drives home the limits of human comprehension, and it suggests that there is an intrinsic goodness in created things that shines through their strangeness and beauty. But it does not provide an answer to Job's question about why the innocent often suffer and the wicked often thrive. Indeed, it calls into question the possibility of applying human moral categories to creation; it suggests a mismatch, a disproportion, between human moral notions and the natural order in which we find ourselves. The Voice from the whirlwind seems bent on proliferating forms of life, on unleashing self-perpetuating vitalities in the world, and not on maintaining an order among them that we can recognize as just or fair. This God is holy, but we may wonder whether such a creator is morally good.

The ancient author who presented this picture of the natural order would find a vast storehouse of supporting material in modern ecological and evolutionary biology. The world of living things described by these sciences has just the characteristics recognized by the poet of Job. Organisms have proliferated with a wild variety, taking forms that often seem bizarre and unlikely, even whimsical. Unlike the poet, however, modern biologists are confident that they can explain, at least in general terms, how such creatures have come to be. Since Darwin, evolutionary theory has provided increasingly sophisticated and thoroughgoing explanations of how new patterns of organization in living organisms can emerge within the dynamic processes of nature. The modern synthesis of genetics and evolutionary theory provides a powerful account of the mechanisms by which life elaborates new forms. Biologists spin out this story of life, of course, without reference to divine action. Indeed, given the prominent role of chance events in shaping the course of evolution, it has been argued (most famously by Jacques Monod) that the development of life on Earth and the emergence of the human species can no longer plausibly be viewed as processes that reflect the purposes and activity of God.[6]

Evolutionary biology, then, might be viewed as posing a double challenge to theology. First, it presents a set of puzzles about God's relation to evolutionary processes. Can the haphazard course of evolutionary history be an enactment of God's creative purposes? If so, then how might we conceive of God's action in and through the long history of variation, mutation, and selection that powers evolutionary change? Second, if we succeed in constructing a theological interpretation of evolution, it appears that we will be left with a powerful version of the problem of evil. Natural selection operates through differential reproductive success, which is determined in part by patterns of competition and premature death. In this process of evolutionary change, lives are generated and destroyed in a vast impersonal lottery of genetic trial and error. This process may appear too accidental, too wasteful, and

---

[6] For Monod's views see *Chance and Necessity* (London: Collins, 1972).

too cruel to be the work of a God who possesses perfect knowledge, power, and goodness.

I want to address both of these questions, briefly surveying an array of options for conceiving of divine action in evolutionary processes, and then turning to the problem of evil as it arises in the context of evolution.

2 Evolution and Divine Action

How might we understand God to act in or through evolutionary processes? In asking this question we are inquiring about the conceptual resources available to a theological interpretation of evolution. I am not supposing it is possible to begin with close attention to the history of life and infer that God has been at work within that history (for example, to argue from evidences of design to a divine designer); my concern is with the theology of nature rather than with natural theology. Nor am I claiming that evolutionary theory is in need of theology to complete or correct it. It has done theology no good (and considerable harm) to seize upon apparent inadequacies in the biologists' naturalistic explanations and then insist upon special divine action as a competing hypothesis. Having acknowledged the impressive explanatory power of evolutionary biology, however, we face a distinctive *theological* task of interpretation. Any contemporary theology concerned with divine action and providence needs to consider God's relation to evolutionary history.

According to classical understandings of God in the theistic traditions, God determines whether a universe will exist at all and what sorts of creatures will populate it. God calls all finite things into being and continuously sustains (or "conserves") their existence. In this creative action, God is universally at work throughout the world at every moment of its history, and God bears the most intimate possible relationship to each individual creature. These basic affirmations form the background against which we will consider the concepts of providence and particular divine action in history. There are, I will contend, several ways in which we can conceive of God acting through evolutionary processes.

2.1 Divine Action in a Deterministic World

In establishing the laws of nature and the boundary conditions under which these laws operate, God sets the parameters of the world's history. If the laws of nature were strictly deterministic, God could specify at the "outset" the entire course of finite events. In such a universe, we could both regard the whole of created history as a single act of God and take every discrete event as an indirect particular divine action.[7] This reflects a familiar feature of our concepts of agency and action; we can describe an agent's action both in terms of the overall outcome that the agent intentionally brings about (for example, producing a landscape painting) and in terms of the particular sub-acts that are the means by which this result is achieved (for example, preparing a canvas, selecting brushes, mixing paint, and so on). On such an account, cosmic evolution and the emergence of life on Earth would be extended indirect divine actions. God's purposes for each event in history would be realized through the lawful operation of the created causes that bring that event

---

[7] William Alston discusses this possibility in "God's Action in the World," in *Evolution and Creation*, ed., Ernan McMullin (Notre Dame, Ind.: University of Notre Dame Press, 1985); see also his "Divine Action, Human Freedom, and the Laws of Nature," in *Quantum Cosmology and the Laws of Nature: Scientific Perspectives on Divine Action*, Robert J. Russell, Nancey C. Murphy, and Chris J. Isham, eds.(Vatican City State; Berkeley, Calif.: Vatican Observatory Publications; Center for Theology and the Natural Sciences, 1993).

about, and the whole of cosmic history would unfold in accordance with the "program" designed into the system at the outset.

2.2 *Indeterministic Chance in the Order of Nature*

This deterministic theology is only one of the ways in which we might develop the classical schema of primary and secondary causation, and it faces a number of significant objections from both science and theology. If the universe in which we find ourselves is not in fact thoroughly deterministic in its structure, then our account of God's action in evolution becomes both more complicated and more interesting. In considering this possibility, it is important to note at the outset that the options for what we are able to say about divine action in this case will depend in part on empirical questions about whether the indeterminacies in nature make the right sort of difference in the development of events at the macroscopic level. If 1) the only points of indeterminism in the order of nature occur at the quantum level, and if 2) these quantum indeterminacies are entirely subsumed within deterministic egularities at higher levels of organization, then the incompleteness of the causal st. actures of nature will be irrelevant to the course of macroscopic events. The world will have an irreducibly probabilistic sub-structure, but this causal openness will disap, ear into the deterministic relations of the entities constituted by these microp. ocesses.

Both of the antecedent clauses in this conditional can be challenged. A number of proposals have been made about the presence of indeterminisms at higher levels in the structures of nature.[8] I want to focus my attention, however, particularly on the second condition. In light of Bell's theorem and the accumulating experimental successes of quantum mechanics, there is an increasingly widespread, though by no means unanimous, acceptance of the view that the probabilistic character of quantum mechanics reflects actual indeterminism in nature, and not merely the limits of measurement or ignorance of underlying deterministic processes.[9] The presence of indeterministic chance within the structures of nature might provide a means by which God could affect the course of events without in any way disrupting or displacing secondary causes (since the finite causes are only necessary, not sufficient, for the quantum effect). This would require, however, that quantum chance not simply be averaged out in deterministic regularities at the macroscopic level.

I have argued elsewhere that there is good reason to think that chance events at the quantum level, over and above their role in constituting the stable properties of macroscopic entities, can affect the course of observable events.[10] The recent development of chaos theory, for example, may be relevant here. Chaotic systems, operating in a perfectly deterministic manner, can produce widely diverging

---

[8] See, for example: Arthur Peacocke, *Theology for a Scientific Age* (Oxford: Basil Blackwell, 1990), Chaps. 3 & 9; John Polkinghorne, "The Laws of Nature and the Laws of Physics," in *Quantum Cosmology and the Laws of Nature*, ed. by Robert John Russell, Nancey Murphy, and Christopher. J. Isham (Vatican City: Vatican Observatory Foundation, 1993).

[9] See, for example, James T. Cushing and Ernan McMullin, eds., *Philosophical Consequences of Quantum Theory: Reflections on Bell's Theorem* (Notre Dame, Ind.: University of Notre Dame Press, 1989); Robert John Russell, "Quantum Physics in Philosophical and Theological Perspective," in Russell, Stoeger, and Coyne, eds., *Physics, Philosophy, and Theology* (Vatican City: Vatican Observatory, 1988).

[10] See my "Particular Providence and the God of the Gaps," in Murphy, Russell, and Peacocke eds., *Chaos and Complexity: Scientific Perspectives on Divine Action* (Vatican City: Vatican Observatory, 1995); see also essays by Murphy and Ellis in the same volume.

outcomes from minutely different initial conditions. In order to predict accurately the future states of such a system, we would need to factor in events at the quantum level.[11] This suggests that quantum events function as "triggers" for chaotic systems, and that chaotic systems function as "amplifiers" for quantum events, elaborating at the macroscopic level the consequences of indeterministic chance at the smallest scales. The science involved here remains uncertain; the relations between the mathematics of chaos and the equations of quantum mechanics remains to be worked out in a satisfactory form. But this represents at least one promising possibility for understanding how indeterministic chance at the quantum level might make a difference in the course of events at the macroscopic level.

Given our interest in evolutionary processes, it is important to note a second way in which quantum indeterminacies may have macroscopic consequences. Robert John Russell has recently argued that the events described by quantum physics play a role in generating variability in individual organisms and populations.[12] Since variability combined with natural selection provides the primary means by which new forms of life emerge, quantum events can in this way affect the direction of evolutionary change. There are many sources of variation in individual organisms. For example, chromosome mutations include translocations, inversions, and deletions, all of which depend upon breaking specific atomic bonds that involve quantum mechanical processes. The same point can be made about mutations at the level of individual DNA molecules, for example, in point mutations and in replication errors. Further, the operation of environmental mutagens may hinge on quantum mechanical processes, as is clear with various forms of radiation damage to genetic material. Finally, the recombination of genetic material in the production of germ cells through meiosis involves the "crossing over," or exchanging, of homologous strands of genetic material between chromosomes. Russell points out that this "is ultimately a *quantum* process at the atomic level initiated by the breaking and making of a *single* hydrogen bond."[13] These genetic changes may be expressed in the phenotype, subjected to selective pressures, and perpetuated or extinguished in the gene pool. In this way, the processes by which genetic information is retained or lost may serve to amplify the results of indeterministic chance at the quantum level. Many scientific questions remain to be answered, of course, about the relative importance of mutation in contributing to variability and about the role of quantum events in the processes of mutation. But this clearly represents another possible route by which indeterministic events at the lowest levels of organization in nature may have significant macroscopic consequences. Indeed, through these mechanisms, indeterministic chance appears to play a fundamental role in the evolution of life on Earth.

2.3 Divine Action and Natural Indeterminacies

1. *The God Who Plays Dice.* In light of these roles for quantum chance in shaping the course of macroscopic events, what might we say about God's action in the world? One possibility is that God creates a world that includes some under-determined events and leaves the outcome up to chance. God would, in effect, make a world that must complete its own creation. More precisely, the design of the

---

[11] See James P. Crutchfield et al., "Chaos," in *Chaos and Complexity*, Russell, Murphy, and Peacocke, eds.; see also Paul Davies, in this volume.

[12] Russell, "Theistic Evolution: Does God Reall Act in Nature?" *Center for Theology and the Natural Sciences Bulletin*, 15.1 (Winter 1995); see also Russell in this volume.

[13] Ibid., 27.

universe would incorporate a means of trying out novel possibilities not rigidly prescribed by the causal laws that give the macroscopic world its consistent structure. If God "turns the world loose" to work out these aspects of its history, then we have an interesting variant on the strategy that explains God's providential governance of history exclusively in terms of God's initial creative action. On this account, God calls the universe into existence and sustains it in existence, but God does not act within the world's history to affect the course of events. All of God's providential guidance is built into the creative act that establishes the laws of nature and sets the boundary conditions of cosmic history. Because some of these laws are irreducibly probabilistic, however, the precise development of events will be left open by God's initial creative action. Given the organization of chance in precise stochastic patterns and the emergence of deterministic laws at higher levels, the course of events must flow in the general channel God has established. But at least some significant details may be decided by particular underdetermined events whose consequences happen to be amplified and then selectively retained in the course of the world's history.[14] On this account, God does indeed play dice with the universe, but since God designs the dice, they generate a distribution of outcomes that expresses God's creative purposes.

It is a matter of scientific dispute as to what extent the dice are weighted to certain results, that is, to what extent the laws of nature and the values for key constants (such as the total mass of the universe, the gravitational constant, the value of the strong nuclear force) make the emergence of life in the universe a probable event.[15] Given the theological view we are currently considering (namely, that God incorporates structured chance into the design of creation), this empirical question will have an important bearing on how strong a claim we can make about God's providential direction of history. For unlike the deterministic view, according to which every event can properly be regarded as God's act, this position entails that much of what transpires in cosmic history can be attributed to divine action only in a qualified sense. If God chooses not to determine events that are left under-determined by the order of nature, then these events (and the causal series that flow from them) cannot be regarded as particular divine acts, even if God perfectly foreknows what those events will be. God brings it about that there is a world that can unfold in this way, but God's plan of creation will not specify that the world's history pursue just the course it does rather than some other. This will be true of many features of evolutionary history, perhaps including the emergence of a particular primate species as the form of life that happens (at least for a brief evolutionary moment on Earth) to have carried farthest the development of intelligent, self-conscious agency.

2. *Particular Providence in an Indeterministic World.* Is there an alternative to this entirely "hands off" view of God's relation to the created world? Theologians have grown accustomed to avoiding talk about God "intervening" in the world to

---

[14] When I speak of the roles of indeterministic chance and natural law in creatively shaping the course of the world's evolution, I have in mind the vast bulk of cosmic history during which these laws have applied (and especially the processes by which life evolves) rather than the very early history of the universe when the world as a whole was a quantum object.

[15] See, for example, John Barrow and Frank Tipler, *The Anthropic Cosmological Principle* (Oxford: Oxford University Press, 1986); Paul Davies, "Teleology Without Teleology."

bring about particular effects. The claim that God performs special actions in history (over and above establishing and sustaining the order of nature) tends to strike the sophisticated modern ear as naive both theologically and scientifically, a retreat from the world described by the sciences into a premodern realm of miracle (even if only very subtle ones). It may also appear to treat God as just another cause that competes for influence with secondary causes, pushing them aside in order to have an effect in the world. These cautionary theological instincts may lead us to be wary of the claim that God acts at the quantum level, lest this become simply the latest version of the perpetually clumsy effort to insist upon a place for God in the causal structures of nature.

It is important to recognize that these familiar objections lose their force if we understand God to act at points of genuine indeterminacy in nature. If the world is put together in such a way that probabilistically structured indeterministic chance can have a significant impact on the course of events, then God could act in history without disturbing the lawful regularities of secondary causes. By acting to determine natural indeterminacies, God could turn events in genuinely new directions without any "violation" of natural law. These novel developments in the world's history would be conditioned but not determined by the past; that is, they would have necessary but not sufficient causal conditions in the events that precede them. As long as God's actions at the quantum level remain within the probabilistic regularities that the sciences observe, they would neither contradict any law of nature nor displace any finite cause. Rather, God has made a world with an open structure, a world that displays a reliable system of causal relations, but one in which the causal net is not drawn so tightly as to require God to disrupt it in order to act within history. In such a world, God's creative action could be continuous and ubiquitous, not only in the sense that God sustains all finite things in existence at every moment, but also in the sense that God acts in the world to affect the ongoing course of history, perhaps including evolutionary history.

A question remains for views of this kind about whether God determines all or only some of the events that are otherwise left to chance within the order of nature. If God determines every such event, then God will be universally and intimately involved in shaping the course of events from "the bottom up." The probabilistic laws of quantum mechanics will, in effect, simply summarize the regular pattern of God's action.[16] By acting in this way, God both establishes the reliable causal structures of the macroscopic world and sets particular causal chains in motion that may profoundly affect the direction of, for example, evolutionary development. This provides for a strong understanding of God's providential sovereignty over the unfolding course of history. Such a view faces important puzzles, however, about whether and how it can avoid a thoroughgoing divine determinism, which would collapse this account back into the first position we considered. If determinism is to be avoided, indeterminacy must be introduced at some higher level in the order of nature, and this causal openness must be compatible with divine determination at the lowest levels of nature. It is not enough to appeal to a strong (in other words, indeterministic) human freedom; the challenge is to explain how such freedom can

---

[16] See Nancey Murphy, "Divine Action in the Natural Order: Buridan's Ass and Schrödinger's Cat," in Russell, Murphy, and Peacocke, eds., *Chaos and Complexity*, and Robert Russell, "Special Providence and Genetic Mutation: A new defence of theistic evolution," in this volume. See also William Pollard, *Chance and Providence* (New York: Charles Scribner's Sons, 1958).

be accommodated within a world organized in this way. The idea of "top-down" causation is a promising possibility here, though it remains to be developed in a conceptually satisfying form.[17]

The alternative is to say that God determines some, but not all, of the events that are left under-determined by their finite causal conditions.[18] In this case, God builds into the design of nature a structured spontaneity from "the ground up." God has created a natural order that is open to selective divine influence without in any way "violating" its integrity. To the extent that God leaves events up to indeterministic chance, God's providential governance of creation is exercised by setting the boundaries (that is, the structure of natural law and the initial conditions) that constrain and direct the development of natural pro esses. In this respect, this view of divine action concurs with the "hands-off" position th.' we noted above, and it too has a stake in the debate over how precisely these constraint direct evolution toward a particular outcome (for example, toward the generation of liv:ng things in general and rational moral agents in particular). This view goes on to affh.n, however, that God determines at least some events that are left under-determined by secondary causes. In this way God may selectively initiate causal chains that contribute to the fulfillment of God's purposes for creation. This is one means by which particular, or special, divine providence may take effect in the world. We do not, on this view, face the puzzles about divine determinism that arise if we assert that God specifies every quantum event. God creates and continuously sustains a world with a structure that is open to particular divine action, but God acts within that structure selectively, limiting the uses of God's power in order to allow the patterned play of chance to elaborate the possibilities of creation.

2.4 Divine Action and Miracle

A discussion of possible modes of divine action would be incomplete without acknowledging that God might also choose to act outside the ordinary course of nature. Talk of miracles, of course, faces a battery of familiar modern objections. We now approach the world with a very different set of expectations than did, say, Gregory the Great when he wrote his account of the miracles of St. Benedict. It is not my purpose here to claim that we ought to give miracles a particular role in our understanding of divine action, but only to suggest that we do not have good grounds to rule them out altogether. If we affirm that God is the creator of the world *ex nihilo*, we have no reason to deny that God could produce effects in the world that exceed the powers created things. There may be good theological reasons to hold that, by and large, God chooses not to do so. And there may good evidential reasons to be

---

[17] For an argument that top-down causation provides a means to avoid determinism, see section 4.4.3 of Nancey Murphy, "Divine Action in the Natural Order: Buridan's Ass and Schrödinger's Cat," in Russell, Murphy, and Peacocke, eds., *Chaos and Complexity*.

[18] This entails that there are some events that do not have sufficient causal conditions, whether in finite events or in divine action. Nonetheless, these events have their being from God as creator. How is this possible? Consider God's relation to secondary causes generally. God directly causes the existence of both the finite cause and its effect (e.g., the flame and the heated water). But God does not directly cause the water to be hot; that is caused by the flame. So a distinction must be made between the act of causing existence as such (*creatio ex nihilo*) and the act of causing an entity to possess a property at some time. This at least suggests the possibility that God might cause the existence of quantum "entities" (e.g., the wave packet) but leave undetermined which of the structured possibilities is actualized. For a bit more on this, see my "Particular Providence and the God of the Gaps," in Russell, Murphy, and Peacocke, eds., *Chaos and Complexity*, 320–22.

skeptical about any particular claim that God has acted in this way. But the God who calls the world into being, setting the terms of existence and operation for all finite things, may also choose to act in creatures to achieve what they, through the exercise of their natural powers, could not accomplish. Note that this mode of divine action includes not only spectacular events that dramatically disarm our expectations, but also vanishingly minute departures from natural regularities that remain entirely hidden from our view. A miraculous intervention in the natural order will have occurred whether God makes the sun "stand still" or merely diverts an otherwise deterministic causal chain beneath the threshold of human observation; the former, however, is a better means of getting our attention.

2.5 Conclusion: A Diversity of Possibilities for Divine Action

There are, then, a number of ways in which we might understand God to act in the world so as to affect the processes that power evolution. In addition to creating and continuously sustaining the being of all finite things; God may 1) determine the causal history of the world by establishing deterministic natural laws and setting the boundary conditions; 2) build indeterministic chance into the structure of nature in such a way that the created world completes its own creation; 3) determine all events left under-determined by secondary causes at the lowest levels of nature; 4) selectively determine some of these otherwise under-determined events; and 5) act outside the ordinary structures of nature. This is *not* an exhaustive list of the ways in which God can be understood to act. It may be possible, for example, to develop an account of ongoing "top-down" divine action that is distinct both from "bottom-up" action in quantum indeterminacies and from creative action in setting the "initial conditions" for the operation of natural law.[19] For the purposes of this discussion, however, we need not work out a more complete account of the possible modes of divine action, nor do we need to settle the question of which approach (or which combination of approaches) is preferable theologically. Our task has been to see whether a reply can be given to the first of the initial two problems posed for theology by evolutionary theory: namely, the question of whether and how God could be understood to act in evolutionary processes. It is now apparent that this can be done in a variety of ways, drawing on the resources of one or more of these modes of divine action.

It is also clear, however, that any of these ways of understanding divine action in evolution will need to grapple with the second theological problem. To the extent that we succeed in giving an account of divine action in evolutionary processes, we will find ourselves faced with a powerful version of the problem of evil. The challenge to God's goodness will take a somewhat different form for each of these views about divine action. Given a deterministic picture of the universe, for example, we have seen that every event can be regarded as an act of God, even if it occurs far down a chain of secondary causes. On this view it must be said that God not only permits but produces every evil that afflicts the created world. This consequence is avoided by the second view, which emphasizes the role of structured indeterminisms in the world's unfolding history. But here too God's goodness can be called in question; the challenge now takes the form of asking whether a perfectly good and all-powerful creator would leave so much to chance, given the magnitude of

---

[19] On top-down causation see, for example, Arthur Peacocke, *Theology for a Scientific Age*, 53–55, 157–60, and idem, "God's Interaction with the World," in Russell, Murphy, and Peacocke, eds., *Chaos and Complexity*.

suffering and loss involved in the meandering course of evolutionary history. Each of the other accounts of divine action in evolution will also face some form of the question about why God would bring into being a world that contains the volume and intensity of suffering that we see around us. Whether we suppose that God 1) controls the evolutionary process precisely, or 2) permits nature to try out at random a bounded set of possibilities, or 3) combines elements of these strategies, we cannot avoid asking whether this evolutionary deity can be the God whose loving care for creation encompasses even the sparrow's fall (Matthew 10:29).

3 Evolution and the Problem of Evil

In considering evolution and the problem of evil, we will be particularly (though not exclusively) concerned with questions raised by natural suffering, or what has come to be called "natural evil." In contrast to moral evils, which can be attributed to the wrong choices of human beings, natural evils befall us and other sentient creatures simply by virtue of the natural conditions of our lives. It may seem odd to apply the word "evil" to natural misfortune, and in fact such suffering raises moral questions (for example, about justice) only when understood in a particular way, namely, as occurring in a world made by a good creator. Given the affirmation that the world is freely called into existence by a loving God, it is legitimate to ask why the world is so rife with suffering. Traditional Christian (and, more generally, theistic) affirmations about God's perfect goodness invite us to consider God's relation to the world in moral terms, and not only in terms of sovereign power or creative fecundity.

3.1 The Task of Theodicy in the Context of Evolution

The task of theodicy is to state a morally sufficient reason for God to create a world that includes the sorts of evils that we find around us. Any formulation of such a reason will need to include two components. First, it must identify the good (or goods) for the sake of which evil is permitted or produced. Second, it must explain, at least in general terms, the relation of evil to this good. I want to discuss each of these topics in turn.

It is important to acknowledge that we possess only a very limited capacity to comprehend the good that God intends in creation. This is a function of our epistemic finitude. Though human knowledge in all of its forms constitutes a remarkable accomplishment, we cannot claim to grasp very much of what is going on in the universe; we know enough about the world to recognize that there is more in heaven and earth than is dreamt of in any of our philosophies. Clearly, we are in no position to pronounce upon the purposes of the whole. We can speak, in the context of Christian faith, of the good that God extends to us. But we ought to resist the temptation to suppose that our good is *the* good sought by God in creation.

Evolutionary biology, set against the background of cosmic history, can play a salutary role in undercutting our anthropocentrism and encouraging an appropriate humility. Our species has made its appearance after a long history of life on Earth, and evolutionary theory gives us no reason to think that this history has reached its consummation and conclusion in us. Neither are there good theological grounds (though I will not argue this here) for supposing that God's one aim in all creation is the production on Earth of the animal species, *Homo sapiens*. God may have purposes for the world quite apart from our species' brief career. If we affirm the intrinsic goodness of creation, then we must affirm that it is good without us, that is, both before us and after us. This is not to deny, however, that finite persons occupy a distinctive place in God's purposes for creation. Nor is this to deny that we can gain some understanding of God's wider purposes in creation by considering God's purposes for us. If we are to think about the good that God intends for the world, we have little choice but to start with the good in which we participate, and then extrapolate in the direction of the rest of creation.

Discussions of the problem of evil have identified various goods as providing the justifying condition for God's permission of evil. The goods associated with moral life are often cited, for example, and they play a key role in theodicies that appeal to human moral freedom. Moral life, however, is *not* the primary good according to Christianity (or any of the theistic traditions). If human freedom has a value, it is because it is a necessary condition for a certain kind and quality of relationship to God. Christian faith affirms that God intends to share with us the good of fulfillment in fellowship with God, as God meets us in the humanity of Jesus and draws us into relationship through the work of the Spirit. Communion with God is a good that includes and transcends the span of our biological lives, so that a Christian discussion of goods and evils can legitimately take note of the exceedingly great good of life with God beyond death. The immanent and transcendent good of this divine-human relationship will shape any distinctively Christian theological reflection on why God creates a world that includes the evils we lament. We should not, as I have said, suppose that our fulfillment in relation to God is the only good God pursues in creation. But our consideration of God's permission of evil must start here, and then work by analogy toward goods further afield, for example, the goods achieved in the lives of non-human living things.

3.2 "Soul-making" and Evolutionary Evils

What, then, is the relation of this good to "evolutionary evils," namely, the natural evils associated with evolutionary processes? I will suppose that these evils are an inherent part of the natural world God creates, and not a result of creation having gone wrong in sin (though human sin will affect their distribution, intensity, persistence, and the way we experience them). One familiar pattern of argument in theodicy has a broadly transcendental structure; that is, it attempts to identify a set of necessary conditions for the possibility of realizing the good. I want simply to sketch a brief example of this kind of theodicy, without trying to work out the deductive connections fully, in order to display the place of natural evils and their relation to moral evils.

It will be useful here to follow roughly the pattern of John Hick's theodicy.[20] If the good sought for us by God is fulfillment in a distinctively personal relationship to God's own (inner-trinitarian) life of love, then clearly there must be finite persons. An essential aspect of being a person is developing a character of one's own as a self-conscious, knowing, valuing, moral agent in community with other such agents. This development of personal life in all its dimensions (for example, physical, cognitive, affective, aesthetic, moral, spiritual) requires that a number of further conditions be met.

First, we must exist at an "epistemic distance" from God; we need "cognitive room" to try out our powers, to learn and grow on our own, to make mistakes and experience their consequences, to develop as seekers of the good.[21] All of this is

---

[20] John Hick, *Evil and the God of Love* (San Francisco: Harper & Row, 1966), especially Part IV.

[21] The expression "epistemic distance" comes from Hick, *Evil and the God of Love*, 281. Austin Farrer makes the same point in an earlier work, *Love Almighty*. Farrer argues that God creates the physical world as a "screen" between finite minds and God. Without this

necessary if, when the highest Good invites and enables our love, there is to be a finite self who responds and can be brought to fulfillment in this relationship.

Second, persons must be free to make morally significant decisions that can have serious consequences for themselves and for others. If each time we make a wrong moral choice God were to intervene to "change our minds" or to prevent harm to self or others, then autonomous moral life would not be possible.

Third, there must be a stable and impersonal environment in which persons carry out this process of "soul making." If the world around us varied constantly in order to accommodate our particular interests, we could not make enough sense of it to gain knowledge of its workings or to act coherently within it. Further, if the natural world were to operate according to a recognizable principle of moral retribution, so that persons always received from nature just what they deserved, then moral life would once again be undercut. Suppose, for example, that a minor self-serving lie were to merit and receive a head cold (with more serious offenses receiving proportionally more severe punishments). In a cosmic moral nursery school of this sort, a moral virtue like compassion would not make much sense, and a moral motive like unselfish regard for others would hardly be possible. If, however, the world displays a lawful and impersonal structure, then it will be possible for persons to harmed by it. This will be a world in which creatures can thrive, but also perish, a world in which survival will depend upon being well-suited to the structures of the natural environment, a world in which it may be advantageous for life to be malleable in response to the conditions in which it finds itself.

A great deal more could be (and has been) said in working out these necessary conditions for the existence of finite persons.[22] It is clear that evolution can be accommodated within such a theodicy, though it is probably not required. We need to ask, however, about the place of non-human lives in such an account. This theodicy begins with the good for human beings, and considers the conditions for its realization. The procedure is candidly anthropocentric, but it need not remain so. There is no reason that such a theodicy must claim that the whole of evolutionary history is just for the sake of producing human beings. God, as I have said, may have other projects underway, and if we seek some insight into the big picture by contemplating our own case, we should not suppose that God's other creations are simply instrumental to our own good.

On the contrary, Christians typically have affirmed that the Earth and its inhabitants are intrinsic, not merely instrumental, goods. To *be* is to be in relation to

---

distancing, "created minds or wills would be dominated by the object of their knowledge or their love; they would lose the personal initiative which could alone give reality to their knowing or their loving. The divine glory would draw them into itself, as the candle draws the moth" (p. 64).

[22] Of particular relevance here are a number of discussions of the third condition above, which links natural suffering to God's respect for the integrity of the created world. John Polkinghorne contends that the free-will defense must be paired with a "free-process" defense which affirms (as a necessary condition for the existence of persons) that God permits the natural order to operate with its own spontaneity and regularity (*Faith of a Physicist* (Princeton: Princeton University Press, 1994), 85; see also Robert J. Russell's response to Polkinghorne, "The Thermodynamics of Natural Evil," *CTNS Bulletin* 10.2 (1990):20–25). Nancey Murphy and George Ellis argue that "suffering and disorder are necessary byproducts of a noncoercive creative process that aims at the development of free and intelligent beings" (*On the Moral Nature of the Universe* (Minneapolis: Fortress Press, 1996), 247). See also Russell's discussion of "Entropy and Evil," *Zygon*, 19.4 (December, 1984): 449–67.

God, and in this lies the good of every creature of every kind. We cannot know the exact character of this relationship for other living things; perhaps it consists just in the creature being what it is, expressing its capacity for life in its particular way. But certainly the suffering of sentient animals is part of the problem of natural evil. There is no reason to doubt that they share with us a capacity to feel pain and fear. It might be argued that this suffering is unfair to them, since they do not also share in the highest personal good for the sake of which our natural suffering is permitted. On the other hand, they realize forms of experience that we can only dimly imagine (for example, when we watch ravens "playing" in a rising current of air or when we hear the recorded sounds of whales). And if they cannot share in the distinctively personal goods that human intelligence makes possible, neither can they suffer from its ills (for example, our capacity for despair and self-hatred, our anxiety about death, and our uniquely cunning cruelty to one another). Perhaps there is a balance between the opportunity to experience various goods and the cost in suffering of making those goods available. It clearly is not possible for us to settle these questions, however, since we cannot fully assess the balance of natural goods and evils in the lives of other sentient animals. Acknowledging these epistemic limits (to which I will return later), we have here at least the general outline of a theodicy that makes a place for evolutionary evils.

3.3 *Should God Permit Evils Only as the Means to a Greater Good?*

I want to consider one of the most powerful objections to such a position: namely, the insistence that there is just too much suffering in the world. We can imagine many changes in the natural world, both large and small, that appear to be significant improvements. "If it were up to me," we say with the confident tone of those who run no risk of having to act on their words, "there would be no flu viruses, and certainly no ebola or HIV." We can readily imagine ways of making the world a gentler, more accommodating place, of providing more abundantly for the well-being of each sort of creature within it; human beings have often envisioned and longed for a peaceable kingdom, rather than Tennyson's "nature red in tooth and claw." More sweepingly, we might wonder why God bothered with so long and haphazard a history of evolution? This question may seem particularly compelling if we are uncritical about the anthropocentric hubris of supposing that our lives are the *telos* of the whole process; surely it would have been more efficient for God, through special creation, to have brought humanity into existence in the midst of a compatible mix of stable species, without extinction?[23] Whatever the particular form of the question, the objector is saying, in effect, that it looks as though the good purposes God is alleged to be pursuing in the world could be achieved more directly, with less waste, loss, suffering, and destruction.

We can formulate this general type of objection in an argument of the following form.

1. An omniscient, omnipotent and perfectly good God would not create a world that includes pointless natural or moral evil, that is, evil that is not necessary or best as the means of producing a greater good or preventing an equal or greater evil.

---

[23] Just this seems to be the picture presented in Genesis. Among the motives for religious discomfort with evolutionary theory, there is not only its denial of special creation of the human species, but also its disruption of an Augustinian theodicy, which defends God's goodness in creation by linking natural suffering to the fall.

2. There are instances of pointless evil in the world.
3. Therefore, the world is not created by an omniscient, omnipotent, and perfectly good God.

The first premise of this argument is offered as a necessary truth that expresses one of the entailments of God's perfect knowledge, power, and goodness. The second premise states an empirical claim. A reply to this argument might challenge either or both of these premises. Let me start by taking a closer look at the first.

Should we say that God will admit evils like natural suffering into the world only when those evils are the necessary or the best *means* to bringing about a greater good? On this account, each natural or moral evil must lead to an outcome that is better than that which would have resulted had this suffering not occurred. The central objection to this principle hinges on the observation that evils are not always related to goods as means to ends. This is a consideration that is often overlooked, although it is implicit in the familiar pattern of argument in theodicy that we briefly considered above. Evils may arise as a "by-product," or "side-effect," of meeting various necessary conditions for the possibility of a good. In some instances, these evils may also serve as the means by which a greater good is achieved, but they need not always do so. Instead, they may occur simply as a collateral cost associated with making a class of goods possible in the world. If these goods cannot be realized without permitting or producing these evils, then a strong case can be made that God is justified in creating a world that includes such evils.

It is easiest to see this point by considering moral evils in the context of a free-will theodicy. If God's purpose in creating finite persons includes or requires that persons possess an indeterministic, or libertarian, freedom, then it will be possible for persons to make morally wrong choices. When we do so, the moral evils we generate may be neither necessary as a means to a greater good nor the best means to such a good. That is, some of these moral evils *do not need to occur* in order to realize God's good purposes for finite persons. Precisely because the human agent is free in undertaking this action, she has it in her power to prevent this moral evil from occurring. The morally wrong action is not a necessary element in the world's history, and while God's intentions for creation no doubt recognize and accommodate this action, they do not require it. On the contrary, the world might very well be better off without this particular evil deed. But if human beings are to possess the freedom to make morally significant choices, then God must *permit* such actions. Contrary to premise (1), therefore, God's purposes in creation may require the permission of pointless evils, namely, evils that are not necessary or best as a means to the good.

It is worth noting an interesting objection to this claim. It is within God's power to limit the scope of human freedom in various ways, including by simply overriding human freedom and determining action in some instances. It might be suggested that God should permit morally wrong free choices only when these evils are either necessary for or the best means to a greater good.[24] In that case, human

---

[24] Eleanore Stump develops a position of this kind in "The Problem of Evil," *Faith and Philosophy*, 1.2 (1985). She argues for the very strong claim that every instance of suffering that befalls a person as a result of another person's wrong moral choices must in fact provide the best opportunity for the moral and spiritual edification of the sufferer. On this account, every victim of wrongdoing needs to suffer in just the way he does. This claim is motivated by Stump's concern to avoid saying that God allows people to suffer "just for the sake of some abstract general good for mankind" (411), such as the permission of morally significant

beings would possess a morally significant freedom, but our freedom would not result in the production of any pointless evils. William Hasker has pointed out a crippling problem with this proposal; it would undermine moral life.[25] Suppose we knew or believed that God was running the world in this way. If we were making a decision between morally significant alternatives, we would know that either 1) doing the wrong thing would in fact be for the best within God's overall plan for the world or 2) God would not permit us to make the wrong choice or to perform the wrong action. On a consequentialist (for example, classical utilitarian) account of morality, this would mean that the putatively wrong action would in fact be right. On a deontological normative theory, one would still have the motive of duty for choosing the right action, but the identification of duties would be complicated and the seriousness of moral choice would be undercut. It appears, once again, that pointless evils must be permitted if God is to realize the goods associated with finite moral agency.[26]

These problems with the first premise of the objector's argument are enough to justify rejecting the argument in which it occurs. Given our focus on evolution, however, it is important to note that the point we have made with regard to moral evils can also be made about natural evils; in other words, they too may occur as a by-product of meeting the necessary conditions for making a class of goods available in the world. Natural suffering, we have noted, is a causal consequence of meeting the necessary conditions for finite personhood, in particular, and sentient life, in general. The suffering that befalls sentient creatures in nature may often serve a greater good. The pain registered by a nervous system may spur action that preserves the creature's life. Pain may also call forth courage or fortitude or insight in a human sufferer and compassion and other-regarding action in an observer. It hardly needs to be said, however, that natural suffering does not always appear to serve these goods, or any others that we can discern. The helpless suffering of sentient creatures seems not to generate any good at all, much less a good sufficient to outweigh this evil. It is within God's power to prevent such suffering, for example, to remove pain that occurs in circumstances under which the information borne by this pain is useless to the sufferer. Should God do so?

Let us note two considerations. First, this would involve massive intervention within the order of nature, suspending its laws in countless particular cases. The action of eliminating all non-functional pains would disrupt the natural structures within which functional pains are possible. If God's purposes for the created world require a stable, impersonal natural order, then this procedure is ruled out. Note that I am not denying that God might sometimes intervene in the natural order; the

---

freedom. In addition to the problems with this view that Hasker points out, I have argued that it rests on an excessively strong moral prohibition against permitting suffering for the general good. See "Victimization and the Problem of Evil: A Response to Ivan Karamazov," *Faith and Philosophy*, 9.3 (1992).

[25] William Hasker, "The Necessity of Gratuitous Suffering," *Faith and Philosophy*, 9.1 (1992): 27–30. Hasker distinguishes two overlapping definitions of gratuitous evils, which correspond to the distinction I make below between pointless and gratuitous evils. I reserve the term "gratuitous" for evils that are *both* pointless *and* excessive (*i.e.*, they could be prevented by God without undermining the necessary conditions for realizing the good, as I explain below). Hasker argues for the necessity of gratuitous evils in both of the senses he identifies. I argue only for the necessity of pointless evils.

[26] Note, however, that this is not to deny that God can bring good out of these evils. It is only to say that they need not occur in order for God's purposes to be achieved in the world.

problem arises only if God does so excessively.[27] In order to preserve lawful structures of nature, pointless natural suffering must be possible, indeed it will be inevitable.

Second, if God did manage the world in such a way that every instance of natural suffering was necessary or best as the means to a greater good, then we would face a problem about moral life comparable to that which we considered in discussing moral evils. If we believed both that God operates in this way and that God is perfectly good, we could trust God never to impose unwarranted suffering on any individual, whether this individual suffers for her own sake or for that of others. All natural suffering would be for the best. It is not clear, in this case, what moral policy we should take toward such suffering. Should we try to prevent it? It would not make much sense to do so, if we knew that 1) God will arrange things so that our efforts fail if this suffering is in fact for the best, and 2) God will intervene to prevent this suffering directly if we do not act to prevent it and it is not, after all, for the best. This consideration suggests that if the moral lives of finite persons are to include ethically significant responses to natural suffering, then some of that suffering must be pointless.[28] Nature will inevitably hurt us and other living things, and we will belong to a community of suffering that calls for compassionate action.

The initial challenge from the objector to evil was that there is too much evil in the world to serve the purposes that the theodicist ascribes to God; the world apparently includes pointless evil. It now appears, however, that one consequence of meeting the necessary conditions for finite animal and moral life is that natural suffering and moral evils will occur which do not, in every instance, serve as the necessary or best *means* to a greater good. A world that makes possible the full range of goods that God intends for creatures, including the good of personal relationship with God, must include pointless evils.

3.4 Does the World Contain Gratuitous Evils?

This is not the end of the matter, however. I have argued that pointless evils must be tolerated by God because the price of eliminating them is to destroy morally significant freedom and the integrity of the natural order. So pointless evils do, after all, have a point, at least in the sense that they are permitted for a reason. It seems fair to ask how *much* pointless evil must be permitted for this reason. The original objection can be rephrased: granted that there must be some pointless evils, why are there so many? Surely there are morally wrong human actions that could be prevented by God without significantly undercutting human freedom? And surely there are pointless natural evils that could be eliminated by God without undercutting the reliable structures of the natural order? These questions suggest that there is some quantity of pointless evils that represents an unavoidable minimum (a "floor" level) that God must permit in creation if the good of creaturely spiritual life is to be made possible. Any pointless evil beyond this minimum will be excessive, and therefore not just pointless but gratuitous. We can make use of this notion of "gratuitous" evil in reformulating the objector's argument in the following way:

1'. An omniscient, omnipotent and perfectly good God would not create a world that includes gratuitous (that is, pointless *and* excessive) natural or moral evils.

---

[27] The question about what frequency of intervention would be excessive is not one human beings are in a position to answer, as I will argue below.

[28] Cf. Hasker, "The Necessity of Gratuitous Suffering," 37–40.

2'. There are instances of gratuitous evil in the world.
3. Therefore, the world is not created by an omniscient, omnipotent, and perfectly good God.[29]

The first premise now asserts that a God perfect in knowledge, power, and goodness would eliminate any pointless evils that could be eliminated *by God* without undercutting the conditions necessary to realize the goods that God intends for the world. There will be many pointless evils that could be eliminated *by human beings* (for example, by making a morally right rather than wrong choices or by acting to relieve pointless suffering); these evils will not be gratuitous, however, if they must be permitted in order to preserve significant moral freedom and a lawful natural order.

Once again, we may reply to this argument by challenging either of its premises. I want to make some exploratory remarks about the first premise, before offering a more decisive criticism of the second. The first premise supposes that there is a determinate level of pointless evil that is both 1) the minimum (the "floor") that must be allowed in order make certain goods possible and 2) the maximum (the "ceiling") that it is morally permissible for God to permit or produce. This picture is surely too simple. It is far from clear that there will be any single optimal level of pointless evils, such that even a relatively small departure from this level would either disrupt the necessary conditions for realizing the good or result in gratuitous evils. Consider natural evils. God could intervene in the natural order to prevent at least some instances of natural suffering, but if God were to do so too often, the integrity of the natural order would be undermined. It is artificial to suppose that there must be a precise balance point at which any further intervention would compromise the natural order and any less would result in gratuitous evils. Furthermore, any single instance of pointless natural suffering could be eliminated by God, but not all can be; therefore, each will appear to be gratuitous.

The same points can be made about God's permission of moral evils, and here there is a further consideration. Suppose for a moment that there were a world (or a class of possible worlds) that generates the moral goods that God seeks in creation at the lowest level of pointless moral evil. It may be doubted whether it would be within the power of God to create such a world intentionally, even though God is omniscient and omnipotent. The production of moral goods and evils in this world will depend upon the free choices of the finite persons it contains. In order to select a world (from among all possible worlds) that achieves the goods associated with morally significant freedom at the lowest cost in pointless moral evils, God would need to know what each possible free creature would choose to do in each circumstance in which it might find itself. Only if God knew this would God be in a position to decide which set of free creatures to create and what circumstances to place them in so that their freedom would be preserved with a minimum of pointless moral evil.[30] God would need to know what finite free agents who never actually

---

[29] Cf. William Rowe's argument in "The Problem of Evil and Some Varieties of Atheism," *American Philosophical Quarterly* 16.4 (1979): 333–41.

[30] Simple foreknowledge (namely, knowledge of all free acts in the actual world alone) would allow God to act so as to minimize pointless moral evil in this world. But in choosing to create this world from among all possible worlds, God would not know whether it was one of those in which the minimum would be the lowest possible in relation to the goods achieved (*i.e.*, there might be another possible world in which this same level of good is achieved at a lower cost in evils).

exist would freely choose to do in circumstances that never in fact obtain. We may legitimately doubt whether there are such truths to be known.[31]

Finally, there is a general objection that can be raised to the suggestion that God must generate goods in the most "efficient" way possible, measured in terms of the "cost" at which these goods come in evils permitted. Calculations of this sort are possible only if goods and evils can be quantified, compared on a single scale, and averaged out. This way of thinking about goods and evils does not hold up well on close examination, however. What, for example, is the net intrinsic good/evil generated by adding to or subtracting from a world a stand of redwoods, a sprained ankle, a free act of kindness, one species of beetle, and another year of life for Mozart? If goods and evils cannot all be measured against each other, then there cannot be, on conceptual grounds, such a thing as *the* best of all possible worlds.[32]

There are good reasons, then, to be cautious about agreeing to the first premise of the argument. Nonetheless, the moral intuition at work here does have considerable plausibility. Although there are conceptual difficulties in trying to articulate this intuition in terms of minimums and excesses, one implication of God's perfect goodness would seem to be that God does not permit readily preventable profound suffering in the world. We must take seriously the charge that the world around us ∠nes in fact contain suffering that could be prevented by God without undercutting the ¡ ⁿssibility of realizing the goods that God intends for the world. We need to turn, thereſo.e, to the second premise, the empirical claim that the world does in fact contain graʻuitous evils.

As we nc·ed earlier, this claim is often put forward by pointing to various evils that strike us as easily eliminated without serious consequences. It clearly will not be possible to reply to this objection case by case, explaining in each instance how this or that evil fits into the scheme of things. On the contrary, we do not and cannot know enough about the world or about God's purposes to offer such an explanation. But precisely this limitation points to what is wrong with the second premise; it supposes that in at least some instances we know (or at least are in a position justifiably to claim) that an apparently pointless evil is gratuitous (that is, excessive). For the argument to succeed, it must show not simply that the theodicist has no explanation for certain pointless evils, it must establish that we are justified in asserting that there is no explanation of the right type. It must show that the world

---

[31] Propositions about what an agent would freely choose to do under circumstances that do not obtain have come to be called "counterfactuals of freedom." God's putative knowledge of such truths has traditionally been called "middle knowledge," because it lies between God's "natural" knowledge of possibility and necessity (a set of necessary truths independent of God's will) and God's "free" knowledge of what is actual (a set of contingent truths dependent upon God's will). There is a longstanding debate among philosophical theologians about whether God has middle knowledge. If all counterfactuals of freedom are false, then there is nothing of this form to be known, and omniscience will not include propositions of this type. On this debate see, for example, Alfred J. Freddoso's translation of Louis de Molina, *On Divine Foreknowledge* (Ithaca: Cornell University Press, 1988), especially the introduction, section 4, 46ff; and Robert Adams, "Middle Knowledge and the Problem of Evil," in *The Virtue of Faith and Other Essays* (New York: Oxford University Press, 1987).

[32] On rather different grounds, Alvin Plantinga also calls into question the concept of a "best of all possible worlds." He observes that "no matter how marvelous a world is—containing no matter how many persons enjoying unalloyed bliss—isn't it possible that there be an even better world containing even more persons enjoying even more unalloyed bliss?" (*God, Freedom, and Evil* (Grand Rapids, Mich.: Eerdmans, 1989), 34). Plantinga's point is well taken, though note that he is working with a quantitative notion of goods.

would be better off if God prevented the instances of evil to which objector points. Failing this, the argument cannot reach its a-theological conclusion. Let me note several reasons why this is something we cannot know.[33]

First, we noted above that pointless evils are bound to look like they are preventable by God. *Ex hypothesi*, they do not serve as the necessary or best means to a greater good, but are tolerated by God as collateral costs of meeting the general conditions for other goods. There must be some evils of this class (for example, God must permit pointless natural suffering if there is to be a lawful natural order). But any single instance of the class could be prevented by God without the loss of a greater good. Furthermore, there may be no principled reason for God to prefer one instance to another (for example, from within a class of comparably severe natural evils). This means that there may be *no reason at all* for God's permission of one such evil rather than another. There will be a reason for permitting instances of this class of evils, but not for permitting this specific instance. We cannot expect to give a specific explanation when in fact there is none to be given. The observation that we cannot think of a reason for God's permission of an evil, therefore, does not warrant the conclusion that the evil is gratuitous, or excessive.

Second, we are not in a position as knowers to say whether or not there could be fewer instances of a class of evils. This would require that 1) we be able to identify which evils are in fact pointless and then 2) determine that some of these need not be permitted in order to preserve the general conditions that make the good available. Both of these two tasks lie beyond our epistemic powers. A simple way to illustrate this point is to conduct a thought experiment in the elimination of evils. Suppose that we identify an evil that is a candidate for being gratuitous (that is, both pointless and excessive). In order to conclude that this evil is in fact gratuitous, we would need to tell a comprehensive story about how things would go in the world without this evil. This would require not only that we perfectly forecast the consequences of this change for the future. It would also require that we retell the history of the world in such a way that this event does not occur. We do not, of course, know enough about how the world works in order to do this. Moreover, what we think we *do* know about the world indicates that we cannot *in principle* know enough to tell such a story.

Here it is helpful once again to take note of chaos theory. Under certain conditions, a non-linear but deterministic system in nature may be exquisitely sensitive to changes in its initial conditions. Two starting points for the system may be infinitesimally close together and yet generate dramatically different outcomes. In order to predict over time the outcome of even a simple system of this sort, we would need to have a knowledge of its initial conditions that includes, for example, the gravitational effects of each electron in the galaxy. It is beyond our epistemic reach, therefore, to reconstruct in full detail the causal history of systems of this sort, yet these systems appear to be ubiquitous in our world (for example, weather systems involve chaotic processes, hence the often noted "butterfly effect").

The result of this for our topic is clear. Adjustments in the operations of nature that strike us as eliminating pointless evils may in fact have major and unpredictable consequences further along in the course of events. These consequences may be

---

[33] For a very helpful and more extensive development of an epistemic reply to the evidential, or empirical, problem of evil see William P. Alston, "The Empirical Argument from Evil and the Human Cognitive Condition," in *The Evidential Argument from Evil*, ed. Daniel Howard-Snyder (Bloomington: Indiana University Press, 1996).

good or bad, and so this argument cuts in both directions on the problem of evil. On the one hand, it is possible that the elimination of a particular evil may lead to the loss of a greater good or the generation of a greater evil. On the other hand, it may be that a relatively minor adjustment would lead to the prevention of a great evil or the production of a great good. Both the theodicist and the objector to evil, then, can appeal to chaotic effects in telling a "just so" story about the world. The only conclusion that emerges with any certainty from such an exchange is that, in considering chaotic systems, we do not and cannot know what will result from our hypothetical changes in the course of events. As a result, we simply are in no position to say whether any particular evil is pointless or gratuitous.

Third, we noted earlier that we have no reason to think that we know or can comprehend the full range of goods that God is pursuing in creation. This will be true even of the goods that God intends for human beings; theologians generally are quite diffident about claiming to understand and describe beatitude, the fully realized communion with God that is the highest good of any creature. Without a complete understanding of all the relevant goods, however, we once again are in no position to assess evils with regard to whether they are pointless or gratuitous.

3.5 Conclusion: The Limits of Theodicy

The cumulative result of these considerations is to induce a degree of epistemic humility in contemplating the place of evil in the design of the world. The sarcastic words of the Voice from the whirlwind may now seem appropriate as a reminder of our epistemic limits: "Where were you when I planned the earth? Tell me if you are so wise." The objector to evil cannot show that there are evils in the world that are incompatible with the existence of a God of perfect knowledge, power, and goodness. But neither can the theodicist give a detailed explanation of why a world made by such a God should include all of the evils that our world contains. The most theodicy can hope to do is to provide a general account of why the sorts of goods and evils characteristic of our world must be permitted or produced in order to achieve certain goods. A detailed explanation of the distribution, relationships, and overall balance of goods and evils cannot reasonably be expected. The "problem" of evil is not one that we are going to "solve."

Trust in God's goodness, then, cannot be rooted in a confidence that we have tallied up the world's goods and evils and found a net advantage on the side of good. Such trust no doubt reflects the experience of what is good in life, and it certainly is challenged by the pervasive presence of suffering, loss, and grief. But it must be rooted elsewhere. For Christians this trust has arisen in response to the enactment of God's love in the life of Jesus Christ. Here it is affirmed that God's own life is made available to us in the midst of our world, with its physical dangers and moral complexity. The God who approaches us in this way is not the cosmic bureaucrat who, from the safe distance of a disengaged eternity, sets up the system of nature and calculates the balance and distribution of goods and evils, assuring that the cosmic ledger show a favorable bottom line in the end. Rather, this God enters into the suffering of creatures, bearing the wounds of the physical creation, and transfiguring these sorrows through incorporation within the life of God. The highest good consists in relationship to this God, a relationship that does not exclude the creature's suffering, but rather includes and transforms it by placing it in relation to God's own suffering in the world. This does not provide an explanation for the world's evils, but it has given rise to faith that none of the world's evils, neither those we suffer nor those we do, can separate us from the divine love.

CONTRIBUTORS[1]

Francisco J. Ayala, Donald Bren Professor of Biological Sciences, University of California, Irvine, California, USA.

Ian Barbour, (now retired) Professor of Physics, Professor of Religion, and Bean Professor of Science, Technology and Society, Carleton College, Minnesota, USA.

Charles Birch, Emeritus Professor and previously Challis Professor of Biology, University of Sydney, Australia.

Camilo J. Cela-Conde, Catedratico (Senior Professor), Facultad de Filosofia y Letras, Universidad de las Islas Baleares, Palma de Mallorca, Spain.

Julian Chela-Flores, Professor, Abdus Salam International Centre for Theoretical Physics, Trieste, Italy, Instituto de Estudios Avanzados, Universidad Simon Bolivar, Caracas, Venezuela.

Anne M. Clifford, Associate Professor, Department of Theology, Duquesne University, Pittsburgh, Pennsylvania, USA.

George V. Coyne, S.J., Director, Vatican Observatory, Vatican City State, Italy.

Paul C. W. Davies, formerly Professor of Natural Philosophy, University of Adelaide, Australia, currently Visiting Professor, Imperial College, London, England.

Willem B. Drees, Nicolette Bruining Professor of Philosophy of Nature and of Technology from a Liberal Protestant Perspective, University of Twente, Enschede, The Netherlands, and Bezinningscentrum (Center for the Study of Religion, Science and Society), Vrije Universiteit, Amsterdam, the Netherlands.

Denis Edwards, Lecturer in Systematic Theology, Catholic Theological College, Adelaide College of Divinity, Flinders University, Adelaide, Australia.

George F. R. Ellis, Professor of Applied Mathematics, University of Cape Town, Rondebosch, South Africa.

John F. Haught, Landegger Distinguished Professor of Theology, Georgetown University, Washington, D.C., USA.

Philip Hefner, Professor of Systematic Theology, Lutheran School of Theology at Chicago, and Director, Chicago Center for Religion and Science, Chicago, Illinois, USA.

---

[1] Also participating in the conference were R. David Cole, Stephen Happel, Michael Heller, Clifford N. Matthews (observer), Bernd-Olaf Küppers, and Fraser Watts (observer).

Gisele Marty, Department of Psychology, Universidad de las Islas Baleares, Palma de Mallorca, Spain.

Nancey Murphy, Associate Professor of Christian Philosophy, Fuller Theological Seminary, Pasadena, California, USA.

Arthur Peacocke, Director, Ian Ramsey Centre, Oxford, England, Warden Emeritus of the Society of Ordained Scientists, formerly Dean of Clare College, Cambridge, England.

Ted Peters, Professor of Systematic Theology, Pacific Lutheran Theological Seminary and the Graduate Theological Union, Berkeley, California, USA.

Robert John Russell, Professor of Theology and Science in Residence, Graduate Theological Union, and Founder and Director, The Center for Theology and the Natural Sciences, Berkeley, California, USA.

William R. Stoeger, S.J., Staff Astrophysicist and Adjunct Associate Professor of Astronomy, Vatican Observatory, Vatican Observatory Research Group, Steward Observatory, University of Arizona, Tucson, USA.

Thomas F. Tracy, Professor of Religion and Chair, Department of Philosophy and Religion, Bates College, Lewiston, Maine, USA.

Wesley J. Wildman, Assistant Professor of Theology, School of Theology, Boston University, Boston, Massachussetts, USA.

Name Index

VATICAN PRESS